# CHOCOLATE, COCOA, AND CONFECTIONERY:

## SCIENCE AND TECHNOLOGY
## THIRD EDITION

# CHOCOLATE, COCOA, AND CONFECTIONERY:

## SCIENCE AND TECHNOLOGY
## THIRD EDITION

### Bernard W. Minifie, Ph.D.
Consultant to the Confectionery Industry
Richardson Researches, Inc.
Hayward, California

An **aVi** Book

CHAPMAN & HALL
I(T)P An International Thomson Publishing Company

New York • Albany • Bonn • Boston • Cincinnati • Detroit • London • Madrid • Melbourne •
Mexico City • Pacific Grove • Paris • San Francisco • Singapore • Tokyo • Toronto • Washington

Chapman & Hall
115 Fifth Avenue
New York, NY 10003

Chapman & Hall
2-6 Boundary Row
London SE1 8HN
England

Thomas Nelson Australia
102 Dodds Street
South Melbourne, 3205
Victoria, Australia

Chapman & Hall GmbH
Postfach 100 263
D-69442 Weinheim
Germany

Nelson Canada
1120 Birchmount Road
Scarborough, Ontario
Canada M1K 5G4

International Thomson Publishing Asia
221 Henderson Road #05-10
Henderson Building
Singapore 0315

International Thomson Editores
Campos Eliseos 385, Piso 7
Col. Polanco
11560 Mexico D.F.
Mexico

International Thomson Publishing - Japan
Hirakawacho-cho Kyowa Building, 3F
1-2-1 Hirakawacho-cho
Chiyoda-ku, 102 Tokyo
Japan

5  6  7  8  9  XXX  01  00  99  98  97

**Library of Congress Cataloging-in-Publication Data**

Minifie, Bernard W.
      Chocolate, cocoa, and confectionery: science and technology / Bernard W. Minifie--3rd ed.
            p.                                  cm.
      Bibliography: p.
      Includes index.
      ISBN 0-412-99541-7
      1. Confectionery. 2. Chocolate. 3. Cocoa. I. Title.
      TX791.M55    1989
      664'.153--dc19

*Visit Chapman & Hall on the Internet http://www.chaphall.com/chaphall.html*

To order this or any other Chapman & Hall book, please contact **International Thomson Publishing, 7625 Empire Drive, Florence, KY 41042.** Phone (606) 525-6600 or 1-800-842-3636. Fax: (606) 525-7778. E-mail: order@chaphall.com.

For a complete listing of Chapman & Hall titles, send your request to **Chapman & Hall, Dept. BC, 115 Fifth Avenue, New York, NY 10003.**

# Contents

# Preface to the Third Edition

The second edition of this book achieved worldwide recognition within the chocolate and confectionery industry.

I was pressed to prepare the third edition to include modern developments in machinery, production, and packaging. This has been a formidable task and has taken longer than anticipated.

Students still require, in one book, descriptions of the fundamental principles of the industry as well as an insight into modern methods. Therefore, parts of the previous edition describing basic technology have been retained, with minor alterations where necessary.

With over fifty years' experience in the industry and the past eighteen years working as an author, lecturer, and consultant, I have collected a great deal of useful information. Visits to trade exhibitions and to manufacturers of raw materials and machinery in many parts of the world have been very valuable. Much research and reading have been necessary to prepare for teaching and lecturing at various colleges, seminars, and manufacturing establishments.

The third edition is still mainly concerned with science, technology, and production. It is not a book of formulations, which are readily available elsewhere. Formulations without knowledge of principles lead to many errors, and recipes are given only where examples are necessary.

Analytical methods are described only when they are not available in textbooks, of which there are many on standard methods of food analysis.

## Acknowledgments

I am still indebted to many of the persons mentioned under "Acknowledgments" in the second edition.

I am especially grateful to the following.
The late Herb Knechtel of Knechtel Research Sciences Inc., Chicago, Ill.

Terry Richardson of Richardson Researches Inc., Hayward, Calif. They have given me continuous support for the past eighteen years. Yvan Fabry, Karlheinz Weiss (now retired), and other members of the German College of Confectionery, Solingen, Germany. By participating in teaching at the College courses and lecturing at seminars, I have met many students, scientists and production personnel from factories worldwide.

Prof. Joachim von Elbe, University of Wisconsin, Madison. During my seven years helping with the NCA vocational course, Professor von Elbe and his staff gave me much encouragement, both in course programming and in providing other useful contacts.

Cadbury-Schweppes, Bournville, Birmingham, England. I wish to renew my thanks to the directors of this company and particularly to Sir Adrian Cadbury, who gave me his support in launching the first edtion in 1970. Besides retaining contact with my friends in the company, I am grateful for the help of F. J. Stanley and his technical library.

Various journals and their staffs have provided good publicity:
Confectionery Production, U.K.
    Dennis Buckley, Barrie Cassey.
Manufacturing Confectioner, U.S.A.
    Allen Allured.
Candy Industry, U.S.A.
    Don Gussow, Walter Kuzio, Patricia Magee.
I have had much help from friends in other colleges, universities, and industrial organizations.

Acknowledgment, where appropriate, is given in the chapter references and appendices.

# Cocoa and Chocolate

Part 1

# 1

# History and Development

The development of cocoa, chocolate, and confectionery over the centuries has been a remarkable phenomenon.

The cocoa tree is an unusual tree, with its cultivation confined to limited areas and climatic conditions. The processing required both in the areas where it is cultivated and in the factory is a complex example of human persistence and ingenuity.

Several developments have been responsible for progress in the industry, as the chocolate products as originally prepared by the natives of Central America would hardly be acceptable today. Until the early 1800s the only product was a very fatty chocolate drink prepared from the whole cocoa bean, sugar, and spices. In 1828, Van Houten of Holland invented the cocoa press, which removed a part of the cocoa fat from the bean, resulting in a powder with about 23 percent fat. This made the drink easier to prepare and digest. At the same time, the natural fat, cocoa butter, was released, making it possible to produce a fluid chocolate that could be molded and also used to cover other confectionery products. In England, during the 1840s, Fry, and later Cadbury, made chocolate bars.

Another major development was the invention of milk chocolate by Daniel Peters of Switzerland in 1876. The ground cocoa nib (the bean cotyledon) was processed with sugar and milk solids and the result was a product that today is the mainstay of the chocolate industry.

Cadbury's Dairy Milk chocolate was developed in the early 1900s and similar products by many other manufacturers followed. Since that time, the popularity of milk chocolate has increased astronomically, and with the development of mass-production molding machines that have helped to reduce its manufacturing cost, it is now available to almost everyone.

In the early days, the processing of milk for chocolate needed a method that would ensure the development of flavor and good shelf life. The American, Milton Hershey, used a special cultured-milk process and Cadbury developed the Crumb process. (Milk chocolate manufacture is discussed in a later chapter.)

In the United States, Hershey and chocolate are synonymous. Hershey established the Hershey Chocolate Company around 1900, after selling his caramel-manufacturing business. After much experimenting, he developed his own method for making milk chocolate, and a factory was built in the rich Pennsylvania countryside where there were plentiful supplies of fresh milk.

The Swiss have also had a close association with chocolate. Chocosuisse, the Association of Swiss Chocolate Manufacturers, which was established in 1945, in the publication *Chocologie* trace the history and development of chocolate through the ages. Many of the early Swiss names cited there are still associated with chocolate products, including Suchard (1797–1884), Cailler (1796–1852), Sprungli (1816–1897), Lindt (1885–1909). Henri Nestlé (1814–1890) came into the chocolate industry later and was more concerned with milk processing. Without Nestlé's development of condensed milk, Daniel Peters would not have invented milk chocolate.

It is interesting to note that the chocolate industry has been associated with many philanthropists and humanitarians. Fry, Cadbury, and Rowntree were Quakers. Elizabeth Fry, the great social reformer, was a descendant of the family that built the first factory in England in 1728. Milton Hershey, a Mennonite, in 1909 established a school for orphan boys that today provides foster care for both boys and girls.

## BOTANY, GROWING AREAS, CULTIVATION

The cocoa tree (Theobroma cacao) is a native of the dense tropical forests of the Amazon where it grows in conditions of semishade, warmth, and high humidity.

The genus Theobroma consists of over twenty species (family Sterculiaceae), but only T. cacao is of commercial value. T. cacao is assumed to have spread naturally westward and northward to Guyana and Mexico, and later to the Caribbean islands. In so doing, two distinct subspecies developed. Morris (1882) classified these as Criollo and Forastero, and later Forastero was divided into several varieties. A third group called Trinitario is essentially a cross between Criollo and Forastero and is not found in the wild.

The Mayas of the Yucatan and the Aztecs of Mexico cultivated cocoa long before its introduction to Europe, and Montezuma, emperor of the Aztecs, is stated to have regularly consumed a preparation called "chocolatl" made by roasting and grinding the cocoa nibs, which were then mashed and mixed with water, maize, and spice. The richness of this mixture no doubt had some connection with the Aztec belief that the cocoa tree was of divine origin, and later led the Swedish botanist Linnaeus to give the name "Theobroma"—Food of the Gods—to the genus, including the cacao species.

The Aztecs also considered the drink to have aphrodisiac properties. Historical illustrations show cups of chocolate being consumed at wedding ceremonies, and in the court of Montezuma, the drink was held in high esteem as a nuptial aid.

Columbus first brought cocoa beans to Europe but only as curiosities. His fellow countryman Don Cortes was the first to recognize their commercial value as a new drink, and he sent back to Spain cocoa beans and recipes for preparing chocolate.

The Spaniards added sugar to the recipe, and the drink gained in popularity. They then introduced cacao into Trinidad, trying to keep the methods of its cultivation and preparation a secret. Eventually, however, the trees were grown in other West Indian Islands and in the Philippines, whence the Dutch probably introduced cocoa into Indonesia and Ceylon (now Sri Lanka).

South America and the West Indies remained the major suppliers of cocoa beans until around 1900, when the cocoa tree was introduced to West Africa where growing conditions were so favorable that production soon reached a very high level. To this day, the countries of West Africa supply most of the world's requirements, although Brazil is now rapidly becoming a major producer. Malaysia and New Guinea have also moved into the realm of important producers and considerable expansion is expected. In West Africa, the Ivory Coast has greatly increased production.

Table 1.1. indicates the trend of the expansion of the cultivation of the cocoa tree, and the map of Fig. 1.1 shows the main growing areas of the world.

After being introduced into Europe by the Spanish, the popularity of the chocolate drink spread to Italy, Holland, and France in the middle 1600s; a short time later, it became known to the aristocracy of England, being mentioned in *Pepys' Diary* in 1664. At that time, chocolate was very costly and was beyond the reach of all but the wealthy. In the 1700s, some improvement occurred and chocolate or, rather, cocoa products, began to be made on a factory scale. The Fry factory in Bristol, England, started production in 1728 but chocolate bars and other chocolate-covered confectionery did not appear until later in the nineteenth century.

Margaret Lang (1978) describes the social conditions in those early days.

TABLE 1.1. WORLD PRODUCTION OF RAW COCOA 1972/73–1984/85 × '000 METRIC TONS

| | 1972/73 | 1973/74 | 1974/75 | 1975/76 | 1976/77 | 1977/78 | 1978/79 | 1979/80 | 1980/81 | 1981/82 | 1982/83 | 1983/84 | 1984/85 (Estimate) |
|---|---|---|---|---|---|---|---|---|---|---|---|---|---|
| **Africa** | | | | | | | | | | | | | |
| Cameroun | 107 | 110 | 118 | 96 | 82 | 107 | 106 | 124 | 120 | 122 | 106 | 198 | 120 |
| Gabon and Congo | 6 | 8 | 7 | 7 | 6 | 6 | 8 | 6 | 5 | 6 | 4 | 3 | 5 |
| Ghana | 418 | 350 | 377 | 397 | 320 | 268 | 250 | 285 | 258 | 225 | 178 | 159 | 174 |
| Equatorial Guinea | 10 | 12 | 15 | 12 | 6 | 7 | 5 | 5 | 8 | 8 | 10 | 8 | 8 |
| Ivory Coast | 181 | 209 | 242 | 231 | 230 | 304 | 312 | 379 | 403 | 457 | 355 | 405 | 545 |
| Liberia | 2 | 3 | 3 | 3 | 3 | 3 | 4 | 4 | 4 | 4 | 5 | 6 | 6 |
| Nigeria | 241 | 215 | 214 | 217 | 166 | 207 | 137 | 172 | 156 | 183 | 156 | 115 | 150 |
| São Tomé and Principe | 11 | 10 | 7 | 8 | 6 | 6 | 8 | 7 | 8 | 8 | 8 | 7 | 3 |
| Sierra Leone | 6 | 8 | 5 | 6 | 6 | 7 | 7 | 7 | 10 | 8 | 10 | 9 | 10 |
| Togo | 19 | 16 | 14 | 18 | 14 | 17 | 13 | 11 | 16 | 11 | 10 | 16 | 11 |
| Zaire | 5 | 5 | 5 | 4 | 4 | 5 | 4 | 4 | 4 | 5 | 4 | 4 | 4 |
| Other Africa | 22 | 7 | 4 | 4 | 5 | 4 | 7 | 9 | 3 | 3 | 6 | 9 | 9 |
| **Total Africa** | **1,028** | **953** | **1,011** | **1,003** | **848** | **941** | **864** | **1,021** | **995** | **1,040** | **852** | **849** | **1,045** |
| **Central and South America** | | | | | | | | | | | | | |
| Bolivia | 2 | 2 | 3 | 3 | 3 | 3 | 3 | 3 | 3 | 3 | 3 | 3 | 3 |
| Brazil | 162 | 246 | 273 | 258 | 234 | 283 | 314 | 294 | 349 | 314 | 336 | 302 | 405 |
| Colombia | 23 | 23 | 25 | 26 | 29 | 29 | 31 | 30 | 39 | 42 | 40 | 41 | 40 |
| Cost Rica | 5 | 6 | 7 | 6 | 8 | 9 | 10 | 8 | 4 | 5 | 5 | 5 | 5 |
| Ecuador | 43 | 72 | 78 | 63 | 72 | 75 | 90 | 98 | 81 | 85 | 55 | 55 | 100 |
| Mexico | 30 | 28 | 32 | 33 | 24 | 33 | 38 | 34 | 30 | 41 | 42 | 40 | 42 |
| Panama | — | — | 1 | 1 | 1 | 1 | 1 | 1 | 1 | 1 | 1 | 1 | 1 |
| Peru | 2 | 2 | 2 | 3 | 4 | 5 | 4 | 4 | 5 | 9 | 12 | 10 | 10 |
| Venezuela | 19 | 17 | 19 | 15 | 15 | 18 | 17 | 17 | 14 | 17 | 17 | 16 | 15 |
| Other America | 2 | 2 | 2 | 2 | 2 | 2 | 2 | 2 | 2 | 3 | 3 | 3 | 3 |

**Total Central and South America**

| | 288 | 398 | 442 | 410 | 392 | 458 | 510 | 491 | 528 | 520 | 514 | 476 | 624 |
|---|---|---|---|---|---|---|---|---|---|---|---|---|---|
| **West Indies** | | | | | | | | | | | | | |
| Cuba | 2 | 2 | 1 | 1 | 1 | 2 | 3 | 1 | 2 | 1 | 2 | 3 | 4 |
| Dominican Republic | 28 | 30 | 27 | 30 | 30 | 30 | 33 | 27 | 32 | 40 | 40 | 39 | 33 |
| Grenada | 2 | 2 | 3 | 3 | 2 | 2 | 3 | 2 | 3 | 2 | 2 | 2 | 2 |
| Haiti | 3 | 3 | 3 | 3 | 3 | 3 | 4 | 2 | 2 | 3 | 3 | 3 | 3 |
| Jamaica | 2 | 2 | 2 | 1 | 2 | 1 | 2 | 1 | 2 | 2 | 3 | 3 | 3 |
| Trinidad and Tobago | 5 | 4 | 5 | 3 | 4 | 4 | 3 | 2 | 3 | 2 | 2 | 2 | 2 |
| **Total West Indies** | 42 | 43 | 40 | 41 | 42 | 42 | 48 | 35 | 44 | 50 | 52 | 52 | 47 |
| **Asia and Oceania** | | | | | | | | | | | | | |
| Indonesia | 2 | 3 | 3 | 3 | 3 | 4 | 7 | 7 | 8 | 10 | 12 | 15 | 16 |
| Malaysia | 9 | 10 | 13 | 17 | 21 | 23 | 26 | 32 | 43 | 60 | 65 | 80 | 100 |
| Papua New Guinea | 22 | 31 | 33 | 32 | 27 | 30 | 30 | 31 | 28 | 30 | 29 | 28 | 30 |
| Philippines | 3 | 4 | 3 | 3 | 3 | 3 | 4 | 3 | 4 | 4 | 4 | 4 | 4 |
| Sri Lanka | 1 | 2 | 2 | 2 | 2 | 2 | 2 | 2 | 2 | 3 | 2 | 2 | 2 |
| Vanuatu | 1 | 1 | 1 | 1 | 1 | 1 | 1 | 1 | 1 | 1 | 1 | 1 | 1 |
| Western Samoa | 1 | 2 | 1 | 2 | 2 | 1 | 2 | 2 | 2 | 2 | 1 | 1 | 1 |
| Other Asia | — | — | — | — | — | — | 1 | 1 | 5 | 5 | 6 | 7 | 6 |
| **Total Asia and Oceania** | 39 | 53 | 56 | 60 | 59 | 64 | 73 | 79 | 93 | 115 | 120 | 138 | 160 |
| **WORLD TOTAL** | 1,397 | 1,447 | 1,549 | 1,514 | 1,341 | 1,505 | 1,495 | 1,626 | 1,660 | 1,725 | 1,538 | 1,515 | 1,876 |

**Notes:**

[1] Production statistics relate to the International Cocoa Crop Year, which runs from 1st October to 30th September, except in the case of Mexico prior to 1973/74 and Venezuela where statistics relate to the calendar year.

[2] The Brazilian Cocoa Crop Year runs from 1st May to 30th April: production statistics, however, have been adjusted to the International Cocoa Crop Year.

[3] In the case of some smaller producers, production has been assumed to equal net exports.

—Production less than 500 metric tons.

*Source: Cocoa statistics.* Oct. 1985. Gill and Duffus, London.

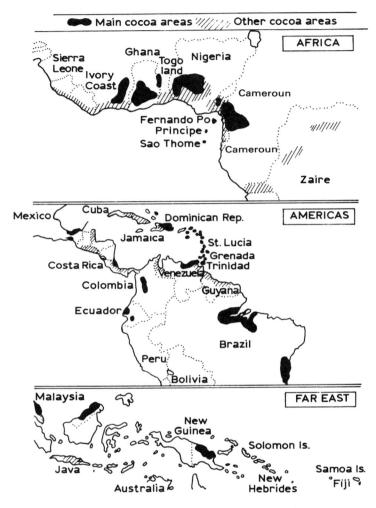

Fig. 1.1.  The Cacao Growing Areas of the World

*Cadbury Schweppes, England*

## THE COCOA TREE

Botanically, the term "cacao" refers to the tree and its fruits (pods and seeds). Cocoa describes the bulk commercial dried fermented beans, as well as the powder produced from the beans (see later).

The tree can only be cultivated within fairly narrow limits of altitude, latitude, and humidity. Wood (1985) notes that the cultivation of 75 percent of the world's cocoa lies within eight degrees of either side of the equator, with exceptions in some areas to about 18 degrees north or south. The optimum growing temperature is between 18 degrees and 32°C. In some areas, temperatures may fall to lower levels, but then lower yields and damage to the trees are likely.

Rainfall is preferably between 1,500 and 2,000 mm per year. Thus, a hot, moist climate favors growth. Relative humidity under these conditions will be 70 to 80 percent during the day, increasing to the saturation point at night. Trees grow mostly in the lower altitudes and it is exceptional for them to flourish above 3,000 ft.

The cacao tree will grow to a height of 20 to 30 ft and requires some shade from larger forest trees. From the time the seedling reaches a height of 3 to 5 ft it throws out three to five fan branches, and later vertical "chupons" from points below "jorquettes" where branches fork. This pattern of growth is repeated until the height of maturity is reached.

The flowers are about half an inch in diameter and are formed in small groups on the trunks and lower main branches of the trees. They are bisexual, have no nectar or perfume, and the pollen is too sticky to be dispersed by the wind. It is only in recent years that it was found that the main agent of pollination is a small midge (*Ceratopogonidae*).

Many flowers are produced but only a small number become

Fig. 1.2.   Drawing of Cacao Tree Showing Pods on Trunk and Lower Branches

*Cadbury Schweppes, England*

pollinated and develop into pods. These pods mature in five to six months during which many wilt and drop off, and this constitutes a further thinning process.

The pod is, botanically, a "drupe" and attains a length of 6 to 10 in. by 3 to 4 in. diameter. It normally contains twenty to forty seeds surrounded by a mucilaginous pulp when the pod is ripe. It is an unusual fact that the ripe pods do not open and scatter the seed, nor do the pods drop off and rot, and the seeds have no dormancy when removed from the pod. In uncultivated areas, the seed is spread by small mammals such as monkeys and squirrels, which open the pods, suck off the sweet pulp, and spit out the seeds.

It is traditional to grow cacao under shade trees, and although such conditions resemble those in the natural habitat, the degrees of shade needed vary considerably. The young plant certainly needs protection but the larger trees may need only a thin canopy, and in some conditions of rich fertile soil, none at all.

Propagation by seed is the most economical way of increasing stock but vegetative methods can also be used, and these provide a more consistent and reliable method of reproducing trees of particular strains. In West Africa, seeds are often pressed into the ground, but it is a better practice to raise the seedlings in a nursery to be planted when they are four to six months old.

The seedlings or cuttings are usually planted 7 to 10 ft apart, and in the early stages of growth, require special precautions against weeds and pests; temporary shade usually is provided by such plants as maize, cassava, and plantain.

A tree should begin bearing pods after three years, and the yield will increase up to the eighth or ninth year, although, considering the size of these trees, the quantity of beans produced per year is quite small.

Yields vary greatly, depending on the method of cultivation, the spacing of the trees, and the nature of the hybrid.

Wood (1985) quotes the following example with Brazilian hybrids:

| Age of tree (years) | 3 | 4 | 5 | 6 | 7 |
|---|---|---|---|---|---|
| Yield (kilograms dry beans per hectare)* | 300 | 1,400 | 2,000 | 2,500 | 2,250 |

* 1 kg = 2.205 lb; 1 hectare = 2.47 acres

The pods are harvested over a period of several months as the trees simultaneously bear mature fruit, flowers, and growing pods. The

Fig. 1.3. Mature Pods on Trunk of
Cacao Tree

*Cadbury Schweppes, England*

ripe pods are removed from the branches by cutlass and delivered to
a suitable central point for opening and removal of the beans and
adhering pulp. These beans with pulp are then taken to the
fermentary; (see Figs. 1.2, 1.3 and 1.4).

Fig. 1.4. Drawing of Section of Cacao
Pod Showing Beans and Surrounding
Pulp

*Cadbury Schweppes, England*

## DISEASES

It is not possible in this chapter to describe in detail the numerous
diseases and pests that attack the cacao tree, although these have
resulted in huge losses. In one survey, by Padwick (1956), losses
attributable to disease were stated to be approaching 30 percent.

One of the most serious cases was the swollen shoot disease that
swept through Ghana and, to a lesser, extent, Nigeria in the years

following World War II. As a result, surveys of other cacao-growing areas were made and virus diseases discovered. In no instance has devastation on the Ghana scale been recorded but large outbreaks are always possible. Strict quarantine in the movement of cacao seeds and seedlings must be rigidly observed.

The disease exists in various strains and many produce the characteristic swelling on shoots and roots. Leaves have disease symptoms such as "mosaic" and pods have a smooth and round appearance. Trees generally die about three years after infection. Viruses are present in many host trees and are transmitted to cacao trees by mealy bugs, of which there are several species. The adult insect moves little but the nymphs and young females can move from tree to tree.

The most effective methods of control were to remove infected trees and to breed resistant strains. Many insecticides were used but were not effective. Systemic insecticides were costly, dangerous to handle, and in some instances caused cacao beans to have an undesirable taste.

The removal of infected trees has had the greatest beneficial effect but this has involved the wholesale destruction of trees in some areas. The acceptance of this procedure by the farmers was made more difficult by the fact that trees in the early stages of the disease still bore pods that yielded good cacao beans. To persuade farmers that such destruction was necessary needed diplomacy, as well as monetary compensation.

Although swollen shoot has caused great havoc in some areas the most widespread disease of cacao is black pod caused by a fungus (*Phytophthora palmivora*). Leaves, young shoots, and pods are attacked, but in the pod, infection, starting as a brown spot, rapidly spreads to the whole pod, which becomes a total loss. The disease is most rampant where high humidity prevails for long periods, and total destruction of the crop can result. It has been successfully controlled by the use of copper fungicides where very damp conditions prevail, as in the West Cameroun.

Another fungus disease specific to cacao is witches' broom (*Crinipellis perniciosa*), which originated in the Amazon area and spread to the West Indies. It causes distorted shoots or "brooms" and affects flower cushions as well as leading to underdeveloped pods and decreased yield. Witches' broom can be controlled by systematic cutting out of the broom and burning, and of recent years resistant clones have been used in plant-breeding experiments.

Other fungal diseases of importance are Monila, Diplodia pod rot, and Ceratocystis wilt. An up-to-date survey of cacao diseases is given by Lass (1985).

## PESTS

Entwistle (1972) has recorded that more than 1,500 different insects feed on cacao. The most damaging and widespread pests are the insect group known as capsids, the two most important species of which are Sahlbergella singularis and Distantiella theobroma. These insects are about half an inch long, speckled or dark brown in color, and attack young shoots and pods, injecting a poisonous saliva while feeding on the sap. The punctures cause lesions that are readily attacked by fungus and the young shoots are killed. More young shoots then are produced and attacked by capsids, until the tree is so weakened that it will die. Very successful control was obtained by the use of BHC sprays applied by knapsack sprayers or, for large-scale operation, by gasoline-driven machines.

In Ghana, where capsid damage was extensive, systematic application of insecticide almost doubled the crop over a period. Other insects attack cacao and of these thrips are widespread. Ants and leaf-eating and stem-boring beetles can cause damage, but none equal to that caused by the capsids. Insects such as mealy bugs, which are sap-sucking, are vectors of virus disease.

For detailed information the reader is referred to Wood, Lass, and Entwistle (1985).

## FERMENTATION AND DRYING

The correct fermentation and drying of cacao are of vital importance as no subsequent processing of the bean will correct bad practice at this stage. A good flavor in the final cocoa or chocolate is related closely to good fermentation, but if the drying after fermentation is retarded, molds develop and these also impart very unpleasant flavors even if fermentation has been carried out correctly.

The chemical processes involved are not entirely fermenting reactions and the changes that occur in the combined processes of fermentation and drying are sometimes referred to as curing.

After the pods are cut from the trees, the beans with the adhering pulp are removed and transferred to heaps, boxes, or baskets for fermentation to take place. Small farmers tend to use the heap method whereas the box method is employed in larger plantations, and is also used on a smaller scale in South America and the West Indies. In Nigeria, a basket method has been used, with the baskets lined with leaves.

With heap fermentation, 500 to 600 lb of beans are formed into a flat cone on banana or plantain leaves, which are also used to cover the heaps. Better results are obtained by using boxes, which give more even fermentation. Fermentation lasts from five to six days. Forastero beans take rather longer than Criollo, and during the first day, the adhering pulp becomes liquid and drains away, with the temperature rising steadily. By the third day, the mass of beans will have fairly evenly heated to 45°C (113°F) and will remain between this temperature and about 50°C (122°F) until fermentation is complete. It is necessary to mix the beans occasionally to aerate them and to ensure that those initially on the outside of the heap are exposed to the temperature in the interior.

Fermentation boxes are made in a variety of shapes and sizes. One that will hold about a ton of wet beans has dimensions of 4 by 4 ft and 3 ft, 6 in. deep. It is filled to a depth of 2 ft, 6 in. Holes are provided in the base to allow the liquefied pulp to drain away. Boxes are sometimes arranged in tiers and the turning of the beans is replaced by transferring them from one box to the other. It is important in the fermentation to have sufficient beans in the heap or box to prevent heat loss, but for experimental purposes, much smaller quantities can be fermented under laboratory-controlled conditions where heat can be conserved. This is particularly useful where chocolate or cocoa needs to be prepared from cacao harvested from specially bred trees when they first begin to bear.

A considerable contribution to the knowledge of the processes was made by Howat, Powell, and Wood (1957) in a series of small-scale fermentations on viable pods transported by air to the Cadbury research laboratories. This not only enabled the fermentation conditions to be closely observed, but it was also possible to prepare samples of chocolate for tasting purposes.

## Chemical Changes During Fermentation

During the fermentation and drying processes, the unfermented wet beans taken from the pod lose about 65 percent of their weight, assuming the final optimum moisture content of 6 percent is attained. The chemistry occurring in the processes may be summarized as in the following.

**Pulp**  In the process of fermentation in the heaps or boxes, the pulp adhering to the beans provides a medium at the correct temperature and pH, not only to liquefy the pulp, but to create the reactions within the cotyledon so vital to the production of good chocolate flavors. The pulp is composed of approximately 85 percent water and 11 percent sugars with small amounts of citric acid, pentosans, and proteins.

During the first two days of fermentation, the sugars are broken down to about 2 percent, forming small amounts of ethyl alcohol and lactic acid, and the temperature rises to about 45°C (113°F).

For the remaining period of fermentation, the temperature is maintained between 45 and 50°C and the quantity of beans must be sufficient to avoid heat loss. The pH rises from 3.5 initially to 4.5 then to 5.0, and the presence of oxygen during turning results in the formation of acetic acid.

The cotyledons absorb some of the acetic acid, and they also attain a pH of about 5.0.

If fermentation is extended beyond five or six days, spoilage organisms take over, with a detrimental effect on the final flavor of the beans. Residual acidity due to acetic acid is a problem with beans from some sources, such as Malaysia and New Guinea.

**Cotyledon (Nib)**  During the fermentation, the cotyledons gain moisture and the texture changes from a cheesy coherent mass to a fissured structure. When dried, they are friable and readily break into pieces called nibs.

There is also a significant color change, and with the Forastero types, the unfermented bean is slaty gray, passing through purple and purple-brown to rich dark brown. With Criollo types, similar changes occur, but the final color is a light brown because the purple anthocyanins are absent.

Three important factors contribute to good flavor and color in the beans:

1. Germination of the live bean must take place during the first stages of fermentation, followed by death of the beans in 30 to 40 hr.

2. Beans must be maintained in the region of 50°C (122°F) for several days after the initial germination. Temperatures as low as 45° to 46°C (113 to 115°F) give a preponderance of purple instead of brown beans, and still lower temperatures give no true chocolate flavor.

3. The carbon dioxide surrounding the beans must be removed; in commercial fermentation, this is accomplished by turning the heap.

## VARIATION, PROBLEMS, EXPERIMENTAL PROCEDURES

While fermentation is largely carried out satisfactorily by the pile or box method, variations do occur that can lead to an unsatisfactory final product. In some areas, the beans, after removal from the pods, are transported wet to the fermentary. If this journey is prolonged, an uncontrolled fermentation may start.

In Ecuador, such fermentation has been observed when wet beans with pulp are dumped in small piles on concrete, in the sun, and remain there, partly drying and partly fermenting.

In Mexico, it is stated that many beans are unfermented. Examination has shown that they are more likely to be underfermented. Certain chocolate products produced there seem to require this treatment.

Experiments with washing off the pulp or removing it by pressing have been made. Much research is still needed to find alternative methods to the traditional fermentation procedure.

### Drying

After fermentation, the beans are placed in shallow trays to dry. In some growing areas where the main harvest coincides with the dry season, sun drying is adequate as long as the seeds are covered during rainstorms or at night, and under these conditions they are dry enough for transport and storage.

The sun-drying trays may be movable on rails so that they can be pushed under canopies, sometimes one above the other, or the canopies can be moved on rails over the trays.

Many authorities are of the opinion that sun drying, properly controlled, gives the best results. The beans are spread out to a depth of about 2 in. and must be raked periodically to expose fresh surfaces.

In some areas, for example the Camerouns, the rainfall and humidity do not permit sun drying and artificial means become necessary. Adequate drying is vital to the preservation of the good flavor of the beans, otherwise molds will develop that will impart an unpleasant flavor that no subsequent process will remove.

Artificial dryers were probably first used in Trinidad in the early part of the century and have since been used in other growing areas. There are numerous types of dryers but an essential feature of all must be that the products of combustion do not come into contact with the beans or the final product will be tainted. Smoky flavors are most objectionable in delicate-flavored milk chocolate and can arise from cocoa butter expressed from contaminated beans.

There are many designs of dryers and in small farm areas economy in construction and fuel usage become paramount. Wood (1985) describes various types of dryers of which the following are probably the best known.

**Samoan Dryer** This is simple in construction and is used in small farming communities. It consists of a flue sloping slightly upward and ending in a chimney. It can be fabricated from oil drums, from which the ends have been removed and the cylinders butted together. To make them smokeproof, the joints are sealed with fireproof material.

The flue is laid down in a trench, preferably on bricks, and above this a drying platform is constructed using timbers, extended to provide support for a roof. Cross-struts are used to support mats on which the beans are placed for drying and the space between the platform and the flue is closed by a wall to conserve heat. Drying conditions of 60 to 70°C to (140 to 158°F) can be maintained in this type of dryer.

**Büttner Dryer** This is the only continuous dryer that has been used for cocoa beans. It consists of a vertical cylindrical chimney about 4 ft in diameter through which a large number of trays pass on an endless chain from top to bottom. These trays, loaded at ground level with wet beans, travel up an incline to the top of the tower and then slowly downward against a current of hot air provided by a turbine and heated tubes. On emergence from the base of the cylinder, the dried beans are unloaded and replaced by wet beans.

The capacity of the dryer is about nine tons with a 16-hour drying cycle. It is economical to run but requires some engineering skill for correct maintenance.

**Lister Dryer** This machine, consisting of a large diesel engine and fan, is used for drying hay and grain. It produces a large volume of air—35,000 ft$^3$ per minute at a temperature of 6 to 12°C above ambient. It had to be modified to deal with wet cocoa beans because the lower temperatures and long times required produced bad odors. The modification involved a preliminary drying in shallow layers at a higher air temperature followed by final drying in the deeper bin. A dry weight of five tons of beans can be handled, and it is stated that this machine is relatively low in capital cost.

**Platform Dryers**   Dryers of this type may employ standard machinery used for other products. Basically they consist of a long narrow platform, 30 ft by 8 ft, 6 in., constructed of perforated aluminum sheet or similar mesh material beneath which is a closed chamber through which warm air is passed. The hot air is produced by a heat exchanger and electric fan.

With all platform dryers, the efficiency is related to the depth of the layer of beans combined with adequate mixing of the beans. With deep layers, mechanical agitation is desirable, but this procedure is not in general use.

## STORAGE AND TRANSPORT OF CACAO BEANS

Great emphasis has already been placed on the necessity for good fermentation and drying but of no less importance are correct transport and storage.

In many cacao-growing areas, methods of transport are still primitive but it is the aim of the producer to get the beans from the farm to the boat as quickly as possible. In areas of high humidity, well-dried beans will soon pick up moisture again, and, for example, in the Camerouns, it has been shown by Powell and Wood that the inside of buildings used for temporary storage had an average relative humidity of about 82 percent. Under these conditions, the beans were in equilibrium at a moisture content of over 8 percent, and 8 percent is the critical upper limit above which molds will develop. The optimum moisture content is 6 to 6.5 percent and the relative humidity in equilibrium with beans at this level will not promote molds. See Table 1.2. These workers recommended using polythene liners for the sacks, and trials showed that these greatly retarded moisture gain and permitted transport and limited local storage under conditions of high humidity.

During World War II, much cacao was stored in tropical-port warehouses, and to prevent damage by damp and insects, rigid rules regarding construction of stores and methods of stacking were formulated. Similar care was taken to see that ships carrying cacao were clean, free from rodent and insect infestation, and adequately ventilated during shipment from tropical to cooler climates. Great

TABLE 1.2. EQUILIBRIUM RELATIVE HUMIDITY OF COCOA
BEANS

| Percent relative humidity | Percent moisture content of beans in equilibrium |
|---|---|
| 65 | 6.4 |
| 70 | 7.0 |
| 75 | 7.5 |
| 80 | 8.5 |
| 85 | 9.5 |

damage to stored beans can result from condensation on the inside of
the ship's plates in the holds because beans loaded at a tropical port
at 29 to 32°C (85° to 90°F) and a moisture content of 7 percent will
deposit considerable amounts of condensed water when arriving in
the colder waters of the north Altantic at 10°C (50°F) and below. It is
the responsibility of the vessel's captain to maintain adequate
ventilation through the hold at all times, but occasionally storms
make this difficult unless the boat is fitted with forced-air ventilation.
The bags of beans are protected from the metal sides of the boat by
sisal dunnage, which minimizes the effect of sweating. However,
proper ventilation of ships' holds virtually eliminated damage due to
condensation.

More recently, bulk container transport has been tried. Here,
again, condensation problems arise, and special construction with
ventilation grilles is used. When these containers are stacked in a
ship's hold, ventilation of the hold is necessary. Also, ventilation of
the containers taking the beans from the ship to the user is required.

It is likely that the basic Hessian "bag" of cacao will remain as
long as present farming methods continue. Traditionally, the bag
contained 140 lb. The main cacao producers are currently shipping at
slightly different weights; Ghana and Nigeria, 137.5 lb; Ivory Coast,
143 lb; Brazil, 132 lb.

On arrival at the port of discharge, the holds should be inspected.
Sometimes, this is done jointly by representatives of the purchaser
and of the port authority.

In any case, an assessment of damage or infestation of the
consignment must be made. It must be decided whether fumigation is
necessary, either in the hold or in the warehouse. At one time, raw
cacao arriving in port often was heavily insect infested. Removal to
the warehouse initiated the migration of Ephestia larvae, which
resulted in many thousands of insects contaminating everything

in the locality. Fortunately, as a result of vigilance in the growing areas, in the warehouses, and during transport, these manifestations are now a rarity.

In many factories, particularly in the United States, consignments of beans on arrival are put into large fumigation chambers and treated with methyl bromide. It is thus ensured that no infested beans get to the production areas. As the first process of roasting destroys all infestation, if the bean cleaning and roasting departments are away from the main production area, this will take the place of fumigation.

## CHEMICAL COMPOSITION OF COCOA BEANS

The general composition of cocoa beans from various sources is similar and the primary analytical composition is shown in Table 1.3.

With properly fermented mature beans, the variation is relatively small, but immature, badly fermented, and sometimes intermediate beans have higher shell and lower cocoa butter contents. The available nib content, and hence the cocoa butter, is the factor that should be used to compare the real values of different consignments of beans.

The cleanliness of the outer shell affects the ash content and the microbiological counts. The shell is virtually a waste material as far as the manufacture of cocoa and chocolate is concerned. Cocoa butter, the natural fat of the nib, is the most costly constituent, although at times powdered cocoa has attained a high value, especially for the manufacture of coatings.

The quality of cocoa beans is determined by the "cut test" (see later), together with the flavor of the chocolate made from the beans.

The alkaloid constituents of the cocoa beans, theobromine and caffeine, have been subjects of considerable publicity of recent years because of their supposed connection with migraine, hypertension, and other health problems. West (1983) has collected much data on this subject. Zoumas, Kreiser, and Martin (1980) have published analyses of cocoa liquors from various sources. They quote the figures presented in Table 1.4. Nevertheless, the stimulating effect of these alkaloids (including theophylline in tea) has to be recognized.

## CHOCOLATE FLAVOR AND AROMA

The essential role played by correct fermentation and drying of the cacao beans in developing the ultimate flavor of the chocolate has

TABLE 1.3. CHEMICAL COMPOSITION OF COCOA BEANS

| | Knapp and Churchman (1937) | | Fincke (1965) | | Jensen (1931) | | Pearson (1981) | | | |
|---|---|---|---|---|---|---|---|---|---|---|
| | | | | | | | Nib, % | | Shell, % | |
| | Nib, % | Shell, % | Nib, % | Shell, % | Nib, % | Shell, % | Max. | Min. | Max. | Min. |
| Water* | 2.1 | 3.8 | 5.0 | 11.0 | 3.9 | 8.1 | 3.2 | 2.3 | 6.6 | 3.7 |
| Fat (cocoa butter, shell fat) | 54.7 | 3.4 | 54.0 | 3.0 | 53.2 | 3.0 | 57 | 48 | 5.9 | 1.7 |
| Ash | 2.7 | 8.1 | 2.6 | 6.5 | 3.1 | 7.6 | 4.2 | 2.6 | 20.7 | 7.1 |
| Nitrogen | | | | | | | | | | |
| Total nitrogen | 2.2 | 2.8 | 2.1 | 2.6 | — | 2.6 | 2.5 | 2.2 | 3.2 | 1.7 |
| Protein nitrogen | 1.3 | 2.1 | — | — | — | — | — | — | — | — |
| Theobromine | 1.4 | 1.3 | 1.2 | 0.8 | 1.3 | — | 1.3 | 0.8 | 0.9 | 0.2 |
| Protein | — | — | 11.5 | 13.5 | 13.9 | 15.9 | — | — | — | — |
| Caffeine | 0.07 | 0.1 | 0.2 | — | — | — | 0.7 | 0.1 | 0.3 | 0.04 |
| Carbohydrates | | | | | | | | | | |
| Glucose | 0.1 | 0.1 | — | — | — | — | — | — | — | — |
| Sucrose | 0 | 0 | }1.0 | — | — | — | — | — | — | — |
| Starch | 6.1 | No true starch | 6.0 | — | 6.0 | — | 9 | 6.5 | 5.2 | 3.4 |
| Pectins | 4.1 | 8.0 | — | — | — | — | — | — | — | — |
| Crude fiber | 2.1 | 18.6 | 2.6 | 16.5 | 2.7 | 14.8 | 3.2 | 2.2 | 19.2 | 12.8 |
| Cellulose | 1.9 | 13.7 | 9.0 | — | — | — | — | — | — | — |
| Pentosans | 1.2 | 7.1 | 1.5 | 6.0 | 1.4 | 8.0 | — | — | — | — |
| Mucilage and gums | 1.8 | 9.0 | — | — | — | — | — | — | — | — |
| Tannins | | | | | | | | | | |
| Tannic acid | 2.0 | 1.3 | 5.8 | 9.0 | — | — | — | — | — | — |
| Cacao purple and cacao brown | 4.2 | 2.0 | — | — | — | — | — | — | — | — |
| Acids (organic)* | | | | | | | | | | |
| Acetic (free) | 0.1 | 0.1 | 2.5 | — | — | — | — | — | — | — |
| Citric | — | 0.7 | — | — | — | — | — | — | — | — |
| Oxalic | 0.3 | 0.3 | — | — | — | — | — | — | — | — |

*Water and organic acids content can vary according to the degree of drying or roasting.

21

TABLE 1.4. ANALYSIS OF COCOA FROM VARIOUS SOURCES

|  | Percent theobromine | Percent caffeine |
|---|---|---|
| Average 22 samples | 1.22 | 0.214 |
| Maximum | 1.73 | 0.416 |
| Minimum | 0.82 | 0.062 |

been emphasized. The true chocolate flavor is finally developed by the roasting process described in a later chapter. Early attempts by Bainbridge and Davies (1912) to isolate chocolate "aroma" used the process of steam distillation of roasted cacao beans followed by hexane extraction. Linalool was identified as a major component of the extract together with a variety of acids and esters. Later, Schmalfuss and Bartmeyer (1932) and Steinmann (1935) identified many more compounds. Mohr (1958) applied the relatively new gas chromatography technique and added to the list. Dietrich (1964) published the results of seven years' work, and although 72 components were detected, it was not possible to reconstitute the true chocolate aroma. Still later, Marian (1967), van der Waal (1968), and Flament (1968) applied mass spectrometry combined with chromatography and the volatile compounds identified rose to 200.

The fundamental stages in the formation of chocolate aroma begin with the production of flavor percursors during the tropical fermentation, and these are changed at the bean-roasting stage into compounds typical of the true chocolate flavor.

Knapp (1937) noted that cocoa butter from well-fermented beans produced no chocolate flavor on roasting—the precursors were, therefore, not in the fat phase. He also found that the precursors were soluble in methanol, and both of these facts formed a very useful basis for later experimental work. However, in spite of Knapp's observations on cocoa butter, this fat does contribute certain flavor characteristics to chocolate, particularly milk chocolate, and the nature of the flavor depends on the method of extraction, the degree of roast, and whether it is obtained from alkalized or unalkalized nibs.

For milk chocolate, deodorized or partly deodorized cocoa butter is often used.

Later research work on the precursors produced evidence that amino acids and sugars are concerned in the formation of final "aroma" compounds and in the roasting process it has been shown that degrading of both free amino acids and reducing sugars occurs.

## Nonvolatile Components of Chocolate Flavor

If the volatile flavors are removed from cocoa liquor by distillation methods and the residual material made into chocolate, the resultant product still has the main characteristics of chocolate, though perhaps somewhat milder.

Papers by Rohan (1967, 1969) summarized in considerable detail the then current knowledge obtained by various workers.

The following is a brief description of his conclusions

*Flavonoids.* A number of polyphenols were identified, including various catechins and anthocyanins.

*Amino acids.* Many of these have high taste intensity and can contribute bitter and sweet tastes.

*Organic acids, phenolic acids.* A large range of these acids has been isolated, and although they are important constituents, their individual and collective influence is uncertain.

*Carbohydrates.* Glucose and fructose are present, with small amounts of sucrose. Flavoring substances develop from the degradation of sugars on heating. The reaction of sugars with amino acids (Maillard) is well known as a contributing factor in the flavor of foods.

A review of the chemicophysical aspects of chocolate processing (Dimick and Hoskin, 1981) summarizes the extensive literature on the subject.

The importance of pyrazine compounds is emphasized and the determination of these substances has been proposed as a method of controlling the degree of roasting. Also, Foster (1978) has given a detailed summary of the Maillard reactions that produce flavors of a chocolate nature.

The research on chocolate flavor will continue. No true imitation of the flavor has, to date, been synthesized.

## SUMMARY

An attempt has been made to explain the complexities associated with the varieties of cocoa beans and the various processes involved.

The following summarizes and, it is hoped, will simplify the understanding of the most important aspects.

Raw beans in the pod exhibit no semblance of chocolate flavor.

Fermentation is the first essential process, and initiates chocolate flavor. This process must take place in the growing area on freshly harvested pods. The beans scooped from the pods are fermented in heaps or boxes. Temperature rises spontaneously to 50°C (122°F) due to yeast/bacterial/enzyme action. In five to six days, if there is no germination, the bean has been "killed."

First is an anaerobic action, in which there is some protein decomposition.

Second is an aerobic action, in which pulp is liquefied and acetic acid is formed. Astringent polyphenols are removed. Flavor precursors formed.

Roasting is the second essential process, which develops the chocolate flavor from the precursors.

The following reactions occur:

1. Some destruction of the amino acids—varies with time.
2. Changes in reducing sugars—reaction with amino acids (Maillard reaction).
3. Nonreducing sugars (sucrose) hydrolyzed to reducing sugars (for Maillard reaction).

The variety of bean provides flavor variations irrespective of processing.

*Criollo*—The original "wild" variety—very small proportion of world supply. Found in Samoa, Java, and Sri Lanka.

*Forastero*—The basic type ("bulk" or "ordinary") accounts for most of world's supply. Found mainly in West Africa (Ivory Coast, Ghana, Nigeria, Camerouns) Brazil (Bahia) and Malaysia.

*Trinitario*—The name given to describe various hybrids (Nacional). In limited use for special chocolates. Found in Ecuador (Arriba), Trinidad, Costa Rica, and Mexico (Tabasco).

An indication of how the origin of the bean can affect the final chocolate flavor is given by the amino-acid contents as shown in Table 1.5.

TABLE 1.5. AMINO ACID CONTENTS

| | Type of bean | | | |
|---|---|---|---|---|
| mg/100 g | Ghana | Ivory Coast | Arriba | Samoa |
| Aspartic acid | 14 | 28 | 8 | 27 |
| Glycine | 4 | 6 | 2 | 5 |
| Lycine | 20 | 24 | 9 | 23 |

## International Cocoa Standards

These standards, which have been adopted by all the main growing countries in Africa, are given in the following; they consist of a Model Ordinance and a Code of Practice. They were formulated as a result of meetings of producers and customers in conjunction with the United Nations Food and Agricultural Organization.

Other standards apply in the world market, and are summarized by Wood (1985). Comments are included on the terms "slaty," "purple beans," "smoky," and "moldy."

## Model Ordinance

### 1. Definitions

*Cocoa bean*: The seed of the cocoa tree (*Theobroma cacao*, Linnaeus); commercially and for the purpose of this Model Ordinance the term refers to the whole seed, which has been fermented and dried.

*Broken bean*: A cocoa bean of which a fragment is missing, the missing part being equivalent to less than half the bean.

*Fragment*: A piece of cocoa bean equal to or less than half the original bean.

*Piece of shell*: Part of the shell without any kernel.

*Adulteration*: Alteration of the composition of graded cocoa by any means whatsoever so that the resulting mixture or combination is not of the grade prescribed, or affects injuriously the quality or flavour, or alters the bulk or weight.

*Flat bean*: A cocoa bean of which the cotyledons are too thin to be cut to give a surface of cotyledon.

*Foreign matter*: Any substance other than cocoa beans, broken beans, fragments, and pieces of shell.

*Germinated bean*: A cocoa bean, the shell of which has been pierced, slit or broken by the growth of the seed germ.

*Insect-damaged bean*: A cocoa bean the internal parts of which are found to contain insects at any state of development, or to show signs of damage caused thereby, which are visible to the naked eye.

*Moldy bean*: A cocoa bean on the internal parts of which mold is visible to the naked eye.

*Slaty bean*: A cocoa bean which shows a slaty colour on half or more of the surface exposed by a cut made lengthwise through the centre.

*Smoky bean*: A cocoa bean which has a smoky smell or taste or which shows signs of contamination by smoke.

*Thoroughly dry cocoa*: Cocoa which has been evenly dried throughout. The moisture content must not exceed 7.5 percent.[1]

## 2. Cocoa of Merchantable Quality

(a) Cocoa of merchantable quality must be fermented, thoroughly dry, free from smoky beans, free from abnormal or foreign odors and free from any evidence of adulteration.
(b) It must be reasonably free from living insects.
(c) It must be reasonably uniform in size,[2] reasonably free from broken beans, fragments, and pieces of shell, and be virtually free from foreign matter.

## 3. Grade Standards

Cocoa shall be graded on the basis of the count of defective beans in the cut test. Defective beans shall not exceed the following limits:

Grade I— (a) moldy beans, maximum 3 percent by count;
          (b) slaty beans, maximum 3 percent by count;
          (c) insect-damaged, germinated, or flat beans, total maximum 3 percent by count.

Grade II— (a) moldy beans, maximum 4 percent by count;
          (b) slaty beans, maximum 8 percent by count;
          (c) insect-damaged, germinated, or flat beans, total maximum 6 percent by count.

*Note*: When a bean is defective in more than one respect, it shall be recorded in one category only, i.e. the most objectionable.

The decreasing order of gravity is as follows:

— moldy beans;
— slaty beans;
— insect-damaged beans, germinated beans, flat beans.

---

[1] This maximum moisture content applies to cocoa in trade outside the producing country, as determined at first port of destination or subsequent points of delivery. The Working Party reviewed the ISO method for determination of moisture content and agreed that it could be used, when recommended by ISO, as a practical reference method.

[2] *Uniform in size*: As a guide not more than 12 percent of the beans should be outside the range of plus or minus one-third of the average weight. It is recognized, however, that some hybrid cocoa may not be able to meet this standard although fully acceptable to the trade.

## 4. Substandard Cocoa

All dry cocoa which fails to reach the standard of Grade II will be regarded as substandard cocoa and so marked (SS), and shall only be marketed under special contract.

## 5. Marking and Sealing

(a) All cocoa graded shall be bagged and officially sealed. The bag or seal shall show at least the following information:

Producing country, grade or 'SS' if substandard, and whether light or mid-crop,[1] and other necessary identification marks in accordance with established national practice.

(b) The period of validity of the grade shall be determined by governments in the light of climatic and storage conditions.

## 6. Recheck at Port of Shipment

Notwithstanding paragraph 5 (b) above, all cocoa so graded shall be rechecked at port within seven days of shipment.

## 7. Implementation of Model Ordinance

Method of sampling, analysis, bagging, marking and storage applicable to all cocoa traded under the above International Standards are set out in the following Code of Practice.

## Code of Practice

### A. Inspection

1. Cocoa shall be examined in lots, not exceeding 25 tons in weight.
2. Every parcel of cocoa shall be grade-marked by an inspector, after determining the grade of the cocoa on the basis of the cut test (see paragraph C below).

Grade marks shall be in the form set out in, and shall be affixed according to, Section ... of ...[2] and shall be placed on bags by means of a stencil or stamp (see also paragraph E below).

### B. Sampling

1. Samples for inspection and analysis should be obtained:
   (a) from cocoa in bulk, by taking samples at random from the beans as they enter a hopper or from the top, middle and

---

[1] Absence of a crop indication means main crop.

[2] i.e., the appropriate reference in national regulations.

bottom of beans spread on tarpaulins or other clean, dustfree surface, after they have been thoroughly mixed;

(b) from cocoa in bags, by taking samples at random from the top, middle and bottom of sound bags using a suitable stab-sampler to enter closed bags through the meshes of the bags, and to enter unclosed bags from the top.

2. The quantity of samples to be taken should be at the rate of not less than 300 beans for every ton of cocoa or part thereof, provided that in respect of a consignment of one bag or part thereof, a sample of not less than 100 beans should be taken.

3. For bagged cocoa, samples shall be taken from not less than 30 percent of the bags, i.e. from one bag in every three.

4. For cocoa in bulk, not less than five samplings shall be taken from every ton of cocoa or part thereof.

5. In importing countries samples for inspection should be taken from not less than 30 percent of each lot of 200 tons or less, i.e. from one bag in three. Samples should be taken at random from the top, the middle and the bottom of the bag.

## C. The Cut Test

1. The sample of cocoa beans shall be thoroughly mixed and then 'quartered' down to leave a heap of slightly more than 300 beans. The first 300 beans shall then be counted off, irrespective of size, shape and condition.

2. The 300 beans shall be cut lengthwise through the middle and examined.

3. Separate counts shall be made of the number of beans which are defective in that they are moldy, slaty, insect damaged, germinated, or flat. Where a bean is defective in more than one respect, only one defect shall be counted, and the defect to be counted shall be the defect which occurs first in the foregoing list of defects.

4. The examination for this test shall be carried out in good daylight or equivalent artificial light, and the results for each kind of defect shall be expressed as a percentage of the 300 beans examined.

## D. Bagging

1. Bags should be clean, sound, sufficiently strong and properly sewn. Cocoa should be shipped only in new bags.

## E. Sealing and Marking

1. After grading, each bag should be sealed with the individual

examiner's seal. The grade should be clearly marked on each bag. Bags should also be clearly marked to show the grading station and period of grading (week or month).

For these purposes the following measures shall be carried out:

(a) suitable precautions will be taken in the distribution and use of examiners' seals to ensure that they cannot be used by any unauthorised person;

(b) parcels shall be numbered consecutively by the official examiner with lot numbers from the beginning of each month. The parcel number or lot number will be stencilled on each bag in every parcel examined, in the corner nearest the seal;

(c) grade marks will be stencilled near the mouth of the bag.

## F. Storage

1. Cocoa shall be stored in premises constructed and operated with the object of keeping the moisture content of the beans sufficiently low, consistent with local conditions.

   Storage shall be on gratings or deckings which allow at least 7 cm of air space above the floor.

2. Measures shall be taken to prevent infestation by insects, rodents and other pests.

3. Bagged cocoa shall be so stacked that:

   (a) each grade and shipper's mark is kept separate by clear passages of not less than 60 cm in width, similar to the passage which must be left between the bags and each wall of the building;

   (b) disinfestation by fumigation (e.g. with methyl bromide) and/or the careful use of acceptable insecticide sprays (e.g. those based on pyrethrin) may be carried out where required; and

   (c) contamination with odors of flavors or dust from other commodities, both foodstuffs and materials such as kerosene, cement or tar, is prevented.

4. Periodically during storage and immediately before shipment, the moisture content of each lot should be checked. Molds can develop on beans with a moisture content higher than 8 percent. It is essential, therefore, that the drying process should reduce the moisture content of the beans to below 8 percent and the average moisture content should be brought down to between 6 percent and 7 percent. At moisture contents below 6 percent the beans tend to become brittle.

## G. Infestation

1. Cocoa beans may be infested with insects which have not penetrated the beans and whose presence is not revealed by the cut test which is employed for grading purposes. Such insects may subsequently enter beans or they may be involved in cross infestation of other shipments.
2. Therefore, when the cocoa is rechecked at port before shipment, as provided under paragraph 6 of the Model Ordinance, it should also be inspected for infestation by major insect pests. If it is found to be infested it should, before shipment, be fumigated, or otherwise treated to kill the pests. Care should be taken to avoid cocoa beans becoming infested in ships and stores from other commodities or with insects remaining from previous shipments.
3. If the use of insecticides or fumigants is necessary to control infestation, the greatest care must be exercised in their choice and in the technique of their application to avoid incurring any risk of tainting or the addition of toxic residues to the cocoa. Any such residues should not exceed the tolerances prescribed by FAO/WHO Codex Committee on Pesticide Residues and the FAO/WHO Expert Committee on Pesticide Residues and by the government of the importing country.
4. Rodents should as far as possible be excluded from cocoa stores by suitable rodent-proof construction, and where direct measures are necessary to control rodents the greatest care must be taken to prevent any possibility of contaminating the cocoa with substances which may be poisonous.

## Comments on Definitions

**Slaty Beans** These can be recognized by their characteristic slate color and cheesy texture. Chocolate made from them is dark gray in color, extremely bitter and astringent, and lacks chocolate flavor. Such beans have been dried before any of the processes of fermentation have been initiated. Slaty beans generally result from inadequate mixing, which allows beans at the surface to dry out. This defect is more likely to arise in heap fermentation where the quantities are small, the heap has not been turned over, and the beans are dried before fermentation is complete. The defect is rarely found when the recognized methods of large-scale fermentation are employed.

**Purple Beans**   While fully purple beans are objectionable because, like slaty beans, they produce bitter and astringent flavors, they are not usually classified as defective because there is a wide range in degrees of purple color and it is not practicable to define the various categories precisely.

Fully purple beans exhibit a bright purple color, which may be associated with the cheesy texture of slaty beans. Such beans are insufficiently fermented and are rarely found when normal large-scale methods are used with adequate quantities of beans.

Normal fermentation will, however, give a proportion of beans described as "partly brown and partly purple." These represent various degrees of fermentation between the inadequate level of "fully purple" beans and the full fermentation of the brown beans. Such beans are not defective; in fact, it is desirable that the samples should contain at least 20 percent of beans in this category. As the proportion increases, the bitter and astringent flavor of inadequately fermented cocoa will tend to predominate, and 50 percent should be regarded as the upper limit for this class of beans.

Overfermented beans are undesirable because they do not give a full chocolate flavor and may introduce off-flavors. They are not classified as defective because of the difficulty of definition. Some overfermented beans are dark brown in color; others tend to a pale yellow-brown color.

**Smoky Beans**   Beans contaminated by smoke have a most objectionable flavor, which is virtually impossible to remove from chocolate. The contamination can arise from the use of crude methods of drying or from defect in dryers that allow smoke to reach the beans. Methods of artificial drying have already been described and proper maintenance is required to prevent smoke damage.

**Moldy Beans**   The development of mold inside the bean is a most serious defect, because even a small percentage of moldy beans will cause the chocolate to taste musty. Moreover, it is now known that the growth of some molds can result in the presence of mycotoxins.

Molds will often develop on the outside of beans during drying if the beans are dried too slowly, and in storage if they are dried insufficiently.

## Consumption

The consumption of chocolate and confectionery products varies considerably from country to country. In recent years, the industry

TABLE 1.6. CHOCOLATE CONFECTIONERY CONSUMPTION (KILOGRAMS PER PERSON PER YEAR)

| Country | 1970 | 1975 | 1980 | 1981 | 1982 | 1983 | 1984 |
|---|---|---|---|---|---|---|---|
| Australia | 3.9 | 4.7 | 4.0 | 4.1 | 4.2 | 4.2 | 4.5 |
| Austria | 4.4 | 4.6 | 6.3 | 6.5 | 6.5 | 7.0 | 7.5 |
| Belgium | 5.3 | 5.2 | 6.0 | 5.8 | 5.7 | 6.0 | 6.0 |
| Denmark | 4.4 | 4.4 | 4.8 | 5.1 | 4.9 | 5.4 | 6.0 p |
| Finland | 2.0 | 2.2 | 2.4 | 2.7 | 3.0 | 2.9 | 3.2 |
| France | 2.7 | 3.2 | 4.0 | 4.0 | 4.1 | 4.0 | 4.1 |
| Germany (F.R.) | 4.8 | 5.7 | 6.6 | 6.5 | 6.1 | 6.2 | 6.8 |
| Greece | N.A. | N.A. | 1.5 e | 1.4 e | 1.4 e | 1.4e | N.A. |
| Ireland (Rep.) | 4.9 | 4.4 | 5.9 | 5.8 | 6.0 | 5.8 | 5.7 |
| Italy | 0.9 | 0.9 | 0.9 | 1.0 | 1.1 | 1.2 | 1.2 |
| Netherlands | 5.1 e | 4.0 | 4.9 | 5.0 | 5.1 | 5.4 | 5.0 |
| New Zealand | 3.5 | 3.9 | 3.3 | 3.3 | 3.7 | 3.8 | 3.8 |
| Norway | 5.0 | 5.4 | 6.7 | 7.0 | 7.1 | 7.5 | 7.6 |
| Sweden | 4.3 | 4.6 | 5.3 | 4.4 | 5.1 | 4.9 | 5.6 |
| Switzerland | 7.7 | 7.5 | 8.4 | 8.5 | 8.3 | 8.4 | 8.6 |
| United Kingdom | 5.7 | 5.9 | 6.5 | 6.7 | 7.3 | 7.6 | 8.0 |
| U.S.A. | 4.2 | 3.7 | 3.7 | 4.0 | 4.0 | 4.3 | 4.5 |

p = preliminary
e = estimated
*Source*: International Office of Cocoa and Chocolate (IOCC); International Sugar Confectionery Manufacturers' Association (ISCMA); Joint International Statistics Committee.

TABLE 1.7. SUGAR CONFECTIONERY CONSUMPTION (KILOGRAMS PER PERSON PER YEAR)

| Country | 1970 | 1975 | 1980 | 1981 | 1982 | 1983 | 1984 |
|---|---|---|---|---|---|---|---|
| Australia | 4.6 e | 4.8 | 3.8 | 4.0 | 4.1 | 4.0 | 4.2 |
| Austria | 2.3 | 2.3 | 2.4 | 2.5 | 2.5 | 2.4 | 2.3 |
| Belgium | 4.2 | 3.8 | 3.8 | 3.8 | 4.1 | 4.2 | 4.3 |
| Denmark | 4.2 | 4.4 | 4.6 | 4.9 | 4.9 | 5.0 | 5.1 |
| Finland | 4.5 | 3.6 | 4.1 | 3.7 | 4.2 | 3.8 | 4.0 |
| France | 3.3 | 3.1 | 2.8 | 2.8 | 2.9 | 2.8 | 2.7 |
| Germany (F.R.) | 4.4 | 4.8 | 5.7 | 5.8 | 5.7 | 5.6 | 5.8 |
| Greece | N.A. | N.A. | 0.7 e | 0.7 e | 0.6 e | 0.5 e | 0.5 e |
| Ireland (Rep.) | 5.2 | 5.6 | 5.3 | 6.2 | 6.3 | 6.4 | 6.3 |
| Italy | 2.0 | 2.1 | 2.2 | 2.0 | 1.9 | 2.0 | 2.0 |
| Netherlands | 5.6 p | 5.8 | 5.3 | 5.3 | 5.1 | 5.4 p | 5.4 p |
| New Zealand | 3.0 | 3.6 | 3.6 | 3.2 | 3.0 | 3.0 | 2.8 |
| Norway | 2.7 | 3.0 | 3.9 | 3.7 | 3.8 | 3.9 | 3.9 |
| Sweden | 5.2 p | 4.6 | 3.8 | 3.6 | 4.1 | 3.8 | 4.5 |
| Switzerland | 2.9 | 2.5 | 3.0 | 3.0 | 2.9 | 2.9 | 2.9 |
| United Kingdom | 5.4 | 5.5 | 5.2 | 5.1 | 5.0 | 4.8 | 4.9 |
| U.S.A. | 4.9 | 3.6 | 3.4 | 3.4 | 3.6 | 3.8 | 4.1 |

p = preliminary
e = estimated
*Source*: International Office of Cocoa and Chocolate (IOCC); International Sugar Confectionery Manufacturers' Association (ISCMA); Joint International Statistics Committee.

has been subjected to a great deal of adverse publicity, much of it founded on incomplete evidence.

Tables 1.6 and 1.7 show the amounts per person consumed in different countries.

## ACKNOWLEDGEMENTS

In the preparation of this chapter, I am indebted to friends who are specialists in the agronomy of the cocoa bean. I would particularly mention G. Ross Wood of Cadbury Schweppes, who also helped with the first and second editions.

More recently, I have been assisted by Dr. John West, professor of botany at the University of California, Berkeley.

For the provision of statistics, I am grateful for the help given by the Cocoa, Chocolate and Confectionery Alliance, London, England, which has supplied IOCC and ISCMA information, and Messrs. Gill and Duffus, PLC, London, England, for cocoa statistics.

## REFERENCES

Bainbridge, J. C., and Davies, S. H.   1912.   *J. Chem. Soc. London* 101, 2207.

Chocosuisse, Union of Swiss Chocolate Manufactures Münzgraben 6, Bern, Chocologie.

Dietrich, Lederer, Winter, and Stoll   1964.   *Helv. Chim. Acta.* 47(6), 1581–1590.

Dimick, P. S., and Hoskin, J. M.   1981.   Chemico-physical aspects of chocolate processing. *Can. Inst. Fd. Sci. Tech. J.* 14, 4.

Egan, H., Kirk, R. S., and Sawyer, R.   1981.   *Pearson's Chemical Analysis of Foods.* Churchill Livingston, Edinburgh.

Entwistle, P. F.   1972.   *Pests of Cocoa.* Longman, London.

Fincke, A.   1965.   *Handbuch der Kakaoerzeugnisse.* Springer, Berlin.

Flament, I., Willhalm, B., and Stoll, M.   1968.   *Helv. Chim. Acta.* 50, 2223.

Foster, H.   1978.   What is chocolate flavor? Manf. Conf., May.

Howat, G. R., Powell, B. D., and Wood, G. A. R.   1957.   Experiments on cocoa fermentation in W. Africa. *J. Sci. Fd. Agric.* 8.

Jensen, H. R.   1931.   *Chemistry, Flavoring, Manufacture of Chocolate, Cocoa, Confectionery.* Blakiston, Philadelphia.

Knapp, A. W., and Churchman, A.   1937.   *J. Soc. Chem. Ind.* (London) 56, 29.

Lang, M.   1978.   Two hundred and fifty years on. Conf. Manf. & Mktg., Oct. 27.

Marian, J. P., et al.   1967. *Helv. Chim. Acta.* 50, 1509.

Mohr, W.   1958.   *Fette, Seifen, Anstrichmittel.* (Germany) 60, 661.

Mohr, W.   1977, 1978.   *Rohkakaobohnen-Untersuchungen.* Gordian, Hamburg.

Morris, D.   1882.   Cacao. How to grow and how to cure it. Jamaica.

Padwick, G. W.   1956.   Losses caused by plant diseases in the tropics. Commonwealth Mycological Inst., Kew, England.

Rohan, T. A., and Stewart, T.   1967.   The precursors of chocolate aroma. *J. Fd. Sci.*

Rohan, T. A.   1969.   *The Flavor of Chocolate, Its Precursors and a Study of Their Reaction.* Gordian, Hamburg.

Rohan, T. A.  1969.  The flavor of chocolate. *Fd. Proc. Mktg.*, Jan.

Schmalfuss, H., and Bartmeyer  1932.  *Z. Lebensmitt. Untersuch.* (*Germany*) 115, 222.

Steinmann, A.  1935.  *Z. Lebensmitt Untersuch* (Germany) 69, 479.

Van der Waal, B., Sipma, B. G., Ketenes, D. K., and Semper, A. T.  1968.  *Rec. Trav. Chim.* 87, 238. Pays Bas (Holland).

West, J. A.  1983–84.  Private communications.

Wood, G. A. R., and Lass, R. A.  1985.  *Cocoa.* Longman, London.

Zoumas, B. L., Kreisler, W. R., and Martin, R. A.  1980.  Theobromine and caffeine content of chocolate. *J. Fd. Sci.* 45(2), 314–316.

# 2

# Cocoa Processes

## RAW-BEAN CLEANING

Raw cocoa beans, in general, are received in user countries in a reasonably clean condition and free from anything but minor amounts of surface grit, bag fiber, and occasional small stones. Insect infestation has been greatly reduced as a result of much better control in the country of origin and in warehouse storage elsewhere.

It is the custom in many chocolate factories to fumigate all incoming beans whether or not this might have been done previously, and this may lead to abnormal pesticide residues. Nevertheless, the first process that must precede the manufacture of chocolate or cocoa is that of cleaning.

The machinery consists of a series of operations, which—by means of screens of varying meshes, brushes, air lifts, and magnetic separators—removes fiber (from the jute sacks), stones and grit, metal, bean clusters, and immatures (Fig. 2.1).

Fine silliceous grit on the shells of some beans can be a great nuisance if it passes beyond the cleaning process, as it can appear in the finished product and increase the insoluble ash by as much as 0.3 percent, as well as contributing to machinery wear.

Immature beans and clusters should not be allowed to proceed to roasting and subsequent manufacturing processes, but may be included in bean blends with other winnowing products for expeller pressing to produce cocoa butter.

The dust from the cleaning operation will have very high microbiological contamination, and even though most of it will be extracted by an efficient cyclone, it is best for the equipment to be located in a separate area away from subsequent production processes.

## ROASTING

Roasting is a process applied to many foodstuffs as a means of developing flavor; it is really a form of cooking, although the latter

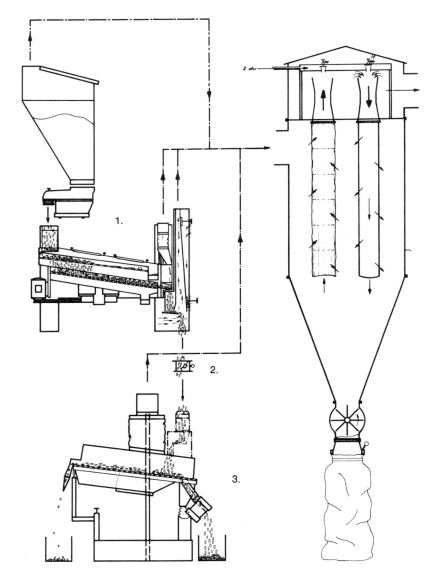

Fig. 2.1. Sequence of Operations in the Cleaning Process
1. Separator with sieves and aspiration channel. Sieves extract large pieces (e.g., bean clusters) and sandy matter. Aspirator removes fiber, string, and other light material.
2. Magnetic separator—removes iron particles.
3. Destoner—fluidized bed with air aspiration. Removes stones and other metal and heavy particles. Process air is also subject to cyclone separation.

expression is more appropriate if water in quantity is also involved. In some modern systems of roasting, however, a type of cooking does occur.

Cocoa beans are roasted to develop further the true chocolate flavor, which should already exist in the form of precursors arising from the correct fermentation and drying of the original beans. The effect of fermentation has already been discussed.

During the roasting of the fermented and dried beans, the following changes take place:

- The bean loses moisture.
- The shell is loosened.
- The nib (cotyledon) becomes more friable and generally darkens in color.
- There is some degradation of the amino acids, as shown by Rohan and Stewart (1966), and proteins are partly denatured. Akabori (1932) has shown that natural reducing sugars are almost completely destroyed during the degrading of the amino acids. These nonenzymatic "browning" reactions associated with cocoa roasting are reviewed by Foster (1978).
- There is a loss of volatile acids and other substances that contribute to acidity and bitterness. A large number of compounds has been detected in the volatile products, including aldehydes, ketones, furans, pyrazines, alcohols, and esters. The substances that undergo least change are the fats, polyphenols, and alkaloids.

The degree of change is related to the time and temperature of roasting and to the rate of moisture loss during the process. Roasting conditions vary greatly according to the machine and product required. Heat treatment, as an alternative to roasting, may proceed in two stages, see Fincke (1965). The first is a drying process in which the bean is subjected to low-temperature heating. This dries and loosens the shell but has virtually no roasting effect on the nib as the temperature is unlikely to exceed 100°C (212°F). This initial heating is followed by higher-temperature treatment, when the temperature may reach 125 to 130°C (257 to 266°F). This temperature is related to the requirements of subsequent processing, that is, whether the manufacture of chocolate or cocoa powder or the expression of cocoa butter. In some batch roasters, temperatures may rise to 150°C (302°F), or higher.

Some chocolates, particularly milk chocolate, require lower roasting temperatures, and some cocoa powders, if red shades are required, also need low-temperature roasting. To produce mild-flavored cocoa butters, very low temperatures are best, and when using the expeller press, unheated raw beans may be used.

As an alternative to whole-bean roasting, some members of the industry prefer roasting the nibs. This entails the drying or very-low-temperature roasting of the bean, which dries and loosens the shell, followed by winnowing. Then the nib is roasted separately, and as the nib particles are much smaller than the beans, they are more rapidly and evenly heated. The NARS process (described later) uses this principle.

The development of roasting processes and machinery is of interest. On some primitive equipment, metal trays containing the product to be roasted were suspended by chains over a fire.

Later, rotating drums heated externally were used, with provision for the escape of unwanted volatiles. Some manufacturers still believe that these give the best results from the standpoint of flavor.

Then it was considered that indirect heating had advantages, as it gave a more even roast and avoided burning the shell. However, these systems were often very extravagant in fuel use; one such machine, long since abandoned, consisted of enclosed water-filled steel tubes passing through a rotating cylinder. The ends of the tubes projected into a coke furnace, and the beans were roasted by the heat conducted along the tubes to the rotating cylinder.

Later, the Sirocco-type roaster was developed, which consists of a bowl or rotating drums through which preheated air is blown, causing turbulence and thus consistent heating of the individual beans. After roasting, the beans are discharged onto a perforated metal plate through which cool air is passed. This type of roasting is still in use in many companies, not only for cocoa beans but also for nuts and coffee (Fig. 2.2).

Many designs of continuous roasters are now available, and these have shown great economies in fuel consumption, and also result in fewer broken beans and less transfer of cocoa butter from the nib to the shell. Kleinert (1966) investigated various roasting systems and concluded that batch or drum roasting has no superiority over continuous methods. Most of these machines have a built-in cooling system that comes into play after the roasting cycle. Figures 2.3, 2.4, and 2.5 illustrate the principle of some modern roasters.

The determination of the optimum end point of the roast for the best results in subsequent processes is still subject to personal judgment.

Beans of different origin—e.g., Ghana, Ivory Coast, Malaysia, Brazil—but of similar type (Forastero) will vary in size, moisture

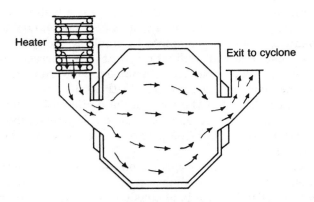

*Diagram of air flow*

Fig. 2.2.  Sirocco-Type Batch Roaster

Roaster relies on the principle of drawing hot air through the beans, thereby reducing the risk of burning the outer shell. The air is heated by gas or oil burners. Cyclones are used to extract the dust from the exit air.

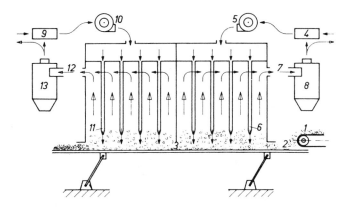

Fig. 2.3.   Operation of a Fluid Bed Roaster, Where the Beans Are Heated by Convection in a Current of Hot Air

1. Raw Bean Feed
2. Vibrators
3. Fluid Bed Zone
4. Air Heater
5. Air Blower
6. Hot Air Inlet Nozzle
7. Spent Air Outlet
8. Dust Separator
9. Air Cooler
10. Air Blower
11. Cold Air Inlet, Nozzle
12. Return Air
13. Dust Separator

*Kleinert, 1966*

content, and shell content. Therefore, it is not practical to establish exact conditions in roasting even with one type of roaster.

In the operation of a roasting plant, it is still customary to decide when the roasting process is complete by smelling a crushed sample, following the instrumental indication that the correct time and temperature have been reached. Obviously, this requires skilled judgment on the part of the operator. With skilled blending of the raw beans to the extent that the contents of a silo with a capacity of, say, fifty tons can be regarded as of similar quality throughout, then, after the initial roast, a time/temperature program may be established for that quantity.

Reade (1983) has made observations on methods of roaster control and has concluded that an estimation of the moisture/equilibrium relative humidity (ERH) ratio is of considerable value. A hygrometer probe is used to determine the ERH. Samples of beans taken at intervals from the roaster are quickly crumbled in a mechanical grinder, transferred to a closed container, and the probe is inserted. He has established an approximate scale as given in Table 2.1.

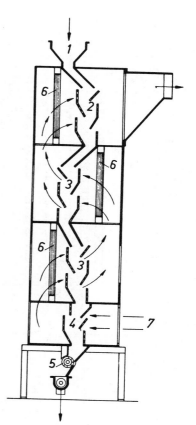

Fig. 2.4. Section Through a Probat-Roaster, Where the Beans Are Prewarmed, Roasted, and Cooled Entirely by Convection
1. Feed Hopper
2. Prewarming Stage
3. Roasting Stage
4. Cooling Stage
5. Outlet
6. Air Heater
7. Cooling or Cold Air Inlet

*Kleinert, 1966*

An ERH of 40 percent or above indicates a very light roast, of 28 to 30 percent is an average roast common to many operations, and below 20 percent means a very dark roast.

The table will only apply as a measure of control of a given process. For a different machine or process the effect on the flavor of the roasted beans would be different.

## The NARS Process

Nibs, alkalizing, roasting, and sterilizing (NARS) is a more recently developed method of heat treating of cocoa beans. The process is described fully in a paper by Mayer-Potschak (1983).

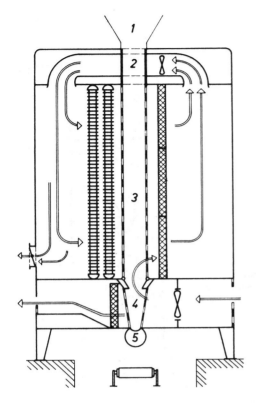

Fig. 2.5. Diagrammatic Section Through a Buhler STR 2 Roaster, Where the Beans Are Prewarmed, Roasted, and Cooled Entirely by Convection
1. Feed Hopper
2. Preheating Stage
3. Roasting Stage
4. Cooling Stage
5. Off-take Mechanism

*Kleinert, 1966*

It consists of two main operations. First, the whole beans are treated by infrared and radiant heat using a machine called a Micronizer. This dries and loosens the shell of the beans without significantly affecting the nib, and is followed by a normal winnowing process. In the second part of the process, the separated nib is

TABLE 2.1. MOISTURE/ERH RATIO

| ERH, % | Moisture content, % |
|--------|---------------------|
| 70 | 7 |
| 48 | 4 |
| 38 | 3 |
| 24 | 2 |
| 12 | 1 |

ERH readings taken at 25°C (77°F).

subjected to a treatment in which the nib first is heated at a relatively low temperature and then roasted at a higher temperature. Finally, a fine water spray is applied to the hot nib, which induces moist air capable of destroying all microorganisms, including heat-resistant spores.

## Micronizer

This Micronizer is illustrated (Figs. 2.6 and 2.7). During this treatment, the raw cocoa beans are subjected to radiant heat for an interval ranging from 60 to 120 sec, the energy being supplied either by gas or by electricity. The beans are conveyed in a thin layer beneath the heaters along a vibrating steel trough, with the motion causing the beans to be continuously turned so that all surfaces are exposed equally.

With this method, the shells of the beans are rapidly heated and they dry, expand, and detach themselves from the nibs. The nibs are heated and dried to a much lower degree but slight expansion assists in loosening the shells. Because the nib remains tough, there is less breakage and less dust is formed in contrast to traditional roasting processes in which the nib is more intensely heated and dried. This

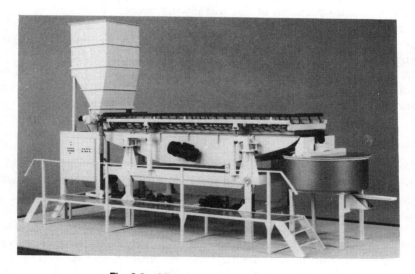

Fig. 2.6.   Micronizer

*Micronizing Co. (U.K.) Ltd., Suffolk, England*

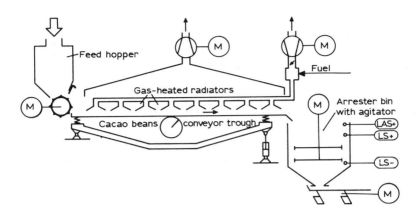

Fig. 2.7. Diagram of Micronizing

*Micronizing Co. (U.K.) Ltd., Suffolk, England*

property helps in the winnowing process as it is well recognized that the smaller the percentage of fine particles or dust in the broken beans, the higher will be the yield of pure nib.

The efficient operation of the winnowing process is described later.

The micronizing treatment also contributes to the reduction of microbiological contamination, including the destruction of rodent hairs and insect fragments.

A further advantage of micronizing is the decreased transfer of fat from the nib to the shell, which, in some roasting processes, can represent a considerable loss of valuable cocoa butter.

## Roasting of Nib

Fundamental to the NARS process is the initial drying of the nib, which has a high moisture content. This condition in the nib results from the micronizing treatment, which leaves the nib containing 5 to 6 percent moisture.

Mohr (1970) had shown that cocoa beans should be dried at a relatively low temperature before roasting if the best flavors are to be obtained. In the initial drying, the natural moisture distributed evenly throughout the nib provides an ideal condition for the removal of volatile substances detrimental to the final chocolate aroma.

In the NARS Tornado roaster (Fig. 2.8), heat is applied externally

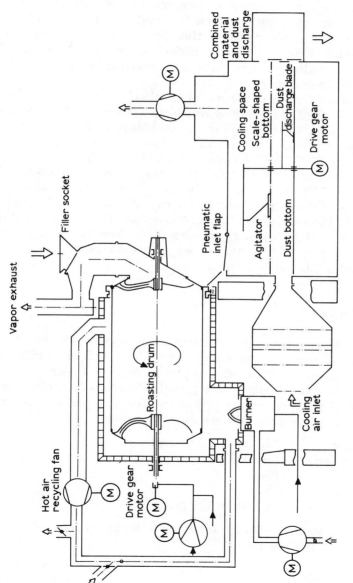

Fig. 2.8. Tornado Roaster—2600RS

G. W. Barth, Ludwigsburg, W. Germany

45

to the rotating drum and the nib is heated and dried by a combination of conduction and natural convection. The drying proceeds evenly, allowing the moisture and volatile substances to diffuse from the inner parts of the nib pieces. With forced air roasting, which is found in many machines, the surface of the nibs dries rapidly, resulting in a type of case hardening, which retards the drying of the internal parts.

In the Tornado roaster, the initial heating is for about 10 min at just below 100°C (212°F), during which time the moisture is reduced to between 2 and 3 percent. In the second stage of the roasting, the temperature is raised according to the flavor development required, and may reach 130°C (266°F) over a period of 15 to 20 min.

Having determined the roasting conditions required to provide the correct aroma, the machine can be set to the specific time and temperature. These are calculated by measuring the product temperature, the heating air temperature outside the drum, and the temperature of the outgoing air. The roast does not change with variations in size of the nib pieces as would be the case with conventional machines using beans or nibs.

## STERILIZATION

Although most microorganisms will have been destroyed in the early stages of roasting under moist conditions, a final sterilization ensures the destruction of heat-resistant bacteria and spores.

The procedure is to inject, over a period of 20 sec, a fine water spray into the roasting drum at the end of the roasting period. This will guarantee the following conditions in the final nib:

| | |
|---|---|
| Plate count | Less than 100 per gram |
| Entero bacteria | Less than 10 per gram |
| *E. coli* | None |
| *Salmonella* | None |
| Spores | Less than 100 per gram |

The roasted nibs are cooled by introducing sterile air that has passed through a bacteria filter.

## ALKALIZATION

Alkalization is used mainly to produce changes in the color of the nibs, and hence in the liquor or cocoa powder subsequently produced. Solutions of potassium carbonate are usually employed.

Alkalization is discussed in detail later, but in the NARS process, the solution of alkali, preheated, is sprayed into the drum after it is charged with the nibs. The alkalized nibs are then slowly dried at a temperature below 100°C (212°F). This develops color and is followed by roasting at higher temperatures, as previously described. Slow drying in the initial stages is an important factor.

## COCOA LIQUOR TREATMENT

It is possible to heat-treat liquor, a process that is regarded by some chocolate manufacturers as an alternative to or partial replacement of roasting. Also, it is used to reduce or eliminate conching (discussed later in the chapter).

## WINNOWING (CRACKING AND FANNING)

The valuable part of the cocoa bean is the nib, the outer shell being a waste material of little value. To separate the shell as completely as possible is the aim of every manufacturer of winnowing machinery.

A typical composition obtained by hand separation of raw cocoa beans containing 6.5 percent moisture is: nib (cotyledon), 87.1 percent; shell, 12.0 percent; germ, 0.9 percent. The moisture of the raw shell may be 8 to 10 percent and the raw nib 4 to 5 percent.

After roasting, the nib moisture will be reduced to 1.5 to 3 percent (depending on the degree of roast), and the shell moisture to about 4 percent. The overall moisture loss from the original bean may average 5 percent, and assuming the shell is wholly removed by winnowing, the total loss is 17 percent. Thus, the yield of the roasted nib should be 83 percent. Because the separation in the winnow is not perfect, the nib is likely to contain 1 to 1.5 percent shell, which means that pure nib yield is under 82 percent.

The germ is mostly present in the finer nib fractions from the winnow. In the large nib fractions, the shell content is likely to be as low as 0.5 percent. Where there is very good control of roasting and winnowing, yields of nibs with 1.0 percent shell content and 1.5 percent moisture have been reported to range between 82.5 percent and 83.5 percent.

Because of the high cost of raw cocoa, a great deal of research has been carried out to improve yields. The mechanics of winnowing machines have been improved and methods of pretreatment and roasting of the beans have been investigated with the objective of reducing the amount of fine particles at the cracking stage and a reduction of the transfer of cocoa butter from the nib to the shell.

These improvements have already been mentioned in the description of the Micronizer. This machine is capable, in most cases, of increasing nib yields by 1 to 1.5 percent, and considerably higher than this with some grades of beans where the shells are firmly adhering.

## Winnowing Machines

The principle of separation by the winnowing process depends on the difference in the apparent density of the nib and shell. Winnowing machines make use of the combined action of sieving and air elutriation. The shell is loosened by roasting or partially drying the beans, which are then lightly crushed with the object of preserving large pieces of shell and nib and avoiding the creation of small particles and dust. The older winnows used toothed rollers to break up the beans but modern machines are fitted with impact rollers. These consist of two hexagonal rollers running in the same direction that throw the beans against metal plates. There has also been a change in the design of the sieves and the method of air separation.

The older winnows used either rotating cylinders or flat vibrating sieves, both having graded mesh sizes. Thus, a cascade of broken beans of various sizes was produced that met air flows from several channels, each capable of velocity adjustment appropriate to the size of the falling particles. This gives maximum efficiency of separation, provided adjustments are made correctly.

The new machines use a multilayer sieve frame with meshes of different sizes, one above the other, the largest mesh at the top. The shell pieces are removed by pneumatic suction at the overflow of each sieve, with the remaining nib pieces being directed to chutes at the side of the machine. The sieves are kept free from blockage by means of the vibratory movement and a system of raking. The fine dusts are collected in a cyclone system following the winnow. The object of the winnowing process is to produce two basic fractions—the nib, which contains the minimum of shell and germ, and the shell portion, which in itself can be divided into various grades that should be free from significant amounts of nib.

It was previously mentioned that nib as produced commercially may contain 1.5 percent of shell, and probably a small amount of germ. In the manufacture of superfine cocoa, shell and germ produce particles that settle rapidly in a drink.

Germ separators have been fitted to some winnows but generally they are not very efficient. To obtain a nib with minimum shell and germ content, it is possible to divert the streams of nib and shell from the different winnow exits. The largest nib has the least shell and germ. The germ is a very hard substance, but with modern refining machinery, it can be well ground and does no harm in most chocolate formulations.

By adjusting the winnow setting or by diverting coarse nibs from the first delivery chutes, it is possible to obtain nibs with shell contents of less than 0.5 percent and very little germ. This is very suitable for the manufacture of superfine cocoa powder. When the older type of winnowing machine was in use, it was a frequent practice to use another winnow with a different adjustment to separate further the fine material discharged from the winnow exits.

Figure 2.9 shows the general construction of a Bauermeister winnow.

It is obvious from the foregoing description that a winnow needs regular and careful maintenance. Checks on shells in nibs and nibs in shells should be made several times daily. Also, it is useful periodically to check the fat contents of the material from the different outlets, and from these figures a good assessment of the proportions of shell to nib can be obtained on the basis of 54 percent fat in the nib and 2.5 percent in the shell.

In a badly maintained winnow, the nib yield can fall below 80 percent (instead of 83 to 84 percent) and it has been calculated that for each loss of 1 percent nib, the cost of the original bean is increased 1.25 percent. In addition, shell contents of nibs much above 1.5 percent can have serious effects on quality.

An excess of fine particles of nib and shell will result in bad yields. When the particle size is 2 mm or less, air separation will fail to distinguish between nib and shell. Therefore, the roasting method and the breaking of the beans become very important. In a well-managed roasting and winnowing plant, the nib pieces over 3 mm in size should account for 85 to 90 percent.

The following representative analyses of fractions from different winnow exits are of interest and, as mentioned above, regular monitoring of samples is useful to judge the performance of the winnow and preceding processes.

**Large nib:**

| | |
|---|---|
| Moisture | 2.0 to 3.5 percent (depending on degree of roast) |
| Cocoa butter | 52.5 to 55.5 percent |
| Shell | 0.2 to 1.5 percent (depending on winnow setting) |
| Germ | 0.1 to 1.5 percent |

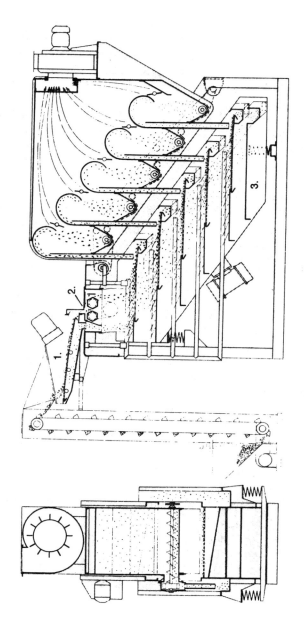

Fig. 2.9. Principle of Operation of Winnowing Machine
1. Vibratory Sieve to Separate Broken Beans
2. Impact Roller Breaker
3. Vibratory Sieves and Air Lifts

*Reproduced by permission of Bauermeister Maschinen Fabrik, Hamburg, W. Germany*

This is the clean pure cotyledon of the bean used for the best-quality chocolate and cocoa manufacture.

**Small nib:**

Moisture            3.5 to 6.0 percent
Cocoa butter     35.0 to 48.0 percent

Small nib mixed with some germ and shell dust. Sometimes subjected to further separation when the clean nib "smalls" obtained can be mixed off with large nib for second-grade chocolate or cocoa, but is often used in blends for expeller pressing and solvent extraction.

**Nib dust (fine nib):**

Moisture            3.8 to 7.5 percent
Cocoa butter     30.0 to 36.0 percent

Not suitable for use in chocolate or cocoa manufacture. Contains fine shell and germ and is likely to contain gritty matter up to 1.5 percent. Mixed in blends for expeller pressing and solvent extraction.

**Shell dust, small shell, cyclone dust:**

Moisture                                        7.0 to 9.0 percent
Cocoa butter (including shell fat)     5.0 to 17.0 percent

Very variable product, mostly fine shell and nib dust, likely to contain gritty matter.

Not suitable for human consumption but can be further separated into large shell and shell/nib-dust mixtures, the latter being suitable for mixing into blends for expeller pressing, and solvent extraction.

**Large shell:**

Moisture     8.0 to 10.00 percent
Fat              2.0 to 3.0 percent

Shell fat is of different composition from cocoa butter. The properties of these fats are described later.

**Cocoa Shell** Shell is of little commercial value but various proposals have been made to make some use of this by-product. On the basis of an annual world production of 1,200,000 tons of cocoa beans, about 140,000 tons of shell are available.

Knapp and Coward (1935) showed the presence of vitamin D in cocoa shell, and this was followed by Kon and Henry (1935), who proved an increase in vitamin D in milk when a proportion of shell was included in fodder. Its use in food for animals is severely limited by the presence of theobromine, which is poisonous to many animals and particularly to poultry, and it has been stated that theobromine intake should not exceed 0.027 g/kg body weight, see Bartlett (1945).

Shell is used in some cocoa and chocolate factories as an auxillary fuel and has been proposed as a filler for plastics. It also has some value as a constituent of fertilizers because of its nitrogen, potassium, and phosphorous content and its organic bulk, which provides humus.

Shell has generally been considered of no value in the human diet and because of its high cellulose content, it is virtually indigestible. Recently, dietary fiber has come into prominence as a necessity to provide bulk in the intestines.

A patent by Kleinert (1983) describes a method utilization of cocoa shell to produce dietary fiber. The alkaloid content is reduced by washing. Because of its high cellulose content, the crude fiber is some three times that of nib—calculated on the fat-free material. This is an important factor in the determination of cocoa shell in cocoa powder and mixtures where it is considered an adulterant. The method is described by Pearson (1981). Another method for the determination of shell is the "stone cell" count by Van Brederode and Reeskamp. This is described in detail by Meursing (1983).

**Expeller Cake**   Some of the products of winnowing previously described are subjected to expression by expeller (screw) presses, which removes a part of the cocoa butter. An expeller cake can be produced with a fat content as low as 8 percent.

The residual fat, which contains a small proportion of shell fat, may be solvent extracted and used with cocoa butter for certain chocolates or for confectionery. The defatted cake, which contains some shell residue, is usually destroyed but sometimes is used in fertilizer or for the extraction of theobromine. Expeller or extrusion presses are described later.


## NIB GRINDING, COCOA LIQUOR, HEAT TREATMENT

There are many machines for reducing cocoa nibs to liquor. The nib has a cellular structure with the fat (cocoa butter approximately 55 percent) in solid form locked within the cells. In the grinding process, the cell walls are ruptured. Frictional and applied heat melts the fat, the particle size of the nonfat constituent is reduced, and the paste becomes progressively more fluid. The viscosity of the liquor is related to the degree of roasting preceding the grinding and to the moisture content of the nib.

Proteins in the nib are degraded to a varying degree in the roasting process and the moisture present is likely to be reduced in the grinding process.

For many years, the machines consisted of stone mills, with three pairs of horizontal stone disks set in tiers. Each pair consisted of a lower stationary stone and an upper rotating stone, and the distance setting between the stones controlled the fineness of grinding and output.

The nib, or partly ground liquor, was fed to the center of the stones and grinding took place as it worked toward the periphery. Grooves were cut in the face of the stones to improve grinding, and these needed constant renewing. An improvement was the introduction of aloxite stones, which were subject to less wear, but there was a gradual adoption of grooved steel disks, and these machines were called liquor mills.

Early models had two vertical disk mills, the first a breaker, trickle fed with nibs, and the second a finisher which produced the fine liquor. Each mill had a fixed and a rotating steel disk and the gap between them was adjustable to give the required output and fineness.

A later model has two vertical disk mills, each with two fixed plates and one rotating disk, both gaps being adjustable.

The original stone and disk mills subjected the liquor to very high temperatures, which gave strong flavors to the liquor and expressed cocoa butter. This was detrimental in the preparation of delicately flavored chocolates, particularly milk chocolate.

Modern liquor mills are equipped with horizontal millstones in three tiers. Whole or partly ground nibs are fed to the center of the top mill and the first liquor issues from the periphery. This is fed to the center of the second mill. Liquor from this mill is fed to the third mill for the final grind. Provision is made to cool the mills to prevent overheating of the liquor and the millstones are capable of very precise adjustment. The mills have outputs up to 2,600 lb/hr and are capable of producing very fine liquor (99 percent through 325-mesh petroleum sieving). A diagram of one of these mills is shown in Fig. 2.10.

Another method of grinding is by the use of pin or hammer mills

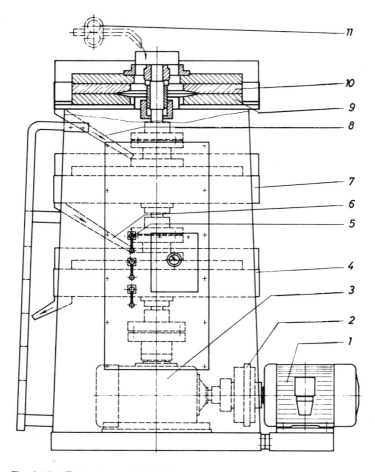

Fig. 2.10. Triple Liquor Mill CMA.

1. Drive Motor
2. Turbo Coupling
3. Reduction Gear
4. Lower Pair of Millstones
5. Pneumatic Distancing Mechanism
6. Mass Outlet Channel

7. Center Pair of Millstones
8. Mass Outlet Channel
9. Lower Millstone
10. Upper Millstone
11. Mass Pump

mounted above three-rolls refiners. A machine of this type is shown in Fig. 2.11.

The pin mill consists of two disks, each fitted with hard steel pins. These disks rotate in opposite directions and are fed with nibs from above by means of a controlled speed screw conveyor. The feed is equipped with a magnetic device to extract any ferrous material that might be in the nibs.

The speed of the disks and pin distribution are such that all nib particles are pulverized, and in this stage of the grinding, the nib is reduced to a mobile paste with particle sizes of the order of 90 to 120 $\mu$m.

Fig. 2.11.   Cocoa Mill Type MPH 411

*Carle and Montanari, Milan, Italy*

The paste then passes into a distribution chamber, which has several outlets and thereby spreads the liquor evenly along the length of the feed roll. The roll pressures are controlled hydraulically and the cooling water to each roll is thermostatically regulated. Instrumentation on the side panel indicates the performance of these controls.

The technical aspects of liquor grinding are discussed at length by Boller (1980). In this paper, the advantages of a combination of impact and shear mills followed by a three-roll refiner are described. It is stated that very low viscosities are obtained, which is not likely with single-stage grinding.

## Ball or Bead Mills

A modern development in the grinding of nibs to liquor is a machine that may be described as a vertical ball mill. The basis of the system consists of a cylinder that contains a rotor and grinding balls of special steel.

The nib is first partly ground to a mobile paste in one of the systems already described; a single shear mill is usually sufficient. This is pumped into the base of the cylinder. During its passage, it is subjected to the grinding action of the balls set in motion by the rotation of the spindle and agitators. There is a progressive reduction in particle size as the liquor traverses the cylinder, and at the top it is separated from the grinding balls by means of a screen.

**The Wieneroto** This ball mill is used extensively where cocoa is grown and processed to liquor, cocoa butter, and kibbled cake, and much research has gone into its development. The refining action that occurs inside the mill is a function of movement of the balls, their size, and the speed of the agitator arms. The particles that pass between the balls are subjected to frictional and impact forces.

The revolution of the agitator is kept relatively low as high speeds cause wear on the balls, agitator arms, and tank. The balls are of specially hardened steel, and with the present speed of operation and the design, metallic contamination is negligible.

A diagram of a Wieneroto mill is shown in Fig. 2.12 and the machine and process are described by Tadema (1983). These mills are now incorporated in a chocolate-making process that is described later (Wiener process).

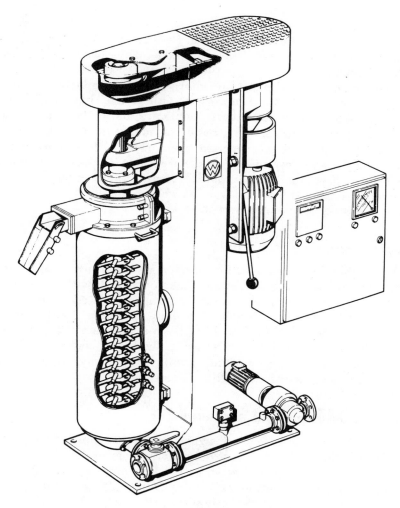

Fig. 2.12.   Wieneroto W45C
*Wiener and Co, Apparatenbouw b.v., Amsterdam, Holland*

## PROCESSING OF COCOA LIQUOR

### Flavor Development Processes

The alkalization of nib, liquor, and cocoa cake is described later. This section deals with the treatment of liquor to remove unwanted flavors and to reduce subsequent conching time.

The production of liquor is carried out to an increasing extent in countries where the beans are grown. Because of this, the purchaser of the liquor generally has little knowledge of the origin and quality of the beans used.

Some liquors are acidic in character, particularly those of Malaysian and South American origin. Others may have a smoky or hammy character and, regrettably, some are distinctly moldy. Little can be done to improve liquor that is obviously moldy. Also, liquor made from inadequately fermented beans shows little improvement from the treatment described. Various proposals for the deflavoring of liquor have been made and some processes have resulted in the removal of some of the good flavors as well as the bad.

The following is a summary of the methods used.

## Simple Heating of Thin Films

This method has been used for many years and consists of heating a thin film of the liquor in a steam-jacketed trough with an internal scroll or by spreading it on an internally heated drum. This treatment has also been applied to dark chocolate. Temperatures range from 80 to 110°C (176 to 230°F). An adaptation of this process is to disperse 2 to 3 percent moisture in the liquor before treatment. The evaporation of the moisture will carry away some of the unwanted flavors, which are volatile in steam.

## Thin-Film Roasting and Air Scrubbing

Several systems have been designed on the principle of subjecting thin films of liquor to streams of hot air and moisture. A development by the Petzholdt Company known as the Petzomat is a machine that can apply various treatments to thin films of cocoa liquor. The method, which is applied to the liquor in three separate towers, is as follows.

Liquor is continuously pumped into the top of column 1, where the paddles of the internal rotor create a turbulence in the thin liquor film as it progresses down the column. As the liquor passes down the tower, it is swept by a stream of hot air and the process is repeated in columns 2 and 3. The liquor is transferred mechanically from one column to the other.

Each tower is jacketed and heat may be applied by steam or hot water.

In addition to the application of hot air, the Petzomat has now been

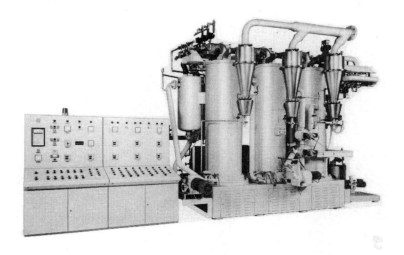

Fig. 2.13.   Petzholdt Injection Unit Type PIA

*Petzholdt GmbH, Frankfurt am Main, W. Germany*

applied to alkalizing and sterilization processes and to the treatment of liquor/milk powder/water systems. Also, when sugar syrups are included, flavors associated with the Maillard reaction are possible. The plant and process are fully described in Petzholdt brochures and in technical presentations, see Obering and Schmitt (1983). Figures 2.13 and 2.14 show the three-tower machine and the internal construction of one tower.

## Thin-Film Roasting with Vacuum

A system that uses a climbing film evaporator, with or without vacuum, is described by Bauermeister (1981). More recently, a vacuum process has been applied to the well-known beater blade mill (Bauermeister), so that the process of grinding can be combined with a roasting procedure similar to that described above.

Bauermeister has also discussed the relationship between the degree of roast and the methylpyrazine content of the liquors. This has been investigated by a number of workers. It is stated that the

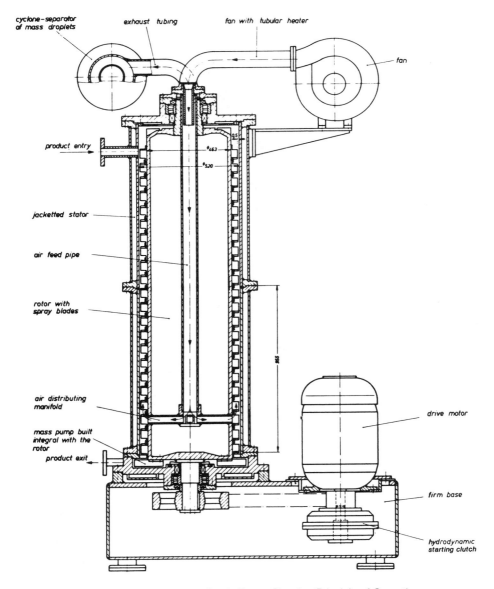

cyclone-separator of mass droplets

exhaust tubing

fan with tubular heater

fan

product entry

jacketted stator

air feed pipe

rotor with spray blades

air distributing manifold

mass pump built integral with the rotor

product exit

drive motor

firm base

hydrodynamic starting clutch

Fig. 2.14.   Petsomat—Single Tower Showing Principle of Operation

*Petzholdt GmbH, Frankfurt am Main, W. Germany*

roast flavor is indicated by the tetra- and 2,5-dimethylpyrazine content, which increases as the roasting temperature increases.

Another liquor-processing system is ,the PDAT/I deodorizing and pasteurizing plant by Carle and Montanari. It uses a vacuum process but, in addition, the liquor, in spray form, is treated with an inert gas (nitrogen or carbon dioxide), which removes oxygen from the system. In this plant, the liquor is pasteurized and deacidified but may also be subjected to flavor development by the addition of sugar syrups (Maillard reaction).

Roasting in spray form takes place and alkalization may be included. By adjusting the alkalization process, it is claimed that a range of colors may be obtained.

## General Comments

Methods of liquor roasting and other flavor development have been described. It must be appreciated that their application requires experimentation on the part of the chocolate manufacturer in relation to the particular raw materials and final flavor objectives. In most cases, there will be a reduction of moisture content, acidity, and microbiological count. Loss of good flavors is also a possibility, and some workers say this may occur with vacuum treatment.

## ALKALIZATION

The method was first used by Van Houten in 1828 and cocoa so treated may be described as "alkalized," "Dutched," or "soluble," but the last term is incorrect as there is no increase in the solubility of alkalized cocoa, but some evidence of improved dispersability or suspension in water is apparent.

Alkalization is used mainly to change color, and the process consists of treatment of the beans, liquor, nibs, or powder with solutions or suspensions of alkali, usually potassium or sodium carbonate.

Other alkalis may be used, such as sodium or potassium bicarbonate or hydroxide, calcium hydroxide, or ammonium carbonate or hydroxide. Of these, ammonium compounds can be used for powder treatment and excess ammonia is driven off by heat—the other alkalis have little to commend them for normal cocoa manufacture.

Alkalization affects flavor as well as color, but it is dubious whether there is any improvement. Some claim the flavor is strengthened; others think the true chocolate flavor is lost. The difference of opinion probably lies in how the alkalized product is to be used—whether as a drink or as an ingredient in a coating or a cookie.

In many countries, the amount of alkali and the chemical compounds to be used are controlled by law, but the way in which they are used is not. The permitted maximum is normally 2.5 to 3 parts of potassium carbonate (or equivalent alkali) per 100 parts of nib. Sometimes, a maximum ash content in the fat-free matter is specified. Some countries permit the use of food acids, added after treatment, to control pH and to assist in the production of reddish shades, but this effect is negligible compared with that produced by dilute alkali solutions.

The chemical reactions occurring during alkalization are not precisely known—there is certainly more than neutralization of the free acids but no saponification. That polyphenolic substances are modified is shown by the color variations that can be obtained and there is physical swelling of the cocoa particles. Some protein breakdown also occurs.

The alkalization process obviously raises the pH, which, for most fermented beans, ranges between 5.2 and 5.6. The change resulting from alkalization depends on the quantity of alkali used, but most alkalized cocoas or liquors have a pH between 6.8 and 7.5. Certain special cocoas, such as the so-called black cocoa, have a pH as high as 8.5, and are used for pigmentation. They usually have a very harsh flavor.

## Alkalization of Nibs

This method of alkalization, which has been the subject of much research, uses whole nibs and was supposed to have been developed in Holland.

The original process used tanks in which the nib was soaked in warm alkali solution until complete penetration was achieved. After soaking, the wet nib had to be handled into driers and roasters—an expensive, cumbersome, and variable process.

Today, equipment is available that alkalizes, dries, and roasts in a continuous process and the machinery is so designed that the fragile wet nib is not broken down to give sludge and fat separation, which can cause serious trouble in conveying systems.

Figure 2.15 shows the Buhler LBCT system of continuous alkalizing. A nib subjected to alkali solutions requires considerable time for the liquid to be absorbed, but eventually will take up half its own weight of liquid. The Buhler system treats the nib under pressure,

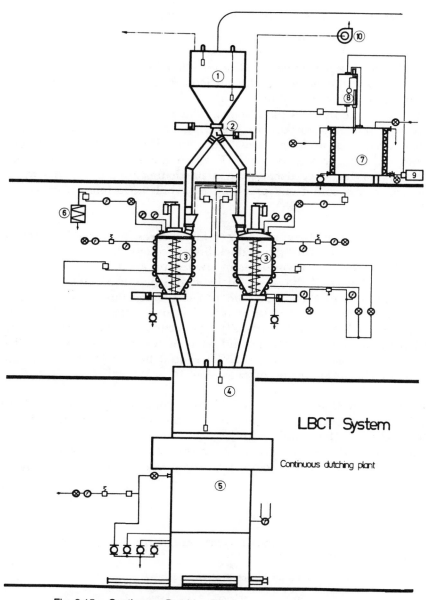

Fig. 2.15. Continuous Dutching Plant

1. Prestorage Bins for Nibs
2. Pneumatically Operated Discharge Device
3. Reaction Vessel for the Treatment of the Nibs
4. Feeding Bin for Drier
5. Drier "STR-2 Special"
6. Steam Condensate
7. Dissolving and Storage Tank for Alkali Solution
8. Dosing Device for Alkali Solution
9. Pump for Alkali Solution
10. Waste Air Exhaust

*Reproduced by permission of Buhler-Miag, Switzerland*

63

thereby getting better penetration of alkali in a short ime. It is also possible to use stronger alkali solutions. The conditions of treatment ensure a low bacteria count.

Certain fundamental conditions in the process control the color and flavor of the final cocoa. The following are results of experimental work carried out under ambient conditions. With pressure treatment, these will not necessarily apply. They do, however, serve to emphasize that having once decided on a particular process, it is vital to adhere to the conditions precisely to avoid variations in color and flavor, see Minifie (1968).

**Degree of Roast** Low roasting temperatures give reddish colors, whereas high roasting gives dark-brown colors. It has been mentioned that there are advantages in low-temperature roasts carried out on nibs that still have a fairly high moisture content. These conditions aid in the production of red shades during alkalization, and the micronizing process or other similar methods that help the separation of shell from virtually raw beans will be advantageous. Alkalization temperatures should be kept low to ensure red shades and a temperature of 80°C (176°F) should be considered optimum. This applies to the drying process also.

**Quantity and Concentration of Alkali** Both the quantity of alkali and its concentration in water have a marked effect on the color of the final cocoa (See Table 2.2).

It is difficult to compare colors accurately by description but it should be noted that the colors 4, 5, and 6 are richer than 1, 2, and 3,

TABLE 2.2. EFFECT OF QUANTITY AND CONCENTRATION OF ALKALI

|   | $K_2CO_3$ lb per 100 lb nib | Water lb per 100 lb nib | % Concentration of $K_2CO_3$ solution | pH of Cocoa | Color |
|---|---|---|---|---|---|
| 1 | 1.7 | 20 | 8.5 | 7.3 | Dull brown |
| 2 | 1.7 | 30 | 5.6 | 7.1 | Darker brown |
| 3 | 1.7 | 50 | 3.4 | 7.2 | Reddish brown |
| 4 | 2.5 | 20 | 12.5 | 7.6 | Rich brown |
| 5 | 2.5 | 30 | 8.3 | 7.7 | Deep red brown |
| 6 | 2.5 | 50 | 5.0 | 7.6 | Deep red |

and there is progressive reddening as dilution increases $1 \rightarrow 3$ and $4 \rightarrow 6$.

The duration of alkalization is determined by the time taken for the alkali solution to penetrate the nib completely. This is usually about an hour, which ensures time for the mixture to reach 80°C (175°F).

Pressure or vacuum, or both, to aid penetration will affect the length of time required.

## Alkalization of Liquor

Liquor alkalization has been widely practiced, but suspensions or solutions of alkali are used with much less water than with nib alkalization, and as a result sandy-brown cocoas are produced. If dilute alkali solutions are used as for nibs the quantity of water to be driven off becomes a problem. To overcome this, the alkalized liquor can be passed through a film cooker that subjects the liquor in a thin layer to a temperature sufficient to evaporate the water. Temperatures of alkalization of liquor are generally higher than for nibs and may reach 115°C (240°F) to ensure evaporation of water.

Typical results obtained by liquor alkalization are shown in Tables 2.3 and 2.4.

Red colors of the final cocoa are not obtained with these processes because of the higher concentration of alkali and higher temperatures. The type of brown color is related to the bean type, roasting conditions, and temperature of alkalization.

TABLE 2.3. BATCH PROCESS—ALKALI SOLUTION ADDED TO TANK OF LIQUOR SLOWLY STIRRED FOR 8 HR

|  | $K_2CO_3$ lb per 100 lb nib | Water lb per 100 lb nib | % Concentration of $K_2CO_3$ solution or slurry | pH of Cocoa | Colour |
|---|---|---|---|---|---|
| 1 | 0.14 | 2.8 | 5.0 | 6.1 | Yellow brown |
| 2 | 1.7 | 5.0 | 25.0 | 7.0 | Yellow brown |
| 3 | 4.0 | 4.4 | 47.0 | 8.5 | Orange brown |

TABLE 2.4. "FLASH" PROCESS—ALKALI SOLUTION FED CONTINU-OUSLY INTO A THIN STREAM OF LIQUOR

| $K_2CO_3$ lb per 100 lb nib | Water lb per 100 lb | Concentration of $K_2CO_3$ solution or slurry | pH of Cocoa | Color |
|---|---|---|---|---|
| 4 | 4.0 | 4.4 | 47.0 | 9.5 | Dark brown |
| 5 | 1.7 | 16.0 | 10.0 | 7.0 | Grayish pink |

## Drying, Roasting, Pressing, and Grinding

Having observed the various conditions required for production of specific cocoa colors, especially from nib alkalization, it must be recognized that subsequent drying, roasting, grinding, and pressing of the nib can also affect color.

Low-temperature processing seems to be necessary to preserve red colors, this is further discussed in the following.

## Alkalization of Cocoa Cake

Cocoa cake, preferably in kibbled form, may be alkalized. It should have been produced from well-fermented beans using low-temperature roasting and pressing. Alkali solution is rapidly absorbed by the cake, and although this helps the chemical action, the amount of water present must be controlled to prevent a wet sludge or slurry from forming. Slurries are difficult to handle and to dry. Alkalization of cake is best carried out in a heated rotating vacuum drum, which, in addition to providing good mixing at the alkalizing stage, ensures that efficient drying takes place at low temperatures. The red colors disappear at high temperatures in the presence of air.

In this equipment, it is possible to make black or "Blackshire" cocoa, used, as previously mentioned, for pigmentation.

The conditions required are much higher alkalization tempera-tures, more alkali (5 to 8 percent), and longer treatment times. The pH of this cocoa is about 8.5, compared with 6.8 to 7.5 for normal alkalized cocoa.

## Alkalization of Whole Beans

It is possible to alkalize whole raw beans by applying the alkali solution in a drum roaster. Low temperatures may be used to obtain reddish shades. Much of the alkali is absorbed in the shell, which retards penetration of the internal nib.

This process is not often used commercially.

## Water Treatment

Before leaving the subject of alkalization, a reference should be made to water treatment. If liquor, or preferably cocoa cake, is subjected to the action of water without dissolved alkali, some reaction still takes place without any marked change of pH. There is a tendency to produce slightly redder shades than would normally be the case with a natural or unalkalized cocoa. This process can be used for obtaining improved colors for coatings, which usually require unalkalized cocoa for their preparation.

## General Remarks

From this summary of alkalization methods, it is obvious that although a wide variety of colors is available, it is very difficult for the manufacturer to produce consistently to an exact standard. Precise control of every part of the process is required, but variations still occur and most manufacturers resort to the blending of powders to maintain consistent colors.

The effect of variations in the raw beans is not discussed here as there seems to be little relationship between the type of bean, the degree of fermentation, and the ultimate color obtained using different methods of alkalization. The answer to this is the same as applies to flavor. The cocoa manufacturer must select beans according to the growing area and maintain a standard that will at least give reasonably consistent results.

## LIQUOR PRESSING

Good-quality cocoa powders are prepared by hydraulic pressing of finely ground liquor—which must have been manufactured from well-winnowed, high-grade cocoa beans.

Like other machinery, hydraulic presses have been modernized in design and are now much more automatic and accurate in obtaining the required fat content in the cocoa cake.

The original presses were vertical with press pots, usually four or six, arranged one above the other. Each pot consisted of a heavy steel shallow cylinder perforated with many small holes at the base. Over this base was placed a filter pad, usually cloth, and the pot was then filled with liquor and a further filter pad placed on top. A ram was then applied over each pot and the butter was expressed through the filter cloths into channels around the pots and then into containers.

The pressure applied in these presses was up to 6,000 lb in.[2] and the larger presses could take a charge of 250 lb of liquor. The control of fat content was largely on a basis of time and about 15 min was required to reduce the cake to 22 percent fat and 30 min to 14 percent. To obtain lower fat contents on these presses was generally unreliable and uneconomic. Control by measuring the volume of the expressed fat was used in some instances but accuracy could not be achieved.

With modern horizontal presses, up to 14 pots are mounted in a horizontal frame and each pot is provided with a metal filter screen supported on the pot plates (Figs. 2.16 and 1.17).

The press, when closed, is filled automatically with hot liquor under pressure and a proportion of the free butter is removed during

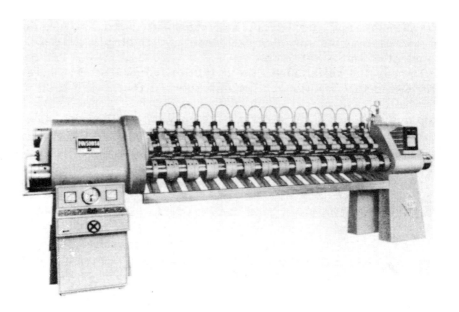

Fig. 2.16.   POV.540.B2.14 Liquor Press

*Carle and Montanari, Milan, Italy*

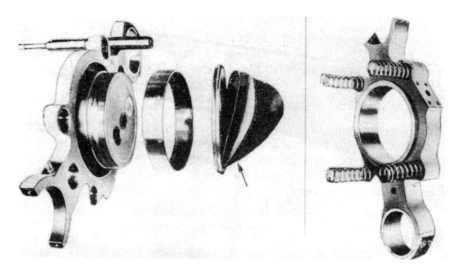

Fig. 2.17. Section of Single Pot (Arrow Indicates Filter Screen)

*Carle and Montanari, Milan, Italy*

the filling operation. A higher pressure is then applied. The larger presses have an output of up to 3,500 lb/hr with cocoa cake at 24 percent cocoa butter. At lower fat contents, outputs are correspondingly lower, and although claims of cake with 8 percent fat content are made, the normal minimum is 10 to 12 percent with outputs of 1,200 to 1,600 lb/hr.

The operation of these presses is completely automatic and the ultimate fat content of the cocoa cake is controlled by the time cycle, weight of cocoa butter, and distance of ram travel. At the end of the pressure/time cycle, the direction of the ram is reversed, and the pots open up and allow the cocoa cakes to drop into a bin or onto a conveyor from which they are removed for grinding. With the older vertical presses, the cakes had to removed manually and the whole operation in the high temperatures of the press rooms was very arduous.

With modern presses, to obtain optimum yields in a minimum time, various factors have to be taken into account, including the following.

## Temperature of the Liquor

In modern practice, the best results are obtained at 95 to 105°C (203 to 221°F). Higher temperatures result in inferior, strongly flavored cocoa butter without any improvement in yield.

## Moisture Content

Moisture content of the liquor is critical. Most processors operate in a range of 1 to 1.5 percent but one press manufacturer recommends 0.8 percent. According to Reade (1983), the optimum can be as high as 1.8 percent, which results in maximum yields of cocoa butter. But it is important to recognize that errors can arise in moisture determinations—a simple method usually recommended is by toluene distillation (Dean and Stark method).

Moisture content and some of the other factors mentioned here are interrelated, including the way in which the moisture is dispersed in the liquor.

## Degree of Roast/Protein Coagulation

It is generally recognized that liquor prepared from high-roast beans presses best and the coagulation of the protein seems most likely when the roasting is done in two stages—at a low temperature with fairly high moisture followed by the true high-temperature roast (see NARS process).

## Homogenizing

The inconsistent dispersion of moisture and other constituents that may be present in liquor causes many problems in pressing. Sometimes, small concentrations of moist gelatinous sludge are found in a tank of liquor that may have resulted from the alkalization process or low-temperature roasting. These can cause blockage in the pipelines feeding the press pots.

A combined homogenizing, heating, and pumping equipment for feeding the press is manufactured by Carle and Montanari (GDO dosing-homogenizing).

## Particle Size

With modern methods of liquor grinding, particularly where the final grind is with a bead mill, liquors with ultrafine particles are likely. Sieving test results of 98 percent through a 400-mesh sieve have been recorded, and such liquors are almost certain to cause pressing troubles, with a buildup of back pressure and blinding of the sieves. These fine liquors are good for chocolate or superfine cocoa manufacture, but more efficient pressing is likely with a coarser liquor, for example 98 to 99 percent through a 200-mesh sieve.

## Pressures

Older presses worked under a pressure at the filter of about 6,000 lb/in.$^2$ but modern presses are designed for 12,000 lb/in.$^2$. The main effect is greater speed of pressing but the final yield of cocoa butter is not significantly increased.

More detailed information is available from manufacturers' brochures (see references) and Kooistra (1981) has described in more detail many of the items already mentioned.

## EXPELLER PRESSING

Extrusion, expeller, or screw presses are used extensively in the oil-seed industry where the oil or fat yield is of principal importance. Similarly, they are used in the cocoa industry for the production of cocoa butter from whole beans, and blends of winnowing products consisting of fine nib dusts, small nibs, and immature beans. Less frequently, fully winnowed whole nib is pressed when the expeller cake is used for the manufacture of coatings and therefore must be free from shell and as low as possible in cocoa butter content.

The principle of the expeller press (Fig. 2.18) is to force the material to be extracted into a tapering tube by means of a rotating screw. The tube is perforated along its length by narrow slits and terminates in an adjustable cone that gives a variable gap between the cone and the tube exit. The material traveling through the press is subjected to a combination of shearing and increasing pressure and the fat is forced out through the tube slits. The cake is extruded through the gap at the end of the press in the form of thick flakes or "corns". The fat contains some fine cocoa and must be filtered or centrifugally separated.

The material being fed to the press is subjected to a steaming process to soften it and help the release of fat in shearing action.

With the extrusion process, it is possible to reduce the fat content of the cake to 8 to 9 percent without great difficulty.

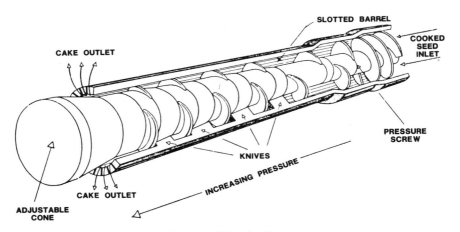

Fig. 2.18.   Diagram of Expeller Press

*Reproduced by permission of Loders and Nucoline, London*

Expeller cake may be ground to a fine powder and used as a cocoa ingredient in coatings or cookies, in which case it would be made from good-quality winnowed nibs. Alternatively, if expeller cake is made from whole beans or from winnowing products containing shell, it may be solvent extracted.

The shearing action in the expeller press results in a cocoa powder with different physical properties. The cell structure is such that the powder shows much greater absorbency than hydraulic pressed powder. This must be borne in mind if it is to be incorporated in drinks.

## COCOA GRINDING

The methods of producing cocoa cake have been described, and having obtained the various types of cake, it is necessary to reduce it to a fine powder.

During the hydraulic pressing, the particles of cocoa matter and the residual fat have become very densely compacted. The shallow cylindrical press cakes, particularly if low in fat content, can be very hard and the first operation is to put them through a breaker, which usually consists of rotating rollers with intermeshing teeth; these break the cake into small lumps, the largest of which should be about the size of a pea (kibbled cake).

This lumpy powder is further reduced by passing it through hammer mills or peg disintegrators that work in conjunction with wind sifters, rotating selectors, and cyclones.

The reduction of cocoa cake to fine powder presents several special problems not associated with the reduction to powder of mineral substances, or even pure organic materials like sugar.

1. The true particle size of the cocoa material is that which was achieved during the liquor grinding process and the cake pulverizer is likely only to reduce the size of the compacted aggregates formed during the hydraulic pressing. Hence, the importance of fine grinding of the liquor is emphasized as the first step toward fine cocoa production.

2. The presence of cocoa butter in the powder necessitates the use of cooling air during the grinding; if the temperature of the powder rises above 34°C (93°F), the fat will melt, and this causes adhesion of the particles and clogging of the disintegrating machinery.

Furthermore, at temperatures below 34°C, some of the glyceride fractions of the cocoa butter will melt (see Cocoa Butter and Chocolate Tempering) with the result that even though the powder can pass through the mills, the fat will set in the powder in an unstable state, give a fugitive or grayish color, and cause the powder to cake.

These effects are most noticeable with the higher-fat cocoa powders. When they have less than 20 percent fat content, the cooling conditions need not be so precise. The color of cocoa powder is mainly of importance for cocoa to be used as a beverage, where a good appearance in the package is wanted. For bulk supply where the cocoa powder is to be used as an ingredient of coatings, cakes, or cookies, the *powder* color is less important. The *inherent* color derived from the method of alkalization then becomes the dominant factor. When it is necessary to obtain good powder colors, the best conditions for running a cake pulverizer are those that set the cocoa butter from the liquid to the solid state, in a stable form, during the grinding process. This necessitates controlling within fairly narrow limits the temperature and rate of feed of the cake being delivered to the pulverizer. The cake temperature should be above the melting point of cocoa butter—e.g., 43° to 45°C (109° to 113°F)—and the cooling and feed rate adjusted to give a powder temperature at the exit of the mill of 21° to 24°C (70° to 75°F). Some variations of the feed temperature conditions quoted may be necessary with different mills, but the powder temperatures mentioned should be fairly precise to ensure good permanent powder colors.

3. The cooling air used must be dry with a relative humidity of not more than 50 to 60 percent in the mill, otherwise cocoa of high moisture content will be produced and microbiological troubles can arise due to mold growth in sections of the mill ducts and conveyors. On the other hand, air of very low relative humidity may give rise to static electricity charges on the powder particles, leading to bulkiness and difficulty in the packing machines. This trouble is even more likely where ambient conditions are very dry.

The air must obviously be free from foreign odors as cocoa powder is very prone to contamination by traces of aromatic substances.

In a modern plant the cocoa cake is fed by a remote controlled, variable delivery electromagnetic doser which incorporates magnetic pickup mechanism for extraction of ferrous foreign matter.

The cocoa cake is subjected to the combined action of two disks in the grinding mill. One of these disks rotates at very high velocity.

The disks are fitted with grinding pins made of special steel and are arranged in such a manner that particles of cocoa cannot escape their grinding action. The housing of the grinding mill incorporates hollow walls in order to ensure an efficient cooling system. After the grinding process, the cocoa is delivered by air stream into the cooling pipes.

These are hollow wall pipes which circulate a cooling liquid. The length of piping from the mill to the cyclone separator is for air cooling and also for the cocoa powder delivered inside the piping. The length of delivery tempers the powder and fixes its color.

A terminal cyclone separates the finished product from the air; the air returns into circulation.

There are many types of powder mill that may be adapted to cocoa powder production, but the cooling conditions previously mentioned must be observed. These mills have built-in systems for automatically returning the coarser particles to the mill before reaching the cyclone separator. It must be recognized, however, that in mills suitable for inorganic powders many problems can arise if they are used for a product like cocoa with some 20 percent fat content.

A recent development to ensure good colors and prevention of compacting of cocoa powder is the Carle and Montanari ISC stabilizer. It is basically a cylinder provided with cooling air from the bottom. The air fluidizes the powder, ensuring contact with every particle.

A line diagram showing the design of the latest powder grinding, cooling and stabilizing system is given in Fig. 2.19.

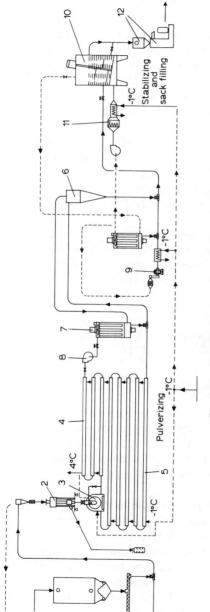

Fig. 2.19.    Powder Cooling and Stabilizing System

 1. Kibbled Cake Feed
 2. Metal Detector
 3. Pin Mill
 4. Air Cooling
 5. Cocoa Cooling
 6. Cyclone
 7. Self-Cleaning Filter
 8. Fan
 9. Pneumatic Conveyor with Cooled Air
10. Fluid Bed Stabilizer
11. Fluid Bed Air Cooler
12. Sack Filling and Weighing

*Carle and Montanari, Milan, Italy*

75

## Cocoa Fineness

Assuming the flavor and color quality desired are achieved, the most important property of cocoa powder is its fineness, for its solubility is low and in a water or milk suspension coarse particles will rapidly settle, giving an objectionable rough sediment. Large particles will also give a speckled appearance in a milk drink or ice cream.

In cakes and cookies, where the cocoa is also required to impart a pleasant chocolate color to the products, a very fine particle size will give improved dispersion and pigmentation.

Some fine cocoas tend to produce flocculation in milk or water suspensions. It has previously been mentioned that expeller press cocoas are prone to doing this, due to cell rupture in the process—it can also be related to pH (natural cocoa pH 5.5) or to calcium content, especially if calcium-sensitive stablilizers are used to help suspension of the cocoa.

There are two basic methods for determining fineness: (1) the sediment test, and (2) the sieving test, which may be either dry or wet (see Cocoa Composition and Analysis and Appendix).

Meursing (1983) describes many of these tests.

## COCOA FOR DRINKING, INSTANT COCOAS, DRINKING CHOCOLATES, MANUFACTURING COCOA

### Drinking Cocoa

The powder prepared by one of the methods mentioned previously is usually flavored by the addition of vanillin, cinnamon, cassia, and other powdered spices or oleo resins. Salt is also frequently added to the extent of about 0.5 percent, and this can be mixed with the alkalizing materials.

The flavoring spices or vanillin are blended with the cocoa as pulverized powders, with a particle size no greater than that of the cocoa, and these ground powders give better results than flavor essences or essential oils, which tend to float to the surface when the drink is made and are more prone to change by oxidation than when they are partially locked in the cell structure of the spice powder. Cocoa powders can make a very rich, pleasantly flavored drink if prepared correctly. This entails making a paste of the powder and a small amount of milk or water, and then adding boiling water or milk and, preferably, whisking. Alternatively, the whole mixture can be boiled in a saucepan. For convenience, and to save time, in recent years the so-called "instant" chocolates have become popular.

## Instant Cocoa

In addition to the traditional hot cocoa, cold chocolate drinks have increased in popularity. It is extremely difficult to make a cold drink with ordinary cocoa powder and the addition of an edible wetting agent is essential to get proper dispersion. Sometimes the powder is "instantized" (see below).

Soya lecithin has been used to aid the wetting of cocoa powder but this is subject to the development of off-flavors. Specially prepared lecithins and substituted lecithins are now available (Meyer, 1983); these are described in Chapter 4.

A synthetic phospholipid is also available that does not result in off-flavors (Bradford and Harris, 1968), which has been used in the manufacture of milk chocolate. In the process of lecithination each particle should be completely coated to obtain the maximum effect. To do this, liquid lecithin may be injected into the cocoa-cake pulverizer. If the cocoa has already been pulverized, lecithin, preferably as a water emulsion, may be sprayed into a rotating tumble mixer. The cocoa in this mixer will be in turbulent air suspension, which provides a maximum opportunity for the particles to become coated. The treated cocoa is then dried.

Even so, 1.5 to 3 percent lecithin may need to be added to obtain a full wetting effect with cocoas with a fat content of 22 to 23 percent, and this is considerably more than is ever used in chocolate manufacture where the limit on the addition of lecithin is usually 0.5 percent. It thus is necessary that the lecithin has little flavor.

## Drinking Chocolate

Chocolate has largely replaced cocoa powder for beverages. These drinks are easier to prepare and most of these chocolates can be used to make cold drinks.

Drinking chocolates generally consist of 70 percent sugar and 30 percent cocoa powder. Some are just simple mixtures with some added flavor, but more often a process is used that cooks and agglomerates the particles of sugar and cocoa. This, it is claimed, improves the chocolate flavor.

One process for the manufacture of drinking chocolate uses a sugar syrup boiled to a supersaturated state whereupon it is rapidly mixed with cocoa powder and then dried. In another recipe, only part of the sugar is used in the above process, the remainder of the sugar added being of a graded size, free from very fine dust, and this mixture is "instantized."

The principle of instantizing is to cement the sugar and cocoa particles together by subjecting the sugar/cocoa powder to an atmosphere of wet steam, which wets the sugar just sufficiently to form a film of syrup on the particles, and this brings about adhesion to, and partial wetting of, the cocoa particles. The grading of the sugar size helps this process as the presence of very fine sugar prevents the proper wetting of the separate larger particles and causes aggregation into solid masses. After the steaming process, the mixture is dried, and usually sieved through a fairly coarse mesh.

Plant and processes employed for instantizing are discussed by Jensen (1973) and in Chapter 10.

Some drinking chocolates incorporate the milk ingredient as well as the sugar. Nonfat milk solids are usually employed as full cream milk solids limit the shelf life of the powder. Milk fat tends to impart off-flavors to the drink after a period, and in some recipes "filled milk" is used.

Filled milk is nonfat milk emulsified with a vegetable fat or oil such as ground nut oil or hardened palm kernel oil. This is then spray dried alone, or with cocoa and sugar added, and then instantized.

Drinking chocolates, like cocoa powders, have added flavors, and because of the high proportion of sugar, the added flavors are usually stronger than in pure cocoa powder.

The wetting of an instant powder is related to the capillary channels in the particle aggregates and it is important, when using an instantizing process, to ensure that the final product will retain its instant properties after the normal processes of packing and transport. If the particle strengths are weak, some fine powder may be formed during movement, and this can destroy a lot of the quick wetting properties.

## COCOA POWDER—MICROBIOLOGY, COMPOSITION, ANALYSIS

The larger manufacturers who use cocoa as an ingredient in their products are becoming increasingly aware of its potential microbiological activity.

The raw fermented cocoa beans arriving in the factory have very high bacteria counts (e.g., $5 \times 10^6$), and during the cleaning, roasting, and winnowing processes, dust is produced that can contaminate finished products. It is very necessary to isolate these first processes from subsequent production. Roasting does not necessarily reduce the bacteria count to acceptable levels but the subsequent processes of alkalization, liquor grinding, and pressing do give a product in press-cake form that is virtually free from bacterial contamination.

However, in the process of converting the press cake to powder, there is ample opportunity for it to become reinfected and further cross-infection can occur in the user's factory. Another way in which the bacteria count can be increased is through the flavor addition. Raw spices can be very heavily contaminated, but manufacturers have overcome this by a process that eliminates the bacterial contamination yet retains the property of a spice to give up its flavor slowly by retention in the cell structure.

It has already been mentioned that infection can occur in the duct work and conveyors of cocoa powder plant, and should moisture condensation take place locally, mold patches can grow and very quickly a reservoir of spores is built up that can contaminate all production in the plant. From similar local infection, ferments and bacteria can be introduced but the amount of such infections is usually insufficient to affect the flavor of the cocoa itself.

Even in a well-serviced plant, appreciable counts can be found in the finished powder and the figures shown in Table 2.5 have been recorded by the author.

Many cocoas purchased on the open market have much higher counts ($50 \times 10^3$ and above). When counts show appreciable increases, the plant is examined thoroughly for sources of contamination—condensation, water leaks—and then treated with ethylene or propylene oxide gas, which provides sterilization. The

TABLE 2.5. INFECTIONS IN COCOA POWDER

| | Plate count. Nutrient agar 32°C/72 hr colonies per gram | E. Coli in 0.1 g | Yeasts and molds. Wort agar pH 3.5 32°C/5 days colonies per gram |
|---|---|---|---|
| Cake from hydraulic press | 100 to 500 | Absent | Absent |
| Cake after breaker (kibbled) | 1 to $6.5 \times 10^3$ | Absent | 0 to 45 |
| Finished cocoa powder | 1.3 to $16.5 \times 10^3$ | Absent | 0 to 2 |

equipment must be thoroughly cleaned afterward and these gases must *not* be used to treat the cocoa powder as toxic by-products are likely to be formed.

When cocoa is used as an ingredient in confectionery coatings or cake fillings, the presence of microorganisms can lead to considerable trouble unless adequate sterilizing precautions are taken. In baked goods, temperatures are usually sufficient to destroy any organisms and many of these products are expected to have a short shelf life and would be sold quickly.

As a result of bacterial or mold activity, cocoa powder can contain active lipolytic enzymes. Confections such as nougats, pastes, and pralines (as well as cake coatings) often use cocoa as a flavoring material and, at the same time, vegetable fats of the lauric glyceride type. If lipolytic enzymes are present, very objectionable soapy rancidity can develop, sometimes several months later. In high-concentration substrates or in dry powders, high temperatures are required to bring about destruction. The addition of cocoa to fondant can also be responsible for fermentation, particularly if the syrup phase concentration is low.

Meursing (1983) has discussed at length the implications of contamination and has proposed very rigid specifications for cocoa powder. He states that, with the ever-increasing use of vending machines, instant foods, and large-scale catering where adequate conditions for sterilization are not used, strict bacterial control of the basic food product is essential.

## Cocoa—Composition and Specifications

In Table 2.6 representative analyses of various cocoas are given, but in constructing a specification for a cocoa powder, the following are the fundamental requirements:

1. Preparations. The cocoa shall be prepared from good-quality, properly fermented cacao beans free from mold, infestation, or extraneous flavors.

2. Winnowing. The beans shall be properly winnowed to produce nibs, commercially free from shell, for the preparation of the cocoa. The shell content of cocoa is estimated by the determination of crude fiber but more recently a method for counting the stone cells, after defatting has been developed. See Meursing (1983).

The shell content of cocoa powder shall be a maximum of 1.75 percent calculated back to unalkalized nib.

TABLE 2.6. ANALYSES OF COCOA POWDER (VARIOUS QUOTED FIGURES)

| | Natural | Alkalized | | | Remarks |
| --- | --- | --- | --- | --- | --- |
| | | 1 | 2 | 3 | |
| Moisture, % | 3.0 | 3.5 | 3.5 | 4.3 | Should not exceed 5.0% |
| Cocoa butter | 11.0 | 10.0 | 23.5 | 21.5 | Depends on pressing |
| pH (10% suspension) | 5.7 | 7.1 | 6.7 | 6.8 | |
| Ash, % | 5.5 | 8.5 | 6.3 | 7.7 | |
| Water soluble ash, % | 2.2 | 6.3 | — | 5.8 | Depends on alkalization |
| Alkalinity of water soluble ash as $K_2O$ in original cocoa, % | 0.8 | 2.9 | 1.8 | 2.5 | |
| Phosphate (as $P_2O_5$), % | 1.9 | 1.9 | 1.4 | 2.0 | |
| Chloride (as NaCl), % | 0.04 | 0.9 | 0.7 | 1.1 | Salt occasionally added as flavor |
| Ash insoluble in 50% HCl | 0.08 | 0.01 | 0.06 | 0.09 | High figure shows bad bean cleaning |
| Shell, % (calculated to un-alkalized nib) | 1.4 | 1.0 | 0.5 | 1.0 | Depends on efficiency of winnowing |
| Total nitrogen | 4.3 | 3.9 | 3.5 | 3.7 | |
| Nitrogen (corrected for alkaloids), % | 3.4 | 3.0 | 2.8 | 3.0 | |
| Protein Nitrogen corrected for alkaloids × 6.25), % | 21.2 | 18.7 | 17.5 | 18.7 | |
| Theobromine, % | 2.8 | 2.7 | 2.3 | 2.3 | |

3. Cocoa butter content. The cocoa butter content shall be within the following limits:

$9\frac{1}{2}$ to $11\frac{1}{2}$ percent, $11\frac{1}{2}$ to $13\frac{1}{2}$ percent ... 22 to 24 percent

In quoting these ranges, the manufacturer and supplier would expect the average fat content to be at the middle of the range, but in commercial pressing, a variation of 2 percent is likely.

Regulations exist in many countries regarding the description of a cocoa in relation to its fat content. Examples of those in the United Kingdom and the United States are

U.K. (*Code of Practice*)    Cocoa to be sold as a beverage must have a cocoa butter content of not less than 20 percent.

Cocoa with less than 20 percent cocoa butter is described as "manufacturing cocoa."

*United States*          "Breakfast cocoa" or "high-fat cocoa"—not
                         less than 22 percent cocoa butter.
                         "Medium-fat cocoa"—10 to 22 percent
                         cocoa butter.
                         "Low-fat cocoa"—less than 10 percent cocoa
                         butter.

Full details of all countries' current regulations on composition are available from the literature and trade organizations.

Methods for determination of fat are given in the Appendix and Meursing (1983).

4. Alkalization. This may be specified in several ways. The type of alkali to be used and quantity may be stated, but it is more usual to quote pH and ash contents, for example,

Alkalized cocoa—pH 6.8 to 7.2. Ash content 6.5 to 8.5 percent or similar ranges according to degree of alkalization.

Unalkalized (natural cocoa)—pH 5.4 to 5.8. Ash content 5.0 to 5.5 percent.

5. Fineness. Two tests are considered adequate to assess the fineness of a cocoa powder:

a. "Wet" sieving. This may be done with petroleum or water.

b. Sedimentation. The conical tube using water (Imhoff test) is a simple empirical test. For more information on the size distribution, the Andreasen pipette or Coulter counter, may be used.

Further notes and a description of the methods are given in the Appendix.

6. Microbiological. A typical specification for the cocoa powder is:

|  |  |
|---|---|
| Standard plate count | Less than 5,000/g |
| (Note; High-quality cocoas are now less than 2,000) | |
| Yeast, molds | Less than 50/g |
| Entero bacteria | Negative in 1 g |
| E. coli | Negative in 1 g |
| Salmonella | Negative in 10 g |
| Lipolytic activity | Absent |

Reference should be made to Meursing (1983) and international methods.

7. Flavor and color. The methods of production to give variations in color and flavor have already been discussed and precise specification for these is a matter of agreement between the supplier and the user. It is not reliable to judge color on powder appearance alone, and it is necessary for the user to test it in a product or agree with the

supplier on a test that represents what will happen in the product. It has been shown how color and flavor can vary with slight deviations in process, and process quality control is of great importance at all manufacturing stages.

Further details of composition may be derived from the analyses of cocoa beans and nibs, with the exception that allowance should be made for the addition of alkalizing chemicals and flavoring salt. Many tables of composition are given in textbooks on food analysis and considerable variation is shown in the published figures. The recent publication by Meursing (1983), together with various bulletins by the same author, gives accurate and up-to-details of the composition and analysis of many types of cocoa powder.

## Contamination and Adulteration of Cocoa Powder

Cocoa manufactured from substandard beans may be contaminated with insect fragments, mycotoxins (from molds), and pesticide residues. In times when prices have been high, cocoa powder has been adulterated with other substances, such as starch, carob powder, cocoa shell, and even iron oxide. These dangers arise mostly when cocoa is purchased from unknown sources.

## REFERENCES

Akabori, S. 1932. Amino acids and their derivatives. *J. Chem. Soc. Jpn* 52, 606.

Bartlett, S. 1945. *Proceedings of Cocoa Research Conference,* HMSO, London.

Bauermeister, P. 1981. Cocoa liquor roasting. Manf. Conf. Oct.

Bradford, L., and Harris, T. L., British Patent 1,032,465.

Boller, G. 1980. Technological aspects of liquor grinding. Manf. Conf. May.

Carle and Montanari, Milan. Cocoa presses, PDAT/I process.

Chang, S. S. 1966. Reversion flavor in soya bean oil. *Chem. Ind.* (Nov. 12). London.

Durr, H. 1973. Refining of cocoa mass on high speed agitation bead mills. *Int. Choc. Rev.* 3. Switzerland.

Egan, H., Kirk, R. S., and Sawyer, R. 1981. *Pearsons Chemical Analysis of Foods* (new ed). Longman, Harlow, England.

Fincke, H. 1965. *Handbuch der Kakaoerzeugmisse,* Berlin.

Foster, H. 1978. What is chocolate flavor? Manf. Conf. May.

Jensen, J. D. 1973. Methods of instantizing powders for food drinks. *Manf. Conf.* 46. Oct.

Kleinert, J. 1966. *Rev. Int. Choc.* 21 May. Lindt and Sprungli, Kilchberg, Zurich.

Kleinert, J. 1983. Lindt and Sprüngli, Switzerland. German Patent D. E. 3125144C1. Feb.

Knapp, A. W., and Coward, K. H. 1935. *Biochem. J.* 29, 2728. London.

Kon, S. K., and Henry, K. M. 1935. *Biochem. J.* 29, 2051. London.

Kooistra, P. C. 1981. Problems of pressing ultra-fine imported liquors. Manf. Conf. Aug.

Lucas Meyer, GMBH 1983. *Lecithin and Additives*. Hamburg, Germany.

Mayer-Potschak, K. J. 1983. Roasting in humid atmosphere. PMCA Conference, Lancaster, Pa.

Meursing, E. H. 1983. *Cocoa Powders for Industrial Processing* (3rd ed.). Cacao-fabriek de Zaan, Holland.

Minifie, B. W. 1968. Special cocoas—their manufacture and uses. Manf. Conf. June.

Mohr, W. 1970. *Fette Seifen Anstr. M.* 695–703.

Newton, D., Micronizing, U. K. Framlingham, Suffolk, England.

Obering, H., and Schmitt, A. 1983. *Schoko-Technik*. Petzholdt GMBH, Frankfurt A-M, West Germany.

Reade, M. G. 1983. Private communications. Checkenden, Reading, England

Rohan, T. A., and Stewart, T. 1966. *J. Fd. Sci.* 31, 202, 209. London.

Regulations. Cocoa, Chocolate, and Confectionary Alliance, 11 Green St., London W.1. British Food Manufacturing Industries Research Association, Leatherhead, Surrey, England.

Tadema, J. K. 1983. *A Unique Way of Producing Chocolate and Coatings*. Wiener & Co. B. V., Amsterdam, Holland.

# 3

# Cocoa Butter and Replacement Fats

## COCOA BUTTER

Cocoa butter is essentially the natural fat of the cocoa bean but definitions in some countries limit it to the natural fat obtained from well-winnowed cocoa nib (cotyledon) by hydraulic or expeller pressing.

The U.S. Food and Drug Administration has defined cocoa butter as "the edible fat obtained from sound cocoa beans (*Theobroma cacao* or closely related species) before or after roasting." Dictionary definitions are very inconclusive and one even describes cocoa butter as the fat obtained from the coco-palm! That is totally wrong.

International arguments on definition have intensified as a result of discussions at conferences of Codex Alimentarius and the European Economic Community (Caobisco). At the U.S. Codex Seminar in 1981, Weik summarized the situation on cocoa products and chocolate and it is necessary for all manufacturers and users of cocoa products to keep in touch with Codex developments. As cocoa beans are now grown in many parts of the world, and products, including cocoa butter, are produced in many of these growing areas, the variations that are likely are being increasingly investigated. A summary of recent findings is given later.

The very high cost of cocoa butter has also increased the possibility of more sophisticated production methods. Since the first edition of this book, the literature published on cocoa butter could fill many new pages. It is not proposed to deal with the legal arguments that surround this unique fat but merely to describe the variations that arise through different methods of extraction.

It is likely that companies that press cocoa butter for their own use have a different outlook from those that make cocoa butter for sale or those that purchase cocoa butter on the open market. The following comments on the different grades of cocoa butter are a useful guide in deciding the policy to be adopted.

## Prime Pressed Cocoa Butter

This is defined as the fat obtained from good-quality cocoa nib commercially free from shell by means of mechanical (hydraulic) pressing. No subsequent refining other than filtration is employed.

## Expeller-Pressed Cocoa Butter

In the expeller, extrusion, or screw press, slightly different conditions prevail during extraction. The nib is normally steamed to assist the separation—chemically, the cocoa butter is the same as prime pressed. Expellers are also used to extract the fat from whole beans.

The flavor of the cocoa butter obtained by this process is different from prime pressed—it can be very mild and floral if raw beans are used. Often, expeller pressing is used to extract the fat from substandard beans (e.g., immatures) and some winnowing products, in which case the resultant cocoa butter is usually subjected to a refining process.

## Solvent-Extracted Cocoa Butter

This is the fat extracted from the cake residues after the expeller process, or it may be extracted from cocoa or chocolate residues. Solvent-extracted cocoa butter must always be subject to a refining process.

As a commercial article, there is a great divergence of opinion as to its acceptance as "cocoa butter." Some authorities claim it should be extracted from residues or expeller cake that have been produced from cocoa nib only.

In relation to the extraction of whole beans, the following is of interest: assuming the composition of the whole bean to be nib 88 percent, shell 12 percent, and the fat contents of roasted nib and shell to be 55 percent and 3 percent respectively, the *calculated* fat content of the whole bean is 48.76 percent. This comprises nib fat 48.40 percent and shell fat 0.36 percent, and, therefore, the percentage of shell fat in the total fat extracted from the whole bean is calculated as 0.74 percent. The composition of shell fat is given later.

The presence of this small amount of shell fat has no significant effect on the properties of the nib fat and no evidence has been produced to show that it is unsuitable as an edible fat. Some authorities argue that the shell, being the outer part of the bean, is liable to contamination and so is unacceptable from a hygienic point of view, and, also, that spray residues would be concentrated on the shell. There is very slight evidence for the latter but the roasting of the bean to some extent distributes the residue throughout the bean.

Those who argue against the inclusion of shell butter will, nevertheless, use "smalls" in the production of second grade chocolate and cocoa. These smalls, which are an intermediate product of winnowing between the large nib and dusts, have a fat content of 40 to 45 percent, and on this basis would be expected to contain shell butter approaching that of whole bean butter.

Because of the great popularity of milk chocolate, which requires the use of excess cocoa butter in relation to cocoa material when compared with dark chocolate, many manufacturers must produce cocoa butter from cocoa beans by the most economical means.

Expeller presses are generally used and the cake is then subjected to solvent extraction to remove the residual fat. By the use of whole beans, very light roasting or even no roasting is needed, and this gives the mild-flavored cocoa butter that is desirable for milk chocolate.

Solvent extraction, which is used for other edible fats as well as cocoa butter, probably had its difficulties in early days because of the use of bad solvents that left traces of residue, but these troubles have been overcome and solvent-extracted cocoa butter is an accepted commercial fat. Highly refined petroleum solvents are generally used for extraction.

Solvent extraction removes certain gums and phosphatides as well as the fat, but extracted fats are usually subjected to deodorization and degumming processes so that these cocoa butters have a bland flavor. They also have the reputation of being softer and with less "snap" than expressed butters.

It should be noted that it is unlikely that solvent-extracted cocoa butter would alone account for the added cocoa butter in a chocolate. Generally, it would be incorporated in a butter blend at the rate of 2 to 5 percent.

The processes for production and properties of these butters are given later.

## Constitution of Cocoa Butter

Cocoa butter, obtained by hydraulic expression of cocoa nib, is a light yellow fat, exhibiting a distinct brittle fracture below 20°C (68°F), a fairly sharp complete melting point about 35°C (95°F), with an incipient fusion or softening around 30–32°C (86 to 90°F). The completely liquid fat displays a marked tendency to supercool, a fact that must be taken into account in the processes of chocolate enrobing and molding.

Cocoa butter is composed of a number of glycerides of stearic, palmitic, and oleic fatty acids with a small proportion of linoleic. Hilditch and Stainsby (1936) and Meara (1949) established that percentages of the constituent glycerides were as follows, with approximate variations:

| | |
|---|---|
| Trisaturated | 2.5 to 3.0 |
| Triunsaturated (triolein) | 1.0 |
| Diunsaturated | |
| Stearo-diolein | 6 to 12 |
| Palmito-diolein | 7 to 8 |
| Monounsaturated | |
| Oleo-distearin | 18 to 22 |
| Oleo-palmitostearin | 52 to 57 |
| Oleo-dipalmitin | 4 to 6 |

The configuration of these diglycerides has been the subject of much research by a number of workers using modern analytical techniques, including Chapman (1957), Lutton (1957), Savary (1957), Schofield (1959), and Steiner (1961). Kattenberg (1981) has analyzed cocoa butters from various cocoa-growing areas and his findings are summarized below. See Table 3.1 and Fig. 3.1.

## Properties of Cocoa Butter

Many analyses of cocoa butter have been published. Table 3.2 gives typical published figures. The main physical and chemical constants show little variation but modern methods of analysis have shown

TABLE 3.1. TRIGLYCERIDE COMPOSITION OF COCOA BUTTERS FROM MAIN GROWING AREAS (KATTENBERG 1981)

| Bean Origin | Fraction, % | | | |
|---|---|---|---|---|
| | Trisaturated | Monounsaturated | Diunsaturated | Polyunsaturated |
| Ghana | 1.4 | 77.2 | 15.3 | 6.1 |
| Ivory Coast | 1.6 | 77.7 | 16.3 | 4.4 |
| Cameroun | 1.3 | 75.7 | 18.1 | 4.9 |
| Brazil | 1.0 | 64.2 | 26.8 | 8.0 |

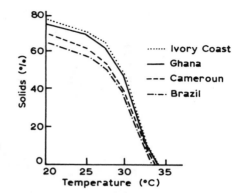

Fig. 3.1.   Graph of Percent Solid Phase
Versus Temperature of Cocoa Butters

*Kattenberg, 1981*

some differences in glyceride structure and there has been considerable study of the physical properties, including rate of crystallization, cooling curves, hardness (penetration), and contraction. These properties show significant variation.

## New Sources of Supply—Natural Variations

Users of cocoa butter from some of the newer important growing areas have reported noticeable variations in the physical properties of these butters.

Many users of Brazilian butter have indicated that it has a softer texture, than West African cocoa butter but it should also be noted that there is appreciable variation in the cocoa butters from any one area. Butters from Ghana and Nigeria, however, have shown consistent quality in the past.

Malaysia is becoming an important supplier of cocoa beans and other cocoa products. An article by Som and Kheiri (1982) compares cocoa butters from Malaysia and Ghana and no abnormal characteristics were found in the Malaysian product.

## Supercooling, Cooling Curves

Cocoa butter has unique supercooling properties, which means that the liquid fat in an undisturbed condition will remain in the liquid state well below its melting point. When cooled and stirred under rigidly controlled conditions, time and temperature can be plotted graphically.

TABLE 3.2. PROPERTIES OF COCOA BUTTER

| Data | Author | Fincke (1965) | Jensen (1931) | Pearson (1970) |
|---|---|---|---|---|
| Specific gravity | 0.8957 (40°/15.5°C) | 0.910–0.912 (15°/15°C) liquid 0.976–0.978 (15°/15°C) solid | — | 0.950–0.975 (15.5°C/15.5°C) |
| Refractive index | 1.4560–1.4580 (40°C) | 1.4565–1.4578 (40°C) | 1.4565–1.4575 (40°C) | 1.456–1.458 (40°C) |
| Iodine value | 35.4(35–40) | 33.5–37.5 | 33–39 | 35–40 |
| Saponification value | 195(188–198) | 192–197 | 191–198 | 188–195 |
| Unsaponifiable matter | 0.8% | 0.3–0.4% | 0.5–1.1% | less than 1.5% |
| Iodine value of unsaponifiable matter | 80–96 | — | — | — |
| Melting point Complete fusion | 33.0°C (32.0–34.0°C) | 32.8–35.0°C | 32.5–34.5°C | 31–34°C |
| Incipient fusion | 32.0°C (31.2–32.7°C) | 31.8–33.5°C (Flow point) | 30.0–32.5°C | — |
| Free fatty acids (as oleic acid) | 1.5% (maximum permitted) | 0.8–3.0% | 0.4–1.05% | 0.5–1.4% |
| Titer point | 49.0°C | Fatty acids—complete fusion 51.5–53.5°C Fatty acids—Flow point 49.0–51.0°C | 49–50°C | 48–51°C |
| Reichert Meissl value | 0.65 | 0.1–0.5 | — | 0.2–1.0 |
| Polenske value | 0.3 | 0.5–1.0 | — | less than 0.5 |

Most companies manufacturing fats and producing cocoa butter have fully mechanized this test. The shape of the curve gives very useful information on the purity and quality of cocoa butter. (See Appendix for details.) An adaptation of the cooling curve is an instrument called the tempermeter, used to measure the degree of chocolate "temper" in an enrober or molding plant.

## Crystallization, Polymorphism

Cocoa butter exhibits a very complex crystallization system as a result of the different glycerides present. It is polymorphic, which means it will crystallize in several different forms according to how the liquid fat is solidified. There are four main polymorphic crystalline forms, with the following properties:

| | |
|---|---|
| $\gamma$ form | is produced by very rapid cooling of the liquid fat. Its melting point is approximately 17°C (63°F). It is very unstable and transforms |
| $\alpha$ form | quickly, even at low temperatures, to the $\alpha$ form (melting point 21 to 24°C (70°–75°F). |
| $\beta'$ form | The $\alpha$ form changes at normal temperatures to the $\beta'$ form (melting point 27 to 29°C (81 to 84°F). |
| $\beta$ form | and then ultimately to the $\beta$ form, which is stable and has a melting point of 34 to 35°C (95°F). |

The stability of the different forms increases from $\gamma$ to $\beta$, and in all chocolate processes, the objective is to set the cocoa butter and chocolate in the most stable form. Failure to do this will result in bad colors, bloom, and delayed setting. These problems are discussed later. Various workers have discovered other polymorphs, but from a practical standpoint the four mentioned are those of importance. See Vaeck (1960).

The speed of crystallization is a factor that has become increasingly important. In the early days of development of equivalent fats, it was soon noted that this property was independent of the glyceride structure. The temperature and cooling curve give some information, but recently developed instruments measure the increase in viscosity or resistance to mixing when the fat is allowed to crystallize under specific conditions of temperature and movement.

The apparatus used is a thermo-rheogram (TRG) as described by Baenitz (1978). Experimental work with the same system is explained in three articles by Kleinert (1982).

The principle is as follows:

The instrument consists of a kneader rotating at a constant speed within a thermostatically controlled trough. The liquid fat can be held at specific temperatures and allowed to crystallize while the kneader is in motion. The crystallization of the fat creates a resistance to the kneader and the reaction force is measured and plotted graphically against time.

The rate of crystallization is a most important factor in the handling of chocolate or vegetable fat coatings in tempering, enrobing, and molding. The TRG method indicates a property not shown by viscosity or cooling curve measurements. An interesting publication by Davidson and Crespo (1979) using a visco-corder indicates similar variations in the crystallization pattern of cocoa butter and alternative fats.

## Hardness, Penetration

It is well known that cocoa butters from different origins or from different methods of extraction vary in hardness. This is related to crystal pattern and is analogous to the TRG measurement.

The instrument used is called a penetrometer and consists of a cone suspended over a specially prepared slab of fat. The cone may be loaded with different weights and the time taken for the cone to penetrate the fat under load and at different temperatures is measured.

Its application is described by Kleinert (1982) and in other parts of this book.

## Contraction

Cocoa butter has the valuable property of contraction on solidification, which enables the molding of chocolate blocks and bars into the attractive confections displayed in shops and stores.

Proper contraction depends on correct seeding of the liquid fat or tempering of chocolate. The solidification of cocoa butter or chocolate to bring about this contraction, and also to give a smooth crystalline solid that will keep satisfactorily without discoloration (fat bloom), depends on the production of the stable polymorphic form of the fat during cooling and setting.

This has been studied by the author by determining linear and volumetric contraction on seeded cocoa butter under different cooling conditions. See Fig. 3.4.

Published works by Lovegren and Feuge (1965) also trace the variation of contraction with time under different conditions of cooling. These workers, for the purpose of experiment, produced three polymorphic forms:

*Form 1.* Obtained by heating cocoa butter to 50°C (122°F), then immersing the sample alternatively in ice water and a water bath at 24°C (75°F) a number of times and ultimately ageing the sample for 65 hr at 25°C (77°F).

*Form 2.* Obtained by heating as in form 1 and chilling in ice water.

*Form 3.* Obtained by six successive heatings to 50°C (122°F) interspersed with coolings, the temperature dropping in stages from 20°C (68°F) to 6°C (43°F).

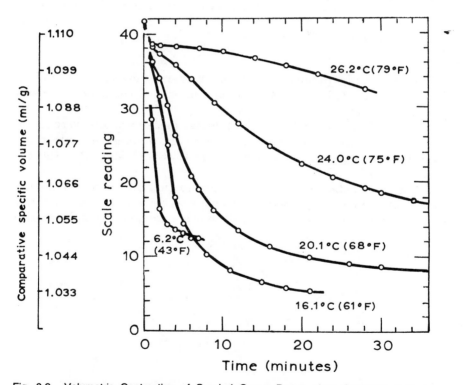

Fig. 3.2. Volumetric Contraction of Seeded Cocoa Butter when Solidified at Various Temperatures
This is interesting in relation to the use of moderate temperature cooling of enrober covered chocolates where excessive contraction is not required.

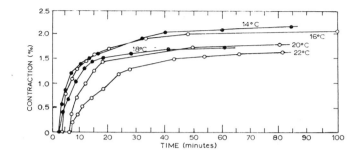

Fig. 3.3. Linear contraction of Seeded Cocoa Butter When Solidified at Various
Temperatures
Linear contraction is important in chocolate block molding where good contraction is
required for demolding. At the same time, stable cocoa butter forms are required.

*Reproduced by permission of USDA, New Orleans, U.S.A.*

The two graphs published by these workers show volumetric (Fig.
3.2) and linear (Fig. 3.3) contractions of seeded cocoa butter when
solidified at various temperatures.

In the author's experiments, dilatometer measurements were made
on well-seeded cocoa butter at 31°C (88°F) and volumetric contraction
measured over periods of cooling at 18°C (65°F) and 10°C (50°F).
(See Fig. 3.4.)

These curves show the volumetric contraction over a period of time
under different cooling conditions and it is very apparent that the
contraction is much more rapid at the lower temperatures. From the

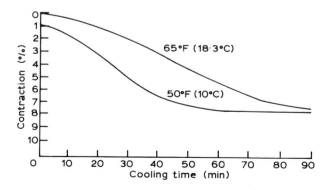

Fig. 3.4. Volumetric Contraction of Seeded Cocoa Butter While Cooling

TABLE 3.3. PERCENTAGE CONTRACTION

| Time of cooling, min | Contraction of volume, % | |
| --- | --- | --- |
| | 18°C (65°F) | 10°C (50°F) |
| 25 | 1.8 | 4.1 |
| 50 | 4.1 | 7.4 |
| 100 | 7.4 | 7.7 |

several experiments, the percentage contraction was calculated and the results are shown in Table 3.3.

Total linear contraction was also measured in rectangular and circular molds and an average figure of 1.9 percent contraction was obtained. Both volumetric and linear results are close to those obtained by Lovegren and Feuge.

In considering these contractions in relation to practical use in enrobers and molding plant, another factor must be taken into account.

Dilatometric measurements on cocoa butter show that at a temperature of 18°C (65°F) an appreciable quantity of the cocoa butter is still in the liquid state dispersed among the solid-crystal phase. Temperatures above or below will have higher or lower quantities of liquid phase present (Table 3.4).

The presence of this liquid phase has an effect on the texture of the cocoa butter or chocolate containing it and at the higher temperatures it has a somewhat plastic consistency. (see graph, Fig. 3.4).

TABLE 3.4. PERCENTAGE OF LIQUID PHASE COCOA BUTTER VARIES WITH TEMPERATURE

| Temperature | | % Liquid phase | | |
| --- | --- | --- | --- | --- |
| | | Lovegren and Feuge well tempered cocoa butter | "Solidified" cocoa butter (Average of various workers) | Cocoa butter/Butter fat 82%/18% as in English milk chocolate (Author) |
| °C | °F | | | |
| 0 | 32 | 0.6 | — | — |
| 5 | 41 | 1.9 | — | — |
| 10 | 50 | 4.1 | 11 | 19 |
| 15 | 59 | 6.8 | 14 | 25 |
| 20 | 68 | 10.8 | 15 | 30 |
| 25 | 77 | 16.7 | 20 | 37 |
| 30 | 86 | 36.1 | 38 | 75 |
| 34.1 | 93.4 | 100 | — | — |

These properties are important in the setting of chocolate, as at the higher temperatures (18°C/65°F) the chocolate coating on a confectionery bar is able to "stretch" or "relax" while cooling, whereas if it is cooled and set completely at temperatures of 10°C (50°F) or below, the chocolate is less likely to relax and can rapidly become brittle and fracture. It may also cause fracture of a fragile center. At the same time, lower cooling temperatures will cause more unstable forms of cocoa butter to be produced, with a risk of discoloration or bloom on storage.

In milk chocolate, the presence of butter fat, which is a low melting fat, increases the proportion of liquid phase at a given temperature.

In the cooling of chocolate, the crystallization and contraction factors mentioned above must be taken into account. These are discussed under "Enrobing," "Chocolate cooling," and "Bloom."

## Solvent Extraction of Cocoa Butter

The solvent extraction of cocoa butter has been developed and perfected over many years. In Holland, Dutch-extracted cocoa butter obtained a reputation as a reliable article of commerce in the chocolate industry, as a result of the attention paid to the selection and processing of raw materials used for extraction, the quality of solvents, and the refining of the extracted fat.

There are many types of solvent-extraction equipment, batch and continuous, with modifications to suit the material being treated, whether seeds, nut residues, offal, or bones. With cocoa, the material for extraction is preferably in the form of corns from expeller presses, or cubes, as powder does not readily allow percolation of the solvent.

Figure 3.5 is a diagram of a continuous extraction plant using pure hexane. This plant consists of a large vertical bucket elevator. The extractor in normal use has thirty buckets, each about 5 ft long by 2 ft wide and 2 ft deep, and each is perforated like a colander. These buckets are filled automatically with the expeller cake when they reach the top of the downward moving side of the elevator and sprayed with the solvent, plus a portion containing some dissolved fat (the source of this mixture, which is known as miscella, will be apparent later). The mixture percolates through the cake and drips from bucket to bucket, dissolving more and more fat as it drips to the bottom of the extractor vessel. It is now approximately a 15 percent mixture of fat and solvent and is pumped to the stills, where the fat is recovered. The buckets on the upward moving side containing partially extracted cake are washed with pure solvent. This, as

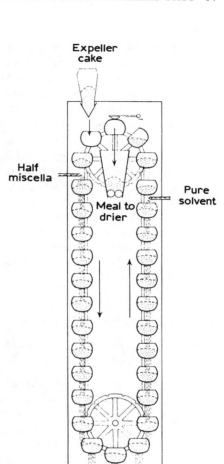

Fig. 3.5. Continuous Cocoa Butter Extractor
Using Pure Hexane

*Reproduced by permission of Messrs Loders Croklaan, Holland*

before, drips from bucket to bucket and eventually falls into a tank at the bottom and forms the spray, which is pumped up to give the washing previously described in the downward-moving side.

After passing the point where the clean solvent has been sprayed on it, the soaked extracted cake continues on its upward journey until it reaches the top, where the bucket is turned upside down, emptying the contents into a hopper, from which they pass to a dryer.

**The Batch Method** Another semicontinuous method of solvent extraction is the batch method, for which the equipment consists of a series of tanks or pots, usually linked in groups of five. These pots are charged with the expeller cake and filled with the solvent. To extract the maximum amount of fat and at the same time economize in operations, every new batch of cake is washed for 20 min each in five solutions of solvent, the first being highly charged with fat from four previous washings, the next having been used three times previously, and so on until the last washing is clean solvent, which in turn is pumped through to the other pots to repeat the process. The solution of fat in solvent is then distilled to recover the fat.

As with many processes unique to an industry, development is continuous. Engineering design is constantly improving to give greater efficiency and economy with regard to the extraction and refining of fats; the student should refer to more detailed books on fat and oil technology.

## Refining of Solvent-Extracted Fat

After extraction, the cocoa butter obtained following evaporation of the solvent is subjected to lye refining, bleaching, and deodorizing.

1. The lye refining consists of degumming with hot water and centrifugal separation followed by alkali washing (neutralization) and further separation. The final stages include washing and vacuum drying. The de Laval short-mix process is a continuous method of carrying out these operations.

2. Bleaching takes place under vacuum with fuller's earth as the bleaching agent, and after reaction, the slurry is pumped to filter presses—still under vacuum—and then to the deodorizer.

3. Deodorizing is carried out with superheated steam under vacuum (see below).

## Deodorization of Cocoa Butter

Cocoa butter extracted by solvent is always refined and deodorized and the result is a relatively bland fat.

It is also the practice of many chocolate manufacturers partially to deodorize their cocoa butter to make it more suitable for the manufacture of milk chocolate and for certain confectionery coatings. Deodorized cocoa butter also has some uses in the pharmaceutical industry because it is a fat that melts just below body temperature and it is preferable to have the strong cocoa aroma removed. When cocoa butter is expressed from alkalized liquor from high-roast beans, it has a very strong flavor that is imparted to delicately flavored milk chocolate, overriding the true milky character. This strong-flavored cocoa butter can be used in dark chocolates, but because of the large demand for milk chocolate, the strong butter is deodorized.

The deodorization of cocoa butter is carried out by the use of superheated steam under vacuum and the aim is mainly the reduction, not complete removal, of flavor.

Figure 3.6 illustrates the principle of the equipment used for this type of deodorization. The essentials are:

1. Dry steam. On no account must condensation take place in the liquid fat at any stage of the process.

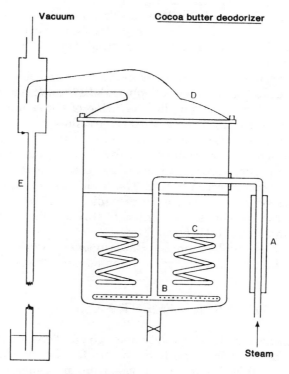

Fig. 3.6.  Cocoa Butter Deodorizing Plant
A. Superheater
C. Heating Coils
E. Drain with Barometric Seal
B. Steam Distributor
D. Vacuum Vessel with Wide Take-off

2. A large wide "take-off" vessel. This is necessary so that condensation does not occur in the neck, causing run-back of odorous substances into the bulk of fat.

3. A vacuum of 3 cm or slightly less.

4. Steam control *before* the superheater.

5. A glass-lined deodorizing vessel. It should include stainless steel heating coils and steamer.

6. Steam produced from clean potable water. Boilerhouse steam from factory mains is unsuitable.

7. Regular cleaning. Bad flavor in deodorized butter has been found to arise from polymerized volatiles accumulating in the take-off vessel from the deodorizer and these can seep back into the bulk of fat.

8. Cooling of the cocoa butter after deodorizing. The process takes place at 104 to 110°C (220 to 230°F) and it is wrong to deliver to storage tanks at that temperature. The fat should be cooled to approximately 70°C(160°F) for storage for short periods before use, but if storage for longer periods is required, then 45 to 50°C (113 to 122°F) should be used.

9. Deodorization under the temperature conditions mentioned above and with adequate steam supply. Both temperature and steam supply depend somewhat on the character of cocoa butter required and each user must experiment with the available equipment.

Pure expressed cocoa butter contains natural antioxidants and these are destroyed or reduced in the deodorizing process. Most manufacturers, therefore, use deodorized cocoa butter as part of a blend with expressed cocoa butter. The inferior keeping properties of deodorized cocoa butter should be recognized and, even if solidified into blocks, storage should be limited to three or four months under cool conditions 13 to 15°C (55 to 60°F).

An interesting publication by Bauermeister (1966) describes in some detail the process for deodorization of cocoa butter.

## COCOA-BUTTER REPLACEMENT FATS

Some explanations have already been given of the need to produce cocoa butter as an independent commodity for the chocolate industry and because of this, its price has risen during times of cocoa crop shortages.

For many years, scientists worked to produce a satisfactory substitute that could be used to replace, at least in part, cocoa butter in high-class chocolate or to replace it fully in coatings.

A description of replacement fats is given in Chapter 6, and the reader is also referred to Chapter 9, on fats.

## Equivalent Fats (CBE)

An equivalent fat is a fat that has all the physical and chemical properties of cocoa butter but has its constituent glycerides derived from sources other than the cocoa bean. It need not have the flavor characteristics of cocoa butter.

The first development in this field was the invention of Coberine by the Unilever Co. working in conjunction with a leading company in the chocolate industry. It was covered by a patent (1961) that described the fractionation of palm oil from acetone solution. The fat has a glyceride composition almost identical to that of cocoa butter and can therefore be mixed with it in any proportion and used in chocolate processing without any alteration in melting, tempering, or cooling. The data given in Table 3.5 and cooling curves show the similarity of the fats. Cooling curves (Fig. 3.7) indicate the eutectic effect when a noncompatible stearine is mixed with cocoa butter, whereas cooling curves (Fig. 3.8) show not only the complete similarity of Coberine to cocoa butter, but also that there is no eutectic effect on mixing.

The legal opinion in a number of countries is that this fat must not be used in chocolate described as such, but in the United Kingdom small additions of Coberine have been made to chocolate for some years. When the United Kingdom entered the European Economic Community (Common Market), it was disclosed that cocoa butter

TABLE 3.5. TYPICAL PHYSICAL AND CHEMICAL CHARACTERISTICS OF COBERINE AND COCOA BUTTER

| Fat | Iodine value | Saponification value | Softening point (Barnicoat) | Dilatation | | Penetration (Hutchinson units) |
|---|---|---|---|---|---|---|
| Coberine | 34 | 196–197 | 35.2°C | $D_{20}$ | 1,950 | 24 at 70°F |
| | | | | $D_{25}$ | 1,840 | 35 at 75°F |
| | | | | $D_{30}$ | 1,380 | |
| | | | | $D_{32.5}$ | 380 | |
| | | | | $D_{35}$ | 0 | |
| Cocoa butter | 37 | 193–194 | 35.4°C | $D_{20}$ | 2,010 | 25 at 70°F |
| | | | | $D_{25}$ | 1,850 | 36 at 75°F |
| | | | | $D_{30}$ | 1,390 | |
| | | | | $D_{32.5}$ | 320 | |
| | | | | $D_{35}$ | 0 | |

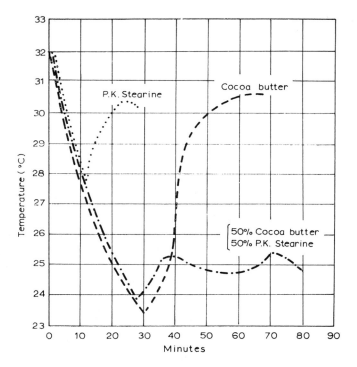

Fig. 3.7. Cooling Curves for Compatible and Noncompatible Fats

*Reproduced by permission of Messrs. Loders Croklaan, Holland*

equivalents were used in chocolate up to a level of 5 percent (approximately 15 percent in the fat phase).

The United Kingdom petitioned the EEC to include a clause in its definition of chocolate to allow up to 5 percent of cocoa butter equivalents. At the same time, the United Kingdom established its own legislation limiting the addition to 5 percent, the law becoming effective in May 1977. The same legal situation exists in Ireland and Denmark.

The advisory agency for the EEC is Caobisco, and if the addition is finally agreed upon, it is likely that the type of vegetable fat will be rigidly defined.

Caobisco established a working party to investigate and recommend methods of analysis to identify the fats that would be permitted in the 5 percent addition. The United Kingdom, with Ireland and

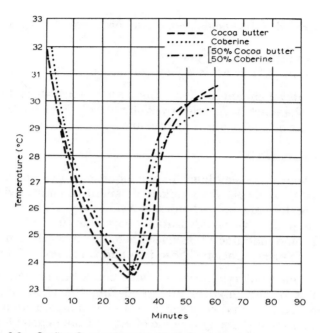

Fig. 3.8.   Cooling Curves Showing Similarity of Cocoa Butter and Coberine

*Reproduced by permission of Messrs. Loders Croklaan, Holland*

Denmark, opposed some of the proposals and the investigations continued.

Reports of the General Assembly of the International Sugar Confectionery Manufacturers' Association (ISCMA)/International Office of Cocoa and Chocolate (IOCC) at Hershey, Pa., in May 1981, and in Brussels, in September 1981, discuss this subject in detail.

The difficulties in determining the amount of fat other than cocoa butter in a mixture are complex, although not impossible, to resolve, according to work carried out by Fincke and by Padley and Tims.

The problem is discussed by Errboe (1981) and an EEC definition of CBE is presented in his article as follows:

1. Level of triglycerides type SOS ≧ 65 percent (S = saturated fatty acids, O = oleic acid).

2. Fractions of the 2 position of triglyceride occupied by unsaturated fatty acids ≧ 85 percent.

3. Total content of unsaturated fatty acids ≦ 45 percent.

4. Unsaturated fatty acids with two or more double bonds $\leqq 5$ percent.

5. Level of lauric acids $\leqq 1$ percent.

6. Level of *trans* fatty acids $\leqq 2$ percent.

Since the development of Coberine, other equivalent fats have appeared on the market. Articles by Wolfe (1977) have dealt with the subject in relation to Codex Alimentarius.

## Substitute Fats (CBS)

Fats that bear no resemblance to cocoa butter but can be used with some degree of success when mixed in small amounts with cocoa butter or chocolate are termed substitutes.

Many cheap fats were advocated years ago for diluting cocoa butter, and some manufacturers used them, to the detriment of their product and their reputation. These fats had three main defects:

1. They formed eutectic mixtures with cocoa butter, lowering the melting point of the mixture and rendering the chocolate soft at normal temperatures and very soft at summer temperatures.

2. They increased the polymorphic effect and made good tempering of the chocolate difficult. This made the chocolate very susceptible to discoloration and bloom formation.

3. They were subject to microbiological or oxidative change that produced rancidity or off flavors.

These fats were produced from coconut, palm kernel, and other seed and nut oils. Their constitution is very different from that of cocoa butter.

Cocoa butter substitutes may be classified in two groups—CBS laurics and CBS nonlaurics.

**CBS Laurics** This is a range of fats with different physical properties but all have triglyceride configurations that make them incompatible with cocoa butter. They are short-chain fatty-acid glycerides. Compound coatings prepared from these fats must use low-fat cocoa powder as the chocolate ingredient.

These fats are mostly based on palm kernel oil and the stearines physically separated from this oil have characteristics similar to cocoa butter and are much superior to hydrogenated or hardened palm kernel oil. The latter fat is still used as a confectionery ingredient.

CBS laurics have good stability and shelf life, and very good texture and flavor release. They do not need tempering as does cocoa butter, they contract and mold well, and, not to be overlooked, they are much cheaper than cocoa butter.

The drawbacks are:

1. Because they are incompatible with cocoa butter, all equipment that has been used for chocolate must be completely cleaned before being used for lauric coatings.
2. When exposed to fat-splitting enzymes (lipase), particularly if in contact with moisture or moisture-containing confections, the free fatty acids (lauric acid) liberated have a strong soapy taste (see Chapter 9).
3. They have a low tolerance to milk fat.

**CBS Nonlaurics**   These fats consist of fractions obtained from hydrogenated oils—usually soya, cottonseed, peanut, and corn (maize) oil. These oils are selectively hydrogenated, with the formation of *trans* acids, which increases the solid phase of the fat.

Their chain lengths and molecular weights are similar to those of cocoa butter, and because of this, CBS nonlaurics will tolerate cocoa butter in a mixture of from 20 to 25 percent.

Many of the nonlauric fats and compound coatings made from them do not have the fracture of cocoa butter, chocolate, or the lauric fats. They may have a waxy texture and do not mold well, and so are frequently used for covering cakes, cookies, and other baked products where this texture is an advantage. Coatings with a brittle fracture will chip away from the centre.

## Other Fats Associated with Cocoa Butter

**Shell Fat**   In the earlier discussion of shell fat, it was indicated that, in the normal extraction of fat from whole beans, the amount of shell fat present in the extracted fat is calculated at 0.74 percent.

If pure shell is extracted with solvent, the resultant fat bears no resemblance to cocoa butter. It is very soft, often with a very high fatty acid content and a poor flavor. It normally would not be used as an ingredient in a greater amount than that which would be present in whole-bean fat. It is regarded by some as an adulterant of cocoa butter.

*Detection in Cocoa Butter*   Fincke (1963) developed a test known as the Blue-Value test based on the presence of behenic acid tryptamide,

TABLE 3.6. PROPERTIES OF SHELL FAT

| Data | Author | Fincke | Other published figures—range |
|---|---|---|---|
| Acidity as oleic acid | 3–7 | 3–64 | 13.7–47.1 |
| Iodine value | 52 | 39.9–41.4 | 39–61.4 |
| Saponification value | 183 | 189–202 | 168–181 |
| Unsaponifiable matter | 1.9% | — | 7.1–14.5% |
| Refractive index | — | Zeiss butyro (40°C) 47.3–57.2 | Zeiss butyro (40°C) 54–61 |
| Melting point (complete) | 28.5°C | 38°–50°C | — |
| Melting point (fatty acids) | — | 50.6°C–51.5°C | — |

which is found only in the shell. The suspected fat, after an alkaline extraction, is redissolved in pure cyclohexane and the ultraviolet spectrum of this solution is measured. The very high unsaponifiable content of shell butter should be noted.

An analysis of shell fat is presented in Table 3.6.

## Allied Fats

Assuming that ultimately the addition of 5 percent equivalent fat is permitted, will there be enough raw material to supply the needs?

The following fats, in addition to palm oil, contain the symmetrical triglycerides required for CBE fats. Some are already used in admixture with the main ingredient—palm oil stearine.

**Illipe Butter** This natural fat bears a very close resemblance to cocoa butter. It is Borneo tallow, known commercially as illipe butter, and is extracted from seeds of the Shorea species. Its melting point is a little higher than that of cocoa butter (complete melting 37 to 38°C/100°F) and the fat imparts slightly different properties to chocolate containing it, particularly the conditions of tempering for enrober use. Organoleptically, it is impossible to detect when added to chocolate in amounts of the order of 5 to 10 percent.

Because of the similarity in glyceride composition to that of cocoa butter it is miscible with it in all proportions and eutectic mixtures are not formed. It is useful, therefore, to raise the softening point of chocolates required for hot climates but the value is marginal only.

*Properties of Illipe Butter* The description 'illipe butter" covers various groups of oil seeds.

Smythies (1957) summarizes them as follows:

1. Illipe nuts from India (sometimes called the true illipe), derived from Bassia longifolia.

2. Mourak nuts from India, derived from Bassia latifolia. These trees belong to the family Sapotaceae, and the fats have different melting points but are often blended and sold as illipe butter.

3. Siak nuts from Sumatra, derived from Palaquium oleosum and P. oblongifolium, which also belong to the family Sapotaceae.

4. Illipe nuts from Sumatra and Borneo, derived from certain species of Shorea in the family Dipterocarpaceae. These are big forest trees producing the common utility timber known as meranti or seraya. The nuts are known as tengkawang in North Borneo and engkabang in Sarawak. The dried kernels are exported under names that denote origin, size, and color, such as large black pontianak.

The nuts are collected as they fall from trees and should not be picked. After collecting, they are put into bamboo crates and immersed in water until the outer coat bursts, which takes several weeks. They are then sun dried to less than 6 percent moisture, and in this condition they are exported for extraction. The fat has a natural pale-green color but is also available commercially in bleached, refined form.

A comparison of the fatty acid and glyceride constituents according to Hilditch (1941) is given in Table 3.7.

Table 3.8 summarizes the physical and chemical constants presented in various publications.

The most significant difference between illipe and cocoa butter is the supercooling limit, which denotes the temperature at which crystals first appear when the liquid fat is slowly cooled under controlled conditions. If illipe fat is used in chocolate in appreciable quantities, the tempering procedure must be altered to a higher temperature range, and preferably with mixing for longer periods, otherwise unstable "seed" will be formed and the risk of fat bloom on storage increased. The use of illipe fat was rife in the late 1920s and early 1930s and there were some disastrous results from lack of knowledge of the special properties of the fat.

Illipe, like cocoa butter, is very stable toward oxidative rancidity.

**Shea Butter** Shea butter is obtained from the nut of an African tree (Butyrospermum parkii) and grows in the vast savannah area from the

TABLE 3.7. COMPARISON OF FATTY ACIDS AND GLYCERIDE CONSTITUENTS

| Fatty acids | Borneo illipe, % | Cocoa butter, % |
|---|---|---|
| Palmitic | 19.5 | 24.3 |
| Stearic | 42.4 | 35.4 |
| Oleic | 36.9 | 38.2 |
| Linoleic | 0.2 | 2.1 |
| Arachidic | 1.0 | — |
| Glycerides | | |
| Trisaturated | 1 | 3 |
| Dipalmitostearin | 2 | — |
| Palmitodistearin | 1 | — |
| Oleodipalmitin | 8 | 4 |
| Oleopalmitostearin | 32 | 57 |
| Oleodistearin | 40 | 22 |
| Palmitodiolein | 3 | 7 |
| Stearodiolein | 13 | 6 |
| Triolein | — | 1 |

Ivory Coast to Nigeria. It is essentially a wild tree, and because it takes fifteen to twenty years to mature and bear crops of nuts, it has not been, so far, successfully cultivated in plantations.

Collection of the nuts is erratic and depends to some extent on the price obtained from alternative crops. However, it is a very useful fat to augment the supply of CBE fats, and ultimately supplies should be stabilized.

TABLE 3.8. PHYSICAL AND CHEMICAL CONSTANTS OF BORNEO ILLIPE AND COCOA BUTTER

| | Borneo illipe | Cocoa butter |
|---|---|---|
| Incipient melting point | 34–36°C | 31.8–33.5°C |
| Complete melting point | 37.5–39°C | 32.8–35.0°C |
| Solidifying point | 28–32°C | 27–29.5°C |
| Refractive index at 40°C | 1.456–1.457 | 1.456–1.458 |
| (Zeiss butyro) | 45–47° | 46–47.5° |
| Saponification value | 188–197 | 191–198 |
| Iodine value | 29–38 | 33–39 |
| Unsaponifiable matter | 0.7–2.0% | 0.5–1.1% |
| Specific heat (temperature range 4–24°C) | 0.51 | 0.49 |
| Latent heat of fusion (from room temperature) | 32 cal/g | 31–33 cal/g |

**Mango and Sal Fats** The trees producing these fats (Mangifera indica and Shorea robusta) grow in India, and it is considered that, with proper harvesting, considerable quantities of fat should be available.

**Antibloom Fats** Bloom is a discoloration occurring mainly on enrobed dark chocolates, particularly if the chocolate is covering fatty centers in which the fat is low melting and in a relatively free state. Coconut paste is a good example of such a center. In some circumstances, bloom will also form on milk chocolate.

At elevated storage temperatures, 21 to 24°C (70 to 75°F), cocoa butter polymorphs in the chocolate coating will melt, diffuse to the surface, and recrystallize in the form of a white bloom. Liquid fat in the center can diffuse into the covering and produce a similar effect. This has already been mentioned when explaining the polymorphic forms of cocoa butter—the causes and prevention of bloom are discussed in detail later in the book.

Certain fats when added to chocolate greatly inhibit bloom formation; of these, the best known is butter oil, and its addition is accepted in countries where substitute fats are not permitted both as an ingredient and for antibloom purposes. The addition of approximately 4 percent in the chocolate produces maximum protection and a minimum of 2 percent is necessary to give an effect of any significance.

Butter oil is dehydrated, curd-free dairy butter, and however it is prepared, it is advisable to heat to 95°C (205°F) for 15 min to destroy lipolytic activity. The 4 percent addition has an appreciable softening effect on the chocolate and some manufacturers prefer to use the lower quantity of 2 percent and retain snap in the texture of their chocolates.

An extensive investigation was carried out by Kleinert (1961) on the relative value of various antibloom additives. From twenty additives tested, only three were really effective: butter oil (dehydrated cow butter), biscuitine (arachis oil plus hardened fat), and hardened arachis oil (Special 36). Note: Biscuitine is a product of the SAIS Company of Switzerland (Unilever). There are now several grades and their respective properties should be checked.

## REFERENCES

Alfa Laval Co. Ltd. Brentford, Middlesex, England.
Baenitz, W. 1978. *Fette, Seifen Anstrich-mittel* 79, 476.

Bauermeister, H. 1966. *House Mag.* 139/66. Hamburg, W. Germany.

British Patents 1961. Nos. 827, 172, and 925, 805. Unilever Co., England.

Chapman, D., Crossley, A., and Davies, A. C. 1957. *J. Chem. Soc.* 1502.

Davidson, R., and Crespo, S. 1979. Cocoa butter crystallization behaviour. Manf. Conf. May.

Errboe, J. 1981. *Bull. ISCMA/IOCC,* Sept., and Official report May.

Fincke, A., and Sacher, H. 1963. *Süsswaren* 428.31.

Fincke, A. 1965. *Handbuch der Kakaoerzeugnisse* (2nd ed.). Springer-Verlag, Berlin.

Fincke, A. 1980. *Deutsche Lebens.* Rund 76, 162.

Hilditch, T. P., and Stainsby, W. J. 1936. *J. Soc. Chem. Ind.* 55, 95T.

Jensen, H. R. 1931. *Chemistry Flavoring and Manufacture of Chocolate, Confectionery and Cocoa.* Blakistons, Philadelphia.

Kattenberg, H. R. 1981. The quality of cocoa butter. Manf. Conf. Jan.

Kleinert, J. 1961. Studies on the formation of fat bloom. *Int. Choc. Rev.* May.

Kleinert, J. 1982. *Review Chocolate, Confectionery and Bakery.* June.

Kleinert, J. 1982. *Review Chocolate, Confectionery and Bakery.* June-Dec.

Lovegren, N. V., and Feuge, R. O. 1965. *J. Am. Oil Chem. Soc.* 42 308.

Lutton, E. S. 1957. *J. Am. Oil Chem. Soc.* 34.521.

Lutton, E. S., and Wille, R. L. 1966. *J. Am. Oil Chem. Soc.* 43.491.

Meara, M. L. 1949. *J. Chem. Soc.* 2154.

Padley, F. B., and Timms, R. E. 1978. *Chem. Ind.,* London, 918.

Pearson, D. 1970. *Chemical Analysis of Foods.* Churchill, London (new edition, 1981).

Savary, P., and Flanzy, J. 1957. *Desnuelle Biochem. Biophys. Acta.* 24, 414.

Schofield, C. A., and Dutton, H. J. 1959. *J. Am. Oil Chem. Soc.* 36, 325.

Smythies, B. E. 1957. *The Illipe Nut.* Government Printing Office, Kuching, Sarawak.

Som, N. H., and Kheiri, M. S. A. 1982. Manf. Conf. Mar.

Steiner, E. H., and Bonar, A. R. 1961. *J. Sci. Fd. Agr.* 12, 247.

Vaeck, S. V. 1960. Manf. Conf., 40(6)35.

Weik, R. W. 1981. Manf. Conf., Jan. 1982.

Wolfe, J. 1977. C.B.E's and codex. Manf. Conf.

# 4

# Emulsifiers in Chocolate Confectionery Coatings and Cocoa

Lecithin is the familiar name for nature's supreme emulsifier and surface active agent. Since its commercial introduction about fifty years ago, it has had a great impact on the food industry, particularly in the manufacture of chocolate. It occurs naturally in all living matter, animal and vegetable, with the highest content in egg yolk (8 to 10 percent). Butter contains 0.5 to 1.2 percent and soya bean oil, at present the main and cheapest source of vegetable lecithin, yields about 2.5 percent.

Vegetable lecithin in its present commercial form is recognized as a nutritious food additive as well as having great practical value. In foods, it is used in chocolate, margarine, vegetable fats, instant powders for drinks, and baked goods, but application has been extended to other industries, such as paint, rubber, plastics, and cosmetics.

With the development of a process for extracting lecithin from soya bean oil, the use of this emulsifier increased manyfold—as lecithin from this source proved to be over 100 times less costly than that from egg yolk.

Because of its molecular structure, commercial lecithin exhibits both lipophilic and hydrophilic properties, and this is responsible for its exceptional value as an emulsifier and wetting agent.

## VEGETABLE LECITHINS

### Soya Lecithin

Soya lecithin is extracted from the beans by continuous leaching with solvent. The solvent is evaporated and the lecithin precipitated

111

from the crude oil by the introduction of steam and water. This precipitate is separated by centrifuge and the residual moisture is removed by vacuum drying. This process results in the light brown commercial product containing about 65 percent acetone insoluble phosphatides, with the remainder mostly soya oil. By choice of solvent and degumming, an improved quality is obtained and the bitter soya flavors reduced, but a still more purified form is prepared by acetone treatment. This solvent will remove residual soya oil and further unwanted flavors and sterols, but the phosphatides remain insoluble and are redissolved in cocoa butter or other vegetable oil.

For some uses, a product light in color is required and this is obtained by treatment with bleaching agents such as hydrogen peroxide and benzoyl peroxide. There is also increasing use for oil-free phosphatides, especially as a diet supplement, but these must contain 2 to 3 percent oil; if completely devoid of oil, they will rapidly deteriorate, oxidize, and become insoluble.

Although these purified and bleached forms of lecithin are available, most of the commercial product now used contains the soya oil carrier. It was originally plastic in consistency but is now mostly in a fluid form, which is more easily dispersed and is suitable for mechanical dispensing.

The approximate composition of commercial lecithin is as follows:

| | | | |
|---|---|---|---|
| Soyabean oil, percent | 35 | Inositol, mg/g | 14 |
| Chemical lecithin, percent (phosphatidyl choline) | 18 | Choline, mg/g | 23 |
| | | Tocopherol, mg/g | 1.3 |
| Cephalin, percent (phosphatidyl ethanolamine) | 15 | Biotin, µg/g | 0.42 |
| | | Folic acid, µg/g | 0.60 |
| Inositol phosphatides, percent | 11 | Thiamin, µg/g | 0.115 |
| Other phospholipids and polar lipids, percent | 9 | Riboflavin, µg/g | 0.33 |
| | | Pantothenic acid, µg/g | 5.59 |
| Carbohydrates (sterol glucoside), percent | 12 | Pyridoxine, µg/g | 0.29 |
| | | Niacin, µg/g | 0.12 |

The main analytical data (AOCS method) are, according to Eichberg (1980):

| | |
|---|---|
| Acetone insoluble | 62 to 65 percent |
| Iodine value | 95 |
| Saponification value | 196 |
| Phosphorus | 2 percent |
| Specific gravity (25°C) | 1.0305 |
| pH | 6.6 |

Moisture                1 percent maximum
Acid value              30 maximum
Benzene insoluble       0.3 maximum
Peroxide value          5 maximum

Manufacturers to also include purity specifications, see Meyer (1983):

| | | | |
|---|---|---|---|
| Lead | 10 ppm maximum | Total plate count/g | 5,000 maximum |
| Arsenic | 3 ppm maximum | Salmonella/25 g | negative |
| Iron | 40 ppm maximum | Yeasts/molds/g | negative |
| Other metals | 15 ppm maximum | Enterobacteriaceae/g | negative |

The chemical structure of the most important phosphatide (phosphatidyl choline) is shown in the following.

$$CH_2-O-\overset{\overset{\displaystyle O}{\|}}{C}-R$$
$$CH-O-\overset{\overset{\displaystyle O}{\|}}{C}-R^1$$
$$CH_2-O-\overset{\overset{\displaystyle O}{\|}}{\underset{\underset{\displaystyle OH}{|}}{P}}-O-CH_2-CH_2-\underset{\underset{\displaystyle OH}{|}}{N}-(CH_3)_3$$

Phosphatidyl Choline

Modern analytical methods have indicated that this substance varies appreciably in extracted "natural" lecithin, mainly in the fatty acid linkages R and $R^1$, which can be any of the higher fatty acids—e.g., palmitic, stearic, oleic, linoleic, or linolenic.

Commercial soya lecithin is soluble in hydrocarbons, fatty acids, and hot animal and vegetable oils. It is not soluble in polar solvents (e.g., acetone) or in water, but water in small amounts will disperse in lecithin and can then be diluted with more water to produce an extended emulsion. This property is very useful when it is required to disperse a water-soluble substance—a color, for example—in a fat medium.

The phosphatides away from the soya oil are less stable and without tocopherol will soon become rancid. It is stated that with soya oil present, the lecithin per se will keep for long periods, but for delicately flavored chocolates such as milk chocolate, a flavor reversion may be detectable after storage. This reversion is recognized in papers by Chang (1966) who has isolated a substance from soya oil, 2-pentyl furan, which is stated to give the undesirable flavor.

TABLE 4.1. PROPERTIES OF VEGETABLE LECITHINS

|  | Cottonseed | Peanut |
|---|---|---|
| Acetone insoluble, % | 54 | 72 |
| Phosphorus, % | 1.9 | 2.4 |
| Moisture, % | 1.0 | 1.0 |
| Appearance | Dark brown | Lighter brown |
| Consistency | Viscous liquid | Plastic solid |
| Flavor, odor | Strong, sometimes offensive odor and flavor | Very slight odor Sour flavor |

## Other Vegetable Lecithins

Vegetable lecithins have been produced commercially from peanut, cottonseed, and corn oil. Representative properties are shown in Table 4.1.

These lecithins were variable and generally had lower viscosity-reducing power. They were not available in sufficient quantities to meet demands of the chocolate industry. Peanut lecithin, when obtainable, did not show the reversion properties of soya lecithin and was suitable for use in delicate milk chocolate.

Two other natural lecithins have been manufactured. Lecithin from safflower oil was produced in the United States as a superior grade, but there was no demand over soya lecithin. Also, lecithin was made from rapeseed oil in Germany during World War II because of the shortage of tropical oils but it has not since become a readily available product. See Markley (1945).

## SYNTHETIC PHOSPHOLIPIDS AND MODIFIED VEGETABLE LECITHINS

Chemical lecithin is phosphatidyl choline, the main constituent of vegetable (soya) lecithin.

Various synthetic phospholipids have been manufactured. One, developed by Cadbury and called "YN", has achieved considerable importance. It does not impart reversion flavors to milk chocolate and shows greater viscosity-reducing power than soya lecithin.

YN is prepared from rapeseed oil by a series of reactions, which involve:

1. *Glycerolysis* under an atmosphere of nitrogen.
2. *Phosphorylation* of the above with phosphorus pentoxide.
3. *Neutralization* with ammonia, filtration and mixing with a proportion of cocoa butter.

This produces a substance with the following composition:

(a) *Tris*-phosphatidic acid (P content, 1.7 percent).
(b) *Bis*-phosphatidic acid (P content, 2.49 percent).
(c) *Bis*-phosphatidyl-*mono*-phosphatidic acid (P content, 3.28 percent).
(d) *Bis*-phosphatidyl-lyso-phosphatidic acid (P content, 3.77 percent) or perhaps the equivalent 1:2 cyclo form.
(e) *Mono*-phosphatidic acid (P content, 4.62 percent).
(f) Lyso-phosphatidic acid (P content, 7.35 percent) or equivalent 1:2 cyclo form.

| | |
|---|---|
| Triglyceride (inactive) | 40 percent |
| Neutral phospholipid—(a) above | 15 percent |
| Mixed phosphatidic acids—(b) to (f) above—as $NH_4$ salts | 40 percent |
| $NH_4$ salts of phosphoric acids mainly (with some organic content) | 5 percent |

There is no evidence of polymeric organic compounds.


## Toxicity Checks on YN

During the development of this product, toxicity checks were made in conjunction with the British Industrial Biological Research Association; Gaunt (1967) and Fouer (1967).

Since 1962, its use in the United Kingdom has been permitted under the Emulsifiers and Stabilizers Regulations no. 720. More recently, in the European Economic Community Directive no. T422, June 30, 1980 (80/608/EEC), it has been approved for cocoa and chocolate products. To date, the following countries have approved YN legally—the United Kingdom, West Germany, Ireland, Iceland, the Netherlands, Switzerland, Australia, Canada, New Zealand, Ghana, Kenya, and Nigeria.

It is now supplied by other companies specializing in lecithin products; Meyer (1983).

TABLE 4.2. TYPICAL MODIFIED LECITHINS

| | Oil-free natural lechithin, % | Phosphatidyl choline + cephalin concentrate (alcohol soluble), % | Inositol Phosphatides + cephalin concentrate (alcohol insoluble), % |
|---|---|---|---|
| Chemical lecithin | 26.8 | 55 | 10 |
| Chemical cephalin | 22.4 | 25 | 30 |
| Inositol phosphatides | 16.4 | 7 | 40 |
| Soy bean oil | 3.1 | 4 | 4 |
| Miscellaneous | 31.3 | 9 | 16 |

## Fractionated and Modified Vegetable Lecithins

These lecithins find particular application in the formation of water-in-oil or oil-in-water emulsions, and in the coating of powders where instant wetting is required for drinking chocolate and cocoa powders.

Lecithins modified to improve their hydrophilic properties are used in baking processes. The fractionation of lecithin is carried out by extraction of natural lecithin with alcohol. The alcohol-soluble fraction is water dispersible and readily forms oil in water emulsions, and the insoluble fraction forms water in oil emulsions.

Typical lecithins have the analyses and properties shown in Table 4.2. Eichberg (1980).

Carriers are added to these concentrates depending on their use—cocoa butter or other vegetable oil and propylene glycol are examples where the lecithin is to be added to food. Hydroxylated lecithins are prepared by treatment with hydrogen peroxide and lactic and acetic acid, which improves the hydrophilic properties. Used in conjunction with mono- and diglycerides in cakes and bread, they improve texture and handling.

## USE OF VEGETABLE LECITHIN AND OTHER PHOSPHOLIPIDS IN CHOCOLATE

Chocolate is a dispersion of very fine solid particles in a fat phase. The solids are composed of sugar and ground cocoa material in the case of dark chocolate. In milk chocolate, milk solids particles are present in addition to milk fat, the latter being included in the fat phase.

In the first processes of chocolate production, the fat is all in the liquid state, but in the later stages the chocolate used for molding or enrobing is in a tempered condition. The solid phase then includes a proportion of fat crystals, normally cocoa butter, which influence the fluidity of the chocolate, in addition to the particles of sugar, cocoa, and milk.

## Viscosity

Because of the presence of these solids, chocolate does not behave as a true liquid but exhibits non-Newtonian properties. The viscosity of the liquid chocolate is, therefore, very much greater than that of the liquid fats alone (70 poise, for example, compared with 0.4 poise) and the degree of flow of chocolate is greatly dependent upon the ease with which the solid particles are able to move over one another within the liquid phase. It is clear that the addition of a surface active agent will have a great effect on fluidity, and this is what happens when lecithin is added. Chocolate with a working viscosity suitable for molding or enrobing can be prepared with a much lower cocoa butter content if lecithin is present, and since cocoa butter is an expensive ingredient, the economic value of lecithin is obvious.

Figure 4.1 shows the effect of lecithin addition on the fat content of

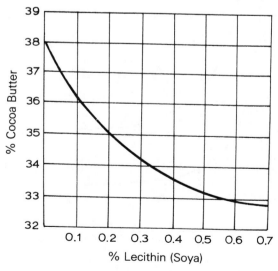

Fig. 4.1.   Effects of Lecithin on Fat Content in Dark Enrobing Chocolate

dark enrobing chocolate. There is a reduction of 5 percent fat due to the lecithin, and this constitutes about 13 percent of the total fat present.

**Moisture Content. Effect on Viscosity** Chocolate normally contains 0.5 to 1.5 percent of moisture, and if further minute amounts are added and mixed in as "free" moisture, the viscosity greatly increases. If the same amount of water is added to the liquid fat alone, no similar viscosity change occurs but with a fine sugar and fat mixture, the viscosity effect is the same as with chocolate.

The addition of lecithin to chocolate or a sugar/fat mixture brings about a marked reduction in viscosity, as shown by the graphs in Fig. 4.2. The effect on a cocoa/fat mixture is very much less.

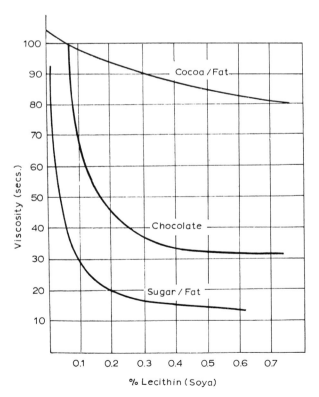

Fig. 4.2. Influence of Lecithin on Viscosity as Determined by Redwood-Type Viscometer

Lecithin displays both hydrophilic and lipophilic properties. Although the exact function of lecithin in chocolate has not been elucidated, T. L. Harris (1968) has made the following observations.

Moisture at the surface of the sugar particles increases the friction between them. This results in a greater resistance when the particles move among themselves and produces the effect of increased viscosity.

When lecithin is added, the hydrophilic groups of the molecules attach themselves firmly to the water molecules on the surface of the sugar particles. This reduces friction, increases particle mobility, and lowers viscosity.

Some support for this theory has been obtained by experiments that show that some of the lecithin is tightly bound to the chocolate particles:

1. If chocolate, to which lecithin in a known quantity has been added, is extracted with warm petroleum ether or similar solvent, all the lecithin does not appear in the extracted fat even if repeated extractions are made. Lecithin estimations using an extraction technique tend to give results that are about 70 percent of the theoretical.

2. A fine sugar/cocoa butter/lecithin mixture is made and the viscosity checked. The mixture is then extracted with petroleum ether until the sugar is fat free. This sugar is then again made up into a mixture with the same amount of fresh cocoa butter alone and it will be found that the viscosity is very similar to the original mixture, indicating that the activity of the lecithin was retained on the sugar particles.

From these experiments, it would appear that only a proportion of the total lecithin added to a chocolate is effective in reducing viscosity. This, however, is not the case, as will be seen from Figs. 4.1 and 4.2 where a steady decrease in viscosity is obtained with increasing lecithin addition up to about 0.5 percent. The lecithin molecules not attached to the particles obviously play some part in decreasing viscosity but the mechanism of their action is not clear.

## Other Physical Effects of Lecithin Addition

In addition to the marked viscosity reduction obtained by the addition of lecithin to chocolate, other physical changes are noticeable.

**Temperature**  When chocolate without lecithin is heated, a marked increase in viscosity occurs above certain temperatures. With dark chocolate, the critical temperature is about 90°F (194°F), a temperature that is rarely reached in chocolate processing, but with milk chocolate, a large increase in viscosity occurs at about 60°C (140°F). Although much milk chocolate processing is carried out below 52°C (125°F), sometimes conching at 60°C (140°F) is used to develop flavor. When lecithin is added, higher temperatures are possible without viscosity change, and with milk chocolate, the temperature can then be taken to 80°C (176°F).

However, with milk chocolate, because of variations in the properties of milk powders, particularly whole-milk powder, the higher conche temperatures sometimes produce a type of granulation in the chocolate. This is less likely if nonfat milk is used (see "Chocolate Manufacture").

**Tempering**  The tempering of chocolate ensures the development of stable cocoa butter crystals within the liquid chocolate. Properly tempered chocolate will set in a condition in which it is not susceptible to discoloration or bloom under normal storage conditions. The addition of lecithin brings about a change in the conditions for tempering, and supercooling to slightly lower temperatures is observed.

Because of this change of crystallization pattern, many claims have been made that lecithin improves bloom resistance, gloss, and latitude in handling tempered chocolate. The evidence for these improvements is very slight and the important point is to recognize the slight difference in tempering and adjust the process accordingly.

If by error, or for any other reason, the lecithin content in chocolate is increased much above the accepted normal limit of 0.5 percent, a very marked difference in tempering conditions occurs. Whereas the normal temperature reduction to induce "seed" in chocolate is 27 to 29°C (82 to 85°F), with a lecithin content of over 1 percent, the seeding temperature may be as low as 21°C (70°F). This problem is seen occasionally in automatic lecithin-dispensing systems.

**Viscosity-Reducing Power of YN Compared with Soya Lecithin**  The viscosity-reducing power of YN continues with additions up to 0.8 percent and over. It also shows a greater effect at the lower additions, e.g., 0.1 to 0.5 percent of the chocolate. With soya lecithin, viscosity reduction ceases at about 0.5 percent addition, and further additions result in increased viscosity.

Tests were made with various additions to milk chocolates and the viscosity determined on a Brookfield viscometer at different rates of shear. It is recognized that the viscosity of chocolate varies with the degree of movement to which it is subjected—or, in technical terms, rate of shear.

Graphs indicating the comparative viscosity reductions of increasing additions of YN and soya lecithin at different rates of shear are shown in Fig. 4.3.

Milk chocolate with a total fat content of 34.0 percent was used with lecithin additions of 0.1 to 0.5 percent. Cocoa butter was added to all the chocolates with less than 0.5 percent lecithin addition so that the total fat, including lecithin, of all samples was 34.5 percent.

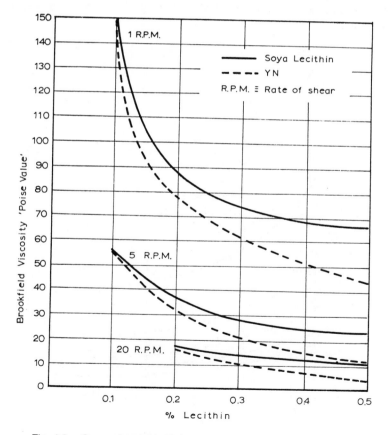

Fig. 4.3.   Comparison of Influence of Lecithin and YN on Viscosity

Viscosity determinations were made at 40°C (104°F), at which temperature the fat was free from seed.

From these graphs, it will be noted that YN has a viscosity-reducing power about 5/3 that of soya lecithin.

The slope of the soya lecithin curves is flattening considerably at 0.5 percent addition compared with YN, and a second series of curves (Fig. 4.4) shows the viscosity effect of additions of YN up to 0.9 percent and a significant decrease in viscosity occurs between 0.5 percent and 0.9 percent additions. For these experiments, a thicker milk chocolate was used with 32.5 percent total fat content.

In various parts of the text, terms related to viscosity are mentioned (e.g., rate of shear, yield value, Newtonian and non-Newtonian liquids). Later in the chapter, these are discussed in relation to the determination of viscosity of chocolate and vegetable fat coatings.

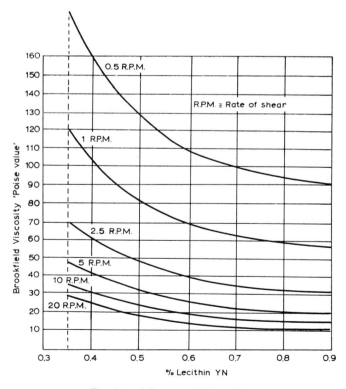

Fig. 4.4. Influence of YN on Viscosity

## Other Surface Active Compounds

In addition to YN, other phosphatides and complex glycerides have been prepared for use in chocolate and other foodstuffs.

Chovis compounds, made originally by the Emulsol Company of America, were used years ago in considerable quantities in chocolate coatings but were found to be too costly in comparison with lecithin. They are approved in the United States [U.S. FDA (1944)] for use in food and bear a close relationship to YN. YN consists of ammonium salts of phosphatidic acids, whereas the Chovis compounds are sodium salts as follows: 1,2-diglyceride monosodium phosphate, 1-monoglyceride-3-monosodium phosphate.

Emargol, an American product, has been proposed for use in margarine; it is 1-mono stearin-3 monosodium sulphoacetate.

Several phosphated monoglycerides are manufactured under the name of Emcol (Witco Chemical Co.), but there is little evidence of their use in chocolate.

Sucrose esters have been tried experimentally in chocolate; Osipow (1956). Sorbitan stearates (Span 60, Tween 60) have been found ineffective in chocolate but are useful in vegetable fat coatings.

Polyglyceryl ricinoleate, the partial polyglyceryl ester of inter-esterified castor oil fatty acids, is synergistic to lecithin and has proved very useful in modifying the yield value of high-viscosity chocolates.

TABLE 4.3. DARK CHOCOLATE

| Lecithin, % | Plastic viscosity (poise) | Yield value (dynes/cm$^2$) | PGPR % | Plastic viscosity (poise) | Yield value (dynes/cm$^2$) |
|---|---|---|---|---|---|
| 0.3 | 18.5 | 155 | 0.1 | 12.5 | 151 |
| 0.7 | 17.1 | 221 | 0.2 | 14.8 | 82 |
| 1.3 | 12.4 | 285 | 0.5 | 14.9 | 13 |

TABLE 4.4. MILK CHOCOLATE CONTAINING 0.5% LECITHIN

| PGPR added, % | Plastic viscosity (poise) | Yield value (dynes/cm$^2$) |
|---|---|---|
| Nil | 15.3 | 72 |
| 0.1 | 15.2 | 64 |
| 0.2 | 15.6 | 56 |
| 0.3 | 17.4 | 30 |
| 0.4 | 16.4 | 26 |

The effects are described by Bamford et al. (1970) and the results from their paper are indicative of the effect of this substance (Tables 4.3 and 4.4).

## RHEOLOGY, VISCOSITY, DEFINITIONS, MEASUREMENT OF VISCOSITY

The terms "rheology" and "viscosity" are used frequently in the food industry to describe the flow properties of various products.

*Rheology* is defined as "the study of the deformation and flow of matter."

*Viscosity* is the name given to the internal friction of fluids.

Energy must be applied to start and maintain flow. The mathematical treatment of viscosity is highly complex and is not considered here. References at the end of the chapter give more detailed information.

However, it is necessary to have a basic understanding of some of the terms used in the measurement of viscosity, particularly in relation to the flow properties of chocolate.

There are essentially two types of liquid—Newtonian and non-Newtonian.

The viscosity of Newtonian liquids is independent of the rate of shear (mixing) but the viscosity changes with temperature. Liquids in this category are water, alcohol, thin oils, liquid fats, and glycerol. Liquids such as chocolate, paint, printing ink, which do not have this simple relationship, are known as non-Newtonian liquids. Their viscosity is affected by the presence of solids in suspension, as well as by temperature.

It is characteristic of these liquids that they only start to flow when their yield point (see later) has been reached, and then their viscosity decreases with increasing rate of shear.

These flow properties were studied by Casson (1959) in relation to printing ink and his findings were corroborated by Steiner (1959) for chocolate.

It was then recognized that the Casson values, as they are known, were much more likely to define the flow properties of chocolate than the figures obtained by the single-speed rotational viscometer (e.g., McMichael) or orifice viscometer (Redwood). Casson values are now defined as:

*Plastic viscosity*—the force required to maintain constant flow in a fluid mass.

*Yield value*—the force required to initiate flow in a fluid mass.

The value of these figures has become much more significant since the introduction of lecithin and the use of high-viscosity (less fluid) chocolates. High-viscosity milk chocolates are particularly susceptible to variation because of the milk proteins present.

The practical effects of yield value and plastic viscosity may be illustrated as follows:

Chocolate molding is aided by a low yield value. Molding chocolates are often made to a formulation that is low in fat and consequently are very viscous and resistant to flow and require considerable energy to make them flow out into the molds. It has been noted that use can be made of synergistic emulsifiers (polyglycerol polyricinoleate) to decrease yield value.

In chocolate enrobing, it is necessary to have a high enough yield value to prevent decorations from collapsing and to avoid the chocolate flowing off the centers, causing flanges on the base edges.

By using single-speed rotation viscometers or orifice flow viscometers, it is possible to have two chocolates with identical viscosity values but which in machine performances are quite different because of different yield values. In the past, this has caused frequent arguments between control and production personnel.

## VISCOMETERS

These instruments come in many designs. Some are simple and inexpensive, and the viscosity values obtained are empirical only. Others are more complicated and capable of obtaining accurate and complete information on the flow properties of both Newtonian and non-Newtonian liquids.

### Simple Viscometers

In earlier times, chocolate was made much more fluid, but with the ever-increasing cost of cocoa butter, this has been reduced and the viscosity consequently increased.

With fluid chocolate, it was possible to measure flow properties by means of simple viscometers of the Redwood type used for lubricating oils. Many companies constructed their own instruments and formed their own internal standards for production control. Such instruments can be made inexpensively and are still used by small companies working with fluid compounds or ice cream coatings.

The Redwood type of viscometer consists of a jacketed cylindrical container fitted with a small tube at the base. The tube is standardized with respect to length and orifice diameter.

The jacket is filled with water at 38°C (100°F), preferably thermostatically controlled. Chocolate at this temperature but cooled from 49°C (120°F) is filled to a given mark on the container, with the orifice at the base plugged by a small rod with a ball fixed at the lower end. Release of the ball allows the chocolate to flow into a small pot, and the time taken to fill this pot is recorded. Normally, the flow time should be 25 to 60 sec and the dimensions of the orifice arranged accordingly.

This is an empirical instrument but is quite reliable for thin coatings (Fig. 4.5). For compound coatings, viscosity is taken at 49°C (120°F). An instrument for thicker coatings is based on the falling-ball viscometer, but instead of the ball a graduated cone is used. The chocolate under test is adjusted, as previously, to the correct temperature. The cone is clamped at a controlled distance above the chocolate, and to make the determination, the clamp is released and the depth to which the cone sinks is measured by the graduations on the surface (Fig. 4.6).

This, again, is an empirical viscometer, but it is useful as a check on the coating viscosity.

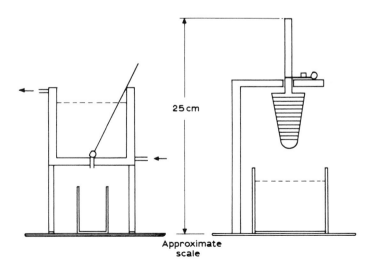

25 cm

Approximate
scale

Fig. 4.5.  Redwood-Type Viscometer        Fig. 4.6.  Falling Cone Viscometer

With modern high-capacity equipment and the continued use of low-fat, high-viscosity chocolates, more precise instruments are required.

## Rotational Viscometers

Rotational viscometers in general use in the chocolate industry include the MacMichael, a single-speed instrument, and the Brookfield and Haake multispeed instruments.

**MacMichael Viscometer (Fig. 4.7)** This rotational viscometer is the instrument accepted by the American Association of Candy Technologists and has been adopted by the National Confectioners Association and most chocolate manufacturers in the United States. The principle of the instrument is as follows. A metal cylinder is suspended on a torsion wire and this, in turn, is immersed in a cup that contains the chocolate to be tested. Chocolate at a given temperature is poured into the cup to a given level on the cylinder and the cup is rotated at a standard speed. The twist in the wire caused by the movement of the rotating chocolate is measured by a scale attached to the torsion

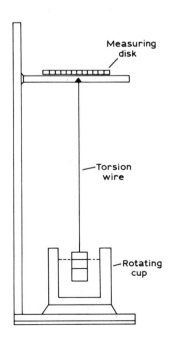

Fig. 4.7. MacMichael Viscometer

wire. For accuracy, the whole instrument should be kept in a thermostatically controlled cabinet.

Standard temperatures and dimensions are as follows:

Chocolate
Cup—internal diameter 6.9 cm, rotation 15 rpm.
Cylinder—diameter 2.0 cm, immersion level 3 cm.
Torsion wire—no. 26.
Chocolate temperatures—100°F (38°C) (cooled from 120°F (50°C)).
Compound coatings
As chocolate but temperature 120°F (cooled from 125°F) (52°C)).
Ice cream coatings
Temperature—38°C (100°F).
Disk 5.7 cm (2¼ in.) in place of cylinder with 4 cm immersion.

The MacMichael instrument needs regular checking for rotation speed, and the torsion against a standard wire.

In recent years, the instrument has been criticized because of its failure to provide full information on the flow properties of different chocolates.

In addition to the fact that it is a single-speed instrument, another failing is the incorrect dimension of the annular space between the cup and the cylinder.

The ratio of the diameter of the cylinder to the cup is so small that shearing of the chocolate does not occur across the whole gap. Nevertheless, Robbins (1979) has compared the performance of the MacMichael instrument with the Brookfield. He has shown that by using a single speed of "bob" on the Brookfield (20 rpm), there is a constant conversion factor of 3.40 from Brookfield to MacMichael viscometers.

**Brookfield/Haake Viscometers**  With these instruments, plastic viscosity and yield value can be accurately determined. The principle is shown diagrammatically in Fig. 4.8 and Robbins has described the method of operation. A descriptive brochure by Haake (Germany) gives full details of the company's instruments and methods of use.

The general procedure is as follows:

The chocolate sample is carefully prepared by melting and mixing at 50°C (122°F), taking care to avoid including air. This is then cooled to about 43°C (110°F) and transferred to the cup of the viscometer. The outer cylinder is a water jacket controlled accurately by means of a circulating thermostat at 40°C ± 0.10. The inner cylinder is set in motion as soon as it is seen that the chocolate temperature has

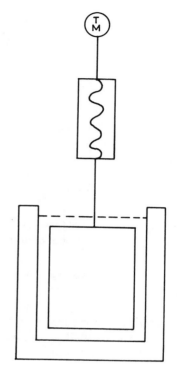

Fig. 4.8. Principle of Haake and Brookfield Viscometers
The Sample is filled into the annular gap between the concentric cylinders. While the center cylinder rotates at a specific speed, the torque is measured.

stabilized at 40°C (104°F). Viscosity readings are then taken at increasing rates of shear, ranging from 1 to 50 rpm, followed by similar readings at decreasing speeds. For the purpose of calculation, the average figures for increasing and decreasing speeds are taken.

Following Casson's theory, to obtain plastic viscosity and yield value, it is necessary to plot graphically the square root of shear rate (rpm) against the square root of the shear stress (viscometer readings). A straight-line graph is then obtained, as illustrated in Fig. 4.9. Complicated calculations may be avoided in routine testing by using a computer system by Richardson (California).

Having obtained the figures so far mentioned, how can they be applied to investigate production variations?

The flow properties of chocolate are influenced by moisture content, fat content, particle size, temperature of the process, and the nature of the ingredients. With the means to measure the flow properties, a tabulation of conditions prevailing in production can be prepared, and the formulations used tabulated (Fig. 4.10). Against each set of

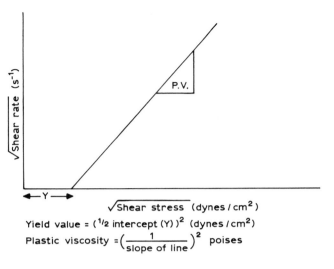

P.V.
$\sqrt{\text{Shear rate}}$ (s$^{-1}$)
Y
$\sqrt{\text{Shear stress}}$ (dynes / cm$^2$)
Yield value = ($^1/_2$ intercept (Y))$^2$ (dynes / cm$^2$)
Plastic viscosity = $\left(\frac{1}{\text{slope of line}}\right)^2$ poises

Fig. 4.9. Plastic Viscosity and Yield Value Determination

readings, a record of any changes in ingredients, formulations, or the process, as well as other observations, should be made.

| Date | Type of Chocolate | Plastic Viscosity | Yield Value | Fat Content | Moisture Content | Particle Size | Remarks |
|---|---|---|---|---|---|---|---|
| | | | | | | | |

Fig. 4.10. Recommended Tabulation of Data

## The Gardner Mobilometer

This instrument was originally recommended to the American Newspaper Publishers Association (1932) to measure the consistency of printing ink.

Martin and Smullen (1981) have investigated its application to chocolate, and in their report recommend it as a reliable and inexpensive instrument for the determination of plastic viscosity and yield value.

The instrument consists of a vertical cylinder, supported on a level base plate and vertical post, through which a plunger assembly (mounted in a bracket) moves downward for a definite distance under a constant load in a measured time. The gravitational driving force of the plunger causes the fluid sample to stream upward through four orifices in the disk of the plunger.

The plunger assembly consists of a piston rod, the lower threaded end of which is attached to a disk (fiftyone-hole, four-hole, or solid) while the upper hollow end supports a weight pan on which additional loads may be placed. The weight of the complete piston assembly is 100 g when it consists of any disk, the piston rod, and the empty, unloaded weight pan.

The chocolate to be tested is brought to 40°C and the tube filled to within 1 cm from the top. The plunger is moved up and down once to remove air bubbles. The plunger rod is marked in centimeters, and for the viscosity test, it is released and the time for it to sink 10 cm is taken.

A weight is then placed on the upper pan and the test repeated. Further tests with other weights will give a series of readings from which plastic viscosity and yield value may be calculated.

The workers found good correlation with the Haake viscometer and further experimental work is taking place.

## USE OF LECITHIN IN CHOCOLATE, COCOA POWDER, CHOCOLATE DRINKS

### Chocolate

From the foregoing, it will be understood that the action of lecithin is purely a surface phenomenon, and because of this, it is essential that the maximum amount of the lecithin added be active on the particle surfaces.

The amount of lecithin that is effective in chocolate is within fairly narrow limits, i.e., 0.2 to 0.6 percent for soya and other vegetable lecithins, or up to about 1 percent for the synthetic phospholipid YN. The figures for vegetable lecithins refer to commercial natural lecithins and normally the soya product contains 65 to 70 percent of active phospholipids. The remainder is oil natural to the source material, or it may be replaced by cocoa butter or a refined vegetable oil. In countries where legislation controls the amount of lecithin that may be added, active phospholipid content is sometimes specified. Other countries will not yet allow lecithin substitutes.

Viscosity control applied to chocolate making is a complicated procedure and the mere addition of, say, 0.5 percent lecithin with all the other ingredients will not give maximum viscosity reduction. In actual practice, lecithin is added primarily to save cocoa butter at all stages of the process; therefore, it is sometimes advantageous to split the addition between the refining and the conching stages. In the final conching stage, lecithin should be added as late as possible to get maximum viscosity reduction, and this is related to the necessity of keeping the lecithin on the surface of the particles. A mixture that is too dry or too liquid will not feed to the refining rolls at its optimum speed and will give uneven films on the rolls. If, however, the refiner paste can be made into a suitable physical condition with low cocoa butter content, then the effect of this fat saving is continued throughout the whole making process.

Thus, if the total lecithin addition is 0.5 percent it is probable that it will be advantageous to add 0.15 percent to 0.2 percent at the mixing stage prior to refining, and the remainder late in the conching stage. If, at the refining stage, the paste without lecithin has a fat content of, say 27 percent to ensure correct feeding to the rolls, then with lecithin the same consistency may be obtained with a fat content 1 to 2 percent less. The actual fat contents depend on the formulation and particle size of the paste but the lower fat contents achieved with the addition of lecithin will be reflected in lower fat contents in the final chocolate.

The next stage in the procedure is to reduce the flake from the refiner rolls to a mobile paste suitable for the commencement of conching *without* the addition of more lecithin. This is achieved by mechanical churning or beating, either in a separate mixer or in the first stage of the cycle of operations in a rotary or continuous conche. This part of the process is termed dry conching (see "Chocolate Manufacture"), and is followed by mixing at a higher speed, extra cocoa butter sometimes is added at this stage.

Near to the end of the conching period, the remainder of the lecithin is added together with any flavor, and after sufficient time for dispersion, the viscosity is checked and adjusted by adding further cocoa butter if necessary. Viscosity should never be adjusted by adding more lecithin.

It is sometimes possible to get a further viscosity reduction by rapid mixing after conching, but this depends on the type of chocolate and the cocoa butter content. The extent of this reduction can only be found by experiment.

## Cocoa and Drinking Chocolate Powders

Fine powders, particularly if they contain fat such as cocoa powder, and chocolate powder for beverages are not easily wetted or dispersed in water or water-based liquids such as milk.

The incorporation of lecithin or a modified lecithin as a surface active agent will bring about a change in the physical condition of the powder, enabling dispersion to take place.

Often, the addition of lecithin is combined with a process of instantizing (see "Milk Powder"). This cements the fine particles into agglomerates that contain capillary channels, and these allow liquid to be drawn in, thereby producing a wetting action. Agglomerates also affect the bulk density so that the material has a greater volume for a given weight.

Modified lecithins are now produced specifically for use in the manufacture of wettable powders (Lucas Meyer, West Germany); they consist of liquid products that can be sprayed at normal temperatures on powders in tumbler mixers or pulverizers, or during a spray-drying process. Alternatively, where a powder has already been agglomerated, a modified lecithin in powder form can be added in a simple dry mixing process.

## REFERENCES

Bamford, H. F., Gardiner, K. J., Howat, G. R., and Thomson, A. F.   1970.   The use of Polyglycerol Polyricinoleate in Chocolate. Confectionery Production U.K.

Cadbury Bros. Ltd.   1966.   British Patent no. 1,032,465.

Casson, N.   1959.   *Flow Equation for Pigment-Oil Suspensions of the Printing Ink Type. Rheology of Disperse Systems.* Pergamon Press, London.

Chang, S. S., and Wilson, J. R.   1964.   Soya bean oil in our foods. *Ill. Med. J.*

Chang, S. S.   1966.   Reversion flavors in soya bean oil. *Chem. Ind.* London.

Eichberg, J. American Lecithin Co., Atlanta, Ga.

Fouer, G.   1967.   Metabolic fate of $^{32}$P labelled emulsifier YN in rats. *Fd. Cosmet. Toxicol.* 5(5), 631. London.

Gaunt, I. F., Grasso, P., and Gangoli, S. D.   1967.   Short term toxicity study of emulsifier YN in rats. *Fd. Cosmet. Toxicol.* 5(5), 623. London.

Haake Inc., Karlsruhe, West Germany.

Harris, T. L.   1968.   Surface active lipids in foods. Monograph no. 32, Society of Chemical Industry, London.

Markley, K. S.   1945.   Oilseeds and related industries of Germany. U.S. Dept. Comm. Office Tech. Serv. P.S. Rept. 18,302.

Martin, R. A., and Smullen, J. F.   1981.   Simplified instrumentation for the measurement of chocolate viscosity. *Manf. Conf.* May.

Meyer, Lucas, 1985. Hamburg, W. Germany.

Meyer, Lucas, *Metarin-Fractionated Lecithin.* Hamburg, West Germany.

Osipow, L., Snell, F. D., York, W. C., and Finchler, A. 1956. *Ind. Eng. Chem.* 48, 1459. London.

Richardson, T. W. 1979. Richardson Researches Inc., Hayward, Calif.

Robbins, J. W. 1979. A quick reliable method for measuring yield value, plastic viscosity and "MacMichael" Viscosity of Chocolate. *Manuf. Confect.*

Steiner, E. H. Rheology of disperse systems. In Casson 1959.

U.S. Food and Drugs Administration. 1944. *Definitions and Standards for Foods.* Title 21. Pt 14, Cacao Products, Sec. 14.6 (a) and 14.7 (a).

U.S. Patent No. 2,629,662.

Witco Chemical Co. Inc., Chicago, Ill.

# 5

# Chocolate Manufacture

## INGREDIENTS

The basic ingredients required for chocolate manufacture are cocoa nibs, cocoa liquor, sugar, other sweeteners, cocoa butter, butter fat (oil), milk powder, milk crumb, and emulsifiers. Some of these ingredients are described in detail elsewhere in the book but their main properties relevant to chocolate are described in the following.

## Cocoa Nibs, Cocoa Liquor

These are prepared as described under "Roasting, Winnowing and Nib Grinding." Lower-temperature roasts are usual for milk chocolate and for some dark chocolates.

The types of beans used for both dark and milk chocolate are primarily the "Bulk" or "Ordinary" grades, mostly Forastero (described in Chapter 1).

For special high-grade dark chocolate, Trinitario, and occasionally Criollo, beans may be employed and the formulations and processes are confined to particular companies.

These "fine" or "flavor" cocoas show great variation and many companies buy their cocoa beans on "mark," which means from the relatively small area where they are grown.

It is appropriate here to mention imported liquor, which means that it has been produced in the area where the beans are cultivated. It is regrettable that its quality is often poor, sometimes contaminated, and the user company has no real check on the quality of the beans used to make the liquor. With the beans themselves, it is comparatively easy to make quality checks and to reject consignments, if necessary.

Methods of liquor deflavoring have been described elsewhere and have met with a measure of success, but some extraneous flavors are not easily removed.

## Sugar and Other Sweeteners

High-grade sugar should be used in manufacturing chocolate; it must be dry and free from invert sugar. Color is less important than in the manufacture of fondant, for example. Certain grades of off-white sugar, particularly for chocolate manufacture, are available.

Washed raw sugars are sometimes used in health-food chocolates but they usually contain some invert sugar and moisture. They cause refining problems (see later) because, whereas fully refined sugar crystals fracture readily, the presence of moisture and invert causes the sugar to roll into "plates," which is detrimental to the chocolate's texture, and presents difficulties in the subsequent conching and enrobing processes.

Dextrose and anhydrous corn syrup (glucose syrup) are used as partial replacements for sugar. Anhydrous dextrose, not dextrose hydrate, should be used. Anhydrous corn syrup is very hygroscopic and can cause trouble with moisture absorption during refining.

Both of these substances reduce sweetness and can be less expensive. Nonsugar substances such as sorbitol, mannitol, and xylitol are occasionally used as sweeteners in dietetic chocolates (see later).

## Cocoa Butter

The properties of cocoa butter are described in Chapter 3. Physical properties are important; milk chocolate requires cocoa butter with a mild flavor. Fracture and contraction also are important to afford good molding and texture. Some butters, including refined butters, may have poor textures.

## Milk Products

Whole-milk powder and nonfat-milk powder are the primary milk products used. In some countries, whey (usually demineralized) is permitted. Milk crumb is described later. Butter fat (or oil) is derived from unsalted dairy butter by dehydration and removal of curd. It is used in conjunction with nonfat-milk powder to make less expensive milk chocolate.

It also is used as an antibloom agent in dark chocolate, and occasionally to replace some of the cocoa butter constituent of either dark or milk chocolate when butter fat is cheaper than cocoa butter.

## Emulsifiers

The most popular emulsifier is lecithin, used to reduce viscosity and save cocoa butter. Other emulsifiers are described in Chapter 4.

## Other Fats

Some countries now permit the addition of small quantities of other fats (5 percent) with the understanding that it can still be called chocolate. Such fats are called equivalent fats and should have the same chemical and physical properties as cocoa butter, except for flavor.

## FLAVOR

Some of the flavor of chocolate, particularly dark chocolate, comes from the blend of beans used. With milk chocolate, milk caramelization plays an important part. Flavors also may be added, including vanillin, cinnamon, cassia oil, essential oils of almond, lemon, and orange, varieties of balsams, and resins, as well as manufactured combination flavors.

The art of flavoring is the prerogative of the chocolate or confectionery experimentalists, but again, it must be emphasized that no skill in flavoring will overcome poor flavors derived from badly prepared or moldy cocoa beans. Certain countries stipulate that "no flavors which simulate chocolate, milk or butter may be added."

## REWORK

This is the name given to reprocessed chocolate bars and confectionery. It is possible to reclaim misshapen chocolate units in the form of pastes, syrups, or crumb, and utilize them as part of the basic ingredient of new batches of chocolate. This practice has been opposed by members of the EEC and Codex on the grounds that it opens the way to adulteration with nonchocolate ingredients used in the fillings or centers of chocolate candies. Reclaiming methods are described later.

## CHOCOLATE PROCESSES

Manufacturing processes, whether for dark or milk chocolate, involve certain basic operations: preparation of ingredients, mixing of ingredients, refining of the mixture, pasting or partial liquefaction of the refined mixture, conching (or an alternative process), and adjustment of viscosity and flavoring.

### Preparation of Ingredients

The two main ingredients, cocoa nibs and sugar, must be pulverized either before mixing or by using a machine with a combined grinding and mixing action. The methods for preparing liquor from nibs have been described.

The grinding of sugar from granulated to a fine powder is now recognized as an explosion hazard and the machinery is usually extremely noisy. Therefore, equipment is now being developed that avoids sugar milling. Cocoa butter and other fats are liquefied and care must be taken to see that they are not overheated when melting and are not stored as liquids for long periods, particularly butter fat.

Milk powders should not be stored in open hoppers and should be used as soon as possible after delivery. Moisture content should not exceed 3 percent; if over 4 percent, staling will occur. In some cases, milk powder and cocoa powder may be further dried before mixing, but this is more likely with compound coatings.

### Mixing

In most chocolate plants, the basic ingredients are dispensed by automatic methods—punched card or computer—which deliver the correct quantities according to any given formula.

In some instances, the ingredients are metered and mixed continuously; in others, they are fed into batch mixers. Considerable controversy has arisen over the merits of continuous versus batch mixing ingredients.

There is no doubt that continuous processes have many advantages and they are being adopted more and more in all branches of the food industry. However, continuous mixing means continuous metering of ingredients to a high degree of accuracy over long periods. Small errors in the bulk ingredients—sugar, liquor, milk powder—may not have a significant effect during production or on the final product but errors in the dispensation of emulsifiers, flavors, or fats can have very serious consequences. The delivery of lecithin, part of which is often added in the preliminary mix, has been known to be subject to variations, resulting in local changes in viscosity of the paste to the refiners. This will cause problems with output and particle size. Batch mixing is considered more positive and the quantities per mix may range from 1,500 to 6,500 lb. Errors in delivery of the raw materials to the mixer would have less effect on the condition of the whole batch, which would still be constant throughout. In any batch mixing process, it is important to establish a precise mixing time to be sure of obtaining a constant consistency. This is normally 12 to 15 min.

How can the batch system be made continuous? The mixers can be duplicated so that while one is discharging, the other is mixing the next batch. Alternatively, a storage system may be used between the mixer and refiners.

One method uses an automatic hopper feeding each refiner. The combined hoppers have the capacity to take the contents of a complete mixer batch. Each hopper has a discharge device that continuously replenishes the feed roll.

Nevertheless, continuous mixing is used by many chocolate manufacturers. The Bus-Ko Kneader combines an oscillation and mixing movement and discharges well-processed paste onto a continuous steel belt to the refiner hopper. Other kneaders are produced by Werner Pfleiderer, Baker Perkins. Another system that combines both systems is the Bauermeister Beetz-Kneader.

Diagrams of the automatic hopper system and batch mixing are shown in Figs. 5.1 and 5.2. Other mixing systems are shown later in discussing alternative methods of chocolate making.

The mixing process prior to refining should produce a chocolate paste of somewhat rough texture and plastic consistency. If the texture is too slack, the paste will not climb the rolls correctly, if too stiff, passage between the rolls will be retarded or will be erratic. With any given formulation, the consistency initially is obtained by experiment, after which a standard recipe can be established for the mixing process.

## Refining

The refining of chocolate paste is an important operation and produces the smooth texture so desirable in modern chocolate confectionery.

Exactly what constitutes smoothness is debatable, as it is clear

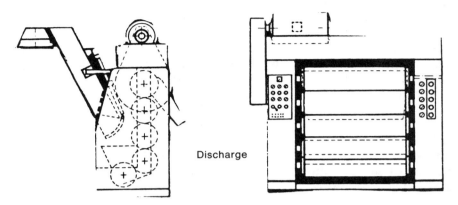

Fig. 5.1. Buhler Automatic Hopper System

*Buhler, Uzwil, Switzerland*

that if refining is carried to an extreme, producing chocolate with maximum particle sizes of less than 25 μm (0.0010 in.), the texture becomes slimy, particularly with milk chocolate.

Research on particle-size distribution seems to indicate that a small proportion of particles up to 65 μm (0.0026 in.) gives better textures for milk chocolate, whereas good dark chocolate requires a maximum of 35 μm (0.0014 in.). These figures are arbitrary and are greatly affected by the nature of the confection and added flavors; as an example, one of the most popular cream bars on the market, and which has sold in large quantities for many years, has a chocolate particle size of the order of 75 μm (0.0030 in.).

Another factor concerning particle size is whether the large particles are sugar, cocoa material, or milk crumb aggregates—each produces a different sensation on the palate. Large sugar crystals are obviously gritty but disperse, cocoa particles give a persistent sensation of roughness, and crumb aggregates soften and disperse in the mouth after a short while and, it is thought, this overcomes the sensation of sliminess.

Having once decided on the degree of refining for any particular chocolate, the problem is to maintain it consistently. Methods used in the factory and in the laboratory are given later in the book.

Particle-size distribution should also be known because whereas

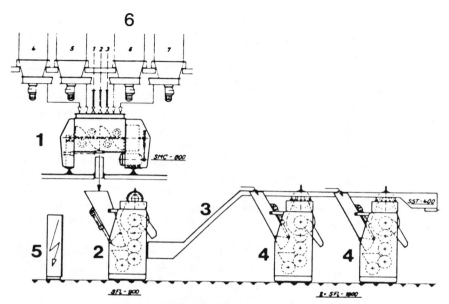

Fig. 5.2. Mixing and Double Refining System
1. SMC Chocolate Mixer for The Blending of the Ingredients to Obtain the Raw Chocolate Mass
2. SFL-900 Five-Roll Refiner for the Prerefining of the Chocolate Mass
3. SST Steel Belt Conveyor to Convey the Prerefined Chocolate to the Refining Line
4. SFL-1800 Five-Roll Refiner for the Final Refining of the Chocolate Mass
5. Electric Control Panel with EMW Program Controls for the Accurate Feeding of Ingredients Into the Mixer and for the Automatic Operation of the Entire Plant
6. Raw Material Hoppers

*Buhler, Uzwil, Switzerland*

the "feel" of the chocolate is related to the larger particles, the presence of a high proportion of very fine particles will result in the use of extra cocoa butter, which is costly.

Today's roll refiners are precision-made machines consisting of five rolls mounted vertically with the bottom feed roll offset. These steel rolls are centrifugally cast by pouring liquid steel into permanent molds. By rotation of the molds, the steel is forced outward, giving a perfect surface free of lighter material, such as slag. The steel is of special composition to give the hardest possible surface. The rolls are slightly "crowned" so that when the machine is working, the pressure exerted will ensure that a film of chocolate of even thickness is spread over the entire roll surface.

The speed of rotation of the rolls increases from bottom to top, and this is known as the differential. It allows the chocolate film to be transferred from one roll to the next and results in a shearing action in the nip of the rolls.

In modern refiners, the pressure between the rolls is controlled hydraulically and each roll is cooled internally with water jets, with the water to each roll being thermostatically regulated. Correct cooling is essential as the refiners generate a great deal of frictional heat, and if this is not dissipated evenly, roll distortion will occur with consequent wear and uneven grinding.

In old refiners without thermostatic water control, it was a problem to obtain correct working conditions. The upper rolls sometimes became overheated and it was not unknown for rolls to be overchilled, particularly when natural water supplies were used, resulting in moisture condensation on the surface, which adversely affected the chocolate film.

Checks on the roll cooling should be made, and although some variation is likely due to the paste formulation and fineness of the final product, the following figures are a useful guide: top roll, 80 to 95°F (26 to 35°C); fourth roll, 100 to 120°F (38 to 49°C); third roll, 85 to 105°F (29 to 40°C); second roll, 80 to 95°F (26 to 35°C); feed roll, 80 to 95°F (26 to 35°C).

Surface temperatures of the rolls should also be ascertained and compared with the water temperature. Grinding performance can be assessed by taking samples of refiner paste from the top roll and measuring the particle size.

The samples should be taken from the middle and the ends of the roll, and if pressure and cooling are set correctly, the figures obtained should be the same.

Another part of the machine that needs great care and attention is the scraper knife, which removes the film of chocolate from the top roll. This must be set at the correct angle and pressure to ensure that the film is completely removed. If chocolate passes the blade, it will contaminate the chocolate on the roll below. A diagram of a modern refiner system is shown in Fig. 5.3.

Machines of the type described when operating correctly have an output some three times that of the older refiners and, depending on the degree of fineness required, will produce 1,800 to 2,200 lb/hr. Some of the latest refiners have even greater outputs.

The main purpose of a roll refiner is to grind the paste fed to it, but it also acts as a dispersion machine in which agglomerates are rubbed out and the particles fully wetted with the liquid-fat ingredient.

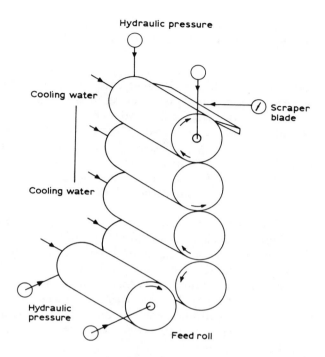

Fig. 5.3.  Diagram of Modern Refiner System

*Buhler, Uzwil, Switzerland*

A recent development in refining has been the reintroduction of double refining. This uses granulated sugar and avoids the use of sugar mills, which are an explosion hazard. See Fig. 5.2. The system was used many years ago but with less efficient machinery.

## Milk Chocolate Processes

The processes described so far apply to both dark and milk chocolate, but with milk chocolate, only after the milk solids have been included.

The essential part of the milk chocolate process is the method used to incorporate the milk ingredient. Milk chocolate is virtually moisture-free (0.5 to 1.5 percent). Full cream milk contains about 12.5 percent of milk solids, including fat, and it is these solids that constitute the milk in milk chocolate. There are several methods of removing the 87.5 percent of water from milk to obtain the milk solids.

**Milk Powder** Liquid milk is concentrated by evaporation under reduced pressure in equipment designed to give maximum efficiency in the consumption of heat (e.g., triple-effect evaporators). This concentrated milk is dried to powder by the spray process or by means of drum evaporators. (These methods and the quality of powder obtained are discussed later in the book.) This milk powder is used with cocoa nibs or liquor, sugar, cocoa butter, lecithin, and flavor in processes similar to those described previously.

**Milk Crumb** The development of this process revolutionized the manufacture of milk chocolate, resulting in greatly increased sales of milk chocolate, far in excess of those of dark chocolate. In countries where milk powder is more often used, milk chocolate has been less popular, possibly because much of it lacks the rich, creamy, partly caramelized flavor obtained by the crumb process. It must be recognized that flavor acceptance varies from country to country, and even within one country, and the extra sweetness of milk chocolate does not appeal to everyone.

Because of the quantities of fresh milk required, the large chocolate companies established their own crumb factories in various rural areas of England and Wales, and in later years similar factories were established in the Irish Republic where milk is abundant. These factories are, in fact, huge dairies in the milk-receiving sections where strict hygiene is observed and bacteriological and composition checks are made continuously on milk from the farms.

Various crumb recipes and processes have been devised and have been the subject of published papers, but all are derived from one basic process. See Koch (1961), Headon (1964), and Powell (1970).

The essence of the crumb process is the Maillard reaction between the milk protein and sugars, which produces a particular flavor. It is a time/temperature/water reaction. Without the presence of water, the typical flavor is not produced and the higher the temperature, the quicker is the reaction.

*Manufacture of Milk Crumb* Milk from the farms is delivered to the factory in churns or tank trucks and is subject to strict quality-control examination for composition and microbiological contamination.

It is then filtered, cooled to 4.5°C (40°F) and pumped to temporary storage in large insulated stainless steel tanks or silos with capacities up to 15,000 gallons.

*Evaporation.* After a preheat stage at 75°C (167°F), it is concentrated to around 30 to 40 percent total solids in a continuous evaporator.

*Condensing.* The total solids content of the evaporated milk is checked and sugar weighed in, according to the recipe. These operations are usually carried out with automatic weighers in large plants, and from this proportioning equipment, the sugar/milk mixture is delivered to vacuum pans for dissolving and condensing.

Condensation takes place under vacuum with rapid boiling at a maximum temperature of 75°C (167°F) until a concentration of around 90 percent solids is reached. At this stage, some of the sugar should be showing signs of crystallization.

*Kneading.* A melangeur or other type of heavy mixing equipment is charged with cocoa liquor (quantity again as required by the recipe) and the batch of condensed sugar/milk is gradually run in. The mixture of liquor and crystallizing condensed milk makes a very stiff paste and this continues to crystallize during kneading, which may last 20 to 30 min.

*Drying.* The paste from the kneaders may be transferred to shallow trays, which are placed in vacuum ovens where the paste is dried. The drying temperature may vary from 75°C (167°F) to 105°C (220°F), according to whether hot water or steam is used for shelf heating, and time of drying may vary from 4 to 8 hr. Alternatively, the paste may be dried in a continuous-belt vacuum oven. The flavor of the crumb is related to the times and temperatures, and for a given quality of product, these must be precisely controlled.

The moisture content of the crumb from the ovens is usually under 1 percent. The dried crumb is finally broken into lumps by toothed rollers and may be stored in this condition in sacks or bins.

The traditional vacuum oven method of crumb manufacture has been described, but various continuous processes have been developed in which all the operations take place automatically and continuously.

Technically, the most difficult part of this process is the continuous vacuum drying and, with this, the development of the flavor obtained by the batch process. Development of the correct flavor when transferring from batch to continuous processes is a problem that must be faced throughout the food industry. Unfortunately, it too often is ignored in the interests of mechanization and cost reduction.

Since the first edition of this book, other continuous processes for the production of crumb have been developed. Powell (1970) describes a process designed to mechanize the traditional vacuum oven process.

After milk concentration by the multiple-effect evaporators, the sugar and concentrated milk are metered into a continuous vacuum concentrator. This sweetened condensed milk meets a metered supply of cocoa liquor and from there to a continuous kneader where sugar crystallization takes place. This is followed by a drum drier constructed inside a vacuum chamber. Final caramelization and drying are accomplished by means of a flow tower.

A development by Minifie (1977), in conjunction with the Groen Company of Chicago, produces milk crumb without the use of vacuum. Sweet condensed milk, made by any of the recognized processes, or a similar milk reconstituted from milk powder, is mixed with cocoa liquor, and extra sugar and water if required. This is passed through a specially designed heat exchanger where the total solids are increased from about 70 percent to 95 to 96 percent. At the same time, caramelization and flavor are developed. After concentration, the thick paste is discharged into a crystallizer, which rapidly transforms the amorphous paste into a crumb with fine sugar crystals. The equipment is described in detail in the published literature. It has several advantages over the vacuum processes:

1. The cooking time in the evaporator is of the order of 2 to 5 min, depending on the degree of caramelization required.
2. Space requirements are much less.
3. Capital and running costs are less.
4. Accurate control of process conditions is possible.
5. Plant sanitation is easily controlled.

Figure 5.4 is a diagram of a pilot plant used for development work. The process can be used for a variety of crumb formulations, including white crumb (without liquor).

*Storage and keeping properties.* Crumb is also sold as a commercial article in coarse powder form, and for transport and storage in uncontrolled atmospheres, it is essential to use polythene-lined bags.

For bin or silo storage, it is necessary to have the crumb in lump form and with a low dust content. Bins or towers are in use that hold 400 to 500 tons, and powder will consolidate under pressure in tall towers.

In storage bins, crumb will retain its low moisture content for long periods and will keep satisfactorily for six to nine months.

In polythene-lined bags, up to six months' storage may be expected

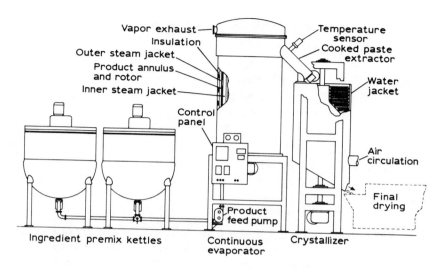

Fig. 5.4. Groen Crumb System Pilot Plant

in uncontrolled atmospheres, but if no lining is provided, crumb will readily pick up moisture and the flavor will deteriorate rapidly when the moisture content reaches 2.5–3.0 percent.

Provided these conditions are met, milk crumb will keep better than whole-milk powder, and this enables manufacturers to build up stocks during the periods when milk is plentiful.

*Composition and bacteriological specification.* A typical composition for milk crumb in the United Kingdom and the Irish Republic is:

| | |
|---|---|
| Cocoa liquor | 13.5 percent |
| Sugar | 53.5 percent |
| Milk solids | 32.0 percent |
| Moisture | 1.0 percent |

giving an analysis of:

| | | |
|---|---|---|
| Total fat | 16.5 percent | { milk fat 9.2 percent<br>cocoa butter 7.3 percent |
| Fat-free cocoa | 6.2 percent | |
| Sugar | 53.5 percent | |
| Milk solids less milk fat | 22.8 percent | |
| Moisture | 1.0 percent | |

As examples of other crumbs the following may be quoted:

*Low milk crumb*

| | |
|---|---|
| Cocoa liquor | 13.5 percent |
| Sugar | 63.0 percent |
| Milk solids | 18.0 percent |
| Added cocoa butter | 4.5 percent |
| Moisture | 1.0 percent |

*High liquor crumb*

| | |
|---|---|
| Cocoa liquor | 18.0 percent |
| Sugar | 55.0 percent |
| Milk solids | 26.0 percent |
| Moisture | 1.0 percent |

White crumb is a product made from milk and sugar by one of the crumb processes. It can be used to make white chocolate or as an ingredient for milk chocolate in which the manufacturer can choose the particular type of beans or liquor.

Because of the heat treatment during manufacture and low moisture content, milk crumb is in a more favorable position to meet a strict bacteriological specification compared with many other milk products on the market. The Groen process is particularly suited to the production of crumb of high microbiological standards. The following specification can be met without difficulty.

| | |
|---|---|
| Total count (three days at 30°C milk agar) | 95 percent under 5,000/gram none over 20,000/gram |
| E. coli | Absent in 1 g |
| Yeasts, molds (three days at 25°C malt extract agar pH 5.4) | Less than 50/g |
| Salmonella | Absent in 100 g |
| Staphylococcus aureus | Absent in 1 g |

## Conching

Conching may be regarded as the last process in the manufacture of bulk chocolate, whether dark or milk. It is certainly an essential process for the development of the final texture and flavor. Over the years, it has been the subject of much investigation by trial and error, and more recently by scientific research. Efforts have been made to substitute other processes and machinery for what is a somewhat cumbersome operation.

**The Conche** The origin of the name of this machine is supposed to be from the Latin word meaning "shell," and the traditional conche does bear some resemblance to a shell. This is the longitudinal conche, which consists of a flat granite bed upon which heavy granite rollers attached to robust steel arms move backward and forward. The shape of the ends of the conche pot is such that the rollers slap the chocolate against these ends and cause it to splash back over the rollers into the main body of the chocolate.

To assist mechanical arrangement, the rolls are usually made to reciprocate from a crank on the driving rod so that several tanks can be operated from one drive (Figs 5.5 and 5.6).

Although this seems a very inefficient piece of machinery, it has the effect of producing the desired final flavor in the chocolate, free from all harshness, which is very apparent in unconched chocolate. In the longitudinal conche, the chocolate eventually must reach a fluid state. Since the paste from the refining rolls is in a friable, flake form, the required fluidity is only obtained by the addition of extra cocoa butter and emulsifier, combined with prolonged mechanical movement. This process was long, labor, and power consuming, and charging and discharging was an arduous task. Nevertheless, the majority of chocolate companies (including the very large ones) adhered to the use of these machines for many years and many conches are still in use. However, the need for improvement was evident when raw material and labor costs escalated.

The first investigations of the process showed the value of what is now known as dry conching. If the flake from the refiner is

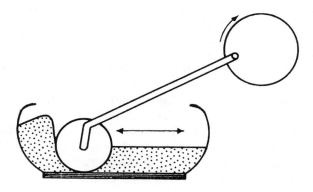

Fig. 5.5.  Diagram of Conche Pot

Fig. 5.6.   An Original Four-Pot Conche

*Baker Perkins, Peterborough, England*

mechanically agitated *without* the addition of extra fat, moisture and unwanted volatiles are removed more rapidly and extra fluidity is developed. It was soon found that the traditional conche had insufficient power to deal with this modified process but the principle was proved in spite of bent cranks, burnt-out motors, and stripped gears.

To assist the efficiency of these conches, mechanical devices for transforming the refined flake into a more fluid paste were used before transferring to the conche. Kneaders similar to those described under "Mixing" have been used successfully.

Roller conches are all very similar in design but vary enormously in size. In one large manufacturer's factory, whole batteries of small single-roller conche tanks have been seen—the rollers reciprocating fairly rapidly and the tank capacity only 300 lb.

At the other end of the scale, another, smaller manufacturer uses a double-roller conche working more slowly with a capacity of 10 tons! Another large manufacturer uses a multiple-roller conche about 80 ft long with continuous feed.

**Rotary Conches**   In the development of conches to replace the longitudinal conche, it was obvious that much more power was required to handle the flake chocolate from the refiner.

The first developments were in the form of rotary conches. In these, the refiner paste is first turned over vigorously and, by means of powerful mixer blades, the paste assumes the condition of small lumps. These turn over and over, giving maximum opportunity for moisture and other volatiles to evaporate. During this part of the process, known as dry conching, the paste becomes progressively softer and, in some cases, reaches a thick fluid condition. Ventilation or a forced-air circulation may be applied to the interior of the conche over the chocolate but *not* through it.

After dry conching, the paste is made more fluid by the addition of some cocoa butter, and conching is continued with the chocolate in a much more liquid state. The speed of mixing is increased—in one conche (Petzholdt, Fig. 5.7) the chocolate is sprayed to the inner side of the vessel. In another (Carle and Montanari, Fig. 5.8), the chocolate is lifted by auger through granite cones and returned to the main body of the conche.

The Frisse is a typical example of the overbeating conche, which consists of a large tank with three powerful intermeshing mixer blades (Fig. 5.9). The intermesh provides a shearing as well as a mixing action.

Fig. 5.7. Petzholdt Superconche Type PVS

1. The Design of the Arms Ensures Intensive Mixing as Well as Scraping of the Inner Walls.
2. The First Stage of Operation Consists of Dry Conching to Remove Moisture and Other Volatiles and at the Same Time the Paste is Progressively Liquified.
3. The Second Stage Completes the Conching Process at a Higher Speed with the Chocolate Sprayed from the Top Center to the Side Walls. Air Circulation over the Chocolate may also be Applied.
4. A Control Panel with Recorder Regulates the Temperature of the Conche Jacket and Contents.

*Petzholdt, Frankfurt am Main, W. Germany*

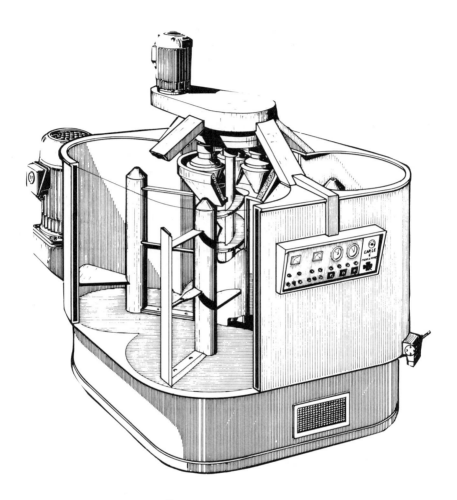

Fig. 5.8.   Rotary Conche

Method of operation:

Dry conching. The chocolate paste from the refiner is introduced into the external vat of the conche without the addition of extra cocoa butter. The vat walls are heated and the planetary stirrers agitate the mass vigorously. A small amount of cocoa butter may also be added at this stage. Mixing usually continued for about 4 hr.

Final conching. At the termination of the dry conching phase, the three shutters located in the bottom of the central vat are opened, and the chocolate flows through these openings into the central vat. At this position, the chocolate is subjected to the action of a screw propeller that forces the paste into truncated granite rollers, after which it is again returned to the external vat and the cycle is continuously repeated. At the end of this stage, lecithin is added and when this is completely incorporated, the viscosity is adjusted to the required value by the addition of cocoa butter.

*Carle and Montanari, Milan, Italy*

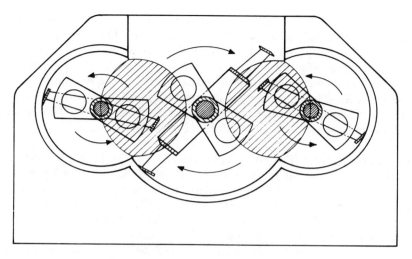

Fig. 5.9.   Frisse Double-Overthrow Conche

*Richard Frisse, Herford, W. Germany*

All conches have heating/cooling jackets that can be thermostatically controlled and are also usually provided with a time/temperature record.

## Other Conches and Chocolate-Making Systems

In recent years, a great deal of research and development has gone into simplifying the process of chocolate making. Probably the most significant development to date involves changes in the conching process. Additionally, there has been considerable improvement in the performance of roller refiners, with increased output and automatic control.

The following machines and processes have been adopted by many chocolate companies.

**The Tourell Conche**   This is a continuous conche that has been in use for a number of years by leading chocolate companies. It was developed mainly for the manufacture of crumb milk chocolate, but with some modifications has been used for other chocolates.

It has also been used as a liquefier or as a means of dry conching refiner paste, in which case much less time is required subsequently in one of the rotary conches described previously.

The conche consists of a series of jacketed troughs that contain specially designed paddles. The paste from the refiners travels through the troughs and is subjected to very vigorous mixing and shearing. The viscosity is gradually reduced and moisture and volatiles removed. Lecithin and extra cocoa butter are added in the final troughs. The operation is continuous once the troughs are full; with crumb milk chocolate, conching is complete in 4 hr. The capacity of the conche is up to 8,000 lbs/phr (Fig. 5.10).

**Macintyre Refiner/Conche** This consists of a large heavy-gauge horizontal drum, the inner surface of which is fitted with inclined steel ribs. An internal rotor, also fitted with steel ribs, can be adjusted by means of an external control. The gap between the rotor and the internal ribs of the drum is fairly wide during the initial period when ingredient mixing is taking place.

As mixing proceeds, the gap is closed to obtain a full grinding effect. The jacket is provided with water circulation for heating or cooling and air circulation through the mixer carries away unwanted volatiles and moisture.

Chocolate may be made starting with nib or liquor and granulated sugar, with cocoa butter and lecithin being added later to give the required fluidity. For milk chocolate, milk powder or milk crumb is added with the sugar and liquor.

Machines of similar design are now made by several companies.[1] Conching time is normally 12 to 24 hr, but depending on the fineness required by the user.

The internal mechanical construction must meet exceptionally high standards; if the rotor becomes out of balance, the result is metal contamination of the product. Figures 5.11 and 5.12 show the exterior and interior construction respectively. These machines are increasingly being used in the manufacture of chocolate, compound coatings, and nut spreads.

**The Wiener Process** The Wieneroto ball mill was mentioned in the section on liquor refining. This mill is also incorporated in a system that has been developed in recent years for the production of dark and milk chocolate, compound coatings, and similar products. The process has found application in factories in many parts of the world.

---

[1] Brierley, Collier, Hartley, Rochdale, England; Lloveras SA, Tarrasa, Spain; Low and Duff, Carnoustie, Scotland.

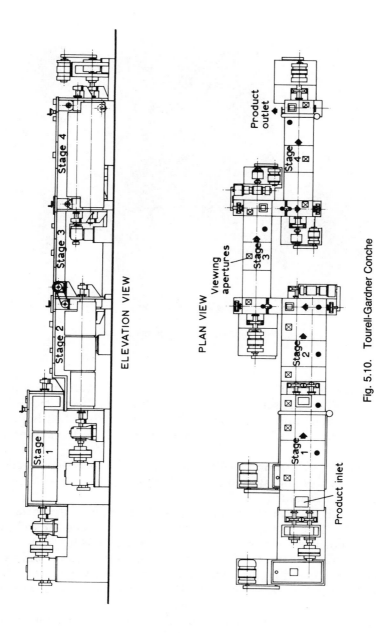

ELEVATION VIEW

PLAN VIEW

Fig. 5.10.   Tourell-Gardner Conche

*Tourrell-Gardner, Cornwall, England*

Fig. 5.11. Macintyre Refiner Conche

*Brierley, Collier and Hartley, Ltd., Rochdale, England*

The basic equipment consists of a mixer conche, a ball mill, a taste improver, and an automatic system to add lecithin. These are linked by pipes, pumps, and a thermostatically controlled heating and cooling system. The whole is operated from an electronic control panel.

For chocolate production, the Wieneroto ball mill has been specially adapted as part of the system in which the ingredients are mixed, refined, and continuously conched.

The mill is filled with approximately 3,500 lb of specially hardened steel balls 0.375 in. in diameter. This charge is suitable for refining products of which the larger particle size is about 0.04 in.

The basic ingredients are fed into the conching/mixing tank and after adequate blending are transported to the Wieneroto. After passage through this mill the paste is returned to the conche tank and the process repeated. In the course of circulation, two other intermediate processes are included: adding lecithin and improving flavor.

In the process of milling, the particle size is reduced and thus the total surface area is increased. This has the effect of greatly

Fig. 5.12. Macintyre Conche—Internal Construction

*Brierley, Collier and Hartley Ltd., Rochdale, England*

increasing viscosity, which causes pressure in the pipelines and an associated increase in the amperage of the motor driving the refiner agitator. To compensate, lecithin is automatically injected and the quantity recorded, and this procedure continues throughout the entire production cycle. Toward the end of the cycle, the remaining lecithin, up to 0.5 percent, may be added.

The flavor improver uses a system whereby thin streams of the product in circulation are exposed to heated air injected across the surfaces. This has the effect of reducing the moisture content, removing of unwanted acidic substances and other volatiles, and reducing viscosity and yield value.

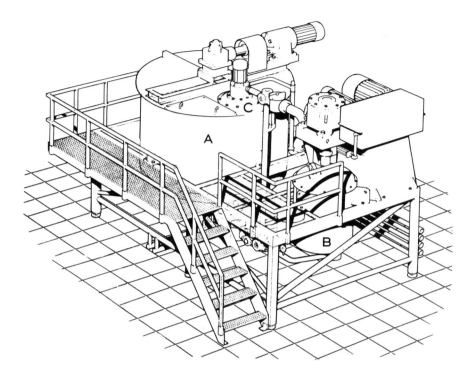

Fig. 5.13. Wiener Process Installation
A. Mixing/Conching Tank
B. Wieneroto
C. Taste Changer

*Wiener and Co., b.v. Amsterdam, Holland*

Diagrams of the equipment are shown in Figs. 5.13 and 5.14 and complete descriptions given during presentations at technical conferences are available.

The latest complete Wiener chocolate making system, having a capacity of 9 tons of finished product per day, consists of the following:

Premixer and conche with double, contra rotating stirrers.
Two variable speed vane pumps.
Double walled pipe system with ball filters and valves.
Two Wieneroto Ball Mills type W-105-SCG.
Two taste changers with ventilators and temperature control.
Automatic lecithin dosing system.

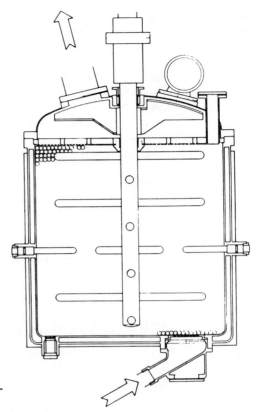

Fig. 5.14. Cross Section of Wieneroto Ball Mill

**Melangeurs, Edge Runners** This chapter on chocolate making would
be incomplete without a mention of these machines. At one times,
melangeurs were used for the complete manufacture of chocolate and
incorporated the processes of mixing, grinding, and conching.

The machine consists of a rotating pan, usually with a granite bed,
and on this bed rollers, also granite, rotate. Scrapers continuously
direct the mixture of ingredients between the bed and rollers. The
rollers can be raised or lowered to control the thickness of the paste
layer—in the initial stages of the mix, they will be higher than when
the paste is well ground and more fluid.

Heat is applied to the base of the machine by steam or hot-water
coils—originally gas jets were used.

In the past, loading and discharge were always manual but batch weighers are now used to load with recipe weights of each ingredient, and after mixing, the machine is mechanically emptied by scoop, Archimedean screw, or central discharge.

The melangeur has limited capacity for its size but machines with automatic discharge are generally larger.

Edge runners work on a similar principle but with a single roller. Smaller machines find extensive use in experimentation and, in addition to chocolate, various confectionery products, such as nut pastes, can be made.

Larger melangeurs (such as shown in Fig. 5.15) are often used for handling scrap confectionery. For example, chocolate-covered creams may be ground with sugar or chocolate refiner paste and, with heat applied, dried so that the final mixture can be reincorporated into chocolate, compound coatings, or other fatty formulations.

**Conching conditions, the effect of conching** With the traditional longitudinal conche conching, periods were generally very long

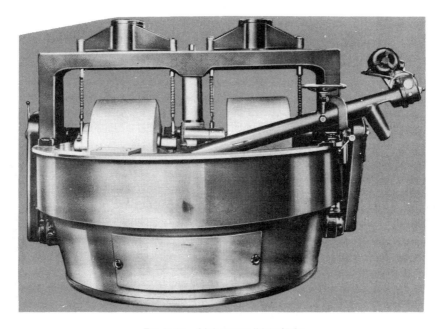

Fig. 5.15. Melangeur (M.22/RC)

*Carle and Montanari, Milan, Italy*

compared with present-day practice—for dark chocolate, 96 hr was not uncommon. With the change to more powerful conches and the introduction of dry conching, the times have been considerably reduced. With crumb milk chocolate 10 to 16 hr is frequently the conching time employed at a temperature of 120 to 125°F (49 to 52°C). With milk powder chocolates, 16 to 24 hr are more likely and temperatures up to 140°F (60°C). If nonfat-milk powder plus butter oil is used instead of full-fat-milk powder, temperatures up to 160°F (70°C) may be used. These higher temperatures give a partly caramelized flavor but not the same as that produced from milk crumb. Dark chocolates are usually conched at higher temperatures, 160°F (70°C), and sometimes up to 180°F (82°C).

Occasionally, with milk chocolates, a type of granulation occurs during conching that is very difficult to disperse unless the chocolate is subjected to a type of milling. Granulation is more likely to take place with chocolates made with whole-milk powder when higher conche temperatures are used. Conching processes that include a form of milling are the Carle and Montanari (granite cones), Loveras (refining roll), and Wiener (ball mill) processes.

Conching conditions may be modified by pretreatment of the chocolate liquor in thin films (see Chapter 2). Dark chocolate may also be heat treated as a thin film on a drum or in a trough at temperatures of approximately 100°C (212°F).

The LSCP process (Kleinert-Lindt and Sprungli) does not use a normal conching process.

Many statements have been made regarding the effects of conching and Table 5.1 summarizes these statements and assesses their validity.

**White Chocolate**   This confection has had bouts of popularity but tends to be sickly and can have a poor shelf life.

White chocolate is milk chocolate without the cocoa matter and is prepared in a similar way, using cocoa butter instead of liquor, with sugar and milk powder. Alternatively, it can be made from white crumb, which is manufactured by the crumb process previously described but with cocoa butter instead of liquor.

With white chocolate, even more than with milk chocolate, a mild-flavored cocoa butter is required, and this favors deodorized cocoa butter, or that pressed from very lightly roasted beans. The latter assists from a keeping standpoint and white chocolate made by the crumb process also has better shelf life. However, white chocolate made from white crumb is likely to have an off-white color due to

TABLE 5.1. VIEWS OF EFFECTS OF CONCHING

| Statement | Validity |
|---|---|
| Removal of moisture from refiner paste | Yes, especially dry conching |
| Removal of undesirable volatiles that give poor flavors | Yes, especially dry conching |
| Reduces viscosity of refiner pastes and throughout the process | Yes |
| Reduces particle size, removes particle edges | Negligible |
| Complete dispersal of solids in liquid fat. Cocoa solids, sugar, milk solids in cocoa butter or cocoa butter and milk fat | Yes. Fat spreads over the particle surfaces—lecithin greatly assists this |
| Dispersion of aggregates of ingredients from previous processes or formed during conching process | May not be properly dispersed in some conches. May require partial milling, e.g., Carle conche, Wiener process |
| Flavor development due to prolonged mixing at elevated temperatures—mellows flavor | Yes, but long periods not necessary |
| Development of flavor in longitudinal conche due to frictional heat between roller and base | Not proved, unlikely |

A publication by Dimick and Hoskin (1981) illustrates a number of these factors.

caramelization in the crumb process but the flavor is probably more attractive.

A feature of white chocolate, which also applies to coatings without added cocoa, is its susceptibility to oxidative rancidity if exposed to light, and it readily picks up foreign flavors.

In normal chocolate or coatings, some protective action arises from the cocoa material, and white chocolates should be closely wrapped in foil to assure best shelf life. White chocolate should, according to some authorities, not be called chocolate as no cocoa solids are present. The legal implications are usually avoided by giving the confection a proprietary name.

## Viscosity of Chocolate

The subject of viscosity and its determination has been dealt with at length in Chapter 4 on emulsifiers. However, it is appropriate at this stage to reiterate some of the points specifically concerned with chocolate making.

With the development of highly automatic chocolate molding and enrobing machinery, the accurate control of fluidity or viscosity has become very important. Furthermore, it has become necessary to conserve the expensive cocoa butter ingredient by using emulsifiers and mechanical methods to get increased fluidity in lower-fat-content chocolates. The fluidity or viscosity of chocolate cannot be defined by a single figure as, because of the presence of many solids (sugar, cocoa material, crumb), it behaves as a non-Newtonian liquid and its actual viscosity depends on the speed at which it is mixed, or rate of shear.

The solids can be further increased by the presence of cocoa butter crystals, which must, of necessity, be present in tempered chocolate, and since this can be a variable factor depending on the method of tempering and the amount of seed, a true viscosity of chocolate in use in a process is difficult to obtain. In practice, it is usually sufficient to control the viscosity of the untempered chocolate and to standardize precisely the tempering procedure in the machine using the chocolate. The process of dry conching and newer types of conche have contributed to increased fluidity (viscosity reduction). This is partly due to an overall reduction of moisture content, which at one time was of the order of 1 to 1.5 percent and is now frequently down to 0.5 percent. The more vigorous mechanical action of the conches has helped with viscosity reduction.

In relation to the viscosity of chocolate, it is not often realized that *free* moisture in minute amounts will increase viscosity greatly and moisture determination on the chocolate will not necessarily give any useful information. It is more appropriate to inspect the equipment for minute water or steam leaks or to examine the ingredients being used. Sugar or crumb may contain local patches of damp material, but this is less likely in bulk supplies. It may occur occasionally in bulk storage bins due to a faulty or prised seam. Rework can cause trouble, particularly if invert sugar is present, and very occasionally, under conditions of extreme atmospheric humidity, moisture can be absorbed on the thin layers of chocolate on the refining rolls.

Something must be said again about the method of adding lecithin to chocolate. Lecithin (and some other emulsifiers) are surface-active substances and the processes should be such that the lecithin is kept on the surface of the chocolate particles.

The correct procedure is to add the lecithin as late as possible in the final process (e.g., conching), allowing sufficient time for proper dispersion.

However, it has been shown that in some formulations it is an advantage to add about one-fourth to one-third of the total lecithin at the mixing stage before refining. Here, the moisture content of the mixture is probably higher than in the final product, and the correct viscosity of the paste suitable for feeding the rolls can be obtained with a lower fat content if lecithin is added. The savings achieved are apparent throughout the process.

## REFERENCES

Czyzewski, T. S., and Minifie, B. W. Groen Division (Dover Corp.), Chicago. Patent Application SN. 660,846.
Dimick, P. S., and Hoskin, J. M. 1981. Chemico-physical aspects of chocolate processing—Review. *Can. Inst. Fd. Sci. Technol. J.*, Canada.
Headon, T. A. 1964. *Proceedings, Pennsylvania Manufacturing Confectioners Assocn.*
Koch, J. 1961. *Mfg. Confect.* 41(12), 23.
Minifie, B. W. 1977. Conf. Manf. & Mkt, London.
Powell, B. D. 1970. Manf. Conf. 48.

## Chocolate-Processing Machinery

Bauermeister, Hamburg-Altona, W. Germany
Buhler, Uzwil, Switzerland
Carle and Montanari, Milan, Italy
Frisse, Herford, Germany
Groen Division, Dover Corporation, Elk Grove, Ill.
Lehmann GmbH, Aalen, Württ, W. Germany
Lloveras S.A., Tarrasa, Spain
Low and Duff Ltd., Carnoustie, Angus, Scotland
Petzholdt GmbH, Frankfurt, W. Germany
Tourell-Gardner, St. Blazey, Cornwall, England
Wiener and Co., Amsterdam, Holland

# 6

# Confectionery Coatings, Chocolate Replacers, Dietetic Compounds

## CONFECTIONERY COATINGS

The manufacture of vegetable fat coatings has developed into an industry almost as large as that of chocolate itself. These coatings are of many types. Some closely resemble true chocolate, dark or milk. Others—made from cocoa powder, nonfat-milk powder, sugar, and one of the many vegetable fats—may have an appearance resembling dark or milk chocolate but a texture that is quite different. Then there are the pastel coatings that usually consist of milk powder, sugar, and a vegetable fat, with added flavors and colors.

Other coatings, which have a "health food" image, may contain such ingredients as carob, soya protein isolates, and sorbitol or fructose. Many international arguments have arisen around the use of the word "chocolate" to describe a product that actually is not chocolate. It may be relatively easy to define the composition of chocolate for eating or for covering candies, but when the word is applied to cakes, cookies, and ice cream there is real difficulty.

A chocolate cake may contain 3 to 4 percent fat-free cocoa material as an ingredient of the flour mix, or it may be coated with a compound prepared from cocoa, sugar, and vegetable fat. Chocolate ice cream, similarly, may only have some cocoa and artificial flavor.

In a number of countries, chocolate must conform to certain standards of composition. It must be made from pure cocoa nibs, cocoa butter, and sugar, or, in the case of milk chocolate, must contain a specified amount of milk solids, including the related quantity of milk fat.

This does not deter manufacturers in these countries from making substitutes that may be sold under some proprietary name, without the word "chocolate" included in the list of ingredients. The description "chocolate flavored" is often used. However, the customer will always recognize good quality, and well-manufactured "real" chocolate is certainly a quality item.

Coatings are manufactured for two reasons; (1) for economy, where the expensive cocoa butter is partly replaced by a vegetable fat; and (2) there are many uses for which pure chocolate is unsuitable, such as for coating ice cream and cakes where the cocoa butter ingredient of chocolate sets too hard, is brittle, and flakes off. A soft vegetable fat is much more suitable for these purposes.

## Ingredients

At one time, confectionery or compound coatings had the image of poor-quality products. In many cases, this was fully justified because the policy of the manufacturers was solely to get down to the lowest price. Apart from inferior ingredients, processing time was reduced and, in many cases, the shelf life of the product was very poor. As time progressed, the value of improved quality was recognized and today there are many good coatings on the market.

**Fats**  The properties and the chemistry of fats generally, including cocoa butter, are discussed in other chapters. Here, the main factors required for the production of fats for compound coatings are given.

Fats as expressed from various seeds and nuts are not suitable as replacements for cocoa butter, the natural fat of the cocoa bean. Usually, they are too liquid and contain substances that afford poor shelf life; they must be refined and hardened. Refining removes free fatty acids, gums, and other extraneous substances. Hardening is carried out by physical and chemical means.

*Stearines*  The physical method called fractionation produces fats known as stearines. The refined natural fat is subjected to a process of cooling and/or selective solvent extraction, which causes the higher melting fractions to separate from the liquid fractions called oleines. These stearines are removed by filtration, pressing, or centrifuge, and the fats so produced have higher melting points and harder texture.

*Hydrogenation*  The other method of hardening is by the chemical process of hydrogenation in which hydrogen is introduced into the unsaturated molecules of the fat. Unsaturated fats usually have a lower melting point; some are liquid at ambient temperatures.

There are other methods of modifying fats, of which interesterification is an example.

**Glycerides**  Chemically, all fats are called glycerides and they are a combination of glycerol with the various fatty acids. Since there are three positions in the glycerol molecule to which the fatty acids can be attached, also in different order, it will be seen that the variation in the properties of these fats can be very large.

$$
\text{Glycerol} \quad
\begin{array}{l}
CH_2OH \,\text{---}\, HOOC.R_1 \\
CHOH \,\text{---}\, HOOC.R_2 \\
CH_2OH \,\text{---}\, HOOC.R_3
\end{array}
$$

$\downarrow$

H$_2$O removed by
Esterification

*Fatty acids*
R$_1$ R$_2$ R$_3$

Stearic
Oleic
Palmitic

**Lauric and Nonlauric Fats**  Natural fats may be classified according to the length of the chain of molecules of their fatty acids. The lauric fats are the shorter chain acids and the main commercial ones are derived from coconut and palm kernels. The nonlauric fats are composed of the longer chain fatty acids; some of the important ones are cocoa butter, palm oil (from the flesh of the palm fruit), soya bean oil, and cottonseed oil.

Certain basic differences exist between the lauric and nonlauric fats, and these must be fully understood when using coatings made from these fats.

Fractionated lauric fats (e.g., palm kernel) have physical characteristics close to those of cocoa butter and coatings made from these have a texture similar to that of chocolate—good snap and eating qualities. However, these fats are *not* compatible with cocoa butter. Therefore, coatings must be made with low-fat cocoa powder (see later).

The presence of cocoa butter over 5 percent in the fat portion will result in bloom formation on the coatings, and probably some softening of texture.

Lauric fats are susceptible to decomposition by the enzyme lipase, but so are nonlauric fats. However, with lauric fats, the enzyme liberates free lauric acid, which has a soapy taste even at very low concentrations. The fatty acids liberated from nonlauric fats do not have these properties.

With lauric coatings, great care must be taken in the selection of the milk powder and cocoa ingredients, and also to ensure that processing reduces the moisture content of the coating to a low level.

Nonlauric fats are more compatible with cocoa butter, and cocoa liquor may be used to obtain stronger flavors in the final coatings. Compatibility varies with the type of fat, and with some nonlauric fats, the texture of the coating is poor.

**Classification of Fats**   Cocoa butter alternatives may be classified as follows:

*Cocoa Butter Equivalent (CBE)*   These are fats with a composition similar to that of cocoa butter both chemically and physically, except for flavor. They can replace cocoa butter in a coating in any proportion. They are made from special nonlauric fats.

*Lauric Cocoa Butter Replacers (CBR)*   Physical properties resemble those of cocoa butter, but they are not compatible with cocoa butter.

*Nonlauric Cocoa Butter Replacers (CBR)*   Their cocoa butter tolerance is of the order of 20 to 25 percent of the fat phase. Their texture is somewhat waxy.

Replacers are sometimes called substitutes. Legally, coatings made from any of these fats cannot be called chocolate, but some countries now permit the inclusion of 5 percent of equivalent fat (i.e., about 15 percent of the fat phase) and the coating can still be classified as chocolate.

The chemical and physical properties of fats are dealt with elsewhere but the parameters in relation to coatings may be defined as to melting point, SFI (solid-fat index), and hardness (penetration). Fats with melting points above body temperature have a waxy texture in the mouth. The correct texture occurs when the fat has a narrow melting range; in other words, the fat starts to melt and is completely melted within a range of two to three degrees just below body temperature. Cocoa butter exhibits this property.

SFI indicates the proportion of liquid fat at different temperatures, and there are several ways to measure this. Originally, the SFI was measured by an analytical determination called dilatometry, but the modern method uses nuclear magnetic resonance (NMR), which gives a more precise result.

Hardness is related to the snap when breaking a bar of chocolate or coating. It is determined by means of a penetrometer that measures the rate at which a rod or a cone penetrates a block of fat or coating at different temperatures.

**Cocoa Powder and Liquor**   Cocoa powder and liquor determine the flavor and quality of the final coating. Cocoa powder must always be purchased from a reputable manufacturer who will guarantee to supply to a high-grade specification. To buy cocoa powder "spot" on the open market without a guarantee is courting disaster.

One specification quoted, Meursing (1983), is as follows:

*Unalkalized (natural) cocoa powder*

| | |
|---|---|
| Moisture content | 5 percent maximum |
| Shell content | 1.75 percent maximum (calculated on the nib) |
| Standard plate count | 5,000 maximum (normally much lower) |
| Molds per gram | 50 maximum (normally much lower) |
| Yeasts per gram | 50 maximum (normally much lower) |
| Enterobacteriaceae (1 g) | Negative |
| E. coli (1 g) | Negative |
| Salmonella | Negative |
| Lipase activity | Negative |

Cocoa liquor and powder should be manufactured with well-winnowed nibs from properly fermented and dried cocoa beans. Low-fat cocoa powders are normal for coatings (10 to 12 percent cocoa butter) in order to reduce the proportion of noncompatible cocoa butter in the formulation. This is particularly important with lauric fat coatings.

Occasionally, alkalized cocoa powder may be used in coatings to give darker or redder colors. The pH will be higher than with natural cocoa (pH 5.5 to 5.8) and it is preferable, if such cocoa is used, to limit the pH to 7.0. pH values above this will encourage the activity of lipase, should it be present, particularly if the moisture content is high.

Some companies use kibbled cake instead of cocoa powder as an ingredient. Freshly manufactured cake usually has a low microbiological count.

**Milk Powder**   Nonfat-milk powder is normally used for compound coatings. It has better keeping properties and avoids the inclusion of a third fat (milk fat), which may cause compatibility problems.

Roller process powder is used by some coating manufacturers but spray process powder usually gives a better flavor. Pretreatment of the milk by heating before spray drying will ensure low bacteria counts. Low-heat and high-heat powders are available; high-heat powder is less fat absorbent. With all types of powder, if the moisture constant rises above 4 percent, off-flavors will occur.

Whole-milk powder has poor keeping properties unless nitrogen or vacuum packed, because the milk fat is so finely divided. A high moisture content and bacteria count will accelerate deterioration. Specification for both types of powder:

Moisture content       3 percent maximum

Microbiological specification:

| Total bacteria count | Not greater than 5,000/g (three days milk agar 30°C) |
|---|---|
| E. coli | Absent in 1 g |
| Staphylococci aureus | Absent in 1 g |
| Salmonella | Absent in 100 g |
| Lipolytic activity | Absent |

Demineralized whey powder may be used as a partial replacement for nonfat-milk powder.

Milk and cocoa powders may be predried before making the coatings, and this helps to provide low moisture contents in the finished product. The dried powders must be used immediately as they are hygroscopic until incorporated in the fat. Alternatively, the coatings may be conched at a temperature sufficient to reduce moisture (see later).

**Sugar** The quality of the sugar ingredient should be similar to that required for chocolate: high grade and free from moisture and invert sugar.

Some special dietetic coatings use washed raw sugar or brown sugar, but these contain some invert sugar and some problems arise in the refining process. The presence of invert sugar imparts a plastic nature to the mixture, and also some hygroscopicity.

**Emulsifiers, and Other Minor Ingredients** Lecithin may be used in the same way as in the chocolate process to reduce viscosity and save fat. It is important, however, to study the implications of this, as saving fat in some vegetable-fat coatings may not be justifiable economically.

*Antibloom Agents* Like chocolate, vegetable-fat coatings are susceptible to fat bloom under certain conditions of storage but the antibloom effect of butter fat in chocolate does not apply to lauric-fat coatings.

The usual antibloom substances employed are the sorbitan and polyoxyethylene sorbitan fatty esters known commercially as Spans and Tweens.

Recommended amounts are 0.5 percent Tween 60 plus 0.5 percent Span 60, but experimental work has indicated that 2 percent of Span 60 alone is more effective. Other proportions of Spans and Tweens are also used. Span 65 sorbitan tristearate is preferred by some manufacturers.

There is considerable conflict of opinion on the effectiveness of these additives in coatings and chocolate for antibloom purposes and the general opinion is that they are of little use for chocolate but are effective in coatings in some recipes. It is very important that they are well dispersed by heating the coatings to 60°C (140°F) and well mixing for 30 min. These additives can improve the texture of a coating as well as retard bloom, and they will also alter viscosity, which must be checked after their addition.

Some users of these substances claim that their chemical constitution is not consistent and varies with different suppliers. It is, therefore, essential to make tests using compounds from a particular supplier and, if effective, to stay with the same source until an alternative has been approved.

**Crystallization Starters** These substances are high melting triglycerides, with a melting point of 55 to 70°C (130 to 158°F), which, when included in a fat in a proportion of 2.5 to 3.0 percent aid the tempering procedure. This is discussed under "Tempering."

## Manufacturing Processes

**General Process** The precise process is largely determined by the quality of final product required. The accepted chocolate process may be used with a preliminary ingredient mix followed by roll refining and conching. Alternatively, the ingredients less a proportion of the fat can be charged into a MacIntyre mixer/refiner, or a similar machine and the whole process completed at a temperature of about 50°C (122°F). Viscosity is adjusted by adding extra fat. For large outputs, continuous kneaders and mixers may be used.

The Wiener process incorporating a vertical ball mill may also be used for these coatings. This machine is described under "Chocolate Making."

**Tempering** In the first edition of this book, a system of tempering was described that required the coating to be mixed for 12 to 24 hr at a specific temperature, depending on the type of fat used in the coating. The object of this was to produce a seed of the higher melting fractions of the fat and this promoted the formation of stable fat crystals in the final coating, thereby preventing bloom during subsequent storage. This procedure is largely historic and was extremely inconvenient—it required considerable storage capacity.

Coating fats as now produced do not require tempering, except for very large or thick blocks. In some fats, 2 to 3 percent of crystallization starters are added, or they may be incorporated during the manufacture of the coating, making sure that the temperature is high enough to dissolve and disperse them—that is, not less than 60°C (140°F).

With large blocks, the center will cool much more slowly than the outer layers, which act as an insulation. Without tempering, the effect of the latent heat of crystallization will be so great that the center overheats and becomes completely liquid, with no seed.

Slow cooling finally results in very large fat crystals and granulation and discoloration of the center of the block.

**Conching** At one time, conching was considered an unnecessary expense for compound coatings. Today, the best results are obtained by some form of conching, as it is claimed that the chocolate flavor is developed by a prolonged heating of the cocoa ingredient with the bland vegetable fat. This is particularly so when cocoa powder is used as the ingredient with lauric fat coatings.

With nonlauric fats where cocoa liquor can be used, the chocolate flavor is naturally more pronounced. Conching may be carried out using normal conching machinery or other suitable equipment where the coating can be mixed, heated, and ventilated. High-temperature conching, up to 80°C (176°F), is beneficial for dark coatings.

### Conditions to Be Observed During Production

*Tempering and cooling* With compound coatings, for best results, the conditions for molding, enrobing, and cooling should be adjusted according to the properties of the fat phase in the coating.

The three basic groups of fats (i.e., cocoa butter equivalents, lauric and nonlauric replacers) are responsible for influencing the fat crystallization in the coating.

*Cocoa Butter Equivalents* These require the same tempering and cooling procedures as chocolate. They are polymorphic.

**Lauric Replacers**  These are based mainly on coconut and palm kernel oils and are not polymorphic. They crystallize spontaneously, on cooling, in one stable form, and this takes place within a narrow temperature range and is very quick. These coatings do not need tempering. Molding of thin bars and the enrobing of confectionery centers can be done with fully liquid coatings at 40 to 45°C (104 to 113°F). Preferably there should be quick cooling with maximum air velocities and this gives good contraction. With large blocks, some tempering, by reduction of the temperature of the coating during mixing until seed is formed, is preferable. In these circumstances, rapid depositing is necessary.

**Nonlauric Replacers**  These fats are also not polymorphic and crystallize in stable form on cooling. Generally, they crystallize more slowly and their contraction is less pronounced. Coatings prepared from these fats do not need tempering and can be used at 40 to 45°C (104 to 113°F). They do, however, need to be cooled slowly with lower air velocities.

**Condition of Centers, Molds**  Molds should be warmed to about the temperature of the deposited coating, and centers to 75 to 85°F, but if large cakes, cookies or aerated centers are being coated, temperature of these centers will be higher.

**Cooling**  It is necessary to emphasize the differences in cooling procedure between chocolate, cocoa-butter equivalent fat coatings, and lauric and nonlauric fat coatings.

Cocoa butter, being polymorphic, needs moderate cooling, and in a chocolate cooling tunnel, the conditions may be as follows:

|  | Entry | Center | Exit |
|---|---|---|---|
| ⟶ | 15 to 17°C | 10 to 12°C | 15 to 17°C |
| From enrober | (59 to 62°F) | (50 to 54°F) | (59 to 62°F) |

The same conditions will apply with a cocoa-butter-equivalent coating. With lauric fat coatings, rapid cooling is needed at the outset with lower temperatures and/or higher air velocities.

|  | | | |
|---|---|---|---|
| ⟶ | 10 to 12°C | 10 to 12°C | 15 to 17°C |
| From enrober | (50 to 54°F) | (50 to 54°F) | (59 to 62°F) |

Additionally, the covered centers should enter the cooling tunnel directly from the enrober.

Nonlauric replacers generally may be cooled under the conditions used for chocolate, but since there is a wide range of nonlauric fats, some variation is likely and the user must consult the supplier.

*Heat Treatment*  Some coatings benefit from heat treatment after covering to encourage the formation of stable forms of the fat in the solid mass. Covered centers may be stored in a room at 75° to 78°F for 24 to 48 hr with beneficial effects on shelf life and appearance.

Heat treatment must be applied immediately after enrobing. Superficial heating with radiant heat immediately after leaving the enrober is advocated by some manufacturers.

**Sanitation of Machinery, Tanks, Pipelines**  Because of the incompatability of lauric fats and cocoa butter, enrobers and other equipment used for chocolate should not be used for lauric fat coatings unless they are completely cleaned of chocolate residue.

With enrobers, this is an exceptionally difficult task, and it is best, if possible, to use separate enrobers for each type of coating.

**Storage**  Storage conditions for lauric-compound-covered confectionery are different from those for chocolate or cocoa-butter-equivalent coatings. It is best to store dark chocolates at a temperature of about 12°C (54°F) or below but lauric coatings should be stored at 20 to 22°C (68 to 72°F). Nonlauric coatings are generally best under conditions similar to those for chocolate.

Whatever the conditions are, they should be fairly constant. Wide fluctuations are bad and the relative humidity should be 50 to 55 percent.

## Colored Coatings and Pastel Coatings

The fats mentioned previously can be used for colored coatings and the processes for manufacture are similar to those for light and dark coatings.

**Flavor**  The fat and milk powder should not impart any significant amount of flavor of their own to the product, but with some coatings caramelization may be derived from the milk powder by the use of high mixing temperatures. Spray process powder is preferred for these coatings.

There is a variety of flavors that can be added and these are preferably oil soluble. Fruit flavors combined with citric acid (0.1 percent) help to reduce sickliness, in fact, some slight acid addition is desirable with most flavors. Flavors should generally be added in the last stage of processing.

**Color** Obviously, the best colors to use would be oil soluble, but few are in the "permitted lists" of various countries. Water-soluble colors can be ground into the coating at the refining stage but the best method is probably to add them in the form of an emulsion. Most of the certified water-soluble colors are also soluble in the permitted solvents glycerol and propylene glycol. To a solution of the dye in one of these solvents at the required concentration, lecithin is added with continuous stirring, maintaining the mixture at 50°C (120°F).

About 2 to 3 percent lecithin is required and a small quantity of water may be necessary to produce a good emulsion. This emulsion is added as an ingredient with the flavor late in the process.

With water-soluble colors, it is often necessary to use fairly high concentrations to get the desired shade. This can result in staining of the mouth upon eating the sweet.

"Lake" colors are now available. These are permitted colors precipitated on an inert inorganic base that renders them insoluble. They are very finely ground and so have intense coloring power. They are best added before the refining stage, but with rapid mixing can be added to the final melted coating.

## FORMULATIONS

A number of companies now manufacture cocoa butter equivalents and replacers and many formulations are given in their trade brochures (see the References). More information on the properties of these various fats is given in Chapter 9.

Tables 6.1, 6.2, and 6.3 show some typical formulations that will serve as a guide to the technologist new to the industry.

Nonfat-milk powder may partly or wholly replace the whole-milk

TABLE 6.1. COCOA BUTTER EQUIVALENT

| Percentage of | Dark compound | "Milk" compound | "White" compound |
|---|---|---|---|
| Cocoa liquor | 40 | 10 | — |
| Cocoa butter equvalent fat | 9.5 | 21.5 | 26.5 |
| Whole-milk powder | — | 20.0 | 25.0 |
| Sugar | 50.0 | 48.0 | 48.0 |
| Lecithin | 0.5 | 0.5 | 0.5 |

TABLE 6.2. COCOA BUTTER REPLACER—LAURIC FAT

| Percentage of | Dark compound | "Milk" compound | "White" compound |
|---|---|---|---|
| Low-fat cocoa powder (10–12% CB) | 14.0 | 5.0 | — |
| Lauric fat | 29.5 | 31.0 | 31.5 |
| Nonfat-milk powder | 8.0 | 17.5 | 20.0 |
| Sugar | 48.0 | 46.0 | 48.0 |
| Lecithin | 0.5 | 0.5 | 0.5 |

powder, in which case adjustment of the proportion of added fat will be necessary.

Nonfat-milk powder has a fat content of 1 percent, whereas whole-milk powder has a fat content of 25 to 28.5 percent.

Since some formulations include a proportion of whole-milk powder this introduces a third fat (milk fat), and care must be taken to ensure that the texture of the final product is acceptable.

Nonlauric fat compounds are generally used for enrobing. The texture of the fats and their poorer contraction on solidifying make them less suitable for molding.

The presence of whole-milk powder may render the final product too soft for warm climates, and in such cases wholly nonfat milk should be used.

Because of their elastic texture, these compounds are very suitable for coating cakes and cookies.

TABLE 6.3. COCOA BUTTER REPLACER—NONLAURIC FAT

| Percentage of | Dark compound | "Milk" compound | "White" compound |
|---|---|---|---|
| Low-fat cocoa powder (10–12% CB) | 12.5 | — | |
| Cocoa liquor | 10.0 | 10.0 | |
| Whole-milk powder | — | 5.0 | 20.0 |
| Nonfat-milk powder | — | 13.0 | 5.0 |
| Nonlauric fat | 30.0 | 27.5 | 29.5 |
| Sugar | 47.0 | 44.0 | 45.0 |
| Lecithin | 0.5 | 0.5 | 0.5 |

# DIETETIC COATINGS

People prone to obesity, who are calorie conscious, or who suffer from certain illnesses such as diabetes must limit their intake of chocolate and confectionery. In order that they not be deprived of these attractive foods, a number of modified confections have been developed.

Confectionery generally is not a well-balanced foodstuff, being very rich in carbohydrate and, in the case of chocolate, also high in fat. Protein is somewhat deficient, as are vitamins.

In addition to products that have been developed for some therapeutic reason, there are numerous confections on the market that claim some health-giving properties or perhaps to aid in weight control. Sweeteners as alternatives to sugar have come into prominence, including sorbitol and fructose, which are useful for diabetics, and xylitol, which is noncariogenic. Others are discussed in more detail in Chapter 8.

Chocolate has received some criticism because of its caffeine and theobromine content, and the cocoa ingredient may be replaced by carob or defatted wheat germ. Soya protein isolates are increasingly used to enhance the protein content of candy bars.

## Diabetic Chocolate

For persons suffering from diabetes, the forbidden ingredients in chocolate and confectionery are obviously sugar, dextrose, invert sugar, and starch-conversion products.

Many ingredients have been tried as replacements for the sugar in chocolate, which constitutes 40 to 50 percent of the final product's weight. If intense artificial sweeteners such as saccharin are used, the bulk normally occupied by the sugar has to be replaced by some other ingredients, otherwise the bitter flavor of the excessive liquor content is objectionable.

Nuts, either whole or ground, are acceptable, and for a period soya flour was also used, but even the best product from these ingredients was not a good replacement.

The use of sorbitol as an acceptable sweetener for diabetic foods has been a useful development as this substance provides bulk as well as sweetening power.

More recently, crystalline fructose has become a commercial article and, like sorbitol, requires no insulin for ingestion.

The sweetness of sorbitol is about half that of sugar whereas fructose is considerably sweeter than sugar. It is very hygroscopic and, as with sorbitol, this presents some problems in production.

With sorbitol formulations, the addition of an artificial sweetener such as saccharin is necessary to enhance sweetness but fructose does not require this.

**Manufacture of Diabetic Chocolate**  The ingredients—cocoa liquor, sorbitol, artificial sweetener (saccharin), nuts, and fats—are mixed to a suitable paste for refining.

The refining stage presents some unusual features since the sorbitol crystals are different from sugar, being elongated in shape and relatively soft, and tend to grind into flat plates instead of roughly rounded particles. This fact, together with the hygroscopic nature of sorbitol, can result in moisture absorption from the air under conditions of high humidity, and this can cause trouble at the conching stage.

After refining, the paste is conched with additional fat and flavor is added. It is best to complete the conching process and use the chocolate without delay in the subsequent molding or covering processes as this type of chocolate exhibits considerable thixotropy and will thicken or partially set if held without mixing for any length of time. This effect is aggravated by the presence of moisture, either in the original sorbitol, in other ingredients such as soya flour or milk powder, or by absorption during refining.

In the conching process, a temperature of 46°C (115°F) must not be exceeded, and above this, viscosity increases very rapidly and the chocolate becomes almost gelatinous and unmanageable.

**Manufacture of Diabetic Assortments**  The assorted centers that can be manufactured from sorbitol are necessarily limited as the physical properties of sorbitol solutions and different from those of sugar. However, certain confectionery pastes and fondants may be made. Sorbitol syrup (70 to 80 percent) is available commercially, as well as crystal sorbitol, and is slightly less expensive. Sorbitol will not crystallize from supersaturated syrups as readily as sugar, and a mixture of sorbitol powder and syrup is used to prepare confectionery pastes and fondants. Boiled toffees can be made from sorbitol syrup, nut pieces stirred in and the mixture poured onto cooling tables for cooling and cutting (see "Confectionery Processes").

Chocolate covering of these centers is usually carried out by hand or fork dipping but enrobing is possible provided precautions are taken to prevent thickening due to moisture absorption or storage of the liquid chocolate.

## Carob Coatings

Carob, also known as Saint John's bread and locust bean, has been recognized as a valuable foodstuff since biblical times. In spite of this, until recent years, its products were little known outside its normal area of growth—the shores of the Mediterranean. The product, which has achieved publicity of late, is obtained from the deseeded pod after roasting and grinding, and is known as carob flour.

Carob is well described by Blenford (1983) and numerous confectionery and food products are now manufactured from it, including drinks, cookies, coatings, and snack bars. The coatings have a composition similar to those already described, with the cocoa ingredient replaced by toasted carob powder either wholly or in part.

An analysis of toasted carob flour is as follows:

| | |
|---|---|
| Protein (N × 6.25) | 4 to 5 percent |
| Carbohydrates—total | 33 to 50 percent |
| Sucrose | 10 to 30 percent |
| Dextrose | 2 to 7 percent |
| Fructose | 5 to 10 percent |
| Fat | 0.5 to 1.0 percent |
| Crude fiber | 3.5 to 10 percent |
| Ash | 2.3 to 3.5 percent |
| Theobromine ⎫<br>Caffeine     ⎬ | none |
| Moisture | 4.0 percent maximum |

*Particle size*:

| | (1) | (2) |
|---|---|---|
| percent through 200 mesh | 97.5 | 99 |
| percent through 325 mesh | 75.0 | 90 |

Note particularly the absence of caffeine and theobromine.

As a result of roasting, carob flour has a good bacteriological analysis. Coatings can be produced using 100 percent carob flour without any cocoa powder, as well as mixtures of the two.

Although subsequent processing is exactly the same as for cocoa or chocolate coatings, care must be exercised in the choice of fat and the choice of carob flour to ensure that the best-quality carob coating is obtained.

## Defatted Wheat Germ

Processed defatted wheat germ, developed by Kovacs (1976–1981), is a highly nutritious substance available in powder or expanded forms.

By roasting, using specific processes, a product that has properties allied to cocoa is obtained. From this, compound coatings may be made that have good flavor and color, are nutritious and are competitive in cost. In an expanded form, they can usefully replace nuts and cereals in chocolate bars.

Defatted wheat germ is a natural material and is a good source of high-quality protein, vitamins, and minerals. See Table 6.4.

Defatted wheat germ may be used for the manufacture of coatings using the processes and formulations previously mentioned. As with carob, allowances must be made for the low fat content and carbohydrate.

TABLE 6.4. TYPICAL ANALYSIS

| Chemical and physical properties | Percent | Essential amino acid content Per 100 g of pure protein | | Vitamin values mg./100 g | |
|---|---|---|---|---|---|
| Protein (N × 6.25) | | Arginine | 3.7 | Thiamin | 2.26 |
| (min.) | 30.0 | Cystine | 1.0 | Riboflavin | 0.79 |
| Carbohydrate | | Histidine | 1.0 | Niacin | 9.75 |
| (by difference) | 53.0 | Isoleucine | 3.1 | Vitamin B6* | 1.8 |
| Fat (max.) | 1.0 | Leucine | 5.6 | Pantothenic | |
| Fiber (crude) | | Lysine | 4.9 | acid† | 1.74 |
| (max.) | 5.0 | Methionine | 1.3 | Choline‡ | 220.0 |
| Ash | 6.0 | Phenylalanine | 3.2 | Folic acid | 0.37 |
| Moisture | 5.0 | Threonine | 3.5 | Inositol | 573.0 |
| Calcium | 0.104 | Tryptophane | 0.9 | Biotin | 0.05 |
| Phosphorus | 1.0 | Valine | 4.8 | | |
| Iron | 0.01 | | | | |
| Sodium | 0.004 | | | | |
| Potassium | 1.1 | | | | |
| Cobalt | 0.00001 | | | | |
| Copper | 0.0008 | | | | |
| Zinc | 0.0007 | | | | |
| Magnesium | 0.3 | | | | |

* As pyridoxin.
† As d-calcium pantothenate.
‡ As choline chloride.

## "Slimming" Chocolates

There has been much discussion in medical and legal circles regarding the validity of the term "slimming" or "nonfattening" as applied to various foodstuffs, and chocolate is no exception.

There are two basic methods of helping to make a food less fattening:

1. Incorporating ingredients rich in protein and low in fat and carbohydrate, such as soya and ground nut protein.
2. Add materials that have no food value at all. There is obviously great opposition to many possible ingredients, such as chalk and similar inorganic substances that were used to adulterate foods in the early nineteenth century.

More readily acceptable are substances such as certain cellulose esters (e.g., carboxy-methyl cellulose), which, while added to the foodstuffs in quite small amounts, are prone to create considerable bulk on the absorption of moisture.

The claim is that with foods containing these materials, bulk is created in the stomach, giving a feeling of fullness that allays further appetite. No difficulty arises when incorporating these ingredients in the normal chocolate processes.

A third method, some times advocated by spurious advertizing, is worthy of mention. This applies to certain aerated products. There is nothing wrong with aerated confections and there are many delectable candies on the market in which the appeal, pleasant texture and flavor are related to the fact that they are aerated and are accordingly of considerably greater bulk.

It is wrong to claim that these are less fattening when the only support for this is that there is less of them by weight in a given volume.

When judging such foodstuffs, the criteria are the composition and the caloric value of the quantity consumed.


## Medicated Chocolates

Chocolate and cocoa, and to a lesser extent cocoa butter, are often an ingredient of tablets and pills used for therapeutic purposes. Chocolate is a suitable and pleasant carrier for many medicines with an objectionable flavour, and is particularly applicable to children's medicines. It can also be used for coating capsules.

## REFERENCES

Blenford, D. 1983. *Carob "Food."* Bentec Products Ltd., Oxford, England

Kovacs, L. 1976–81. Wheat germ nuggets. *Food Processing* (Apr. 1978). Vitamins Inc., Chicago, Ill.

Meursing, E. H. 1983. *Cocoa Powders for Industrial Processing.* Cacaofabriek de Zaan B. V. Koog aan de Zaan, Holland.

The following companies publish informative technical brochures.

Aarhus Oliefabrik A/c, Aarhus, Denmark

Capital City Products, Columbus, Ohio

Durkee Foods, Cleveland, Ohio

Dynamit Nobel A.G., Troisdorf, W. Germany

Friwessa B.V., Zaandam, Holland.

Friwessa Inc., East Hanover, N.J.

Loders Croklaan, Wormerveer, Holland

Loders Croklaan, Burgess Hill, England

Loders Croklaan, New York, N.Y.

Sais, Zurich, Switzerland

Sorbitol-fructose:

Roquette Freres, Lestrem, France

# 7

# Chocolate Bars and Covered Confectionery

## PRODUCTION METHODS

The methods of manufacture of bulk chocolate have been described and there are certain well-defined processes for using chocolate to produce a variety of confections. These are summarized in the following.

## Molding

This is the casting of liquid chocolate into molds (metal or plastic) followed by cooling and demolding. The finished chocolate may be a solid block, a hollow shell, or a shell filled with a confectionery material such as fondant, fudge, or soft caramel.

## Enrobing

This is the mechanical method of coating confectionery centers with chocolate by putting them through a curtain of liquid chocolate followed by cooling.

Under this heading, hand covering and fork dipping may be mentioned. The former is the traditional method of coating centers by immersion in tempered liquid chocolate, removal by hand, usually one finger, although most of the hand is covered in chocolate, and then depositing on a glossy surface. The finger is detached from the liquid chocolate in such a way as to leave a decorative twirl. This is a highly skilled process, though not entirely hygienic.

Fork dipping avoids the immersion of the hand in the chocolate by the use of a thin pronged fork. The result is not quite so good in appearance.

## Panning

This process, sometimes called the Volvo process, employs a rotating pan in which the centers rotate and cascade over one another. The chocolate is applied by hand or spray during rotation and is set by cooling air applied to the pan as it revolves. In this manner, layers of chocolate are built up around the center to any desired thickness and the shape of the final article approximates that of the center and is quite smooth. A polish or glaze is usually applied to this type of chocolate unit.

## Tempering

This process is necessary as a preliminary to all the other chocolate processes. It ensures that the cocoa butter constituent is seeded and that the chocolate will set in a stable condition with a good permanent color and gloss.

# OTHER CHOCOLATE PROCESSES

## Chocolate Drops (Chips)

Chocolate drops or chips are made in very large quantities. They are sold in small bags for household use and in bulk for inclusion in cookies and other flour confectionery products.

Viscous chocolate, properly tempered, is deposited continuously from a battery of small nozzles onto a moving metal belt. The belt passes through a cooler and, after setting, the drops are discharged either directly into bulk packages or to a bag packaging line.

A necessary precaution with bulk packing is to see that the drops are completely cooled. If they are not, latent heat evolved in the bulk pack will raise the temperature of the drops well above ambient and ultimately severe fat bloom will form.

Drops are produced usually in small pyramid shapes but also may be deposited as small flat domes. These can be coated with colored nonpareils before the chocolate is set. Packed in attractive colored transparent bags, they make a popular children's product.

## Roller Depositing

This production system has been designed for the manufacture of small solid articles using chocolate or other fat-based compounds. The articles may be "lentil" centers, coffee beans, small eggs, balls, and so on, and they are formed in a pair of stainless-steel rollers in which the two halves of the article have been engraved.

The hollow rollers are equipped with a special coolant circulation system. This cools the rollers in such a way that when the liquid-tempered mass is fed between them, it forms a partly cooled continuous belt containing the shaped articles. This continuous belt is then passed through a cooling tunnel for the final cooling process. The cooling tunnel length can be designed to suit requirements, that is, if space is a problem, then a multipass cooler can be arranged that will reduce the length of the cooling tunnel.

After cooling, the web is transferred by conveyor to a rotating perforated drum. The speed of rotation and the angle of inclination of the drum are adjusted to separate the shaped articles from the flanges connecting the articles during forming. This process also smooths the surfaces for further processing, such as sugar or chocolate panning or polishing. The pieces passing the drum perforations are reprocessed.

The capacity of the roller line will depend on the width of the rollers and the number of pairs installed and can vary between 275 and 1,300 lb per hour.

Design of the roller assembly facilitates complete replacement and change of product (Figs. 7.1 and 7.2).

## Aerated Chocolate

Aerated chocolate in bar or tablet form has been a very popular product for many years. The basic method of preparation is to subject well-beaten tempered chocolate to vacuum. The small bubbles will expand, producing a cellular product, and in this form it must be set by cooling. If the vacuum is released while the chocolate is still warm, the aeration will collapse.

A similar product may be made by mixing chocolate with carbon dioxide under pressure. Under these conditions, the gas is partially soluble in cocoa butter.

This chocolate may be released through nozzles to normal atmospheric pressure whereupon a cellular product, very similar to that formed by vacuum, is produced.

A third method relies on the addition of certain fatty substances and vigorous mixing. This gives a fine aeration that is not prone to collapse under normal chocolate handling processes.

Fig. 7.1.  Roller Depositing Machine

*Aasted Mikroverk, Bygmarken, Denmark*

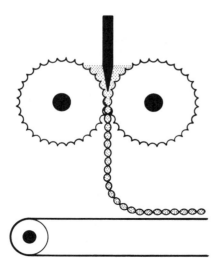

Fig. 7.2.  Principle of Operation of the Roller Depositing Machine

*Aasted Mikroverk, Bygmarken, Denmark*

## Chocolate Flake or Bark

"Flake" is the trade name of a very popular English chocolate bar, but similar chocolate pieces are sold in Europe and in the United States. The idea of this form of bar arose from the physical condition of chocolate paste as it came from the refiner rolls, but because it is produced at considerable speed when making normal chocolate, it is broken into many small flakes. If the roll speed is reduced, the film on the roll surface can be removed in a continuous sheet and, depending on the angle of the scraper blade, can be collected as a wrinkled, partially compressed bar of a length equal to the width of the roll.

The width of the bar, or the quantity scraped from the roll, can be mechanically controlled by intermittent stopping and starting of the roll, and a take-off device, also mechanical, removes the bar from the scraper blade in unison with the halting of the roll. The length of the separate bars is controlled by means of a small sharp protrusion at intervals along the scraper blade.

Refiner paste for flake has a higher fat content than that for chocolate manufacture (usually about 31 percent) and is subjected to a certain amount of physical kneading.

Another method of getting the paste into a suitable condition for flaking is to add a small amount of water, which must be very adequately distributed through the chocolate.

During the kneading process, some cooling of the mass is necessary to temper the chocolate and to produce a stable cocoa butter seed. In this condition, at about 30 to 32°C(86 to 90°F) for milk chocolate, it is fed to the flaking rools, and after producing the paste in bar form, it is tranferred automatically from the scraper to a tunnel cooler where it is set solid in its crinkled form.

The pasting process and the slightly higher fat content than is normal for refiner paste ensures greater rigidity in the finished bar, which, because of its structure, is somewhat fragile. It also has a low weight for its bulk—a useful sales point.

Of recent years, this chocolate bar has achieved additional popularity as a sweet to be served with ice cream. It has also been marketed as a thinly chocolate covered bar, which gives the flake more rigidity.

A diagram of the mechanical principle of flake manufacture is given in Fig. 7.3.

There are alternative methods of transferring the flake scraped from the rolls. One uses a grab device that not only does the transfer but slightly compresses the flake. Another uses a system where the

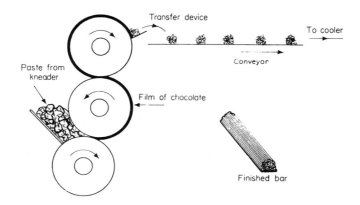

Fig. 7.3. Manufacture of Chocolate "Flake"

take-off plate slopes downward intermittently and pushes the flake bars off onto the cooler conveyor.

## Chocolate Vermicelli or Streusel

Vermicelli is derived from the Italian, meaning "little worms," which it resembles during manufacture. This chocolate product was very popular at one time for coating truffles and is probably used more today for flour products than for chocolates.

Chocolate vermicelli is usually manufactured by the principle of extrusion through a perforated plate. The plate is fed with chocolate paste by a worm and rotating blade as in a mincer or by rollers rotating under pressure on the top of the plate. To produce vermicelli from pure chocolate is not easy as the paste has to be tempered to obtain the correct texture, and to maintain it in this condition in the extruder requires careful cooling control to prevent frictional heat from destroying the temper. This is helped by rapid transfer of the chocolate paste through the perforated plate, and to assist this, the holes are drilled with recesses (Fig. 7.4). In this way, the strength of the plate is maintained against the pressure of the paste, but friction during passage is reduced.

A recent machine by Lloveras S.A. uses a thin section sieve supported by crossbars. The paste is pressed through the perforations by Teflon rollers, and very little fractional heat is developed and there is negligible buildup on the rollers. The paste must be fed

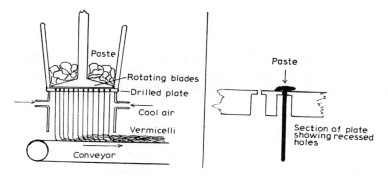

Fig. 7.4. Chocolate Vermicelli Manufacture

consistently and in keeping with the optimum throughput of the extruder. Cooling must be provided immediately below the sieve so that the vermicelli strands are rapidly set and retain some rigidity by the time they reach the conveyor belt. Otherwise, the strands will curl and solidify in the tunnel cooler, making it difficult to break them up during panning.

The extruded paste is carried away on a belt for cooling, and emerges as a series of strands about 6 to 9 in. long. This is transferred to a comfit pan and on rotation is broken into pieces ranging from $\frac{1}{8}$ in. to $\frac{1}{4}$ in., after which it is glazed with syrup. To give richer shades, the syrup may be colored with a "permitted" food color.

An easier method of producing vermicelli from chocolate is to mix about 6 percent of water with the chocolate paste before extrusion and this gives a suitable texture that is permanent irrespective of changes of temperature in the extruder.

The vermicelli produced, however, requires some drying, and the final texture is inferior to that of the pure chocolate material.

A third method of manufacture that gives a product suitable for flour confectionery where long shelf life is not required uses a plastic vegetable fat. The manufacture of the paste is similar to that described under "Coatings," with cocoa, sugar, and vegetable fat (and milk powder for light colors), a vegetable fat being chosen that has a pasty consistency at 31 to 32°C (88 to 90°F), but which does not have a complete melting point above 40°C (104°F). This paste can be extruded over a fairly wide range of temperature, but even after glazing, may be susceptible to the development of fat bloom.

## Laminated Chocolate

Tempered chocolate may be laminated by means of rollers and, while in a plastic condition, can be cut into bars or pieces.

Laminations consist of dark, milk, or white chocolate.

## Chocolate Tempering

It has already been noted that cocoa butter exists in a number of polymorphic forms and the nature of the crystalline form depends on the method of cooling the liquid fat.

If chocolate is solidified from the liquid state without any attention to seeding of the liquid cocoa butter constituent or to the method of cooling, it will be granular in texture and of poor color or blotchy in appearance.

To obtain chocolate tablets or covered confectionery of good texture, color, and in a stable condition such that bloom will not develop on the surface on storage, good tempering and correct cooling are essential.

The process of tempering consists of cooling down the chocolate with continuous mixing to produce cocoa butter seed crystals and distributing these throughout the mass of liquid chocolate.

Originally, with hand processes, this was carried out by turning over the chocolate repeatedly on a marble slab with a flexible pallet knife. This chocolate was then transferred to a small covering bowl that was warmed from underneath and could be topped with unseeded chocolate from time to time, and this well mixed in. The slab process is still used for introducing the first seed into small enrobers such as are employed in some retail establishments.

On a still smaller scale, a bowl of chocolate containing about 4 to 6 lb of chocolate is the starting point. This chocolate, if taken from a storage tank, will be at about 45°C (113°F) and it is preferable to cool this to about 35 to 38°C (95 to 100°F) before starting the tempering procedure.

About one-third of this chocolate is mixed on the marble slab until it thickens, taking care that lumps are not formed by repeatedly scraping from the outside to the center.

The thickened chocolate is then returned to the bowl and mixed thoroughly with the contents of the bowl. The mixture is then fully tempered. A skilled operator can tell quickly whether the chocolate is tempered by applying a small amount to the lips. A very small sample spread on a piece of foil and placed in a cooler should set rapidly. A thermometer is a useful guide and the final temperature should be about 31°C (88–89°F), and a degree or so lower for milk chocolate.

A warning: use the thermometer only as a final check. The essence of hand tempering is visual judgment of the condition of the mixed chocolate on the slab.

With mechanical and large-scale tempering, temperature control is employed but visual checks are always useful. The Tempermeter, described later, is a precise instrument for determining exactly the degree of temper.

Hand covering and fork dipping require skill and practice but correctly done produce very attractive chocolates with good gloss and shelf life.

There is an ever-increasing number of small retail shops that use these methods. Illustrated books giving details with formulations are available (Wilton, 1983).

**Machine Tempering**  The simplest and first large-scale temperers to be constructed were cylindrical steel tanks, water jacketed and provided with an internal scraper mixer (Fig. 7.5). Water is circulated through the jacket at 13 to 15°C (55 to 60°F) and the action of the stirrer is to sweep the liquid chocolate over the cool surface, thereby forming cocoa butter seed and distributing it through the liquid mass.

The scraper must be in intimate contact with the interior surface of the kettle to avoid building up a layer of overcooled chocolate that will contain unstable cocoa butter seed.

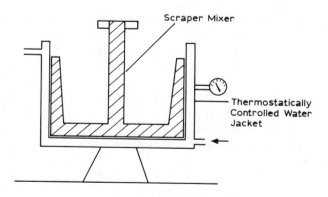

Scraper Mixer

Thermostatically Controlled Water Jacket

Fig. 7.5.   Tempering Kettle

The temperature of the chocolate from the storage tank is steadily reduced from about 46 to 49°C (115 to 120°F) to about 28 to 29°C (83 to 85°F) for dark chocolate or about 1°C lower for milk chocolate because of the milk fat present.

At this temperature, the chocolate should have sufficient cocoa butter seed in stable form to ensure that the chocolate after cooling will also contain stable forms of cocoa butter. Moderate forms of cooling are essential (see later).

[Note: In Chapter 3, on cocoa butter, it was pointed out the unstable crystal forms have melting points of

$\gamma$ 17°C (63°F), $\alpha$ 21 to 24°C (70 to 75°F), $\beta'$ 27 to 29°C (81 to 84°F):

therefore, during the kettle process described above, unstable forms should not be formed.]

During the seeding process, chocolate viscosity increases as the proportion of seed increases, therefore, it either must be used quickly or heated to remelt some of the seed.

With molding machines with high output, quick usage is possible and duplicate tempering kettles are often used, but for enrober use it is necessary to raise the temperature of the seeded mass to 32 to 33°C (90 to 92°F)—lower for milk chocolate (88 to 90°F, depending on milk fat content)—and this has the effect of decreasing viscosity while still retaining sufficient seed to allow the cocoa butter to set in a stable form.

Raising the temperature will also ensure that any unstable seed is melted. This is important for the enrobing process as enrobed units are more susceptible to fat-bloom formation than molded bars, which have a highly glossy compacted surface.

Tempering kettles are now usually supplied with thermostatically controlled water circulation, which considerably simplifies maintenance of the correct degree of temper. A second kettle that contains untempered chocolate at about 92 to 94°F (34°C) is useful to supply small quantities to the tempering kettle if the viscosity should become too great.

If this is done, care must be taken to see that it is adequately mixed in or streaks will appear on the enrobed pieces. This same kettle may supply an enrober by drip feeding chocolate to maintain correct viscosity and temper in the enrober tank.

Another method of tempering is to use solid chocolate shavings that are mixed into liquid chocolate at 32 to 33°C (90 to 92°F), and this avoids the cooling and reheating process. This is particularly useful for small confectionery manufacturers who buy their chocolate in block form. It has one disadvantage in that sometimes the shavings are difficult to disperse or will aggregate, and lumps of chocolate traveling round a pipe system can cause serious trouble as, even if screens are provided, these may become blocked. It is worth noting that certain chocolate technologists advocate this method of tempering and even go the the length of preparing their chocolate seed by cooling well-grained chocolate in very thin layers on an enrober band. They claim that the nature of this seed encourages the formation of a maximum of stable forms of cocoa butter in the chocolate.

**Automatic Tempering** Modern chocolate plants for molding, enrobing, or shell production with large outputs are provided with automatic temperers. These are constructed in the form of tubular or plate heat exchangers with each section controlled thermostatically. Chocolate is delivered from the storage tanks and is subjected to the cooling and warming cycles, as previously indicated.

An example of the plate type is the DMW temperer by Aasted International of Denmark. These heat exchangers have the advantage of large cooling surfaces with very efficient mixing and scraping of the surfaces. This produces a fine-grain, stable seed in the tempered chocolate. The principle of operation is described in Fig. 7.6.

The whole system is enclosed in a module with external controls and capacity may vary from 650 to 9,000 lb per hour.

**Principle of Operation** By means of a pump, the chocolate is transported through the temperer. Continuous scraping of the surfaces ensures moderate cooling without any shock treatment.

The temperer is equipped with up to seven cooling zones, each with electronic controllers set at the required temperature. The temperer operates fully automatically.

To prevent the tempering machine from freezing when stopped, all tempering elements may be circulated with hot water from a built-in hot water system.

**Technical Description** Consider Fig. 7.6. The built-in drive motor (1), drives over V-belts a strong worm gear (2). The main shaft (3) with its scrapers (4) is driven by this worm gear. The scrapers remove chocolate from the elements (5) which are water jacketed. The various tempering sections are equipped with digital electronic controllers, which, by means of probes (6) placed in the chocolate flow, control the solenoid valves (7). At (8) cooling water is intro-

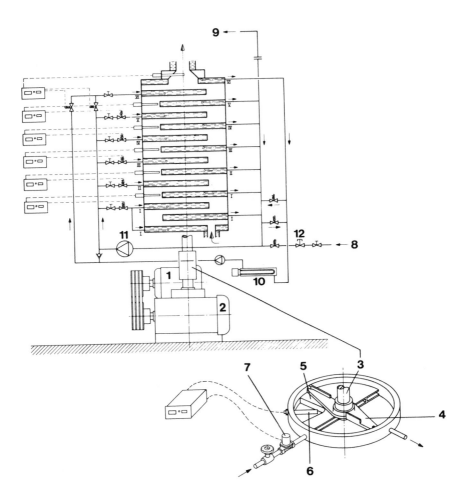

Fig. 7.6. DMW Temperer

*Aasted Mikroverk, Bygmarken, Denmark*

duced through a thermostatic water valve, and at (9) is returned to the water cooler. In the latter section, cooling or reheating can take place as required.

A cooling water pump (11), and hot water supply (10), together with a thermostatic valve, forms part of the cooling water stabilization. The stabilizer ensures that the cooling water is supplied at a suitable temperature and at a sufficient water pressure. The required cooling temperature is set by a thermostatic water valve (12).

## Compound Coatings

It should be mentioned here that the conditions for tempering and cooling of chocolate do not apply to compound coatings. Lauric fat coatings, for example, require other conditions, and these fats are not compatible with cocoa butter.

These factors are described in Chapter 6 on confectionery coatings.

## Measurement of Degree of Temper

Well-tempered chocolate contains approximately 3 to 8 percent cocoa butter crystals.

It is necessary with large molding machines and enrobers to know quickly whether chocolate being used is properly tempered. Experienced chocolate workers can assess the degree of temper fairly well from touch and by setting a small sample on metal foil. Temperature recording gives some confirmation. However, the precise degree of temper needs rapid scientific confirmation if large machines are to work consistently.

One instrument developed to accomplish this is called the Tempermeter. It is based on the principle that under controlled conditions of cooling chocolate there is a relationship between the cooling curve and the degree of temper.

The instrument consists of a temperature recording probe that is immersed in the sample of chocolate contained in a narrow tube. The tube is immersed in ice water and the probe is connected to an automatic recorder that prints the temperature on a moving chart at equally spaced intervals of time. Thus, a cooling curve is formed and the shape and slope of the curve depend on the degree of temper. The curve is produced in about 4 min.

The latest instrument by Sollich uses a transparent inclinometer, which greatly facilitates the measurement of the slope of the cooling curve.

In sampling, it is essential that the chocolate be taken directly from the molding machine or enrober to fill the tempermeter tube. Figures 7.7, 7.8, and 7.9 show the instrument, typical cooling curves, and the inclinometer.

Apart from measuring the degree of temper of normal chocolate,

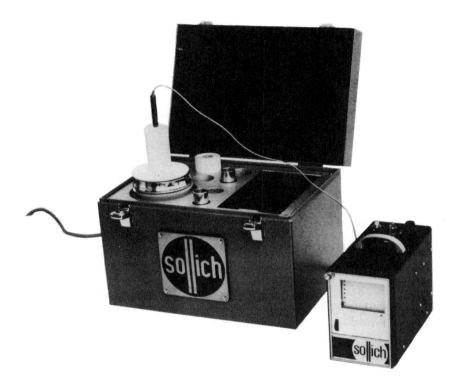

Fig. 7.7.   Portable Control Unit for Determining the Degree of Temper in Chocolate

*Sollich KG, Bad Salzuflen, W. Germany*

the instrument is useful in estimating the effects of additions to chocolate.

Fats claimed to be cocoa butter equivalents, which, in all chemical and most physical respects, are identical to cocoa butter, sometimes retard the tempering process when mixed in the chocolate. By taking samples from a tempering kettle at prescribed intervals and determining the degree of temper, this effect can be measured. Lecithin added to chocolate in amounts much above the normal of 0.5 percent will severely retard the tempering process. Occasionally, an error in dispensing in a continuous mixing process may produce this effect.

Richardson (1986) has shown that the principle of the Temper-

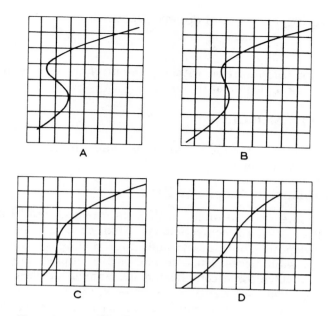

Fig. 7.8.   Representative Curves Produced by Tempermeter
A. Undertempered          B. Slightly Undertempered
C. Correctly Tempered     D. Overtempered

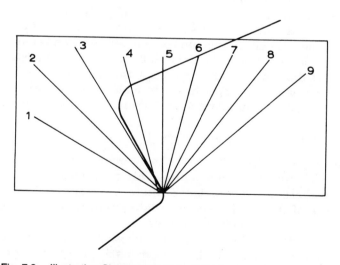

Fig. 7.9.   Illustration Showing the Use of "Inclinometer"
1. Very Undertempered          6. Very Slightly Overtemperred
2. Undertempered               7. Slightly Overtempered
3. Slightly Undertempered      8. Overtempered
4. Very Slightly Undertempered 9. Much Overtempered
5. Correctly Tempered

197

meter may be computerized with a printout of the curves and ready assessment of the degree of temper.

## CHOCOLATE MOLDING

The main types of molding machine make solid tablets, hollow figures, and filled shells.

### Molding Chocolate Blocks or Bars

The chocolate may be mixed with nuts, raisins, small cookies, or small pieces of hard confectionery. This chocolate or chocolate mixture is deposited into molds and, after cooling, is knocked from the molds as solid pieces. These machines are, in principle, relatively simple but involve some complicated mechanical devices, especially the depositor. They consist of the temperer, the depositor, the mold cycle carrier, the cooling tunnel, and the packing belt. The temperer has already been described. The depositor has multiple valves that will take chocolate or a mixture with nuts, raisins, and other ingredients from a hopper and deliver a row of deposits of correct weight into the mold impressions.

The design of molds is an expert's job; although the general appearance of the finished block or bar may be decided by the sales staff, the exact design of the impression requires knowledge of the optimum angles for the mold sides and the type of embossing that will facilitate demolding and give a good appearance.

Certain companies are specialists in mold manufacture and must work closely with the chocolate manufacturers' engineers and pattern makers.

The following comments are relevant:

1. Steep angles at the sides of the mold impression obviously cause more difficulty with demolding. This problem may be emphasized if there is a slight flange where the mold impression joins the face of the mold. Milk chocolate products are more difficult to produce in these circumstances.
2. Areas of flat surfaces without embossing should be avoided. Force marks, which look like finger marks, may appear on these areas.
3. Do not have deep embossing. Intricate design causes problems with the occlusion of air bubbles and small particles of chocolate.

Metal molds were used originally and were carried on an endless chain. Depending on the size of the molded article, the mold frame contains many identical impressions, and many hundreds of these molds may be transported on the chain.

At the start of the cycle, a warm tunnel heats the molds to a temperature approaching that of tempered chocolate. The molds then receive deposits of chocolate and the whole chain and molds are subjected to violent vibration or tapping to distribute the chocolate evenly over the molds and to remove air bubbles from the bottom of the mold, which will be the face of the block when demolded. After this operation, the molds pass into the cooler. The design of coolers and the air conditions used for all types of chocolate plant are discussed elsewhere. At the end of the cooler, the molds are reversed and the molded blocks tapped out onto a conveyor belt that takes them to the wrapping machines. The empty molds return to the mold heater and then to the machine to receive the next deposit of chocolate.

Metal molds were once the only type available, but plastic molds are now in almost universal use.

Thermal conductivity, which is inferior to metal, was originally a problem when the molds were of very solid construction. Designs have been improved, and by means of special bracings, the mold impressions themselves are now sufficiently thin to give good heat transfer.

Plastic molds are considerably lighter in weight than metal molds and a recent development in this direction was the introduction by Aasted International of a carrierless molding plant. This new design eliminates the use of the heavy metal mold carrier. When using plastic molds, the physical load on chains, rails, and turning points is reduced substantially, with a consequent increase in the life of these units. In addition, the power rating for the motors is reduced, and energy consumption, both in the mold heater and the cooling chamber, is reduced to a minimum. This new mold is designed with guide pins on each end and a long hole in the middle for the insertion of the suspension pins. Only the guide pins from one side are required to turn the molds over when producing hollow articles or when conveying molds from one level to another.

The long hole in the middle—that is, the central suspension—provides a more efficient shaking of the mold. In addition, the noise level of the plant is considerably reduced by the elimination of any metal-to-metal contact.

## Hollow Goods

Hollow chocolates include Easter eggs and rabbits, Santa Clauses, and the like. The molds are in two halves, usually hinged to open and close like an oyster. The requisite amount of tempered chocolate is deposited in one half of the mold, and the mold is closed and clipped or magnetically attached to a machine that rotates the mold through a variety of directions, thereby distributing the chocolate evenly on the inside of the mold. While rotating, the chocolate is partially cooled until it ceases to flow. The molds are then removed and setting is completed in a tunnel cooler, after which the molds are opened and the articles removed.

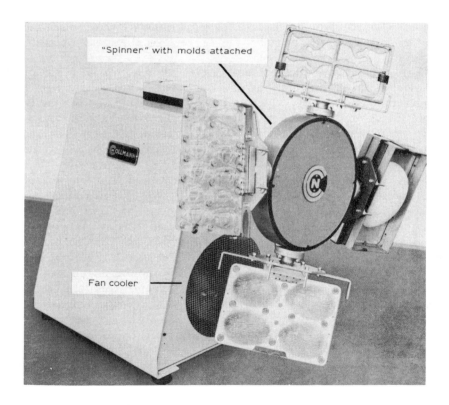

Fig. 7.10.   Minispinner for Hollow Chocolate Articles

*Collman GmbH, Lubeck, W. Germany*

Fully mechanized equipment is available in which chocolate is deposited into the molds, followed by closure, complete cooling, and opening of the molds.

The chocolate must be well tempered but sufficiently fluid to flow evenly around the inside of the molds. There must be no delay between filling and placing on the spinner. Uneven chocolate will result in splitting of the finished article as a result of uneven contraction.

Figure 7.10 shows the principles of operation of the spinner.

## Foiled Articles

It is possible to mold solid or hollow chocolate articles in printed foil that is pressed into the mold shape prior to depositing the chocolate. Hinged molds are used and the final article, in such shapes as Santa Claus or the Easter Rabbit covered with brightly colored foil, are very attractive.

## Shell Forming Equipment

These are very complicated machines and are now fully automatic. Chocolate is deposited into molds and, by a reversal operation, a lining of liquid chocolate is left coating the inner surface of the mold. This is set by cooling and the hollow shell then receives a deposit of confectionery center followed by backing with more liquid chocolate. After cooling, the chocolates with centers are removed from the mold by reversing and tapping. See Fig. 7.11.

Fig. 7.11.  Shell Plant Production Sequence
*Note*: Liqueur leveling applies to liqueur centers only.

*Gebr. Bindler, Bergneustadt, W. Germany*

The original shell equipment made small shells for assortment units and these were usually filled with a liquid type of center at a temperature below the melting point of chocolate, and then backed with a deposit of liquid chocolate. A great deal of trouble was experienced with this type of assortment, from oozing of the center through imperfect seals between the shell and back and from fermentation of the center through the use of recipes with a low concentration syrup phase.

Shell equipment developed in recent years have been capable of receiving deposits of hot fondant into the chocolate shells, and this has meant very special design of the cooling section immediately before and after deposit of the fondant. This remarkable advance in technique has greatly increased the variety of confections that it is now possible to make on such equipment. Large fondant-cream-filled bars, as well as small assortment chocolates, can be made by this method and a still further spectacular development is the production of assorted-center-filled blocks. The chocolate shell is made by the same method but the indentations in the shell receive small deposits of different confections, such as fondant, fudge, soft caramel, Turkish delight, and coconut paste. These various centers are ejected in hot liquid form from multiple depositor nozzles fed from a series of hoppers. These hoppers are kept supplied from separate kettles where the different recipes are usually made batchwise.

This equipment has eliminated much of the production of centers by the starch-deposition method.

In addition to the sheer engineering of these large machines, a profound knowledge of handling the chocolate and centers is required. This includes standardization of viscosity, control of tempering, the nature of the fat phase of the chocolate, and particularly the type of centers. When casting centers into starch some moisture loss occurs. In a chocolate shell this cannot occur and it is vital that the syrup phase concentration of the confection be maintained at a safe level.

A diagram of modern shell equipment is shown in Fig. 7.12. It is capable of depositing hot centers. The direction of flow of the molds is shown by the continuous dotted line.

Where depositing of hot centers such as fondant is involved, reduced cooling of the shell before deposit is necessary in addition to extra cooling after the deposit. By this means, the hot center will melt only a superficial layer of the chocolate on the inside of the shell.

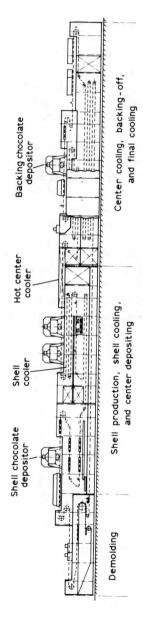

Demolding

Shell chocolate depositor

Shell cooler

Hot center cooler

Backing chocolate depositor

Shell production, shell cooling, and center depositing

Center cooling, backing-off, and final cooling

Fig. 7.12. Diagram of a Modern Shell Plant

*Aasted Mikroverk, Bygmarken, Denmark*

## The Westal SCB Process

This process is for the production of filled items in one depositing sequence.

The basic principle of operation has been used for over fifty years by, notably, Cadbury and Toms of Denmark. The machine was developed and made commercially available in 1978 by Westal Ltd., now a division of Baker Perkins.

The principle of the process is the simultaneous depositing of liquid chocolate and confectionery into a mold. Deposits are shaken down into the mold impression and, after cooling, a complete shell with filling is formed.

Filled bars and double items may also be made, and multiple deposits of different chocolates and centers in the same mold are possible. A version is available for marbled chocolate and fillings.

Figures 7.13 and 7.14 show the method of formation of the separate units and the sequence of operations in the molding equipment. Discharge of the molded units can be on a continuous belt or on plaques.

The process is especially suitable for soft or semifluid centers but it is essential to ensure that the syrup phase concentration of 75 percent minimum is maintained. Where alcoholic liqueurs are permitted as ingredients, a reduced concentration is possible.

The equipment is compact, reliable, and economical of energy, labor, and space.

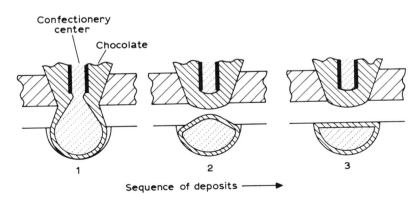

Fig. 7.13. Formation of Separate Units in a Molding Plant

*Westal Company, Redditch, England*

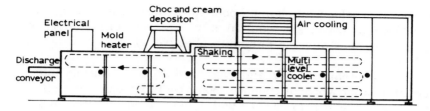

Fig. 7.14.    Diagrammatic Sequence of Operations in a Molding Plant

*Westal Company, Redditch, England*

## COMPOSITION OF CHOCOLATE FOR MOLDING

Chocolates made as previously described, either dark or mill, are suitable for molding, but by reason of the different methods of manufacture and inclusion of lecithin, fat contents are much lower than they were some years ago.

There has been a steady reduction in the total fat contents of all chocolates, whether for molding or enrobing purposes. With the higher cocoa butter contents that prevailed at one time, the contraction of the chocolate was considerable and simplified the demolding. With the lower fat contents, which have been obtained by the use of emulsifiers and the additional mechanical aids to reduce viscosity, the effect of contraction is minimal. Good-quality, finely ground chocolates with total fat contents down to 28 percent can now be produced and satisfactorily molded. Lower-quality chocolates with coarser particle size can be produced down to 25 percent. Particularly with milk chocolates, quality suffers with fat contents below 30 percent, the texture becoming pasty with a lack of snap.

Demolding of small chocolates from deep molds can be difficult with milk chocolate with high milk solids (25 percent including milk fat) and an improvement results from reducing the milk solids to 15 percent. If it is not required to call the confectionary block by a name that implies that the covering is whole milk chocolate, a proportion of nonfat milk or whey powder may be used.

The addition of rework may also cause demolding trouble due to the presence of soft fats in the centers used for the rework—such chocolate is best not used for chocolate shells. Dark chocolate with butter fat addition for antibloom purposes can present difficulty if the maximum of 4 percent is used. Molded surfaces are much less prone to bloom development than enrober-covered chocolates and generally antibloom additives are required only to prevent bloom on the backs of molded lines and the minimum effective addition of 2 percent is adequate.

Many chocolate blocks and bars contain nuts (hazels, almonds, brazils) either whole or chopped. Some contain raisins as well, which must be dry and firm or they will adhere to one another and cause thickening of the chocolate in the depositing hopper. Raisins are purchased in large cartons in a slightly compressed condition and it is common practice to dust lightly with fine sugar or maize starch when breaking up the blocks for use.

Other popular ingredients in chocolate blocks are small Rice Crispies or other cereals, and small pieces of boiled sugar, preferably aerated, to avoid a gritty texture when eating.

Some of these additions, particularly cookies, will absorb cocoa butter from the chocolate, and to avoid the use of excess cocoa butter, rapid mixing and depositing are desirable. Devices are in use for the continuous feeding of an ingredient into a stream of tempered chocolate. They are mixed in the recipe proportions and fed directly to the depositing hopper.

An extremely popular type of molded line is the wafer biscuit bar. These generally consist of a laminated wafer with a flavored fat-cream filling. The machinery for manufacture of these lines is complicated and the general principle is to fill a multiple-section mold with chocolate and to press wafer fingers into the sections. The excess chocolate is backed off by means of a roller or scraper.

## MOLDING AND SHELL PLANT COOLERS

Problems arise in the cooling of chocolate in molds and shells that differ from those in enrobers. The following are important points.

### Tablets

In the molding of tablets, blocks, and shells, good contraction is required to help the demolding process. That means good tempering and fairly rapid cooling, but the cooling is best achieved by high-velocity air rather than low temperatures so as to avoid the formation of unstable cocoa butter seed. In enrobing, rapid contraction is undesirable as it causes splitting of the chocolate covering, and possibly oozing from weak points in the covering.

## Shells

After deposition of the chocolate and reversal of the mold, the liquid chocolate must be cooled without disturbing the chocolate. Therefore, high-velocity air should not be directed onto the chocolate initially.

The shell, being thin, will cool quickly even with moderate air flow, and if the cooling is too vigorous and uneven, distortion of the chocolate will occur. This is particularly noticeable with shells for filled bars or blocks and the distortion may result in the bars becoming loose in the molds and detached at the ends, which are bent upward. If this occurs and the center deposit is, for example, hot fondant, the shell may be softened and settle again in the mold. This will probably cause defacing of the embossings. Following this, when the backing chocolate is applied and cooled, contraction will occur and this can emphasize the distortion problems. Bent bars create packaging problems and result in breakage during packing.

# GOOD MANUFACTURING PRACTICE

In all molding plants, efficiency depends on obtaining 100 percent ejection of units from the molds because, if some of the molds become blocked, they have to be removed from the cycle. The chocolate that has stuck to the mold must be removed by hand and the mold cleaned. With equipment for shells and filled articles, this can be a very expensive operation and the disposal of the rejected material poses another problem.

Good demolding depends on proper tempering of the chocolate, adequate and even cooling, and starting with clean molds. Good tempering will result in good contraction, but milk chocolate is more difficult to demold than dark chocolate because of the soft milk fat it contains; this applies particularly to small shell chocolates, for which it is often desirable to use a chocolate with reduced milk or milk fat content.

Mold cleaning is done periodically. Special washing equipment is used for this purpose.

High-pressure hot-water jets with some detergent are used, followed by a rinse, preferably with soft water, and hot-air drying.

Care must be taken in the washing of plastic molds and information should be obtained from the manufacturers regarding suitable detergents.

## CHOCOLATE ENROBING

Chocolate assortments, confectionery bars coated with chocolate, and chocolate cookies and cakes are all manufactured by the coating process known as chocolate enrobing. Reference has been made to hand covering, but the coating of chocolates by this process is reserved for the very expensive assortments mostly sold in exclusive shops.

The principle of the enrober may be summarized as follows:

The confectionery centers to be covered with chocolate are placed on a conveyor belt—usually made of plastic-coated canvas. With small units for assortments, this is mainly done by hand, but with larger bars, mechanical placing is used, particularly if enrobing is preceded by cutting from a large slab or extrusion from a series of nozzles. From the canvas belt the centers are transferred to a wire net specially designed for enrobers, which first passes over a bottom-coating device. This part of the machine consists of a system of pumps or rollers that produce a shallow surge of tempered chocolate through the net, lifting the centers slightly and at the same time coating the bottoms. These bottom-coated units are transferred automatically to another belt, where the bottom chocolate is cooled by passing over a cold table, and from this to the main enrober net.

The prebottomed centers are carried on this net through a curtain of tempered chocolate, thereby coating the tops and sides as well as providing an additional thin coating on the bottoms. After emerging from the curtain, the covered chocolates are subjected to blowing by air from fans and to vibration of the net, both of which are adjustable, and are used to control the amount of chocolate remaining on the centers. From the enrober net, the chocolates are again transferred to a conveyor belt, which takes them to the cooler, but at the transfer an "antitailing" device consisting of a small-diameter spinning roller takes off the drips of chocolate at the ends and edges of the bars.

An antitailing device is also necessary at the discharge end of the prebottomer, or the tails formed will be partly solidified on the cooling slab and will be enlarged when passing through the enrober.

Many enrobers are worked without prebottomers. This may be satisfactory for small centers, but it certainly is not for large bars. In the enrober machine, an upward surge of chocolate will provide some bottom coating, but with the heavier pieces, the vibration of the net will cause them to sink through the chocolate, resulting in very thin bottom coating. Very little protection thus is given, and subsequently

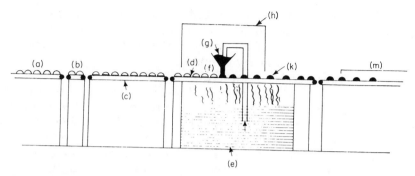

Fig. 7.15.  Enrober System
a. Conveyor-Uncovered Centers
c. Bottom Cooling Conveyor
e. Tank of Tempered Chocolate
g. Reservoir of Tempered Chocolate
k. Covered Centers

b. Bottom Coater
d. Enrober Net
f. Chocolate Curtain Supplied From
h. Shield
m. Cooler

Decorating by hand or mechanically is carried out at position k. There are ingenious machines available for applying chocolate designs to the enrobed chocolates.

the center will dry out or pick up moisture, depending on the nature of the center and ambient conditions.

A diagram of the enrober system is shown in Fig. 7.15. Enrobers have, during the past fifty years, developed from small machines requiring a great deal of skilled attention and continuous adjustment to very big machines with automatic tempering of the chocolate. The original machines had a net width of 15 to 18 in., whereas modern machines may be 54 in. wide and use great quantities of chocolate.

With the larger enrobers that are not automatically operated, a constant supply of tempered chocolate is necessary. This may be obtained from tempering kettles used to supply the enrober reservoir tank.

A skilled enrober worker is required to maintain the quantity of chocolate in the enrober tank at the right degree of temper to prevent overthickening or loss of seed by adjusting hot and cold water circulation. The same operator is required to keep the chocolate kettles filled with tempered chocolate, or, if the enrober tank is kept full enough, it can be supplied with cooled but untempered chocolate, with the tempering taking place in the enrober tank itself.

As a development of the last principle, drip feeding of the enrober tank with untempered chocolate equal in quantity to the chocolate demand is done in some factories. From this description, it is readily understood that it takes skill to work an enrober efficiently and to maintain correct weights of chocolate on the center, at the same time keeping the chocolate properly tempered.

With the latest high-output enrobers, automatic tempering has been introduced. This is a complicated procedure as the enrobing process differs from molding in that a proportion of the tempered chocolate from the chocolate curtain returns to the enrober reservoir tank. This chocolate will increase in viscosity as a result of the continuous formation of more seed of cocoa butter crystals. It is necessary, therefore, to remelt these, either by returning the chocolate to the storage tank or by mixing it with untempered chocolate.

Automatic enrobing systems are described later.

## Mechanical

Whether an enrober is fully automatic or hand controlled, it is necessary to have some understanding of the mechanical processes involved. The following is a brief summary of the various operations.

**Feeders** This is the preliminary process to deliver the centers in a uniform manner to the enrober (or prebottomer).

The bulk centers are transferred by vibration and oscillation into channels, which space out the lines across the width of the belt. The lines are then transferred from the feeder belt to the wire belt (prebottomer or enrober), which is moving faster, thereby spacing out the centers longitudinally. This prevents them from sticking together at the enrobing stage.

At the transfer stage, there should be a gap with a deflector system to remove center tailings.

**Prebottomers and Cold Tables** As previously mentioned, prebottomers are necessary when enrobing large articles. The prebottomers should be separate mechanical units, and generally it is not satisfactory to supply them with tempered chocolate from the enrober except during the startup period.

The tempering system should be thermostatically controlled, and because the unit is more exposed than the enrober, it is usual to run the chocolate at maximum temperature but obviously avoiding loss of seed.

After coating of the bottoms, the pieces pass over a de-tailing rod to a cold plate, which partially sets the chocolate. Thus, it will not melt on the enrober net when passing through the enrober.

Cold plates should be thermostatically controlled to avoid moisture condensation and the belt conveying the bottom-coated pieces should always be maintained in close contact with the plate. Sometimes, slight doming of the upper surface is used to achieve this. About 2 min of bottom cooling usually is enough to allow sufficient setting of the chocolate.

If condensation occurs on the cold plate due to high humidities in the enrober room, a cover can be provided to form a type of tunnel.

**Coating Section**   This part of the enrober consists of the following:

1. Reservoir tank. This provides tempered chocolate for the curtain and has a slowly moving agitator.
2. Pump and mixer. These lift chocolate to the flow pan and curtain of chocolate. Heated jackets are necessary and the reservoir tank must never reach a chocolate level where the pump discharges air as well as chocolate. Air bubbles in the chocolate curtain are a great nuisance.
   The flow pan is adjustable and the curtain discharge should be as close as possible to the tops of the units being covered.
4. Blower. This controls the thickness of the chocolate coating and can be adjusted for angle and height. Air temperatures must be watched, as overheating will cause some of the chocolate to be detempered; 82 to 85°F (28 to 29.5°C) may be considered optimum but air speed is also a factor.
4. Vibrator. After the curtain, the transport net is subject to vibration by either mechanical or electrical means. It should be possible for the amplitude and the speed of vibration to be controlled separately, but in some enrobers this is not the case. Ripples on the chocolate surface are controlled by vibration and blowing, but if the chocolate becomes overseeded, these operations may not be effective.
5. Wire belt extension. This serves as a decorating table and may be warmed to ensure that excess chocolate is removed and returned to the reservoir tank.
6. De-tailing device. This has already been mentioned but correct adjustment is important. It should not be too high or chocolate will be whipped off the bottoms with the tails. It should be just touching the wire belt and not the tunnel belt.

## Chocolate Cooling

A complete understanding of the principles of chocolate cooling is essential if chocolate products with a good appearance and shelf life are to be obtained.

Regrettably, the construction of coolers is often at fault, and those properly constructed are frequently used incorrectly. Little attention is given to the relative humidity of the air passing through the cooler, and in older installations, icing and deicing of the cooling units are often responsible for the introduction of damp air.

## Enrober Coolers

**Tunnel Coolers** The following description mainly concerns enrober coolers but the conditions also apply to other chocolate equipment.

The requirements for good chocolate tempering have already been described. It is necessary to ensure that chocolate in the enrober is well seeded with stable forms of cocoa butter crystals, but it must be realized that the chocolate on the enrobed centers still contains a high proportion of liquid cocoa butter that must be solidified in the stable crystal form during the cooling process.

This liquid chocolate *must not* meet very cold air, which will make the remaining cocoa butter unstable. The air in the first stage of the cooler should be at a temperature of 15 to 17°C (59 to 62°F), and under these conditions the chocolate covering will cool from about 30°C (87°F) to 20°C (68 to 70°F) and the formation of stable crystals will be ensured. In the next stage of the cooler, colder air may be used, and for enrober coolers, this is preferably not lower than 10 to 13°C (50 to 55°F), but it is undesirable for the chocolates themselves to reach this temperature. A third stage of the cooler is useful in which the chocolates again meet warmer air so that they emerge from the cooler at about 15°C (60°F). It is not generally understood by the designers of coolers that *the important factor in a cooler is the temperature gradient of the chocolate and not only the air temperatures*. With similar air conditions, chocolate pieces of different sizes can have greatly different rates of cooling and very small chocolates can soon reach the lowest temperature of the air in the cooler (in the example above, 10°C). This can produce unstable crystals, and also, in some chocolates, promote cracking of the covering due to rapid contraction.

Deposition of moisture may also occur if there is no warming section of the cooler. If air conditioning is not provided in the packing room, it is very likely that chocolates coming from a cooler at 12.5 to

14.5°C (55 to 58°F) will, on occasion in warm weather, be below the dew point.

Another disadvantage of a low-temperature section is that if the conveyor belt stops with the cooler loaded, the chocolates will soon reach the lowest air temperatures and cause the troubles mentioned above. A safeguard is to provide a means for directing the cooling air out of the tunnel when the belt stops.

**Design of Coolers**  Many types of cooling systems have appeared over the years. Some have been designed at minimum cost, and with these many problems arise in trying to maintain the air conditions outlined above.

When a cooler, particularly a tunnel cooler, is used for a variety of chocolate articles—such as small enrobed pieces, large candy bars, or molded blocks—adjustments have to be made to air temperatures and velocities according to the load in the cooler.

**Multizone Coolers**  Air control becomes very difficult with large enrober coolers where a single-section tunnel is used. A development that has resulted in a great improvement in temperature control is the multizone cooler. This system uses a series of separate cooling sections, each with its own air flow and cooling unit. The tunnel consists of any number of these sections in line, depending on the load and type of product to be cooled.

There are various designs of the sections of these coolers. In some, air travels horizontally at right angles to the direction of the belt movement. In others, air is directed vertically onto the top of the chocolates. (An example is shown in Fig. 7.21.) Adequate cooling of the bottoms of the covered articles is necessary or the pieces will not readily detach from the belt at the end of the tunnel.

With candy bars, uneven cooling will cause distortion as a result of the chocolate on the top of the bar setting and contracting while the bottom chocolate is still soft. Modern coolers have extra cooling applied to the bottoms by means of a separate water or air-cooled bed beneath the conveyor belt.

**Radiant Cooling**  Radiant cooling has been the subject of much argument in the chocolate industry. The system makes use of the fact that black surfaces are capable of absorbing heat rays, and tunnel coolers were constructed in which the ceiling of the tunnel was provided with a series of matte black platens through which coolant was circulated. Radiant heat from the chocolate, theoretically, would

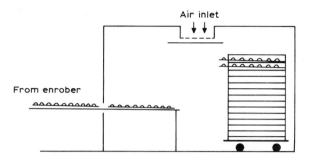

Fig. 7.16.  Simple Room Cooler
Chocolate units are transferred from enrober to plaques, which are placed in racks on a
trolley. The room is cooled with moving air to a temperature of 15 to 16°C (59 to 61°F), and
the relative humidity is less than 60 percent. After a cooling time of 20 to 40 min, the trolley
is removed from the room for packing.

be absorbed by the platens. In fact, there are several associated
factors that arise. The coolant temperature must be fairly low to be
effective and moisture condensation may occur on the platens, with
detrimental results to the contents of the cooler. The fact that the
platens are in the top of the tunnel results in a convection current of
cold air falling on the chocolates. This contributes to the cooling and,
therefore, the effect is not entirely one of radiant heat absorption. The
system may be used in conjunction with convection air cooling and
bottom cooling using water jackets.

Figures 7.16 to 7.21 illustrate some chocolate-cooling systems, and
many modifications of these are available.

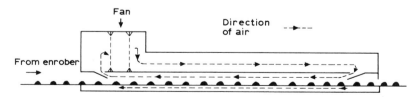

Fig. 7.17.  Simple Tunnel Cooler
Cool, dry air is supplied from freon-type refrigeration with a proportion going both above
and below the conveyor belt. The tunnel is 8 to 12 m (26 to 39 ft) long for small units—extra
length is needed for larger tablets or candy bars. Air temperatures are critical and the
velocity is variable. Care is needed to prevent condensation on the chocolates at the exit.

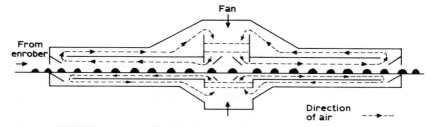

Fig. 7.18. Tunnel Cooler with Central Air Outlet

There is a separate air cooling system to the bottom of the conveyor belt. Divided air stream gives better control of the temperature at the entry and exit of the cooler. The air velocity is variable. The tunnel length is 8 to 12 m (26 to 39 ft for small units—extra length is needed for larger tablets or candy bars).

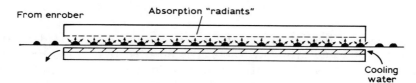

Fig. 7.19. Radiant Cooling Module
Length is 4 m (13 ft).

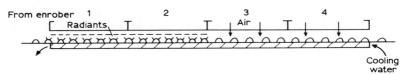

Fig. 7.20. Modules Placed in Series to Give Multizone Cooling

The first zones are radiant cooling, and the final zones are air convection cooling.

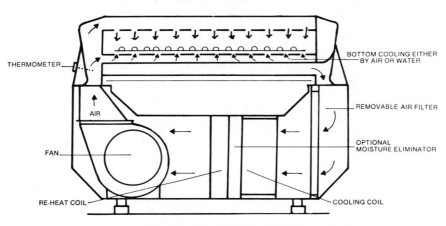

Fig. 7.21. Cross Section of Module of Multizone Cooler

*Baker–Perkins, Peterborough, England*

## Automatic Enrober Systems

Modern enrobers are now available in which the chocolate is automatically maintained in a constant degree of temper irrespective of the quantity of chocolate passing through the system.

The Sollich Temperstatic TSN is an example and operates with enrober net widths ranging between 25 and 50 in. The throughputs are 770 lb/hr for the "62" and 1540 lb/hr for the "130."

The built-in tempering system is described below and diagrams of the chocolate circulation and enrober are shown in Figs. 7.22 and 7.23.

**Tempering**   Chocolate from the main storage tank is fed to the heating tank and cylinder, where it is ensured that it reaches 40°C (104°F) and is completely free from cocoa butter seed. It then flows to the tempering cylinder cooling stage 1, where it is cooled to 28°C (83°F) (milk chocolate) or 29°C (85°F) (plain chocolate). Precise sweeping of the cooling surface together with intense mixing ensure full development of cocoa butter seed. The chocolate then passes to cooling stage 2, where cooling takes place at about 1°C (2°F) above stage 1.

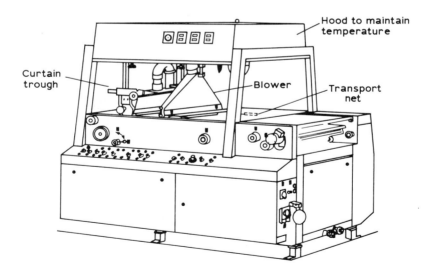

Fig. 7.22.   Temperstatic ® Enrober

*Sollich GmbH., Bad Salzuflen, W. Germany*

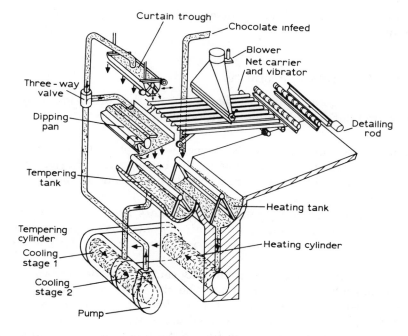

Fig. 7.23. Diagram of Chocolate Circulation

*Sollich GmbH, Bad Salzuflen, W. Germany*

At the end of cooling stage 2 is a pump that feeds chocolate to the curtain trough or dipping pan. The throughput of this pump is purposely well above that traversing cooling stage 1, and the difference is made up by a feed from the tempering tank to between cooling stages 1 and 2.

The tempering tank acts as a reservoir and is kept at a temperature just above the chocolate temperature at the end of cooling stage 1. The retention time in this tank is permanently preset and is not adjustable by the operator. It is here that any unstable cocoa butter crystals are remelted.

Since there is a continuous flow of chocolate through the tempering cylinder/tempering tank circuit, there is always a surplus of chocolate in the tempering tank. This either overflows into the heating tank or is incorporated into the chocolate from the enrober. The ratio between these flows does not affect the retention time in the tempering tank.

Temperature control is by means of a special water-circulation system.

## Chocolate Enrobing Problems

In the foregoing descriptions, some of the problems may be anticipated but the following remarks are additional.

**Tempering**   Chocolate *must* be properly tempered. Undertempered chocolate causes delayed setting in the cooler and adhesion to the conveyor belt, and ultimately bad chocolate color and fat bloom.

**Centers**   Many kinds of confectionery centers are chocolate covered and these are fed by hand or mechanically onto a conveyor band and then to the enrober net, which carries them through the chocolate curtain.

One common fault arises from cold centers. If the centers have been made independently, they may be stored for a period in a cool room, particularly caramels, to retain their shape. These *must* be warmed before covering or the covered pieces will have a dull grayish appearance due to premature cooling of the chocolate. It is customary to bring trays of centers into the enrober room to warm up before enrobing, but here another trouble may arise, because if the centers are below the dew point of the room, condensation of moisture on the centers will occur. In its worst state, this can cause an increase in viscosity in the surplus chocolate that drains from the enrober curtain, and may promote bloom formation.

The enrober room and the centers should be at a temperature in the region of 24 to 27°C (75 to 80°F), as centers that are too warm will delay cooling and cause the chocolate to run off, giving flanges on the bottoms and bad control of unit weights.

When centers are made mechanically (e.g., by extrusion), these are fed directly from the extruder to the enrober. To warm these, a radiant heater over the conveyor is used. Centers that have been cast in starch must be very thoroughly brushed or air-blown free of dust. If this is not done, the chocolate will not coat evenly and many pinholes will form. The vicinity of the enrober must be free from drafts, which, in addition to causing bad chocolate colors, may cause setting of chocolate on the enrober net.

Fatty centers such as fudge, marzipan, and nut pastes will cause fat bloom to develop on dark chocolate coatings upon storage. The

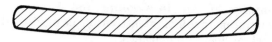

Fig. 7.24.   Distortion

addition of 2 to 4 percent of anhydrous butter fat to the chocolate coating will prevent this.

**Contraction on Cooling**  This is a factor often completely misunderstood but may have serious consequences.

In Chapter 3 cocoa butter, contraction properties on cooling were explained in detail, and it will be noted that the degree of contraction and rate of contraction are related to cooling. If chocolate-covered centers are cooled too rapidly or unevenly, two problems are likely to occur:

1. Distortion. Figure 7.24 shows the effect of rapid cooling of the top of a chocolate-covered bar while the bottom is still soft.

2. Oozing, splitting. Figures 7.25 and 7.26 show the effects of contraction pressure causing the center to be forced through pinholes or weak patches in the covering. In some cases, the covering cracks and may separate from the center.

Distortion and oozing are worse with soft centers such as fondant or fudge. Splitting or cracking is more likely on light, aerated centers, for example, wafers or aerated hard candies.

Distortion is emphasized if the top chocolate covering is thick and the bottom thin. Oozing is frequently incorrectly diagnosed as fermentation.

Fig. 7.25.   Oozing

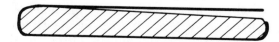

Fig. 7.26.   Splitting

**Relative Humidity of Air in Coolers and Packing Rooms** It is surprising how often insufficient attention is paid to the problem of air humidity when designing coolers.

The most obvious cause of trouble is the condensation of moisture on chocolates when they arrive in the packing room from the enrober cooler. At one time, many factories worked without air-conditioned packing rooms and these same factories usually had badly designed low-temperature coolers from which the chocolates often emerged at temperatures below 13°C (55°F). During summer periods, dew points frequently rose above these temperatures and production had to stop. In recent years, air conditioning of packing rooms has increased steadily as it was uneconomic to stop production.

A less obvious cause of trouble is high-humidity air in the cooler itself. In some designs of chocolate coolers using air at, say, 13°C (55°F), there is no accurate control of the relative humidity of the air entering the cooler. Air from outside the factory is drawn into the refrigeration system, and for economy purposes, it is merely cooled to the minimum temperature required in the cooler. The relative humidity of this air depends entirely on the dew point of the supply air and the result is that the air in the cooler can rise to 80 to 90 percent relative humidity. Since chocolate will pick up moisture from the air at 85 percent and higher (dark chocolate) or 75 percent and higher (milk chocolate), chocolate in these coolers becomes damp and will ultimately develop sugar bloom.

Another cause of trouble is high humidity during defrosting of cooling coils, or temporary failure of the cooling system. This can be very serious as air at close to 100 percent relative humidity may be introduced into the coolers and, apart from damage to the chocolates, molds can develop on chocolate residues in the cooler and on cooler surfaces. In this way, mold spores will be distributed over manufactured goods and musty odors will arise in the coolers. In any cooling system, to avoid these troubles, air intake should be partly dehydrated by reducing its temperature to well below that of the air used for circulation in the cooler. For cooling air to be used at 13°C (55°F) to 14.5°C (58°F), the intake of air should be drawn through cooling coils that reduce it to 7°C (45°F), and this air when warmed to 55 to 58°F will have a relative humidity of about 63 to 70 percent [dew point 7°C (45°F)]. At this relative humidity, no chocolate or cooler damage will occur. There are two methods of raising the air from 7°C (45°F): (1) by passing through heating coils, or (2) by mixing with the original air supply. The former method is the most positive and is recommended for enrober coolers.

As with many processes in the confectionery industry, there have been great improvements in cooling systems. The older brine circulation has been replaced by compact freon refrigeration units that can be used locally were cooling is required.

## Chocolate Panning

Panning is a unique process of coating a center with successive layers of chocolate or sugar in a rotating pan known as a Volvo or comfit pan. Sugar panning is dealt with elsewhere in the book. The following description concerns only chocolate panning.

Chocolate is frequently applied to nuts. Almonds, hazels, and peanuts are usually roasted, but brazils are not.

Other centers may be raisins, malted milk balls, preserved ginger or cherries, some pastes, or nougat. The center must have some solidity or it will break up in the pan.

Usually the process consists of three separate operations:

1. Preglazing of the center.
2. Engrossing, which means the building up of the layer of chocolate (or sugar with sugar panning)
3. Final glazing to give a glossy appearance and provide some protection from the effects of dampness and abrasion.

The pans (Fig. 7.27) are made of copper or stainless steel and particular pans are generally retained for the specific purposes of sugar coating, chocolate coating, or glazing. They may be tilted and operated at different speeds.

The pans are provided with ducts that blow air onto the units while they are rotating. It is essential for the temperature and humidity of this air to be controlled to fairly fine limits. Cool dry air is required to set the chocolate when it is being applied, but warmer dry air is required when the syrup glazes are used. Originally, the syrups and chocolate were applied by hand, but in present-day large installations, spray guns are used. These are fitted to each pan and are adjustable to direct the spray where required onto the rotating product. They are supplied by pipeline from bulk tanks, and one person can supervise a battery of pans by this method.

An example of the process stages for nut panning is as follows:

**Preglazing** This is used to seal the nut surface and with a thin coating of chocolate prevents transfer of nut oil to the chocolate and

Fig. 7.27.   Volvo Pan

*Norman Bartleet Ltd., London*

subsequent development of fat bloom or softening of the chocolate. It is possible to dispense with this process with thick chocolate coatings.

To glaze the nuts, they are rotated in the pan and a syrup of 50 percent concentration is poured on in small quantities. Rotation of the pan causes this to be distributed evenly over the surface of the nuts and the air blowing onto the syrup layer will cause it to dry out as a glaze. Successive applications of syrup will increase the thickness of the glaze, but it is necessary to dry out each layer before the next application.

Various recipes have been proposed for this glazing, but for the best results, the syrup should be viscous and not crystallize from the dried glaze. To increase viscosity, gum arabic is added or a suitable substitute gum, and to prevent crystallization, glucose syrup. The actual quantities of each should be found by experiment to suit the confectionery center.

Some confectioners build up the glazing layer by the application of dusting powders consisting of cocoa/powdered sugar mixtures. It is particularly useful when preserved fruits or ginger are being panned as it serves to neutralize the stickiness of the adhering preserving syrup. The syrup is dried off by application of air from the duct at 21 to 24°C (70 to 75°F) and 50 percent relative humidity.

**Chocolate Coating** Dark or milk chocolate may be used and should be fluid enough to distribute readily over the surface of the centers being coated. With hand panning, it is better to use tempered chocolate to encourage quick setting and to prevent streaky layers and discoloration of the finished chocolates. With spray panning, good results are obtained with untempered chocolate provided the temperature has been reduced to 34 to 35°C (93 to 95°F).

The mechanical process of spraying induces a seed in the chocolate that helps the setting of the chocolate.

Since finishing glazes are usually applied to panned goods, slight discoloration of the chocolate is acceptable.

The chocolate layer is built up by even application and setting is achieved by using duct air cooling at about 13°C (55°F) with humidity controlled at 60 percent or below. Care is required in this panning process to prevent agglomeration of units, and this rests largely with the operator's skill in the application of chocolate.

**Finishing Glaze** There are various methods of obtaining the final glaze and for the best results two distinct layers are applied—first, a syrup glaze, and second, a wax or edible shellac glaze.

The chocolate-coated centers are preferably allowed to set completely in shallow trays before glazing. This ensures a rigid unit; if they are glazed straight from the chocolate coating pan, they may distort.

The syrup glaze is similar to that used for preglazing or an edible dextrin may be used. "Crystal gums" may be used instead of gum arabic, and recipe quantities of ingredients and strength of syrup may be adjusted to suit the line being glazed—the object of the glaze being to create a polish that does not crack, flake off, or crystallize, and to serve as a base for the final glaze.

The syrup glaze is applied by pouring into the rotating mass and will quickly distribute over the surface of all the chocolate-covered pieces in the pan. Spraying is unnecessary and the entire recipe quantity of syrup may be added in two or three successive applications with drying in between.

Air from a duct is again used for drying at a temperature of about 18°C (65°F) for milk chocolate, and 21°C (70°F) for dark chocolate, with a 50 percent relative humidity. The drying is continued until all signs of stickiness have disappeared.

After this glazing, the chocolates are best left in shallow trays at about 18°C (65°F) and 50 to 60 percent relative humidity for at least 12 hr—this stabilizes the glaze and creates a film with an equal moisture content throughout. After this period of drying, a wax or varnish glaze is applied. If this glazing is to be done in a rotating, pan, a wax is usually employed and the pan itself is coated internally with a wax layer—the wax may be beeswax or carnauba wax, the latter giving a harder glaze. Compositions containing acetylated glycerides have also been proposed for these glazes. Whichever wax is used, a relatively dull polish is obtained, but for a brillant surface, it is necessary to use a varnish prepared from edible shellac.

It is preferable to use shellac containing the natural wax as this gives a more flexible film and seems to spread more evenly. The resin is dissolved in ethyl or isopropyl alcohol and a thick opaque solution results because of the presence of wax, but the glaze obtained appears quite clear. The strength of solution depends on the confectionery being glazed but is usually 25 to 30 percent. Proprietary shellac glazes are obtainable and these are generally applied in a rotating pan. The method used needs some skill and the pan containing the pieces to be glazed is rotated and the varnish solution applied in a quantity just sufficient to distribute it evenly over the surface of each piece. As soon as this has occurred, the pan is stopped and the solvents allowed to evaporate. *The pan must not be rotated while this is happening,* or the glaze will be torn off. The solvent evaporation can be accelerated by blowing air into the pan. These solvents present an explosion hazard, and in some countries this operation must be carried out in a separate room by remote control. As soon as the glaze is dry, the pan is rotated two or three times to free the pieces and the contents emptied onto trays for final drying and hardening.

Zinsser U.S.A. has introduced a range of confectionery shellac glazes, including one that contains small amounts of beeswax and acetylated monoglyceride. This addition is claimed to reduce "tack" during rotation in the pan.

Kaul GmbH of Germany produces a range of water and alcohol-soluble confectionery glazes under the name "Capol." These are suitable for both chocolate and sugar panning.

Shellac coating, in addition to giving a brilliant finish, provides very good protection against dampness, and because of this, glazed panned chocolate lines are particularly popular in tropical export markets.

In addition to chocolate panned goods, sugar glazing on top of chocolate is now very popular. Sugar glazing, in effect, is similar to the syrup glaze applied to chocolate units, but it is built into much thicker layers to form a hard shell. This process is described in Chapter 6.

### Zein Glazes

Considerable interest has been shown in the confectionery industry in the use of zein as a protective glaze. Zein is a corn protein and there are no official restrictions or limits on its use in foodstuffs.

It is soluble in ethyl alcohol or isopropyl alcohol in proportions of 5 to 20 percent of the total weight of the solution. Plasticizers such as glycerol and propylene glycol are also used in zein solutions.

It is claimed that zein glazes may be applied in a rotating pan. They have also found considerable application in the manufacture of compressed tablets.

Descriptive brochures are available from Freeman Industries Inc., Tuckahoe, N.Y., and Benzian A.G., Lucerne, Switzerland.

## REFERENCES

### Manufacturers, Research Establishments

Aasted International Bymarken, Denmark—tempering, molding
Baker Perkins, Peterborough, England
Norman Bartleet Ltd., London, England—comfit plans
Bindler GmbH, Bergneustadt, W. Germany—shell plants
Collman GmbH, Lubeck, W. Germany—hollow figure machines
Freeman Industries Inc., Tuckahoe, N.Y.—zein glazes
Kaul GmbH, Elmshorn, Germany—general glazes
Lloveras S. A., Tarrasa, Spain—vermicelli
Richardson Researches, Hayward, California
Sollich GmbH, Bad Salzuflen, W. Germany—enrobers, coolers, Tempermeter
Westal Company (Baker Perkins), Redditch, England—shell equipment

Zinsser and Co., Somerset, N.J.—shellac glazes

Wilton Book Division, Woodridge, Illi. (*The Complete Wilton Book of Candy*)—Hand covering, etc.

## Equipment

Many companies manufacture molding machines, enrobers, coolers, comfit pans, and the like.

Readers are advised to consult the *Directory of Equipment and Supplies*. Manf. Conf. 1986 (July) and later editions.

Many companies also advertise in the trade journals listed in the Appendix.

# Part 2

# Confectionery: Ingredients and Processes

# 8

# Sugars, Glucose Syrups and Other Sweeteners

The terminology used in different parts of the English-speaking world to describe candy, chocolates, cakes, and other sweet foods varies greatly. If other countries, in which English is not spoken, are included, the literal translations can be very misleading. Sometimes journals publish articles that have been translated from another language by someone who may be a good linguist but does not know the technology of the industry.

A few examples are worth quoting: *Eiweiss* is often translated as "egg white" when it means albumen or protein. *Stärke-sirup* is translated as "strong syrup" when it means starch syrup or better glucose syrup. *Hart Karamel* is translated as hard caramel when it means hard candy. Caramel has many meanings, including the color. *Weich Karamel* usually means all caramels, as well as soft caramels. *Praline* refers to boxed chocolates as well as nut pastes. *Humidité* is often translated as "humidity" when it means "moisture."

A recent dictionary by Silesia-Essenzenfabrik (1984) gives the international equivalent of these terms.

Commercially, in the United Kingdom, the following usually applies:

*Sugar confectionery*. This includes hard candy, toffees, fudge, fondants, jellies, pastilles, and others not covered with chocolate. It does not include cookies or cakes.

*Chocolate confectionery*. This includes much of the sugar confectionery covered in chocolate, and usually chocolate bars and blocks.

*Flour confectionery*. Baked fancy cakes, iced or chocolate covered, come within this group. Cookies may also be included but their production is often closely associated with chocolate confectionery.

In the United States and some other countries, all sweets are known by the general name of candy, and this as a rule includes chocolate. In the United Kingdom, candy has a particular meaning and usually indicates that a sweet is coated in sugar or sugar crystals, or "candied." It is also applied to candied fruit and fruit peel.

The art of confectionery goes back a very long time and Egyptian writings show its existence 3,500 years ago. Excavation of the ruins of Herculaneum, Italy revealed a complete confectioner's workshop with utensils similar to many used today. Most of the sweets of ancient times were based on honey, but sugar cane juices, crudely evaporated, were used in India and China.

The cultivation of sugar cane spread and it was about the seventh century in Persia when sugar refineries came into being.

The Venetians were the sugar brokers of Europe in the Middle Ages and sugar cane growing spread to Sicily, Spain, Madeira, and the West Indies where production increased greatly. Later it has been cultivated in most of the tropical and semitropical parts of the world. (See also "Sugar")

It was not until the sixteenth century, when sugar refining became a commercial process, that sugar confectionery began to develop. As time went on the introduction of other ingredients resulted in the various forms of confectionery that fill the candy shops of today.

Cocoa beans, with sugar, made chocolate, and with milk as well, milk chocolate. Fats and milk with sugar made caramels and toffee, and later came invert sugar and glucose syrup and a great many other ingredients, including flavors, emulsifiers, aerating agents, and stabilizers, all of which have transformed a craft into a science. Along with this, engineering developments have made continuous processing possible and the craftsman has gradually given way to the scientist and the engineer.

A series of articles by Lees (1967) describes in detail the development of the craft and industry and makes interesting reading for those wishing to know more of the historical side of the business.

## CONFECTIONERY INGREDIENTS

This section deals mainly with the ingredients used for sugar confectionery. Some indication of the distinction among chocolate, sugar, and flour confectionery has already been given but there are many ingredients that are common to all three. The main chocolate ingredients have been considered separately.

The products described under "Sugar Confectionery" are frequently used as centers for chocolate. Today, some uncovered confectionery is included in boxed chocolates, together with centers coated with attractively colored pastel coatings.

Cookies, and particularly wafers, are finding increasing use in chocolate and confectionery bars. In these circumstances, compatibility is a critical factor (see "Multiple Confectionery Bars").

## Carbohydrates—Chemistry

The carbohydrates are a large group of chemical substances that occur widely in the vegetable and animal kingdom and conform to a general formula $C_x(H_2O)_y$, $x$ being generally a multiple of 6.

Sugar is $C_{12}(H_2O)_{11}$, dextrose $C_6(H_2O)_6$ and starch $\{C_6(H_2O)_5\}_n$.

These represent the three main groups—the monosaccharides (dextrose, fructose), the disaccharides (sucrose (sugar), lactose, maltose), and the polysaccharides (starch, cellulose).

The monosaccharides are sugars that cannot be broken down by acid hydrolysis to simpler sugars, and as far as the confectionery technologist is concerned, dextrose and fructose, which are called hexoses, are the most important. Unfortunately, both sugars have the synonyms glucose and levulose and the former is confusing as it is often associated with confectioners' glucose syrup, which, although containing dextrose, also contains other complex carbohydrates.

Dextrose and fructose are the constituents of invert sugar and arise from the inversion of sugar (sucrose). They are known as isomers—the same chemical formula but a different molecular arrangement. A characteristic property of the natural sugars is that when dissolved in water they rotate the plane of polarized light passing through the solution. The hexoses, dextrose and fructose, turn the plane in opposite directions—dextrose is dextrorotatory, turning the plane to the right, and fructose (levulose) is levorotatory, turning the plane to the left.

These facts are used in the analysis of confectionery containing several sugars.

These sugars are also capable of oxidation to corresponding acids by mild oxidizing agents, and this is also made use of in sugar analysis in which the sugar is added to Fehling's solution (copper sulfate, tartrate, and sodium hydroxide) until the blue of the cupric salt is replaced by red cuprous oxide. Sugars that have this reaction with Fehling's solution are known as reducing sugars and this property is evidence of the aldehydic structure of certain sugars.

The disaccharides are sugars in which two monosaccharides are linked, and this may or may not be through their aldehydic group. If not linked through this group, the sugar will have reducing properties; for example, sucrose is nonreducing whereas maltose and lactose have a reducing action.

The complexity of sugar technology is indicated later in the description of "glucose syrup." Special analytical techniques such as high-performance liquid chromatography (HPLC) are now used to determine constituents of sugar mixtures and to monitor production.

Some disaccharides are readily broken down by dilute acids and by certain enzymes, and with sucrose this is the principle of manufacture of invert sugar.

$$C_{12}H_{22}O_{11} + H_2O \xrightarrow[\text{invertase}]{\text{acid}} \underset{\text{dextrose}}{C_6H_{12}O_6} + \underset{\text{fructose}}{C_6H_{12}O_6}$$

Inversion also occurs in many confectionery processes where acid conditions prevail, as in fruit jellies.

Polysaccharides all have a much higher molecular weight than the mono- and disaccharides, and their configuration is complex.

Starch is the most important substance in this group as far as the confectioner is concerned, and it occurs naturally in corn, rice, wheat, potatoes, and many other plants. Starch in its purified form is used for confectionery center molding and in various modified forms as an ingredient. It can be hydrolyzed by acid or specific enzymes or by a combination of these. The range of glucose syrups so formed is described later. Modified starches are being used to an increasing extent in confectionery, especially for jellies and associated products.

Dextrin is also a product of starch and is used for confectionery glazes as well as for adhesives.

For further information on the molecular configuration and chemistry of the carbohydrates, reference should be made to books on organic chemistry, but where necessary scientific explanations will be given in the references to particular sugars, starches, and derivatives.

## Carbohydrates Used in Confectionary

**Sugar (Sucrose, Saccharose)** Sugar is the main ingredient of confectionery and chocolate, and something has already been said about the history and development of cane sugar.

Beet sugar was of much later development and its extraction met with many failures before it became a commercial proposition. The beet originated in the Mediterranean area and the German chemist

Marggraf (1747) was the first to extract a quantity of sugar from this plant.

Its commercial development was energized in the Napoleonic era when a decree was issued in France to advance the production of beet sugar to break the British monopoly on cane sugar, but the defeat of Napoleon, together with colonial opposition, caused the industry to decline. It was as late as 1912 before it could be said that a beet sugar industry existed in England. During World War II, Britain depended greatly on the beet sugar industry to provide the nation's sugar.

There has been great controversy over the relative merits of cane and beet sugar and confectioners and jam makers at one time would have nothing to do with beet sugar if they happened to know the origin.

Everything rests with the refining, and a highly refined sugar whether of cane or beet origin is practically pure sucrose, being of the order of 99.9 percent. Modern methods of refining make highly refined beet and cane sugar virtually indistinguishable. However, the difference is very noticeable when the raw sugars are compared. Raw cane sugar has a pleasant flavor and smell, whereas raw beet sugar is positively unpleasant if of low purity.

Occasionally, even in present times, beet sugar appears on the market that does not compare favorably with well-refined cane sugar. These sugars have a slight odor when boiled in a syrup and exhibit strong foaming tendencies, which are very detrimental in making jam and confectionery. This is related to the presence of traces of proteins and breakdown products, saponins, and mucilages that are not likely to exist in raw cane—in fact, poorly refined cane sugar may contain traces of cane wax, which will act as a foam inhibitor. The purity and foaming of a sugar are also related to its ash content and a very low ash is indicative of good refining.

Methods of foam value determination and the effect on manufacture of aerated confections are discussed later.

Cane sugar is grown in all tropical countries, especially where rich soil is abundant and irrigation is available. The Central American countries and Caribbean islands are very large producers. Australasia also has been an important producer of recent years.

Beet sugar is grown in temperate climates, and all European countries and the U.S.S.R. are large producers. The development of beet sugar production has been associated with politics, as it is a source of sugar that is independent of tropical countries and of the effects of blockade during wartime.

In the United States, both cane and beet sugars are grown and produced—cane in the southern states and Hawaii and beet in the north. Beet sugar was grown experimentally in the Minnesota/North Dakota area as early as the 1880s but it was not until 1920 that commercial quantities were produced.

## Production of Sugar

A detailed description of sugar production is beyond the scope of this book but the following summarizes the main processes.

**Cane Sugar**  Sugar cane is really a gigantic tropical grass. The stalk is 1 to 2 in. in diameter and matures from the root stock in twelve to eighteen months. The stalk has a hard casing with interior pulp, and when cut, the cane contains 14 to 17 percent of sugar. After cutting, the canes are delivered to the factory, where they are crushed and then subjected to high pressure. This is followed by water washing to get maximum extraction and the juice so obtained contains up to 20 percent sugar. The cane residues, called bagasse, are waste material and are generally used as boiler fuel.

The extracted juice is limed to precipitate impurities and the clear juice concentrated by evaporation until sugar crystals form. This is centrifuged, giving raw sugar crystals and residual syrup, which is called cane molasses.

Normally, the raw sugar is exported to consumer areas where the final refining is carried out. This consists of a series of washings and recrystallizations.

The cane syrups and molasses are useful flavoring materials whereas those from beet sugar are unsatisfactory as food ingredients.

**Beet Sugar**  The beet seed is sown in early spring, the exact time depending on climatic frost. Harvesting begins in September, and may extend through the winter in some climates.

The roots are lifted mechanically, and on arrival at the factory, are thoroughly washed to remove soil residues. They are then sliced and transferred to a continuous diffuser. In this machine, water contacts the slices in such a way that the freshest water meets the most exhausted slices first. The sugar in solution diffuses through the root cells by a physical process known as osmosis so that the sugar solution becomes increasingly concentrated. The solution is limed, filtered, concentrated, and crystallized in a manner similar to that for cane sugar.

Brown sugars cannot be made from beet and the molasses are unsuitable for human consumption. They are used in cattle feed or for fermentation to produce alcohol.

However, it should be emphasized once again that properly refined beet and cane sugars are indistinguishable and supplies for commercial or household use may be of either cane or beet origin.

## Types of Sugar

Besides the highly refined sugars used for chocolate, hard candy, and fondant, various grades of brown sugar are used for flavor in caramels, fudge, and similar products. Various syrups are used for similar purposes, such as golden syrup, treacle, and molasses.

Invert sugar, described later, is also prepared and sold in different forms.

An interesting development of recent years is fondant sugar, a specially prepared fine sugar powder containing about 3 percent invert sugar or up to 10 percent of a modified glucose syrup (maltodextrin). It is used to prepare a type of fondant without using complex boiling and beating equipment (see "Fondant"). There is also a new type of microcrystalline sugar (microtal, see below) that has the special characteristics of free flow, rapid solubility, low bulk density, and absorptive properties.

Solutions of sugar, invert sugar, and glucose syrups of varying composition are supplied commercially, often to a specification required by the users. They are delivered by tank truck and are used mostly by jam and soft drink manufacturers, but are increasingly employed in confectionery factories where there is large-scale production of products with similar formulations.

While tank storage effects useful economies in handling, care must be taken to avoid microbiological problems and crystallization in pipelines. The installation of equipment for bulk handling should be entrusted to experts in the field.

The solubility of sugar (sucrose) in water is 67 percent (at 20°C) and above this level, crystallization will occur in storage in a relatively short time. The addition of invert or glucose syrup increases the total solubility and also prevents microbiological spoilage. The accepted minimum concentration to resist this spoilage is 75 percent. Tank storage of syrups with concentrations of less than 75 percent needs special precautions.

Air drawn into a storage tank during emptying must be filtered; otherwise microorganisms can settle on the surface of the syrup. In

TABLE 8.1. CRYSTALLINE AND POWDERED SUGARS

| Type of sugar | Method of manufacture | Characteristics and appearance | Reserve description (U.K. Specified Sugar Prod. Regs) | Uses |
|---|---|---|---|---|
| Mineral water sugar | Extra refined product using granulated sugar process | Very pure white crystal of granulated size but of superior analysis | 'Extra-white sugar' ('white sugar/sugar) | Pharmaceutical and high-quality drinks and foods |
| Granulated (ordinary sugar) | Raw sugar has its surface film of impurities removed by washing and centrifuging. The raw sugar crystal is then dissolved and filtered with the aid of lime treatment. Natural colouring matter is removed by bone charcoal, activated carbon, and/or ion exchange resins. Purified sugar liquor is then recrystallized, dried, sieved and graded | White crystals | "White sugar" (sugar) | General manufacturing purposes where a pure white product is required |
| Preserving, fine, extra fine, vending and caster sugar | Prepared by the granulated sugar process, but processed and sieved to give the appropriate crystal size | White crystals | "White sugar" (sugar) | Various manufacturing purposes where a pure white product is required and size is critical |
| Superfine powdered (pulverized) and finest icing sugar | White sugar is milled and sieved to the appropriate size | White powder | Powdered sugar, icing sugar (white sugar, sugar) | As for preserving, etc. |
| Icing CP sugar | White sugar is milled together with a free flow agent | White powder containing up to 1.5% free flow agent, usually tricalcium phosphate E341 | "Icing sugar" | Special manufacturing requirements, e.g., dusting, cake toppings, etc. |
| Fondant icing sugar | White sugar is milled with 10% maltodextrin to the appropriate size | White powder containing 8–10% milled glucose syrup solids | — | Used as an ingredient in an alternative method for making fondant |

| Name | Process | Appearance | Alternative name | Uses |
|---|---|---|---|---|
| Industrial sugar (formerly known as 'SMX') | Produced as per granulated sugar but to a lower refining standard | Low grade of white granulated sugar | None | Where final color is not important |
| London Demerara | Produced by mingling a coarse-grained refined white cane sugar with cane molasses that has undergone a clarification process | Large-grained white sugar with cane molasses, analytically comparable to raw Demerara but with extraneous matter removed | None | Sweetening coffee and specialized baking and confectionery products |
| London fourths, London Primrose | Produced by mingling a fine-grained refined white cane sugar with blends of intermediate-cane refinery sugar liquors to give the required flavor and color. Mingling syrups undergo a clarification process and evaporation before mingling | Moist brown sugars | "Soft sugar" | Used in confectionery and baking, etc., where flavor and color are required |
| Dark pieces | Made by mingling cane caster sugar with clarified cane molasses | Moist dark brown sugar | "Soft sugar" | Used in confectionary and baking etc., where flavor and color are required |
| Microcrystalline Sugar (Microtal)* | | | | |
| Microtal DCE (Direct compression excipient sugar) | Microcrystalline agglomerates prepared by rapid crystallization at high temperature of a solution of white sugar and 3% maltodextrin | Granular white powder | None | Direct compression pharmaceutical and confectionery tabletting |
| Microtal Fine "100" | As for DCE but feed liquor is T2000 (see later) | Off-white granular white powder | None | Dry mixes requiring rapid solubility Vending sugars |
| Microtal Light soft 100 | As for DCE but the input liquor is a blend of sugar liquors from the cane-refining process | Flavorsome, nonsticky, free flowing, granular brown sugar | Brown "soft sugar" | Various uses requiring flavorsome but free flowing brown sugar |

*Microtal is a registered trade name of Tate & Lyle Industries.

TABLE 8.2. LIQUID SUGARS, INVERTS, SYRUPS, TREACLES

| Type of sugar | Method of manufacture | Characteristics and appearance | Reserve description (U.K. Specified Sugar Prod. Regs.) | Uses |
|---|---|---|---|---|
| T1000 superfine liquid | Mineral water sugar, dissolved in water and microbiologically filtered; 67% solids | Water-white sugar solution | Sugar solution | Pharmaceutical and high-quality drinks and foods (Syrup B.P.) |
| T1001 gran liquid | As for T1000 but using white sugar (granulated) | Water-white sugar solution | Sugar solution | Pharmaceutical and high-quality drinks and foods (Syrup B.P.) |
| T10004 gran liquid | As for T1001 but delivered hot | Water-white sugar solution | Sugar solution | Pharmaceutical and high-quality drinks and foods (Syrup B.P.) |
| T2000 fine liquid | High-grade liquor from cane refining process, which has been decolorized and filtered | Pale straw-colored sugar solution | None | Pharmaceutical and high-quality drinks and foods (Syrup B.P.) |
| T2001 fine liquid | As for T2000 but delivered hot | Pale straw-colored sugar solution | None | Pharmaceutical and high-quality drinks and foods (Syrup B.P.) |
| T2800 T3000 T3302 T7000 | Examples of other liquid sugars prepared from intermediate sugar liquors of the cane sugar refining process. These are bright liquids having increasing color, flavor, and reducing sugar | Bright liquid sugars of increasing color | None | Various food uses depending on the specific flavor and properties required |
| Lyle's Golden Syrup | Processed from selected cane refinery sugar liquors, partially inverted | Distinctive color and flavor produced by traditional process | None | Used for flavoring purposes Confectionery and baking |

| Product | Description | Appearance | Type | Uses |
|---|---|---|---|---|
| N4314 glycerine extender | T1000 partially inverted at high solids | Bright, water-white, syrup | Invert sugar syrup | Pharmaceutical, confectionery, baking, and other uses requiring high purity |
| N3222 | Partially inverted blend of cane refiners intermediate sugar liquors (Industrial Golden Syrup) | Characteristics of golden syrup | None | Flavoring purposes in baking and confectionery |
| N3212 N3231 N3233 | Partially inverted blend of cane refiners intermediate sugar liquors (Industrial Golden Syrup) | Dark partially inverted syrups | None | Flavoring purposes in baking and confectionery |
| Treacles N9002 N9004 N9005 N2101 | Partially inverted and filtered cane molasses which may be blended with low-purity sugar liquors from the cane refinery process | Thick black viscous liquid having the characteristics of cane molasses | None | Used for special flavoring |
| Puratose | Fully inverted solution of white sugar (granulated) in water | Water-white invert syrup | Invert sugar syrup | Pharmaceutical, confectionery, baking, ice cream, and other uses requiring high purity |
| Inverts (liquid) N1200 (No. 1 invert) N1201 (No. 2 invert) N1202 (No. 3 invert) | Fully inverted intermediate sugar liquors from the cane refining process of increasing color and flavor and decreasing sugar purity | Fully inverted syrups | None | Confectionery, baking, brewing, and other uses |
| Water white invert | Fully inverted solution of white sugar seeded and crystallized | Water-white crystallized invert | Crystallized invert sugar syrup | Mostly used as for liquid inverts but with more convenient handling |
| No. 1/2/3 invert Block invert | Liquid Inverts N1200–N1202 are seeded and crystallized to give solid inverts of increasing color and flavor and decreasing sugar purity | Crystallized fully inverted syrups | None | Mostly used in brewing but also confectionary and baking |

an enclosed humid environment, mold and fermentation will soon develop. The base of the tanks must ensure complete emptying and all pipelines should be constructed to prevent accumulation of syrup in bends and joints.

To prevent condensation in the air space above the syrup, a useful precaution is to have a ventilation fan in addition to the filter.

After each emptying, the tank should be sterilized. If the syrup usage is high, an interval of a week is permissible. Live steam or chemical detergent may be used for sterilization after rinsing with hot water. Special spray systems are available for tank cleaning and sterilization.

The subject of liquid sugars is discussed at length by Hoynak and Bollenback (1966), Diefenthäler (1976), and Tate and Lyle (1984).

The U.K. sugar refiners Tate and Lyle have prepared charts (Tables 8.1 and 8.2) describing the comprehensive range of sugars and syrups available.

In the United States, commercial standards exist for various grades of sugar and those shown in Tables 8.3 through 8.8 are typical for confectionery uses. (California and Hawaiian Sugar Co. (1983))

Confectioners A.A. and sanding sugars (Table 8.3) are exceptionally well refined sugars suitable for white fondant and to produce sparkling coatings on gums and jellies.

Granulated (all-purpose) sugar (Table 8.4) is suitable for most commercial and retail uses.

TABLE 8.3. CONFECTIONERS A.A. AND SANDING SUGARS

| Average chemical analysis, % | |
|---|---|
| Sucrose | 99.98 |
| Moisture | 0.015 |
| Invert | 0.001 |
| Ash | 0.002 |
| Organic non-sugars | 0.002 |

| Screen analysis | | | |
|---|---|---|---|
| A.A. | % | Sanding | % |
| On U.S. 12 | 3.1 | On U.S. 20 | 11.0 |
| On U.S. 16 | 72.0 | On U.S. 30 | 50.3 |
| On U.S. 20 | 18.2 | On U.S. 40 | 35.0 |
| On U.S. 30 | 6.5 | On U.S. 50 | 3.4 |
| On U.S. 40 | 0.1 | Through 50 | 0.3 |
| Through 40 | 0.1 | | |

TABLE 8.4. GRANULATED—ALL-PURPOSE SUGAR

| Average chemical analysis | % |
|---|---|
| Sucrose | 99.94+ |
| Moisture | 0.02 |
| Invert | 0.015 |
| Ash | 0.01 |
| Organic non-sugars | 0.01 |

| Screen analysis | % |
|---|---|
| On U.S. 30 | 0.4 |
| On U.S. 40 | 20.6 |
| On U.S. 50 | 35.1 |
| On U.S. 80 | 34.8 |
| Through 80 | 9.1 |

Powdered sugar (Table 8.5) is specially prepared from all-purpose granulated sugar for dusting, use in icings, and so on. It contains 3 percent of predried edible starch to retard caking.

"Easy fond" (Table 8.6) is a free-flowing fondant sugar containing about 3 percent invert sugar and no added starch. It can be reconstituted with syrup, frappé, fats, and flavors to produce a fondant especially suited for extrusion and hand-rolled cremes (see "Confectionery Processes").

TABLE 8.5. POWDERED SUGAR

| Screen analysis | % |
|---|---|
| On U.S. 100 | 0.1 |
| On U.S. 140 | 1.2 |
| On U.S. 200 | 2.3 |
| On U.S. 270 | 4.2 |
| On U.S. 325 | 4.8 |
| Through 325 | 87.4 |

TABLE 8.6. EASY FOND

| Analysis | % |
|---|---|
| Invert | 3.2 |
| Moisture | 0.44 |
| Particle size | 5.4 μm (0.0002 in.) |

TABLE 8.7. SOFT BROWN SUGARS

| | Analysis | |
| --- | --- | --- |
| | Golden "C" % | Yellow "D" % |
| Sucrose | 89.3 | 87.9 |
| Moisture | 2.7 | 2.8 |
| Invert | 4.2 | 4.6 |
| Ash | 1.4 | 1.7 |
| Organic non-sugars | 2.4 | 3.0 |

TABLE 8.8. WASHED RAW SUGAR

| Analysis | % |
| --- | --- |
| Sucrose | 98.6 |
| Moisture | 0.12 |
| Invert | 0.45 |
| Ash (sulfated) | 0.43 |
| Other natural substances | 0.40 |

Soft brown sugars (Table 8.7) are prepared from Hawaiian raw cane sugar, which contributes a special flavor. Two types are available—Golden C (light brown) and Yellow D (dark brown).

Washed raw sugar (Table 8.8) is a "turbinado"-style sugar. It is a natural raw sugar that has been washed to remove impurities, and is suitable for health foods, including candy bars. It meets U.S. Food and Drug Administration (FDA) requirements.

## INTERNATIONAL STANDARDS FOR SUGAR AND SUGAR SYRUPS

Standards for quality of sugars have been agreed on within the EEC and are described in the following.

### Reserved Descriptions for Sugar Products

The EEC Directive 73/437/EEC of December 1975 for grades of sugar has now been adopted by all member nations.

The United Kingdom version is the Specified Sugar Products Regulations 1976 (SI 509) and amendment 1982 (SI 255). These regulations designate certain prescribed categories of sugar and sugar products and their compositional requirements.

A product described by one of the Reserved Descriptions must conform to certain compositional requirements, the most important of which are color (in solution), ash, and visual color.

In some cases, the results are converted to "points", that is,

| | |
|---|---|
| Color (in solution) | 7.5 International Commission for Uniform Methods of Sugar Analysis Color units = 1 point |
| Ash | 0.0018 percent = 1 point |
| Visual color (Brunswick Institute Scale) | 0.5 unit = 1 point |

The analytical requirements for the most important specified sugar products are given in Table 8.9. Analytical methods are described in the amendment regulations of 1982.

Composition requirements for powdered sugar and icing sugar are the same as for white sugar, except that up to 1.5 percent of a permitted anticaking agent may be added.

**Mesophilic Bacteria**  Average of last twenty samples during routine testing equals not more than 100 colonies per 10 g of dry sugar equivalent. Not more than 5 percent should have a count greater than 200 per 10 g, although this sample may be omitted from the average.

**Molds**  Average of last twenty samples during routine testing equals not more than ten per 10 g of dry sugar equivalent. Not more than 5 percent should have a count greater than eighteen per 10 g, although this sample may be omitted from the average.

The standards for granulated sugar are twice those for liquid sugar.

## Invert Sugar

It has already been mentioned that disaccharides are broken down by acid hydrolysis or enzymic action and this is the principle of manufacture of invert sugar used in confectionery. The original confectioners' method was to make sugar into a syrup and boil it with citric, tartaric, or acetic acid for 30 to 45 min. The quantity of acid required may be as much as 1 percent and the process will darken the color appreciably.

TABLE 8.9. ANALYTICAL REQUIREMENTS FOR SPECIFIED SUGAR PRODUCTS

| | Extra-white sugar | White sugar | Soft sugar | Sugar solution | Invert sugar solution | Invert sugar syrup |
|---|---|---|---|---|---|---|
| Color (in solution), max. | 3 points = 22.5 ICU | — | — | 45 ICU | — | — |
| Ash (conductivity), max. | 6 points = 0.0108% | — | — | 0.1% on solids | 0.4% on solids | 0.4% on solids |
| Visual color, max. | 4 points = 2.0 units | 12 points = 6.0 units | — | — | — | — |
| Total points, max. | 8 points | — | — | — | — | — |
| Polarization, min. | 99.7 | 99.7 | — | — | — | — |
| Loss on drying, max. | 0.1% | 0.1% | 4.5% | — | — | — |
| Invert | 0.04% max. | 0.04% max. | 0.3–12.0% | 3% max. on solids | 3–50% on solids | 50% min on solids |
| Dry matter, min. | — | — | — | 62% | 62% | 62% |
| Ash (sulfated), max. | — | — | 3.5% | — | — | — |
| Total sugars, min. | — | — | 88% as sucrose | — | — | — |
| SO$_2$, max. | 15 ppm | 15 ppm | 40 ppm | 15 ppm on dry matter | 15 ppm on dry matter | 15 ppm on dry matter |

A much more satisfactory method is to use hydrochloric acid, which will invert high-grade sugar at a much lower concentration of acid. Two processes are available: (A) is a 2-hr reaction and gives a syrup that is practically as colorless as the sugar syrup used; (B) is shorter but gives a slightly brown syrup.

Invert sugar from process (B) is suitable for use in many confections, but for white fondants or where pale clear colors are required, it is better to use process (A) syrup.

**Process (A)** A cylindrical tank with a capacity of about 2,000 lb is provided with a stirrer and steam heating coil, and it is an advantage to fit thermostatic control for ease of operation. A head space of about 12 in. above the 2,000-lb mark should be available (Fig. 8.1). Five hundred pounds of water are run into the tank, the mixer set in motion, and the heat applied. The requisite amount of sugar (about 1,400 lb) is added slowly to produce a syrup concentration of 70 to 72 percent.

The thermostat is set at 71°C (160°F), and when the syrup has

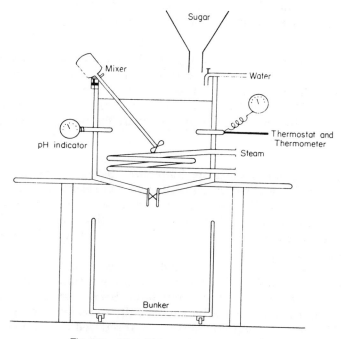

Fig. 8.1. Manufacture of Invert Sugar

reached this temperature, all the sugar should have dissolved, but for quicker solution, the sugar may be dissolved at a higher temperature—88°C (190°F) and cooled to 71°C (160°F). This necessitates water circulation through either the steam coil or a jacket round the tank, and when no solid sugar is visible, 0.1 percent of hydrochloric acid (S.G.1.18) is added as *free acid*. This point is important because commercial waters usually have a natural "hardness" that will neutralize some of the acid. This neutralizing power of the water can be determined by titration, and the acid used for inversion increased accordingly. Alternatively, the pH of a sugar syrup prepared in neutral distilled water can be determined by laboratory experiment, and by fitting a pH indicator to the inversion tank, acid can be added until the correct pH is obtained. For efficient inversion, pH 2.15 is required, determined on a solution of the syrup in an equal amount of distilled water. When the correct acidity and temperature are achieved, the heating is continued for 2 hr, at the end of which time inversion should be complete. A laboratory check should indicate that the residual sucrose is less than 4 percent.

The residual acidity is neutralized by the addition of a calculated quantity of sodium bicarbonate. This results in a mild effervescence but the head space should be sufficient to contain it.

**Process (B)** Syrup is prepared as in (A) but the thermostat is set at 96°C (205°F). At this temperature, acid is added as before but inversion will be complete in about 15 min, after which the syrup is neutralized. Extra acid may be required with this shorter inversion and it is possible to use lower-grade sugars followed by decolorization with active carbon.

**Invert Sugar from Low-Grade Sugar or "Scrap"** Syrups prepared from low-grade sugars or confectionery scrap often contain small amounts of dissolved mineral salts, and these have marked buffering action towards acid inversion. The result is that inversion of sucrose by the methods just described is retarded or much higher concentrations of acid are required.

A much more satisfactory method of inversion is to use the enzyme invertase. Invertase concentrate is a preparation used mainly to soften fondant centers after they have been cast and covered in chocolate, but it has been used successfully to reclaim scrap as invert syrup. Methods of scrap reclaiming are discussed in another chapter, and although much scrap is reused as syrup, conversion to invert sugar means that the syrup can be stored and used as an ingredient and no invert sugar need be purchased or made by the foregoing methods unless light-colored syrup is required.

**Invertase Process** The efficiency of inversion of sugar by the enzyme invertase is related to syrup concentration, pH, and temperature, and also to the quantity of invertase used. These factors are described in detail under "Invertase Concentrates."

The following is a typical process used in scrap reclaiming.

Scrap is made into a clear decolorized syrup by one of the methods described under reclaiming of scrap, and adjusted to 50 percent concentration. A tank similar to that used for invert sugar manufacture is suitable, but thermostatic control is essential and the pH must also be accurately adjusted.

The thermostat is set at 60°C (140°F), and when the syrup has reached this temperature, the pH is adjusted to 5.0 ± 0.1. Invertase concentrate is then added in a quantity equal to 0.15 percent of the syrup, but since the activity of invertase preparations varies somewhat according to the supplier, reference should be made to the relevant instruction brochures. Under these conditions, inversion should be complete in 8 hr and the syrup can then be used at once or concentrated to 75 percent soluble solids for storage.

To preserve color, it is preferable to concentrate in a vacuum evaporator, in any case the syrup must be heated to destroy invertase activity unless the subsequent manufacturing process involves boiling.

**Decrease of Moisture Content During Inversion** It often is not realized that during the inversion of sugar, the absorption of one molecule of water has an effect on the syrup concentration.

$$\underset{\text{sucrose}}{C_{12}H_{22}O_{11}} + H_2O \rightarrow \underset{\text{dextrose}}{C_6H_{12}O_6} + \underset{\text{fructose}}{C_6H_{12}O_6}$$

$$\text{Mol. wt.} \rightarrow \quad 342 \quad + 18 \rightarrow \quad \underset{\diagdown \qquad \diagup}{180 \quad + \quad 180}$$
$$360$$

This amounts to over a 5 percent increase in sugars" if there is complete inversion, and this should be allowed for in inversion processes and in the use of invertase in fondants. However, while sucrose will crystallize from solutions of 75 percent concentration, when dextrose and fructose are present (as in invert sugar), no crystallization will occur at normal temperatures (see "Solubility—Mixed Sugars").

If invert syrups are concentrated to about 80 percent soluble solids, dextrose will eventually crystallize out, particularly if cooled and seeded. A paste is then formed that resembles crystalline honey, but for most confectionery recipes, the clear syrup is more convenient to use.

## Honey

Honey is used essentially for flavor purposes in modern confectionery recipes. At one time it also served the purpose of a humectant as well as contributing to the recipe a noncrystallizing sugar and a means of raising the soluble solids. Invert sugar, which is cheaper, now provides these properties.

A typical composition of honey is invert sugar, 74 percent (consisting of fructose 39 percent; and dextrose 35 percent); sucrose, 1.8 percent; ash, 0.2 percent; dextrin, 1.5 percent; proteins and wax, 0.4 percent; and moisture, 18 percent.

The composition of a large number of American honeys has been described by White (1962) and the following average figures have been given: moisture, 17.2 percent; fructose, 38.2 percent; dextrose, 31.3 percent; sucrose, 1.3 percent; maltose, 7.3 percent; other sugars, 1.5 percent; and pH, 3.9.

The fructose content is always higher than the dextrose in genuine honey and the ratio, varying from about 1.15 to 1.35, has been a suggested standard. Manufactured invert sugar has a ratio of approximately 0.9. The composition can vary slightly according to the origin, but the flavor is greatly dependent on the flowers from which the bee obtains the nectar. The moisture content is important and should not exceed 20 percent or fermentation may occur through the activity of osmophilic yeasts. Natural honey contains active enzymes (amylase, invertase) and ferments, and high moisture contents favor their activity, so it is important to ensure that these microorganisms do not affect any manufactured product containing the honey. Heat treatment at 71°C (160°F) will inactivate most organisms present, and temperatures up to 88°C (190°F) can be used if the honey is a minor ingredient (less than 10 percent).

Pure honey is supposed to lose its medicinal properties and some of its flavor if heated alone above 50°C (122°F).

Honey is used for its special delicate flavor. Even so, the stronger flavored honeys are preferred, such as those of Central and South

American origin (Mexican, Chilean, Guatemalan, Jamaican). Table honeys are generally too weak. Because of the natural origin of honey, there is considerable variation in flavor and quality, even from the same source, and it is the policy of the main brokers to blend their imports to a standard flavor.

With nougat, montelimart, coconut pastes, and fondants, the flavor of honey comes through well and may be reinforced by the addition of honey flavor. This is one of the true artificial flavors and many are available. Honey has a therapeutic reputation and is recognized for easing throat troubles, and because of this it is an ingredient of pastilles, gums, and lozenges.

The presence of genuine honey is also useful in advertising, for even without any specific medicinal claims, its presence in a list of ingredients or in a wrapper illustration conveys the effect of quality associated with natural products.

**Adulteration of Honey** For the detection of added artificially prepared invert sugar, reliance is placed on Fiehe's test for furfural, but this is not entirely reliable as overheated honey can give a reaction, and invert sugar prepared by the invertase method does not react to the test. However, it is quoted in the absence of other methods.

*Fiehe's Test* Fiehe's test for the detection of invert sugar in honey depends on the formation of a red color when the aldehyde reacts with resorcinol. The modified test attributable to Lampitt (1929) is as follows:

Dissolve 20 g honey in 20 ml cold water and shake gently with 40 ml ether. Decant off the ether and evaporate at room temperature. Dissolve the residue in 10 ml ether and to 2 ml of this ether extract add 2 ml of a fresh 1 percent solution of resorcinol in concentrated hydrochloric acid. A positive reaction is indicated by the immediate appearance of a pink color in the acid layer. This darkens and after 20 min a deep cherry-red color appears at the junction of the layers. Then for a confirmatory test, allow the rest of the ether extract from the honey (8 ml) to evaporate in a porcelain dish at room temperature and to the residue add 2 ml of fresh aniline acetate solution (1 ml redist. aniline + 4 ml glacial acetic acid). A positive reaction is indicated if a pink to orange color appears within 15 min. Lampitt et al. state that both the resorcinol and aniline tests will give positive reactions with honey containing 5 percent of commercial invert sugar.

**Other Tests to Be Applied to Honey**  *Specific Rotation* The specific rotation of honey ranges from +5° to −15°, but those that are dextrorotatory (+) are rare. The specific rotation changes only slightly on inversion. This is determined on a solution prepared as follows:

Dissolve 10 g of the sample in distilled water in a 100-ml graduated flask, add 0.5 ml of ammonia solution (0.880), 5 ml neutral lead acetate solution, and 5 ml of alumina cream. Fill flask in these proportions to 100 ml. Shake well and filter.

*Sucrose* To the solution used for specific rotation add 0.5 g of potassium oxalate, allow to dissolve, mix, and filter. Determine the sucrose by the Herzfeld method.

*Reducing Sugars* Dilute 10 ml of the solution used for sucrose determinations to 250 ml with distilled water and total reducing sugars determined by the Lane and Eynon method.

*Microscopic Examination* Dilute a sample of the honey with twice its volume of water and allow to stand for 24 hr. Remove the sediment by pipette and examine microscopically. Genuine honey contains pollen grains, wax particles, and, frequently, insect fragments.

Further information on the composition and analysis of honey is given by Pearson (1981) and Lees (1975).

## Maple Sugar

This is another natural sugar that is claimed to have health-giving properties. Maple syrup has a very pleasant flavor and is a great favorite in the United States and Canada where it is produced by tapping the bark of the sugar maple tree (Acer saccharum) when the sap is rising in the spring. The tapping extends over a period of three to four weeks and the syrup produced contains up to 3 percent of sugar. One maple tree yields syrup equal to about 4 lb of maple sugar in a season. The syrup is evaporated and a crystalline paste is obtained containing about 83 percent of sucrose, but maple syrups that retain their fluidity are used for domestic purposes. They have a concentration of 70 to 75 percent soluble sugars.

**Use of Maple Syrup** Maple fondant can be prepared using maple sugar and about 10 percent of glucose syrup, and this syrup can be concentrated and creamed in a beater as a normal fondant syrup. This fondant paste can be used as a filling or ingredient in other confections and is particularly popular with walnuts as in walnut fudge.

As with honey, very good artificial maple flavors are available.

## Malt Extract

Malt extract may be regarded as a sweetening ingredient used in the confectionery industry. It has particularly useful flavor characteristics and some products, such as malted milk balls, are quite popular. Malt is also used in caramels and chocolate drinks. Commercially, it appears as an extract that is a tenacious heavy syrup of various colors and flavor strength, and as a powder of about 2 percent moisture that is very hygroscopic.

**Manufacture of Malt**  In the manufacture of malt extract, high-grade barley is wetted and allowed to germinate under controlled conditions of temperature and humidity for about seven days.

During this process, the starch is solubilized and the proteins are broken down to peptides and amino acids. This is followed by kiln drying and some sifting to remove dust and rootlets, but the natural enzymes are retained.

The resultant product is then extracted with water, followed by filtration and evaporation until the extract reaches a concentration of 80 percent total solids. The extract may be vacuum dried to produce the powder.

**Composition**  Malt extract has a total solids content in the region of 80 percent. The predominant sugar is maltose 55 percent. Other sugars are approximately sucrose, 4 percent; dextrose and fructose, 2 percent each; and dextrins, 13 percent.

A range of malt extracts is commercially available and it is well to obtain a stated composition from the supplier for a given grade. Since malt is incorporated in a formulation in significant amounts to obtain a flavor level, adjustment of quantities of other sugars (e.g., glucose syrup) becomes necessary.

## GLUCOSE SYRUP, CORN SYRUP, LIQUID GLUCOSE, CORN SWEETENERS

Some 160 years ago, the Russian chemist Kirchoff discovered that a sweet substance was formed when starch was heated with dilute acid. Later work by other chemists showed that starch was a polymer of D-glucose and that this polymer could be broken down by hydrolysis. Thus, the beginning of the corn sweetener industry was established. Development was stimulated during the Napoleonic wars in Europe as a result of the blockade of cane sugar imports by the British.

Sweeteners described as corn sugar were produced in the mid-nineteenth century from potato starch in Europe and from corn (maize) in the United States. After World War II, there was a rapid advance in technology owing to the introduction of enzyme conversion and the range of conversion products was greatly expanded. At one end of the range was crystalline dextrose hydrate and at the other very low conversion products—the maltodextrins. But the corn sweeteners were at a disadvantage in some industries, such as those producing soft drinks, because of their low sweetness compared with sugar.

In the 1970s, there was a major breakthrough that offset this disadvantage—the commercial development of enzyme catalyzed isomerization of dextrose to fructose, which is a sweeter sugar. High-fructose corn syrups are now available commercially, and still more recently, crystalline fructose has become a marketable product.

## Definitions

Considerable argument arose concerning the name to be given to the conversion product of starch. This happened largely because of the medicinal associations of glucose, synonymous with dextrose hydrate, which achieved the reputation of a sugar readily assimilated, with quick energy-producing qualities and therefore able to allay fatigue.

The following have been adopted by the FDA and agree with the proposals of the Codex Alimentarius.

*Corn syrup* (glucose syrup) is the purified concentrated aqueous solution of nutritive saccharides obtained from edible starch and having a dextrose equivalent of 20 or more.

*Dried corn syrup* (dried glucose syrup) is corn syrup from which the water has been partially removed.

*Dextrose monohydrate* is purified and crystallized D-glucose containing one molecule of water of crystallization with each molecule of D-glucose.

*Dextrose anhydrous* is purified and crystallized D-glucose without water of crystallization.

*Maltodextrin* is a purified concentrated aqueous solution of nutritive saccharides obtained from edible starch, or the dried product derived from said solution and having a dextrose equivalent less than 20.

The EEC "Sugars" directive (27.12.73) definition is "the purified and concentrated aqueous solutions of nutritive saccharides, obtained from maize or potato starch having the following characteristics: dry substance, not less than 70 percent by weight [normally 80 to 82 percent]; dextrose equivalent, not less than 20; sulfated ash, not more than 1.0 percent by weight of dry substance; sulfur dioxide, normally less than 20 ppm, but for confectionery use, up to 400 ppm is permitted subject to the regulations of each country."

Dextrose is one of the constituents of glucose syrup, and because this syrup is a useful ingredient in sugar and flour confectionery, soft drinks, and some alcoholic drinks, the advertising agencies were quick to seize on the value of its medicinal reputation to promote the sales of the various foods and drinks containing it.

Nevertheless, glucose syrup is a major and valuable constituent of confectionery apart from its food value. It has a greater solubility than sucrose, and when in solution with it, retards crystallization or graining. Because a mixture of sucrose and glucose syrup has a greater solubility than sucrose alone, the soluble solids can easily be maintained above the 75 percent which is necessary to ensure resistance to activity by microorganisms. For the same reason, the equilibrium humidity can be maintained at a low level, which prevents drying out, but this effect varies according to the degree of conversion of the glucose syrup.

## Manufacture of Glucose Syrup

Glucose syrup is manufactured principally from corn, but some European manufacturers use potato starch as the raw material. The starch, after separation from the raw material, is changed to glucose by acid conversion, and this may be followed by additional enzyme conversion. The acid conversion is carried out under pressure and the degree of conversion is controlled by variations in temperature, time, pH, and pressure. More recently, enzyme/enzyme conversion has been developed.

As with many present-day commodities, the manufacture of glucose syrup has developed into a mechanized technology, and control of conversion is precise to the extent that all types of syrup can be produced to within quite narrow ranges of composition. Not many years ago, one glucose only was made, known as 42 dextrose equivalent (DE), and this varied appreciably from one manufacturer to another.

Control of composition today is greatly aided by the modern analytical techniques available; see Jackson (1984). By means of

HPLC, a spectrum of the proportions of the various saccharides in a syrup is obtained in as little as 15 min.

A summary of the process of manufacture is as follows:

Corn, as received, contains foreign matter in the form of dust, stalks, chaff, and stones, and this is removed by sieving and air separation. The cleaned corn is soaked in warm water containing sulfur dioxide for about 48 hr, which swells and softens the corn—the $SO_2$ preventing microbiological action—and this part of the process removes soluble proteins and mineral salts. The softened corn is then coarse ground wet to remove the germ without damaging it and this process produces a slurry of loose starch, gluten, germ, and some fibrous matter. From this the germ is separated by centrifugal means and is separately processed for extraction of the corn oil—now an important oil of commerce that is also finding increasing popularity for domestic use.

The slurry now contains fiber, starch, and gluten, and by means of a series of screens, the fiber is removed, leaving the starch and gluten. High-speed centrifuges then remove the gluten, and the remaining starch slurry is further purified and concentrated, automatically, by continuous centrifuge and elutriation.

The purified starch suspension is transferred to the converters, which, originally, were large pressure vessels, but continuous converters are now in operation.

It is very important that proteins are removed from the starch before conversion, as the presence of protein in the final glucose syrup can result in foam formation on boiling, a great nuisance in some high boiled confections.

In the converters, by the catalytic action of acid under pressure, the starch is changed to dextrose, maltose, maltotriose, maltotetrose, and a complex variety of oligosaccharides. This is known as the acid conversion process—the original method.

There are now, in addition, the acid/enzyme and multiple-enzyme processes.

In the acid/enzyme process, the starch slurry is partly converted by acid to a given DE, which indicates a low dextrose content. This is then treated with the appropriate enzyme to complete the conversion. $\beta$-Amylase is normally employed, which produces a high-maltose syrup. In the multiple-enzyme process, the starch granules are first gelatinized and the starch polymer broken down by $\alpha$-amylase enzyme. By use of these various systems, a range of syrups may be produced that have different viscosities, sweetness, hygroscopicity, and fermentability.

For the manufacture of high-fructose syrups, high DE syrups prepared by one of the above processes are further treated by an enzyme isomerase that converts a proportion of the dextrose to fructose.

The maltodextrins are produced in the same way as the corn syrups but with the conversion process stopped to keep the DE below 20.

All syrups after conversion are filtered, decolorized, and concentrated. Some are further purified by exchange resins.

**Commercial Grades of Glucose Syrup**  It is well understood from the brief description of the manufacturing processes that the chemical and physical properties of the different syrups vary considerably. The variation largely determines their method of use.

The main properties of the grades of glucose syrups are given below, together with an explanation of terms used in the industry.

Certain accepted trade definitions are associated with glucose syrup grades, and these must be understood by the confectionery technologist.

*Dextrose Equivalent*  DE is the percentage of reducing sugars on a dry basis, calculated as dextrose, or the pure dextrose percentage that gives the same analytical result as is given by the combined reducing sugars in the glucose syrup. The higher the DE, the further the conversion has been taken, resulting in less of the higher carbohydrates and a lower viscosity.

*Baumé Density*  For the purpose of simplification, the glucose syrup industry uses the Baumé scale in preference to specific gravity. To the newcomer this is a confusing method of describing the density of the product. The relationship with specific gravity is given by the equation

$$°Bé = 145 - \frac{145}{\text{True spec. grav. } 60°F/60°F(15.5°C)}$$

In practice, the Baumé reading is taken at 140°F (60°C) using a special hydrometer having a range of 10 Bé graduated in 0.1°. It is usual to report "commercial Baumé," which is related to observed Baumé at 140°F (60°C) as follows:

Commercial Baumé = Observed Baumé (140°F/60°C) + 1

In some countries, the use of Baumé density has been abandoned.

*Specific Rotation*  The specific rotation of a liquid is obtained by dividing the observed optical rotation by the length of the column of liquid and by the specific gravity of the liquid.

The specific rotation $(\alpha)_D^{20}$ of a sugar is defined as follows:

$$(\alpha)_D^{20} = \frac{100 \times \alpha}{c \times l}$$

$\alpha$ is the observed rotation at 20°C (68°F) using the illumination of the $D$ line of sodium

$c$ is the concentration of the sugar in solution expressed in grams per 100 ml of solution

$l$ is the length of the column of solution

The specific rotation of glucose syrup varies from about 90 for the high conversion syrups to 130 for the very low conversion syrups.

**Classification and Properties (Corn Refiners Association, 1979)** A relatively simple method of classification of glucose syrups is according to their DE.

1. Maltodextrin—less than 20
2. Low conversion—20 to 38
3. Medium conversion—39 to 58
   42 DE is known as "Regular" or "Standard"
4. High conversion—59 to 65
5. High fructose—75 to 96

Conversion Type of Glucose Syrup

| Property | Low(30DE) | Regular(42DE) | High(65DE) |
|---|---|---|---|
| Sweetness | → | | → |
| Control of crystallization | ← | | ← |
| Viscosity | ← | | ← |
| Humectancy | → | | → |
| Vapor pressure | ← | | ← |
| Osmotic pressure | → | | → |
| Fermentability | → | | → |
| Depression of freezing point | → | | → |
| Browning | → | | → |

Direction of arrow indicates the trend of the properties in relation to degree of conversion.

Fig. 8.2. Conversion Type of Glucose Syrup
Direction of arrow indicates the trend of the properties in relation to degree of conversion.

All the syrups are wholly carbohydrate and are fully digestible and nutritive. The degree of conversion affects the physical properties considerably.

As conversion increases, the syrups become sweeter and less viscous, are more readily fermented, and are more hygroscopic. The lower conversion glucoses are more viscous with greater body, and therefore retard crystallization as well as acting as stabilizers toward foams. Figure 8.2 indicates these properties.

**Composition of Glucose Syrups**  Different companies manufacturing glucose syrups use proprietory names for the various types and there are small differences in the composition of these syrups of similar dextrose equivalent.

Table 8.10 is representative and indicates the relation between DE, method of conversion, and saccharide composition (Roquette Frères, 1984).

## Maltodextrins (Roquette Frères, 1984)

These are products with a low DE from 3 to 20, obtained by enzyme conversion (usually α-amylase). They are generally spray-dried powders (less than 5 percent moisture) as the corresponding liquids have extremely high viscosities and tend to become cloudy on storage.

The saccharide composition of the commercial maltodextrins is given in Table 8.11. The very low DE product is prepared from waxy corn starch, a natural starch that is almost entirely the amylopectin polymer.

## Dried Glucose Syrups

Glucose syrups with DEs between 20 and 65 may be spray dried and powders with less than 5 percent moisture are obtained. These powders may replace a proportion of the sugar (sucrose) in low-moisture confections or in chocolate, thereby reducing sweetness, and possibly the cost.

These powders are very hygroscopic, which may hinder their handling in production processes.

## Uses of Glucose Syrups and Maltodextrins

*Low DE syrups and maltodextrins* have high viscosities and low sweetness. They inhibit sucrose crystallization and act as stabilizers in aerated products such as marshmallows.

TABLE 8.10. GLUCOSE SYRUP—DEXTROSE EQUIVALENT AND METHOD OF CONVERSION

| Dextrose equivalent | Method of conversion | Dextrose, % | Fructose, % | Disaccharides (maltose), % | From tri- to heptasaccharides, % | Polysaccharides i.e. above hepta-saccharides, % |
|---|---|---|---|---|---|---|
| 28 DE | Acid | 9 | — | 8 | 32 | 51 |
| 37 DE | Acid | 15 | — | 12 | 39 | 34 |
| 37 DE | Acid + β-amylase | 5–10 | — | Approx. 40 | Approx. 25 | Approx. 25 |
| 42 DE | Acid | 18 | — | 14 | 42 | 26 |
| 45 DE | α-amylase + β-amylase | 5–10 | — | Approx. 50 | Approx. 20 | Approx. 20 |
| 50 DE | Acid | 26 | — | 17 | 42 | 15 |
| 60–65 DE | Acid + β-amylase | 30–35 | — | 35–40 | 10 | 20 |
| 75 DE | Acid + β-amylase + isomerase | Approx. 35 | Approx. 20 | Approx. 20 | Approx. 10 | Approx. 15 |
| 96 DE | α-Amylase + amyloglucosidase + isomerase | 48–50 | 42–44 | 7–8% | | Nil |

TABLE 8.11. MALTODEXTRIN

| Dextrose equivalent | Dextrose, % | Maltose, % | Saccharides from tri- to hepta, % | Polysaccharides, i.e. above the hepta-saccharides, % |
|---|---|---|---|---|
| 5 DE | less than 1% | 1–2 | 5–7 | 92 |
| 10–13 DE | 1–2 | 3–4 | 20 | 75 |
| 17 DE | 2–3 | 5–6 | 24 | 68 |
| 20 DE | 4–5 | 8–10 | 28 | 58 |

Roquette Frères (1984).

They have low hygroscopicity and may be used as protective glazes on hard candies and other confections. Because of their viscosity, they impart body and a chewy texture. In some instances, maltodextrins may replace gum arabic.

*42 DE syrup* is the Regular grade, which is used for all general purposes. Its properties may be modified by blending in high or low conversion syrups.

Where bulk storage of the 42 DE syrup is employed, blending is useful for certain small-volume formulations. The dried syrups or maltodextrins are convenient for this purpose.

*High conversion syrup (50 DE)* is sweeter and more fluid than the regular grade but retains its crystallization-retarding properties. The greater fluidity is useful, for example, in the formulation of fondant for depositing into chocolate shells, thus preventing tailing.

*High maltose syrup (45 DE)* is an acid/enzyme conversion syrup in which the enzyme has been selected to produce the disaccharide maltose rather than dextrose.

Its sweetness compares to that of 42 DE Regular syrup and it exhibits high moisture retention, low browning, and a neutral flavor. It is useful in high boilings where high humidities might be expected.

# DEXTROSE, GRAPE SUGAR, POWDERED GLUCOSE, D-GLUCOSE

When starch is completely hydrolyzed by acid treatment, the resulting solid obtained by crystallization is the monosaccharide dextrose, which is chemically known as D-glucose—a confusing anomaly.

It is the most abundant sugar in nature and most fruits and berries, as well as honey, are rich in dextrose.

## Dextrose Manufacture

The main commercial product today is dextrose monohydrate, $C_6H_{12}O_6 \cdot H_2O$. Prior to 1960, dextrose was produced by hydrolysis of starch with acid, but now practically all is manufactured by the enzyme process. Although the enzyme process was patented as long ago as 1942, it did not achieve commercial success until the development of submerged culture techniques for the production of fungal amylases. Along with this a method had to be worked out for the selection of the correct enzyme preparation and removal of those that interfered and prevented high yields.

Before treatment with the converting enzyme, the starch slurry is thinned, which can be done by acid or enzyme and a partial change to complex intermediate dextrinlike substances occurs.

The final conversion takes place after adjustment of the pH to 4.0 to 4.5 and up to 72 hr is required at a temperature of 55 to 60°C (130 to 140°F). The converted syrup is filtered to remove small residues of protein, fat, and unconverted starch and then decolorized with activated carbon, and again filtered. The solution is concentrated and crystallization is controlled by seeding, stirring, and cooling so that wholly monohydrate is obtained.

**Properties of α Dextrose Hydrate** Moisture content—theoretical: 9.1 percent; commercial product: 8.5 percent approx.; specific rotation $(\alpha)_D$ 20°C, $+52.6°$; melting point 85°C (185°F); solubility 10°C (50°F) 41 percent, 15.6°C (60°F) 45 percent, 20°C (68°F) 48 percent, 30°C (86°F) 55 percent, 50°C (122°F) 70 percent.

Crystal structure—six-sided plates in nodular masses. The heat of solution is minus 25 cal/g at 25°C (77°F). This, together with the property of rapidly going into solution, causes the cool sensation experienced when dissolved in the mouth.

**Forms of Dextrose** Dextrose exists in three crystalline forms described chemically as:

α-D-Glucose hydrate (α-dextrose hydrate)—This crystallizes from concentrated solutions below 50°C (122°F).

Anhydrous α-D-glucose—This crystallizes from concentrated solutions above 50°C (122°F) and below 110°C (230°F).

Anhydrous β-D-glucose—This form separates if the solution is allowed to crystallize at temperatures above 110°C (230°F) [more specifically, above 115°C (239°F)].

**Solubility of Dextrose**  When in solution dextrose can exist in $\alpha$ and $\beta$ forms. Dextrose hydrate will dissolve readily in water at 25°C (77°F) until a concentration of 30 percent is reached. After this, further quantities will dissolve at a much reduced rate until a saturated solution of 51 percent concentration is obtained. This peculiarity arises from the gradual transition of the original $\alpha$ form to the $\beta$ form, which is more soluble, and ultimately a saturated solution of the mixed $\alpha$ and $\beta$ forms is obtained in the presence of solid $\alpha$ dextrose hydrate. Anhydrous $\alpha$- and $\beta$-D-glucose have initially solubilities greater than 51 percent but will, in solution, eventually crystallize and reach equilibrium with $\alpha$ dextrose hydrate as in the previous instance.

The solubility of dextrose increases with temperature and at 50°C (122°F), 70 percent concentration is obtained, which is nearly 20 percent above that at 25°C (77°F). Above 50°C (122°F), the solubility increases still further but the anhydrous $\alpha$ dextrose then becomes the stable crystal form that will deposit from solution when cooling a saturated solution from, say, 93°C (200°F) to 50°C (122°F). Further cooling will deposit crystals of the $\alpha$ hydrate form and the $\alpha$ anhydrous form will transform to the hydrate.

It will be seen, therefore, that to prepare dextrose fondants or any other confectionery product where crystallization of dextrose is required, temperature control will be an important factor. This is discussed under the use of dextrose in confectionery.

**Specific Rotation of Dextrose**  Because of the slow transformation of $\alpha$ dextrose to the $\beta$ form in solution the phenomenon known as mutarotation occurs. When a solution of $\alpha$-dextrose hydrate is made, it originally has a specific optical rotation of 112 and this will gradually reduce until a constant figure of 52.5 is reached. Similarly, $\beta$-D-glucose when first dissolved has a specific rotation of 19 and this will gradually rise to the same constant figure of 52.5. Equilibrium is quickly obtained by the addition of alkali, and in analytical procedures, ammonia is added for this purpose.

## Fructose (Levulose)

Fructose has already been mentioned as a constituent of honey, invert sugar, and high-fructose glucose syrups. It is abundant in nature, being present in fruits, vegetables, and honey. Fructose, originally, was available only in solution but a process has been developed of recent years for the production of crystalline fructose with a moisture content of less than 0.1 percent.

It has achieved importance as an ingredient of dietetic (diabetic) foods.

**Sweetness—Caloric Value** Fructose has a sweetening power of from 1.3 to 1.7 times that of sugar (sucrose), the variation being related to other constituents of the food containing it. Therefore, for a given sweetness compared with sucrose it has a lower caloric value.

The type of sweetness is generally regarded as very pleasant and blends well in most foods.

**Diabetics** Fructose provides a readily assimilable carbohydrate without creating a demand for insulin.

**Hygroscopicity** Crystalline fructose is very hygroscopic, as also are concentrated solutions. Handling of pure fructose as an ingredient, in chocolate production, for example, needs special precautions to prevent moisture pickup. Inclusion of fructose in high-moisture confectionery centers will retard drying, but with low-moisture products such as hard candies, the reverse applies. Inversion of the sucrose in hard candies, which produce some fructose, is detrimental and induces stickiness.

**Browning** Fructose, being a reducing sugar, will develop brown pigments on heating and it is very active in the Maillard reaction with milk proteins. This is a factor that affects flavor intensity in caramels.

## Sorbitol

Sorbitol is a polyhydric alcohol and is widely distributed in nature, the richest source being the rowan or mountain ash berry, but no natural supply is important commercially. It is now produced by the chemical reduction of glucose (dextrose). Sorbitol is a valuable substance in the manufacture of diabetic chocolate and confectionery because, unlike some intense sweeteners, it has bulk as well as sweetness. In this respect, it is roughly equivalent to dextrose. Sorbitol is available commercially in crystalline and syrup form. The crystalline form is used for chocolate manufacture and the purity and moisture contents are critical or processing difficulties will arise.

The liquid form is used in various foods, including confectionery, as a softener, and it will also help to prevent drying out. It is used in the manufacture of sugarless confectionery and chewing gum.

Sorbitol is polymorphic and exists in three crystalline states. Only one form, the gamma, is stable, and the other unstable forms will, under the influence of moisture or heat, transform to the stable form.

**Equilibrium Relative Humidity of Sorbitol (Water Activity)**  Many claims are made for sorbitol as a humectant, and although it finds use in the paper, textile, tobacco, baking, and other trades, its value in normal confectionery recipes must not be overestimated. Tests with formulations of the nut paste and fudge type indicate that its humectant properties differ little from those of invert sugar, which is less expensive. About half the sugar of invert sugar is fructose, which has very good humectant properties. Sorbitol syrup, because of its viscosity, has an effect in some formulations of retarding crystallization. In such cases, it could convey extra humectant properties to the confection as well as softening the texture.

Manufacturers of sorbitol publish data on the equilibrium humidity of solutions of different strengths, but when using any ingredient with humectant properties, it is necessary to examine its effect in the finished product, and this applies to sorbitol. Sorbitol has a marked cooling effect when taken into solution, and also has a high solubility.

**Sorbitol in the Diet**  It has been shown that sorbitol is utilized in the body and 98 percent of that taken in food is digested, with the other 2 percent being excreted.

It exhibits no toxic symptoms. According to the medical literature, sorbitol in a diabetic food is a precursor of glycogen and hence fructose, but because of the delay in converting sorbitol to fructose, the enfeebled pancreas does not become overworked.

Sorbitol produces a laxative effect and the limit of intake recommended is about 3 oz per day.

## Mannitol

Mannitol occurs widely in nature, for example, in celery, larch, and especially the manna ash (Fraxinus ornus). The dried juice of the last is "manna."

Mannitol is produced synthetically by the hydrogenation of fructose and is a polyhydric alcohol. Commercially, it is a crystalline powder, odorless and nonhygroscopic with low solubility.

Its sweetness is about 0.6 that of sucrose. Its low solubility limits

its use but it has special value in sugarless chewing gum and, because of its low hygroscopicity, in effervescent powders. It is also used in pharmaceutical tablets.

## Lycasin 80/55

Lycasin is the proprietary patented product of Roquette Frères and is a hydrogenated glucose syrup, which is obtained by enzyme hydrolysis of starch. It is a liquid with 75 percent dry substance. Approximately 7 percent of sorbitol is present.

It has interesting properties that make it useful in confectionery and chewing gum.

*Crystallization.* It will not crystallize even at low temperatures.

*Hygroscopicity.* It is hygroscopic and useful as a humectant.

*Sweetness.* It has about 0.75 the sweetness of sucrose.

*Viscosity.* The low viscosity enables it to be easily worked in confectionery formulations.

*Dental caries.* Exhaustive tests have indicated that it has valuable effects in the prevention of dental caries.

**Use of Lycasin** Hard candies can be made with lycasin with moisture contents as low as 1 percent and no inversion or browning will occur. In sugarless chewing gum, lycasin can replace the glucose syrup and the sugar ingredient by sorbitol or a mixture of sorbitol and mannitol.

Soft panned coatings may be prepared using a mixture of lycasin and sorbitol, and sorbitol alone for hard panning.

## Xylitol

Xylitol is a pentahydric alcohol with a 5-carbon chain, and thus differs from dextrose and fructose, which are 6-member. It is present in many natural substances, such as fruits, vegetables, and mushrooms, and may be extracted from birch wood.

The substance has been known for many years but recently it has been the subject of many biological studies and is now available in substantial commercial quantities. It is a white crystalline powder, hygroscopic, and resistant to heat.

Xylitol, when consumed, is slowly but completely absorbed in the intestine. It does not require insulin for metabolism and, therefore, does not change the blood sugar level in diabetics. It has exceptional value as a noncariogenic sweetener and is not metabolized by cariogenic bacteria. It has a sweetening power very similar to that of sugar, being about twice that of sorbitol and glucose syrup. It has not shown any toxic properties but abnormalities arise when exceptionally high quantities are taken. Like sorbitol, it has laxative properties (30 to 40 g per single administration). In confectionery, it has special applications as a noncariogenic sweetener in chewy products such as chewing gum and caramels.

## NONNUTRITIVE (SYNTHETIC) SWEETENERS

A number of these sweeteners have been produced. Generally speaking, they are intensely sweet substances without bulk and some have had a doubtful medical history. Some of the more important of these are discussed in the following.

### Saccharin

Saccharin was first discovered over 100 years ago and has been in general use for over eighty years. It was first banned in the United States in 1912, without real reason, but was reinstated during World War I. It was again subject to some doubt in relation to carcinogenicity in 1977 and a ban was proposed by the FDA. However, there was such an uproar that the ban has, to date, not been imposed, but it has led to a search for other sweeteners.

$$C_6H_4 \underset{CO}{\overset{SO_2}{\diagdown}} NH$$

*Saccharin*    "ortho-sulpho-benzimide" is the ammonia derivative of O-sulpho-benzoic acid

Saccharin is a white crystalline powder about 300 times as sweet as sucrose, and not very soluble in water. The imide hydrogen may be replaced by sodium and this forms the soluble salt.

Saccharin, if used to replace sugar entirely, results in a rather bitter effect—some people find it objectionable.

### Cyclamates

"Cyclamates" is a general term of the group of substances that includes sodium and calcium cyclamates and cyclamic acid.

Cyclamates have a very clean sweet taste of high intensity, very closely resembling sugar sweetness. There is no bitter aftertaste. Their sweetening power varies from thirty to sixty times that of sugar, depending on the medium and flavor. Citric acid, other citrus products, and saccharin have synergistic effects on the sweetness.

## Sodium Cyclamate $C_6H_{12}NSO_3Na$

HN·SO$_3$Na

Sodium cyclohexyl
sulphamate

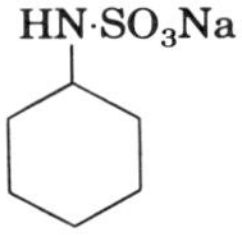

Sodium cyclamate is a white crystalline powder, odorless, and intensely sweet.

Cyclamates were developed commercially between 1942 and 1950, mainly for use in pharmaceutical products. Later they were approved for general use (FDA, 1950; United Kingdom Food Standards Committee, 1955).

In 1970, as a result of further evidence of possible carcinogenic effects the use of cyclamates in foods was banned by the U.S. Department of Agriculture and other countries followed suit.

## Acesulfam K (Acesulfam Potassium)

This was first manufactured in Germany in 1967 at the laboratories of Hoechst. It is a white crystalline powder and is the potassium salt of 6-methyl-1.2.3 oxathiazine 4(3H) 1,2,2 dioxide. It has from 130 to 200 times the sweetness of sucrose, depending on its application, and from a flavor standpoint, it is most acceptable when mixed with other sweeteners such as sucrose. Health tests have shown no unfavorable effects.

## Aspartame (Nutrasweet, Canderel)

This was developed in the United States by the Searle Company and its chemical name is 3-amino-N (a carboxyphenyl) succinic acid methyl ester. It is a white, odorless, crystalline powder about 200 times as sweet as sugar.

It was granted approval by the FDA in 1980 and is being used by large American companies. It is officially approved by many other countries.

## Talin

This substance, a relatively new product, is manufactured by the British Tate and Lyle Company. It is classified as a wholly natural sweetener and is extracted from the fruit of an African plant known as Thaumatto-coccus danielli. It is the sweetest substance known, with an intensity about 2,000 times that of sucrose. It is very soluble in cold water but is unstable to heat if the pH is less than 5.5.

It is permitted in foods generally.

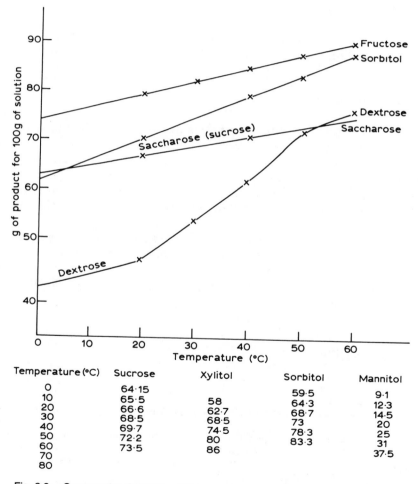

Fig. 8.3. Comparative Solubility of Fructose, Sorbitol, Dextrose, and Saccharose

| Temperature (°C) | Sucrose | Xylitol | Sorbitol | Mannitol |
|---|---|---|---|---|
| 0 | 64·15 | | 59·5 | 9·1 |
| 10 | 65·5 | 58 | 64·3 | 12·3 |
| 20 | 66·6 | 62·7 | 68·7 | 14·5 |
| 30 | 68·5 | 68·5 | 73 | 20 |
| 40 | 69·7 | 74·5 | 78·3 | 25 |
| 50 | 72·2 | 80 | 83·3 | 31 |
| 60 | 73·5 | 86 | | 37·5 |
| 70 | | | | |
| 80 | | | | |

*Roquette Frères (1984)*

## LEGISLATION

With all the sweeteners so far described, only brief notes on legislation have been mentioned. The U.K. Food Research Association (1983) has given an up-to-date appraisal of the situation but potential users should always obtain the latest information.

## SOLUBILITY OF SUGARS AND REPLACEMENT SUBSTANCES

Comparative solubilities of sucrose, dextrose, fructose, xylitol, sorbitol, and mannitol are given in Fig. 8.3 (Roquette Frères, 1984).

## SWEETNESS OF SUGARS

The sweetness of synthetic sugar substitutes was included in the descriptions of these substances.

The relative sweetness of the various sugars present in confectionery has been the subject of much conjecture. Tastes vary greatly and precise measurements are not possible. Concentration affects relative sweetness and sugars may be affected by the synergistic effect of other substances.

The following figures have been collected from various sources and represent the relative sweetness of the more common sugars: sucrose, 100; invert sugar, 115; fructose (levulose), 130; glucose syrup (42 DE), 45; glucose syrup (55 DE), 55; glucose syrup (65 DE), 65; maltose, 33; lactose, 16.

## REFERENCES

California and Hawaiian Sugar Co. 1983. *The Industrial Sugars of C and H*. San Francisco, Calif.

Corn Refiners Association, Inc. 1979. *Nutritive Sweeteners from Corn*. Washington, D.C.

Egan, H., Kirk, R., and Sawyer, R. 1981. *Pearson's Chemical Analysis of Foods*. Churchill-Livingstone, Edinburgh, Scotland.

Hoynak, P., and Bollenback, G. 1966. *This is Liquid Sugar*. Corn Products Co., Tenefly, New York, U.S.A.

Jackson, E. B. 1984. *Glucose Syrup*. Confectionery Production, London.

Lampitt, L. H. 1929. *Analyst*. London.

Leatherhead Food Research Association 1983. *Users Guide to Newly Permitted Sweeteners*. Leatherhead, Surrey, England.

Lees, R. 1967. *Books of Historic Significance to the British Sweet and Chocolate Industry*. Confectionary Production, London.

Lees, R. 1975. *Honey and Its Uses*. Confectionery Production, London.

Lees, R., and Jackson, E. B. 1973. *Sugar Confectionery and Chocolate Manufacture*. Specialized Publications Ltd., Surbiton, Surrey, England.

Roquette Frères 1984. Hydrolysis Products, Lille, France.

Silesia 1984. *Biscochoc Dictionary. Handbook of the Confectionery Industry*, Vol. 1/11. Silesia-Essenenfabrik, Gerhard Hanke K.G., Neuss, W. Germany.

Tate & Lyle Refineries 1984. *Industrial Sugars, Liquid Sugars, Brewing Sugars*.

White, J. W. 1962. Composition of American honeys. *Tech. Bull.* 1261, U.S. Department of Agriculture.

# 9

# Confectionery Fats

Vegetable fats generally are used in great quantities in the production of all kinds of confectionery, such as caramels, fudge, nougat, truffles, and pastes for wafer and cooky fillings. The only animal fat normally used in these products is butter, with the other animal fats employed in the general baking and other food industries.

Many trees and plants supply natural oils and fats but these oils are not usually suitable for use without further physical and chemical treatment. The oil expressed from the fruit or seeds is refined, and then subjected to hardening either by glyceride separation or by hydrogenation followed by deodorization. The degree of hardening is determined by the way the fat is to be used subsequently.

The description "oil" or "fat" is often applied indiscriminately. The explanation often given is that an oil is a liquid fat and a fat a solid oil with regard to normal temperatures. When seeds or nuts are expressed, an oil is mostly produced, which is then treated further. Several hundred varieties of oil-bearing plants are known but only a few of these are of commercial importance.

Vegetable fats are usually described as domestic and nondomestic, the former obtained from crops growing in moderate climates and the latter of tropical origin.

| *Domestic oils* | *Main producing countries* (1984) |
|---|---|
| Soybean oil | United States, Brazil, China (P.R.) |
| Cottonseed oil | United States, U.S.S.R., China (P.R.) |
| Peanut oil | India, China (P.R.) |
| Sunflower seed oil | U.S.S.R., United States |
| Rapeseed oil | Canada, China (P.R.), India |
| Olive oil | Mediterranean countries |
| *Tropical oils* | *Main producing countries* (1984) |
| Palm oil | Malaysia, Indonesia, Nigeria |
| Coconut oil | Philippines, Indonesia |
| Palm kernel oil | Malaysia, Nigeria |
| Cocoa butter | Brazil, Ivory Coast, West Africa, Malaysia |

Additionally, there are the wild crops often called the exotic oils, and among these are illipe and shea butter. World production varies greatly but they have acquired importance in relation to the production of cocoa butter equivalent fats.

## CHEMISTRY OF FATS

Chemically the fats and fatty substances are more correctly termed lipids. These consist of (1) the true fats or oils called triglycerides, (2) phospholipids, and (3) sterols.

### Glycerides, Fatty Acids

Glycerides are combinations of the trihydric alcohol glycerol with various fatty acids. Glycerol is represented chemically as

$$\begin{array}{l} CH_2OH \\ | \\ CHOH \\ | \\ CH_2OH \end{array}$$

*Fatty acids* are mostly chain compounds. They include palmitic acid and stearic acid, which are the most abundant in natural fats.

| | |
|---|---|
| Palmitic acid | $CH_3(CH_2)_{14}COOH$ |
| Stearic acid | $CH_3(CH_2)_{16}COOH$ |
| Lauric acid | $CH_3CH_2)_{10}COOH$ |

Lauric acid is abundant in coconut and palm kernel oils.

A saturated fatty acid may be represented by the following:

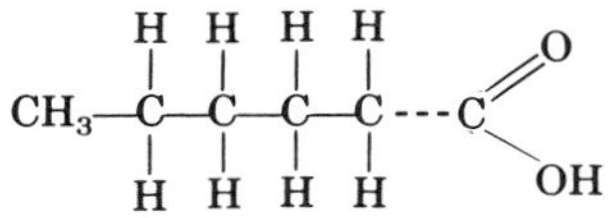

An unsaturated acid becomes

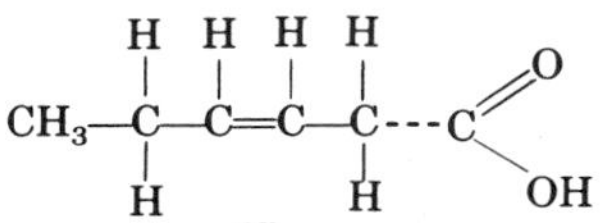

(*cis*)

or

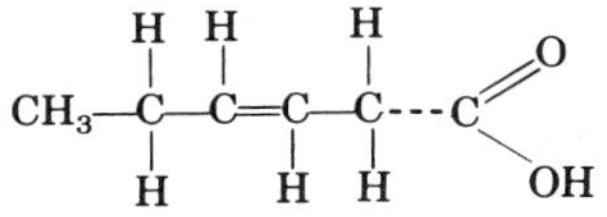

(*trans*)

These have links termed 'double bonds' and different fatty acids may have one, two, three, or occasionally four double bonds. The presence of these bonds will greatly increase chemical reactivity and susceptibility to rancidity of the fat formed from them. Also, the melting point is generally lower. The chemistry of hydrogenation is related to the presence of double bonds (see later).

In the formation of a glyceride, the glycerol molecule becomes esterified with the fatty acid, losing three molecules of water in the process. Alternatively, a glyceride may be hydrolyzed into glycerol and the free fatty acid.

The following chemical formulas indicate the reversible reaction:

$$
\text{Esterification} \rightarrow
$$

$$
\begin{array}{l}
\text{CH}_2\text{—OH} \quad \text{R}_1\cdot\text{C}\overset{\displaystyle O}{\diagup}\text{OH} \\[4pt]
\text{CH—OH} + \text{R}_2\cdot\text{C}\overset{\displaystyle O}{\diagup}\text{OH} \\[4pt]
\text{CH}_2\text{—OH} \quad \text{R}_3\cdot\text{C}\overset{\displaystyle O}{\diagup}\text{OH}
\end{array}
\qquad
\begin{array}{l}
\text{CH}_2\text{—O·C}\overset{\displaystyle O}{\diagup}\text{R}_1 \\[4pt]
\text{CH—O·C}\overset{\displaystyle O}{\diagup}\text{R}_2 + 3\text{H}_2\text{O} \\[4pt]
\text{CH}_2\text{—O·C}\overset{\displaystyle O}{\diagup}\text{R}_3
\end{array}
$$

$$
\leftarrow \text{Hydrolysis}
$$

$R_1$, $R_2$, $R_3$ represent fatty acids and it can be seen that with different acids and positional interchange, many variations of glyceride are possible.

In plant life, glycerides are built up by the complicated reactions of photosynthesis. Thus, fats are formed in the seeds and fruit as a means of providing food for the new plant in its early stages of development.

Here, hydrolysis occurs usually by enzyme action (lipase). In the production of fats from natural sources, premature action by enzymes must be avoided to prevent the fat from containing free fatty acids. The presence of free fatty acid can have a profound effect on the flavor of a fat as is exemplified by lauric ($C_{12}$) fat.

## Phospholipids, Sterols

These substances occur in fats, usually less than 0.5 percent. The best known phospholipid is lecithin.

Sterols are fat-soluble ring compounds and cholesterol is well known, which is mostly present in animal fats. The fat-soluble vitamins are closely related to the sterols.

## PRODUCTION AND PROCESSING OF FATS

Fats are extracted from the natural seeds or fruits by combined processes of expeller pressing and solvent extraction. The raw fat so obtained is unsuitable for use in food and must be refined.

Refining is carried out in three stages:

1. Neutralization. The fat is washed with alkali solution, which removes residual fatty acids in the form of a soap.
2. Bleaching. The hot liquid fat is mixed with an absorbent substance (fuller's earth) and decolorizing carbon, followed by filtration. This removes odorous substances, color, and slime.
3. Deodorizing. This process is described under "Cocoa Butter" and removes the last traces of undesirable volatiles. The refined fat so produced is practically unchanged physically and for most confectionery uses must be hardened.

### Hardening Fats

Two basic processes are used—one physical and the other chemical.

**Physical Glyceride Selection**  Removal of some of the lower melting components in a fat usually results in a simpler mixture of glycerides and widens the scope of available fats.

The technique originally applied to coconut and palm kernel oils was to allow the liquid fat partially to solidify under controlled conditions so that it contained a mixture of solid fat crystals and liquid fat. It was then submitted to hydraulic pressing, thus separating the fat into stearines (higher melting points) and oleines (liquid). This method of crystallizing and pressing is not sufficiently selective for many requirements and new techniques of separation are now used. These are based on fractional crystallization from suitable solvents such as acetone, which enables a much greater control to be exercised over separation of required glycerides. by this technique it is possible to isolate fractions in which single glycerides predominate.

Stearines from coconut and palm kernel oils are hard brittle fats with melting points similar to that of cocoa butter, and find a ready use as fats for confectionery coatings. Their properties arise from a small number of similar glycerides present, in which myristodilaurin predominates. These stearines, although used considerably as cocoa butter substitutes, unfortunately give rise to eutectic effects when some cocoa butter is present, and the resultant mixture of fat components of a coating may be too soft. To produce a satisfactory cocoa butter alternative fat, free from eutectic effects when mixed with cocoa butter, a more specific selection of glycerides similar in chemical constitution to those present in cocoa butter itself is required (see "Equivalent Fats").

Coconut and palm kernel stearines are also subject to soapy rancidity in circumstances where breakdown of the fat occurs, liberating the free lauric acid (see Lipase Activity).

**Chemical Hardening (Hydrogenation)** Generally speaking, the saturated fats have higher melting points and are harder than unsaturated fats. The glycerides of the various fatty acids may be saturated or unsaturated.

The unsaturated acids, which have lower melting points, may be converted to saturated acids by combining with hydrogen. Therefore, the degree of hardening that may be effected is governed by the amount and type of unsaturated acids present in an oil.

For most edible-oil requirements, partially hardened oils are used and the relationships among texture, plasticity, and melting point are of great importance. In order to understand how these properties are manipulated, some knowledge of the chemistry of the reaction is necessary.

**The Chemistry of Hydrogenation** Fatty acids have long chains of carbon atoms (each having four bonds) to which are attached hydrogen atoms.

An unsaturated acid chain has the following structure:

$$\underset{\text{H } cis \text{ (together)}}{\overset{\text{H H H H}}{-C-C=C-C-}} \quad or \quad \underset{\text{H } trans \text{ (opposite)}}{\overset{\text{H H--}}{-C-C=C-C-}}$$

These double bonds between carbon atoms combine with hydrogen to form saturated acids:

$$\underset{\text{H H H H}}{\overset{\text{H H H H}}{-C-C-C-C-}} \quad \text{and} \quad \underset{\text{H H H H}}{\overset{\text{H H H H}}{-C-C-C-C-}}$$

which are identical, whether from "cis" or "trans" unsaturated acids.

A diunsaturated fatty acid contains two such double bonds and a triunsaturated acid three.

When hydrogenation commences, the triunsaturated acids are converted to diunsaturated acids, and then monounsaturated acids, and finally to saturated acids. At the same time, the existing diunsaturated acids are converted to monounsaturated and then saturated, whereas the existing monounsaturated are converted to saturated acids.

If all these processes occur together, the process is nonselective, whereas if the triunsaturated acids are hydrogenated preferentially, followed by the diunsaturated, etc., the process is selective.

Depending upon the type of reaction taking place in the intermediate stages, a semihardened oil will have different characteristics, but a completely hardened oil will have all acids saturated and will finish up with the same characteristics.

In addition, the *"cis"* and *"trans"* acids may change formation from one to the other, and the double bonds in unsaturated acids may change their relative position in the chain, forming "iso" unsaturated acids.

These changes also affect the physical characteristics of any semi-hardened oil, but to a lesser degree than selective and nonselective hardening.

Vegetable oils are hydrogenated by bringing them into intimate contact with hydrogen gas at suitable temperature and pressure.

The oil is heated to 120°–180°C (248° to 356°F) in a closed vessel containing hydrogen under pressure, and contact is brought about by vigorously stirring the oil or by passing the hydrogen through the oil in the form of fine bubbles. The reaction takes place in the presence of a catalyst, usually nickel, which is deposited on *kieselguhr* in a very finely divided state.

The method of preparation and deposition of the catalyst is very important because it influences both the activity of the catalyst and the selectivity of the reactions.

The science of hardening an oil to obtain the desired relationship between melting point and texture for any given degree of hardening depends upon precise knowledge of the temperature and time of hydrogenation and the activity of the catalyst.

## COMMERCIAL EDIBLE OILS

Table 9.1 gives a summary of the structure of the common edible oils. A brief description of those important in the confectionery industry follows.

## Coconut oil

The coconut is a well-known product of the palm of the same name, which grows on the islands and in the coastal areas of tropical countries. The nut has a hard outer husk and inner kernel and it is this kernel that is dried to give the article of commerce known as copra from which the crude coconut oil is expressed.

**Characteristics** The average oil content of copra is 66 percent, and, dependent upon its qualities, yields a white to brownish-yellow fat. This fat is somewhat crystalline in appearance and possesses a marked degree of brittleness at low temperatures, but as its melting point is 25°C (77°F), it becomes soft and almost liquid at normal summer temperatures. Before refining the fat has a flavor that can vary from that of fresh coconut to a disagreeably harsh taste, but after refining the oil is odorless and tasteless, of good white appearance, and stable.

**Chemical Nature** Coconut oil is distinguished chemically from many other fats by the presence of large quantities of glycerides of lower saturated fatty acids, for example, lauric and myristic acids, as well as small quantities of the short-chain volatile acids—caproic, caprylic, and capric. Unsaturated fatty acids are present in small amounts only, thus giving excellent stability against the development of oxidative rancidity.

The rather unusual fatty acid composition of coconut oil accounts for its high saponification value, high Reichert and Polenske values, and low iodine value.

These fatty acids are combined in coconut oil to give different triglycerides with the lauric and myristic acid triglycerides predominating. These particular triglycerides are present in large quantity and account for the fact that coconut oil does not soften gradually with increasing temperatures, but passes rapidly from a hard solid to a liquid state within a range of a few degrees.

The small content of glycerides of unsaturated fatty acids is sufficient for hydrogenation to be applied and the sharp melting coconut oil may be hardened to a product of longer plastic range and higher melting point.

## Palm Oil, Palm Kernel Oil

The oil palm is a native of tropical West Africa but is now also cultivated in Malaya and Indonesia. The fruit of the palm is unusual

TABLE 9.1. TYPICAL COMPOSITIONS AND CHEMICAL CONSTANTS OF COMMON EDIBLE FATS AND OILS (Fatty acid compositions were determined by gas–liquid chromatography and are expressed as mean average weight percent compositions on a fatty acid basis. Trace acids (less than 0.1%) are excluded.)

| Fatty acid | Carbon atoms* | Butterfat | Cocoa butter | Coconut oil | Corn oil | Cottonseed oil | Lard | Olive oil | Palm oil | Palm kernel oil | Peanut oil | Rapeseed oil | Rapeseed oil (low erucic) | Sesame seed oil | Soybean oil | Sunflower seed oil | Tallow (beef) | Tallow (mutton) |
|---|---|---|---|---|---|---|---|---|---|---|---|---|---|---|---|---|---|---|
| Butyric | 4:0 | 3.8 | | | | | | | | | | | | | | | | |
| Caproic | 6:0 | 2.3 | | 0.5 | | | | | | 0.3 | | | | | | | | |
| Caprylic | 8:0 | 1.1 | | 8.0 | | | | | | 3.9 | | | | | | | | |
| Capric | 10:0 | 2.0 | | 6.4 | | | 0.1 | | | 4.0 | | | | | | | 0.1 | 0.2 |
| Undecanoic | 11:0 | 0.1 | | | | | | | | | | | | | | | | |
| Lauric | 12:0 | 3.1 | | 48.5 | | | 0.1 | | 0.3 | 49.6 | | | | | | 0.5 | 0.1 | 0.3 |
| Tridecanoic | 13:0 | 0.1 | | | | | | | | | | | | | | | | |
| Myristic | 14:0 | 11.7 | 0.1 | 17.6 | | 0.9 | 1.5 | | 1.1 | 16.0 | | | | | 0.1 | | 3.3 | 5.2 |
| Myristoleic | 14:1 | 0.8 | | | | | | | | | 0.1 | 0.1 | | | | 0.2 | 0.2 | 0.3 |
| Pentadecanoic | 15:0 | 1.6 | | | | | 0.2 | | | | | | | | | | 1.3 | 0.8 |
| Pentadecenoic | 15:1 | | | | | | | | | | | | | | | | 0.2 | 0.3 |
| Palmitic | 16:0 | 26.2 | 25.8 | 8.4 | 12.2 | 24.7 | 24.8 | 13.7 | 45.1 | 8.0 | 11.6 | 2.8 | 3.9 | 9.9 | 11.0 | 6.8 | 25.5 | 23.6 |

278

| | | | | | | | | | | | | | | | | | |
|---|---|---|---|---|---|---|---|---|---|---|---|---|---|---|---|---|---|
| Palmitoleic 16:1 | 1.9 | 0.3 | | 0.1 | 0.7 | 3.1 | 1.2 | 0.1 | | 0.2 | 0.2 | 0.2 | 0.3 | 0.1 | 0.1 | 3.4 | 2.5 |
| Margaric 17:0 | 0.7 | | | | 0.1 | 0.5 | | | | 0.1 | | | | | | 1.5 | 2.0 |
| Margaroleic 17:1 | 0.2 | | | | | 0.3 | | | | | | | | | | 0.7 | 0.5 |
| Stearic 18:0 | 12.5 | 34.5 | 2.5 | 2.2 | 2.3 | 12.3 | 2.5 | 4.7 | 2.4 | 3.1 | 1.3 | 1.9 | 5.2 | 4.0 | 4.7 | 21.6 | 24.5 |
| Oleic 18:1 | 28.2 | 35.3 | 6.5 | 27.5 | 17.6 | 45.1 | 71.1 | 38.8 | 13.7 | 46.5 | 23.8 | 64.1 | 41.2 | 23.4 | 18.6 | 38.7 | 33.3 |
| Linoleic 18:2 | 2.9 | 2.9 | 1.5 | 57.0 | 53.3 | 9.9 | 10.0 | 9.4 | 2.0 | 31.4 | 14.6 | 18.7 | 43.2 | 53.2 | 68.2 | 2.2 | 4.0 |
| Linolenic 18:3 | 0.5 | | | 0.9 | 0.3 | 0.1 | 0.6 | 0.3 | | | 7.3 | 9.2 | 0.2 | 7.8 | 0.5 | 0.6 | 1.3 |
| Nonadecanoic 19:0 | | | | | | | | | | | | | | | | 0.1 | 0.8 |
| Arachidic 20:0 | | 1.1 | 0.1 | 0.1 | 0.1 | 0.2 | 0.9 | 0.2 | 0.1 | 1.5 | 0.7 | 0.6 | | | | 0.1 | |
| Gadoleic 20:1 | 0.2 | | | | | 1.3 | | | | 1.4 | 12.1 | 1.0 | | 0.3 | 0.4 | 0.1 | |
| Eicosadienoic 20:2 | | | | | | 0.1 | | | | 0.1 | 0.6 | | | | | | |
| Arachidonic 20:4 | 0.1 | | | | | 0.4 | | | | | | | | | | | |
| Behenic 22:0 | | | | | | | | | | 3.0 | 0.4 | 0.2 | | 0.1 | | 0.4 | 0.4 |
| Erucic 22:1 | | | | | | | | | | | 34.8 | | | | | | |
| Docosadienoic 22:2 | | | | | | | | | | | 0.3 | | | | | | |
| Lignoceric 24:0 | | | | | | | | | | 1.0 | 1.0 | | | | | | |
| Iodine value range | 25–42 | 32–40 | 7–13 | 110–128 | 99–121 | 53–68 | 76–90 | 45–56 | 14–24 | 84–102 | 97–110 | 110–115 | 104–118 | 125–138 | 122–139 | 33–50 | 35–48 |
| Saponification value range | 210–240 | 190–200 | 248–264 | 186–196 | 189–199 | 192–203 | 188–196 | 195–205 | 243–255 | 188–196 | 168–183 | — | 187–196 | 188–195 | 186–196 | 190–202 | 192–198 |

* Number of double bonds shown by second figure, e.g., 18:3.
Reproduced by permission of Durkee Foods, Cleveland, Ohio.

in that it yields oil from the fruit and the kernel—the fibrous layer of pulp on the outside gives palm oil and the hard kernel supplies palm kernel oil and the yields of crude oil are approximately 56 percent and 50 percent respectively.

**Palm Oil** Refined palm oil is a pale yellow fat with good keeping properties. It is soft in consistency and melts completely at about 40°C (104°F).

*Glyceride Composition* The glycerides of palm oil consist of approximately 10 percent trisaturated glycerides, 50 percent monounsaturated glycerides, 30 percent diunsaturated glycerides, and 10 percent triunsaturated glycerides. The trisaturated glycerides consist mainly of tripalmitin, which is not a common feature of natural fats where mixed glycerides are the rule rather than the exception.

This glyceride composition, ranging as it does from triunsaturated glycerides through di- and monounsaturated glycerides to fully saturated ones, gives refined palm oil its soft texture and reasonably long plastic range and makes it satisfactory for so many purposes.

*Hydrogenation* Owing to its relatively high iodine value, palm oil may be hydrogenated to any desired melting point, usually to between 40 to 42°C (104 to 107.6°F) and 46–48°C (114.8 to 118.4°F). Complete hydrogenation gives a melting point of about 58°C (136°F). Hydrogenated palm oil is a useful ingredient of fat blends incorporating any desired degree of stiffness; the consistency of palm oil may be increased by incorporation of the hydrogenated oil.

**Palm Kernel Oil** In many respects, refined palm kernel oil bears a close resemblance to coconut oil. It is a white solid fat, somewhat less brittle than coconut oil but with the slightly higher melting point of 28 to 29°C (82.4–84.2°F). Its iodine value is higher than that of coconut oil, owing to a higher unsaturated fatty acid content.

The oil is composed mainly of glycerides of lauric and myristic acids. Short-chain volatile acids are present, but in a lesser quantity than in coconut oil.

This fatty acid composition accounts for the high saponification value and for the high Reichert and Polenske values, which are, however, lower than those for coconut oil. The unsaturated fatty acid content, although higher than that of coconut oil, is still quite low, and thus does not favour the development of oxidative rancidity.

*Glyceride Separation* It is possible to separate suitable glycerides from palm kernel and coconut oil, and these form the basis of the well-known palm kernel stearines, which have physical properties resembling the more expensive cocoa butter, being hard brittle fats with melting points significantly below body temperature.

The higher content of unsaturated fatty acids renders palm kernel oil a very suitable fat for hydrogenation, and a useful range of hardened palm kernel oil products may be obtained according to the degree of hydrogenation.

## Peanut Oil, Arachis Oil

The nut from which the oil is obtained is now of great commercial importance and is grown widely in tropical and subtropical countries.

Peanuts come from a small annual plant that reaches maturity in about four months. The yellow flowers change into pods that bury themselves in the ground, and there they enlarge and ripen. They then have rough ribbed shells around the kernels, the latter containing about 45 percent of oil.

**Physical and Chemical Nature of the Oil**  Peanut oil is used for almost every edible purpose.

The color of the unrefined oil varies from a light brown to a water white, and has a distinct nutty flavor. It is liquid at normal temperatures, but at lower temperatures deposits a crystalline stearine. As can be seen from the fatty acid composition, its degree of unsaturation makes it a very suitable fat for hydrogenation.

**Fatty Acid Composition**  See Table 9.1. It has been suggested that differences in climatic conditions might account for the variations in the linoleic/oleic acid ratios found in nuts received from different sources. One look at Table 9.2 might suggest that the cooler temperatures give rise to a greater linoleic content but a lower oleic content, and vice versa.

TABLE 9.2. PEANUT OIL—CLIMATIC CONDITIONS AND FATTY ACID COMPOSITION

|  | Saturated | Oleic | Linoleic |
|---|---|---|---|
| Spain | 22 | 53 | 25 |
| Philippines | 18 | 55 | 27 |
| West Africa | 18 | 65 | 17 |
| Senegal | 15 | 66 | 19 |

The fatty-acids composition of peanuts of different origin could well be related to the great differences in keeping properties of roasted nuts. This has been the subject of research and is discussed elsewhere.

### Soya Oil

The soya bean had its origin in eastern Asia but enormous expansion of its cultivation has occurred in the United States during this century. As a result, soybean oil has become one of the world's leading sources of vegetable oil in spite of the fact that the bean contains about 20 percent. Along with this, however, it is very rich in protein (40 to 50 percent), which makes the extracted meal a valuable animal fodder.

As it is a cheap source of protein a great deal of research has been done to make it suitable for human consumption. Soya flour has a peculiar earthy flavor that must be removed if it is to be fully acceptable as a major constituent of human food.

The soya bean plant has clusters of small purple flowers that develop into two to five having seed pods. The beans are oval and range in color through yellow, green, and black. The plant grows best in warm, damp climates but adapts itself to a variety of conditions, provided the soil is rich and well drained.

### Cottonseed Oil

It was long after the plant was cultivated for cotton that the seed became an important source of vegetable oil.

The mature plant produces fluffy white seeds called cotton bolls with the fiber adhering. The seeds are oval, about $\frac{3}{16} \times \frac{3}{8}$ in., and yield from 15 to 25 percent of oil.

It is a plant of tropical or subtropical regions.

### Sunflower Seed Oil

The plant is very tall (5 to 8 ft), although there are dwarf varieties. The flower has a dark-brown center with yellow petals. It is native to Central America but is now cultivated in many parts of the world.

The U.S.S.R. is the main producer. Originally the seed contained 20 to 30 percent of oil but the cultivation of new strains has led to an increase to 40 percent.

## Sesame Seed Oil

Sesame seed originated in China and India. Today it is also grown extensively in Africa and Mexico. It is a crop that grows in poor soil and is easily cultivated. The seed contains about 50 percent oil, which has uses similar to those of olive oil.

## Rapeseed Oil

Rapeseed can be grown in colder climates and in recent years countries like Sweden, Denmark, Poland, and Canada have increased production considerably, thereby reducing consumption of imported tropical oils. The plant is of the Brassica (cabbage) family and fields when in flower are brilliant yellow. Rapeseed contains 35 to 40 percent oil. Rape oil from original seed contained a high proportion of erucic acid that has been shown to be dietetically undesirable. New genetic varieties of seed have been developed giving oil of low erucic acid content.

## Olive Oil

The olive tree has been a source of edible oil for many centuries and olive oil, although perhaps now not so important commercially, is the highest-quality vegetable oil and greatly prized for table use. The tree grows in the countries around the Mediterranean, with Spain and Italy the main producers. The fruit contains about 15 percent of oil.

## Corn Oil

This has been an important edible oil of recent years, produced as a by-product of the vast starch, glucose syup, and dextrose industry. Corn oil is pale yellow in color, liquid at normal temperatures but deposits a small amount of stearine at lower temperatures. The oil is almost entirely in the germ, which is separated in the early stages of wet milling and starch extraction.

Corn is a major crop in the United States and as a result of intensive scientific development there is an ever-increasing supply of derived products—not only in the food industry but also such materials as adhesives and paper.

## CHEMICAL AND PHYSICAL PROPERTIES OF OILS AND FATS

In the examination of oils and fats, specific chemical and physical tests are used. The significance of the figures obtained by the analyst is of importance to the technologist and the meaning of the more important tests is summarized in the following descriptions.

### Saponification Value

This is the number of milligrams of potassium hydroxide (potash) required to saponify 1 g of oil or fat.

The results of this test are indicative of the nature of the combined fatty acids present, for example, values above 200 indicate the presence of fatty acids of low or fairly low molecular weight, and values below 190 indicate the presence of high-molecular-weight fatty acids. Coconut and palm kernel oils consist of glycerides of fatty acids with low molecular weights and hence have saponification values between 240 and 265. On the other hand, rapeseed oil containing large amounts of a high-molecular-weight fatty acid (erucic acid) has an average value of 175.

### Acid Value

This is the number of milligrams of potassium hydroxide required to neutralize the free fatty acid of 1 g of oil or fat.

Normally, the free acidity is expressed as a percentage of the major fatty acid present, for example, coconut and palm kernel oils are expressed in terms of lauric acid, molecular weight 200; palm oil as palmitic acid, molecular weight 256; and liquid oils (peanut, cottonseed, etc.) as oleic acid, molecular weight 282. It is interesting to note that there is evidence to show that the mean molecular weight of the free acids in an oil or fat may differ somewhat from the mean molecular weight of the combined fatty acids. The free fatty acids if calculated as oleic acid (mol. wt. 282), as it often is, is *half* the acid value.

Acid values, however expressed, are of value to the refiner of crude oils as indication of the amount of free acid to be removed; applied to neutral oils, they show how well this has been carried out. To users of refined products, low acid values indicate purity, but better still, if applied to stored products at regular intervals they will indicate deterioration if regular increases occur.

## Unsaponifiable Matter

This refers to material present in an oil or fat, that remains nonvolatile on drying to constant weight at 80°C (176°F), after the saponification of the oil or fat by alcoholic caustic potash and extraction by a specific solvent.

Unsaponifiable matter as defined above includes, among other things, hydrocarbons, higher alcohols, and the sterols, cholesterol and phytosterol. The method of determination has been designed to exclude free fatty acids, soaps, and mineral matter; readily volatile substances will be removed during the period of drying.

The majority of oils and fats have unsaponifiable contents below 2 percent; many have values below 1 percent. A few, however, have much higher values, up to 10 percent, and in these cases, the normal method has been modified to avoid the formation of troublesome emulsions; shea nut oil is a typical member of this group. With substances like wool fat (lanolin), which contains a high proportion of wax esters, saponification is sometimes difficult to effect in one operation. In such instances, the unsaponifiable matter is obtained in the normal way and then subjected to a further saponification with caustic alkali; reextraction of the saponified product should then yield the unsaponifiable matter free from contamination.

As stated above, the majority of oils and fats have low contents of unsaponifiable matter that consists mainly of sterols. It has been found that cholesterol is the sterol present in the unsaponifiable matter of animal oils and fats, whereas vegetable oils and fats contain phytosterol. If the acetates of these two sterols are prepared and melting points determined, it is found that cholesterol acetate has a much lower melting point than phytosterol acetate and this fact has been used to distinguish between animal and vegetable products.

## Iodine Value

This is a measure of the degree of unsaturation in oils and fats and denotes the percentage by weight of halogen, calculated as iodine, absorbed under standard conditions.

Where the proportion of saturated acids is high, as in coconut oil and similar fats, iodine values are low, but with liquid oils the values are high and range from 80 to 200. Oils of the highest value, such as linseed oil, will absorb oxygen from the air and find use in surface coverings, but oils used for edible purposes have lower iodine values; these run from 80 to 130, of which peanut oil (85 to 95) and cottenseed oil (105 to 115) are the most common.

Iodine values are useful in deciding purity, but their main function is in plant control of hydrogenation. As partial hydrogenation is usually carried out, determination of iodine value is very important, as the decrease in this value will indicate the degree of saturation attained.

## Volatile Fatty Acids

The Reichert value is a measure of the water-soluble steam volatile fatty acids present in an oil or fat. The Polenske value is a measure of the water-insoluble steam volatile fatty acids present.

The Kirschner value is a measure of the water-soluble steam volatile fatty acids present that form water-soluble silver salts.

These processes do not determine the total quantities of steam volatile fatty acids present and hence are merely empirical values, but by strict adherence to the dimensions of the apparatus used and details of procedure, they afford useful information as to the presence or absence of certain fats in a mixture.

All the values are concerned with the presence of short-chain fatty acids in a fat and are applicable to butter fat and coconut and palm kernel oils. Cow butter and other milk fats are unique in possessing glycerides containing butyric acid $CH_3 \cdot CH_2 \cdot CH_2 \cdot COOH$, and as this acid is water soluble, high Reichert values are obtained. Insoluble volatile acids are practically absent from these fats and hence Polenske values are very low.

With coconut and palm kernel oils, however, the picture is somewhat different. These oils contain both soluble and insoluble acids and therefore significant Reichert and Polenske values are obtained; they do not, however, contain butyric acid.

The Kirschner value is practically specific for butyric acid and is of value in indicating whether other fats containing volatile acids (e.g., coconut, palm kernel) have been added to butter and other milk fats.

It must be emphasized that the tests are empirical and the results, unlike the saponification and iodine values, are not additive, but a fair approximation of the presence of coconut and palm kernel oils in a mixture in which butter fat is absent may be obtained by considering the sum of the Reichert and Polenske values.

## Peroxide Value

The peroxide value is a measure of the peroxides contained in the oil or fat expressed as milliequivalents of peroxide oxygen per kilogram of sample.

When oils and fats are subject to oxygen absorption there is a small but steady increase in the oxygen uptake as measured by peroxide value until a point is reached when the rate of oxygen uptake is materially increased. The time to reach this point is referred to as the induction period and is considered to indicate the stability of the product under test. Determinations of induction periods are carried out under conditions of accelerated oxygen absorption (i.e., at high temperatures) and there is considerable argument as to whether they are truly indicative of shelf life. The author has shown that the test can be of great value in the assessment of the stability of certain fats and fatty foods but it is desirable to run experimental parallel shelf-life tests in the first instance on the products being investigated.

## PHYSICAL TESTS

### Specific Gravity and Apparent Density

The specific gravity at $t$ [15.5°C (59.9°F)] in air of an oil is the ratio of the weight in air of a given volume of the oil at $t$°C to that of the same volume of water at 15.5°C (59.9°F). The apparent density (grams per milliliter) in air of an oil at $t$°C is the weight in air, expressed in grams, of 1 ml of the oil at $t$°C.

These definitions have been in use in the fatty oil industry for many years; they are specific for the industry. In other industries, these properties may be defined in a different way. These are useful as much fatty oil is delivered in bulk form these days.

It is interesting to note that hydrogenated oils have lower specific gravities than the unhydrogenated oils from which they were produced. It is found that the reduction in specific gravity is approximately proportional to the degree of saturation applied.

Specific gravity is used in conjunction with other figures in assessing purity of an oil.

### Melting or Fusion Point, Slip Point, Complete Fusion

A great deal of argument has arisen in the fat industry and between fat suppliers and users concerning the method to be employed for the determination of the melting and softening points of fats.

The capillary method used in many laboratories is given in the Appendix and defines the test as follows:

Fusion point—the temperature at which softening is first seen.

Slip point—the temperature at which the fat starts to rise in the tube.

Clear point (complete fusion)—the temperature at which the fat clarifies.

By negotiation between the chemists of suppliers and users it is usually possible to agree on the precise details of the tests so that any variability of fat deliveries can be detected. It is best to agree to use recognized standard methods such as British Standards, International Union for Pure and Applied Chemistry (IUPAC), or Wiley.

## Softening point

The Barnicoat point is a more precise and reliable test than the slip point and is advocated by the fat manufacturers. See Appendix.

## Flow and Drop Point

The method was devised by Ubbelodhe.

A special apparatus is required, one determination only may be carried out at a time and the time taken is rather long. The method briefly is as follows: A sample of the fat, in fine shavings, is placed in a small cup with a standard-size orifice. This cup fits round the bulb of a special thermometer arranged so as to be in contact with the sample and heat is applied. The temperature at which the sample is observed to move through the orifice is recorded as the flow point, and when the first drop of melted fat falls from the cup this is recorded as the drop point.

## Dilatation Test (Solids/Fat Index)

The dilatation of fat may be defined as the isothermal expansion from solid to liquid, of a fat that due to change of state has previously been solidified under carefully prescribed conditions.

The majority of fat products consist of intimate mixtures of solid and liquid glycerides and it is possible to estimate the percentage of solid fat present at any given temperature by means of dilatometry.

It has been observed experimentally that 100 g of completely solidified fat expand in volume approximately 10 ml (10,000 microliters) on melting. The dilatation is usually expressed in microliters per 25 g of fat and represents the difference between the volume of solid fat and the volume of the liquid at the same temperature. A dilatation expressed in microliters expansion per 1 g is often referred to as the percentage of crystal fat present at a given temperature. It is useful to plot dilatation values against temperature readings; the shape of the resultant curve yields useful information about the product and also shows precisely the point of complete fusion. Dilatometry is more objective in application than melting point methods and can determine the ratio of solid and liquid phases at any temperature below that of complete fusion.

Dilatation results give useful information on the texture of fats at temperatures between ambient and complete melting but penetrometer values should also be determined (see "Hardness").

In companies where large quantities of fat are being produced or used, the solid fat content is measured by nuclear magnetic resonance. It gives a more accurate picture of the physical nature of a fat than dilatation.

Preparation of the sample must be standardized and is described in IUPAC method 2.141 "Determination of the Dilatation of Fats." Solids content are expressed as $N_t$ values and the properties of a fat may be indicated graphically (Figs 9.1 and 9.2).

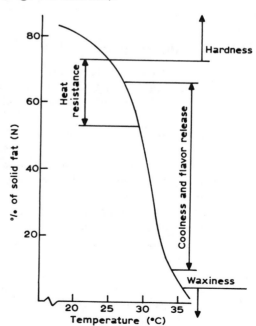

Fig. 9.1. Interpretation of an $N_t$ Curve

*From "The Characteristics of Chocolate," Loders Croklaan b.v., Wormerveer, Holland*

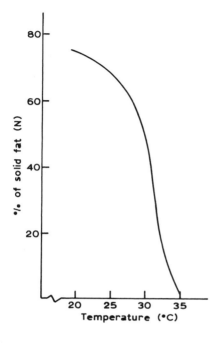

Fig. 9.2.  The $N_t$ Curve of Cocoa Butter

*From "The Characteristics of Chocolate," Loders Croklaan b.v., Wormerveer, Holland*

## Cooling Curve — Setting Point

The cooling curve determination is of most value in assessing the supercooling quality of fats of the cocoa butter type where tempering or seeding of that fat or coating containing the fat is required, as in enrobing.

With fats that exhibit supercooling, a point is reached when the latent heat of solidification overcomes the cooling effect, and the temperature begins to rise, finally reaching a maximum point. This is called the setting point. If temperature readings throughout the test are plotted each minute, the shape of the resultant curve can yield useful information regarding the likely behavior of the fat in use.

Cooling curves are discussed under "Cocoa Butter". It is essential that standard procedures be observed to obtain reproducible results.

## Hardness

When a fat or mixture of fats is used in a coating or filling, the texture of the products at different temperatures is often a very important property. The hardness is not necessarily indicated by the results obtained from dilatation and melting point and a penetrometer test is a useful additional value to obtain.

The method uses the standard penetrometer as employed for the testing of bitumen (IP 49 and ASTM D5). Either the standard penetrometer needle or cone may be used, depending on the consistency of the fat. Tests are normally made at temperatures ranging between 15.6°C and 29.4°C (60°F and 85°F), and by plotting penetrometer values against temperature a very good picture of texture change with temperature is obtained.

The penetrometer is of particular value as a routine check on deliveries of fat and on samples of any standard product where, for reasons of economy or supply, it has become necessary to substitute, wholly or partly, one fat by another. The test is empirical and the procedure depends on how the fat is to be used, but it is important to prepare the blocks of fat for testing by a standard procedure of tempering (seeding) and cooling. (See Appendix.)

### Refractive Index

This is a measure of the extent to which an oil or liquid fat will bend a beam of light as determined by refractometer. The refractive index is allied to the iodine value in that it also indicates the degree of unsaturation, oils of high iodine value having a high refractive index. Owing to the rapidity with which a test may be carried out, it is a useful guide during hydrogenation, but generally it is less informative than iodine value.

### Fatty Acid Composition of Glycerides

The traditional method of separating and determining the component fatty acids of the various glycerides was extremely complex and time consuming.

Since the development of chromatography in its various forms (including thin-layer chromatography and HPLC), the problem of analysis has been greatly simplified. It is speedy, more accurate, and uses less material.

## PACKING AND STORAGE OF FATS

Fats are usually received into a factory in three ways—in solid blocks in fiberboard containers, in solid form in drums, or in liquid form in tank trucks. Fat is an expensive ingredient and needs care in storage and melting. In the solid form, it will keep for three to six months in cool conditions (i.e., 15°C/60°F or less) with relative humidity of the order of 55 to 65 percent.

## Fiberboard Cartons

These must be stored in dry conditions—damp packing will cause the surface fat on the blocks to deteriorate. When unpacking and transferring the fat for melting, great care must be taken to avoid packing material, particularly polythene, getting into the liquid fat.

## Drums

Fat in drums presents a considerable problem when it comes to emptying and melting. To dig the fat out manually is expensive in labor and may result in foreign matter contamination from the drum interior.

Melting can be achieved by means of electrically heated jackets or steam coils, but local overheating must be avoided at all costs and temperatures up to 60°C (140°F) for short periods only are permitted. If melting is carried out slowly, as in a hot room. 52°C (125°F) should not be exceeded.

Solid fat (from cartons) is often subjected to gross mishandling during melting, particularly if large pieces are involved.

Fat so melted in a steam-jacked pan can reach temperatures of 100°C (212°F) and over very quickly, and sometimes the fat is left at this temperature for a considerable period awaiting use in a production department.

The fat is best melted at a maximum of 60°C (140°F), but preferably 52°C (125°F) if time of melting is extended. Stainless-steel pans should be used, *never* copper. Stainless-steel metal pipe grids heated internally with low-pressure steam or hot water are best and blocks of fat can be rested on these.

Fat that is overheated may not show immediate deterioration due to a delayed chemical process known as the induction period. Such fats, which may have been included in a product, will develop rancidity and off-flavors in due course.

Fat that has been chipped for production use is often transported through the factory in trolleys. Exposure of these to bright light near windows will promote rancidity. Fat residues in the trolleys must be removed regularly.

## Bulk Storage

Refined oils should not normally be stored in bulk for more than ten days or hydrogenated oils, fourteen days.

The tank site should be chosen to minimize the length of pipeline, especially from the tank to the point of use. Tanks should be housed away from dust that may carry yeasts or enzymes. Either cylindrical or rectangular tanks are suitable and, although a steel tank is adequate for most vegetable oils and fats, stainless-steel or glass-lined tanks are preferred. Tanks should be covered and have dished or similar bottoms to facilitate complete drainage, the take-off pipes being at the lowest point. The tanks should be properly lagged and heated so as to avoid local overheating. The use of hot water jackets or coils is satisfactory but where a number of storage tanks is used, it is often the practice to have them sited in a thermostatically controlled hot room, thereby avoiding the need for jacketing and lagging. Air filters to remove dust and associated microorganisms should be fitted so that air entering the tanks as fat is withdrawn does not cause contamination.

The air may be drawn over ultraviolet sterilizers and nitrogen can be drawn in in place of air.

It is desirable to maintain the temperature of the oil in the storage tanks as low as possible, commensurate with keeping the oil completely liquid in the tanks and associated pipelines. The delivery pipe should be heated and lagged. A maximum of 50°C (122°F) is desirable.

## Cleaning of Tanks

An accumulation of oxidized oil, pipe scale, or other foreign materials in tanks or pipelines may impair the keeping properties of fresh deliveries. Periodic examination of tanks by technical staff should be maintained and, when cleaning is found to be necessary, the inside surfaces should be completely freed from oil with steam and a suitable detergent. It is most important that the tanks and pipes be thoroughly flushed clean of all traces of detergent, and properly dried before refilling.

All pipelines should be installed with sufficient fall to allow easy and complete drainage. Since copper is a strong pro-oxidant, and contact with it will reduce the stability of an oil, the use of bronze fittings, such as valves and pump impellers must be avoided, steel being preferred.

Fresh oil should be discharged into an empty tank and not mixed with old supplies.

## REFERENCES

Egan, H., Kirk, R. S., and Sawyer, R. 1981. *Pearsons Chemical Analysis of Foods.* Churchill Livingstone, Edinburgh, Scotland.

Hudson, B., Gurr, M., Kirtland, J., Patterson, H. Thomas, A., and Paulicka, F. 1976. Recent advances in chemistry and technology of oils and fats. *Chem. Ind.,* London.

International Society of Fat Research—World Congress. 1978. Review. *Chem. Ind.,* London.

Landon, J. W. 1975. Palm oil. *Chem. Ind.* London.

Padley, F. B. 1984. New developments in oils and fats. *Chem. Ind.* London.

*Vegetable Oils and Fats, The Chemistry of Glycerides.* Unilever Educational Series, Unilever House, Blackfriars, London, England.

Weiss, T. J. 1970. *Food Oils and Their Uses.* AVI Publishing Co., Westport, Conn.

Information Brochures are available from:

Loders Croklaan b.v. (Unilever), Burgess Hill, W. Sussex, England

Loders Croklaan b.v., Wormerveer, Holland

Friwessa b.v., Zaandam, Holland

Witocan (Dynamit Nobel) D5210 Troisdorf, W. Germany

Durkee Foods, Cleveland, Ohio

# 10

# Milk and Milk Products

Milk products are major ingredients in the chocolate and confectionery industry and uses are described in many parts of this book.

Milk chocolate is increasingly popular, and although milk powder is still used in great quantities in the manufacture of this chocolate, milk crumb is steadily gaining ground because of its quality and particular flavor.

In confectionery, sweet condensed milk and concentrated (evaporated) milk are used extensively. Reconstituted milk powders are also employed but special care in the preparation of their dispersions is needed. Whey products are finding increased use. Liquid milk is rarely used because of the large amount of water required for its evaporation but some companies claim that the best caramels are made from liquid milk. Many manufacturers use fresh cream.

Other milk products used in the industry are butter and butter fat (oil), lactose, and modified milk protein.

The confectionery technologist should know something of the properties and composition of the various milk products used in the industry and the following summaries should be of some assistance. For greater details about the various materials, reference should be made to the bibliography at the end of the chapter. Also, there are entire libraries concerned with the production of milk and milk products.

## LIQUID MILK

Cow's milk is the dairy product in most countries but other milks are produced elsewhere. Buffalo milk, for example, is used in India and some Middle Eastern countries. Goat's milk is a domestic product in many countries.

### Composition

The base components of all milk are described as fat, solids not fat (SNF), and water. The proportions vary considerably not only in

TABLE 10.1. AVERAGE COMPOSITION OF COW'S MILK IN UNITED KINGDOM

| | | | Solids not fat (SNF), % | | |
| | Water, % | Fat, % | Lactose | Casein/albumin (protein) | Ash |
|---|---|---|---|---|---|
| Davis (1955–75) | 87.61 | 3.62 | | 8.77 | |
| | 87.70 | 3.53 | 4.30 | 3.67 | 0.80 |
| Pearson (1970–1981) | 87.70 | 3.61 | 4.65 | 3.29 | 0.75 |

different animals but among different breeds of the same animal. Some figures are given in Tables 10.1, 10.2, and 10.3; a search of the literature will reveal many variations. There are seasonal variations and differences arise attributable to the lactation period, age and physical condition of the animals, and location.

TABLE 10.2. MILK FAT ACCORDING TO BREED OF COW

| Breed | Fat, % | SNF, % |
|---|---|---|
| Ayrshire | 3.7, 4.3 | 8.9, 8.8 |
| Friesian | 3.5 | 8.6 |
| Guernsey | 4.5, 5.1 | 9.1, 9.3 |
| Jersey | 5.5 | 9.5 |
| Holstein | 3.7 | 8.5 |
| Shorthorn | 3.5 | 8.7 |

TABLE 10.3. AVERAGE COMPOSITION OF OTHER MILKS

| | Water, % | Fat, % | Lactose, % | Protein, % | Ash, % |
|---|---|---|---|---|---|
| Buffalo milk | 82.9 | 7.5 | 4.7 | 4.1 | 0.8 |
| Goat's milk | 85.3 | 4.9 | 4.8 | 4.1 | 0.9 |
| Human milk | 88.2 | 3.3 | 6.8 | 1.5 | 0.2 |

## Cow's Milk

Table 10.1 shows the average composition of cow's milk in the United Kingdom.

Seasonal changes produce milk with the highest fat content in the autumn (average 4 percent) and the lowest in the spring (average 3.5 percent) SNF figures are highest in the autumn (8.8 percent) and lowest in the spring (8.6 percent).

Breeds of cow give significantly different milks. Jersey and Guernsey cows in particular produce milks rich in fat. Table 10.2 gives the fat composition of milk from various breeds.

The average composition of other milks is given in Table 10.3. Published figures, however, vary considerably depending on the environment, food, and, with animals, the quality of the husbandry.

The composition of the components of buffalo and goat's milk differs from cow's milk, particularly in fat and mineral salts.

## MILK STANDARDS

There are regulations governing the composition of milk in most countries and these prescribe minimum figures for fat and for SNF.

Dairy companies that use very large quantities of milk process to a standard composition to meet the required regulations. Regulations also prescribe the different categories of milk, for example, "untreated," "pasteurized," "sterilized," or "ultra heat treated." Large chocolate and confectionery companies often process liquid milk for their own use, manufacturing condensed milk, milk powder, and milk crumb. These companies employ their own dairy technologists, who usually combine a knowledge of milk processing with confectionery technology.

## MILK PROTEINS

The proteins of milk consist of casein, albumin, and globulin, and all are high grade and easily digested, and they provide the essential proteins for nutrition. Because of this, milk is rated very highly as a food.

Casein, which mainly imparts the characteristic white color to milk, is present as a colloidal suspension in association with calcium in the molecule as tricalcium phosphate. It is precipitated from milk by acid or the enzyme rennin (rennet), and whereas with the acid the calcium is separated, with the enzyme the calcium remains linked with the protein molecule. The casein of milk constitutes about 80 percent of the total nitrogen content averaging about 2.85 percent.

Casein is prepared for commercial use in the manufacture of plastics, adhesives, paint, paper, and medicinal products. It is used in confectionery as a stabilizing agent and replacement for egg albumen and as an ingredient indirectly when milk powder or condensed milk is included in a recipe. The remaining 20 percent of the protein is now classified as "whey protein" and consists mainly of lactalbumin and lactoglobulin.

According to Muir (1985), the composition of milk protein is

Casein (80 percent)
$\begin{cases} \alpha \text{ S1 and S2} & 48 \\ \beta & 36 \\ \kappa & 13 \\ \gamma & 3 \end{cases}$

Whey protein (20 percent)
$\begin{cases} \beta\text{-lactoglobulin} & 50 \\ \alpha\text{-lactalbumin} & 20 \\ \text{other proteins} & 30 \end{cases}$

The whey proteins are water soluble and are coagulated by heat, lactalbumin at approximately 100°C (212°F) and lactoglobulin at about 72°C (162°F). This latter temperature is recognized as the minimum heat treatment required for bacteriological stability.

The stability of these proteins is closely related to the properties of milk powders (Sweetsur, 1976). Instantized skimmed milk powder, for example, used in many beverages as a replacement for liquid milk, may be subject to feathering or flocculation in the beverage solution. This is avoided by controlled heat treatment of the liquid milk.

Denaturation has been studied in relation to a property known as the casein number. This is the percentage of total nitrogen precipitated at pH 4.7.

For untreated milk, the casein number is 77.6 percent. As milk is heated, this figure rises to 90 percent and powder prepared from it becomes less suitable for beverage use. Instantizing is discussed later.

## LACTOSE

Lactose is the natural sugar of the milk of all mammals and it does not occur in plant life. Chemically, it is a disaccharide consisting of dextrose (glucose) and galactose and is a reducing sugar.

Lactose is nutritionally a very important sugar and promotes the assimilation of calcium and phosphorus in young animals and is stated to be the source of sugar matter in brain tissue.

It is not sweet and when in solution it is about one-sixth of the sweetness of sucrose, and because of this it has been proposed as a sugar replacement in some confectionery to allay excessive sweetness. This replacement involves some difficulties because of its low solubility and crystallization peculiarities. The solubility of lactose at normal temperatures is about 16 percent, increasing to 60 percent at 90°C (194°F) and in solution it exists in two forms $\alpha$- and $\beta$-lactose. In the crystalline state, lactose can have three forms: $\alpha$-lactose hydrate, $\alpha$-lactose anhydride, and $\beta$-lactose anhydride. $\alpha$-Lactose hydrate (mono) is commercial lactose—the anhydride is obtained by dehydrating at a temperature between 65°C (149°F) and 93.5°C (200°F) at a reduced pressure or it can be produced at normal pressures by heating between 110°C and 130°C (230°F to 266°F). The anhydride is hygroscopic in moist air at temperatures below 93.5°C when it reverts to the hydrate, but above 93.5°C it changes to $\beta$-anhydride. $\beta$-Lactose anhydride also forms when lactose crystallizes from concentrated solutions above 93.5°C.

Lactose is slow to crystallize from solutions unless seeded and slow crystallization produces large gritty crystals. Sucrose depresses the solubility of lactose considerably and these factors must be carefully studied if lactose or high milk solids contents are used in confectionery; otherwise objectionable gritty textures arise on storage.

Commercial lactose is made from the whey arising from cheese making and also from the residual liquid in casein manufacture (see "Whey").

This liquid is acidified, heated with active carbon, and filtered. It is then subjected to repeated crystallizations and the degree of purity is determined by the number of these.

High-quality edible lactose has the following specification, in percent.

| | |
|---|---|
| Lactose | 98.0 minimum |
| Ash | 0.15 to 0.40 |
| Protein | 0.20 to 0.50 |
| Moisture | 0.10 to 0.20 |
| Acidity as lactic acid | 0.05 to 0.10 |

## MINERAL SALTS

The mineral salts content of milk is 0.7 to 0.8 percent and consists mainly of calcium and magnesium phosphates and citrates. Muir (1985) quotes the figures given in Table 10.4. Trace elements are also present (Al, Cu, Mn, Zn, I) and these are recognized as important nutritionally.

TABLE 10.4. MINERAL SALTS CONTENT OF MILK

|  | Millimoles per litre | |
| --- | --- | --- |
|  | Total | Soluble |
| Calcium | 29.5 | 7.8 |
| Magnesium | 4.6 | 2.6 |
| Sodium | 22.1 | 21.9 |
| Potassium | 37.9 | 35.0 |
| Phosphate (inorganic) | 20.5 | 10.2 |
| Citrate | 10.8 | 8.3 |
| Chloride | 33.6 | 33.6 |

Preservation of the mineral balance is important in preventing curdling during processing.

## MILK FAT, BUTTER, BUTTER FAT (OIL), FRACTIONATED BUTTER FAT

Butter and butter fat are important ingredients to the confectioner and chocolate manufacturer.

Milk fat is part of the formulation wherever whole milk products are used—i.e., condensed milk, milk powder, concentrated milk, or cream. Caramels, toffees, and butterscotch rely on these ingredients and butter for flavor and texture. Anhydrous butter fat is an antibloom ingredient of dark chocolate. Butter (dairy butter) is the product of churning cream. Cream is separated from milk by warming and settling or, commercially, by centrifuge.

Churning is a process of beating that causes the fat membranes to be broken up and the fat globules to coalesce. Salt is added to give flavor and help as a preservative. Butter may be made from cultured (sour) cream or sweet cream, the former having the stronger flavor.

The average composition of commercial butter is, in percent

| | |
| --- | --- |
| Fat | 80 to 84 |
| Moisture | 15.3 to 15.9 |
| Added salt | 0.03 to 1.8 |
| Protein (curd) | 0.7 |
| Lactose | 0.4 |
| Ash | 0.15 |

Butter is unique in that it contains an appreciable proportion of glycerides of the lower fatty acids, including butyric acid. Minute amounts of liberated butyric acid give a buttery flavor in cooked confectionery but diacetyl ($CH_3 \cdot CO \cdot CO \cdot CH_3$) formed from cultured cream is mainly responsible for flavor.

Hilditch (1964) quotes the following constituent fatty acids combined as glycerides in butter fat, in percent.

| | |
|---|---|
| Butyric | 3.0 to 4.5 |
| Caproic | 1.3 to 2.2 |
| Caprylic | 0.8 to 2.5 |
| Capric | 1.8 to 3.8 |
| Lauric | 2.0 to 5.0 |
| Myristic | 7.0 to 11.0 |
| Palmitic | 25.0 to 29.0 |
| Stearic | 7.0 to 13.0 |
| Oleic | 30.0 to 40.0 |
| Others | 3.0 to 6.0 |

Dairy butter, whether salted or unsalted, has limited keeping properties at normal temperatures. For long-term storage, it should be kept in cold storage at $-10°C$ ($14°F$).

At normal temperatures, there is a gradual transition of flavor toward rancidity and the liberation of free fatty acids. These changes are brought about by the action of microorganisms, enzymes, and atmospheric oxidation. The degree of deterioration is best assessed in its early stages by taste and smell, but acid value and peroxide value determinations are useful.

Rancidity of butter fat cannot be removed entirely by the normal fat-refining methods, and even if the volatile constituents are removed, tallowy flavors remain.

The presence of enzymes and microorganisms in fresh butter makes it essential that confectionery recipes including the butter are subjected to a process of heating or boiling, or very bad flavors may develop on storage.

Butter oil or dry butter fat is made by removing moisture, curd, salt, and other mineral matter from dairy butter. There are several ways of making this fat:

1. By boiling the butter at a temperature well above $100°C$ ($212°F$); this removes the water and precipitates the curd and mineral matter, which are strained off. This method is likely to give a marked deterioration of flavor—possibly arising from partial charring of the curd.

2. By melting in tanks and draining off the settled water. This is wasteful as the separation is incomplete.

3. By melting, washing with hot water, followed by a centrifuging process.

The original butter or butter fat separated by methods 2 and 3 must be subjected to a heating process, 93°C (200°F), to ensure enzyme destruction. Many enzymes are present in milk, and lipase in

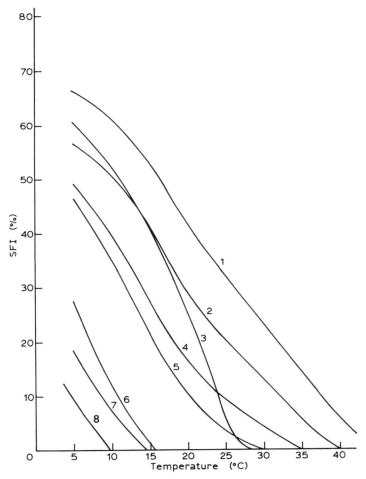

Fig. 10.1.  Typical SFI Curves of Butter Fat and Butter Fat Fractions

Butter Fat:
1. 42°C
2. 38°C
3. 26°C
4. Normal Grade 32°C

5. 28°C
6. 15°C
7. 10 to 13°C
8. 10°C

*S.A.N. Corman* (1987)

particular must be destroyed to prevent soapy rancidity from developing in confections containing lauric fats. Stainless-steel equipment must be used throughout.

Phosphatides (lecithin) are present in butter (0.3 percent) and it is stated that they are partly responsible for better emulsification of butter in confectionery formulations, compared with vegetable fats.

Anhydrous butter (butter fat or butter oil) in solid form has relatively good keeping properties, provided the dehydration has been done correctly. Modern suppliers state that a shelf life of six to twelve months is possible at normal temperatures.

The moisture content of butter fat must be less than 0.05 percent. If higher than this, the water will separate when the fat is liquid and cause rapid deterioration at the interface. This is particularly noticeable if the liquid fat is held in tanks and care should be taken in the design to see that the bottom is dished to give complete discharge.

## Fractionated Butter Fat

The buildup of excessive stocks of butter in some countries can be attributed to various causes. One of the modern methods of finding new uses is to produce fractions with different melting points. Corman (1987) describes a series of products with melting points ranging from 10°C to 42°C.

Solids/fat/index curves determined by nuclear magnetic resonance are shown in Fig. 10.1.

## CONDENSED MILK, EVAPORATED MILK

Some reference is made to the preparation of condensed milk in the manufacture of milk crumb, but for the commercial product the process is briefly as follows: The whole or skimmed milk is preheated at 82 to 85°C (180 to 185°F) for 15 min and this destroys pathogenic organisms, yeasts, and molds, and inactivates enzymes. It should be noted here that lipase inactivation occurs at this relatively low temperature because of the high water content. Flash heating at higher temperatures may also be used.

Sugar is then added to the hot milk and the solution evaporated under vacuum. When the required concentration is reached, the liquid is rapidly cooled with continuous agitation, and to prevent coarse lactose crystals from developing on storage, seeding with finely ground lactose is practiced.

The sugar in condensed milk acts as a preservative and reasonable storage periods can be expected without sterile packing. The best temperature of storage is below 16°C (60°F); at higher summer temperatures 21°C (70°F) and over, thickening will occur with darkening of color and development of stale flavors, particularly with whole milk. If factory hygiene has been properly observed and preheating correctly carried out, microbiological troubles will not arise.

Unsweetened (evaporated) milk is prepared in the same way, but the concentrated whole milk from the evaporator is homogenized while still warm to distribute the fat globules in a fine state and prevent separation.

Unsweetened concentrated milk must be sterilized in sealed containers unless it is to be used within 24 hr of manufacture, but, as with milk, it can be stored longer if cooled to 4 to 7°C (40 to 45°F).

### Composition of Condensed Milks

For the purpose of recipe calculations the compositions shown in Table 10.5 may be used but it is as well to check supplies by analysis and against the manufacturer's declared composition.

TABLE 10.5. COMPOSITION OF CONDENSED MILKS

|  | Whole sweetened, % | Skimmed (nonfat sweetened), % | Unsweetened (evaporated), % |
|---|---|---|---|
| Fat | 9.3 | 0.6 | 10.5 |
| Sugar (sucrose) | 41.0 | 43.0 | — |
| Lactose | 11.4 | 15.0 | 11.8 |
| Protein | 9.3 | 10.2 | 9.5 |
| Ash | 2.0 | 2.2 | 2.0 |
| Water | 27.0 | 29.0 | 66.2 |

### BLOCK MILK

This is a form of solid sweetened condensed milk where the moisture content has been reduced to 8 to 9 percent. It is useful where long-distance transport is involved and it has been imported in this form from Holland. It is no longer made commercially in large quantities.

A typical composition is, in percent,

| | |
|---|---|
| Nonfat-milk solids | 26.5 |
| Milk fat | 18.5 |
| Moisture | 9.0 |
| Sugar | 46.0 |

## MILK POWDER

Almost complete desiccation has long been a means of preserving milk, but with milk powder there are difficulties when it has to be reconstituted into liquid form. The milk powder industry has developed mainly since 1900 and many processes have been tried but they have now resolved into two classes, namely, the drum or roller process and spray process.

### Drum Drying

In the drum process, milk, concentrated to about 23 percent solids and homogenized, is fed onto drums steam-heated internally. The drums rotate slowly and the film of milk loses its moisture in less than a complete revolution. The film is then scraped off and the flakes broken up and sieved or pulverized. The solubility of this product varies between 80 and 90 percent.

There are various patented designs of drum driers that claim improved efficiency and quality. A later development enclosed the drum in a vacuum chamber, thus enabling the drying to proceed at a lower temperature. This resulted in less damage to the proteins and gave greater solubility to the powder.

Roller process powder, as it is called, has found use in vegetable fat coatings. It is less fat absorbent. The manufacture of this powder has steadily decreased, mainly because of the much greater efficiency of the spray-drying process.

### Spray Drying, Instantizing

In spray drying, concentrated milk is atomized by means of a pressure sprayer or centrifugal disk into a large chamber in which turbulent hot air currents are circulating. The milk particles slowly settle in the chamber and are dried in the process. Cyclone separators remove powder carried through with the exhaust air.

In modern installations, dry milk that has been concentrated to 52 percent and air temperatures up to 150°C (300°F) are used in the drying chamber. A further development uses a fluid bed drier integrated into the system. These improvements have resulted in greatly increased efficiency and economy in heat usage.

Spray drying has also been combined with the process of instantizing, which results in agglomeration of the particles. These aggregates are readily wetted and will disperse easily in beverages such as coffee or tea.

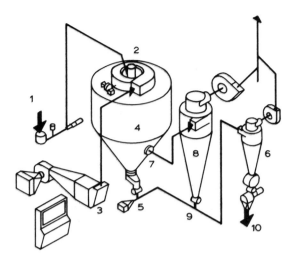

Fig. 10.2. Conventional Spray Drier with Conical Base Chamber (One-Stage Drying)

*A/S Niro Atomizer, Denmark*

Systems developed by A/S Niro Atomizer, Denmark, are described and illustrated in Figs. 10.2 and 10.3.

In Fig. 10.2, the concentrated milk (or other product) is pumped from the feed tank (1) to the atmoizer (2). A fan, heater, and filter (3) deliver clean, hot air to the drying chamber (4).

During the drying process, each droplet is converted into a powder particle, and the more concentrated the liquid before the atomization, the greater is the powder output in relation to the amount of water evaporated. Therefore, it is a great advantage for the atomizer head to be capable of handling liquids with a high solids content.

The drying chamber is designed so as to allow the newly formed powder particles to remain suspended in the air for some seconds before they reach the conical section of the chamber. The dry powder slides down the cone to the vibrator feeder (5) and then through a pneumatic transport system to the small cyclone (6). The hot air that leaves the drying chamber at (7) enters the cyclone (8), where it is rotated at high velocity, thereby separating the suspended particles by centrifugal action.

This powder is transported along the pneumatic system where it joins the stream from the cyclone at (9).

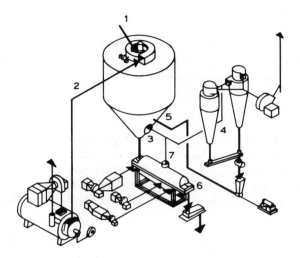

Fig. 10.3.   Spray Drier with Fluid Bed Attachment (Two-Stage Drying)

*A/S Niro Atomizer, Denmark*

Discharge is at (10).

Production economy and quality of product can be improved if powder with a higher moisture content can be handled in the first drying stage. At the the same time, the powder must be free flowing, thus allowing higher inlet air temperatures to be used without degrading the quality.

As shown in Fig. 10.3, the powder with a higher moisture content is transferred directly from the primary drier to a fluid bed drier where final drying and cooling of the powder take place. The first part of the system is similar to the conventional drier: (1) atomizer and (2) hot air supply.

Air is removed from the conical section of the drier at (3). This is fed to the cyclone (4) where suspended powder is removed. This powder is returned to the atomized cloud in the drying chamber (5).

Powder is discharged from the base cone of the primary drier to the fluid bed dryer (6). From this secondary drier, it is delivered to the collecting station.

Air discharged from the fluid bed is also fed to the cyclones (7).

## Instantizing

This process has been mentioned previously under "Cocoa" and "Drinking Chocolates." Instantized nonfat- (skimmed) milk powder has become a widely used commercial article.

## Production

**Multipurpose Re-wet Agglomeration Equipment** The principle of this instantizing unit is shown diagrammatically in Fig. 10.4. For surface agglomeration to be obtained, the powder from the silo (1) is conveyed pneumatically into an agglomerator tube (2). Simultaneously, warm moist air enters the tube tangentially, creating a vortex

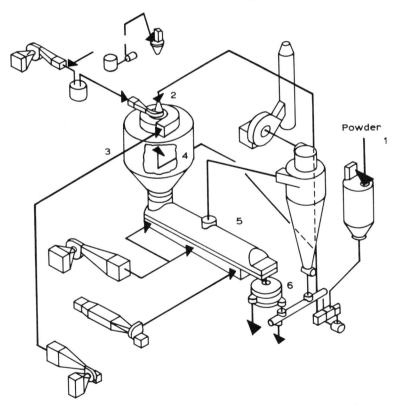

Fig. 10.4. Multipurpose Re-wet Agglomeration Plant

*A/S Niro Atomizer, Denmark*

flow (3). The air velocity, temperature, and humidity can be widely varied. When the agglomerates have reached a certain dimension, they leave the agglomerator tube. The surface of the agglomerates is dried in the chamber (4) shown below the agglomerator tube. The final drying and cooling of the powder take place in the vibro-fluidizer (5). Finally, the product passes to the sieve (6). Powder extracted from the air in the cyclone is returned to the agglomeration section. This procedure converts skim milk powder into an instant product, with bulk density not more than 0.3 g/ml, and which consists entirely of large agglomerates.

**New-Systems**  Today, spray driers with integrated fluid beds (i.e., mounted inside the chamber) are used to produce nonfat- (skimmed) milk powder.

These show appreciably lower energy consumptions. They will eventually phase out the re-wet agglomeration equipment as well as the older straight-through systems where the instant powder is produced directly from liquid skim milk.

The latest methods are described by Hansen (1985).

## Sanitation

Sanitation in spray-drying equipment is of paramount importance. Accumulations of partially dried powder in the upper part of the primary drier are possible and these, if not removed, can develop very high microbiological counts, salmonella included.

Special cleaning, steam, and detergent sterilization must be applied rigorously, as is the case with most other dairy operations.

## Properties of Milk Powder

Spray-dried powders have a greater solubility (99 percent) than roller process powders, and generally the solubility is related to the degree of denaturing of the protein, which is greater with the roller process. A good spray process powder is virtually fully soluble in cold water, whereas a roller powder may be less than 85 percent soluble.

At one time, spray process powder had the reputation of poor keeping properties compared with roller, but preheat treatment methods have largely corrected this difference. The method of heat treatment varies from holding at a temperature of 73.9°C (165°F) for 30 min to flash heating to 87.8°C (190°F) followed by holding at this temperature for 3 min. Additionally, there is the ultra heat treated process. The milk is "flash heated" at about 140°C (204°F) for 2 to 4 sec, either by steam injection or by passing through heat exchangers under pressure.

The keeping properties of milk powder are related to moisture content, storage temperature, and oxygen access. Shelf life, particularly that of whole-milk powder, is extended by vacuum or inert gas (nitrogen) packing.

The moisture content should not exceed 4 percent and for long storage a temperature of 7°C (45°F) is necessary with the powder in containers that are moisture-proof. Gas or vacuum packing reduces the effect of oxygen trapped in the separate particles during the drying operation and this oxygen will diffuse into the surrounding atmosphere in a closed container. If this is an inert gas such as nitrogen, the oxygen level will be reduced and improve keeping quality. Nonfat-milk powders are now classified as "Low Heat," "Medium Heat," and "High Heat" according to the treatment they have had during concentration and drying. Low-heat powders are used where good solubility and reconstitution in water are required. The higher heat powders are used in compound coatings and in baking—they are less absorbent.

## Composition of Milk Powders

The composition of milk powder is obviously related to the milk from which it is made. Standards vary somewhat in different countries. Average compositions are as shown in Table 10.6.

TABLE 10.6. COMPOSITION OF MILK POWDERS

|  | Whole-milk powder, % | Nonfat-milk powder, % |
|---|---|---|
| Moisture | 3.5 | 4.0 |
| Fat | 26.5 | 1.0 |
| Protein | 29.0 | 39.0 |
| Ash | 5.8 | 7.5 |
| Lactose (by difference) | 35.2 | 48.5 |

## WHEY PRODUCTS

Whey is the by-product from the manufacture of cheese or casein. It is not a product of precise composition as, in cheese manufacture,

TABLE 10.7. SEPARATED LIQUID WHEY

|  | Whey, % | Whey solids, % |
|---|---|---|
| Water | 93.4 | 0 |
| Lactose | 5.0 | 75.7 |
| Protein | 0.9 | 13.7 |
| Mineral salts | 0.7 | 10.6 |

sodium and calcium salts may be added. These will affect the mineral salt composition of the whey.

Whey products have in the past been largely diverted to the animal feed market but constituents of whey make it a valuable human food adjunct and at low cost. The approximate composition of separated liquid whey is given in Table 10.7.

It will be noted that the solids comprise about 14 percent protein, which can be a useful ingredient in some confectionery recipes. Whey is sold as whey paste (condensed whey) and this may contain from 60 to 70 percent whey solids. This paste is pasteurized, packed in polythene-lined containers, and should be used within four weeks of manufacture when produced under carefully controlled conditions.

Other whey products that have better keeping properties are sweetened whey paste and whey powder. Sweetened whey is made by adding sugar to whey in an amount equal to the total solids and then condensing. Whey powder is usually spray dried and has keeping properties similar to nonfat-milk powder.

Whey solids have a high mineral salt content, which imparts an unsatisfactory flavor if added in quantity to some confectionery products.

In recent years, demineralized whey has become available. The whey has its mineral salts removed by electrodialysis or ion exchange and products with different degrees of demineralization are available. These can be used in greater proportions as replacements for nonfat-milk powders.

Table 10.8 provides examples of the composition of demineralized whey powders.

Whey proteins, described earlier in the chapter, consist mainly of $\alpha$-lactalbumin and $\beta$-lactoglobulin. These are proteins of high nutritional value and are available commercially. They are used in special diets.

Smith (1980) describes the production and uses of whey products.

TABLE 10.8. DEMINERALIZED WHEY POWDERS

|  | Example 1 | Example 2 |
|---|---|---|
| Protein, % | 11.5–14.0 | 13.0–15.5 |
| Lactose hydrate, % | 73.0–78.0 | 80.0–85.0 |
| Fat, % | 0.6–0.9 | 0.6–0.9 |
| Mineral salts (ash), % | 5.0–5.5 | 0.7–1.0 |
| pH | 6.2–6.6 | 6.2–6.6 |

## RECONSTITUTED MILKS, FILLED MILKS

Milk can be reconstituted from milk powder for use in various confectionery formulations. A method is described under "Caramels." For a truly reconstituted milk, high-grade, soluble, nonfat-milk powder and butter are used.

Filled milks consist of nonfat-milk powder plus a vegetable fat. The milk powder is adequately dispersed in warm water, followed by homogenizing. Heat treatment should also be applied (e.g., pasteurization, ultra heat teatment. These milks may be used in a variety of ways and may be combined with sugar, dextrose, or lactose to produce sweetened condensed milk.

Filled milk powders are manufactured by emulsifying a vegetable fat into a solution of nonfat-milk powder before spray drying. A typical composition is, in percent, as follows:

| Nonfat-milk powder | 70 |
|---|---|
| Vegetable fat | 25 |
| Emulsifier, sugar | 2 |
| Moisture | 3 |

Vegetable fat may be lauric (coconut, palm kernel) or nonlauric (soya). Care to avoid lipase action with lauric fats is essential.

Recombined milk production is described by Alfa-Laval (1985).

## OTHER MILK PRODUCTS

### Malted Milk

True malted milk is manufactured by the heat treatment of a mash of whole milk, barley malt, and meal. The control of temperature, in the region of 65°C (149°F), is vital to this process; otherwise side reactions that result in objectionable products will occur.

Other malted milks are prepared from milk and malt extract or from simple mixtures of milk powder and dried malt.

## Cultured Milks

**Yogurt (Yoghourt)**  This product is now sold in vast quantities, usually fruit flavored. It normally is made from a nonfat-milk dispersion that is first heated for about 30 min at 88°C (190°F). This is followed by homogenizing with about 5 percent of sugar and fruit juice or flavor at 43°C (110°F) and incubation at this temperature for 2 to 4 hr with the culture (L. bulgaricus).

**Enzyme-Modified Milk Fat**  This has received much attention in the United States as a flavor for milk chocolate. Controlled lipase action on milk or milk products produces a flavor that some people like. Farnham (1958) describes the characteristics of these products.

Cultured milk powders are available and may be used as flavors in chocolate, caramels, fudge, and similar confections.

## REFERENCES

Alfa-Laval  1985.  *Recombined Milk Production.* Brentford, Middlesex, England.

Corman, N.  1987.  *Fractionation of Butter Fat.* Corman, S. A., Brussels, Belgium.

Davis, J. G.  1955, 1965.  *A Dictionary of Dairying.* Int. Textbook Co., Aylesbury, England.

Davis, J. G.  1968.  *Dairy Products in Quality Control in the Food Industry,* Vol. 2, p. 29. Academic Press, New York.

Davis, J. G.  1975.  *Dairy Products in Materials and Technology,* Vol. 8, p. 263. Longman de Bussey, London.

Egan, H., Kirk, R. S., and Sawyer, R.  1981.  *Pearsons Chemical Analysis of Foods.* Churchill Livingstone, Edinburgh, Scotland.

Farnham, M. G. 1958. *Twenty Years of Confectionery and Chocolate Progress.* AVI Publishing Co., Westport, Conn.

Federal and State Standards for Composition of Milk Products  1962.  U.S. Dept. of Agiculture, Washington, D. C.

Hall, C. W., and Hedrick, T. I.  1971.  *Drying of Milk and Milk Products* (2nd ed.). AVI Publishing Co., Westport, Conn.

Hansen, O.  1985.  Evaporation, membrane filtration and spray drying in milk powder and cheese production. *N. Europ. Dairy J.,* Copenhagen, Denmark, p. 299.

Harper, W. J., and Hall, C. W.  1976.  *Dairy Technology and Engineering.* AVI Publishing Co., Westport, Conn.

Hilditch, T. F., and Williams, P. N.  1964.  *The Chemical Constitution of Natural Fats.* Chapman & Hall, London.

Hunziker, O.  1940.  *The Butter Industry.* La Grange, Ill.

Hunziker, O.  1949.  *Condensed Milk and Milk Powder*. La Grange, Ill.
Milk Marketing Board  1978.  *Dairy Facts and Figures*. Thames Ditton, England.
Muir, D. D.  1985.  Hannah Research Inst., Ayr, Scotland.
S/A. Nicolas-Corman.  1985.  Metz, France.
A/S Niro Atomizer.  1984.  Soeborg, Denmark.
Smith, G. M.  1980.  *The Technology of Whey Production*. Scottish Milk Marketing
    Board.
Sweetsur, A. W. M.  1976–1983.  *J. Soc. Dairy Tech.*, England.
Webb, B. H., Johnson, A. H., and Alford, J. A.  1974.  *Fundamentals of Dairy
    Chemistry* (2nd ed.). AVI Publishing Co., Westport, Conn.

# 11

# Egg Albumen and Other Aerating Agents

Aeration may be described as a method of introduction of air (or other gas), in the form of very small bubbles, into a liquid or solid. The effects of aeration are as follows:

1. A reduction in density of the product, as low as 0.2.
2. A physical change in texture that results in a completely different "mouth feel" and, in many cases, a modification of flavor.
3. A change in shelf life. The distribution of finely divided air in a food may have far-reaching effects on quality and shelf life. Fat and flavor ingredients may be affected by oxidation and rancidity. Drying of the product or hygroscopicity is affected by the lighter density compared, for example, with a dense hard candy or caramel where surface layers only are affected. In some instances, collapse of aeration may occur due to the presence of fatty ingredients.

## METHODS OF AERATION

The following methods of aeration are used in the confectionery industry. Actual processes and formulations are given in other parts of the book.

1. Beating of an aerating agent with air and other ingredients, for example, marshmallows.
2. Pulling. This process is used to incorporate air into hard or chewable candy.
3. Use of vacuum or pressure change. Fine bubbles distributed through a product may be expanded by pressure reduction, as in aerated chocolate and marshmallows.
4. Chemical change. Some salts, such as sodium bicarbonate,

evolve gas on heating or by the addition of acid. Hard candy may be expanded by the incorporation of sodium bicarbonate in the hot syrup.

The nature of an aerated product depends on a number of factors. In all systems, there are two phases—the continuous phase consisting of the sugar syrup and other "liquids" and the disperse phase, which is the gas in the form of small bubbles. In some instances, solids, such as fine sugar, are present in the continuous phase. The interphase is where the surface of the bubbles meets the continuous phase and the property of the interphase is related to the nature of the surface active ingredients.

The surface active agent imparts strength to the bubbles' "skin" and prevents coalescence.

The viscosity or texture of the aerated product depends on the concentration of the continuous phase and the size of bubbles in the disperse phase.

## AERATING AGENTS

Aerating agents have a molecular structure that possesses hydrophilic and hydrophobic properties and, in a foam, will concentrate at the interphase and reduce the surface tension.

All soluble proteins have this property, and during the beating process the molecular chains become distributed in a very thin layer over the bubble surfaces.

Aerating agents that are used in confectionery, besides solubility, must not be affected by pH changes that are normally expected, or by variable sugar concentrations.

## WHOLE EGG, EGG ALBUMEN, EGG WHITE

### Whole Egg Products

Liquid whole egg and liquid egg white are used a great deal in baking and flour confectionery, but very little in sugar confectionery, (but see frozen egg white later). Composition is given in Table 11.1.

Dried whole egg may be used as an ingredient in drinks prepared from cocoa or chocolate, sugar, milk, or nonfat milk and dried malt. Dried whole egg keeps much better if, before drying, sugar is

TABLE 11.1. AVERAGE COMPOSITION OF WHOLE EGG/EGG WHITE

| Whole egg | % | Egg white | % |
|-----------|-----|-----------------|---------------|
| Shell | 12 | Moisture | 88 |
| White | 56 | Protein | 10.6 Fat ~0.03 |
| Yolk | 32 | Carbohydrates | 0.6 |
| | | Mineral salts | 0.8 |

dissolved in the liquid egg. Commercial powder contains two parts of egg solids to one of sugar with a moisture content of 3.5 percent maximum.

Early experiments by the author showed the effectiveness of glycerol and invert sugar additions in maintaining good shelf life of spray-dried whole egg.

Brooks and Hawthorne (1943) also showed the value of sugar addition.

## Egg Albumen

Dried egg albumen is still the main aerating agent used in confectionery. Although substitutes have been produced and are used in considerable quantity, nothing completely replaces egg albumen in some formulations. Substitutes are frequently used as extenders.

The manufacture of egg albumen was a traditional industry in China before World War II and very large quantities of their product, known as crystal albumen, were exported.

Blomberg (1932) described the Chinese method, which is of interest because it led to the development of modern techniques. It included a fermentation process initiated, apparently, by microorganism contamination from the shell during separation of the white.

The method originally used by the Chinese was to store several hundred pounds of the liquid white in wooden vats for up to six days at a temperature in the region of 27°C (80°F). The progress of fermentation was judged solely by inspection of the consistency and clarification of the liquid and ultimately the sediment and scum formed were removed. The clear liquid remaining was treated with 0.05 percent strong (0.880) ammonia and transferred to shallow trays where it was dried at temperatures not exceeding 125°F in about 48 hr. The trays should have been aluminum, but galvanized trays were also used and this accounted for serious contamination of albumen with zinc at one period.

The dried albumen in the trays was a semiplastic mass and this was broken into large granules for further drying and hardening. This then became the crystal albumen of commerce with a moisture content of 8 to 14 percent.

From this description, it can be imagined that the process could be thoroughly unhygienic and it is not surprising that salmonella contamination was found in a number of imported consignments of albumen.

Similar processes were used in the United States, and because of their objectionable nature, drying without fermentation was tried but the resulting product would not keep and would turn brown, become insoluble, develop a bad odor, and lose its whipping properties.

As a result of these experiments, investigation of the Chinese process was undertaken, and Stewart and Kline (1941) showed beyond doubt that dried albumen produced by this method was stable because the fermentation removed the natural glucose (0.3 to 0.5 percent) from the egg white.

These workers also showed that the correct period of fermentation was that which just completed the change of the glucose to acid, and if fermentation was prolonged, protein breakdown occurred and the albumen became foul.

Hawthorne and Brooks (1944) found that the glucose in egg white could be fermented by yeast with the same effect but that the product had the flavor of yeast. Hawthorne (1950) later removed the yeast cells by centrifuge and the dried albumen then equaled that prepared by the natural process.

Following this work, a commercial process was developed that had the advantage of speed and easy control, and produced a dried albumen of superior bacteriological quality. Much crystal albumen was produced by this method and ultimately a good spray-dried albumen was manufactured. Most albumen on the market now is spray-dried powder of good flavor, microbiological quality, and solubility.

**Pasteurization of Egg Albumen** The discovery of salmonella in consignments of dried egg albumen led to a great deal of research on methods of pasteurization of liquid egg. Processes were developed that could be used by confectionery manufacturers as a safeguard against contamination of their products by the use of supplies that may have contained undetected infection.

A difficulty arises because of the close proximity of the thermal death point of salmonella and the coagulating temperature of egg albumen in water solutions.

The U.S. Department of Agriculture established a plant process that flash heated the egg to 60°C (140°F), followed by 3 min holding at the same temperature. Temperatures are critical, a drop of 3°C (5°F) will not guarantee salmonella destruction and 3°C (5°F) above will lead to films of coagulated egg forming on the pasteurizing equipment.

With spray-dried egg, flash heating to 60°C for a few seconds before spray drying has resulted in salmonella-free products.

**Factory Methods for Pasteurization**   When consignments of albumen were found to contain salmonella, heat treatment of the dry product in its original packages was recommended by the British Ministry of Food. This consisted of storing the original cases in a warm room at 54.5°C (130°F) for six days.

Three days were needed for the cases to reach the required temperature; these were left for a further two days before they were opened. The process therefore took eleven days. The destruction of salmonella was not always complete, but the results were sufficiently good to justify the treatment of all flaked albumen by this process until supplies could be guaranteed.

Some loss of solubility and whipping power resulted from this treatment.

The presence of sugar in solution retards the coagulation of egg albumen and this is very useful as a means of attaining more effective pasteurization temperatures. Albumen dissolved in syrup at 40 to 60 percent concentrations will not begin to coagulate until the temperatures shown in Table 11.2 are reached, irrespective of the concentration of albumen.

A factory process that has been used successfully to pasteurize albumen solutions without any deterioration of whipping properties

TABLE 11.2. COAGULATION OF ALBUMEN

| Concentration, % | Temperature | |
| --- | --- | --- |
| | °C | °F |
| 40 | 65 | 149 |
| 50 | 70 | 158 |
| 60 | 75 | 167 |

is as follows:

A stainless-steel jacketed kettle is provided with hot water circulation and a stirrer that sweeps the sides of the vessel to prevent local overheating. A recording thermometer that traces the time/temperature cycle is fitted to the kettle. To the albumen, previously dissolved in water, is added sugar/glucose syrup to give a concentration of 72 to 75 percent soluble solids. The temperature of the mixture is then raised to 71°C (160°F) and held at this temperature for 20 min.

The need for strict sanitation in handling egg solutions is emphasized in all descriptions of confectionery processes using egg albumen.

**Others Forms of Egg Albumen**  Fabry (1985) describes other commercial albumens that are used in European confectionery. They are frozen egg white and concentrated egg white.

*Frozen Egg White*  Supplies are of good microbiological quality and freezing does not seem to affect its performance. Care must be taken in the defrosting process and a temperature of 40°C (104°F) is recommended. A double-jacket heater should be used to prevent local overheating and coagulation.

Under these conditions, 4 to 7 hr are required to complete the process.

*Concentrated Egg White*  Two processes are used, one under vacuum and the other by ultrafiltration. Sugar may be added to improve keeping qualities. See Table 11.3.

## Factors Affecting Foam Stability

The recipe for and method of production of foams from egg albumen have considerable influence on their stability and volume and the

TABLE 11.3. COMPOSITION OF EGG WHITE

| | Egg white | | | |
| --- | --- | --- | --- | --- |
| | Fresh | Powder | Crystal | Sweet concentrated |
| Water, % | 80–88 | 5–8 | Less than 16 | 40 |
| Protein, % | 10.5–13 | 80–95 | Greater than 75 | 20 |
| Carbohydrates, % | 0.45–1.4 | 7.5 | — | 40 |
| Mineral salts, % | 0.3–1.1 | 2–5.7 | — | — |
| Fat, % | Trace | 0.5–1.8 | 5–6 | — |

confectionery technologist must examine methods critically in relation both to the manufacture of the whip and to its incorporation in a product.

There have been occasions when great efforts have been made to get maximum expansion in the preparation of the whip only to find that the subsequent processes destroy a lot of the aeration.

It is important to follow through experiments with albumen foam to the finished product, making density determinations at each stage of the process. This is very necessary if confectionery units are being made to weight for packing into a box of standard size; if there is density variation, either the units will not fill the box or they may be crushed on closure.

Continuous-pressure whipping equipment developed in recent years makes it doubly important to follow through production after the preparation of the whip.

## Coagulation of Egg Solutions

Coagulation and denaturation describe the changes that occur in egg albumen solutions.

Coagulation means the conversion from a solution to a solid suspension, that is, from a clear liquid to a white, opaque one. Coagulation will occur from the action of heat, but also from mechanical agitation.

The egg portion that is not coagulated has an affinity for water (hydrophilic) whereas after coagulation it becomes insoluble and hydrophobic. It has already been noted that the addition of sugar elevates the coagulation temperature.

**Effect of pH**   The pH of a water solution has some influence on coagulation but is not very significant between pH 8 and 5.5. In very acid solutions (pH 2), a type of gel is formed that does not coagulate in the usual way when heated.

## Extenders

During times of albumen shortage many extenders were proposed and exorbitant claims were made for some of them. A summary of these is given below and in the main they are of little value, but in certain specific recipes they may have uses.

**Phosphates**   Hexametaphosphates, trimetaphosphates, pyrophosphates, and sodium dihydrogen orthophosphate have been tried and very little evidence that they helped was forthcoming.

**Acid Salts**   Acids and acid salts are stated to increase the stability of egg albumen foams. Potassium hydrogen tartrate (cream of tartar) at pH 6 is a recommended addition.

**Calcium Lactate**   The manufacturers of lactic acid and lactates claimed that up to 20 percent albumen could be saved in a recipe by the inclusion of 2 to 3 percent of calcium lactate. No consistent results were obtained to corroborate this.

**Alginates**   These gave some evidence of improved stability but their value was marginal only.

## Microbiological Quality of Egg Albumen

For many years, supplies of albumen were variable and the heat treatment given to eliminate salmonella did not improve matters. The present situation has greatly improved, but deliveries should still be rigorously checked.

Tests are summarized below.

**Appearance, Flavor, and Smell**   Crystals should be very pale yellow; brown shades indicate stale material and insolubility. The albumen, when soaked in water at 21°C (70°F), should dissolve completely to a pale-yellow liquid and be free from any objectionable smell. If mixed with fondant (two parts of albumen in solution to 100 parts of fondant), the flavor of the fondant should be little changed and no fetid odor or bad taste should be detectable. Egg albumen powders (spray dried) should be very pale yellow, free from odor, and wholly soluble.

**Solubility**   A solution is prepared by soaking 10 g of the albumen in distilled water at 21°C (70°F) for 12 hr, occasionally stirring, after which 5 g of this liquid is diluted, with 150 ml of water, well stirred, and then allowed to settle. The clear liquid is filtered through a coarse filter paper and the sediment mixed with a further 50 ml of water and again allowed to settle. The clear liquid is filtered first, followed by the sediment, and this is washed with more water and then transferred, by washing, into an evaporating dish, dried in an air oven, and weighed.

Filtration is somtimes difficult and separation by centrifuge is more satisfactory. Here 5 g of the liquid as before is mixed with 25 ml of water in a centrifuge cup and after centrifuging is similarly treated with two further washings of water. The sediment is washed into an evaporating dish, dried, and weighed.

Insoluble residue by these tests should not exceed 4 percent.

**Whipping Test** A whipping test with the albumen in aqueous solution is often advocated but this test bears little relationship to the performance in sugar confections.

The following test using syrup has been shown to give good correlation with factory performance.

| | |
|---|---|
| Egg albumen | 38 g |
| Water | 176 ml |
| Glucose syrup | 900 g |

The albumen is dissolved in the water by soaking for 12 hr at 70°F. The glucose syrup is then added and the mixture transferred to the jacketed bowl of a Hobart mixer. Water is circulated in the jacket at 27°C (80°F) and the albumen syrup gently stirred until it attains the temperature of the jacket water. The whisk is then set in motion at medium speed and beating continued for 30 min (the wire cage whisk is used). The whipping is observed and the foam should reach a maximum volume before 30 min and not show signs of reducing.

The foam is sampled in a weighed cylinder, preferably open at both ends or with a wire mesh bottom to avoid large air bubbles, the ends leveled off, the foam weighed again, and the density determined.

As this, and any similar test, is empirical, it is as well to retain a good-quality albumen as a control and to test this at the same time as the sample.

With this test, a good-quality albumen will give a foam density of 0.35 to 0.40 and should not show any liquid separation after 3 hr.

**Purity** Tests for heavy metals—arsenic, lead, copper, zinc—should be determined by approved analytical methods.

**Microbiological Quality** Egg albumen should conform to the following standards:

| | |
|---|---|
| E. coli | |
| Salmonella | Absent in 1 g |
| Staphylococcus aureus | |
| Plate count | Maximum in 1 g—10,000 |
| | Milk agar at 30°C (86°F)—three days |
| Lipolytic activity | Negative |

It is very important that no lipolytic activity is present as the process of manufacture of whips and certain confections containing them may not involve temperatures sufficiently high to destroy lipases. Lipolytic activity in some albumens may be high if the processor introduces pancreatic lipase during desugarization to hydrolyze traces of yolk fat. This residual fat is detrimental to the whipping properties of the albumen.

It must be emphasized that lipases are resistant to high temperatures in strong syrups or low-moisture foods.

**Protein Composition of Egg Albumen** The exact nature of the proteins in egg albumen was unknown for many years but the figures shown in Table 11.4 have been recorded.

TABLE 11.4. PROTEIN COMPOSITION OF EGG ALBUMEN

|  | Csonka and Jones (1952) | Fabry (1985) |
| --- | --- | --- |
| Ovalbumin, % | 58.4 | 63 |
| Conalbumin, % | 13.2 | 14 |
| Ovomucoid, % | 14.1 | 9 |
| Ovomucin, % | — | 1.6 |
| Globulins, % | 8.7 | 9.0 |
| Lysozyme, % | 3.8 | 3.4 |

## Albumen Substitutes

During the severe shortage of albumen in the postwar years, many substitutes appeared, some of which were quite good and are still used in considerable quantities. Many others were suggested, some undesirable for inclusion in foods. Blood plasma was among the latter and it is unfortunate that such substances received publicity. Although never used, except for experimental purposes, their infamy persisted for a very long time.

Cellulose esters and fish albumen also staked a claim; however, the former is not digestible and the latter gave reversion flavors.

The most satisfactory substitutes have been based on milk and soya protein.

**Casein** This is the principal protein in milk and the well-known Dutch product "Hyfoama" is prepared from casein by hydrolysis with calcium compounds. This product in many recipes requires less quantity than egg albumen to give the same density of foam but, unlike egg, it requires the presence of sugar or glucose to give foam stability. Many good recipes can be produced with this material and technical details are available in the manufacturer's handbook (1955 et seq.).

"Hyfoama" is commercially available as "S" Standard and "DS" Double strength. The product may be used in conjunction with egg albumen.

**Whey** Modern techniques have been developed that have resulted in the separation of the lactalbumins and, to a lesser extent, the lactoglobulins from whey. These products have very good foaming properties and can be used in conjunction with equal quantities of egg albumen, and in some formulations alone.

**Skimmed Milk Powder** The foaming properties are enhanced by various additions of acid salts, pectin, gums, or cellulose esters. These may be incorporated during the spray-drying process.

**Soya Protein** Soya protein has been the subject of much research in the United States and an enzyme-modified protein developed exclusively for use in confectionery has been patented (Gunther Products (1969). See Fig. 11.1.

*Manufacture* Soy protein whipping agents are generally prepared by enzymatic hydrolysis of native soybean protein as it exists in the oil-free flake or flour, isolated soy protein, or soy protein concentrate.

These products dissolve rapidly in water and the solutions are readily whipped into a foam with syrup. Recipes may use the modified protein wholly or mixed with an equal quantity of egg albumen.

The advantages claimed for these whipping agents are:

1. The whips show excellent stand-up properties and collapse of the foam does not occur over long periods.
2. The whipping efficiency is not impaired by very hot syrup. This is a decided advantage over egg albumen and is useful from a microbiological standpoint.
3. Unlike egg albumen, they do not beat down—in other words, prolonged beating does not reduce foam volume. When mixed with egg albumen, they reduce the beating time.
4. These protein whips show considerable stability in the presence of fat and it has been possible to aerate confectionery such as nut pastes. They have a low flavor profile.

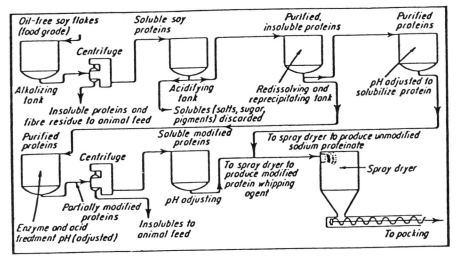

Fig. 11.1. Processing Soy Protein
Schematic diagram of the general processing conditions employed for the production of both an enzyme-modified whipping agent and an unmodified sodium proteinate directly from flakes. In this process, the whipping agent is made directly from a soy protein isolate by treatment with a proteolytic enzyme, such as pepsin, under carefully controlled conditions of time, temperature, concentration, and pH. The degree of hydrolysis of the protein may be controlled by the type of enzyme, concentration of the enzyme, pH, and temperature and time of the reaction.

*Staley Manf. Delatur, Ill., U.S.A.*

**Gelatin** Gelatin is often used as an aerating agent and many formulations incorporate it, such as marshmallows and chewable candies. The texture produced is quite different from that of egg albumen, partly resulting from the fact that, after beating to a foam, gelification takes place.

Gelatin is frequently mixed with other aerating agents. The texture can also be varied by use of gelatin of different "Bloom" values. For example, gelatin of 120 Bloom will give a soft flexible texture whereas 200 Bloom will produce firm or chewy products. (See also "Gelatinizing Agents.")

## REFERENCES

Blomberg, C. G. 1932. *Food Ind.* 4, 100.
Brooks, J., and Hawthorne, J. R. 1943. *J. Soc. Chem. Ind.* 62, 165.

Bryselbout, P. H., and Fabry, Y. 1985. *Guide Technologique de Confiserie Industrielle*, Sepaic, Paris.

Csonka, F. A., and Jones, M. A. 1952. *J. Nutr.* 46, 531.

Hawthorne, J. R. 1950. *J. Sci. Fd. Agric.* 1, 199.

Hawthorne, J. R., and Brooks, J. 1944. *J. Soc. Chem. Ind.* 63, 232.

Lenderink & Co. (N. V.) (Naarden), Hyfoama, Schiedam, Holland.

Stewart, G. F., and Kline, R. W. 1941. *Proc. Inst. Fd. Tech.* 48.

U.S. Patent No. 2,844,468. Gunther Products (Staley Manf.) Galesburg, Ill.

# 12

# Gelatinizing Agents, Gums, Glazes, Waxes

Numerous minor ingredients are used in the confectionery industry and an important group includes substances that form gels and foams or act as stabilizers. Some of these may be used as glazing coatings, but lacquers or waxes are more often employed for this purpose.

In this chapter, the origin and properties of these substances are given. This enables the technologist to acquire a better understanding of the various processes, described elsewhere, that use these ingredients.

## AGAR-AGAR

The name is an Eastern term for seaweed and it has been shortened in commercial use to agar. Japan was the only supplier until 1939, where it was extracted from a red seaweed known as Gelideum. This origin accounted for its becoming known as Japanese isinglass. Isinglass, however, is a glue extracted from fish offal.

With the failure of this supply, investigation of other sources resulted in satisfactory agars being obtained from other seaweeds (Gigartina, Gracilaria, Furcellaria, Chondrus) and these have been grown in Australia, New Zealand, South Africa, Denmark, Spain, and Morocco.

The chemical constitution of agar varies with its origin, but according to Mantell (1947), it is basically the sulfuric ester of a long-chain galactan. The seaweed is extracted by boiling and straining and then extruded into strips that are characteristic of the commercial product, but powdered agar is available and has found favor because it needs very little soaking before dissolving.

The jelly forming power of agar is very high; a 0.2 percent solution will set and a 0.5 percent solution gives a firm jelly.

TABLE 12.1. VARIATION OF AGARS OF DIFFERENT ORIGINS

| Agar | Strength by Bloom gelometer of 1% gel | Melting point of 2% gel, °C | Setting temperature of 2% gel, °C | Ash, % | Acid insoluble ash, % |
|---|---|---|---|---|---|
| Japanese (23 samples) | 260–310 | 89–93 | 33–34 | 2.3–3.6 | 0.02–0.30 |
| Danish (3 samples) | 130–135 | 64–65 | 43–44 | 16.4–18.3 | 0.13–0.80 |
| British (2 samples) | 55–100 | 56–57 | 40–40 | 35.1–37.2 | 0.40–0.31 |
| New Zealand (4 samples) | 610–625 | 90–92 | 35–36 | 0.9–1.2 | 0.06–0.20 |
| South African (5 samples) | 243–306 | 86–88 | 36–36.5 | 2.3–3.0 | 0.10–0.20 |

A detailed description of extraction methods is given by Ingleton (1971).

Jellies are prepared by boiling the strips or powder in water after soaking. Boiling itself has little effect on jelly strength but if flavor acids are present, a considerable reduction of jelly strength takes place. With commercial agars, jelly solutions should be sieved while hot to remove extraneous insoluble matter.

The jelly strength can be determined by the Bloom gelometer (see "Gelatin") on 0.5 percent or 1 percent solutions and it must be recognized that agars of different origin can vary considerably. It is therefore necessary to check supplies regularly to avoid variations in jelly confections made from agar.

Forsdike (1950) gives data that indicate the variation that might be expected (Table 12.1).

Although agar is often used alone to make jellies, in some formulations the texture is not good. Shelf life is also limited and breakdown of the gel, causing syneresis, is possible. Therefore, agar is often combined with other gelling agents, such as starch, gelatin, pectin, and gum arabic.

## ALGINATES, CARRAGEENEN

From the discovery of alginic acid in 1883 by the English chemist Stanford, who was experimenting with seaweed as a source of iodine, the production of alginates has increased to enormous proportions. In the United States alone, vast quantities of a seaweed known as Macrocystis pyrifera are harvested mechanically off the California coast.

The seaweed plant is a perennial and may be collected continuously, with as many as four cuttings per year. If the plant is not harvested, the old fronds break away and float onto the beaches to rot. There, if not removed they cause great problems with infestation by flies.

Off the California coast, the beds of seaweed are under the control of the state of California but private companies employ fleets of boats that have special cutters that operate about 3 ft below the surface. The seaweed is taken to the processing plants where it is washed and milled, followed by treatment with hot alkali solution.

After some clarification, calcium chloride is added, which precipitates calcium alginates, the remaining liquid being separated. The calcium alginate then receives an acid treatment that produces alginic acid. A further treatment with sodium carbonate produces sodium alginate.

The complex constitution of alginic acid, its various salts, and their uses are discussed in publications by "Kelco" (1977).

Alginic acid and its salts of sodium, potassium, ammonium, and calcium are available commercially. More recently, propylene glycol alginate has been produced. In the food industry, it now has many applications and is used as a stabilizer, emulsifier, and thickener for ice cream, chocolate milk suspensions, cake icings and fillings, and chocolate syrups.

One interesting application is the manufacture of "Chellies", which relies on the formation of calcium gels by the interaction of sodium alginate solutions with calcium chloride solutions (see "Fruits").

Alginate gels do not disperse in the mouth in the same way as pectin or gelatin gels do and this must be borne in mind when devising recipes that include sodium alginate.

Alginates have numerous industrial uses other than in the food industry.

# CARRAGEENAN (IRISH MOSS)

This is another product obtained from the seaweeds Chondrus crispus and Gigartina stallata. Chemically, it resembles agar and may be classified as a linear (straight-chain) polysaccharide. There are three main types that are structurally different, namely, kappa, iota, and lambda.

Carrageenan forms gels in water in concentrations as low as 0.5 percent and the property of these is affected by various salts, notably potassium compounds. Brochures describing the carrageenans and their uses are published by Genu Kobenhavns Pektinfabrik (Div. Hercules Inc., U.S.A.).

In the food industry, carrageenans generally are used as stabilizers. In confectionery, they are useful additives in chocolate syrups.

## XANTHAN GUM

Xanthan gum is produced by a process of biopolymerization and is defined as a high-molecular-weight natural polysaccharide. It is formed during an aerobic fermentation process using the microorganism Xanthomonas campestris. The media consist of glucose syrup, phosphate and nitrogen compounds, and certain trace elements.

The rheology of xanthan gum solutions is unique and renders it invaluable as a stabilizer and suspending agent in liquids, pastes, and syrups.

These gum solutions exhibit a yield point. When this is overcome by mixing, the viscosity is reduced, proportional to the rate of mixing (shear), but as soon as mixing ceases, the return to the original viscosity is immediate. This property can be seen as an important asset in the drink and food industries.

In confectionery technology, its use to date has been inadequately pursued.

Xanthan gum can be combined with guar and carob (locust bean) gums to give increased viscosities. This effect is useful in bakery product fillings. As with alginates, publications by "Kelco" (1977) describe its properties and uses.

## GELATIN

The raw materials for the extraction of gelatin are cattle bones and skins, including pork skins. It is obtained by degreasing, liming, and successive extractions with hot water, followed by filtration. The first extracts give the best edible gelatin, with lighter color and stronger gelling power; the later extracts produce the nonedible gelatin and glues.

The first treatment produces insoluble collagen, which is converted to soluble gelatin with hot water. Bones are demineralized with dilute acid to remove calcium phosphate, which then yields bone collagen—known as ossein.

Gelatin is defined in the *United States Pharmacopoeia* as "the product obtained by partial hydrolysis of collagen derived from the skin, white connective tissue, and bones of animals." It is also recognized that there are, basically, two types:

*Type A.* Derived from an acid-treated precursor with an isoelectric point between pH 7 and pH 9.

*Type B.* Derived from an alkali-treated precursor with an isoelectric point between pH 4.7 and pH 4.5.

Commercial gelatin is in the form of sheets, flake, buttons, or powder, and the best grades are colorless, odorless, and without taste. It swells when soaked in cold water and will dissolve on heating, and this solution will set to a jelly on cooling. This gel will melt on warming, and for certain food purposes this melting point is an important property and some grades of gelatin are specified by their melting point.

Because of its method of preparation and the origin of the bones and skins, gelatin may be contaminated with impurities, and tests for metallic contamination should be made, particularly for copper, lead, zinc, and arsenic.

Gelatin is used in a great number of foods including confectionery. Although food-grade gelatin is quite wholesome, it is objected to by some because of its sordid animal origin. It also promotes the growth of microorganisms and great care must be taken during the preparation and storage of solutions.

Gelatin is also partly destroyed by boiling, and as a result of this, it is never boiled in a batch of confectionery but is added in solution or soaked form after boiling is complete. Because of this, there is a small risk that any bacterial contamination may not be destroyed.

Gelatin can be replaced in many recipes by other gelling substances of vegetable origin such as pectin and agar, and often these afford better keeping properties for the product. It should be noted that gelatin is a heat-reversible gel, whereas many other gelling substances are not. This is one of its valuable properties.

It is comparatively easy to make gelatin jellies provided the simple precautions mentioned in their preparation are observed.

In addition to confectionery jellies, gelatin has many other food uses. Table jellies, for example, are a very popular use, and it is utilized to a large extent in ice cream and pie fillings.

## Properties and Testing of Gelatin

*Moisture content.* 9 to 10 percent.

*Solubility.* In addition to water, it is soluble in aqueous solutions of polyhydric alcohols, e.g., glycerol, propylene glycol.

*Protective colloid.* Gelatin is very valuable as a stabilizer in preventing crystallization and separation of emulsions.

*Viscosity.* The viscosity of gelatin solutions can be specified and this factor has many useful applications.

*Strength.* The strength of jelly produced by a given gelatin is classified by the Bloom rating. The jelly strength is determined by an instrument known as a Bloom gelometer. Briefly, the procedure is as follows: A water solution containing 6.67 percent gelatin is carefully prepared in a special wide-mouth bottle, which is then placed in a chill bath maintained at $10 \pm 0.1°C$ for 17 hr. At the end of this time, the firmness of the resulting gel is measured by the gelometer. The instrument delivers the amount of lead shot required to depress a standard plunger (12.7 mm in diameter) a distance of 4 mm into the surface of the gel. The shot is delivered at a controlled rate and the flow is cut off mechanically by an electromagnet. This weight of shot in grams is then known as the jelly strength or Bloom rating of the gelatin. The greater the amount of shot required, the higher is the strength of the gel. Commercial gelatins vary from 50 to 300 Bloom grains. For confectionery purposes, gelatin of Bloom value 180 to 220 is normally used, except for mucilage, where the lower Bloom gelatins are better.

The following publications give further information on the testing of gelatin and its properties:

Publications of Gelatin Manufacturers Institute of America, Inc., New York. *British Standard No.* 757 (1959). Revisions published periodically, British Standards Institution, London. *British Pharmacopoeia,* Pharmaceutical Press, London. *United States Pharmacopoeia.* Publications of the Association of Official Analytical Chemistry, Inc., Washington, D.C.

## GUM ARABIC, GUM ACACIA

There are numerous varieties of acacia trees throughout the world and the exudation from the bark of these is known commercially as gum arabic. The species of acacia tree that grows across the African continent from Senegal to the Red Sea has been responsible for most of the supply of the gum. Political problems of recent years have caused many shortages.

The trees are tapped from incisions made in the trunks and the "tears," which are at first soft with a dry skin, eventually dry to solid lumps. These vary in color from very pale amber to a reddish shade and their size depends on the weather conditions. They are largest after heavy rain, have characteristic cracks and fissures, and fracture fairly readily.

The gum arabic of commerce is graded by its color, the very pale shades commanding the highest prices, particularly for confectionery, as the darker gums usually have an unpleasant taste. According to Mantell (1947), this is due to the presence of tannins. The molecular structure of gum arabic is very complex and Hirst (1942) states that it consists of the calcium, magnesium and potassium salts of D-glycuronic acid, D-galactose, D-rhamnose, and L-arabinose, and the arrangement is intermediate between that of the simple sugars and hemicellulose.

The following are properties of gum arabic:

*Moisture.* The commercial gum should have a moisture content between 12 and 15 percent—outside these limits, the gum will either break down to a powder or be too soft to crush.

*Solubility.* It has remarkable solubility in water, being about 40 percent at temperatures of 24°C (75°F) and over. In other solvents it shows no solubility except for glycerol and ethylene glycol, in which the quantity that will dissolve is small.

In making strong gum solutions, warm water must be used with careful stirring; otherwise a great deal of air is incorporated, which makes handling difficult. Solutions must always be strained, as much foreign matter is present in commercial gum.

*Viscosity.* The high solubility of the gum results in solutions of very high viscosity, and this is of great value in the stabilizing of emulsions and in holding together solids in paste form. This is made use of in confectionery for lozenge manufacture and for the same reason the solution can be applied as a glaze. Its viscosity is retained over a wide pH range and in the presence of other gums and ingredients. The pH of 40 to 50 percent solutions ranges from 4.5 to 5.5 but maximum viscosity is obtained if the pH is adjusted to 6.0 to 7.0.

## Uses of Gum Arabic

Gum arabic has many uses in confectionery but of recent years, because of a general shortage with a consequent increase in cost, substitutes have been developed. Generally, they do not have the special properties of the true gum.

Among its uses, which will be found in other parts of the book, are as a glaze, a binder for lozenges, an ingredient of gums, and a stabilizer to control crystallization.

## GUM TRAGACANTH

This gum is a product of various species of thorny shrubs known as Astragalus that grow in the semidesert areas of Turkey, Iran, Syria, and India.

To obtain the gum, an incision is made near the root of the tree and this is held open by a wedge. The gum that exudes does so in a shape that is related to the incision, so that a narrow slit produces flakes that in good weather dry quickly and are clean and white. This is the best-quality variety.

Its molecular constitution resembles gum arabic in some respects, but its physical properties are different. When treated with water, tragacanth absorbs a great amount and swells, and upon adding more water and warming, a viscous dispersion is formed, but this is not stable and will separate, after standing, into a solution and a sludge.

The solid gum should not be stored for lengthy periods as its solubility decreases with age. Solutions must be freshly prepared as they are subject to bacterial degradation. Gum tragacanth has its main use in confectionery as a mucilage for lozenge paste and for this purpose best results are obtained when it is used in conjunction with gelatin.

It is preferable not to mix it with gum arabic as under certain conditions the tragacanth will precipitate. The gum solution should have a pH of 5 to 6 but maximum viscosities are obtained under slightly alkaline conditions. The viscosities of gum solutions are subject to great change by temperature and mechanical mixing.

## GUAR GUM, LOCUST BEAN (CAROB) GUM

Both of these gums have been developed and modified in recent years. They are employed as stabilizers and thickeners in all branches of the food industry and have numerous other industrial uses, such as in paper making and printing.

In confectionery, they can be used as extenders in starch, agar, and pectin jellies where they prevent syneresis, splitting, and shrinkage. They have a particular value in this respect in licorice recipes.

## Guar Gum

Guar gum is a hydrocolloid—a polysaccharide derived from the seed of the guar plant (Cyamopsis tetragonoloba). The plant, which reaches about 6 ft in height, is grown intensively in India. The purified gum is extracted from the seed after removal of the outer husk and germ. The endosperm is practically all gum.

Chemically, it is a galactamannan with a very specific formulation.

## Carob Gum

Carob gum is obtained from the seed of the locust bean tree (Ceratonia siligna), which grows in the Mediterranean area.

The yield of gum from the beans is small, about 3 to 4 percent, which means that the extraction must be made in the country of origin to effect transport savings. The extracted gum can be improved in solubility and viscosity by chemical modifications to the hydroxylethyl carboxyl esters. Quite strong solutions can be prepared without gelling taking place.

In agar jellies, 0.1 to 0.2 percent may be replaced by carob gum, resulting in an increased rigidity and prevention of syneresis (weeping). This improvement is more noticeable in low-sugar-concentration syrups and is therefore of less interest in sugar confectionery. The gum is valuable as a stabilizer in spreads, chocolate syrups, lozenge pastes, and ice cream.

Wielinga (1976) has described the chemical constitution and properties of these gums and the publications by the Meyhall Company give details of the various modifications and uses.

## PECTIN

Pectin is the gelatinizing agent natural to fruits and it also occurs in many other vegetable products. When extracted, standardized, and in some cases modified by chemical or enzyme treatment, it becomes one of the most valuable gelling ingredients in the confectionery industry. The use of pectin successfully in confectionery necessitates an understanding of a certain amount of the chemistry of its reactions in different media.

Compounds present in plant tissues and derived from colloidal carbohydrates associated with lignins and hemicelluloses are called pectic substances. These consist largely of $\alpha$-D-galacturonic acid residues linked in the 1–4 position. Other sugars are also present, both in side chains and the main chain, but are of secondary importance. Part of a polygalacturonic acid chain without terminal units is shown below.

These polygalacturonic acid chains can be esterified with methyl groups to transform the carboxylic acid groups, for example, $COOH \rightarrow COOCH_3$, and they can also be neutralized by various bases.

The various methylated compounds are described by one collective term "pectin."

Kertesz (1951) classifies pectic substances according to the following descriptions:

*Protopectins.* Pectic substances in their natural state, not soluble in water, but extractable as pectinic acids at elevated temperatures in the presence of acid.

*Pectinic acids.* Colloidal polygalacturonic acids with some of the carboxylic groups esterified. They form gels with sugars at a suitable pH and, if the degree of esterification is low, with metallic salts. These are called pectinates.

*Pectic acids.* Colloidal polygalacturonic acids that have not been esterified with methyl groups. The salts are called pectates.

## Commercial Pectins

The pectins of commerce are almost entirely obtained from citrus fruit and apple residues. After extraction of the juice, and, in the case of citrus fruits, the essential flavoring oils, a pithy material remains. That from apples is called pomace.

The extracted purified pectins are classified into two main groups known as high methoxyl and low methoxyl pectin. A third group now finding increasing application is known as amidated or amide pectins.

**High Methoxyl Pectins** These have 50 percent or more of the carboxylic groups esterified and they need the presence of sugar and an acid to form gels. The degree of esterification above 50 percent determines their behaviour in gel formation, e.g., rate of set.

**Low Methoxyl Pectins** These are the pectins that have less than 50 percent of their carboxylic groups esterified. They require a metallic salt to set them to a gel—usually calcium—but they have a particular value in confectionery because a jelly can be made with sugar, pectin, and a calcium salt, with or without acid, and often from a flavor standpoint this is an advantage, particularly as the high methoxyl pectins require distinctly acid conditions (pH 3) for maximum gel strength.

Sugar is also not essential with low methoxyl pectins and this is useful in the preparation of puddings and sauces.

## Amidated Pectins

These pectins have a proportion of the galacturonic acid COOH groups replaced by $CONH_2$ groups, and usually 15 to 25 percent of the groups are thus substituted.

These pectins tolerate a wider range of calcium concentration than low-methoxyl pectins, and in certain confections will set more rapidly. This is useful where pressure or jet cooking is used.

They are also used in the preparation of heat-reversible bakery jams and jellies.

**Gel Strength** The strength of the jelly produced with a given quantity of sugar and acid is related to the molecular chain length of the pectin molecule and this can vary according to the quality of the raw material and the process of extraction.

Different varieties of the fruit, whether citrus or apple, and the season cause variations in the final product, and to overcome this the gel strength is standardized from the lowest values expected in a commercial extraction—the pectins extracted with high gel strength being diluted with sugar, as this is an ingredient likely to be most common to a variety of recipes.

This standardization necessitates the adoption of a method of testing that is agreed between pectin manufacturer and user, and accepted internationally. The IFT Committee on Pectin Standardization recommended the adoption of the Cox and Higby "Exchange" Ridgelimeter.

The mechanism of gel formation is thought to be due to lowered thermodynamic activity of the water in the presence of a solute such as sugar. The pectin, therefore, starts to come out of solution and a three-dimensional network is built up by the formation of junction zones between adjacent pectin molecules. The exact structure of these zones has not yet been established but it is certain that the galacturonide residues are involved.


## High Methoxyl Pectin—General Properties

This pectin is manufactured in several grades with different rates of set, with the rate and temperature of set being related to the methoxyl content—the higher the esterification, the higher is the temperature of set. If the sugar or soluble solids content is high, as is the case with confectionery jellies, the setting temperature is raised. Another factor that can influence the setting temperature and the rate of set is the presence of certain buffer salts, and, particularly in confectionery jellies, addition of these salts can be of great value in the prevention of premature setting. The salts most frequently used for this purpose are sodium citrate and the various sodium polyphosphates. It must also be recognized that buffer salts of a similar nature can exist naturally in many fruits and when making jam or jellies from fruit pulp this must be taken into account. The water ingredient can also have significant effect if it is "hard".

The setting temperature is the temperature at which gelling commences as the mass cools after it has been boiled, and a confection prepared with any type of pectin *must* be cast or poured above the setting temperature.

High methoxyl pectins form gels in the presence of three other ingredients—water, sugar, and acid—and the part played by each of these may be considered as follows:

Pectin dissolved in water partly dissociates to form $COO^-$ ions resulting in a net negative charge on the molecules that produces a repulsive force between them.

Sugar lowers the solubility of the pectin in the water, and the addition of acid suppresses the ionization of the pectin sufficiently to overcome the coulombic repulsion and allow the formation of junction zones.

## The Pectin/Sugar/Acid Balance

The successful manufacture of pectin jellies depends entirely on a clear understanding of the relative activity and proportions of the ingredients.

The first rule to observe with high methoxyl pectins is that jellification will take place only when the soluble solids—mainly sugar—are between 60 and 80 percent. It must also be recognized that sugar (sucrose) has a limit to its solubility in water; this is about 67 percent at normal temperatures, and it is therefore necessary to use additional, more soluble sugars if jellies with high soluble solids are to be made. In the manufacture of jams or fruit jellies that have a natural acidity from the fruit, some of the sugar will be "inverted" or converted into a mixture of dextrose and levulose (fructose) and the presence of these other sugars creates an increased solubility. With confectionery jellies where no fruit is added, it is often convenient to add the acid at the end of the boil and include in the basic recipe invert sugar or liquid glucose (corn syrup). By this means it is possible to control more precisely the proportion of invert sugar in the final product.

The presence of invert sugar and/or liquid glucose prevents crystallization of the sucrose in a gel of concentration between 67 and 80 percent soluble solids. In confectionery jellies, it is necessary to observe a minimum soluble solids concentration of 75 percent, therefore, the zone of optimum jellification will be reduced to the darker portion of the shaded area in Fig. 12.2. Jams and preserves are now mostly packed in sterilized jars or cans, therefore, the less soluble solids can be tolerated.

The acidity of the mixture is of great importance if a good gel is to be produced and this must be assessed as pH (active acidity), as the presence of "buffer" salts can reduce active acidity. With high methoxyl pectin, the pH range for optimum gel formation is 2.9 to 3.6, and above 3.6 partial gel formation only occurs, which is a waste of pectin. Below 2.9 a phenomenon known as syneresis arises, which results in slow breakdown of the jelly with syrup oozing or weeping out of the gel. In all jam or confectionery production, accurate determination of pH of the gel given by a recipe is essential if the above troubles are to be avoided. With confectionery jellies where the soluble solids range is confined to 75 and 80 percent it should be noted that the pH range for optimum jellification is considerably narrowed.

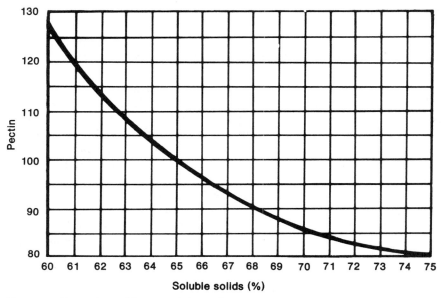

Fig. 12.1. Proportions of Pectin at Different Soluble Solids Giving Jellies of Equal Strength

*Unipectina S.p.A, Italy*

Indications have been given of the permitted variations in solids concentration and acidity but along with these there is also the quantity of pectin required. This varies according to the soluble solids content and in practice is between 1 and 1.75 percent of the final gel weight for optimum gel strength. This ratio between the soluble solids and pectin content is shown in Fig. 12.1.

The relationship between gel formation, pH, and soluble solids content is shown in Fig. 12.2.

## Grades of Pectin

**High Methoxyl Pectin**   Pectin manufacturers make several grades of pectin to suit various recipes and processes and this is exemplified by Table 12.2.

The following are factors that influence gel strength:

***Speed of Cooking***   In any mix that is being boiled to produce a jelly, prolonged cooking will cause (a) excessive inversion of the sucrose and (b) degradation and loss of strength of the pectin. As a result of

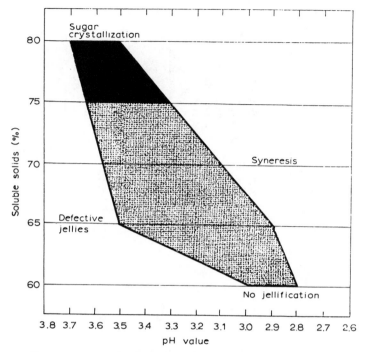

Fig. 12.2.   High Methoxyl Pectin—Zone of Optimum Jellification

*Unipectina S.p.A, Italy*

(a) dextrose crystallization can occur on storage, and (b) will mean loss of gel strength, and in the worst cases no gel at all. The latter fault is all too prevalent in domestic jam making where boiling may continue for interminable periods.

Rapid boiling to the requisite soluble solids should take less than 15 min for best results.

**Depositing the Batch**   After boiling, the batch should be deposited into containers or molds quickly, otherwise the same effect as with slow cooking will arise.

**Premature Setting**   If the temperature in the cooking equipment or depositing hoppers drops below the setting temperature, premature setting will occur, and if depositing is carried out after this, there will be loss of gel strength and the gel will have a granular appearance when cut. Incorrect pH of the jelly will also cause this trouble.

TABLE 12.2. HIGH METHOXYL PECTINS

| Types | % Esterification | % Soluble solids | Setting temperature °F | Setting temperature °C | pH range | Normal uses |
|---|---|---|---|---|---|---|
| Rapid set | 70–76 | 60–70 | 167/185 | 75/85 | 3.1–3.6 | Preserves, jams, jellies—small containers |
| Medium set | 68–70 | 60–70 | 131/167 | 55/75 | 3.0–3.3 | Preserves, jams, jellies—larger containers |
| Slow set | 60–68 | 60–70 | 113/140 | 45/60 | 2.8–3.2 | Preserves, jams, jellies where batches may be held before pouring |
| Confectioners "buffered" | 60–66 | 75–80 | 194/203 | 90/95 | 3.2–3.7 | Confectionery jellies with or without fruit |

These pectins are generally standardized to 150 grade SAG.

*Incomplete Solution of the Pectin* For the pectin to dissolve completely it must be dispersed in water or dilute syrup. The solubility of pectin is repressed by strong syrups and to use these results in great loss of gel strength. It is now general practice to mix the powdered pectin thoroughly with eight to ten parts of sugar. When this mixture is added to water with rapid mixing the pectin particles are well dispersed and there is no balling.

*Water* A factor often disregarded in making pectin jellies is the composition of the water ingredient. Hard waters contain calcium salts that can have appreciable buffering action, but this effect is much more noticeable when using low methoxyl pectins.

Water analysis should be carried out from time to time, and if manufacture is transferred to a factory in another part of the country, this is a very important point to check.

The presence of traces of iron in water is very detrimental to pectin jellies, giving bad colors and flavors.

*Size of Deposit or Container Receiving Deposit* The effect is similar to prolonged cooking or depositing time, if a large container is filled with a freshly boiled pectin jelly, the time of cooling may be sufficiently prolonged to cause loss of gel strength. Some jam makers use heat exchangers to cool the product immediately after boiling to a temperature just above the setting temperature. If this is not done, up to 20 percent extra pectin will be required to get the same set in a large container compared with a small deposit, which will cool quickly in ambient temperatures.

**Low Methoxyl Pectin** Low methoxyl pectins are a comparatively recent development in confectionery technology and because of the flexibility of the conditions under which they set they have wide and increasing applications in foods, cosmetics, and pharmaceutical products. Unfortunately, many confectionery technologists do not fully understand the chemistry of this interesting pectin, particularly its sensitivity to calcium salts, and its use is sometimes abandoned after one or two failures instead of trying to find the exact conditions for preparation of the gel.

In contrast to high methoxyl pectin, sugar and acid are not essential for gel formation because with low methoxyl pectins the reticular structure of the gel is formed with calcium pectinate. Gels with solids content as low as 2 percent and pH close to neutrality can be prepared but it must be understood that gels can still be made with sugar and some acid is required for keeping and flavor purposes provided that some calcium salts are present.

TABLE 12.3. LOW METHOXYL PECTINS

| Type | % Esterification | % Soluble solids | Setting temperature | | pH range | Normal uses |
|---|---|---|---|---|---|---|
| | | | °F | °C | | |
| A | 45–53 | 50–70 | * | * | 2.8–3.3 | Jams, jellies |
| B | 40–50 | 40–65 | * | * | 2.8–3.5 | Low sugar content jams and jellies |
| C | 40–50 | 60–70 | 140/158 | 60/70 | 3.5–4.0 | Jam for "tarts" |
| | | 75–80 | 185/203 | 85/95 | 4.0–5.2 | High sugar confectionery jellies |
| D | 32–37 | 20–50 | * | * | 2.8–3.2 | Low sugar jellies |
| | | Very low | — | — | 6.5 | Milk puddings, creams |
| Polygalacturonic acid | 0 | — | — | — | — | Pharmaceutical uses |

* Setting temperatures can be greatly influenced by calcium salt additions.

To cover all the various uses a variety of low methoxyl pectins is manufactured, modified with regard to molecular weight, degree of esterification, and ability to gel under different conditions of pH and sugar concentration. A range of low methoxyl pectins by the same manufacturer is given in Table 12.3. These, as prepared, contain the necessary calcium and buffer salts to match the degree of methoxylation and method of use.

*Factors Influencing Gel Strength*  The various points discussed under high methoxyl pectin apply to low methoxyl pectin also, but particular attention in all recipes must be given to the calcium balance, otherwise failures will result from pregelation or lack of gel formation.

Although calcium salts and other buffer salts are a constituent of the pectins supplied by the manufacturer, it is sometimes necessary to make adjustments by further additions, particularly with variable water supplies.

*Calcium Salts*  These are reactive towards the low methoxyl pectin according to their solubility—calcium chloride, for example, will give rapid setting, whereas tricalcium citrate or calcium sulfate will give slower setting rates because of low solubility. They are normally added in quantities ranging from 0.05 to 0.10 percent of the final jelly.

*Buffer Salts*  These are added to prevent pregelation and will assist in allowing a jelly, which has been boiled to finality, to be held for a limited period while awaiting to be deposited. It must be remembered, however, that this delay causes some loss of gel strength. The salts that are frequently used for this purpose are sodium citrate and tetrasodium pyrophosphate and may be added in quantities between 0.20 and 0.50 percent of the final jelly.

## Amidated Low Methoxyl Pectins (Amide Pectins)

These have certain advantages over the normal low methoxyl pectins. Buckle (1979) quotes the following:

1. They tolerate a much wider range of calcium levels.
2. Jellies are heat reversible; that is, they melt on heating and set again on cooling.
3. They have thixotropic properties; just below the setting point, their fluidity is maintained by stirring and setting occurs as soon as stirring ceases.
4. Syneresis (syrup separation) is much reduced.
5. They tolerate a wider range of soluble solids.

Chemically, these pectins are produced by ammonia treatment of the citrus or apple residues or high methoxyl pectin. In this process, the methoxyl groups are partially replaced by $CONH_2$ groups. Normally, 15 to 25 percent of the groups are substituted.

## Liquid Pectins

Liquid pectin was used commercially, particularly for jam making, well before pectin powders were on the market. These are always of the high methoxyl type, made from apple pomace, and normally 5 grade. This pectin is usually preserved with sulfur dioxide.

The large bulk due to the water content makes transport and storage inconvenient and expensive and its use is confined almost entirely to jam manufacture. For confectionery purposes, it is less suitable, particularly as it has a light-brown color and a slight flavor.

**Special Pectins** While it has been emphasized that pectin should only be dispersed in water or dilute syrups, more recently pectins have been developed that will disperse and dissolve in high concentrations of syrup. These are particularly useful where continuous, pressure cookers are used and the pectin, water, sugar, and any other ingredients are rapidly heated in thin films—the period of heating being very short.

## The Manufacture of Commercial Pectins

The precise details of the process and plant for the manufacture of pectin are the responsibility of the producers but some knowledge of the principles is of interest to the confectionery technologist.

The raw material is citrus pomace, after extraction of citrus essential oils from the peel and juice from the center, or apple pomace after pressing the juice for cider or apple juice manufacture. A number of other vegetable wastes have also been shown to contain appreciable quantities of pectin but none has yet been exploited economically. The Tropical Products Institute showed that cacao pod waste could yield useful quantities of pectin but there are difficulties in getting the pod waste to the factories in a suitable condition, and small commercial tests gave lower yields than obtained by the Tropical Products Institute. The method of handling and, if necessary, the drying of these pomaces must be carefully controlled to prevent hydrolysis of the pectins by enzyme and microbiological action.

The pectin is extracted from the peel or pomace by dilute acid at a pH of 1.5 to 3.0 and after separation of the insoluble residue, which is normally carried out in filter presses, the extract is decolorized with carbon if necessary before further treatment. Almost all pectins used in the confectionery industry are in powdered form and there are two main methods of separation of the pectin from the liquid:

1. Precipitation with isopropyl alcohol after concentration of the liquid extract, followed by separation of the precipitated pectin centrifugally or on a press.
2. Precipitation of the pectin from a relatively weak extract by metallic salts such as aluminum chloride, followed by removal of the metallic salt by washing with acidified isopropyl alcohol. The pectin is then dried, ground, and blended with sugars and/or buffer salts to give the desired gel strength.

## The Testing of Pectins

While it is true to say most pectin manufacturers have adequate control of their products and make each grade of their own product within narrow specification limits, it is possible for variations to occur between different manufacturers.

In previous tables, the various types of pectin have been summarized but each of these is made to a particular strength or "grade". The "grade" is measured by assessing the strength of a jelly made using a standard recipe and process.

The internationally accepted method is the Cox and Higby "Ridgelimeter" method, which measures the loss of height or sag of an unsupported jelly and this figure can be converted to "grade" by reference to a graph.

Two other typical methods may be mentioned and these have also been used for measuring the strength of gelatin jellies. They are:

*The Tarr-Baker apparatus.* This measures the pressure required to break the surface of the jelly by means of a piston.

*The Owens and Maclay apparatus.* This measures the torsion force required to turn a flat metal blade set in the jelly. There are several modifications of apparatus using this principle.

Both these methods are less accurate than the Ridgelimeter. Various other methods have been proposed.

The Ridgelimeter method applies specifically to high methoxyl pectins. With low methoxyl pectins, while the jellies can be measured by the Ridgelimeter apparatus, the method of jelly preparation for various confections varies so greatly that it best to check supplies of low methoxyl pectin against a known standard using, on a small scale, the production process and recipe.

Grade is defined as "the number of grams of sugar per gram of pectin in a 65 percent pure water solution, at optimum acidity, which can produce a gel of standard strength." The grade is given the description SAG and the method of determination is that agreed by the U.S.–IFT Committee on Pectin Standardization (Method 5.54). Most powdered pectin is now standardized to 150 grade.

## Recipes and Processes

Detailed recipes are not given in this book but it is useful with a specialized ingredient such as pectin to include a few typical processes to illustrate the differences between the various grades.

In the boiling process, apart from the method of adding the ingredients, it is very important to determine accurately the end point of the boil and hence the soluble solids content that controls the set, texture, and keeping properties of the jelly.

Originally, the thermometer was the only means used to decide the end of the boil but this has been superseded almost entirely by the refractometer.

Several types of refractometer are now available for checking purposes (see Appendix):

1. *Prismatic refractometers*. These require the taking of a sample from the boiling vessel and measuring the refractive index by placing between the prisms.
2. *Reflection refractometers*. These also require a sample to be taken but this is put on a glass surface—there are no prisms to close and these instruments are more robust, easier to clean, and suitable for factory use.
3. *Pan and pipeline refractometers*. These are installed as fixtures in the side of the boiling pan or in a pipeline carrying the boiled mixture from a continuous cooker.
4. *Electronic refractometers*. These are for recording or with dial for visual readings, work on an electronic principle, and give constant information of the state of the boil.

With instruments 1 and 2 it is better to cool the hot syrup before placing on the refractometer surfaces. This can be done by spreading the sample on an aluminum plate. After a few minutes, a portion of the jelly is transferred to the refractometer.

## Jelly 1

| | |
|---|---|
| Final soluble solids | 75 to 76 percent |
| pH | 3.2 to 3.3 |
| Type of pectin | buffered confectioner's pectin (high methoxyl) |
| Water | 35.5 parts wt |
| Pectin | 1.65 parts wt |
| Sugar | 48.5 parts wt |
| Glucose syrup | 29.5 parts wt |
| 50 percent citric acid solution ≡ | 0.75 parts wt citric acid |
| Color and flavor | As required |
| Yield | 100 parts wt |

Mix the pectin thoroughly with eight parts of sugar taken from the bulk. Sprinkle this mixture into the water while stirring, then bring to the boil and continue boiling and stirring for 1 to 2 min.

Add the remainder of the sugar, and when this is dissolved, add the glucose. Continue boiling and stirring until the specified soluble solids is reached (refractometer).

Remove the heat, add the acid solution, flavor, and color, stirring vigorously. Then deposit quickly into molds.

*Note*: It is frequently the practice with modern pectins to add the buffer salt and *one-third of the acid* during the dissolving of the pectin. This is done to aid solution of the pectin and to prevent the partial decomposition during boiling that may take place if the hardness of the water and addition of buffers result in the pH of the slurry rising above 5. However, in some circumstances the addition of acid at the dissolving stage may give pH values low enough to start gelation. It is always advisable to check the pH of the preliminary mix.

## Jelly 2

| | |
|---|---|
| Final soluble solids | 77 to 79 percent |
| pH | 3.5 to 3.6 |
| Type of pectin | Slow set (high methoxyl) |
| Water | 31.3 parts wt |
| Pectin | 1 parts wt |
| Tetra sodium pyrophosphate | 0.42 to 0.46 parts wt |
| or Sodium citrate | 0.20 parts wt |
| Sugar | 47.5 parts wt |
| Glucose syrup | 36.0 parts wt |
| 50 percent citric acid solution ≡ | 0.50 parts wt citric acid |
| Flavor and color | As required |
| Yield | 100 parts wt |

Mix the pectin thoroughly with the buffer salt and with about eight parts of sugar taken from the bulk. Sprinkle this mixture into the water while stirring, then bring to the boil and continue boiling and stirring for 1 to 2 min. When the mixture is dissolved, add the remainder of the sugar and the glucose. Continue boiling and stirring until the specified soluble solids is reached. Stop heating, add flavor and color, and finally the acid solution. Mix thoroughly and deposit quickly.

## Jelly 3

| | |
|---|---|
| Final soluble solids | 75 to 77 percent |
| pH | 4.5 to 4.8 |
| Type of pectin | Low methoxyl (containing calcium and buffer salts) |
| Water | 22.0 parts wt |
| Pectin | 1.8 parts wt |
| Sugar | 8.5 parts wt |
| Glucose syrup | 30.0 parts wt |
| Invert sugar (75 percent syrup) | 13.0 parts wt |
| | |
| Thin boiling starch | 3.7 parts wt |
| Water | 22.0 parts wt |
| Sugar | 33.0 parts wt |
| Flavor and color | As required |
| Yield | 100 parts wt |

Mix the pectin with eight and a half parts of sugar, sprinkle this mixture in twenty-two parts of cold water, stir vigorously, and boil for 2 to 3 min. When solution is completed, add the invert sugar and the glucose.

Dissolve separately the starch in twenty-two parts of water, and after a short boil, mix with the pectin solution.

Add the remainder of the sugar and boil until the required soluble solids is reached.

Add color and flavor and immediately mold in starch.

## Milk Pudding

| | |
|---|---|
| Type of Pectin: low methoxyl | |
| Milk | 100 parts |
| Sugar | 12 parts |
| Pectin | 0.8 parts |
| Vanilla or other flavor | As required |

Mix thoroughly the pectin and the sugar, then add the cold milk while stirring. Bring slowly to the boil, remove the heat, and pour into molds. Allow to cool and set.

## LACQUERS AND WAXES

These substances are used in the confectionery industry mainly for the preparation of protective coatings on panned goods. In addition to forming a moisture barrier that improves shelf life, they enhance gloss. In some cases, they are used on other confectionery pieces to prevent sticking or blocking.

### Shellac

Shellac is used as a glaze for confectionery products, particularly for panwork or dragees. Its application to chocolate panning is described under "Chocolate Processes."

Shellac is the refined form of lac, which is the resinuous secretion of the lac insect and has been used for centuries for the preparation of varnishes and polishes of all descriptions. Of more recent years, grades have been prepared specially for the food industry and these are free from metallic contamination and foreign matter. The production of these grades presented considerable difficulties in the early days as it was a native practice to add various substances to the resin to improve color—among these was arsenic!

It is important that shellac for confectionery use should be guaranteed food grade and purchased from a company specializing in the manufacture of edible lac.

Food-grade shellac to meet the U.S. Pure Food, Drug and Cosmetic Act must have an arsenic content of less than 1.4 ppm and be rosin-free. Strict regulations regarding the solvent exist in many countries; some do not permit isopropyl alcohol.

U.S. regulations state that the solvent must be denatured alcohol 35A in which 100 parts of ethyl alcohol are denatured with 4.25 parts of ethyl acetate (by volume). Glazes of less than 28.8 percent solids must be further denatured by the addition of 5 percent acetone (by volume).

## Other Glazes

As an example of the confectionery glazes available, the following by Zinsser (1985) are quoted. They are produced in concentrations ranging from 22 to 45 percent.

**Regular Confectionery Glaze** It is prepared from food-grade bleached shellac containing 5 percent natural wax. The solution has an opaque appearance but yields a transparent film. The wax content provides for a more flexible film; it builds up well during processing of confectionery pieces and is suitable on porous surfaces.

It is widely used in the confectionery industry.

**Wax-free Confectionery Glaze** Made from wax-free bleached shellac; the glaze is less viscous and, therefore, can be used at the higher concentrations. The dried film is very clear. The use of this lacquer is specially recommended as a finishing coat over waxed coatings.

**Orange Confectionery Glaze** Made from orange flake shellac, it is used where color is unimportant or in cases where the orange color is a useful feature.

This company has recently introduced a product called Sparkle Glow 200, which is a shellac glaze containing small amounts of beeswax and acetylated monoglyceride to reduce tack and thereby the "twinning" of units in the pan.

The history of shellac and its many uses, including confectionery, is described in a publication by Angelo Ltd. (1965). See also "Panning."

## Waxes

Several waxes have applications in the confectionery industry, mainly as glazes applied in the chocolate or sugar panning processes.

The comfit pans are usually coated internally with one of the waxes and they may also be applied as an emulsion, sometimes combined with glyceryl monostearate.

**Beeswax** Beeswax is secreted by honey bees for the manufacture of their combs. The crude wax is obtained by melting these in hot water and straining through cloth. Refining is done by melting the wax and filtering, and also by bleaching with fuller's earth and hydrogen peroxide.

*Properties*

| | |
|---|---|
| Density | 0.95 to 0.97—15.5°C (59.9°F)/15.5°C |
| Saponification value | 85 to 107 |
| Acid value | 18 to 22 |
| Iodine value | 7 to 11 |
| Melting point | 61 to 70°C (142 to 158°F) |

Oriental bees secrete another type of wax called ghedda wax.

**Spermaceti**  This is obtained from the head of the sperm whale and is lower melting than beeswax. It is a relatively pure wax consisting mainly of cetyl palmitate.

melting point    42 to 50°C (107.6 to 122°F)

**Carnauba Wax**  Carnauba wax is a plant wax obtained from the leaves of the palm Copernica cerifera. It is a very hard wax of high melting point and imparts a hard gloss to chocolate or sugar dragees.

*Properties*

| | |
|---|---|
| Density | 0.99 to 1.00—15.5°C (59.9°F)/15.5°C |
| Saponification value | 79 to 95 |
| Acid value | 4 to 9 |
| Iodine value | 7 to 14 |
| Melting point | 78 to 85°C (172.4 to 185°F) |

It is an expensive wax in great demand for many commercial polishes, and two other waxes, ouricouri and candellila, are used as partial substitutes.

## Zein

This is an alcohol-soluble protein obtained from the corn endosperm. It has found increasing use as a food glaze and as a natural product is free from restrictions as to its use.

It produces a flexible film free of color, odor, and flavor. (See also "Panning".)

## REFERENCES

Alginates, xanthan gum Kelco (Division Merck Co.)  1977.  Clark, N.J.
Anon.  1962.  Apple pectin production. *Confect. Prod.,* 349 (April), London.
Anon.  1968.  Improved pectin products. *Food Trade Rev.* 35 (April), London.

Buckle, F. J. 1979. *Pectins*. H.P. Bulmer Ltd., Hereford, England.

Carrageenan, Kobenshavns Pectinfabrik 1979. Lille Skensved, Denmark.

Forsdike, J. L. 1950. *J. Pharm.* 2, 796. London.

Gelatin. *Encyclopedia of Chemical Technology* 1966. Wiley, New York.

Gelatin Manufacturers Institute of America. New York.

Hirst, E. L. 1942. *J. Chem. Soc.* 70.8. London.

Ingleton, J. F. 1971. Agar in confectionery jellies. *Conf. Prod.* (Sept.) England.

Kertesz, Z. I. 1951. *The Pectin Substances*. Interscience Pub., London.

Lawrence, A. A. 1973. Guar, carob gums. *Fd. Technol. Rev. no.* 9. Noyes Data Corp., Park Ridge, N.J.

Lees, R. 1974. Natural gums in the manufacture of sugar confectionery. *Conf. Prod.* (Nov.), England.

Mantell, C. L. 1947. *The Water Soluble Gums*. Reinhold Publishing Co., New York.

Minifie, B. W. 1971. Pectin—its use in candy technology. *Manf. Conf.* (Nov.).

Shellac 1965. Angelo Rhodes Ltd., London, (out of print).

Wielinga, W. C. 1976. *Guar, Carob Gums*. Meyhall Chemical A.G., Kreuzlingen, Switzerland.

*Zein—Its Uses* 1986. Freeman Industries Inc., Tuckahoe, N.Y.

Wm Zinsser & Co. Inc. 1985. *Shellac Glazes*. Somerset, N.J.

# 13

# Starches, Soya Flour, Soya Protein

In the confectionery industry, starch has been used for many years, mainly in two ways. First, it is used as a molding medium to make impressions in filled shallow trays, in which liquid confectionery is deposited. Second, starch is used as a gelling agent or stabilizer in many confectionery products. It is an ingredient in jellies, gums, Turkish delight, and various "spreads."

Many of the common starches have been modified chemically or genetically to meet special requirements or to overcome some of the deficiencies of the unmodified starches, such as texture change and syneresis of prepared gels with aging.

## NATURAL STARCHES

The main starch used in the confectionery industry is corn, but wheat, rice, potato, and tapioca are sometimes employed. In plant life, starch is the reserve carbohydrate and is present in the form of cells, the shape and size of which are characteristic of the plant.

Starch granules viewed under the microscope will clearly indicate the origin; Fig. 13.1.

All starches are polymers of anhydroglucose with linkage through alphaglucosidic bonds and generally consist of two types of polymer called amylose and amylopectin. Most commercial starches contain from 20 percent to 30 percent amylose and the remainder amylopectin, based on the dry starch, but starches generally contain 12 to 15 percent moisture. Moisture content depends on the ambient relative humidity.

There are natural starches that show abnormality—waxy cornstarch contains no amylose and wrinkled pea no or very little amylopectin. In recent years, types of corn have been developed genetically, with high amylose fractions (55 percent to 70 percent).

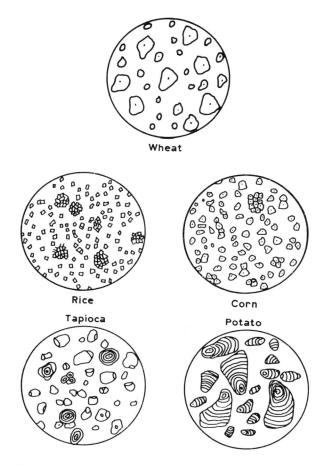

Fig. 13.1.   Granule Size: Rice 3–5 μm, Corn 10–25 μm,
Tapioca 10–25 μm, Potato 15–100 μm (Depends on Origin)
Wheat 10–25 μm

Amylose molecules have a linear structure that contributes to the property of film strength. Amylopectin molecules are much branched, are larger than those of amylose, and are less mobile. Amylopectin forms weak films, but its aqueous collodial solutions have good clarity, are stable, and do not change with age.

High amylose starch has had a very useful effect on the confectionery industry and has greatly increased the production of jellies using starch as an ingredient. Amylose starch needs to be pressure cooked to gelatinize and this can be done rapidly and continuously. The jelly is very quick setting. This is discussed further under "Jellies." Chemically, the normal reaction of starch with iodine is an intense blue but this is entirely due to the amylose fraction. Amylopectin gives a reddish-blue coloration and the reaction can be used for an approximate quantitative analysis.

Starches are insoluble in cold water but when suspended as a slurry and heated they start to absorb water at 60°C (140°F) and this causes the cells to swell very greatly and eventually burst, giving the typical starch dispersions. The nature of this dispersion depends on the amount of water, cooking conditions, and pH, and the temperature of initial absorption varies with the type of starch. In this respect, cereal starches generally cook at a slower rate. With natural starches the suspension in water goes through a series of changes on heating. When the granules are swollen and before they are broken, the liquid has a high viscosity and resembles a smooth paste, but this state is transient and with increased boiling, lowering of pH, or rapid mixing, much thinning takes place. On cooling, these starch solutions tend to thicken and form gels.

The presence of sugars in solution greatly retards the rupture of starch cells and in confectionery boilings, when unmodified starch is used, this has to be cooked for a considerable time in water or dilute syrup before concentration of the batch; otherwise uncooked starch appears in the finished product. In these circumstances, during the period when the starch suspension is very viscous, films form on the surface of the boiling pan and retard boiling.

## Separation of Natural Starches

**Corn** The general principle of starch separation involves steeping in slightly warm water containing sulfur dioxide. This breaks up the protein matter and acts as a bleach. This process is followed by wet grinding, further washing, elutriation, or centrifugal separation, followed by drying. Modern processes achieve very low protein residue, which is necessary if the starch is to be used for glucose syrup manufacture, and it also aids boiling in confectionery recipes.

**Wheat** Wheat flour contains a high proportion of protein (10 percent gluten) and the starch is separated by kneading in running water—the starch being carried away as a slurry and subsequently settled out. Alternatively, the flour suspension is "fermented" to destroy the gluten. The preparation of starch from wheat is prohibited in some countries because of food shortages.

**Potato, Arrowroot, Tapioca**  These are tubers and the separation of starch consists of first cleaning well and then grinding wet to a fluid mash. This is strained to remove fiber and the suspended starch passing the sieve is repeatedly washed and separated by centrifuge.

## MODIFIED STARCHES

### Boiling Starches

In the confectionery industry the first modified starches to be used were the "boiling starches." These are cornstarches that have been "softened" by treatment with dilute acid solutions and a typical method is to suspend a thin slurry of starch in a 0.5 percent solution of sulfuric or hydrochloric acid at 52°C (125°F) for 12 hr. These boiling starches are specified by fluidity, and decrease of pH plus increase of time will give a more fluid starch.

The acid-treated starches are neutralized, separated by filtration, and dried.

The fluidity may be determined using a No. 1 Redwood oil viscometer as follows:

Adjust the temperature in the water-jacket of the Redwood Viscometer to 71°C (160°F) and maintain it at this temperature ±1°F throughout the test. Determine and record the time taken for 100 ml of water at 160°F to be delivered.

Measure 500 ml of water into a one-liter beaker. Into a small beaker weigh 25 g of a sample of the starch and mix to a cream using a portion of the 500 ml of water. Heat the remaining water to boiling and pour into it the starch cream with continuous stirring. Wash out the small beaker with a little of the hot starch solution and add to the bulk. Boil for exactly 5 min stirring continuously, cool to 71 to 72°C (160 to 162°F) in cold water again with stirring, and transfer the requisite amount to the Redwood Viscometer. Determine the volume of starch solution delivered in the same time as that found for the water.

Thin boiling starches have a fluidity of about 65 to 70 whereas medium starches are about 40. Boiling starches do not swell and thicken initially when heated with water and are particularly suitable for use in confectionery recipes such as Turkish delight and gums.

## Oxidized Starches

These are manufactured by a method similar to that for boiling starches but substituting sodium hypochlorite for the acid. This affects the molecular structure of the starch, which stabilizes it against gelling. They produce clear, soft pastes. As with the boiling starches, there are numerous varieties.

## Cross-Linked Starches

The cross-linking reaction is applied to starch when it is required to resist change during boiling in low pH media or if the product is to be heated under pressure.

Cross-linking is brought about by treating starch suspensions with reagents such as phosphorus oxychloride or acetic anhydride, and this is supposed to create a bridge between two separate molecules of starch thereby strengthening the hydrogen bonds and improving the resistance of the granules to disintegration.

Cross-linking also makes starch gels more resistant to damage by mechanical action. Natural starches when swollen by heating are readily broken down by mixing and as a result viscosity is lost. This is a problem when starch is used as a thickener in fillings and spreads, and controlled cross-linking will enable viscosity to be maintained and this is particularly valuable in salad dressings. In confectionery, these starches may be used in chocolate spreads, cordial fillings for shell chocolates, and certain liqueurs where a viscous and unchanging fluidity is required.

The process of cross-linking does not help in preventing the amylose component in starch from subsequent gelling or syneresis. In such cases, the cross-linking of starches with no amylose is used.

A third problem arises when starches are used in foods that are stored for long periods at low temperatures. Colloidal solutions made from the cross-linked starches already mentioned tend to lose clarity and dehydrate to some extent if used in frozen foods.

To overcome this, partial substitution in some of the hydroxyl groups of the starch molecule is resorted to, acetyl or phosphate groups being introduced. These interfere in such a way that alignment of the molecule branches is prevented and this gives greater stability during the freezing/melting process.

## Pregelatinized Starches

These are prepared by treatment of a starch slurry on heated rollers that carry out the combined operations of cooking and drying.

These starches rapidly absorb moisture to form a paste. They are used extensively to make various types of instant foods.

## Zein

This is another interesting substance obtained from natural starches. It is derived from the corn endosperm (tissue inside the developing seed) and is one of the alcohol-soluble proteins (prolamines). It has a special value in confectionery glazes and in the preparation of cozeen, a special antioxidant glaze for coating nuts.

A comprehensive list of the large number of starches used in confectionery is given by Jackson (1973). Brochures published by various companies and the Corn Industries Research Associations provide further information on application of the starches summarized above.

## Moisture Equilibrium of Corn Starch

It is often necessary to know the moisture content of starch in equilibrium with different temperature and relative humidity conditions. This particularly applies when confectionery products such as gums, marshmallows, or fondants are dried in starch in hot rooms.

Table 13.1 indicates the general pattern of equilibrium moisture

TABLE 13.1. PERCENT MOISTURE CONTENT

| %<br>Relative<br>humidity | 25°C<br>77°F | 40°C<br>104°F | 50°C<br>122°F | 60°C<br>140°F | 70°C<br>158°F |
|---|---|---|---|---|---|
| 10 | 6.0 | 5.5 | 5.0 | 4.0 | 4.5 |
| 15 | — | — | 7.0 | — | — |
| 20 | 8.5 | 7.5 | 8.0 | — | — |
| 30 | 10.5 | 9.5 | 9.0 | 7.0 | 7.5 |
| 40 | 12.5 | 11.5 | 11.0 | — | — |
| 50 | 14.0 | 13.0 | 12.5 | 9.0 | 9.0 |
| 60 | 16.0 | 15.0 | 13.5 | — | — |
| 70 | 18.0 | 17.0 | 15.0 | 13.0 | 12.5 |
| 75 | — | — | 16.0 | — | — |
| 80 | 20.0 | 19.0 | — | — | — |
| 90 | 23.0 | 25.0 | 21.0 | 20.0 | — |

contents. The actual values vary somewhat, due to the presence of small amounts of confectionery residues that accumulate in the starch when it has been in use for a period.

In closed drying rooms, air circulation and ventilation are often poor and the local conditions around piles of starch trays may not be the same as those of the air intake. Erratic drying frequently results.

The figures in Table 13.1 are based on the work of Hellman and Melvin (1952), Ubertis and Roversi (1953), and the author. The figures are a useful guide but the following should be noted.

**Oiled Starch**   The addition of 0.5 percent of mineral or vegetable oil to starch to improve the molding properties has no significant effect on the figures.

**Sugars**   The presence of sugars in the starch due to tailings of confectionery affect the equilibrium moisture content most at high ambient humidities.

The technique of depositing confectionery into starch molds is dealt with in the "Confectionery" section.

## SOYA FLOUR, SOYA PROTEIN

The soya bean is an annual leguminous plant (Glycine max) growing about 3 ft high and has been cultivated as a food crop in Eastern Asia for centuries. It thrives best in a continental type of climate where the summers are long and hot. Since World War II, its cultivation in the United States has been greatly increased and much of the world output is now produced there.

The beans or seeds grow like peas, three to six in a pod, and vary in color according to variety. The seeds have two coats enclosing two cotyledons and from these the flour is prepared.

Probably the most important constituent of the bean is soya oil, which is the raw material for many edible fats, including confectionery fats. From this oil vegetable lecithin is prepared, and this is a very important emulsifier in the chocolate and confectionery industry. Its properties are described fully in Chapter 4.

### Soya Flour

Soya flour has been recognized as a valuable foodstuff for many years but its acceptance has been limited in Western countries because of its flavor characteristics. It is regrettable that some of the original flours put on the market had distinctly earthy and bitter flavors that were noticeable in all products in which the flour was used.

Great improvements have been made in the method of preparation and the unpleasant bitter flavor has been removed. There may be still a small residual flavor characteristic of the soya bean but this is imperceptible at the level of use in many food products, which is usually less than 10 percent.

Full-fat soya flour is a very nutritious substance and Tables 13.2 and 13.3 give average composition and comparison with other foods.

Taking into account these compositions it is possible to make "marzipan" pastes in which soya flour replaces ground almonds up to 50%. In so doing it is necessary to observe the sterilization procedure and soluble solids concentration described in the section on marzipan. The definition of "marzipan" varies from country to country and in some the use of this description infers that it is prepared from ground

TABLE 13.2. COMPOSITION OF SOYA FLOUR

| Analysis on a moisture-free basis | % | Mineral content | % |
|---|---|---|---|
| Protein | 44.1 | Calcium | 0.22 |
| Fat | 21.0 | Phosphorus | 0.59 |
| Phosphatides (including lecithin) | 2.1 | Sodium | 0.38 |
| Nonreducing sugars | 11.4 | Chlorine | 0.02 |
| Reducing sugars | trace | Potassium | 2.09 |
| Other carbohydrates (not starch) | 14.5 | Magnesium | 0.24 |
| Fiber | 2.0 | Iron | 0.008 |
| Ash | 4.9 | Copper | 12 ppm |
| | | Manganese | 32 ppm |
| | 100.0 | | |

| Vitamins mg/100 g moisture-free basis | | Protein composition | % |
|---|---|---|---|
| Vitamin A | traces | Arginine | 7.0 |
| Beta carotene | 0.05 | Histidine | 2.5 |
| Thiamine ($B_1$) | 0.60 | Lysine | 6.2 |
| Riboflavine ($B_2$) | 0.31 | Tryptophane | 1.5 |
| Niacin | 1.6–2.2 | Phenylalanine | 5.0 |
| Pyridoxin ($B_6$) | 0.7–1.1 | Methionine | 1.3 |
| Pantothenic acid | 1.1–2.2 | Leucine | 7.5 |
| Choline | 290 | Isoleucine | 5.2 |
| Biotin | 0.08 | Valine | 5.2 |
| Inositol | 220 | Threonine | 4.0 |
| Tocopherols | 21 | | |
| Vitamin K | 0.17 | | |

TABLE 13.3. COMPARISON WITH OTHER FOODS

| | Moisture, % | Protein, % | Fat, % | Carbohydrates, % | Calories per 100 g |
|---|---|---|---|---|---|
| Potatoes | 75.0 | 2.0 | 0.1 | 21.0 | 73 |
| Flour (fine wheat) | 13.0 | 11.0 | 1.1 | 75.0 | 360 |
| Eggs (excluding shell) | 74.0 | 14.0 | 11.0 | 0.7 | 160 |
| Cow's milk | 87.0 | 3.2 | 3.5 | 4.8 | 67 |
| Nonfat-milk powder | 3.0 | 34.0 | 1.0 | 55.0 | 341 |
| Soya flour | 9.4 | 39.9 | 19.0 | 23.5 | 430 |

COMPARISON WITH DRIED EGG AND GROUND ALMONDS

| | Moisture, % | Protein, % | Fat, % | Carbohydrates, % | Phosphatides, % | Ash, % |
|---|---|---|---|---|---|---|
| Dried egg | 5.0 | 45.8 | 32.3 | 3.2 | 9.7 | 4.0 |
| Soya flour | 9.4 | 39.9 | 19.0 | 23.5 | 1.9 | 4.5 |

| | Moisture, % | Protein, % | Fat, % | Carbohydrates, % |
|---|---|---|---|---|
| Ground almonds | 4.0 | 24.0 | 54.0 | 12.0 |
| Soya flour 4: | | | | |
| vegetable fat 3 | 3.6 | 23.3 | 54.8 | 14.1 |

almonds and sugar. In such cases paste containing soya (or other nuts) must not be called "marzipan."

Some bitter almonds are used in soya nut pastes to give the best flavors. Almond essences, because of their benzaldehyde content, tend to oxidize and lose their character.

High-grade soya flour has more recently been used as a partial replacement for condensed milk, particularly in caramels, fudge, and toffees. It is also used in chewable candies. The protein content of the flour imparts body and "stand-up" properties to caramels. The lecithin content aids emulsification and its valuable antioxidant properties preserve the quality of the fat ingredients.

The soya flour should always be dispersed in twice its weight of water before incorporation into a formulation.

There has been a large increase in the use of soya flour in the baking industry. It has been shown that the keeping properties of cookies in particular are improved by the incorporation of 2.5 to 5 percent of soya flour in the flour ingredient.

A general improvement in quality and appearance is also apparent, due probably to the aerating and emulsifying properties of the protein and lecithin constituents.

## Soya Protein

The soya bean (or soybean) is a very valuable source of high-quality protein. Of recent years, much research has been carried out, particularly in the United States to produce soya protein concentrates that have a bland flavor free from the "earthy" character previously mentioned. They have also been structured to have a meat fiber character and good cooking stability. They have the advantage of the nutrition of meat without its cholesterol and fat.

These concentrates have found much application in the development of "nutrition" candies as most confections are high in fat and carbohydrate and low in protein.

Products marketed by the Central Soya Co. may be quoted. An example, sold under the name of Response, has the nutritional composition shown in Table 13.4. Fortified products are also available.

TABLE 13.4 COMPOSITION OF "RESPONSE"

| | | | |
|---|---|---|---|
| Protein (g/100 g) | 70 min. | | |
| Fat (g/100 g) | 0.8 | | |
| Carbohydrate (g/100 g) | 15 | | |
| Moisture (g/100 g) | 8 | | |
| Calories (per 100 g) | 330 | | |
| | | Grams amino acid/ | |
| *Vitamins* | | 100 g protein | |
| Thiamine ($B_1$) (mg/100 g) | 0.2 | *Amino acid* | |
| Riboflavin ($B_2$) (mg/100 g) | 0.1 | Lysine | 6.5 |
| Niacin (mg/100 g) | 0.8 | Methionine | 1.4 |
| Pyridoxine hydrochloride ($B_6$) | | Cystine | 1.5 |
| (mg/100 g) | 0.1 | Threonine | 4.3 |
| Pantothenic acid (mg/100 g) | 0.1 | Leucine | 8.2 |
| Folic acid (mg/100 g) | 0.3 | Isoleucine | 5.0 |
| | | Phenylalanine | 5.2 |
| *Minerals* | | Tyrosine | 3.9 |
| Sodium (mg/100 g) | 50 | Tryptophane | 1.4 |
| Potassium (mg/100 g) | 2,220 | Histidine | 2.7 |
| Calcium (mg/100 g) | 350 | Valine | 5.3 |
| Iron (mg/100 g) | 11 | | |
| Phosphorus (mg/100 g) | 810 | | |
| Magnesium (mg/100 g) | 330 | | |
| Zinc (mg/100 g) | 3 | | |
| Copper (mg/100 g) | 1.3 | | |

Central Soya Co., Fort Wayne, Ind.

# REFERENCES

Central Soya Co., Fort Wayne, Ind.

Corn Industries Research Foundation Inc., Washington, D.C.

Corn Refiners Association Inc., Washington, D.C.

Hellman, N. N., and Melvin E. H. 1952. *J. Amer. Chem. Soc.* 74, 348–50.

Lees, R., and Jackson, E. B. 1973. *Sugar Confectionery and Chocolate Manufacture.* Specialised Publications Ltd., Surbiton, Surrey, England.

Radley, J. A. 1968. *Starch and Its Derivatives,* Chapman-Hall, London.

"Trusoy" in Sugar Confectionery. Spillers Premium Foods Ltd., Cambridge and Puckeridge, England.

Ubertis, B., and Roversi, G. 1953. *Stärke 5,* 266-7. W. Germany.

# 14

# Fruits, Preserved Fruits, Jam, Dried Fruit

Products derived from fruits find wide application in the confectionery industry. Fruits, fruit pastes, and jams are used as ingredients in fondants, jellies, pastilles, and gums. Many natural fruit flavors are delicate and are generally insufficiently strong when used in conjunction with strong chocolate. Strawberries and raspberries are examples and these are often reinforced with artificial flavors. The citrus fruits—lemon, orange, lime, and grapefruit—in conjunction with their natural essential oils are exceptions, and these provide adequate flavor strength in most products.

## COMPOSITION OF NATURAL FRUITS

The variation is very great not only from fruit to fruit but also within any one type of fruit.

Tables 14.1 and 14.2, from McCance and Widdowson (1960) and Macara (1930), give the composition of the more common fruits used in confectionery. Table 14.3 gives the composition of the common citrus fruits. Duckworth (1966) gives a much more extensive list that includes their vitamin, calcium, and iron contents.

## JAMS, PRESERVES, FRUIT PULPS, AND PUREES

Jams sold on the retail market for table use would normally have a soluble solids content of 69 to 70 percent as determined by refractometer, an invert sugar content of 27 percent and a pH of 3 to 3.5. For confectionery use, a soluble solids figure of 70 percent has proved too low, and in the presence of other confectionery ingredients, microbiological spoilage can occur from osmophilic yeasts and molds. This only applies where the jam is used as a filling for chocolate shells or fondants; if the jam is mixed with fondant or other

TABLE 14.1. COMPOSITION OF NATURAL FRUITS

| Fruit | Description | *Water, % | Sugars, % | Protein, % | Calories per 100 g |
|---|---|---|---|---|---|
| Apples | Cooking, no skin or core | 85.6 | 9.2 | 0.3 | 37 |
| Apricots | Dried | 14.7 | 43.4 | 4.8 | 183 |
| Cherries | Raw, no stones | 79.8 | 11.6 | 0.6 | 46 |
| Black currants | No stalks | 77.4 | 6.6 | 0.9 | 29 |
| Dates | Dried, no stones | 14.6 | 63.9 | 2.0 | 248 |
| Figs | Dried, whole | 16.8 | 52.9 | 3.6 | 214 |
| Prunes | Dried | 23.3 | 40.3 | 2.4 | 161 |
| Raisins | Dried | 21.5 | 64.4 | 1.1 | 247 |
| Raspberries | Raw, whole fruit | 83.2 | 5.6 | 0.9 | 25 |
| Strawberries | Raw, flesh and pips, no stalks | 88.9 | 6.2 | 0.6 | 26 |

*The water contents given for some of the dried fruits are higher than would be desirable for use in chocolate or confectionery centers. Raisins, for example, should be within the range 14–17% for chocolate use.
McCance and Widdowson (1960).

370

TABLE 14.2. COMPOSITION OF NATURAL FRUITS

| Fruit | | Insoluble solids (fiber etc.), % | Soluble solids, % | Total sugars, % | Acidity as citric acid, % | Pectin as calcium pectate, % |
|---|---|---|---|---|---|---|
| Apples | Max. | 6.0 | 13.6 | 9.8 | 1.8 | 1.3 |
| | Min. | 1.6 | 9.5 | 4.2 | 0.5 (as malic acid) | 0.5 |
| | Av. | 2.6 | 11.7 | 7.6 | 1.1 | 0.8 |
| Black currants | Max. | 7.9 | 16.7 | 8.3 | 4.3 | 1.7 |
| | Min. | 4.7 | 10.0 | 2.3 | 2.7 | 0.6 |
| | Av. | 5.7 | 14.3 | 6.4 | 3.5 | 1.1 |
| Cherries, stone-free | Max. | 2.7 | 14.8 | 10.6 | 1.7 | 0.4 |
| | Min. | 1.0 | 10.7 | 6.9 | 0.4 | 0.1 |
| | Av. | 1.9 | 12.4 | 8.3 | 0.9 | 0.2 |
| Raspberries | Max. | 9.2 | 11.9 | 7.9 | 2.7 | 0.9 |
| | Min. | 4.4 | 5.4 | 1.3 | 1.2 | 0.4 |
| | Av. | 6.2 | 8.0 | 3.6 | 1.7 | 0.5 |
| Strawberries | Max. | 3.5 | 13.6 | 8.5 | 1.7 | 0.8 |
| | Min. | 1.3 | 5.4 | 3.2 | 0.5 | 0.4 |
| | Av. | 2.1 | 9.0 | 5.5 | 0.9 | 0.5 |

Macara (1930).

TABLE 14.3. COMPOSITION OF CITRUS FRUITS

| | Water, % | Sugar, % | Protein, % | Calories per 100 g |
|---|---|---|---|---|
| Grapefruit | 70–91 | 5–7 | 0.5–0.8 | 22–50 |
| Lemon | 85–94 | 2–3 | 0.6–0.9 | 15 |
| Lime | 86–92 | 0.5 | 0.8 | 36 |
| Orange | 77–92 | 7–11 | 0.8–0.9 | 35–53 |

From Duckworth (1966).

confections for flavor purposes, the quantity used is usually insufficient to affect the soluble solids of the product.

For a filling, the soluble solids content of the jam must exceed 75 percent to be safe from spoilage provided other safeguards are taken in processing. The solids content must also not be too high or crystallization and depositing troubles occur—the usual range is 75 to 78 percent.

Jam suppliers generally use special recipes by arrangement with large confectionery manufacturers and adjustments are made not only in the soluble solids but in the consistency and composition of the sugar ingredients.

Jam is often used as a filling for chocolate shells or as an inner filling for a fondant center, and this jam must have a high fruit content for the best flavor and a fluid consistency to enable it to be deposited easily through mogul or shell plant depositor nozzles.

To make this type of jam, a base syrup prepared from sugar, invert, and sometimes liquid glucose is boiled to a concentration of 85 to 87 percent and fruit or fruit purée mixed in rapidly without additional pectin. This procedure results in the mixture being very close to the ultimate soluble solids of 75 to 78 percent or it will require a very short boil only. Some fruits, such as raspberries, may be partially depipped.

This confectioners' jam may be used for filling as it is or with slight warming, or it may be mixed with pectin syrup to obtain a partial set after it is deposited.

Some confectionery manufacturers make their own jam fillings from canned or frozen purées and by doing this it is possible to obtain higher fruit contents. These purées are concentrated in a vacuum pan and the required amount of sugar/invert/glucose added at the end of the boil. Then vacuum boiling is continued until the correct soluble solids content is obtained.

When high-fruit-content syrups are used for chocolate shells, premature setting is often experienced during depositing because of the pectin in the fruit. This may cause tail formation and an increase in viscosity.

Sometimes a fluid center is required in the chocolate shell. It is possible to use a pectin-reducing enzyme[1] in the prepared fruit mix and this will maintain fluidity at the depositing stage, and subsequently in the unit.

## Handling of Fruit and Fruit Pulps—Preservation

Because of the fibrous nature of a fruit, its viscosity, and the presence of seeds, machinery, valves, depositors, and pipelines may become partly choked, and operators may swill them out with water to clean them. This needs very careful supervision as fibrous and pectinaceous material can adhere to surfaces and entrain moisture or low-concentration syrups.

Experience has shown that resistant strains of microorganisms will build up in these layers and cause serious trouble by infecting subsequent deposits of jam. These troubles are particularly likely if the machinery is used intermittently and sterilization of all equipment with live steam should be standard procedure with complicated machinery. Periodically, sterilization with a proprietary-detergent will also be of value.

Fruit purées or pulps with low solids content are particularly prone to discoloration by contact with metallic iron. Stainless steel is preferred for all equipment.

Fruit that is fresh or preserved in cold storage with, in some cases, the addition of about 10 percent sugar makes the best preserves, but preservation of fruit pulp with sulfur dioxide is a useful means of quickly dealing with fruit from a particularly productive season.

Preservation with sulfur dioxide affects the flavor of some fruits. Citrus fruits, currants, raspberries, and loganberries retain their flavor well but strawberries are less satisfactory in this regard. The fruit may be preserved raw or after cooking, but most fruits with stones are cooked before preserving.

The fruit or pulp is usually sulfited and stored in wooden casks or heavy polythene-lined drums using a 6 percent solution of sulfur dioxide. The final concentration in the fruit is 1,500 to 2,000 ppm. Fruit juices may be preserved using sodium or calcium metabisulfite.

[1] Pectinex 3x, Pectinex Ultra-SP, Novo Enzyme Products Ltd., Windsor, England.

Sulfited pulps are corrosive to metallic iron and should be boiled rapidly in stainless-steel kettles, preferably under vacuum, to remove the sulfur dioxide before making jams or other confection. Sulfur dioxide has a bleaching effect on the natural coloring substances but much of these are restored on boiling.

Pulp may also be desulfited by the addition of sodium percarbonate and this process has been used for apple pulp that was not subjected to boiling.

Many fruit pulps are preserved in large cans that are subjected to pasteurization, a process facilitated by reason of the low pH of the fruit. Much research has been carried out on the long-term storage of fresh fruit in "controlled atmospheres." The metabolism of many fruits can be controlled by storage in atmospheres containing precise contents of oxygen and carbon dioxide. The conditions required for various fruits are summarized by Lipton et al. (1978).

The spray drying of fruit juices is another process that has received much attention. The concentrated juice is dried after inclusion of an additive such as glucose syrup or gum arabic.

Concentrated and dried fruits are becoming important ingredients in the confectionery industry. Apple pulp has a particular value as an ingredient of fruit fillings for hard candies.

The pulp, which is normally sulfited, is first boiled to remove the sulfur dioxide and is then mixed with sugar and glucose syrup and boiled.

| Apple pulp | 2,000 g | Boiled to |
|---|---|---|
| Sugar | 600 g | approximately 112°C |
| Glucose syrup | 400 g | (233°F) |

This "base conserve" may be mixed with other fruit jams and added flavors for the center filling. The center filling should have a syrup phase concentration of 78 to 80 percent (refractometer).

## CANDIED AND PRESERVED FRUITS

### Manufacture

At one time, much of the preserved fruit was prepared in the fruit-growing areas and frequently was incompletely or incorrectly preserved. It did not meet the standards required by the confectioner, being low in soluble solids and often containing osmophilic yeasts. In these circumstance, sterilization and further concentration were necessary before using it in confectionery.

Many fruits, particularly those with a rigid tissue, are preserved whole or in halves. The candied fruit pieces are used in confectionery centers and bars and form very attractive candies in which the nature of the original fruit can be recognized. Cherries and pineapple are probably the most popular fruits for this process, but apricots, pear, apples and plums can also be candied. The peels of citrus fruits, particularly lemons and oranges, also make excellent candied products, and because of the natural essential oil in the peel, retain their original flavor.

The soft fruits, strawberries and raspberries, are difficult to preserve whole but continuous preserving equipment is available that circulates syrup through "baskets" containing trays of the fruit, and by this method, soft fruits will retain their original appearance, but even with this process a lot of the delicate flavor is lost (see later).

With fruits that have a strong cell structure—and this applies to cherries, pineapple, and the citrus peels—preservation must take place in syrups of gradually increasing concentration. This applies whether the syruping is done by the batch or continuous process. If these fruits are immersed in hot syrup of 75 percent concentration, osmosis will cause water to pass outward through the cell walls more quickly than the syrup (which contains high-molecular-weight sugars) passes inward. The effect of this is twofold—first, the difference of the rates of passage of water and syrup will cause the collapse of the cell structure and the fruit becomes very tough in texture and shriveled. Second, the fruit/syrup mixture will continue to transfer syrup and water for a considerable period after cooling, and because of the viscosity of the syrup, areas of low concentration will be formed around the fruit pieces. This can lead to microbiological action in these regions, and if the preserved fruit pieces are drained of free syrup and then coated in fondant or chocolate, the whole piece of fruit may have a low soluble solids concentration.

For this reason, it is essential when preserving fruit to determine the soluble solids in the fruit substance and not in the surrounding syrup. This is done by cutting thin layers from the fruit with a sharp knife or microtome and closing the prisms of the refractometer on these sections.

**Batch Process**   Cherries may be taken as an example. The fruit, which is stemmed, stoned, and washed, is first immersed in boiling water for a short period to soften it and to remove some of the air in the tissues. Scalding with steam is also used for this process. The fruit should remain whole in this operation and is then drained and immersed in low-concentration syrup (30 to 40 percent concentration). The composition of the syrup for this process is preferably a mixture of sugar and liquid glucose in the proportion of 47 sugar to 30 glucose as solids (=36 glucose syrup) but the glucose can be partly replaced by invert sugar to decrease the viscosity of the final syrup at 75 to 78 percent soluble solids concentration. Alternatively, high conversion glucose may be used. Immersion in this syrup should continue for 16 to 24 hr when it is drained off, concentrated to 60 to 65 percent soluble solids, and returned to the fruit for a further period of 24 hr. After this second treatment, it is desirable to warm the syrup as the viscosity has increased and this will accelerate draining. The concentration and immersion are repeated at least twice more until the soluble solids content of the fruit has reached 75 percent. This laborious process was at one time done entirely by hand in vats or barrels, and although mechanization and syrup pumping reduced manual work, this process always involved soaking the fruit in syrup in a static condition in tanks. Because of this, equilibrium was reached fairly quickly in the region of the fruit pieces, and with stronger syrups of high viscosity further diffusion was very slow unless some stirring was done.

**Continuous Process**  Figure 14.1 is a diagram of a plant by Carle and Montanari. Sugar and glucose syrup from storage containers are dissolved to give a syrup of 30 to 40 percent concentration. Equal proportions of sugar and glucose are best to assure freedom from crystallization of the sugar in the preserved fruit.

The fruit, prepared by washing and scalding as in the batch process, is placed in a series of baskets held in a cage supported on a gantry. The preserving vat is filled with syrup, the cage of fruit let down, and the vat closed. By means of the circulatory pump, the syrup is forced between the fruit pieces and slowly concentrated under vacuum.

With this system, the static conditions of the batch process are avoided and the fruit is impregnated by osmosis without shrinkage and in a much shorter time.

The ultimate concentration of the syrup phase in the fruit must be 75 percent minimum, determined by refractometer. With vacuum concentration, the syrup temperature is kept lower and a dark color is avoided.

The cage of processed fruit is lifted from the vat and most of the

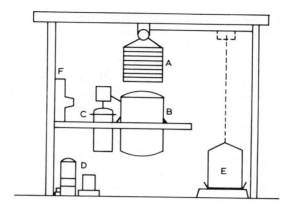

Fig. 14.1.   Diagram of Fruit Candying Plant
A. Cage of Baskets Containing Fruit
B. Processing Vat
C. Heating and Circulation of Syrup
D. Vacuum Pump
E. Position of Cage for Final Draining
F. Control Panel

*Carle and Montanari, Milan, Italy*

syrup drains away. Final draining takes place on a tray which may be done in a hot room to afford complete draining and a slight drying of the surface.

This fruit is now ready for use in various confectionery lines. For chocolate coating by panning, surface syrup is absorbed by rolling in fine sugar and the free dust removed by sieving; alternatively, the preserved fruits may be coated in fondant for chocolate covering. A third method is to deposit the fruit with liquid fondant in a molded chocolate shell. These last two methods are used for the very popular maraschino cherry confections and machines are available for placing and coating the preserved fruit. Cherries and some other fruits are also used in the manufacture of true liqueur-chocolate centers, and genuine alcoholic liqueurs are used for their production.

Candied peels, chopped candied cherries, and pineapple are popular ingredients of soft nougats and pastes.

## Glazed or Glacé fruit

Glacé fruits are popular in many European countries and packed boxes of these mixed fruits certainly look very attractive. Unfortunately, with the exception of perhaps pineapple, cherries, apricots, and peaches, these fruits are oversweet and sickly and have lost a lot of their flavor.

Glacé fruits are prepared by first "candying" the fruit by impregnating with syrup as previously described. Sugar is then dissolved in water at 70 percent concentration and boiled to 80 percent concentration, and this is allowed to cool until the first signs of crystallization occur, when the candied fruit is immersed, drained on wire grids, and then dried in a warm room.

The strength and appearance of the sugar film can be improved if a small amount of gelatin is included in the syrup, or preferably low methoxyl pectin with appropriate setting salts.

## COLORING AND FLAVORING

Many fruits when preserved by the processes already described lack color and it is usual to color them artifically with one of the approved food dyes. This is necessary if the candied fruit is made from fruit preserved in sulfur dioxide solution and it is also necessary, with cherries in particular, to reinforce the flavors with synthetics. Flavor addition is not generally required with pineapple or with citrus peels.

The coloring of preserved fruit is preferably carried out before sugar preservation begins, but if the fruit has been sulfited all sulfur dioxide must be removed first. It is best if the color can be "fixed" in the fruit substance as the bleeding of color from fruit in a confectionery product is unsightly and creates a bad impression on the consumer. Certain approved food colors have this fixative property, and food color manufacturers will advise on the suitability of colors for various fruits.

Flavoring is done in the final syruping process and care must be taken to avoid using a flavor that separates to the surface of the syrup.

A soluble flavor or an emulsion should be used and the fruit and syrup soaked for 24 to 48 hr. When the syrup is drained from the fruit, there is considerable flavor retention in the syrup and it must be used again only for the same fruit. Otherwise, it must be reclaimed by treatment with active carbon (See "Reclaiming").

## Precautions to Be Taken with Purchased Fruits

If it should be necessary to purchase imported preserved fruit, it is important to check these for the presence of fermenting organisms. Some processors using the vat method of preservation obviously do not practice the hygiene recommended previously for depositing machinery and do not raise the syrup fruit mixtures to temperatures sufficiently high to sterilize them. As a result, preserved fruits are often infected with osmophilic yeasts (Zygosaccharomyces torulopsis) and a safe practice is to sterilize these fruits before incorporating them in a confection. To do this, it is sufficient to immerse the fruit in a 75 percent syrup of composition the same as that used for the syruping process, heat to 93.3°C (200°F) for 15 min, and drain. Sometimes these yeasts become active in the candied fruit and produce small bubbles inside the fruit, but more often they cause trouble in the confection made from the fruit, resulting in burst chocolates, which ooze beery-smelling syrup.

## DRIED FRUIT

The drying of food was probably the original method used for preservation, and much fruit is preserved today by this method. For drying to be effective, the moisture content must be reduced until the equilibrium humidity is below 60 percent (water activity 0.6)—the moisture content alone is not a reliable guide (Duckworth, 1966).

Much of the fruit that is dried is grown in the Mediterranean countries or in similar climates in other parts of the world such as California, the Middle East, and parts of Australia. In most of the producing countries, the traditional method was sun drying where generally a period of sunny weather could be guaranteed after the fruit had ripened. The fruits that were dried by this method were dates, figs, and the various types of grape. During this drying process, most of the sucrose in the fruit is converted to invert sugar. Tree fruits—apricots, peaches, pears, and apples—are now similarly dried after cutting into halves and these are exposed to sulfur dioxide for several hours before drying. This prevents attack by microorganisms and improves the appearance. The sun drying of grapes, currants, figs, and dates has proceeded for centuries in some areas under the most unhygienic conditions, particularly on small farms, often on the ground exposed to dirt, dust, animals, and insects. Fortunately, by the combined action of local governments and pressure from food manufacturers who have sent technical advisers to the growing areas, conditions have steadily improved and drying on trays and more hygienic handling áre being practiced.

## Artificial Drying

Artificial drying is developing more and more as this is not only more hygienic but also more reliable. The method usually employed is tunnel drying where the fruit is spread in thin layers and subjected to an air current at temperatures between 60 and 77°C (140 to 170°F). Prunes, which are made from a special type of plum, must be dried this way because, as they are whole fruit, this may take up two days to complete. Dried fruits contain up to 2000 ppm of sulfur dioxide. Other preservatives may be present, such as diphenyl in citrus fruit, benzoic acid, and possibly copper salts from fungicides. Government regulations vary from country to country. Analytical textbooks deal with this subject adequately (Pearson, 1981).

**Raisins, Currants** Small seedless raisins are probably used more than any other dried fruit in the chocolate and confectionery industry.

They should be clean, free from foreign matter, with a moisture content of 14 to 17 percent, and free from crystalline sugars. They should also be easy to separate and in some countries it is the practice to oil the dried fruit with purified mineral oil or vegetable oil to help this. It is important to see that the vegetable oil does not become rancid and the legal limit for mineral oil is 0.5 percent.

**Apricots, Peaches** The dried fruit contains sulfur dioxide but this is easily removed by soaking the fruit and heating to boiling. The fruit can be used for fruit jellies and pastes and, in spite of drying, has a good flavor.

**Figs and Dates** Dates are sometimes covered with chocolate after destoning. Both figs and dates are also used as pastes for fillings in sugar confectionery and cookies. Seeds are usually removed from the fig paste.

**Prunes** These can be used as a paste after destoning, or can be minced to include with other fruit in nougatine and confectionery pastes.

**Apples, Pears**   These have little application as dried fruits in confectionery work.

## Chellies

Chellies (Stewart and Arnold, High Wycombe, England), is the commercial name for small jelly pieces resembling cherries. They are made by an ingenious process covered originally by patents.

The principle of the process is to prepare a solution of sodium alginate in a sugar/glucose syrup solution with, sometimes, other gelling agents, and to eject this solution through nozzles into a coagulant solution of calcium chloride dissolved in a solution of glucose syrup.

The ejection is intermittent, controlled by pistons, so that consistency of size of the drops can be obtained. As soon as the drops meet the calcium chloride solution, an outer skin forms that retains the spherical shape. The hardness of the sphere or "chellie" is determined by the time of immersion in the setting solution. This period may be 1 to 4 min.

Chellies are used in some confectionery lines but more often in cakes and they may be made in any color. Other shapes are possible by a modification of the molding process.

A similar type of jelly piece may be made using low methoxyl pectin instead of sodium alginate. Low methoxyl pectin is set into gel form by calcium salt solutions; the method of producing jellies is described under "Pectin."

## Freeze Drying

This process involves the evaporation (sublimation) of the moisture from a food, including fruit, at temperatures below the freezing point. A high vacuum (less than 0.5 mmHg) is required to effect the drying. It is an expensive process but gives a product of the highest quality with the true flavor of the original fruit.

The dry power is very hygroscopic and its porosity renders it liable to oxidation. It is used only in very special products where cost is less important.

## Ginger

Ginger in its processed forms is a very popular confection and in many ways its preparation is similar to that of various fruits.

Historically, its value as a spice dates back many thousands of years and its medicinal qualities were similarly recognized.

Ginger is the underground stem or rhizome of the herbaceous plant Zingiber officianale roscoe and is a native of the southern provinces of China and India. It is a fibrous rooted perennial related to bamboo, with a pungent aromatic rhizome. Stalks are about one meter high with occasional flower spikes.

Cultivation orginally spread to Malaysia, Africa, and the Caribbean area, but in recent years ginger also has been extensively grown in the Buderim, Queensland, area of Australia. It was introduced there in the nineteenth century and has since been the subject of scientific cultivation and processing. This has resulted in the production of grades of reliable quality and flavor and Australian ginger is now sought worldwide. Figure 14.2 indicates the avenues of disposal of the Australian products but sales of fresh green ginger account for a very small proportion of the crop.

Early harvest ginger, which is picked before the plant reaches full maturity, is tender and relatively free of fiber; late-harvest ginger is more mature and fibrous. As early-harvest green ginger deteriorates on standing, it is provisionally preserved in brine; ginger in brine will keep for several years, but it is normally taken through the next stage of processing within twelve months of harvesting. At this stage, the ginger is drained free of brine, hand cut, hand graded, peeled, boiled, and then impregnated in sugar syrups of progressively increasing strength until it becomes ginger in syrup. Crystallized ginger is obtained after the syrup is drained from the ginger and a coating of sugar crystals applied. Dry ginger, ground ginger, oleoresin of ginger (gingerine), and oil of ginger are obtained and used principally in the ways shown in Fig. 14.2.

Australian ginger is processed to a syrup concentration of 72 percent soluble solids minimum with controlled invert and pH levels. Confectionery manufacturers may use this directly in the process for making chocolate-covered or crystallized ginger. All that is required is to heat the ginger to 85°C (185°F) and drain on a sieve. This treatment should raise the syrup concentration to 75 percent.

For the manufacture of chocolate-covered ginger, the heated, drained ginger is rolled in fine sugar or a mixture of sugar and cocoa. It is then covered with chocolate by enrober or preferably coated with chocolate in a revolving pan and then enrobered—the latter process gives a thick coating that does not ooze. A thick, dark chocolate coating blends very well with preserved ginger.

Crystallized ginger is generally made by the sugar-sanding process

## Green Ginger

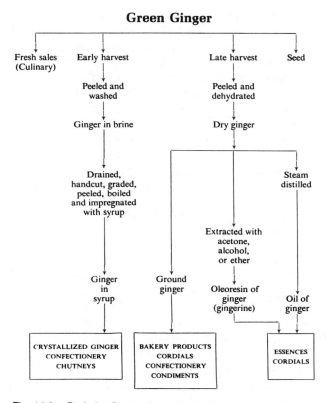

Fig. 14.2.   Buderim Ginger Growers Ltd., Queensland Australia

and the methods used for crystallizing fondants are not applicable to ginger and are unnecessary. The well-drained ginger is rolled in granulated sugar and dried off in a stove until, on cooling, the sugar coating is a layer of fused crystals. Light steaming can be used before stove drying to improve adhesion and some confectioners advocate a light rinse with water before sugar sanding to remove surface syrup. This is a practice to be deprecated, but is mentioned because it is used by many of the older confectioners not only for ginger but for preserved fruits as well, and should be opposed absolutely by the technologist. Similarly, the steaming process must be carefully controlled because, if overdone, low syrup concentrations will arise with all the attendant troubles with fermentation and mold.

The original Chinese "cargo" ginger was imported in barrels and was covered with a sugar syrup of about 65 percent concentration. Because of the practice of using acetic acid in the brine preservation solution, a variable degree of inversion took place that ranged from 20 to 90 percent and pH was found to vary from 3.4 to 4.2.

For use in chocolate centers, the concentration of the syrup phase had to be raised from 65 percent to 75 percent, and this involved treatment similar to that described under "Preserved Fruit." If the inverted cargo syrup were used, dextrose crystallization was likely to occur later in the center of the ginger pieces, detracting from the pleasant texture of good ginger. It was necessary, therefore, to preserve the ginger in sugar/glucose syrup as with fruits, and this ensured a good texture and no subsequent crystallization.

**Ginger Extracts**   In recent years, ginger extracts have been used for flavoring purposes instead of the dried, ground root. They are more uniform in flavor and free of the microbiological contamination so often found in raw spices.

Ginger oil is extracted by steam distillation and is a pleasant aromatic substance without excessive pungency.

The oleoresin is also a very popular flavoring extract. It is a combination of the natural oil and resin and is readily soluble in alcohol. It is used for the preparation of essences and the flavoring of soft drinks.

Apart from confectionery, ginger, in its various forms, is used extensively in general cooking, as well as in cakes, salads, and pickles.


## REFERENCES

Bendall, R. L., and Daly, R.A.   1966.   Ginger growing in the Nambour area. *Q. Rev. Agric. Econ.* Australia.

British Patents Nos. 556,718; 586,157; 708,992. Stewart and Arnold, High Wycombe, England.

Buderim Ginger Growers, Buderim, Queensland, Australia.

Carle and Montanari, Milan, Italy.

Duckworth, R. B.   1966.   *Fruit and Vegetables.* Pergamon Press, London.

Egan, H., Kirk, R. S., and Sawyer R.   1981.   *Pearsons Chemical Analysis of Foods.* Churchill-Livingstone, Edinburgh, Scotland.

Lipton, W. J., *et al.*   1978.   Controlled atmospheres for fresh fruits and vegetables. *Encyclopedia of Food Sciences,* Vol. 3, 182–194.

McCance, R. A., and Widdowson, E. M.   1960.   *Composition of Foods.* H.M.S.O., London. New edition 1985 revised by A. A. Paul and D. A. T. Southgate.

Macara T.   1931.   The composition of fruits. *Analyst,* 56, 39. London.

# 15

# Nuts

Nuts are an important ingredient in the chocolate and confectionery industry. They are compatible in dark or milk chocolate because of their low moisture content. Their flavor, particularly if they are roasted, also combines well with chocolate and reduces sweetness.

Nuts ground to pastes or to small particles are used in such confections as marzipan, persipan, noisette, and praline pastes.

## COMPOSITION

Nuts are a well-balanced nutritive food. Average compositions of the well-known nuts are given in Tables 15.1 and 15.2. As natural products, variation is to be expected, and in some, peanuts for example, composition varies with type and origin. This is discussed in the descriptions of the various types of nuts.

## VARIETIES OF NUTS

### Almonds

The almond (Prunus amygdalus) is probably the most popular of all nuts and its popularity over the centuries has never waned. A huge proportion of the world' supply of almonds is now grown in California. It is stated that the production of shelled nuts in that state increased from 125 million pounds in 1970 to 376 million pounds in 1979. The estimated crop in 1990 is 700 million pounds. Other large producing areas are Spain, Iran, Morocco, and Portugal. Minor quantities are produced in Afghanistan, Tunisia, Cyprus, and Libya. The available supplies in 1981 were estimated in million pounds as United States, 415; Spain, 185; Italy, 56; and other, 64.

The tree requires a deep fertile soil, freedom from frost, rain during

TABLE 15.1. COMPOSITION OF NUTS

| Type of nut | Moisture, % (raw dried nut)* | Protein, % | Fat, % | Total carbohydrates, % (starch, sugars cellulose) | Dietary fibre, %† | Ash, % | Calories per 100 g |
|---|---|---|---|---|---|---|---|
| Almond | 3.5–6.5 | 17.0–24.0 | 52.6–59.0 | 10.0–17.3 | 14.3 | 2.0–3.2 | 560–655 |
| Brazil | 4.6–6.0 | 12.0–17.0 | 61.5–67.0 | 5.7–10.9 | 9.0 | 3.3 | 619–699 |
| Cashew | 4.0–6.0 | 17.2–21.0 | 38.0–48.0 | 22.0–29.2 | 1.0–1.4 | 2.6–3.0 | 531–559 |
| Chestnut (dry)‡ | 6 | 10.5–12.0 | 8.0–10.0 | 70.0–84.0 | 13.5 | 2.5 | 380–404 |
| Coconut (desiccated unsweetened) | 3.1–3.5 | 5.3–9.0 | 57.0–64.0 | 29.0 | 3.9(23.5) | 1.2 | 604–660 |
| Hazel | 4.3–5.8 | 12.7–18.0 | 62.0–68.0 | 12.0–16.4 | 10.0 | 2.0–3.0 | 643–669 |
| Macadamia (oil roasted) | 1.7 | 7.3–8.0 | 73.0–76.5 | 12.9–14.0 | 1.7 | 1.7 | 700–718 |
| Peanut | 5.6–7.0 | 26.0–31.0 | 44.0–49.5 | 10.0–18.5 | 8.1 | 2.4–3.0 | 581–587 |
| Pecan | 3.4–3.5 | 12.0–19.6 | 70.0 | 9.0–14.7 | 3.8 | 1.7 | 675–742 |
| Pistachio | 4.0–5.3 | 19.3–23.0 | 53.7–55.0 | 15.0–19.0 | — | 3.0 | 594–647 |
| Walnut | 3.5–4.0 | 14.7–19.0 | 60.0–63.4 | 13.5–15.7 | 5.2 | 1.7–3.0 | 647–705 |

*Moisture content varies with method of harvesting and preparation. Any nut with moisture content over 7% is subject to deterioration on storage. Roasted nuts may have moisture contents down to 1 to 2% and these, in ambient conditions, will soon pick up moisture. In some tables, moisture contents of raw nuts are quoted and this must be allowed for in calculations.
†"Dietary fiber" in some tables is included in "total carbohydrates." According to McCance and Widdowson (1985), dietary fiber consists of the various polysaccharides (cellulose, hemicellulose, pectic substances) and lignins that are not absorbed in the digestive tract. Fibre in Coconut.—McCance and Widdowson quote 23.5%; Ruehrmund (General Foods) quotes 3.9%.
‡Raw chestnuts contain approximately 50% water.
*Source:* Ranges taken from various published figures.

TABLE 15.2. COMPOSITION OF NUTS—ELEMENTS AND VITAMINS (mg/100 g)

| Nut | Na | K | Ca | Mg | Fe | Cu | P | S | Cl | Thiamin | Riboflavin | Niacin | Biotin | Tocopherol | Ascorbic acid |
|---|---|---|---|---|---|---|---|---|---|---|---|---|---|---|---|
| Almond | 6 | 580–860 | 240–250 | 260–290 | 4–6 | 0.14–0.3 | 440–511 | 37–145 | 2–11 | 0.20 | 0.33–0.92 | 2–7 | 0.4 | 3.0 | Trace |
| Brazil | 1.5–2 | 710–760 | 180–185 | 410 | 2.8–3.5 | 1.1 | 590–680 | 290 | 60 | 1.00 | 0.12 | 1.7 | — | 11.0 | Trace |
| Chestnut | 10 | 490 | 46 | 32 | 1.0 | 0.25 | 74 | 30 | 15 | 0.2 | 0.22 | 0.2 | 1.3 | — | — |
| Coconut (desiccated) | 28 | 680–750 | 22–27 | 90 | 3.4–3.6 | 0.6 | 160–190 | 76 | 196 | 0.03–0.06 | 0.03 | 0.3–0.6 | — | 0.6 | Trace |
| Hazel | 1.0 | 350 | 44 | 56 | 1.1 | 0.21 | 230 | 75 | 6 | 0.10–0.4 | — | 0.8 | — | 1.5 | Trace |
| Cashew | 14 | 462 | 39 | — | 4.0 | — | 371 | — | — | 0.42 | 0.25 | — | — | — | — |
| Peanut | 6.0 420 salted | 670–680 | 61–74 | 180 | 2.1 | 0.27 | 370 | 380 | 7.0 | 0.23–0.90 | 0.10–0.15 | 16 | — | 8.8 | Trace |
| Macadamia | 7 260 salted | 329 | 45 | 117 | 1.8 | 0.30 | — | — | — | 0.21 | 0.11 | 2.0 | — | — | — |
| Pecan | Trace | 540 | 70 | — | 2 | — | 260 | — | — | 0.8 | 0.10 | 1.5 | — | — | Trace |
| Pistachio | — | 972 | 13.0 | — | 17.3 | — | 500 | — | — | 0.67 | — | 1.4 | — | — | — |
| Walnut | 2.7–3.0 | 690 | 61 | 130 | 2.4 | 0.3 | 510 | 104 | 23 | 0.30 | 0.13 | 1.0 | 2.0 | 18.0 | Trace |
| Green Walnut | | | | | | | | | | | | | | | 1,300–3,000 |

Source: Figures from various sources.

blossom time, and dry, sunny conditions during harvesting. The tree will blossom well in the northern temperate countries where it is grown as an ornamental tree known for its early pink flowers, but such trees do not yield good nuts.

All Californian almonds are shelled by machine. Much of the European crop is machine shelled but some is still done by hand. Ideally, almonds are dried naturally rather than artificially. The nuts, after being shaken from the trees, are swept into wind rows, where they dry prior to collection for hulling and shelling. Artificial drying is needed only when the kernels are moist, for example, during bad weather when there is a risk of mold developing.

There are two types of almond—sweet and bitter. Sweet almonds are used by the confectionery industry and for desserts, and bitter almonds are used in small amounts for flavoring almond flours and pastes.

The worldwide production of bitter almonds is small. Bitter almonds are poisonous if consumed in quantity, and contain the glucoside amygdalin, which, if hydrolyzed, yield hydrogen cyanide. Hydrolysis takes place with bitter almonds and apricot and peach kernels if they are pulverized, moistened, and incubated at 30°C (86°F). The natural enzyme, emulsin, in the outer layers of the kernel decomposes the amygdalin into hydrogen cyanide, benzaldehyde, and D-glucose. Benzaldehyde is "almond flavor."

This method has been used to debitter the cheaper apricot kernels for use in marzipans and the reaction may be accelerated by the addition of emulsin enzyme. Genuine marzipan should be made from almonds and not debittered apricot kernels. The product from apricot kernels is usually persipan, and in some countries this is the legal description.

There is a great variety of sweet almonds and these are described by Howes et al. (1945). Probably the most famous is the long Jordan almond of Spain used for decorating choice confectionery. The most prominent variety, however, making up some 60 percent of the California crop and at least 30 percent of the world crop, is the soft-shell "nonpareil" variety.

Almonds may be eaten raw or roasted. They may be blanched by immersion in boiling water; this loosens the brown skins, which are easily removed by squeezing the nuts between rubber rollers or bands. Blanched almonds can be used for decoration after drying or light roasting but more often are ground by being passed through rollers for the manufacture of almond paste or marzipan. The blanching process is also essential for the destruction of enzymes and microorganisms when the nuts are subsequently processed to confectionery pastes.

Almond butter made by grinding the nuts to a smooth paste is similar in consistency to peanut butter but has a better flavor and is much more expensive. This almond delicacy is much favored by vegetarians.

Almonds are an ingredient of toffee "brittles" when they are roasted in high boiled syrups. The brittles may be ground to a paste called praline or nugat, and this paste becomes an ingredient of other confections such as truffles and noisette.

Almonds are easier to store than many other nuts provided they are dried correctly and the kernel skin kept intact, but are liable to become insect infested under warm conditions. Suitable conditions are 4 to 7°C (40 to 45°F) with a relative humidity of 55 to 65 percent. As the nuts are firm, jute bags may be used for their transport, and these can be stacked up to 8 ft without damage to the kernels. Choice Jordan almonds are better boxed. Much of the California crop is packed in fiberboard cases.

## Brazil Nuts

Brazil nuts (Berthollettia excelsa) are a tropical nut greatly favored for dessert purposes and very popular as an ingredient in chocolate bars and blocks, both as a whole nut and chopped. The smaller Brazil nuts are covered in chocolate by enrobing or panning for use in chocolate assortments.

As with almonds, the nuts are very popular with vegetarians and are in great demand in health food shops.

The nuts are native to the Amazonian forests where they grow on very large trees up to 150 ft high that tower above the rest of the forest.

The nuts are contained in woody spherical cases about 6 in. across, that hold up to twenty-four, closely packed together like the segments of an orange. The containers or fruits are on the branches at the top of the tree and when ripe they fall to the ground with great force. This makes collection a dangerous job, particularly in windy conditions. Added to this hazard are the effect on the workers' health of the severe tropical conditions and lack of proper transport, which makes the size of the crop greatly dependent on the prevailing weather and availability of labor. When collected, the fruits are opened locally under temporary shelters and the nuts removed and washed. They are then taken to the local landowner where the collectors receive payment at a nationally agreed upon proportion of the existing market price. From these small centers, the nuts are taken to the large warehouses in Manaos or Para, where they are sorted and cleaned to prevent loss from attack by mold. They are then transported in sacks or in bulk in special boats to user countries where the kernels are removed by mechanical methods. Brazil nut kernels are usually packed in greaseproof paper or polythene-lined cases or tins holding about 28 lb and care must be taken not to bruise the surface of the nuts as this will cause the formation of a gray outer layer and a deterioration in keeping properties and flavor. This also occurs if they are stored in conditions of high humidity. The storage conditions suitable for almonds are also correct for Brazils.

Brazil nut trees have been cultivated in other tropical forests, for example, Malaysia. Commercially, however, this has not been very satisfactory as the trees normally take nine years to bear. In Brazil, there are some jungle areas that contain many trees but transportation is so difficult that the crops cannot be harvested.

For confectionery purposes, Brazil nuts are never roasted. A popular, but expensive, confection is the "buttered Brazil" in which the nut is dipped in hot butterscotch and then allowed to set on a greased plaque.

Brazil nuts are remarkably resistant to infestation as compared with other nuts.

## Cashew Nuts

The cashew nut (Anarcardium occidentale) is a native of Brazil but during the sixteenth century the tree was taken to India as well as countries in Africa, Indonesia, and Southeast Asia.

Figures produced by FAO in 1976 gave world production as shown in Table 15.3.

The cashew tree is an evergreen that grows up to 40 ft high. The

TABLE 15.3. WORLDWIDE PRODUCTION OF CASHEWS

| Country | Production (in 1,000 tons) |
| --- | --- |
| India | 243 |
| Mozambique | 200 |
| Tanzania | 115 |
| Brazil | 27 |
| Kenya | 20 |

nut is kidney shaped and grows at the end of a fruit called the cashew apple.

The shell of the nut is really a tough skin and contains a juice that causes skin blisters. Roasting destroys the blistering compound and the nuts can then be opened by hand cracking. Roasting improves the flavor. This operation used to be carried out by the pickers who heated a few pounds at a time in a pan over an open fire, but mechanical roasters are now established in the larger factories.

The nuts have a soft, mealy texture and have gained greatly in popularity as salted nuts, but their use in confectionery has so far been confined to nut pastes and pralines. The nuts are a little unusual in composition because of their high carbohydrate content.

A remarkable discovery has resulted from the separation of the cashew nut shell liquid, which in the original method of roasting was lost. Modern roasting methods allow the recovery of most of this substance. It is a highly unsaturated compound, and although originally considered a waste product, its main use now is in the production of friction dusts for brake and clutch linings.

The traditional method of packing the nuts was in large sealed cans that had been heated to destroy infestation. At one time, infestation was so bad that it jeopardized the market. In these cans, storage is no problem, but in boxes, cool conditions are required as for almonds.

## Chestnuts

Chestnuts (Castanea sativa), very popular in some countries, have found very little application in the confectionery industry except for marrons glacés. These are made by heating the chestnuts in syrups of gradually increasing concentration until they are thoroughly impregnated, when they can be packed attractively and will keep as well as any other sweet.

In Italy, France, and Spain, where chestnut trees flourish, the nut meats and flour are esteemed as delicacies and used for the preparation of many culinary dishes.

Chestnuts are very high in starch and sugar and low in fat compared with other nuts. They do not keep unless very well dried.

## Coconut

The coconut (Cocos nucifera) is one of the most valuable foods known and the various other non-edible products of the coconut palm only enhance the importance of this remarkable tree. Whole books have been written about the coconut palm. It is widely distributed in the tropics where it flourishes best in sandy coastal areas having 80 in. per year or more of rainfall. The trees take seven years to bear from planting, and may bear for fifty years. Coconut palms attain heights of up to 100 ft. The main producing countries are Sri Lanka and the Philippines. The dried coconut meat is known as copra and is the source of coconut oil, which is used in enormous quantities for making fats for baking and confectionery.

An important product used in the chocolate and confectionery industry is desiccated coconut. The nut is removed from the hard shell and the thin brown rind pared off. This must be very thorough or the finished coconut will be contaminated by flecks of brown. The pared nuts are then washed free of "milk" and skin residues and put through shredding machines, which give a milled product of varying degrees of fineness. After shredding, the ground meat is subjected to a steam or hot water treatment to destroy salmonella and lipases and to reduce bacteria counts to low levels. The pasteurized meat is then dried on wire mesh belts by means of hot air, which reduces the moisture content to less than 4 percent. The critical moisture level of desiccated coconut is 5 percent and the product should be well below this to ensure good shelf life.

After drying, the desiccated coconut is put through a multiple sieving operation to produce the four main grades; Extra Fine, Macaroon (Fine), Medium, and Coarse. In addition, cuts such as Flake, Long Thread, Extra Fancy Shred, Slice, Chip, and Strip are also available for specialized uses. A variety of other coconut products is available to the chocolate and confectionery trade such as; moist white sweetened coconut, toasted coconut, creamed coconut and "tenderized" coconut. The manufacture of various confections with coconut is described under "Confectionery."

Top-quality desiccated coconut should be white in color, free from foreign matter, and uniform in granulation against standard, and the expressed oil should have a free fatty acid content of less than 0.1 percent.

Coconut confectionery at one time had the reputation of developing soapy rancidity. This was generally due to the presence of residual fat-splitting enzymes remaining from unsatisfactory processing of the original coconut. To overcome this, a steam sterilization process of the desiccated coconut was introduced at the user factory. The desiccated coconut, in thin layers, was subjected to steam jets on a conveyor belt, followed by hot air drying. Sterilization may also be carried out in hot syrup but fat separation then becomes a problem.

Coconut and coconut confections will exude oil, and thin films are subject to oxidative rancidity catalyzed by heat, light, and certain metals. Oxidative rancidity, however, is less likely to occur than soapy rancidity. Apart from incorrect processing in the manufacturing of desiccated coconut, the inclusion of lipase-containing ingredients in a coconut confection will cause breakdown of the lauric coconut oil with consequent soapy rancidity.

Cocoa powders, milk powder, egg albumen, and spice flavors occasionally are suspect regarding lipolytic activity.

Desiccated coconut, in the past, has been found to be contaminated in a variety of ways and this was related to the very primitive conditions prevailing in the countries of origin. Unbelievable as it may now seem, there was a practice of lining cases with lead foil, causing contamination of foods containing coconut. Often there were fine pieces of iron present, wood splinters, and frequently infestation with a variety of beetles.

The discovery of salmonella in a number of consignments some years ago initiated an investigation of producing areas and a very great improvement in conditions of manufacture and quality of product resulted.

The majority of desiccated coconut is now packed in 100-lb (or 50-kg) four-ply paper bags with heat-sealed polythene liners. At one time, it was the practice to store coconut in low-temperature cold stores—9.4°C (15°F) but with the prevailing better-quality supplies 4.5 to 7°C (40 to 45°F) is adequate and a shelf life of at least six months may be realized.


## Hazelnuts—Barcelonas, Levant Nuts, Filberts

Hazelnuts (Corylus var.) are very popular as an ingredient of chocolate bars. The roasted nuts have a flavor that combines well with chocolate and the chocolate coating helps to retain crispness. Hazelnuts can also be included in confectionery centers provided the moisture content (or water activity) is not too high. Under moist conditions, the nuts become sour and lose their crispness.

The nuts are cultivated mainly in Spain, Italy, France, and Turkey, and more recently in Oregon in the United States.

The Spanish and Turkish nuts are derived from trees and bushes of the wild hazel or cob nut and different varieties are distinguished by the shape of the husk or "hood" that covers the shell.

The trees yield from 20 to 100 lb of nuts or about half that weight of kernels. The kernels vary in shape, size, and color, and are interchangeable from a flavor standpoint but small sizes are preferred for certain thin chocolate bars or dragees.

Hazelnuts are usually roasted for confectionery and chocolate use but are often eaten raw as a dessert nut. During roasting, the brown skin becomes detached and is removed by sieving and air suction, resulting in a blanched kernel but the color varies from off-white to brown according to the degree of roast.

Fresh crop hazelnuts have a plump appearance with no shriveling and on cutting should show no internal discoloration or mold.

If stored in warm conditions, the nuts gradually shrivel and develop a dark layer beneath the surface, and many of the kernels will have a sour taste. Hazelnuts, even when kept in a cool storage area with controlled humidity, lose some of their quality by the time the new crop arrives. Experiments have shown that if new crop nuts are given a very light roast, thereby reducing the moisture to about 2 percent, they can be kept in closed containers in cool storage, 4.5 to 7°C (40 to 45°F) for at least 18 months without loss of quality. This method is useful to preserve a consistent high quality in a "self" line such as panned roast hazelnuts. Roasting, however, does help to remove slight sourness that may have developed in the raw nut. Fully roasted nuts should never be stored in conditions that allow air access. They soon become rancid.

## Macadamia Nuts

The macadamia (Macadamia ternifolia) is a native of Australia and it is likely that the tree was first discovered in about 1870 growing in the coastal districts of Queensland. It received its name from Dr. John Macadam, former secretary of the Philosophical Institute of Victoria.

Trees were exported to other subtropical countries but only in Hawaii has the nut been developed to a large extent. Here, largely as a result of intensive research by the Mauna Loa Macadamia Nut Corporation in conjunction with the University of Hawaii, the nut has achieved tremendous popularity in a relatively few years. Its texture and flavor are unique and its popularity has been helped greatly by tourism. Visitors have extolled the virtues of the macadamia nut far and wide.

Growing conditions in Hawaii are exceptionally good, with abundant rainfall, porous volcanic soil, and an almost year long growing season. The early research efforts did not meet immediate success because it was found that seedlings from the same parent would often produce nuts very different in quality. This led to twenty years of research by the University of Hawaii. Some 60,000 trees were observed and tested in a painstaking process of selection and grafting before seven varieties were eventually developed that reliably produced a desirable quality of nut. These major strains make up the bulk of Hawaii's crop today.

The macadamia tree is a subtropical evergreen of the Proteaceae family. It grows to a height of 40 ft and has long, dark-green leaves resembling those of the holly. The nuts begins as a series of tiny green buds. These soon develop into sprays of 300 to 400 sweet-scented white blossoms. Each spray produces only about four to eight nutlets. When mature, the fruit resembles small limes. Beneath an outer husk is a hard, brown shell, and inside that the macadamia nut kernel.

Every Mauna Loa macadamia tree today is a product of grafting, which is the only way to be sure of quality. They are carefully tended in a nursery for about two years until they are strong enough to stand in the rough lava earth. After they are transplanted to an orchard, it takes five more years before they begin to bear, and they do not reach full production until they are fifteen years old.

The tree flowers in waves over a period of four to five months and thus there are five to six harvests annually. Mature fruits fall to the ground, but because the tree sheds leaves all year long, the leaves have to be blown from the area before the nuts can be picked up.

The harvested nuts are first dehusked and then must be dried and cured until the nuts' moisture is reduced to about 1.5 percent. This is done to separate the kernel from the shell, making it possible to shatter the shell without damaging the kernel inside. Macadamias have the reputation of being the toughest nut to crack. It takes 300 psi of pressure to break the shell. Today, the nuts are passed between counterrotating steel rollers, precisely spaced to break the shell without disturbing its contents.

The major product is the whole, roasted, lightly salted nut, which is vacuum packed in jars or cans, but other confections are made, such as brittles, sugar-panned nuts, and chocolate-coated nuts.

## Peanuts

The peanut (Arachis hypogaea) is not a true nut, being a member of the pea and bean family. It is unusual in its method of growth. When the flowers wither, the stalks bury themselves in the soil, and at the stalk end the nuts in shell are formed. Because it is an earth nut, it is liable to be contaminated with soil organisms, and when removed from the soil, it has a moisture content of 30 to 40 percent. The methods of harvesting and subsequent drying are important, and the aim is to reduce the moisture as rapidly as possible or molds will develop and toxins are produced.

It is an easy crop to grow in warm climates and peanuts have been a staple food for people living in those areas.

The peanut is also regarded as an oil seed, and large quantities are sold for their oil. Arachis or peanut oil is used for a great number of purposes. The purified product is employed as a salad and frying oil and the hardened oil is an ingredient of bakery and confectionery. It also finds use in such unusual areas as pharmaceuticals and metal polish, and low-grade oils are the raw material for soap manufacture. Vast quantities of peanuts are consumed as dessert nuts and used for salting and in a variety of confections and health foods and uses continue to increase.

Peanuts are particularly popular in the United States. Great progress has been made in their production and improved techniques and seed research have doubled the yields during recent years.

Nearly one-half of the edible consumption in the United States is in the form of peanut butter, a high-protein nutritious product with a unique flavor but, admittedly, it is not liked universally.

The United States has also become an important exporter of edible peanuts.

In Europe, peanuts are available on the open market from the United States, South Africa, India, Argentina, Brazil, and the Sudan and other African countries. Although the popularity of the peanut is increasing in Europe, resistance to its acceptance by the public in the past has been attributable to the development of bad flavors in confectionery containing peanuts and the sale of roasted nuts in nonprotective packages.

Deterioration is also associated with peanuts purchased from countries where harvesting methods are less satisfactory and proper storage facilities are not available. Lack of correct storage conditions, which are needed for a product with such a high oil content, is a reason why peanuts from these countries are used for oil expression.

**Types of Peanuts**   There are three main types of peanuts:

Spanish peanuts consist of varieties with small round kernels. They blanch easily, and are used in peanut brittle, nut clusters, and for other purposes where small nuts are preferred.

Runners have a medium-sized kernel, not as smooth or regular in shape as Spanish peanuts, with a darker skin. These nuts are liable to split in processing and blanch less easily. This type is sometimes classified as the "hardy" variety of peanut. It has general acceptance for confectionery products.

Virginia peanuts are long, large nuts used for in-shell roasting and for salting, they are sold as edible nuts in glass jars, cans, or bags.

Numerous hybrids are available and nuts produced as the edible nuts are usually graded to size. Experience has shown that nuts for human consumption must receive special attention right from the harvesting and any confectionery manufacturer using quantities of peanuts should be able to trace the origin and age of the supplies and to supervise storage. To purchase nuts when their history is unknown is courting disaster.

Work by Minifie and Butt (1968) on peanuts of different origin showed great variation in keeping properties.

The stability of peanuts is also related to unsaturated fatty acids present in the oil. Worthington and Holley investigated the linolenic and linoleic acid contents of peanut oil in different varieties of nuts. They found the relative linoleic acid contents to be: Spanish, 34.2 percent; Virginia, 29.6 percent; Runner, 22.0 percent.

The degree of maturity of the nut had an appreciable effect on the fatty acid composition of the oil. The unsaturated fatty acids present in immature and mature nuts were linolenic acid, 0.3 percent (immature) and 0.02 percent (mature); linoleic acid, 33.0 percent (immature) and 28.0 percent (mature).

Although the type of nut, age, storage, and maturity ultimately affect the keeping, deterioration or rancidity is hardly ever detectable in the raw nut. It is only when nuts have been roasted that the effect of their past becomes apparent.

Apart from the origin, age, maturity, and storage, the keeping properties are greatly affected by chopping the nuts. Roasted nuts are more resistant to oxidative rancidity than are the same nuts chopped into pieces and the release of oil and the greater surface exposed accelerates deterioration.

A recognized method of improving keeping is to coat the nuts, after roasting, in Cozeen, a product developed by Alikonis. This is a solution of zein (corn protein) with acetylated glycerides in alcohol, and is applied by spraying the nuts in a rotating pan.

Roasted nuts should not be stored, but should be used in the chocolate or confectionery product as soon as possible after roasting. If packed for retail sale, the nuts should be vacuum or inert gas packed.

Freshly roasted nuts have an attractive flavor that is soon lost when surface oxidation of the oil occurs.

**Aflatoxin**  At one point, the peanut industry was rocked by the sensational discovery that an epidemic of deaths among turkeys in the United Kingdom was caused by Brazilian peanut meal in their feed. Investigation showed that a toxic substance, called aflatoxin, was formed by the action of the mold Aspergillus flavus and two less common molds A. parasiticus (Spear) and Penicillium puberulum (Bainer).

The mold developed primarily between the digging and the drying of the nuts while the moisture was between 20 percent and 40 percent. As a result, increased importance was placed on artificial drying and improved storage.

The discovery of aflatoxin brought great pressure on the nut processors and users. Large users of peanuts now have all incoming consignments examined for aflatoxin (see later).

## Pecans

The pecan (Carya pecan), which resembles the walnut in appearance, is essentially American, and large quantities of the nut are consumed in the United States. It is a nut that became known comparatively recently when the more remote areas of the United States were opened up.

The pecan tree grows wild in groves in Texas, Oklahoma, and the Mississippi valley, reaching heights of nearly 200 ft with a large branch spread. In some areas, pecan trees line the roads. The yield of nuts from wild trees may reach 800 lb but where orchards have been cultivated, the yield is about 100 lb per tree.

To harvest the nuts, the trees are beaten with long bamboo poles and canvas sheets are spread over the ground to catch the nuts.

The nuts are cracked by machine and it is the practice to wet them beforehand, allowing time for penetration to the kernel. This avoids breakage of the kernel.

Where nuts are to be kept for later use, they are best stored in their shells in cool areas and cracked when required for use.

Pecans are popular with vegetarians, and also are used in confectionery, mainly for topping chocolates and cakes. In the United States, they are a popular ingredient of ice cream.

The crop yield in the United States in the years 1980 through 1982 was assessed at 220 to 256 million pounds per year.

## Pistachio Nuts

Pistachios (Pistacia vera) are probably better known for their appearance than for their eating qualities. They are pale green throughout and are attractive as toppings or may be included in nougat where, with cherries and citrus peel, they make a colorful candy. They are also an ingredient of the original Turkish delight (rahat locoum).

The nut grows on small trees that are native to Turkestan but they are cultivated throughout the Middle East and in Mediterranean countries.

The trees can grow under exceptionally dry conditions in very poor soil. Seedling trees live to a great age—as much as 300 years—but most propagation is by grafting. These trees have a much shorter bearing life.

In India, the nut is regarded as a delicacy and many dishes are prepared from it.

Iran, a large supplier of pistachio nuts some years ago, decided not to export to the United States for political reasons. This accelerated the industry in the United States and the California Pistachio Association was formed. Trees were planted in 1970, and by 1978, 2.5 million pounds of nuts were harvested, and by 1985, crops of 50 million pounds were being anticipated.

## Walnuts

The walnut (Juglans regia) is one of the widely used nuts and has been known as a food since very early times. The variety "regia" is the nut that is generally sold on the European and American markets but other types are available of which the American black walnut (nigra) is the best known. The black walnut is renowned for its exceptionally hard shell but the kernels have a good flavor.

The walnut tree is native to Iran but is now grown over very wide areas. The main nut-producing regions are in France, Italy, California in the United States, and China, but the tree is also cultivated in South Africa, Australia, and Rumania. In some countries, numerous superstitions have grown up around the nut because of the unusual shape of the shell and its two distinct sections.

The tree grows to a height of 100 ft in suitable soil and is cultivated in orchards as well as being grown in the wild state in many areas. The tree is self-fertile but with single trees the male and female flowers may open at different times and it is therefore more successful to grow the trees in groups or in orchards to ensure good yields of nuts.

In Europe, France is the largest producer of walnuts and the main growing areas stretch eastward from Bordeaux with the Isére district producing the best-quality nuts. In the Isére, walnut growing is treated as an institution where grafting, hybridizing, and scientific methods generally are practiced.

In Italy, the walnut areas are concentrated in the area of Naples where soft-shelled varieties are grown.

The walnut industry in California has grown greatly of recent years and much attention has been given to scientific methods of cultivation, preparation, and storage.

Most of the California crop is grown in the San Joaquin and Sacramento valleys, and in the early 1980s the annual crop was of the order of 200 to 220,000 tons (in shells).

In countries where conditions are favorable to walnut growing, most of the nuts separate from the husks and fall from the trees to the ground. They must be gathered quickly to prevent discoloration and damage from mold and insects and are then washed, cured, and dried. This is done in the sun by the small grower but the large orchards use artificial drying with air at 38 to 43°C (100 to 110°F) and this prevents splitting and opening of the shell, which occurs with rapid drying. Without proper drying, the nuts deteriorate rapidly in storage. The fresh kernel has a bitter skin but this bitterness disappears with keeping.

The mechanical methods of cracking, and the cleaning and sorting of the kernels, are complicated because of their intricate shape, and three basic grades of walnut come on to the market—halves, quarters, and pieces.

Halves are used for decoration of the tops of assortment chocolates, walnuts whirls, and cakes; quarters can be used in confectionery centers and cakes where individual nut pieces are required; pieces are used for inclusion in chocolate blocks, cakes, and caramels, and are often reduced further in size for this purpose.

With walnuts it is very difficult to ensure complete freedom from shell, and in spite of careful sorting by the suppliers, it is essential for the confectionery manufacturer to give additional conveyor band inspection.

**Black Walnut** The black walnut (Juglans nigra) is a widely distributed North American forest tree found largely in the hilly areas of

Missouri, Mississippi, and Arkansas. It is almost a national tree in the United States and is also grown in parts of Europe as an ornamental tree.

The kernel is of good quality and flavor but the nut has the drawback of having a fibrous husk, which has to be removed by machine. The remaining shell soon turns black—hence the name.

Some farmers use the tree as a secondary source of income as other crops can be planted on the open ground between the trees.

The overall United States crop is considerably less than the regia nut, at about 12,000 tons per annum.

**Keeping Properties and Storage** Walnut kernels will deteriorate rapidly under warm conditions, particularly if the relative humidity is high, and for this reason it is not advisable to include walnuts in high-moisture confections unless for rapid sale.

A confection should have a water activity not exceeding 0.5 if walnuts are to be included and good shelf life is expected. If this is exceeded, the nuts become sour and later rancid.

A publication by the U.S. Department of Agriculture describing work carried out with the Diamond Walnut Growers of California indicated that walnut kernels should be stored at temperatures below 3°C (38°F) with an optimum relative humidity of 60 percent and the moisture content of the nuts should be maintained at between 2.8 and 4 percent.

Storage properties can be extended by the application of sprays of edible oil containing the antioxidants BHA and BHT. Cozeen, which was found useful for peanuts, was not found to be effective on walnuts.

Walnuts have the reputation of being very susceptible to infestation and topping nuts that have been attacked by the larvae of ephestia or plodia present a very revolting sight. With cold storage, the nuts are likely to arrive at the user's factory free of infestation, but even so many users fumigate them in their cases. If infestation arises later, on confectionery, it is generally due to insect attack in warehouses or shops. Once established, larvae grow rapidly and foul the nut with excreta and webbing. Because of this, some manufacturers refuse to have walnuts in their assortments; if they are included, it is best not to have exposed nuts but to coat them with a thin chocolate covering. This, however, does not ensure freedom from insect attack.

## Other Nuts

Some lesser-known nuts are described briefly. Some may eventually achieve the popularity of the well-known nuts.

*Paradise (Sapucaia) nut.* This is similar to the Brazil. It is native to South America but its use is confined to the local population.

*Butter (Swarri) nut.* A large South American nut about four times the size of a Brazil, it has a very thick, hard shell and soft, buttery kernel with a pleasant flavor.

*Oyster nut.* This is really a gourd seed about 1.5 in. in diameter by 0.5 in. thick. The kernel is pleasantly flavored and has been tried in chocolate and confectionery in place of almonds and Brazils. It keeps well.

*Hickory nut.* These nuts, which are great favorites in North America, resemble the pecan.

*Pignolia nut (pine nut).* A soft nut with an unusual flavor, it is often an ingredient in Italian and Middle Eastern recipes. These nuts have been used for marzipan.

## HANDLING AND STORAGE OF NUTS

In the confectionery industry, nuts are practically always purchased with the shell removed, and before inclusion in a product may be blanched to remove the outer skin of the kernel, then roasted, and sometimes chopped or ground.

The purchase of good quality, sound nuts is by no means an easy task and the big users find it best to be in direct contact with suppliers in the country of origin. By doing this it is possible to exert a measure of control over hygiene, drying, dehusking, and the period and conditions of storage and transport between the growing country and the user's factory.

Serious incidents have arisen through bad handling of nuts; for example, the presence of salmonella in coconut due to very poor hygiene in the tropical factories where it was dried and shredded, and the presence of aflatoxin in peanuts arising from molds growing on the kernels as a result of poor harvesting, drying, and storage.

### Insect Infestation—Rodent Damage

Insect and rodent damage can be very serious when nuts are stored in bulk or in bags in uncontrolled storage.

Even when the nuts are subsequently cleaned and roasted, insect fragments and rodent hairs remain. These can be detected by analytical methods and consignments will be rejected on these grounds (see "Pest Control").

## Harvesting

With the exception of peanuts, most of the important nuts are tree crops, and when growing, the nuts are enveloped in a flesh or husk. As ripening proceeds the husk dries and perishes and usually becomes detached when the "fruit" falls from the tree, leaving the nut in its shell. This has to be dried and the original practice was to expose to the sun on trays, covering at night or during rain, but much drying is now done artificially.

The quality of the nut kernel is largely dependent on efficient and speedy drying after harvesting, especially if the nuts are washed to remove dirt and other foreign matter.

## Cracking

The efficient cracking of nuts in shell depends on good drying with a loose kernel in the shell, but some nuts such as the cashew have a leathery shell that has to be cut off or, alternatively, the nut is quickly roasted when the shell becomes brittle and can be split off.

Cracking at one time was entirely a manual operation but more and more is now done mechanically.

## Storage

The correct storage of nut kernels is of great importance and the lack of care shown by many agents who deal in such an expensive commodity is a very sad reflection on that section of commerce. Nuts are a seasonal crop, which means that some of the kernels must be kept for as long as twelve months unless imports from the northern and southern hemispheres can be alternated.

Cool storage is always to be preferred but many users do not understand the additional need to control the relative humidity of the store and the moisture content of the nuts. Light is also detrimental to nuts and promotes oxidative rancidity in the oil in the outer layers. Many nuts are not even stored in cool conditions and infestation becomes a problem in the warmer months.

## Preparation for Confectionery

**Foreign Matter**   The great problem in the use of nuts for chocolate and confectionery is to eliminate pieces of shell, immature nuts, defective nuts (including moldy and infested), and stones.

There are various mechanical and electronic methods available that can remove a high proportion of the foreign matter but there is always some residue, which necessitates human inspection on a conveyor belt with hand removal.

**Electronic Sorting**   This method, which relies on an electronic eye detecting darker colored pieces, has been used successfully for beans and peanuts. It is less successful for other nuts because of their natural color variation and the presence of skins. With almonds, however, the method has been successfully applied by detection of a "light" spot, thereby detecting broken and chipped kernels. When it can be used, it is a very rapid method of examining each nut.

**Density Methods**   By means of air elutriation, separation of stones and shell can be achieved and this can be combined with vibratory action on screens, conveyors, and spirals. Certain firms specialize in this machinery for many types of food.

A flotation method using a brine solution was found useful for stone removal from hazelnuts but this needed subsequent washing of the nuts. Although roasting followed, water penetration did some damage to the nuts.

**Belt Sorting**   Hand sorting on a conveyor belt becomes tedious, and however conscientious the workers, some foreign matter will be missed. It is desirable to change the inspection teams regularly and a bonus can be paid on the basis that periodic samples taken from sorted material show no foreign matter.

**Metal Detectors**   Electronic detectors have the advantage of finding nonmagnetic metal pieces. With bulk ingredients, these detectors can divert contaminated material immediately from a continuous stream of material.

**Roasting**   Batch and continuous roasters similar to those used for cocoa beans are suitable for nuts. Peanuts are often roasted by heating in vegetable oil. In recent years, there has been a tendency to favor drum dry roasting, which is stated to improve flavor.

Roasting decreases the stability of nuts and renders them prone to oxidative rancidity, and thus it is necessary to use roasted nuts immediately in the product for which they are intended. They should not be stored unless under vacuum or inert gas.

**Chopping and Shredding Nuts**   These processes expose more surfaces to the action of air and release nut oil. Therefore, chopped roasted nuts will deteriorate more quickly. These and any fine siftings must be incorporated in a base such as chocolate, fat, or caramel as soon as possible. This particularly applies to peanuts and hazelnuts.

**Blanching**   Almonds are blanched wet and any nut blanched wet must be roasted or dried immediately or rapid deterioration will set in. Wet nuts also discolor in the presence of iron. Peanuts and hazelnuts are dry blanched, the skins being removed by blowing and suction.

## SUMMARY

The following general comments augment the previous discussion.

Nuts vary greatly in their susceptibility to deterioration in confectionery and this may be due to several factors:

*Type of nut.* Some nuts are much more prone to deterioration than others. Of the common nuts, walnuts and peanuts are probably the worst, hazelnuts occasionally give trouble, and almonds and Brazils rarely.

*Origin of nut.* Certain types of nut vary considerably according to their origin, and this applies particularly to peanuts. Research has shown how difficult selection can be if judged by the usual methods of physical inspection.

*Age and storage conditions.* All nuts contain oils, which are of varying degrees of saturation. These oils are liable to oxidative rancidity, and deterioration is encouraged by warm storage, bruising, or damage to the nuts, and dampness, which increases the moisture content of the nuts.

If molds grow on or inside the nuts, due to dampness, hydrolytic rancidity may occur, imparting a soapy flavor. Coconut is most likely to show this defect. What is more disturbing is that traces of mold growth arising from some temporary dampness and the hyphae having disappeared with subsequent drying, lipolytic enzymes will be left behind, causing rancidity after the nut has been incorporated in a confectionery product. The presence of mold will produce bad flavors apart from hydrolytic rancidity, and mycotoxins may also be formed in the nuts. Aflatoxin in peanuts is a typical example.

*Moisture.* High-moisture contents will accelerate the deterioration of all nuts, and if roasted nuts are allowed to pick up moisture, the effect is even worse. Under these conditions, they turn sour and ultimately rancid.

Mistakes are all too often made in confectionery recipes by including nuts in moist pastes and fondant. The water activity (A/W) of roasted nuts is quite low (0.3 to 0.4) and to include these in a fondant or paste with an A/W of 0.7 will mean fairly rapid transfer of moisture to the nuts. Not only will this bring about flavor changes, but the texture is impaired and the nuts lose all their attractive crispness and become soggy.

*Reclaiming nuts.* It is the practice in many factories to reclaim nuts from misshapen or broken confectionery or chocolate. While this is done for the sake of economy, it can lead to early deterioration and bad flavors if they are included in new production. Reclaiming of nuts from chocolate is reasonably safe if the chocolate is soaked off with cocoa butter, but sometimes vegetable oils are used and this is not recommended.

The reclaiming of nuts from confectionery by making a water slurry and sieving is a dangerous practice unless they can be dried rapidly and roasted. Hazelnuts have been satisfactorily recovered, but walnuts and peanuts should never be treated this way. It is possible to make a paste from confectionery containing nuts by grinding with chocolate, or with sugar and vegetable fat, and if this is done warm, the moisture in the original confection is dried out.

# REFERENCES

Egan, H., Kirk, R. S., and Sawyer, R. 1981. *Pearson's Chemical Analysis of Foods.* Churchill-Livingston, Edinburgh, Scotland.

Howes, F. N. 1948. *Nuts—Their Production and Everyday Uses.* Faber and Faber, London.

McCance, R. A., and Widdowson, E. M. 1960. *The Composition of Foods.* Her Majesty's Stationery Office, London. (New edition, 1976, A.A. Paul, and D. A. T., Southgate.)

Minifie, B. W., and Butt, K. C. 1968. A method of assessment of the shelf life of peanuts. *Manuf. Confect.* (*Oct.*).

Paul Beich Inc., Cozeen. Bloomington, Ill.

Rockland, L. 1960. The keeping properties of walnuts. *IFT Proc.* San Francisco (May).

Tressler, D. K. 1976. *Fruit, Vegetable and Nut Products.* AVI Publishing Co., Westport, Conn.

Woodroof, J. G. 1979. *Tree Nuts.* AVI Publishing Co., Westport, Conn.

Woodroof, J. G. 1983. *Peanuts. Production, Processing, Products.* AVI Publishing Co., Westport, Conn.

Worthington, R. E., and Holley, K. T. 1972. "The Linolenic acid content of peanut oil." J.A.O.C.S., 44, 515–516.

## Additional References

*Peanuts*: National Peanut Corporation (Standard Brands) Norfolk, Va. Goldblatt, L. A. 1969. *Aflatoxin.* Academic Press, New York.

*Coconuts*: General Foods Corp. (Franklin Baker Coconut) Dover, Del. Association of Philippine Desiccators, Makatil, Philippines.

*Almonds*: California Almond Growers, Sacramento, Calif.

*Walnuts*: Diamond Dunsweet Inc., Stockton, Calif.

*Edible nut statistics*: Gill and Duffus Landauer Ltd., London, England.

*Macadamia nuts*: Mauna Loa Macadamia Nut Corporation, Hawaii.

# 16

# Chemical and Allied Substances Used in the Confectionery Industry

Numerous additives are permitted in the food industry, and reference to the lists of substances approved in the legislation of different countries will show that the number is very large indeed.

In this chapter, only those substances that have common use in the confectionery industry are discussed. The technologist should always refer to published legislation with regard to a substance being considered for use. The media are very quick to publicize information concerning additives that have doubtful acceptance. As a result, many food companies try to avoid the inclusion of any substances that may be classified as additives. However, difficulties will always arise; for example, in the manufacture of a jam or jelly, citric acid may be added to improve setting and flavor. It can be considered an additive even though citric acid was present naturally in the fruit constituent of the jam.

Food additives are defined internationally by the Codex Alimentarius Commission and in the U.K. 1984 Food Labelling Regulations. The latter states that an additive is "any substance, not commonly regarded or used as food, which is added in, or used in or on, food at any stage to affect its keeping qualities, texture, consistency, appearance, taste, odour, alkalinity or acidity, or to serve any other technological function in relation to food, and includes processing aids in so far as they are added to, or used in or on, food as aforesaid."

However, vitamins, nutrients, salt, and certain other substances are not included.

## ACIDS IN CONFECTIONERY

The addition of acids to confectionery performs various functions. It provides a flavor effect and is essential in this respect for fruit-flavored confectionery. It alters the pH of the product and in so doing helps to control the setting of pectin jellies. Low pH also brings about some inversion of the sugar (sucrose) ingredient during boiling, sometimes a problem if uncontrolled. Acidity has some preservative action, partly attributable to pH and partly to the nature of the acid radical. Acetic acid and sorbic acid have significant preservative effects.

Originally, tartaric acid was mainly used in sugar confectionery. It has now been largely replaced by citric acid.

In the continuous production of hard candies, acids and flavors are added directly to the boiled syrup. The addition of crystalline citric acid causes difficulties.

Buffered lactic acid, which is a liquid, is now used and is easier to dispense. Buffered lactic acid has a significant lower inversion effect on sugar solutions than citric or tartaric acids.

## Citric Acid

Citric acid ($COOH \cdot CH_2 \cdot C(OH)COOH \cdot CH_2 \cdot COOH$) occurs naturally in lemon juice, from which it was first isolated by Scheele (1784). This was originally the source of commercial supplies but it is now produced by fermentation through the action of certain molds on sugar syrups or molasses. It is available in the anhydrous form and as a monohydrate; it is odorless and colorless and dissolves readily in water (50 percent solution) and in alcohol (rectified spirit) 35 to 40 percent.

It is used in most confectionery as a 50 percent solution, but with hard candies, the powdered acid may be used. In this respect, the relative melting points of the crystalline and anhydrous acids should be noted.

*Anhydrous.* Melting point 153°C (307°F).

*Monohydrate.* No sharp melting point. When heated, loses water of crystallization and finally melts at about 130°C (266°F).

At the temperature of a boiled sugar batch prepared by pouring onto a table, the hydrate will melt whereas the anhydrous acid will not, but the hydrate will add a small amount of moisture to the batch. If anhydrous acid is used, it must be finely powdered to provide good dispersion. Bad dispersion causes objectionable sharp-flavored concentrations in parts of the candy.

*Analytical details*:

Assay $\begin{cases} \text{Monohydrate as } C_6H_8O_7 \cdot H_2O & \text{99.5 to 101.0 percent} \\ \text{Anhydrous as } C_6H_8O_7 & \text{99.0 percent min.} \end{cases}$

Residue on ignition                                      0.05 percent max.
Sulfate                                                  Nil
Oxalate                                                  Nil
Assay { Heavy metals (lead, copper, iron)                10 ppm max
       { Readily carbonizable substances                 Pale brown color only
                                                         (BP test)


## Tartaric Acid

Tartaric acid (COOH·CH(OH)·CH(OH)·COOH) is prepared from potassium hydrogen tartrate (cream of tartar), which separates in the dregs during wine manufacture.

The pure substance is anhydrous in the form of colorless crystals with a melting point of 169°C (336.2°F). It will dissolve readily in water, and 60 percent solutions may be made. It is also soluble in ethyl and isopropyl alcohol. At one time, it was used a great deal in confectionery but has since been superseded by citric acid.

Tartaric acid has a sharper taste than citric acid. Instances have been recorded of bad flavors developing when used in conjunction with some citrus oils, and with these ingredients citric acid should be employed.

*Analytical details*:

Assay anhydrous acid                        Not less than 99.7 percent
Loss on drying at 105°C (221°F)             Less than 0.5 percent
Sulfate                                     Absent
Oxalate                                     Absent
Heavy metals (lead, copper, iron)           10 ppm max


## Fumaric Acid

Fumaric acid (HOOC·CH=CH·COOH) is an anhydrous crystalline substance, only slight soluble in water. It occurs in Fumaria officinalis and various fungi.

It is used in foodstuffs as an acidulant, particularly where noncaking properties are required, as in sherbet powders. It is nonhygroscopic and has a pleasant, strongly acidic flavor.

It is used in gelatin goods to enhance gel strength and to improve the whipping qualities of egg albumen.

*Analytical details*:
Assay as anhydrous acid     99.8 percent (min. 99.5 percent)
Moisture     Less than 0.2 percent
Chlorides     10 ppm max.
Sulfates     25 ppm max.
Malic acid     Less than 0.2 percent
Heavy metals (lead, copper, iron)     10 ppm max.

The acid does not melt but sublimes at about 200°C (392°F).

*Solubility in water*:
  25°C (77°F) 0.6 percent
  40°C (104°F) 1.1 percent
  60°C (140°F) 2.4 percent
100°C (212°F) 9.8 percent

## Malic Acid

Malic acid [$COOH \cdot CH_2 \cdot CH(OH) \cdot COOH$) is a natural acid widely distributed in the vegetable kingdom, particularly in unripe apples, but also in grapes, quince, and rowan berries.

It forms anhydrous, colorless crystals melting at 130°C (266°F) and is not hygroscopic. It is readily soluble in water—nearly 60 percent at 25°C (77°F).

Malic acid is nontoxic and is particularly useful as an acidulant in hard candy manufacture, where it is often used in conjunction with lactic acid.

These acids may be buffered to give a high enough pH to reduce inversion but retain the acid taste. This is very useful when the acids, flavors, and colors are added to the high boiled syrup as with continuous hard candy production. Sodium lactate is used as a buffer. (See "Lactic Acid.")

Assay and purity are similar to tartaric and fumaric acids.

## Lactic Acid

Lactic acid ($CH_3 \cdot CH(OH) \cdot COOH$) is a natural acid prepared by the fermentation of sugars such as lactose, sucrose, and dextrose and allied substances including starch and various gums. The fermentation is induced by lactic bacteria.

Commercial lactic acid is a thick hygroscopic liquid that does not crystallize and is miscible with water and ethyl alcohol. The commercial acid is normally the racemic acid with a slight excess of the dextrorotatory substance. It is a pleasant-flavored acidulant and can be buffered for adding directly to high boilings, thereby inhibiting inversion. The buffer salt used is sodium lactate, which is itself a liquid, and will conveniently mix with lactic acid.

Another useful property associated with lactic acid is the solubility of calcium lactate. Many fruits, gums, and even natural water supplies contain appreciable amounts of calcium that will be precipitated as a haze if citric acid is used as the acidulant. Lactic acid does not give this trouble and this is very helpful in the manufacture of fruit gums to retain a bright, clear appearance.

Lactic acid may also be regarded as a mild preservative, although it does not possess general bactericidal properties.

It has been used to inhibit fermentation in fondants with low syrup phase concentration, and 0.2 to 0.4 percent is effective in syrups of concentration down to 70 percent. In this respect, it is similar to acetic acid but the flavor effect is much less apparent. Acetic acid, however, is effective at much lower concentrations (see below).

*Analytical details*:

| | |
|---|---|
| Melting point | 16.8°C (62.2°F) (racemic acid) |
| Decomposes at 250°C (482°F) | |
| Ash | Less than 0.07 percent |
| Sugars | Less than 0.05 percent |
| Heavy metals (lead, copper, iron) | Less than 10 ppm |
| Arsenic | Less than 0.2 ppm |
| Total nitrogen | Less than 250 ppm |

## Acetic Acid

Acetic acid ($CH_3COOH$) was known to the ancients as crude wine vinegar (4 to 5 percent acetic acid) and the concentrated acid was prepared by Stahl in about 1700. Acetates are found in plant juices, milk, and certain animal matter such as muscles and perspiration. It is the final product of many aerobic fermentations.

The pure substance is a strong corrosive acid, liquid, with a characteristic pungent odor.

It may be regarded as a mild natural preservative and has been employed in the confectionery industry for this purpose in marzipan and macaroon pastes, and in some fondants.

*Analytical details*:

| | |
|---|---|
| Specific gravity 15°C (59°F)/15°C | 1.055 |
| Boiling point | 118°C (244.4°F) |
| Melting point | 17°C (62.6°) |

## PREVENTION OF FERMENTATION

### Acetic Acid

In the "Confectionery" section of this book, reference is made to the constitution of nut pastes and the presence of microorganisms in relation to their keeping properties. Base pastes made with sugar (sucrose) syrups only, have a soluble solids content insufficiently high to inhibit fermentation. At the same time, some nuts contain organisms of the osmophilic type capable of acting in higher concentration syrups.

Fermentation in these pastes may be avoided by attending to:

1. Equipment and process sanitation.
2. Maintenance of a syrup phase concentration over 75 percent.
3. Sterilization of the nut meats.

However, work was carried out on the use of acetic acid as a preservative where, for reasons of texture or the process adopted, these conditions could not all be met.

Experimental work with fondant of a syrup phase concentration between 70 and 75 percent showed that the addition of 0.05 to 0.07 percent of acetic acid made it resistant to fermentation when the fondant was inoculated with yeasts.

In marzipan pastes, higher quantities were required (up to 0.15 percent). Above 0.10 percent most people were able to taste the acid; some were able to detect much lower levels and even found 0.05 percent unpleasant.

The acetic acid must be well dispersed in the syrup ingredient.

Sorbic acid (see below) is now used and is more effective.

### Sorbic Acid

Sorbic acid ($CH_3$—CH=CH—CH=CH—COOH) is an organic acid that has come into prominence because of its antimicrobial properties. It is found naturally in the juice of the unripe sorb apple (Sorbus aucuparia/mountain ash) and in the pure form is a white crystalline powder with a slightly acidic, agreeable taste. It melts at 130 to 134°C (266 to 273°F).

It is now generally accepted as a harmless food preservative and, historically, the preservation value of mountain ash berries has been known for a very long time. They were used to prevent mold in fruit preserves.

TABLE 16.1. SOLUBILITY OF SORBIC ACID

| Solvent | Temperature | % Solubility |
|---------|-------------|--------------|
| Water | 20°C  (68°F) | 0.16 |
|  | 50°C  (122°F) | 0.6 |
| Ethyl alcohol 95% | 20°C | 14.0 |
| Glycerol | 20°C | 0.5 |

The effectiveness of sorbic acid is greatest in an acid medium, which should always be less than pH 6, and under these conditions, it inhibits molds and yeasts, as well as some bacteria.

Most confectionery products do not require preservation if conditions of concentration and sterility, previously mentioned, are observed, but with some nut pastes, soft fillings, and fruit purées, sorbic acid may be usefully employed.

**Regulations for Use**  The permitted use varies from country to country, and reference should be made to the current regulations, but as a guide the level of use in various foodstuffs is as follows:

Jams, conserves (pH 3.5 approx)     0.025 percent
Fruit syrups, cordials              0.02 percent
Fondants, marzipan,
    confectionery pastes            0.10 to 0.20 percent

The sorbic acid is best added to the hot syrup or final product, but a stock solution may be prepared by dissolving in sodium carbonate in the ratio of 1 sorbic acid to 1.25 sodium carbonate. Syrups should not be boiled after addition of the acid. Its solubility is shown in Table 16.1.

## Phosphoric Acid

Phosphoric acid ($H_3PO_4$) is available as a pure grade, suitable for food purposes, at two concentrations:

90 percent $H_3PO_4$     Specific gravity 1.75
67 percent $H_3PO_4$     Specific gravity 1.50

It is used to a limited extent as an acidulant in foods, particularly soft drinks. It has also been used as a partial neutralizing acid after the alkalization process for cocoa, but is not permitted in some countries, tartaric or citric acids being more generally acceptable for this purpose. Phosphoric acid and phosphates are used as plant and machinery detergents in the food industry.

*Analytical details*:

|  | SG 1.75 | SG 1.5 |
|---|---|---|
| Boiling point | 171°C (340°F) | 123°C (253.4°F) |
| Freezing point | 28°C (82.4°F)* | −58°C (−72.4°F) |
| Arsenic[1] | 2 ppm max. | |
| Lead[1] | 10 ppm max. | |

* Phosphoric acid (S.G. 1.75) will crystallize at ambient temperatures.

## BUFFER AND OTHER INORGANIC SALTS

Various approved inorganic salts are added to foodstuffs, and this applies to chocolate, cocoa, and confectionery. When using any of these substances, it is essential that the supplier label every container with the name of the chemical (or a code agreed upon between the manufacturer and user). Disastrous mistakes have been made when the contents of unnamed or incorrectly labeled packages were used without adequate checking.

Assuming that the packages are correctly labeled, the quality control analyst should still check periodically for identity and purity in spite of guarantees by the supplier and a stated analysis on the labels. Arsenic, lead, and other heavy metals may occasionally be present and it is not unknown for a "commercial grade" to be delivered in place of a "food grade." Purity limits are given in the following but reference should also be made to *Pharmacopaeia* standards.

### Sodium Citrate, Trisodium Citrate Dihydrate

Sodium citrate ($C_6H_5O_7Na_3 \cdot 2H_2O$) is a white crystalline powder with a saline taste. It is slightly hygroscopic in moist air but will lose water of crystallization under dry conditions. It therefore should be kept in sealed containers.

It is used as a buffer salt to control the setting of pectin gels. It also has the effect of acting as a synergist to antioxidants and is stated to sequester metals such as copper and iron, which act as catalysts in the development of oxidative rancidity.

[1] Should conform to BP tests for food use.

Monosodium citrate has been used as a buffer salt in crystallizing, enabling syrups to be used a greater number of times in the process.

*Analytical details*:

| | |
|---|---|
| Assay as dihydrate | 99 to 101 percent |
| Acidity/alkalinity | Less than 0.5 ml 0.1 normal NaOH/HCl per 2 g |
| Chloride | Less than 0.03 percent as Cl |
| Sulfate | Less than 0.12 percent as $SO_4$ |
| Heavy metals (lead, copper, iron) | Less than 10 ppm |
| Oxalate | Nil |

## Sodium Lactate

Sodium lactate ($CH_3CH(OH)\cdot COO\ Na$) is used as a humectant and plasticizer and the commercial product is a 70 percent wt/wt viscous solution in water. It has a water activity below glycerol at all dilutions and is a useful plasticizer in water-based foods. It is miscible with water. Lactates occur naturally in foodstuffs and in the body, and are nontoxic. It is used as a buffer salt with lactic acid in sugar boiling.

*Analytical details 70 percent solution*:

| | |
|---|---|
| Specific gravity | 1.380 |
| pH 10 percent solution | 7.0 |
| Refractive index | 1.435 |
| Freezing point | Below 10°C (50°F) |
| Equilibrium humidity | 38 percent A/W 0.38 |
| Purity limits | As lactic acid. |

## Sodium Pyrophosphate, Neutral Sodium Pyrophosphate, Tetra Sodium Pyrophosphate (Tetron)

Sodium pyrophosphate ($Na_4P_2O_7$) a white powder, is a sequestering salt and is slightly alkaline. It is useful in the manufacture of low methoxyl pectin jellies as a means of sequestering calcium which forms a gel with this type of pectin. For this purpose, about four parts of phosphate to every part of calcium are required.

*Analytical details*:

| | |
|---|---|
| Solubility 20°C (68°F) | 5 percent |
| pH 10 percent solution | 10.4 |

| | |
|---|---|
| Assay $P_2O_5$ | 53.5 percent |
| Purity $SO_4$ | 0.25 percent max. |
| Cl | 0.2 percent max. |
| Arsenic | Less than 1 ppm |
| Lead | Less than 2 ppm |
| Iron | 25 ppm |

## Potassium Hydrogen Tartrate—Cream of Tartar

Cream of tartar ($C_4H_5O_6K$) settles out in wine dregs and is purified by recrystallization. The commercial article is a white powder with acid taste.

It is used to produce inversion in boiled sugar products, but has mainly been superseded by the controlled addition of invert sugar and glucose.

Inversion by cream of tartar is unreliable unless boiling times and alkalinity of the water used are rigidly controlled.

*Analytical details*:

| | |
|---|---|
| Solubility (20°C) | One part in 180 parts of water |
| 100°C (212°F) | One part in 15 parts of water |
| Purity | See tartaric acid |

## Sodium Carbonate (Anhydrous)—Soda Ash

Sodium carbonate ($Na_2CO_3$) is mainly used for the alkalization of liquor or nib in cocoa manufacture. Occasionally it is employed for neutralizing acidic foodstuffs, but more often the bicarbonate is used.

*Analytical details*:

| | |
|---|---|
| Assay | 98.0 percent $Na_2CO_3$ min. |
| Arsenic | 2 ppm max. |
| Lead | 5 ppm max. |

Soda ash may absorb up to 2 percent moisture in transport and storage if packed in bags without protective lining.

Impurity in the form of arsenic was the cause of the celebrated British court case of "arsenic in cocoa" in 1921. While such impurity is most unlikely in present times, vigilance is still necessary.

## Sodium Bicarbonate—Baking Soda

Sodium bicarbonate ($NaHCO_3$) is used for neutralizing acid syrups in reclaiming processes (see "Scrap reclaiming").

This chemical is also used as a raising agent due to the evolution of carbon dioxide in hot boiled syrups or in the presence of acid.

$$2NaHCO_3 \rightarrow Na_2CO_3 + H_2O + CO_2$$

$$C_6H_8O_7 \cdot H_2O + 3NaHCO_3 \rightarrow C_6H_5O_7Na_3 \cdot 2H_2O + 3CO_2 + 2H_2O$$
Citric acid                                   Trisodium citrate

The particle size distribution is important in determining the type of aeration in the final confectionery product. This is discussed under "Aerated Confections."

There is a considerable difference in the nature of the aerated high boiling depending on whether or not acid is used. The decomposition by heat alone results in a high pH and alkaline degradation of sugars, particularly dextrose and fructose. This gives the typical caramelized flavor and color. With acid decomposition, the pH is kept lower and a much lighter product with little caramelized flavor is obtained. Such aerated high boilings are more suitable for delicate fruit flavors.

Sodium bicarbonate will slowly decompose in syrup suspensions, and if these are used, they should be made up freshly.

*Analytical details*:

| | |
|---|---|
| Assay | 99.0 to 101.0 percent as $NaHCO_3$ |
| Solubility 20°C (68°F) | One part in eleven parts of water |
| Alkalinity (1 percent solution) | pH 8.6 max. |
| Arsenic | 2 ppm max. |
| Lead | 5 ppm max. |

## Other Chemicals

Sodium chloride (common salt) is frequently used as a flavoring substance. It has synergistic effects on other flavors and is often added to caramels and fudges. At 0.5 to 1 percent it is used as an ingredient in cocoa powder for drinking purposes.

Potassium carbonate, potassium hydroxide, and calcium hydroxide are used as alkalizing chemicals for cocoa liquor. Calcium hydroxide is often employed as a neutralizing substance in syrups and milk products.

## ANTIOXIDANTS

In the section of the book dealing with microbiological problems and rancidity, reference is made to antioxidants. These are substances, some natural, some synthetic, that delay the onset of oxidative rancidity.

Although the study of antioxidants and their mechanism is comparatively recent, use of such processes as smoking, which involves antioxidation, has been recognized since prehistoric times.

Herbs, spices, and oat flour have been known to have preservative action, and it was discovered more recently that vegetable fats mixed with animal fats retarded rancidity of the latter.

## Mechanism of Oxidative Rancidity

The first reaction that occurs is a process of autooxidation when air comes into contact with fat. Activated by some external factor such as light, heat, or traces of metals, peroxide radicals appear in the fat and these enable the fat itself to react with oxygen. Thus a continuous reaction is set up. The decay of quality in fats is measured by determining the peroxide value, and it is generally recognized that rancidity is detectable by taste when the peroxide value reaches 20 in animal fats and 50 in vegetable oils.

The time needed for a fat to reach these peroxide values is called the induction period. An accelerated test used to measure the induction period under artificial conditions is the Swift test in which air is bubbled through a quantity of the liquid fat at 98°C (208.4°F). Peroxide values are taken at intervals plotted on a graph and the Swift value estimated from the curve as the time taken to reach a peroxide value of 20. There are various modifications of this test.

## Action and Limitations of Antioxidants

If the formation of peroxides is retarded, rancidity development in fats will be reduced. This is precisely what antioxidants do—they are usually phenolic substances that interfere with the mechanism of autooxidation. They stop or slow down the chain reaction mentioned above.

The majority of oils and fats contain natural antioxidants and the most important are the tocopherols and lecithins. The amount varies from fat to fat and some fats have glyceride structures that render them more prone to rancidity. Vegetable fats usually have more natural antioxidant than animal fats and for this reason added antioxidants have a greater protective effect on animal fats. Cocoa butter, for example, is so well provided with natural antioxidants as well as a stable glyceride structure that it will keep free from rancidity for very long periods and needs no addition of antioxidants. On the other hand, fish oils are very likely to become rancid quickly and added antioxidants will give maximum protection. Natural antioxidants are only partially effective and in some circumstances are not heat stable, during fat deodorizing or deep frying, for example. The tocopherols themselves are susceptible to oxidation and may give rise to fishy odors. It must be emphasized that antioxidants have no protective effect against hydrolytic rancidity, which generally is recognized by the presence of soapy flavors and an increasing acid value due to the hydrolysis of the fats into glycerol and fatty acids. Oxidative rancidity gives tallowy or fishy flavors and may be detected analytically by peroxide value determination.

Antioxidants have no effect on the prevention of flavor reversion, which sometimes occurs in refined fats due to chemical changes taking place before oxidation sets in. They are most effective when added to fresh fat and are of much less value when the fat has started to deteriorate, for example, when the peroxide value has reached a figure of between 10 and 20. The peroxide value is therefore an important test to be made on a supply of fat of unknown age. Even when added to a fresh fat, they do not prolong its life indefinitely and the antioxidant, in due course, becomes exhausted.

**Sequestering Agents, Synergistic Action** Certain substances are able to reduce the catalytic effect of metals such as copper and iron, which are known to accelerate oxidative rancidity. These are called sequestering agents and consist of citric acid (and some citrates), phosphoric acid, and EDTA (ethylene diamine tetra acetic acid). Synergistic action exists between the primary antioxidants themselves. So the best protection is given by a mixture of antioxidants plus a sequestering agent.

## Permitted Antioxidants

Regulations vary from country to country and from time to time alterations are made. It is not advisable to quote specific limits in a textbook; the food technologist must keep up to date on these matters from current journals and trade association and government publications.

In the early studies of antioxidants, scientists realized that most antioxidants were phenolic in nature and, although many were discovered, only four synthetic substances have been widely employed. They are butylated hydroxyanisole (BHA), butylated

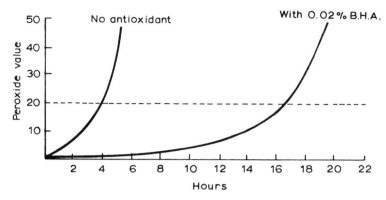

Fig. 16.1.   Effect of Antioxidant on Fat Stability

*Courtesy of N. V. Chemische Fabriek, Naarden, Holland*

hydroxytoluene (BHT), tertiary butylhydroquinone (TBHQ), and propylgallate.

Additionally, octyl gallate, and dodecyl gallate have been considered with ethoxyquin for apples and pears.

The effect of antioxidants is illustrated graphically in Fig. 16.1, in which the change of peroxide value is plotted against time in an experiment where a fat is heated in the presence of a stream of air.

## Natural Antioxidants

Consumer reaction against additives may cause a manufacturer to avoid the use of synthetic antioxidants. Antioxidants are present in many natural products—examples are the tocopherols, lecithin, and ascorbic acid.

Table 16.2 gives the formulas and properties of the main antioxidants.

## Methods of Incorporating Antioxidants

**Solution**   As such small amounts are added, there is no difficulty in dissolving the antioxidant in some of the fat or another solvent and adding to the bulk.

**Spraying or Dipping**   Where protection of an oil-fired product is required, the antioxidant in solution may be sprayed on the product or it may be dissolved in the "glazing" oil often used in the final stage of nut frying.

## Wrapping Materials and Utensils

Antioxidants are embodied in wrapping materials used for fatty foods where fat is likely to soak into or spread as a thin layer over the wrapper surface. Thin films of fat are very prone to oxidative rancidity. There can be no objection to this provided any transfer to the food does not cause the limit of permitted oxidant to be exceeded.

Sometimes a laminate or varnish on a wrapping material will oxidize and produce a contaminating substance. Antioxidants are used to prevent this.

Some plastics used for utensils and food containers may contain antioxidants not permitted for food use. Fats are known to leach out plasticizers and other substances, and when using plastic containers for food, the possibility of extraction of prohibited antioxidants must not be overlooked.

## RELEASE AGENTS, EMULSIFIERS

In the manufacture of confectionery, the product comes into contact with molds, metal, and fabric conveyor belts and side guides, as well as many types of utensil.

The confection is often hot, particularly when hard candy boilings are being made. When the hot melt is poured onto a metal or marble table conveyor belt, or into a mold, strong adhesion will occur unless some substance is interspersed between the confection and the surface receiving it.

This substance must be easily spread, able to maintain a continuous film, not mix with the product, and remain mobile when the confectionery product has cooled. It must also be edible, flavorless, and have good keeping properties.

The simplest release agents are those used by a cook who smears baking tins with butter, vegetable fat, or oil, and these same fats were used originally in the confectionery industry.

They were discontinued as the volume of production increased and the product was required to have longer shelf life—butter was expensive and thin layers on the surface of sweets became rancid. Although vegetable oils are still used, they have to be selected with care as many will give bad flavors on storage.

TABLE 16.2. PROPERTIES OF ANTIOXIDANTS

| Commercial name | Chemical formulae | Remarks |
|---|---|---|
| BHA | 3-tertiary-butyl-4-hydroxy-anisole 2-tertiary-butyl-4-hydroxy-anisole | Readily soluble in fat. Waxy solid. Effective in animal fats, less in vegetable fats. Not decomposed in processing of food. Consists of two isomers. Commercial product 90% 3-tertiary, which is most effective. |
| BHT | 2,6-ditertiary-butyl-4-methyl phenol | Similar to BHA with which it is synergistic. |
| TBHQ | tertiary-butylhydroquinone | The most effective antioxidant for both animal and vegetable fats. Fat soluble. |

424

Propylgallate

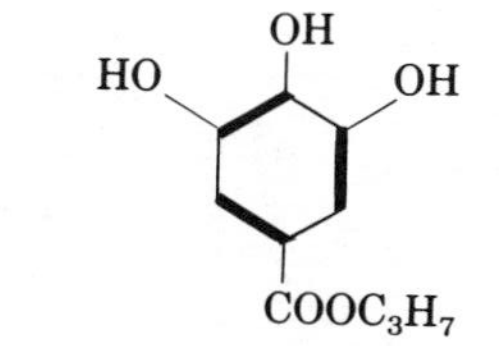

COOC$_3$H$_7$

OH
OH
HO

propanol ester of:
3,4,5-trihydroxybenzoic acid

Synergistic with BHA and BHT. Poor fat solubility. Decomposes at about 150°C. Forms colored compounds with metallic radicals.

Tocopherol

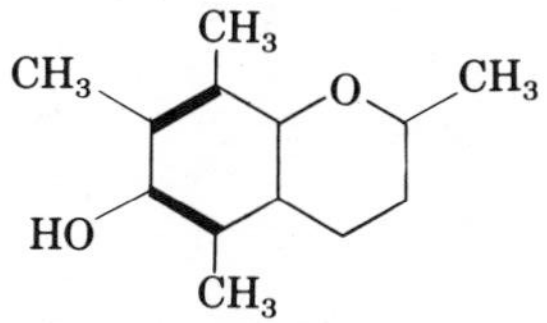

CH$_3$
CH$_3$
CH$_3$
CH$_3$
O
HO
Alpha-tocopherol

CH$_3$
CH$_3$
CH$_3$
O
HO
Gamma-tocopherol

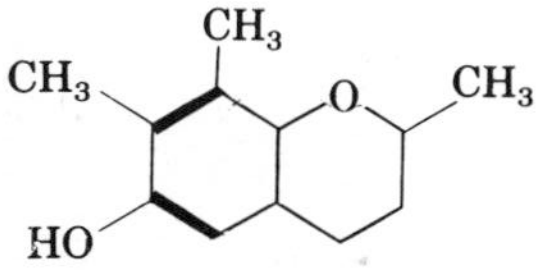

CH$_3$
CH$_3$
O
HO
Delta-tocopherol

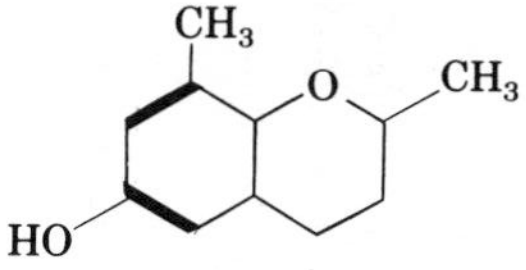

Natural product. Gamma and delta most effective. Sometimes decomposed during processing. Fat soluble.

With the introduction of large-scale continuous production, conveyor belts and large cycles of molds or baking pans have to be treated with release agents. It is important that any substance used to coat them must remain in good condition for long periods as, although it is usual to apply fresh release agent at each cycle of operation, some of the original material remains.

As an alternative to vegetable oil, refined light mineral oil (slab oil) has been used extensively—it is neutral in flavor, chemically inactive, and will not become rancid. Mineral oil is not acceptable universally, and where it is, legislation controls the quantity of residue remaining in the edible product.

Certain release agents are incorporated in the product. This occurs in the manufacture of compressed tablets where good release from the compression dies is required.

## TYPES OF RELEASE AGENTS

### Slab Oils, Mineral Oils

These are low-viscosity oils that have been purified to the extent that they have no flavor and do not develop flavor from decomposition when subjected to the temperature of hard candy boiling.

*Analytical details*:
Specific gravity 15.5°C (60°F)/15.5°C    0.856 to 0.870
Ultraviolet light test                  Not more than a trace of
                                           fluorescence
Carbonizable matter                      Trace only (BP test)
Flavor                                   Bland—no mineral oil
                                           characteristic

These oils have a short life on the coated surface and the film of oil has to be renewed fairly frequently. The incorporation of cornstarch to form a paste gives better film retention and this has been used successfully for some recipes, but some starch pickup will occur on the product.

### Acetylated Monoglycerides

Marketed under various proprietary names, they are monoglycerides in which the hydroxyl groups have been esterified with acetic acid and subsequently distilled. The presence of polar groups is supposed to cause "orientation", which enables the formation of thin continuous films capable of adhering to various surfaces.

They may be represented by the formula

$$CH_2 \cdot O \cdot CO \cdot (CH_2)_{16} \cdot CH_3$$
$$|$$
$$CHOH$$
$$|$$
$$CH_2 \cdot O \cdot COCH_3$$

The substances are also used as protective coatings to retard drying out or to reduce the onset of rancidity (see "Peanuts"). Some of the compounds are applied to panned confectionery to give a final gloss.

There were problems originally with some of these compounds, which developed off-flavors on storage, but manufacturers now guarantee freedom from this defect.

The very thin layers present on the surface of the coated products are exposed to air and thus are susceptible to oxidation, so the constituents of the coating must be resistant.

## Wax Release Agents

This type of release agent consists of the esters of long-chain fatty acids and high-molecular-weight monohydric alcohols. A well-known product of this type is Boeson Trennwax. It is available in solid and liquid forms and as an aerosol for spraying.

It is an anhydrous product resistant to the high temperatures experienced in baking and high boiled sweets and no development of off-flavors has been recorded.

## Lecithin/Fat Release Agents

Lecithin/fat mixtures are mixtures of synthetic lecithin (YN) and cocoa butter and have been used successfully as release agents. The first application consists of YN/cocoa butter (50/50) mixture. Subsequent applications are of cocoa butter alone. A temperature of 38°C (100°F) or over is required for release. Application of cocoa butter alone maintains release properties for at least twelve cycles, after which the lecithin mixture film has to be renewed.

## Silicone Compounds

A very satisfactory material applied to the confectionery industry is silicone rubber. It has excellent nonstick properties and will withstand high temperatures.

It is used for the manufacture of molds for the deposit of caramels and other high boilings. The molds are flexible, which facilitates demolding. It is still an expensive substance but the advantages obtained through mechanization of production methods usually offset material cost.

Silicone varnishes and greases have been used for coating belts, cutting knives, and baking trays. Sometimes they are applied to the interior surface of pans during the concentration of sensitive materials such as fruit pulps. These varnishes, however, break down if mildly abrasive confectionery such as fudge or fondant is in contact with the coated surface.

## Teflon (Polytetrafluoroethylene, Fluon P.T.F.E.)

This substance has an extremely low coefficient of friction and is well known as an internal coating for domestic utensils.

In confectionery, it is used for the manufacture of cutting knives for sticky pastes and caramels and to coat spreading rollers. The side guides on conveyors are often made of this material.

Teflon tape is used on wrapping machines as an intermediate layer to prevent adhesion of hot sealing irons to wrapping materials.

## Other Substances Used as Release Agents

These substances include stearic acid and magnesium stearate.

Spermaceti wax, calcium carbonate, and many other substances are described in the literature as "release agents" or constituents thereof. Manufacturers of emulsifiers and similar substances describe these products and their uses in very comprehensive brochures.

## Emulsifiers

Lecithin, sorbitan (fatty esters), and polyglyceryl polyricinoleate are used in chocolate and compound coatings.

Lecithin is also used in confectionery and another important emulsifier is glyceryl monostearate (GMS). Glyceryl monostearate resembles a fat by being a combination of a fatty acid with glycerol

TABLE 16.3. GLYCERYL MONOSTEARATE—SPECIFICATIONS

|  | GMS 35 | GMS 65 | GMS S/F |
|---|---|---|---|
| F.F.A. (as oleic acid), % | 0.4 max. | 0.4 max. | 0.4 max. |
| "Soap," % | 2.7–3.3 | 5.5–6.5 | 0.3 max. |
| Moisture, % | 2.0 max. | 2.0 max. | 2.0 max. |
| Monoglyceride, % | 32.5 min. | 32.5 min. | 32.5 min. |
| Free glycerol, % | 5 max. | 4–6 | 5 max. |
| Melting point (°C) | 56–60 (132.8°–140°F) | 56–60 | 56–60 |
| Sap. value | 177–183 | 177–183 | 177–183 |

but it differs in that only one of the OH groups is combined:

| Glycerol | Tristearin | Glyceryl monostearate |
|---|---|---|
| $CH_2OH$ | $CH_2OR$ | $CH_2OR$ |
| $CHOH$ | $CHOR$ | $CHOH$ |
| $CH_2OH$ | $CH_2OR$ | $CH_2OH$ |
|  | $(R = C_{17}H_{35})$ |  |

Commercial monostearates are not wholly monostearates but consist of mixtures of mono-, di-, and tristearates, but the active emulsifier is the "mono" substance. They are available in self-emulsifying forms that contain 3 to 6 percent sodium stearate and this enables them to be emulsified by stirring vigorously in hot water at 88 to 90°C (190°F to 194°F).

Specifications for typical commercial products are given in Table 16.3.

Glyceryl monostearate is very useful in caramel and fudge manufacture where it greatly assists the emulsification of the fat ingredient. Where these are subject to extrusion for bar formation, the monostearate prevents fat separation. In nut pastes (e.g., coconut bars), the addition of monostearate emulsifies any oil that separates from the nut during processing.

In fillings, such as lemon curd, monostearate prevents fat separation on subsequent heating. Its emulsifying properties are also very valuable in baking and ice cream processes.

## SOLVENTS

Various solvents are used in confectionery for the solution of flavors and other substances. Solution ensures proper dispersion throughout the food.

Solvents are constituents of compound flavors and of "additive" concentrates. Four solvents have practically universal acceptance. They are ethyl alcohol, glycerol, propylene glycol, and isopropyl alcohol.

Other solvents that have special uses are:

glycerol monoacetate (monoacetin)
glycerol diacetate (diacetin)
glycerol triacetate (triacetin)
diethyl ether and ethyl acetate.

As with other additives, reference should always be made to current regulations in each country. In some instances, a solvent may be permitted only in specific materials and quantity.

Care must be taken in the choice of a solvent, and even though countries have specific regulations regarding solvents for use in foods, variations in purity occur. It is important to test for trace residues that may impart objectionable flavors to a food.

## Ethyl Alcohol, Ethanol, Spirits of Wine

Ethyl alcohol ($C_2H_2OH$) is usually used in the form of rectified spirit, which is 95 percent ethyl alcohol by volume. It is a highly purified form of alcohol, free from any extraneous flavors or odors, and because of the excise duty it carries, it is an expensive solvent.

*Analytical details*:

| | |
|---|---|
| Assay | 94.7 to 95.2 percent vol/vol |
| | 92.0 to 92.7 percent wt/wt |
| Specific gravity (20°C) | 0.8119 to 0.8139 |
| Refractive index (20°C) | 1.3637 to 1.3639 |
| Nonvolatile residue at 105°C | 0.005 percent max wt/vol |

## Glycerol

Glycerol ($CH_2OH \cdot CHOH \cdot CH_2OH$) is available commercially in several forms—crude, pale straw, chemically pure—and only the chemically pure should be used as a solvent.

Flavors, colors, and other ingredients have limited solubility in glycerol and sometimes an increase is obtained by partial dilution with water. It is always necessary to make tests when this solvent is used, and particularly to examine for supersaturation, which may give rise to solid deposits on standing. It is miscible with water and alcohol. Glycerol is used in some confectionery recipes as a humectant and a "softener," usually at a level of 2 to 3 percent. It has also been used in some confectionery that may be frozen—the glycerol causes the product to remain soft at the lower temperatures.

*Analytical details*:

| | |
|---|---|
| Specific gravity (20°C) | 1.255 to 1.260 |
| Refractive index (20°C) | 1.471 to 1.474 |
| BP tests | |
| Purity | Lead: less than 1 ppm |
| | Arsenic: less than 2 ppm |

Copper

Mix 10 ml with 30 ml of water, 1 ml of hydrochloric acid, and 10 ml of hydrogen sulfide solution; no color is produced.

Reducing substances

Mix 5 ml with 5 ml of dilute ammonia solution and heat at 60°C (140°F) for 5 min. Add rapidly 0.5 ml of silver nitrate solution, making the addition from a pipette, the nozzle of which is kept above the mouth of the tube, and allowing the reagent to fall directly into the solution without touching the sides of the tube. Mix and allow to stand in the dark for 5 min; no darkening is produced.

Acidity/alkalinity

A 10 percent w/v solution is neutral to litmus solution.

## Propylene Glycol, 1,2-Propanediol

Propylene glycol ($CH_3 \cdot CHOH \cdot CH_2OH$) is a colorless, odorless viscous liquid resembling glycerol with a slightly sweet taste. It is miscible with water and alcohol and is hygroscopic. It dissolves many food colors, essential oils, flavors and oleoresins, and is becoming a main solvent in the confectionery industry.

*Analytical details*:

| | |
|---|---|
| Specific gravity 15°C (59°F)/15°C | 1.0409 |
| 20°C (68°F)/20°C | 1.0381 |
| Refractive index (20°C) | 1.4326 |
| Boiling point | 187°C [95 percent distils between 187°C and 189°C (368.6°F to 372.2°F)] |

| | |
|---|---|
| Flash point | 107°C (225°F) |
| Acidity as acetic acid, percent | 0.005 max. |
| Ash, percent | 0.005 |
| Moisture, percent | 0.2 max. (note— propylene glycol is hygroscopic) |
| Freezing point (60 percent solution) | −60°C (−76°F) |

## Isopropyl Alcohol, Propanol-2

Isopropyl alcohol ($CH_3 \cdot CH(OH) \cdot CH_3$) is a colorless low boiling liquid with a characteristic odor. It should be 99.5 percent minimum pure isopropyl alcohol by weight. Although the taste and odor are generally not apparent in confectionery if the solvent is used in small amounts, some people seem to be able to detect it and find it objectionable.

Some brands of this solvent, when it was first introduced to the food industry, left strongly flavored residues on distillation, but the pure food grades now available do not suffer from this defect.

| | |
|---|---|
| *Analytical details*: | |
| Specific gravity 20°C (68°F)/20°C | 0.785 to 0.787 |
| Refractive index (20°C) | 1.378 |
| Boiling point | 82°C (179.6°F) range 81.5 to 83°C (178.7 to 181.4°F) |
| Residue on distillation | 2 mg per 100 ml max. (this should have no odor or taste) |
| Acidity as acetic acid, percent | 0.002 max. |
| Freezing point | −89.5°C (−128.2°F) |

## INVERTASE CONCENTRATE

Invertase is a very active enzyme preparation extracted from cultures of yeast. It hydrolyzes sucrose to invert sugar. It is available commercially in liquid form as invertase concentrate under various proprietary names.

There are two enzymes that hydrolyze sucrose: $\beta$-h-fructosidase and $\alpha$-n-glucosido invertase, the former working on the fructose and the latter on the glucose part of the sucrose molecule. Commercial invertase prepared from yeast contains $\beta$-h-fructosidase and this has optimum activity at pH 4–5. $\alpha$-n-Glucosido invertase is obtained from the mold Aspergillus oryzae and an $\alpha$-glucosidase can be prepared from dried yeast, which hydrolyses sucrose at pH 6 to 7.

The organism that secretes invertase is "yeast"; from this, a special strain is cultivated and stimulated to give increased secretion. In commercial manufacture, this special strain is propagated from the culture stage by deep aerobic fermentation. Ultimately, the yeast obtained is conditioned by treatment with molasses under aeration conditions that allow only fractional growth, but the invertase secretion is doubled or trebled. Extraction of the enzyme is by plasmolysis followed by assisted autolysis, that is, digestion by proteolytic enzyme. The plasmolysis is carried out at an elevated temperature by toluene addition followed by papain digestion after which the cell residue is separated by filtration. The filtrate is chilled, the pH adjusted to 4.5, and the enzyme precipitated with chilled industrial alcohol, yielding a product known as invertase gum. The most usual method of making commercial invertase is to dissolve this gum in 55 percent glycerol and after its activity has been determined, the solution is adjusted to standard strength. In addition to invertase solutions, dried commercial preparations are available. In the manufacture of these products, drying is carried out under conditions that destroy the viability of the yeast without reducing invertase activity.

## Use in Fondant Cremes

The prime use of invertase in confectionery is to invert the sucrose in a fondant creme center after it has been cast and covered in chocolate. This inversion softens the texture of the creme and, because of the hygroscopic nature of the fructose (levulose) portion of the invert sugar, assists in the prevention of drying out. The presence of the invert sugar also alters the composition and increases the proportion of the syrup phase of the creme. This reduces the likelihood of further unwanted crystallization and acts as an additional safeguard against fermentation. A creme texture is obtainable by this means that would not be possible to handle directly from starch molds of a Mogul machine.

Fondant creme manufacture is described under "Confectionery Processes" and it will be noted that it consists of a solid phase of minute sucrose crystals dispersed in a syrup phase consisting of all the glucose syrup in the recipe plus some of the sugar.

This fondant creme can be remelted at 60 to 65°C (140 to 150°F) and flavors, colors, and other ingredients such as jam, fruit, or frappé (egg whip) added at this stage, as well as the invertase concentrate. This hot liquid mixture is then deposited into impressions in trays of corn starch, and when cool and set, the cremes are removed by brushing and sieving. These are then covered with chocolate and, after a period of time, the action of the invertase will render the creme semiliquid.

As is the case with any enzyme, the environment has an appreciable influence on invertase activity. With fondant cremes and syrups, the important factors are pH, temperature, and soluble solids together with a constant invertase content. Variation of the speed of action is also related to the quantity of invertase concentrate used, and in the recommendations of different suppliers, the quantity varies from 1 to 3 oz per 100 lb of fondant. The degree of inversion of the sucrose portion of a fondant creme is related to the conditions mentioned above. If the soluble solids concentration is too high, activity is greatly retarded and the inversion of the sucrose is incomplete. It is possible with a creme overdried in starch that very little liquefaction takes place (see later).

Invertase acts best in a slightly acid medium and pH 4.5 to 5 is optimum. Activity persists at a slower rate at pH above this range, but ceases above pH 7. As shown in Fig. 16.2, the optimum temperature for invertase action is 60°C (140°F) at a concentration of 50 percent and below. These conditions are applied, together with pH adjustment in the production of invert sugar by the invertase method. With the higher soluble solids in fondant, higher temperatures can be tolerated but inactivation of invertase accelerates greatly above 65.6°C (150°F.) The loss of activity in an average fondant at various temperatures during 20 to 30 min remelting time is shown in Table 16.4.

At 82.2°C (180°F) and above, at least 70 percent activity is lost instantaneously. It will be seen, therefore, that even at the normal remelting temperatures, 60 to 65.6°C (140 to 150°F), considerable loss of activity occurs if the fondant is held for extended periods before casting. A wide range of ingredients added to fondant cremes—e.g., sugars, egg albumen, milk products, nut pieces, gelatin, agar-agar, fat, and modified starches—have negligible effect on invertase activity. Flavors and colors can have a very detrimental effect if they are mixed with the invertase concentrate in undiluted state. Flavor solvents can be particularly destructive, and it is therefore essential that the fondant and all ingredients be mixed and

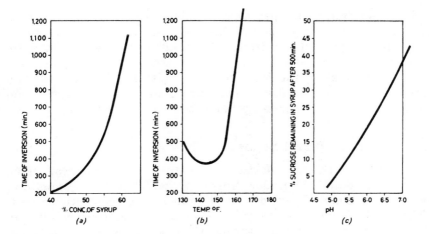

Fig. 16.2.   (a) Effect of Concentration on Rate of Inversion (b) Effect of Temperature on Rate of Inversion (c) Effect of pH on Rate of Inversion. Invertase Concentrate = 0.15 percent Syrup; pH = 5.0 —(a) and (b). Temperature = 60°C (140°F)—(a) and (c).

*Janssen* (1961) *Manf. Conf.*

brought to the remelting temperature before addition of the invertase. This allows just sufficient time for the enzyme to be mixed in thoroughly before casting into starch molds. Many of the heavy metals have a poisoning effect on invertase but copper is the only likely contaminant that may arise in confectionery processing, at a level of 1 or 2 ppm. Copper pans are still used in the industry and the only precaution needed is to see that the invertase concentrate does not come into direct contact with the metal surface.

The rate of inversion depends on the moisture content of the

TABLE   16.4. INVERTASE—LOSS   OF   ACTIVITY   AT DIFFERENT TEMPERATURES

| | | Percentage loss of activity in | |
| --- | --- | --- | --- |
| °F | °C | 20 min | 30 min |
| 140 | 60.0 | 4 | 6 |
| 150 | 65.6 | 8 | 12 |
| 160 | 71.1 | 15 | 20 |
| 170 | 76.7 | 20 | 30 |

fondant and storage temperature. At about 18°C (65°F) maximum inversion occurs after about two months, when the moisture content is 12.5 to 13.0 percent and the soluble solids of the syrup phase are 76 percent.

If, for any reason, the fondant loses moisture, such as delay in enrobing the center or drying out during storage, invertase action will cease at about 80 percent soluble solids concentration.

## Manufacture of Invert Sugar

A process for the manufacture of invert sugar from scrap using invertase is described under "Invert Sugar" but it is sometimes necessary to modify the conditions of temperature and concentration.

The effect of concentration, pH, and temperature are shown in Fig. 16.2. The economics of using syrups of higher or lower concentration depends on how the scrap is dissolved or used after inversion. Similarly, the actual quantity of invertase used in relation to the time of inversion needs careful costing.

## REFERENCES

### Antioxidants

*Antioxidants in Food Regulations.* Her Majesty's Stationary Office, London.
Buck, D. F.   1985.   Antioxidant applications. Manf. Conf. (June).
Eastman Chemical Products Inc., Kingsport, Tenn.
Ostendorf, J.F. Naarden, Bussum, Holland.

### Release Agents, Emulsifiers

Boehringer GmbH, Ingelheim, W. Germany.
Croda Food Products Ltd., Goole, England.
Eastman Chemical Products Inc., Kingsport, Tenn.
Grinsted Products A/S, Brabrand, Denmark.
Imperial Chemical Industries Ltd., London.

### Enzymes, Invertase

Cochrane, A. L.   1961.   Production and application of enzyme preparation in food manufacture. SCI Monograph No. 11. *Soc. Chem. Ind.,* London.
Dixon, M., and Webb, E. C.   1964.   *Enzymes* (2nd. ed.). Longman, Harlow, England.
Janssen, F.   1961.   *Manufac. Confect.* 1, 56, Aug., U.S.A.

Minifie, B. W., and Carpenter, W. J. 1966. Micro-organisms in the confectionery industry. *Proc. Biochem.* (May). London.
Novo Industri, A/S 1986. Bagsvaerd, Denmark.

## Other

*Directory of Equipment and Supplies.* 1986. Manf. Conf. (July).
*Food Additives—The Professional and Scientific Approach.* 1986. Institute of Food Science and Technology, London.
Solvents—Egan, H., Kirk, R. S., and Sawyer, R. 1981. *Pearsons Chemical Analysis of Foods.* Churchill-Livingstone, Edinburgh, Scotland.

# 17

# Colors for Use in Confectionery

Although a prime factor in making a food palatable is its flavor, color plays no mean part. If food looks unattractive, it seems to create a type of nausea in some people and the flow of digestive juices is not stimulated. Color helps a great deal to make all foods pleasing, and candy is no exception.

In a survey by Good Housekeeping Institute (1984), color linked with appearance was ranked second in importance when judging a foodstuff. Freshness was first and taste third.

Color is often accepted as a judgment of quality, particularly of natural foods, such as fruit and vegetables. Sometimes there are anomalies, as with eggs, for example; in some countries brown eggs are preferred, and in others white. Unnatural colors are unacceptable as the bright blues or greens in some foodstuffs indicate that they are synthetic.

The colors used in packaging must not be overlooked. The appeal of an attractively colored and designed package on the supermarket shelf can be responsible, at least, for an initial sale.

The appearance and color of a food under different lighting conditions must also be considered as what might look good in daylight may look quite different under fluorescent lighting.

The solubility of a color may also be important as there are the water-soluble dyes and insoluble pigments. Water-soluble dyes, sometimes used for the surface coloring of dragees, may stain the mouth when eaten. This may create a resistance to the product by parents. Pigments ("lake" colors) are not soluble and the staining effect is very much less.

## ORIGIN AND SUITABILITY OF COLORS

In the early days, serious poisoning occurred from the use of such dangerous inorganic pigments as copper sulfate, copper arsenite, red lead, and cinnabar. Yet how were the ancients to know, without poisoning a few people, when similar colors, such as red oxide of iron, ultramarine, and titanium dioxide (which are still used occasionally in foods) were harmless?

Later, the organic dyes were synthesized and these discoveries led to their application to foods as well as to textiles, plastics, inks, and various other nonedible commodities.

Soon it was found that many were toxic in one way or another. They may have been contaminated by poisonous metals during processing, or contained, as an impurity, some organic by-product that was highly toxic. Some pure dyes are known to be carcinogenic or to have some other biological action.

The coloring of food has become an item of international importance. All the leading countries have laws regarding colors for foods and agree with the principle of positive-list permissibility. These positive lists are based upon the assessment of available biological data derived from food-additive toxicological-type feeding tests in selected species of animals. Variability in the lists has occurred due to variability in available data and the interpretation of the data.

Natural colors would seem to be satisfactory, but often they are weak, fugitive, and affected by a change in pH as well as being limited in color range.

Contrary to general thinking, they are not always nontoxic and may be contaminated during extraction. Toxicological hazards have been recognized and, as a result, purity specifications are being established for natural colors by the regulatory bodies. Like the synthetic colors, the naturals will also be the subject of the positive list system on a worldwide basis. Von Elbe (1986) has described the development of natural pigments during the last twenty-five years.

A permitted list appears to be the right thing for both natural and artificial colors correlating purity and specifications for each individual dyestuff and also those designated foods where colors can be used. A number of countries have "permitted lists" but they are different and an examination of these lists will reveal that there are only a few colors common to the majority. It is also said that legislation through a permitted list prevents the development of new colors. This is hardly likely as new colors will always be developed for textiles and other purposes and there is no reason why some of these should not be tested for application to food manufacture.

It would seem that ultimately an international permitted list must be set up with a limited range of primary dyes that have been shown to be harmless, without biological activity, and capable of being blended to give colors that will satisfy all users.

In the transition period, it is best to examine carefully the permitted lists of the major countries and select those colors that appear in most of them.

Assuming a permitted list is obtained, it seems necessary to specify pure dye content, the nature of diluents used to standardize the intensity of color, and purity in respect of foreign metals and other organic substances.

## THE PRESENT SITUATION

In a textbook, it is not possible to be completely up to date on legislation and the users of colors must always make reference to the latest regulations in specific countries. This is particularly necessary for those companies that have a worldwide export business.

Colors in use today are in three categories:

1. Synthetic—no similar natural color.
2. Synthetic—identical to a natural color.
3. Natural—obtained from plants or animals.

### The JECFA List of Colors

In 1984/85, the joint Food and Agriculture and World Health Organization (FAO/WHO) Expert Committee on Food Additives (JECFA) established a list of synthetic colors that consisted only of the well-known, tested products in general use. This list is shown in Table 17.1.

### U.S. Certified Colors

In the United States, there are certified colors, and the list of these was agreed upon through the FDA and the Color Certification Laboratory in Washington, D.C.

For a color to be certified, the FDA must be satisfied that the manufacturer uses satisfactory processes and keeps adequate records of every batch of color made and sold. A sample of each batch must be sent to the FDA laboratories for approval, whereupon it is given a lot number, and only then can it be sold.

Additionally, the manufacturer's premises may be inspected at any time.

The permitted list as of 1985 is shown in Table 17.2.

TABLE 17.1. JECFA LIST OF COLORS

| | Class | Color |
|---|---|---|
| U.K. approved | Quinoline yellow | Greenish yellow |
| | Erythrosine | Red |
| | Indigotine | Red/Blue |
| | Brilliant blue FCF | Blue |
| | Patent blue V | Blue |
| | Green S | Green/Blue |
| | Fast green FCF | Green |

"AZO" Colors

| | | |
|---|---|---|
| U.K. approved | Tartrazine | Yellow |
| | Sunset yellow FCF | Orange |
| | Ponceau 4R | Red |
| | Red 2G | Red |
| | Azorubine | Red |
| | Amaranth* | Red/blue |
| | Brilliant black BN | Purple/black |
| | Brown FK | Yellow brown |
| | Brown HT | Brown |
| | Allura red AC | Red/yellow |
| | Fast red E | Red |

* Amaranth was subsequently removed from U.S. lists.

TABLE 17.2. U.S. CERTIFIED COLORS

| Color | Status |
|---|---|
| FD and C Red no. 3 | Permanent |
| FD and C Red no. 40 | Permanent |
| FD and C Blue no. 1 | Permanent |
| FD and C Blue no. 2 | Permanent |
| FD and C Green no. 3 | Permanent |
| FD and C Yellow no. 5 | Permanent |
| FD and C Yellow no. 6 | Provisional |

Lake colors prepared from the above are all provisional with the exception of Red no. 40.

## European Economic Community

A list of EEC-approved colors is given in Tables 17.3 and 17.4.

TABLE 17.3. EUROPEAN ECONOMIC COMMUNITY LIST OF COLORS AND THEIR BASIC PROPERTIES

| | | | Red colors |
|---|---|---|---|
| E120 | Carmine | Natural | Water soluble. Resistant to oxidation and reducing agents. Resistance to fading—poor. Maximum dose 50 mg/1 kg. |
| E122 | Carmoisine | Synthetic | Water soluble, not oil soluble. Maximum dose 20 mg/1 kg or 100 mg/1 kg (jellies). |
| E123 | Amaranth | Synthetic | Water soluble. |
| E124 | Ponceau 4R | Synthetic | Water soluble. Maximum dose 50 mg/1 kg. |
| E127 | Erythrosine BS | Synthetic | Limited solubility in water. Avoid using in acid products. Maximum dose 50 mg/1 kg. |

| | | | Yellow and orange colors |
|---|---|---|---|
| E100 | Curcumin | Natural | Stability poor in some media. Slightly soluble in water. Soluble in alcohol and fats. Maximum dose 50 mg/1 kg |
| E101 | Riboflavin (vitamin $B_2$) | Natural and synthetic | Water soluble. Good resistance to other ingredients. Maximum dose 50 mg/1 kg. |
| E102 | Tartrazine | Synthetic | Water soluble. Good resistance to other ingredients and light. |
| E104 | Quinoline yellow | Synthetic | Water soluble. Good resistance to other ingredients and light. |

TABLE 17.3. EUROPEAN ECONOMIC COMMUNITY LIST OF COLORS AND THEIR BASIC PROPERTIES (*Continued*)

| | | | |
|---|---|---|---|
| | | **Green colors** | |
| E110 | Chlorophyll | Natural | Only slightly soluble in water. Subject to fading in light. Maximum dose 500 mg/1 kg. |
| E141 | Chlorophyll derivative | Natural | Soluble in water. Subject to fading in light. Maximum dose 100 ng/1 kg. |
| E142 | Green S Brilliant green BS | Synthetic | Soluble in water. Maximum dose 100 mg/1 kg. |
| | | **Blue colors** | |
| E131 | Patent blue V | Synthetic | Soluble in water. Resistant to other ingredients but requires an acid medium less than pH 5. Maximum dose 50 mg/1 kg. |
| E132 | Indigo carmine Indigotine | Natural and synthetic | Soluble in water. Subject to fading and action of reducing agents. Maximum dose 100 mg/1 kg. |
| | | **Brown color** | |
| E150 | Caramel | Natural | Soluble in water. See description later. Maximum dose 500 mg/1 kg. |

**Black colors**

| | | | |
|---|---|---|---|
| E151 | Black PN | Synthetic | Soluble in water. Has tendency to change color in hot solutions containing glucose syrup. Maximum dose 50 mg/1 kg. |
| E153 | Vegetable black | Natural | See also "Blackshire" cocoa. Not soluble in water. Maximum dose 1,500 mg/1 kg. |

**Colors of other various shades**

| | | | |
|---|---|---|---|
| E160 | Carotenoids | Natural (synthetic) | See further descriptions later. Not soluble in water—fat soluble. Various derivatives of carotene have colors ranging from yellow to purplish-red. Maximum dose 100 mg/1 kg. |
| E161 | Canthaxanthin | Natural (synthetic) | Similar to carotenoids. Maximum dose 500 mg/1 kg. |
| E162 | Betanine (beet red) | Natural | See further description later. Water soluble. Color ranges from red to purple. Maximum dose 500 mg/1 kg. |
| E163 | Anthocyanins | Natural | Water soluble. Red under acid conditions or bluish under alkaline conditions. Maximum dose 500 mg/1 kg. |

TABLE 17.4. INORGANIC PIGMENTS OF MINERAL ORIGIN (CLASSIFIED AS "NATURAL")

The following pigments are used mainly for surface coloring such as dragees, aniseed balls, sugar-coated chocolate balls, and silver cachous.

| | |
|---|---|
| E170 | Calcium carbonate (precipitated chalk) |
| E171 | Titanium dioxide |
| E172 | Iron oxide (ferric oxide) |
| E173 | Aluminium powder |
| E174 | Silver leaf |
| E175 | Gold leaf |

*Source*: Guide Technologique de la Confiserie Industrielle.

## NATURAL COLORS

"Natural" colors are those obtained from plants and animals and there are sufficient variations among these to suit the consumers. The pigment classes are the carotenoids, quininoids, porphyrins, betalaines, and flavonoids.

All of these coloring materials, which are exempt from certification, are either crude extracts or unpurified preparations. von Elbe (1986) has classified the natural colors presented in Table 17.5.

Experiments have shown that the coloring power of these natural pigments is adequate for most foods. Work by Pasch and von Elbe (1977) with betanine showed that with a variety of confectionery products sufficient coloring was obtained at pigment levels between 8 and 80 ppm.

TABLE 17.5.  NATURAL COLORS—CLASSIFICATION

| Class | Coloring extract | Pigmenting substance |
|---|---|---|
| Carotenoids | Annatto | Bixin |
| | Carrot oil | $\beta$-Carotene |
| | Vegetable juice | $\beta$-Carotene, lycopene |
| | Paprika | Capsanthin, capsorubin |
| | Saffron | Crocetin |
| Quininoids | Cochineal | Carminic acid |
| Porphyrins | Vegetable juice | Chlorophylls |
| Betalaines | Vegetable juice | Betanine |
| | Beet powder | |
| Flavonoids | Fruit juices | Anthocyanins |
| | Grape skin extract | |
| Others | Riboflavin | Riboflavin |
| | Turmeric | Curcumin |

## PREPARATION OF COLORS FOR USE

With the limited number of primary colors available, it becomes necessary to make blends to suit each confectionery product. It may seem to be merely a question of mixing the colors until the right shade is obtained but this is certainly not the case because not only do some dyes not mix well, but they may react in an unusual way with the ingredients of the confection.

Most of the water-soluble dyes have an average of about 90 percent pure dye content. The process of manufacture relies on the interaction of water-soluble compounds, and to obtain the solid dye it has to be "salted" out by the addition of inorganic salts such as sodium sulfate or chloride. This practice is well known in organic chemistry; the addition of these salts represses the solubility of the dye and precipitates it, but some of the inorganic salts that are occluded in the dye remain there when it is filtered off and dried.

In addition to the water-soluble dyes, there are "lakes." In the process of dyeing some fabrics a mordant is used to fix the dye; this consists of first soaking the material in an inorganic salt solution, usually of a weak base such as aluminium hydroxide, so that when the dye is applied it forms an insoluble compound in the fiber.

Similar lakes are made with food dyes using aluminum, calcium, or titanium bases. These lakes are very finely ground insoluble powders and their coloring power depends on good dispersion of the particles throughout the product to be colored. They are useful for coloring fat-based coatings where oil-soluble colors are not available to give the tints required. They are also useful for coloring small pieces of jelly, preserved peels, fruit, or other small confectionery items where they are subsequently included in another product such as nougat, montelimar, or baked goods. Water soluble dyes will bleed into the surrounding material, and this looks unsightly, but lake colors, being insoluble, will not.

Lake colors and the inorganic pigments previously mentioned have a special application for the surface coloring of dragees, sugared almonds, aniseed balls, and other "panned" confections. In pastel compound coatings where water-soluble colors are unsuitable or suitable oil-soluble colors are not available, it is best to grind the color into the original paste.

Water-soluble colors are sold in powder or granulated form. At one time, powder was the only type available, and it created many problems during dispensing due to dust that contaminated the surroundings and personnel. Granular colors are much more satisfactory.

Paste colors have been prepared by grinding the powder into a glycerol or glycerol/sugar mixture.

A further development is the manufacture of "jellied" colors that consist of cylindrical pieces ("sausages") of soft gelatin in which the color is dispersed at a given strength. It is only necessary to weigh out or, in some cases, measure a length of the jelly for a batch and add it at the end of the process, and it will disperse without difficulty. Sachets of colors have been made (unit packs) with contents of given weights to correspond with batch requirements.

Commercial suppliers of food colors compound the primary dyes to produce a variety of shades to suit the food manufacturer, and these have been mixed so that there is no unsatisfactory interaction. The supplier will also provide instructions for dissolving and use.

Larger manufacturers make their own color solutions from primary dyes and these are prepared by dissolving in hot water at a strength of 1 to 2 percent and then filtering. These solutions will keep in a cool place for several weeks or longer, but to guarantee longer keeping, 25 percent of glycerol may be added.

Sometimes it may be desirable to dissolve the color in another solvent and propylene glycol is useful when color emulsions are required.

The mixing of dyes in solution generally presents no problems, but if there is obvious precipitation, separate solutions should be made and each added to the batch in the required quantity.

It must not be assumed that a color obtained by adding specific quantities to one confectionery product will give the same sort of color in another product. Experimental work must always decide the types and quantity of colors to be used for each recipe.

Frappé, invert sugar, fruit substance, sulfur dioxide, and pH can have appreciable influences on the ultimate color and on its permanence.

## Change of Color, Fading

Some ingredients of confectionery, especially when associated with certain processes, may cause considerable fading of the colors. Where two or more primary colors are used to make a desired shade, the fading of one of them may cause a complete change of color. One instance of this was a purple color that changed to a pale green when one constituent color faded completely after a few weeks.

Apart from chemical action, the intensity of a color is affected by the physical condition of the confection. A color that appears quite bright in a straight fondant will lose a lot of its brilliance when whipped up to occlude air. The presence of solids such as fine sugars, starch, or titanium dioxide also has a deadening effect on an added color.

Jellies or Turkish delight present a completely altered appearance when a chocolate-covered unit is cut or bitten exposing the colored center. With the uncovered piece, the color is viewed through a semitransparent medium, which gives a different effect.

Table 17.6 shows the effect of light, invert sugar, sulfur dioxide, and pH on some of the colors used in confectionery. This is a guide only, and where fading is experienced, some experimenting is required to find the exact cause.

Invert sugar as a main ingredient in a confectionery line may cause considerable fading. If sulfur dioxide is not removed by boiling this may result in appreciable fading. Light causes fading, particularly when moisture contents are high.

The pH of a confectionery mix affects fading in different ways. An example occurred with an egg albumen substitute, which itself had a pH of 10, and when used as a frappé in neutral sugar/glucose fondant caused rapid bleaching of amaranth. When the pH was

TABLE 17.6. FADING OF COLORS IN DIFFERENT MEDIA

| Color | EEC index no. | Sulfur dioxide solutions | Invert syrup in fondants | Invert + light | Hard candies and sugar/glucose fondants |
|---|---|---|---|---|---|
| *Reds* | | | | | |
| Ponceau 4R | E124 | 2 | 2 | 1 | |
| Amaranth | E125 | 1 | 1 | 1 | |
| Erythrosine | E127 | 1 | 1 | 1 | All colors are reasonably |
| *Yellows* | | | | | stable if |
| Tartrazine | E102 | 3 | 3 | 2 | sulfur dioxide |
| Sunset yellow | | | | | contents are |
| FCF | — | 2 | 1 | 1 | reduced by boiling. |
| *Blues* | | | | | |
| Indigo carmine | E132 | 1 | 2 | 2 | |

1 = Rapid fading.
2 = Slight or slow fading.
3 = Negligible or no fading.

reduced to 6 or below, fading or color change was eliminated but the same results were not obtained with egg albumen, showing that the substitute had some effect irrespective of the pH. Many colors have a greater tendency to fade above pH 6.

## Caramel Color

This is a unique coloring matter worthy of special description. Although it is primarily used in the brewing industry it has found many applications in confectionery recipes. Licorice is a particular example.

Caramel as a coloring matter for foodstuffs has been loosely defined as "burnt sugar," and in its infancy was indeed just that, but the caramels produced by the straight burning of sugar are of limited use when seen purely from the point of view of a coloring agent, as tinctorial values are very low.

The maximum color value of such caramels is of the order of 20,000 EBC (European Brewing Convention) but they can have special qualities that put them much in demand for such materials as coffee extracts, toffee products, and pudding syrups, where the tinctorial value is not the sole property desired—flavor and added solids may be equally important.

High-tinctorial-value caramel is the product of heat plus acid or alkali treatment of sugars, and as a generalization, the treatment of a hexose with ammonia or its compounds is probably the origin of most caramels in general use, where a demand for added color is the primary need. Various other "catalysts" are now employed to accelerate the development of color.

In such caramels, the maximum practicable color value is in the region of 60,000 EBC. The actual constitution of a high colored caramel is of very great complexity and variability, the constituents being colloidal in nature, as a result of the polymerizations brought about by the high temperatures and chemicals used in the manufacturing process. Endeavors to produce materials of higher color value than about 60,000 EBC can result in the production of rubberlike substances, almost completely insoluble and useless as coloring materials.

However, not all the caramels in use have escaped the terms of reference of legislative bodies, that is to permit only those additives with a justified technological rating at levels that present no risk to public health. The major caramels used throughout the food industry are the subject of extensive toxicological investigation in order to assemble additional information to assess their safety in use.

If color were the only requirement of a foodstuffs manufacturer, a maximum-tinctorial-value caramel may appear ideal. Unfortunately, other associated characteristics of such caramels may be influencing factors depending upon the product in which the caramel is to be included. Very high color value is often associated with high viscosity, high total solids, or instability in certain conditions, and lower color values may be advantageous.

In the higher color ranges, the collodial characteristics are more marked, showing isoelectric points of 5 to 5.5, and this factor has to be considered by the user if the caramel is required for coloring products where its inclusion may cause precipitation with other colloidal matter present. Some low-color-value caramels can be prepared where isoelectric points are as low as 2.0 and the use of such a caramel may solve a precipitation problem.

## Isoelectric Point

Caramel, being colloidal in nature, is characterized by its isoelectric point. At a pH above the isoelectric point, the caramel in solution is negatively charged whereas below it is positively charged. The majority of caramels fall into two ranges—those in the neutral range with an isoelectric point around 7.0, and those in the distinctly acid range with an isoelectric point of 3.0. Therefore many caramel colors will not mix without coagulation or precipitation. Figure 17.1 illustrates this point.

The evaluation of caramel depends on its use. Tinctorial power is probably the most important and is measured by tintometer or

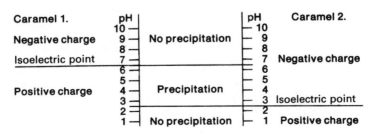

Fig. 17.1

absorptiometer. The importance of other factors, such as viscosity, moisture content and isoelectric point, must be decided by the technologist in relation to the recipe.

## Measurement of Color

Some information has already been given on the importance of color in relation to the acceptance of a food. The necessity of being able to measure color stems from the opinion that color variation in a food (or in the package) is an indication of bad processing or a stale product. Critical examination will decide what color properties are important, for example, reflection, transmission, absorption.

It is not the purpose of this book to discuss color measurement and the reader is referred to publications by Clydesdale (1975, 1984).

## REFERENCES

Ackerman, O. 1968. Manufacture of Caramel Color. U.S. Patent no. 3,385,733.

Clydesdale, F. M. 1984. *Color Measurement in Food Analysis*. Marcel Dekker Inc., New York.

*Coloring Matter in Food Regulations* 1973–1978. Her Majesty's Stationery Office, London.

European Brewery Convention. *J. Inst. Brewing,* London.

Food and Drugs Administration, Washington, D.C.

Francis, F. J., and Clydesdale, F. M. 1973. *Food Colorimetry—Theory and Applications.* AVI Publishing Co. Westport, Conn.

Good Housekeeping Institute 1984. *Consumer Food and Nutrition Study.* GHI, New York.

*Guide Technologique de la Confiserie Industrielle* (in French) 1986. Y. Fabry, Zentralfachschule, Solingen-Grafrath, West Germany.

Institute of Food Science and Technology 1985. Symposium on Food Colors, London.

Von Elbe, J. H. Department of Food Science, University of Wisconsin, Madison. "Natural colors—where are we?" 1986. *Manufac. Confect.* Jan.

# 18

# Flavor and Flavoring Materials

Whatever we consume, be it food, beverages, or confectionery, may be judged from several standpoints, but pleasure rates high as a desirable response. Within the limits of choice, imposed by such external factors as availability and price, we usually select those products that give us most pleasure, and it is most unlikely that we would accept something that is unattractive to look at, has an objectionable smell or is unpleasing to the palate. Of these, the enjoyment provided by the flavor of the product is probably of greatest significance. It will be appreciated that none of these attributes has anything to do with nutritive value of the product, but are all purely sensory and pleasurable, subjective, and personal.

Flavor preference and acceptance/rejection of foods, and confectionery products are no exception, display enormous variability, depending on such factors as ethnic origins, education, upbringing, age, and the environment. A product that is completely acceptable to one consumer may be equally rejected by another. Psychological factors may also be involved, and it is not uncommon for one to be revolted by being served an excess of food, whereas the same food served in a quantity within one's capacity to eat would be consumed with pleasure. It is finding the correct balance of acceptable attributes that largely determines the commercial success of any product.

Flavor is a complex sensation induced by chemical compounds that are present in what we eat or drink and are in equilibrium at the time of ingestion. It may be more accurately defined as the simultaneous appreciation of stimuli on the receptors of taste on the tongue and smell in the nasal cavity, and of general pain, feeling, and temperature receptors located throughout the mouth and throat (Heath, 1981). The stimuli produced by everything we smell, or that enters the mouth, give rise to microelectrical impulses that pass from the specific receptors to definite locations in the brain, where they are translated into recognizable experiences. Just how this occurs is still not well understood, although the nature of the receptors and the nerve pathways have been closely researched and several theories proposed (Wright, 1971; Laffort, 1975).

The first impression one has of any product is its appearance, including, of course, color. This has a direct impact as visual attractiveness makes one product outstandingly different from another. This is particularly the case with children, who tend to select products according to color and anticipated flavor. Appearance is a valuable attribute right up to the point of ingestion. At the time of eating, the volatile components in some products may be inhaled through the nose, thereby stimulating the olfactory receptors; this effect is fortified as the volatiles are released in the mouth during chewing and swallowing. At the same time, the basic taste factors and other superficial stimuli are appreciated so that the complete flavor is recognized. It is at this point that the full pleasurable or unpleasurable response is made and the product judged acceptable or unacceptable.

In the confectionery trade, presentation and appearance are very important in influencing the public to buy one's products. Persuasion by this means and by the various forms of advertising can create the initial interest and lead to a first-time purchase, but unless the product is of an acceptable quality and has a good and enjoyable flavor, repeat sales cannot be expected.

The subject of flavor preference is far from simple and numerous unrelated factors have to be taken into account. Of these, regional preferences are of significance in planning product marketing. The preference for dark chocolate in European countries and for milk chocolate in England and the United States is well known. Dark-chocolate creme-filled bars that are very popular in Europe are being only accepted slowly in the United States. The prestigious after-dinner mint, so widely liked in the United Kingdom, took much expensive advertising before it became the choice of the American hostess. It is important to realize these differences when planning product launches in new countries—sales experiences in one country may not necessarily be repeated in another, although advertising can sometimes significantly affect consumer opinion and tastes. Advertising and popular "trends" within age groups can sometimes lead to the acceptance of flavors that otherwise would not be widely enjoyed. This "acquiring" of taste is difficult to explain and the popularity of products based on it may not be long-lived. Factors contributing to our appreciation of what we eat and drink are many and complex. Although it is generally recognized that the majority of us do not use our senses to the full, they tend to act as an integrated whole to give a conscious response to which we may or may not react. Flavor is a factor that has no chemical or physiological basis for inclusion in the diet, but without it there would likely be no confectionery industry—flavor is that important.

## SOURCES OF FLAVOR

Confectionery products differ from the majority of foods in that the prime component is sugar. Apart from its intrinsic sweetness, sugar has no other flavoring effect and hence the flavor profile of the finished confectionery must be achieved by the deliberate addition of flavoring materials. It is difficult to generalize on the selection and use of added flavorings as there are so many variables that must be taken into account, particularly the composition of the product, its method of manufacture, marketing requirements, and consumer acceptability. Flavorings may be entirely natural in composition, they may be compounded from synthetic chemicals, or they may be a blend of both. Whatever the formulation, the aim in using flavorings is to produce a flavor in the end product that has the maximum level of positive acceptability. In achieving this, it is necessary to recognize that any flavoring must satisfy the following criteria:

1. It must be quite harmless in use and present absolutely no health hazard, particularly when the consumers are children.
2. It must fit the end product technologically and in keeping with the total product concept.
3. It must comply with any legislative requirements in force in the country in which the product is offered for sale.
4. It should be convenient to handle, capable of accurate dosage, and readily incorporated into the product mix, giving a uniformly dispersed flavor.
5. It must be stable before, during, and after incorporation into the final product.
6. It must resist adverse storage conditions.
7. It must be economically viable for both the flavor manufacturer and the confectionery producer.

The nature of available flavorings will be discussed in a later section.

## FLAVOR BALANCE

Many flavors used in confectionery are based on well-known fruits (e.g., raspberry, strawberry). Unfortunately, the natural flavor components in these are in relatively low concentration and, although this enables fruit to be enjoyed when eaten, it makes them a poor source of flavor concentrates for use in confectionery. As a consequence, most flavorings used in confectionery production are imitations based on synthetics that may be identical to those present in the natural fruit. Such flavors have attained a very close similarity of profile to that found in nature and can be suitably formulated to give almost any desired aromatic effect. Flavor manufacturers ensure that their products do comply with the criteria stated above.

Most fruits are characterized by the presence of fruit acids (e.g., citric, malic, tartaric). Without these, the flavor profile would be incomplete and either sickly sweet or unpleasantly insipid. To achieve a really acceptable fruit flavor in confectionery products, it is necessary to pay particular attention to the sugar/acid balance appropriate to the flavor type. This is not quite so straightforward as it may appear. Imitation flavorings are neutral and do not have any acidic components so that this can only be achieved by adding appropriate food acids to the end product. Care must be taken in the levels used as acids control the inversion of sugar as well as fortifying the true fruit flavor. The ratios of sugar to acid content of some ripe fruits are given in Chapter 14. Citric or malic acids are widely used not only to give an acceptable acidic taste, but also to control the pH of jellies, and possibly to modify the consistency of the gel. A level of citric acid that is too high can result in jellies that lack body and have a tendency to sweat (see "Pectin").

## TOTAL PRODUCT CONCEPT

It will be appreciated that the various attributes of a confectionery product cannot be considered in isolation. During product development, it is necessary first to have a concept of the product as a whole, probably based on a novel idea or in response to a known market need. The limitations imposed by the economics of raw materials, processing, packaging, distribution, and marketing must then be recognized and evaluated. Once this has been carried through to everyone's satisfaction, the process of product development can take place, usually involving various sections working concurrently. The flavor of the product is, as we have seen, of vital importance to its ultimate success, and this aspect must be decided upon very early in the development programs. Few confectionery technologists are experts in flavor, and as the flavor manufacturers employ specialists who are available to give advice on the selection and use of all types of flavoring materials, it is very important that their assistance be sought at an early stage. It has to be remembered that consumers are relatively conservative and tend to prefer flavors to which they are accustomed. Most manufacturers are unlikely to risk heavy losses in educating the public to accept an entirely new taste experience. Fortunately, close and confidential collaboration between the technologists of both the flavor and the confectionery manufacturers enables such risks to be assessed and action taken with minimal waste of time or facilities (Clarke, 1972).

## THE FLAVOR SENSATION

### The Sensory Organs

Taste    Taste is detected through the contact of certain water-soluble, mostly nonvolatile, compounds, dissolved in the saliva, with the taste buds on the tongue. It is widely accepted that there are four primary taste sensations: sour, sweet, salty, and bitter. Recent studies, however, suggest that there may be two further well-defined taste sensations—alkaline and metallic. Undoubtedly, sodium bicarbonate and metals do have a distinctive influence on the flavor profile and their particular taste cannot be classified within the four primary taste sensations.

The taste buds, located in specific regions on the upper surface of the tongue, are well supplied with nerves. Recent research has indicated that the different type of taste buds present (Fig. 18.1) may have some preference for recording the various taste sensations. Sweet tastes are recognized mostly by the tip of the tongue, whereas the sides record salty and sour tastes, bitterness being detected at the base of the tongue. This sensitivity is by no means specific and there is some indication that it may change with age.

Over many years, numerous theories have been put forward to explain the recognition of taste sensations. Supporting research work is complex and to date has not been conclusive. There are some 9,000 taste buds on the human tongue, each containing fifteen to eighteen sensory cells. For those who wish to study the subject in detail, there are several standard works for references (Amerine et al. 1965; Meiselman, 1972).

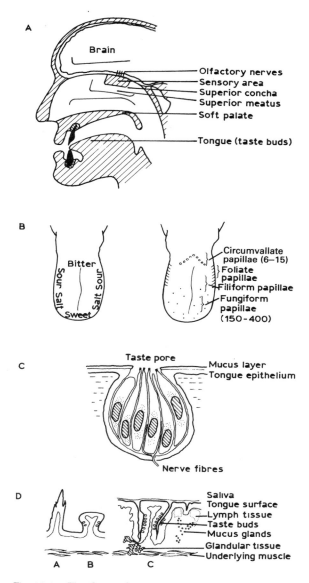

Fig. 18.1.   The Sense Organs
A. Cross Section Through Nasal Cavity and Mouth
B. Distribution of Papillae and Areas of Special Sensitivity
   to the Prime Tastes
C. Cross Section Through a Taste Bud of a Rat
D. Diagrammatic Cross Section of Papillae
   A—Filiform   B—Fungiform   C—Circumvallate

**Odor** For a chemical compound to possess an odor, it must vaporize to enable it to pass into the nasal cavity, where the olfactory sensory region is located on the dividing folds of tissue. The precise mechanism enabling humans to recognize and positively identify thousands of different odors is not yet understood (Schneider, 1968; Kafka, 1975). The vast surface area that is available to receive olfactory stimuli may be judged from the fact that, in the rabbit, some 100 million receptor cells have been shown to exist—each bearing six to twelve sensory hairs.

Sensitivity to odor is much greater than taste, and only molecular quantities of a substance are necessary for detection. No instrument is yet as sensitive as the nose, but, with a complex smell, where the nose detects the overall aromatic profile and can only be used as an analytical "tool" after considerable training, modern equipment can separate and differentiate the individual constituent odorants at extremely low levels of concentration but gives no information on their sensory impact.

**Texture** Texture is very important to product acceptability, particularly in confectionery. Although not yielding any flavor effect, texture does influence the rate at which a flavor is made available. When a hard candy is sucked, it dissolves slowly from the surface inward and the flavorings present are continuously released over the eating period. With fondants, pastes, chocolate, and other soft confectionery, the whole is quickly softened in the mouth and almost the total flavor is available quickly. The type and quantity of flavorings used must be related to these factors to achieve optimum response. Crisp, aerated sweets provide another texture experience, and with these, acceptability is often more related to mouth-feel than flavor. Loss of crispness inevitably results in a lowering of palatability, irrespective of the quality of the flavor.

Other textural characteristics are detrimental to flavor appreciation—chocolate that is badly ground is gritty, or conversely, if it is refined too much, it may be slimy to the palate—a factor that is more noticeable in milk chocolate than in dark chocolate. Texture is related closely to an apparent sickly sweet taste, even where the sugar content is of an acceptable level. All of these effects underline the need to evaluate flavors in the end product made under normal processing conditions so that the interrelated attributes can be correctly assessed.

Many modern snack products depend for their popularity on the crunchy texture given by using shortbread, wafers, cornflakes, peanuts, etc, coupled with smooth caramel or fudge and a smooth chocolate coating. In such products, texture is probably more important than flavor.

## Temperature and Climate

The acceptability of confectionery products is often influenced by climate. In hot weather, fatty confectionery and chocolate are less popular, and consumers prefer the sharper, more acidic and refreshing fruit flavors.

Fatty products become soft and messy to handle under warm conditions, and this undoubtedly contributes to their unpopularity. However a great deal of chocolate is eaten in hot climates where cold storage is readily available to keep the products in good condition up to the time of eating, and iced chocolate drinks are very popular.

Dragees, including chocolate items, are popular in warm climates as these are protected by a hard sugar glaze that preserves their appearance and makes them clean and easy to handle.

The actual temperature of a food certainly affects its apparent flavor even though the ambient temperatures are not sufficient to induce adverse chemical changes in the flavor constituents. Chocolate with ice cream has a taste that is quite different from the same material molded into a block and eaten at normal temperatures. Temperature affects texture so that caramels and similar chewy confectionery will taste a little different according to whether it is hard or abnormally soft. As with texture, the effects of temperature should be determined experimentally during product development and formulations adjusted, as necessary, to maintain an acceptable product quality.

## FLAVOR ASSESSMENT

### Sensory Versus Instrumental Methods

There is still a practice of equating sensory analysis with purely subjective judgments and thereby implying that such methods are unreliable in contrast to the apparent respectability of instrumental or so-called objective test methods. In the light of modern sensory tests using scientifically designed methods involving a correct statistical approach, a full understanding of the purposes for which the test is being used, and precise experiment design of the test conditions, such distinctions are quite unfounded. The sensory assessment of flavors is now well established and of incalculable value in the maintenance of product quality.

## Instrumental Methods

The quality assessment of many flavoring materials, in particular, the essential oils, is amenable to the instrumental determination of physical characteristics. An indication of the quality of essential oils can be obtained from a knowledge of their density, refractive index, specific rotation, and solubility in diluted ethanol. Over many years, such statistics have been carefully recorded and most specifications include limits for these physical characters. They are useful but not absolute, as it is relatively easy to blend materials, often of inferior quality, to give the desired results. The quality of a flavoring material must be directly related to its end use, and although such figures may be of value, the decision on suitability for use must ultimately be based on sensory assessment.

The past two decades have seen an enormous expansion of techniques using gas-liquid chromatography for the quality assessment of many flavoring materials. Again, the essential oils have proved most suitable for such control although the manufacture of compounded flavorings and the detection of adulteration in aromatic chemicals also benefit from these techniques. Most recent test methods, involving combined gas-liquid chromatography and mass spectrometry coupled with the computerized treatment of results, have proved a powerful tool in the hands of researchers and quality controllers, but, again, the results are of little value unless supported by sensory assessment. Research chemists may be able to identify a particular component but its odor and flavor qualities must be determined by sensory means. The quality controller also must support any instrumental findings by sensory judgments made under controlled usage conditions (Sawyer, 1971). The final judgment of any flavoring material must rest with the expert human taster or a panel familar with the ultimate use of the material under test. Instrumental methods have a very valuable part to play in our understanding of flavor chemistry but many can only be applied after careful sensory evaluation. It is doubtful whether the day will ever come when instruments will totally replace the trained nose and palate in assessing flavoring materials and end products.

The literature has many examples of the use of instrumental techniques (Heath, 1981) and full use must be made of the results obtained by such methods. For the technologist involved in product manufacture and control, far greater reliance should be placed on sensory judgments, so long as these are based on well-founded information obtained from fully representative samples.

Much has been written during the past two decades on the subject of sensory analysis, and most confectionery manufacturers have established the necessary facilities and defined sensory procedures as part of their quality assurance systems. But it is amazing how many times one encounters sensory testing of raw materials and finished products where the evaluation is carried out under far from ideal conditions, where judgements are made based on results that have no statistical basis, and where the conduct of the test methods used is open to bias. Three specific aspects of sensory analysis demand close attention: (1) the conditions under which sensory analysis is carried out; (2) the data necessary to establish statistical significance; and (3) the test methodology. Reference should be made to the considerable literature on this subject (Amerine et al. 1965; Goodall and Colquhoun, 1967; ASTM, 1968a, 1968b; Larmond, 1970; Stewart, 1971; Spencer, 1971; Heath, 1981).

## Sensory Test Methods

When one is presented with one or more samples of any aromatic raw material or product and asked to give a reaction, the questioner may have several fairly clearly defined aims in view: to establish and to characterize any changes that may be observed; to distinguish between the samples; to ascertain whether some identified character can be detected and quantified; to establish the degree of pleasure given by the sample; or to establish its value for a stated use. Such questions may be broadly classified into one of four categories:

1. Analytical—applied to differences.
2. Ranking and rating—to establish relative values.
3. Descriptive profile tests—to establish the specific attributes contributing to product quality, etc.
4. Acceptability appraisal—applied to the subjective pleasurable response, preference, consumer, and marketing tests.

**Analytical Difference Tests** Here the aim is to determine whether there is a difference between two samples. Usually, this difference is small, and judgment about the nature of any difference should normally be resolved by post test discussion.

In the single-sample test, which is not widely used, assessors are first familiarized with a standard or reference sample, and then this and the test sample are presented one at a time in random order. Judgment: Is this sample "A" or "not-A"?

The paired-difference test comes in two forms:

1. Assessors receive several pairs of samples that may be the same or different. Judgment: Are the samples the same or different?

2. Assessors receive samples differing in one character only, these being presented in pairs in random order. Judgment: Which sample is stronger in the stated characteristic?

These tests are generally used to compare a production or experimental sample against a control but the method may be adopted for preference determinations. Large numbers are required to achieve any significance as the chance results are 1:2.

The triangular test is the best-known and most widely used difference test method to check samples that differ only slightly from standard. Assessors are presented with three coded samples at the same time. Two of these samples are identical, one is different. Judgment: Which sample is different? Samples can be presented in six ways: ABB : BAA : AAB : BBA : ABA : BAB. If there is a "stronger" flavor, this should be given as the odd sample as this reduces fatigue during the test. Chance probability is 1:3.

The duo-trio test lies between the two previous tests. Assessors receive a designated control sample, then one or several pairs are presented, each pair containing one control and one test sample offered in different positions. Judgment: Which sample is different from control?

**Ranking and Rating Tests**  These are usually employed where there are more than minor differences to be established, where a combination of quality factors are to be evaluated, or where samples have to be sorted to determine relative value or suitability for use.

The aim is to arrange a series of samples in an ascending or descending order of intensity or some other specified characteristic (e.g., flavor strength, pungency). The test may be combined with scoring, in which the samples are evaluated on a predetermined numerical basis covering a single quality factor or a combination of quality factors (Cloninger et al. 1976). Scales may include any number of steps, depending on the degree of precision possible in making a judgment. The method is speedy but its value is very dependent on the experience of the assessors. It can be applied to multiple samples and, by using reference samples, fixed scales can be set or anchored. It is necessary to avoid:

1. Any tendency to be too arbitrary or to give a false sense of exactness that does not exist.
2. A tendency to try to obtain too much information from too many samples at one tasting session.
3. Bias and preconceived ideas to achieve a desired result.
4. Assuming more knowledge about a product than one actually has.

Quality is defined as a degree of excellence and quality factors can be systematically examined and evaluated under the following headings:

1. Appearance
2. Textural factors
3. Odor
4. Flavor

Each can be rated as long as the following precautions are taken to avoid:

1. Attempting to obtain precise quality measurements.
2. Personal bias usually associated with overfamiliarity with a product.
3. Trying to equate products of different characters.
4. Hasty judgments.

Many variations of the scoring system are used a great deal for production quality control and for assessing samples after periods of storage under different conditions. One such system involves presenting samples to a panel of assessors who are asked to place a score against each sample on one of the bases of judgment shown in Table 18.1.

On completion of the test, the results can be examined statistically

TABLE 18.1. SCORING SAMPLES

| Production quality control | Storage tests |
|---|---|
| +2 fully acceptable: equal to control | +2 saleable—equal to finished product |
| +1 acceptable but not equal to control | +1 just saleable |
| −1 quality below control: doubtful acceptance | −1 saleability doubtful |
| −2 unacceptable | −2 unsaleable |

and related directly to the bulk product before making a decision as to its acceptability. It is necessary to maintain an authentic control sample, which should be checked at regular intervals by a panel of assessors (see also Chapter 23).

**Flavor Profile Method** This type of test was initiated by Caincross and Sjostrom in 1948 and consists of a combined qualitative and quantitative descriptive method of flavor analysis that is applicable to total food products as well as to individual components and raw materials. The method depends on the judgment of three factors:

1. The single odor, taste, and mouth-feel characteristics that can be detected and described in order of their perception.
2. The intensity of each characteristic.
3. The overall impact and quality of the test material.

The method depends very much on the specialist training of from four to six panel members. When properly used the method results in a reproducible tabulation of all the describable features of the test subject. Results can be expressed graphically, each component being identified by lines in order of its perception—the distance between the lines giving the relative time intervals between perception and the length of each representing the intensity of the characteristic on a predetermined scale. The overall quality of the product is shown by a semicircle, the radius of which increases with increasing quality. An inner circle may be used to indicate the threshold level of intensity of any particular attribute or character. A full discussion on this test method was published by Sjostrom et al. (1957) and is available in a booklet published by A. D. Little. With a correctly trained panel, the method is reproducible and very informative, particularly for the identification and assessment of product characters that may be responsible for successful sales. It is also used to assess a new product to see whether it has any outstandingly good or bad features that may affect consumer acceptability.

The method does have certain disadvantages:

1. The choosing of panel members must be done with care as selection on a simple basic odor and taste response gives no indication of their ability to judge finished products.
2. The selection, training, and conducting of profile panels is time consuming and very expensive.
3. The delineating of agreed-upon descriptive terms to ensure unformity and understanding is tedious.

4. Results cannot be analyzed statistically.

5. The scales normally used are of limited value and lack precision.

6. Open-ended discussion, which forms an essential part of the test method, is potentially dangerous and does not necessarily reflect the true facts. Comments may mirror the views of a senior staff member or those of the loudest and most assured assessors, who can sway the opinion of the panel.

7. Individual sensitivities cannot be quantified or correlated.

Alternative descriptive test methods are discussed by Stone et al. (1974).

**Methods Based on Pleasurable Response**  These relate to pleasurable and unpleasant stimuli, to likes and dislikes. Preference tests are very widely used, particularly during new product development. They are characterized by the use of large numbers of panelists, evaluation being spontaneous and based purely on personal emotion. It is very important to recognize that trained or semitrained staff no longer represent the general consumer and must not be used where purely subjective, pleasure-giving judgments are required, as judgments by the former may be conditioned and not based on pure emotion.

**Paired Tests**  This method uses the direct comparison of two samples, which may result in a straight preference of one against the other or the degree of preference may be scored on a defined pleasure-response scale. Preference does not necessarily imply overall acceptance and more information than is properly available from the test should not be sought from its results. If additional information is required, then a new test must be established and the appropriate questions asked.

Several scales have been suggested but the following are most widely used:

| | | | |
|---|---|---|---|
| 9 | like extremely | +4 | excellent |
| 8 | like very much | +3 | very good |
| 7 | like moderately | +2 | good |
| 6 | like slightly | +1 | −good + fair |
| 5 | indifferent | 0 | fair |
| 4 | dislike slightly | −1 | −fair + poor |
| 3 | dislike moderately | −2 | poor |
| 2 | dislike very much | −3 | very poor |
| 1 | dislike extremely | −4 | extremely poor |

Multiple tests can be designed to give specific information but the technique is complex and very costly, and the results are difficult to interpret with any degree of confidence. Simple questions give the most direct answers and acceptance can be numerically calculated by using a plus/minus scale as above. A final plus answer would denote acceptability and minus would be indicative of rejection.

The importance of control of all aspects of panel testing cannot be overstressed.

**Taste Panels**   Panels have been mentioned in the foregoing and it could well be imagined that these are produced by just collecting together a group of people who happen to be available at the time. Nothing could be further from the truth, and for work concerned with factory quality control and new product development, panels must be carefully selected, specifically nominated, and trained.

Consumer panels drawn from the public are quite different; these usually involve large numbers of untrained and broadly representative consumers. Panels covering specific groups of the community, based on age, ethnic origins, and the like, may be nominated to establish particular marketing needs, and so on. The only requirement is that such panelists be able to understand and answer questions put to them, intelligently and honestly.

Quality control panels should consist of about ten to fifteen members, but to be able to call on this number at any one time needs the training of many more. From experience, some four times this number is necessary, and these should consist of an equal number of men and women. Age is not significant, although some authorities recommend that persons over 55 should be left out as there is some evidence of loss of sensitivity with increasing age starting at about age 50. Nonsmokers and smokers have been found to be equally sensitive as panel members but the effect of residual smoke odors, inevitably carried by smokers, may seriously affect any nonsmokers. It is strongly recommended that smokers participating in sensory panels should refrain from smoking for at least 30 min before taking part in a test. In addition, it is most desirable that smokers wash their hands well before joining the panel. Female members of a panel should avoid the use of highly perfumed cosmetics and hands should be well washed with nonperfumed soap.

*Panel Selection and Training*   This will depend very much on the available staff and facilities within each company, but the following has been found satisfactory.

Department heads are first approached by the quality control manager seeking permission for some members of their staffs to serve on tasting panels. It is important to get departmental management cooperation for this activity and an agreement on the number of persons likely to be involved, the time they will be away from their normal work routine, and the priority that must be given for this company commitment. Failure to get such agreement will inevitably lead to problems, particularly as there is ever-increasing pressure to reduce staffing.

Possible nominees are then informed and asked whether they would be prepared to serve on a tasting panel and to undertake tests prior to selection. It is no use having unwilling panelists.

Depending on the size of the company one might expect some fifty to eighty nominees, but may have to be content with far fewer than this. Each nominee should be subjected to a series of tests, receiving appropriately coded samples and a questionnaire. The following are examples of the type of test products that can be used in this selection:

1. Two samples of chocolate, one containing a trace of mineral oil.

2. Two samples of chocolate, one badly ground (gritty) and the other normal.

3. Several samples of fondant cream, colored and flavored with recognizable flavorings, such as orange, lemon, rose, or vanilla, but with colors different from those usually associated with the particular flavors.

4. Two samples of fudge or paste, one contaminated with a trace of rancid fat.

5. Two samples of milk chocolate, one of which has been in a wrapper with strong-smelling ink.

6. Two samples of milk chocolate, one having been in contact with a moldy substance.

These tests are fairly searching and inevitably some candidates will be rejected, but those who pass are suitable for inclusion in various panels. It is preferable to select panels for specific training and test routines based on product groups, such as, chocolate, fondants, caramels, or fudges. Often it is desirable to select a special panel to taste one product only. Such panelists, when trained, become experts on the particular product, and can be used to maintain its quality standard within fine limits.

Once selected, panel members are then repeatedly acquainted with the products they are to taste and control, substandard samples being submitted under code until they are proficient in recognizing the required standard of quality.

The panel is then ready to operate as part of the quality assurance scheme, but not indefinitely without specific checks. Tasters are only human and can display normal variability of response and may deteriorate with time.

**Environment for Sensory Assessment** The environment in which sensory assessments are made can significantly affect the value of the results. Two conditions will dictate the minimum facilities required:

1. The need to provide an area in which assessors can apply maximum concentration with the minimum of distraction.

2. The physical conditions necessary for the preparation and presentation of samples.

Probably the most important factor is the need for complete concentration on the part of anyone engaged in sensory evaluation and this, generally, is the least observed of the basic requirements. Assessors, whether working alone or as a panel, are often required to endure all sorts of extraneous and distracting stimuli (e.g., telephone calls, movement and conversation of others, discussion of findings by assessors who have completed the test). All such interruptions detract from the essential calm conducive to critical attention to detail. However simple, sensory tests should be carried out in a separate room, or at least in a room with divided cubicles or a test booth where the assessor can work in relative isolation. The test conditions should be as natural as possible so that the assessor is at ease but not over comfortable, procedures should not impose any haste in reaching a judgment, the recording of results should involve a minimum of writing and be on preprinted questionnaries, and all movement and speech, even by panel members, should be minimized. With these conditions, assessors are given the best chance of producing meaningful results.

Temperature and humidity in a tasting room should be controlled within fairly narrow limits. The conditions that have been found most suitable for chocolate and confectionery evaluation are:

Temperature          17 to 20°C (63 to 68°F)
Relative humidity    55 to 65 percent

The physical requirements for sensory testing are critically examined by Larmond (1973).

**Tasting Sessions**

It is usual for the panel supervisor to summon the panel at a specific time, between meals, with some preference for midmorning. The samples, together with the appropriate tasting instructions and questionnaires, should be laid out in each booth so that the tasting session can start immediately, enabling panelists to concentrate on the work in hand.

It is best not to taste more than three sets of samples at a time (preferably fewer) and these should be presented with each clearly identified by a random three-figure code that has no characteristics that can give any clue to identity.

Sometimes samples have an intense flavor and marked cooling or pungent effects on the palate. In confectionery work, this applies particularly to peppermint oil and ginger. After tasting the first sample, the assessor's palate becomes partially paralyzed so that a judgment on a second sample is likely to be incorrect. Special techniques must be employed to overcome this (Heath, 1978). Such materials should be diluted, if possible, to minimize this effect, comparisons should be made using the triangle test procedure, and the test should be repeated after an interval of 24 hr. No other tasting should be carried out when these products are under test.

Although the above description has concentrated on flavor the odor of products if often of importance. Foreign odors in confectionery will almost certainly be observed as part of the flavor assessment but, occasionally, contamination will be more readily detected by smell. Wrapper solvent or printing ink contamination is best detected by the bottle test described under "Wrapping Materials."

**Test Results** One thing that is of considerable importance in all sensory testing is that judgments be recorded at the time they are made. Preferably, this should be done on preprinted questionnaires. This cannot be stressed too much, as frequently assessors are merely given a piece of paper with nothing on it and are expected to write their comments and findings without guidance. Results should always be written down and not communicated verbally to the panel supervisor or any other member of the panel, or the results of the test could be valueless. It is always worthwile to go to a little trouble to draft a proper test form and for each assessor to be given a copy partially completed with the sample details. The preparation of reports containing sensory evaluation data has been discussed by Prell (1976).

**Significance of Results**  Statistical methods of assessment are particularly valuable for the determination of the validity and significance of sensory test results. The subject lies outside the scope of this book but reference should be made to ASTM 1968B for a more detailed discussion on this method of handling test results.

# FLAVORING MATERIALS

## Classification

Flavoring materials constitute a very important group of substances in confectionery manufacture and, as has been mentioned previously, the character of many lines depends entirely on added flavoring.

All flavorings, whatever their physical character, are composed of highly aromatic constituents that exist naturally or are specifically selected to contribute a particular profile or subtle differences in the overall flavor of any product in which they are used. Flavoring materials include:

1. Naturally occurring plant materials (e.g., herbs, spices, vanilla, fruits, nuts, aromatic vegetables).

2. Products derived directly from such natural materials by physical process only (e.g., extracts, essences, essential oils, oleoresins, fruit juices, and concentrates).

3. Isolates or pure chemicals prepared from natural products (e.g., eugenol from clove leaf oil or citral from lemon grass oil).

4. Synthetics chemically prepared from natural isolates or other natural products (e.g., vanillin from wood lignin).

5. Nature-identical synthetic chemicals.

6. Artificial aromatic chemicals (i.e., those not yet identified in nature) (such as, gamma-undecalactone).

7. Flavor enhancers (e.g., maltol).

8. Taste modifiers (e.g., salt, sugars, and sweeteners, organic acids, or bittering agents).

9. Solvent or carrier.

Only a few of these are of direct interest to the confectioner and chocolate manufacturer, the remainder being provided in the form of compounded flavorings mostly as imitations of well-known fruit flavors. These will be discussed later.

## Natural Flavors

These are obtained from suitable plant sources and are generally subjected to some form of concentration before being used in confectionery products.

## Fruit Flavors

The flavors of natural fruits are delicate, refreshing, and pleasing to most people, but unfortunately the majority are weak and when included in confectionery contribute little of their original character to the product.

The soft fruits—strawberries, raspberries, and currants—contain 85 to 90 percent water and must be concentrated considerably if used as a flavor in a fondant, for instance. Concentration under vacuum is essential, and even so appreciable loss of flavor results. The modern method of freeze drying preserves the original flavor true to type but the process is costly.

Concentrated fruit pulps, if used in confectionery, usually require fortification with some synthetic flavor, particularly if combined with chocolate. Table 18.2 describes the properties and uses of some fruits in confectionery.

## Essential Oils

The essential oils constitute an enormous range of aromatic materials available for use as flavoring or fragrance materials. The nature and value of these are widely known but a definition is appropriate here: An essential oil is a volatile mixture of organic

TABLE 18.2. PROPERTIES OF NATURAL FRUITS

| Fruit | Properties and use |
|---|---|
| Apple (pulp or purée) | This is usually marketed as a sieved purée prepared by cooking the macerated fruit and forcing through a mesh. It is sometimes cooked by blowing steam through the mash—a convenient way of adulteration with water! It may be sulfited or canned but has very little use as a flavoring material. It gives body to fondants, fillings, and spreads and can be used as a fruit base with artificial flavors for bonbon fillings and for fruit jellies. |

TABLE 18.2. PROPERTIES OF NATURAL FRUITS (*Continued*)

| Fruit | Properties and use |
|---|---|
| Apricot (pulp) | This is supplied as canned fruit, stoned, with a fair amount of whole fruit. It has a good flavor that is retained in the process. It makes good jellies, bonbon fillings, and fruit jam sections for fondants and chocolate-covered assortments. |
| Black currant (pulp) | Generally supplied as canned fruit, with plenty of solid berries. Characteristic but weak flavor in confectionery. Gives intense color and the obvious appearance of real fruit substance in a creme center. Black currants have some medicinal reputation due to high vitamin C content, and the juice is used in throat pastilles. |
| Cherries | Used as whole cherries (sulfited) or preserved as glacé cherries—see "Jams, Preserved Fruits" |
| Damson Gooseberry Greengage Plum | These fruits are not often used in confectionery and need reinforcement with artificial flavor. |
| Peach (pulp) | Usually supplied in cans—somewhat similar use to apricot but less attractive and weaker flavor. |
| Pineapple (rings) (cubes) (minced) | Very useful preserved or as an ingredient of fillings and jellies. Needs very little reinforcement, particularly in uncovered cremes. |
| Raspberry (pulp or purée) Strawberry | These are very popular fruits but their true flavors, particularly strawberry, cannot be preserved except by freeze drying. They are always reinforced with artificial flavoring. Used in cremes for assortments and fondants. |
| *Citrus fruits* Orange Lemon Grapefruit Lime | Citrus fruits are different in having a powerful essential oil flavor within the cells of their peels. These fruits can be obtained as pastes or concentrates that include the rind, or as juice or concentrated juices without the rind. Juices are generally used in the cordial industry. Concentrates are occasionally used to flavor fondants and confectionery pastes but by far the greatest demand in the confectionery industry is for citrus essential oils. |
| Reinforced fruit concentrates | These should be mentioned in this grouping as they are largely concentrated natural juices reinforced with "artificials." The artificial flavors are synthesized chemical compounds equivalent to those found in the natural flavor. Flavor companies market some very good products of this type. |

compounds derived by some physical process (i.e., distillation, expression, or solvent extraction) from odorous plant materials. A specific oil is derived from one botanical source with which it agrees both in name and odor. Different parts of the same plant may yield essential oil of a different composition and aromatic quality. The same plant grown in different locations may yield essential oil of different quality.

Early investigators established that essential oils are complex mixtures of several substances, although, in certain instances, one main constituent may predominate (e.g., clove bud oil contains 85 to 90 percent eugenol). The constituents of essential oils may be classified into the following groups:

1. Hydrocarbons of the general formula $(C_5H_8)_n$—known as terpenes. Where $n = 2$, they are called monoterpenes; $n = 3$, sesquiterpenes; and $n = 4$, diterpenes.
2. Oxygenated derivatives of these hydrocarbons.
3. Aromatic compounds having a benzenoid structure.
4. Compounds containing nitrogen or sulfur.

**Production** Essential oils are all volatile in steam, which means that they have a vapor pressure low enough to enable them to be distilled in the presence of water without decomposition. Being insoluble in water, they can readily be separated from the aqueous distillate. Distillation in air at ordinary pressure involves higher temperatures and, in the presence of oxygen, results in decomposition of the heat-sensitive constituents. However, many essential oils can be distilled under reduced pressure, and this technique is used to prepare isolates and de-terpenized oils, which are valuable flavoring agents. The plant used for distillation is described by Heath (1981).

The principal steam-distilled oils used in confectionery include peppermint, cassia, cinnamon, nutmeg, clove, pimento (allspice), caraway, anise, and rose.

The essential oils obtained from the peels of various citrus fruits (e.g., orange, sweet and bitter; lemon; grapefruit; tangerine) are not recovered by distillation, which causes serious decomposition, but by hand or machine expression. Only lime oil is recovered as a by-product of lime juice production, which involves boiling. The old method of hand extraction has largely been discontinued but it did yield the finest oils. The majority of citrus oils are machine expressed as part of the vast citrus juice operation. Special machines have been devised to ensure that the separated oil is exposed to minimum water treatment as this results in a better-quality product. Some oil is recovered from citrus waste by distillation, but this is of very poor quality.

The presence of excessive quantities of terpenes has a detrimental effect on the keeping properties of many essential oils. A cleaner and better flavor is obtained by using terpeneless oils, which are more concentrated than the natural citrus oil. These are made by selective solvent extraction using diluted alcohol, which results in the terpenes separating as an oily layer leaving the flavorful, oxygenated compounds in a clear solution; They are obtained by fractional distillation under reduced pressure; or by a chromatographic process using silica gel.

Because of the complex nature of essential oils and the variability of the source of supply, they can be very susceptible to deterioration by either loss of their more volatile components or the development of off-odors due to oxidation. Some of these defects are observed as resinous, turpentine-like, characteristics. Because of these effects of oxidation, the use of antioxidants is permitted in many countries.

Off-odors and off-flavors can also arise in essential oils if exposed to iron surfaces and direct sunlight. Containers used for handling essential oils must be made of stainless steel, as must stirring rods and funnels. It is preferable to store essential oils in well-filled, amber-colored, glass bottles, in a cool, dark place.

**Use in Chocolate, Cocoa, and Confectionery**   The following essential oils have been found of considerable use in the flavoring of many types of confectionery products. The blends and quantities to be used are best determined experimentally as these are affected by the other raw materials present and the processing conditions used.

*Oil of Almond Origin.* Kernels of bitter almond (Prunus amygdalus), peach (P. persica), apricot (P. armenica).

*Constituents and main constants.* Contains about 97 percent benzaldehyde.

| | | |
|---|---|---|
| Specific gravity | 1.050 to 1.055 | [15.5°C (60°F)/15.5°C] |
| Refractive index | 1.542 to 1.546 | [120°C (68°F)] |
| Optical rotation | 0° to −0.1° | (20°C) |

Badly purified oil may contain small amounts of hydrocyanic acid and this is shown by high specific gravity. If the oil is stored in badly stoppered bottles, some oxidation to benzoic acid will occur.

*Uses.* This flavor and the synthetic product are frequently added to excess in marzipan, noyeau, and other pastes, and the result may be objectionable to many people. In small amounts with other flavors, it is useful for flavoring chocolate, nut pastes, and nougat.

**Oil of Aniseed** *Origin.* True aniseed, Pimpinella anisum (Europe). Star aniseed, Illicium verum (China). The main article of commerce.

*Constituents and main constants.* Contains 80 to 90 percent anethole.

| | | |
|---|---|---|
| Specific gravity | 0.978 to 0.992 | [15.5°C (60°F)/15.5°C] |
| Refractive index | 1.553 to 1.560 | (20°C) |
| Optical rotation | −2 to +1° | (20°C) |
| Melting point | 17 to 20°C (62.6 to 68°F) | |

*Uses.* Correctly employed, this is an attractive flavor. Apart from the famous aniseed balls, it may be blended with other flavors in milk chocolate and light coatings.

**Oil of Bay—Bay Leaves** *Origin.* Leaves of a laurel tree of the eastern Mediterranean and West Indies. Pimento acris and Laurus noblis.

*Constituents and main constants.* Contains phenolic compounds 50 to 70 percent.

| | | |
|---|---|---|
| Specific gravity | 0.965 to 0.985 | [15.5°C (60°F)/15.5°C] |
| Refractive index | 1.510 to 1.520 | (20°C) |
| Optical rotation | −3.0 to −0.3 | (20°C) |

*Uses.* Bay leaves are well known in culinary art. Both the dry leaves and the oil used in trace amounts seem to have a synergistic effect in cocoa and chocolate.

**Oil of Caraway** *Origin.* Distilled from the seeds of Carum carui, grown in Holland and central and southern Europe.

*Constituents and main constants.* Carvone 50 to 60 percent is the main organic constituent together with D-limonene.

| | | |
|---|---|---|
| Specific gravity | 0.910 to 0.920 | (15.5°C/15.5°C) |
| Refractive index | 1.484 to 1.490 | (20°C) |
| Optical rotation | +69° to +82° | (20°C) |

*Uses.* A very useful adjunct in the flavoring of chocolate and coatings. Is also used in liqueurs and liquid centers and in baked confectionery. Flavoring is objectionable if used in strong concentrations.

**Oil of Cardamom—Cardamom Seeds** *Origin.* From the fruits of Alleppy cardamoms—Elettaria Cardamomum, also from Malabar and Mysore cardamoms.

*Constituents and main constants.* A complex mixture of the esters of cineol and terpineol.

| | | |
|---|---|---|
| Specific gravity | 0.923 to 0.945 | (15.5°C/15.5°C) |
| Refractive index | 1.460 to 1.475 | (20°C) |
| Optical rotation | +23° to +48° | (20°C) |

*Uses.* Blends well with many fruit flavors. The ground seeds enhance the flavor of ground coffee and this can be applied to coffee cremes or coffee chocolate.

**Oil of Cassia — Cassia Bark** *Origin.* The oil is distilled from the bark, leaves, and twigs of the Chinese cinnamon—Cinnamonum cassia. The bark is marketed in bales of woody dried lengths (quills), which are dusty to the touch and typically aromatic.

*Constituents and main constants.* The bark contains a variable amount of essential oil (1.5 to 4.0 percent) and the flavoring power should be based on this. The quality is sometimes affected by molds. The oil is mainly cinnamic aldehyde (75 to 90 percent). The following constants refer to the oil.

| | | |
|---|---|---|
| Specific gravity | 1.050 to 1.072 | (15.5°C/15.5°C) |
| Refractive index | 1.600 to 1.610 | (20°C) |
| Optical rotation | −1.0° to +6.0° | (20°C) |

*Uses.* Finely ground cassia bark has been used as a cocoa flavor for a very long time and the oil is frequently added to dark chocolates and coatings. Neither oil nor bark finds much use in confectionery, except perhaps for some throat pastilles, but both are used in cookies and cakes.

**Oil of Cinnamon, Cinnamon Bark** *Origin.* The quills, as they are called, are really the underbark of the tree Cinnamonum zeylanicum, principally grown in Seychelles and Sri Lanka.

*Constituents and main constants.* The bark contains about 2 to 2.5 percent of the essential oil. The oil varies somewhat according to the place of distillation, but normally contains 60 to 75 percent of cinnamic aldehyde with 5 to 10 percent of eugenol. The following are constants for English distilled oil.

| | | |
|---|---|---|
| Specific gravity | 0.995 to 1.040 | (15.5°C/15.5°C) |
| Refractive index | 1.570 to 1.585 | (20°C) |
| Optical rotation | 0° to −1° | (20°C) |

*Uses.* Similar to cassia but is generally considered to have a less harsh flavor. Traces of cinnamon are supposed to enhance fruit flavors and coffee.

*Celery Seed, Oil of Celery Origin.* The celery plant (Apium graveolens) grows widely in northern Europe. The seed used for flavoring is usually of French origin.

*Uses.* The pulverized seed is used as a flavoring for cocoa, chocolate, and various confections such as nougat and pastes. It has the property of leaving a slightly bitter aftertaste, and because of this contributes to neutralizing sickliness.

Oil of celery may be distilled from all parts of the plant but that from the seeds, which contain about 3 percent, is considered the best.

*Constituents and main constants.* The oil contains mainly dextro-limonene. A substance known as sedanolide ($C_{12}H_{18}O_2$) is stated to give the characteristic flavor.

| | | |
|---|---|---|
| Specific gravity | 0.860 to 0.895 | (15.5°C/15.5°C) |
| Refractive index | 1.478 to 1.486 | (20°C) |
| Optical rotation | +60°C to +82° | (20°C) |

*Uses.* Similar to seed but pulverized seed is preferred for powders and the oil for confectionery.

*Oil of Cloves Origin.* Extracted from Zanzibar clove flower buds (Eugenia caryophylata).

*Constituents and main constants.* The oil contains about 90 percent eugenol with associated esters.

| | | |
|---|---|---|
| Specific gravity | 1.044 to 1.069 | (15.5°C/15.5°C) |
| Refractive index | 1.530 to 1.536 | (20°C) |
| Optical rotation | 0° to −2.5° | (20°C) |

*Uses.* Clove oil is a very potent flavor and is used in confectionery in minute amounts with other flavors, particularly vanilla. In cooking, it always seems to be associated with apple dishes and could possibly be applied to some fruit flavors in confectionery.

*Oil of Coriander Origin.* Prepared from the fruits of the Russian coriander—Coriandrum sativum. The plant is also grown in India and Morocco.

*Constituents and main constants.* The main constituent is D-linalool, which is estimated at about 70 percent.

| | | |
|---|---|---|
| Specific gravity | 0.870 to 0.883 | (15.5°C/15.5°C) |
| Refractive index | 1.463 to 1.476 | (20°C) |
| Optical rotation | +7° to +14° | (20°C) |

*Uses.* This pleasantly aromatic flavor is very usefully added to lemon flavors and liqueurs, and as a minor constituent of chocolate. Coriander seeds may be used as the center of dragees or comfits.

*Fennel and Fenugreek* *Origin.* These are prepared from fruits of Foeniculum vulgare and Trigonella foenum graceum, respectively. Sweet fennel is obtained from Foeniculum dulce.

*Constituents and main constants.* In sweet oil of fennel, the main constituent is anethol, but fenchone and dipentene are also present.

| | | |
|---|---|---|
| Specific gravity | 0.964 to 0.976 | (15.5°C/15.5°C) |
| Refractive index | 1.528 to 1.538 | (20°C) |
| Optical rotation | +6° to +20° | (20°C) |
| Congealing point | +3° to +10°C | characteristic of a good oil |

Bitter oil of fennel has quite different constants, for example,

| | |
|---|---|
| Specific gravity | 0.905 to 0.925 |
| Optical rotation | +40° to +65° |

Anethol is practically absent but there is a greater amount of phellandrene.

*Uses.* These flavors have usually been associated with the true spices used in pickles and sauces. Experimental work in chocolate and confectionery products suggests that they have some synergistic action when used with other flavors and contribute to the elusive property that makes a consumer want more of the product containing them.

*Oil of Ginger* *Origin.* This is usually manufactured from Jamaican or African ginger Zingiber officinale, but Australian ginger has become a new source of supply.

*Constituents and main constants.* About 2 to 3 percent of ginger oil is obtained from the dried rhizomes. The main constituents of the oil are dextrocamphene and $\beta$-phellandrene.

| | | |
|---|---|---|
| Specific gravity | 0.874 to 0.886 | (15.5°C/15.5°C) |
| Refractive index | 1.488 to 1.495 | (20°C) |
| Optical rotation | −25° to −50° | (20°C) |

*Uses.* Ginger oil is more often used for cordials and ginger wine than for confectionery. Preserved ginger is a very popular confection.

*Oil of Lemon* *Origin.* Lemon oil is obtained from the peel of the lemon Citrus limonum, one of the citrus trees now grown in many subtropical parts of the world. Methods of extraction of citrus oils have previously been described.

*Constituents and main constants.* The main constituent is the terpene D-limonene (85 to 90 percent) but the essential flavoring substance is the aldehyde citral.

Specific gravity       0.856 to 0.860   (15.5°C/15.5°C)
Refractive index       1.474 to 1.476   (20°C)
Optical rotation       +56° to +62°     (20°C)
Citral                 4.2 to 5.5 percent

*Uses.* Lemon oil probably has a greater use in confectionery than any other essential oil, and because of the enormous demand, the natural product is not infrequently sophisticated. Certain types of oil, particularly Sicilian, are in demand for their excellent flavor and stability but unfortunately it is said that more Sicilian oil comes out of Sicily than was ever grown there. Citral is used to fortify lemon oil and the only way to be sure of the origin and quality of an oil is to deal with a reputable supplier who has factories or control of production in the country of origin.

Chromatographic analysis has helped to establish a measure of checking on the composition of consignments but this does not necessarily match the oil's true flavor or keeping properties.

*Terpeneless oils.* Mention has already been made of these and they are prepared by vacuum distillation during which about 90 percent of the terpenes are removed. Because of the insolubility of the terpenes, terpeneless oils are used in cordials, but although these terpeneless oils are supposed to be stronger in flavor, this is not apparent in confectionery where they are rarely used.

*Keeping properties.* Lemon oil will deteriorate as a result of autooxidation if exposed to light and air, and develops a turpentinelike flavor. This may be retarded by the addition of an antioxidant (see also Oil of Orange).

*Oil of Limes   Origin.* Lime oil is derived from the peel of the fruits of the citrus trees Citrus limetta and Citrus aurantifolio.

The preparation of lime oil is somewhat different from that of lemon or orange oil. The cold hand extraction (hand pressed) gives a mild lemonlike flavored oil not really characteristic of lime flavor, but if the fruit is mashed and the oil steam distilled from this, a strong lime flavor is obtained and some of the citral is decomposed during this process.

*Constituents and main constants.* The pressed oil has 6 to 9 percent citral and distilled oil has 1 to 10 percent.

Lime flavor on its own is not very acceptable to some people and flavoring with it is frequently overdone. A good lime creme can be made by using concentrated lime juice or lime marmalade with a blend of pressed and distilled lime oil.

|  |  | Hand pressed (West Indian) | Distilled |
|---|---|---|---|
| Specific gravity | (15.5°C/15.5°C) | 0.878 to 0.902 | 0.860 to 0.872 |
| Refractive index | (20°C) | 1.482 to 1.486 | 1.470 to 1.472 |
| Optical rotation | (20°C) | +30° to +38° | +33° to +47° |

*Oil of Neroli   Origin.* This is a strongly perfumed oil which in concentrated form is somewhat nauseating. It is steam distilled from the fresh flowers of the bitter orange—Citrus aurantium amara and Citrus bigaradia.

*Constituents and main constants.* Linalyl acetate and L-linalol are the main constituents but the odor is attributed to methyl anthranilate.

|  |  | Bitter— Bigarade | Sweet Spanish |
|---|---|---|---|
| Specific gravity | (15.5°C/15.5°C) | 0.870 to 0.883 | 0.865 to 0.870 |
| Refractive index | (20°C) | 1.467 to 1.475 | 1.473 to 1.475 |
| Optical rotation | (20°C) | +1.5° to +8° | +25° to +45° |

*Uses.* Probably the greatest use is associated with honey flavors or in confections where a floral note is required, such as Turkish delight or violet cremes. It has also been used in coconut pastes. Very small quantities are required—10 to 15 ppm being readily detected.

*Oil of Orange   Origin.* There are two varieties—oil of sweet orange derived from the tree Citrus sinensis and the bitter oil from Citrus aurantium amara or Citrus vulgaris. It is extracted from the peel by methods described earlier in the chapter.

*Constituents and main constants.* Orange oil contains 90 to 95 percent D-limonene and various esters and aldehydes, notably decaldehyde and nonylacetate, which give the main orange character.

|  |  | Sweet oil | Bitter oil |
|---|---|---|---|
| Specific gravity | (15.5°C/15.5°C) | 0.848 to 0.851 | 0.852 to 0.856 |
| Refractive index | (20°C) | 1.473 to 1.475 | 1.473 to 1.477 |
| Optical rotation | (20°C) | +95° to +100° | +91° to +96° |

*Uses.* Orange oil probably is second only to lemon oil in its use in confectionery. Bitter oils are often used in creme centers with dark chocolate coating in preference to sweet oils. Orange oil is very prone to deterioration, particularly in aerated confections such as marshmallows or in a fondant containing egg whip. The oil itself will also deteriorate rapidly if exposed to light in the presence of air, and storage in clear bottles partly full will result in the development of a resinous and caraway type of flavor.

The mechanism of this change is interesting and results from an autooxidation process in which the limonene is converted to carvone: hence the caraway flavor—caraway oil contains 50 to 60 percent carvone.

Limonene       Carvone

This change may be retarded by the addition of an antioxidant. These are described in a separate chapter. Oils that show considerable flavor change in about six weeks when exposed to light and air are preserved for a further four or five months when an antioxidant is added.

When purchasing orange (and lemon) oils, it is important to assess their stability and age. Obviously, oil that has deteriorated badly can be detected by flavor tests but in autooxidation processes the reaction may have proceeded so far without noticeable flavor change.

Some idea of the condition of the oil and its potential keeping properties may be obtained by determination of the peroxide value before and after aeration test as in the following.

**Stability of Orange and Lemon Oil** A 1 in.-diameter-boiling tube is fitted with a stopper carrying two tubes, with each tube bent at a right angle. The longer tube is drawn to a 3/32-in. jet and the horizontal arm is lightly packed with cotton wool.

Essential oil is added to the boiling tube until it is two-thirds full, taking care that none is on the stopper. Air is drawn through the oil at a rate of one bubble a second for 80 hr. The test is carried out in diffused daylight.

At the end of this time, the treated and untreated oils are compared for odor, color, flavor, and peroxide value. A large increase in the peroxide value indicates that the oil has poor keeping properties.

**Other Types of Orange Oil** Terpeneless oils are available as with lemon oil but their value is dubious for confectionery. Tangerine or

Mandarin orange oils have a special type of flavor and are popular for certain assortment cremes.

*Oil of Peppermint*  Peppermint as a flavor in confectionery has increased in popularity. This is probably due to a great deal of research into the growing of various strains of the plant and in the distillation and rectification of the oil.

The peppermint oils on the market today are a variety of blends, and for confectionery purposes, those that have a smooth floral bouquet are most popular. The peppery strong oils, at one time used for mint tablets, are not generally liked.

*Origin.* True peppermint oil is distilled only from the fresh plant, Mentha piperata, and the traditional source of supply is the English Mitcham mint.

Continental- and American-produced oils from the same plant have a slightly different character but many have a very good flavor and are generally less expensive.

Another oil of commerce is derived from the Japanese plant, Mentha arvensis, and is not considered a true peppermint. To produce a palatable product from this plant, it is necessary to fractionate the oil, retaining only the sweeter portions for use.

*Constituents and main constants.* See Table 18.3.

*Uses.* Peppermint oil is used for specific types of confectionery. The best known are peppermint patties, sometimes crystallized, clearmints—glassy high boilings, "mintoes," peppermint crisps— usually aerated high boilings, and compressed mint tablets.

TABLE 18.3. CONSTITUENTS AND MAIN CONSTANTS OF PEPPERMINT OIL

| | Mentha piperata | Mentha arvensis | |
|---|---|---|---|
| Constituents→ | L-menthol 50–65%<br>menthone 9–19%<br>menthyl acetate<br>and iso-valerate<br>5–15% | L-menthol, menthone quantities depending on rectification. Piperitone is stated to be characteristic of Japanese oil. | |
| Constants | | Original oil | "Dementholized oil" |
| Sp.grav.<br>  (15.5°C/15.5°C) | 0.900–0.915 | 0.900–0.910 | 0.895–0.905 |
| Ref. index (20°C) | 1.459–1.467 | 1.458–1.464 | 1.458–1.465 |
| Opt. rot. (20°C) | −18°−−32° | −26°−−42° | −24°−−35° |
| Menthol | 50–65% | 78–92% | 40–60% |

Chocolate covered peppermint cremes have increased in popularity with the development of oils that blend well with chocolate flavor.

It is not always recognized by practical confectioners how volatile peppermint oil can be. Even if they realize that some evaporates when added to a hot batch, few understand that partial fractionation occurs and the flavor remaining in the batch may be different from that which is obtained if the oil is added to the cold product. When adding peppermint oil to a batch, it should always be added as late as possible and mixed in rapidly. This can apply easily to remelted fondants but high boilings are a difficult problem.

When the oil is added to a high boiled batch in a pan, a great deal of flavor is lost, and even when it is folded into the partly cooled plastic batch on the table, appreciable evaporation occurs. Loss is retarded by mixing with a less volatile substance such as propylene glycol, and emulsifying with a small amount of water and lecithin. A malpractice of some old confectioners was to mix the peppermint with an equal amount of refined mineral oil! Powder flavors are also available where the particles enclose minute globules of oil. Evaporation of the oil is retarded by the "skin," which may be dextrin or some similar material, enabling the flavor to be dispersed throughout the batch before it vaporizes. The following method of making an emulsion is specially applicable to peppermint oil and uses an alginate ester. The dispersion is prepared as follows:

| | |
|---|---|
| Propylene glycol alginate | 5 g |
| Water | 800 ml |
| Dissolve by soaking and mixing. | |
| Peppermint oil | 200 ml |

The oil is stirred into the alginate solution and the mixture emulsified.

*Tasting tests.* To make reliable sensory tests on products containing peppermint is extremely difficult as the taste buds seem to be temporarily paralyzed. To form an opinion of a second sample tasted soon after the first is not possible and it is probably best to examine samples by the flavor profile method, or at least by triangular tests at well-spaced intervals. In any case, repeat tests must be made to get any worthwhile results on samples with a marginal difference in flavor.

*Distinction between piperata and arvensis oil.* The following tests may be applied:

1. One milliliter of oil and 0.5 g of a mixture of equal weights of

paraformaldehyde and citric acid are heated on a water bath. Arvensis oil gives no color, whereas the other gives a purple or brownish color.

2. Five drops of the oil and 1 ml of glacial acetic acid are mixed in a test tube and allowed to stand for several hours. Arvensis oil remains colorless whereas the other develops a blue color, reaching a maximum intensity in 24 hr.

**Spearmint Oil**  This oil, which is very popular in the United States for flavoring chewing gum, is distilled from the herb Mentha viridis. It is more often used in vegetable flavoring in the United Kingdom.

*Constituents and main constants.* Spearmint oil contains mainly carvone, phellandrene, and L-limonene. For the American oil:

| | | |
|---|---|---|
| Specific gravity | (15.5°C/15.5°C) | 0.920 to 0.940 |
| Refractive index | (20°C) | 1.480 to 1.489 |
| Optical rotation | (20°C) | −35° to −53° |

**Oil of Rose, Otto or Attar of Rose**  Otto of Rose is one of the perfumery extracts used for special flavoring in confectionery. Its greatest use as a single flavor is probably in Turkish delight, which is the traditional sweetmeat of the Eastern harems.

Like neroli oil, it has an extremely powerful flavor and in excess is nauseating. Very small quantities blended with other flavors are used for liqueurs and the hard candies known as "rose buds," which at one time were very popular with children.

*Origin.* The oil is obtained by steam distillation of fresh rose petals. The original, considered the highest quality obtainable, is from the Bulgarian damask rose, Rosa damascena. Otto is now obtained also from Rosae alba, moschata, and centifolia, and a considerable industry has grown up in the province of Turkish Anatolia. The yield is very small and about 0.03 percent only is obtained from fresh rose petals. A proportion of the oil dissolves in the distillate and this is known as rose water.

*Constituents and main constants.* Geraniol and isomers are the main substances present, with citronellol and isomers constituting a high proportion of the remainder. The hydrocarbon stearoptene is also present and will deposit on cooling. Phenyl ethyl alcohol and N-nonyl aldehyde are important minor constituents, together with complex esters of all these components.

| | | |
|---|---|---|
| Specific gravity | [30°C(86°F)/15.5°C] | 0.849 to 0.862 |
| Refractive index | [25°C(77°F)] | 1.460 to 1.465 |
| Optical rotation | (25°C) | −1° to 4° |

| Melting point | 19 to 22°C (66.2 to 71.6°F) |
| Freezing point | 17 to 21°C (62.6 to 69.8°F) |

If the freezing and melting points are low, alcohol is to be suspected. The oil is washed with water in a separating funnel and the refractive index again determined. If it has increased by more than 0.0001, the presence of alcohol is likely. The aqueous layer is fractionated and the iodoform test applied to the first few milliliters of distillate.

*Stearoptene.* Five grams of otto plus 25 ml 85 percent alcohol are warmed to dissolve, then cooled at 0°C (32°F) for 6 hr. The deposit is collected on a filter paper in a Buchner funnel and washed once with 15 ml 85 percent alcohol at 0°C. The stearoptene is dried *in vacuo* over concentrated sulfuric acid for 24 hr, transferred to a watch glass, and weighed. The stearoptene content varies considerably with the origin of the oil. Figures ranging from 18 to 40 percent have been recorded.

Synthetic rose flavors of very good quality are now available.

***Otto of Violet***   Natural otto is available but more often the synthetic product is used. In confectionery, it has limited use in assortment cremes, fondants, and cachous (lozenges).

## Essences

The term "essence" is now usually applied to alcoholic extracts of fruits, essential oils, and certain aromatic plant materials (e.g., vanilla, cocoa). Many fruits when pulped and extracted with alcohol yield a delicate flavor that is of little value to the confectionery manufacturer as they are too weak and very expensive. Such essences are of more value in cordial and table jelly manufacture.

The use of the word "essence" to describe blended imitation flavors is no longer acceptable, the description "flavoring" is preferred. These products are discussed in a later section.

## Herbs, Spices, and Spice Products

Although herbs and spices play a significant role in the seasoning of savory food products, their use in confectionery is strictly limited. Powdered spices such as cassia and cinnamon bark are used in the production of cocoa and chocolate products and ginger is an acceptable flavor in chocolate fillings. The majority of herbs and spices find no use in sweet products. However, the following spice products are available commercially and may be of value in new product development:

1. Ground herbs and spices
2. Processed spices:
   a. Essential oils
   b. Oleoresins
   c. Spice extractives dispersed on edible carriers (e.g., dextrose)
   d. Spice extractives encapsulated in gum or modified starches
   e. Blended seasonings and flavorings
   f. Emulsions
   g. Solubilized spices

Details of such products can readily be obtained directly from the manufacturers and there are numerous articles in the literature (Heath, 1978, 1981).

## Other Natural Flavoring Materials

**Coffee** This is undoubtedly a very popular flavor in confectionery and chocolate, but since it is generally recognized that coffee is an "acquired" taste, there have been many arguments as to the correct blend to use. The connoisseur usually prefers high roast strong coffee but the popular flavor is the milder light roast and dried water extracts are used a great deal.

*Origin.* The coffee of commerce is obtained mainly from the berries of a small evergreen tree, Coffea arabica. The berries grow in clusters and consist of a fleshy outer section with two inner seeds flattened against one another. These have a casing of hard husk.

By far the biggest coffee producer in the world is Brazil, with Colombia second. Smaller quantities are produced in various African countries.

*Curing and preparation of coffee.* Dry curing consists of sun or artificial drying during which some fermentation occurs. The pods of the dried berries are then removed mechanically.

With wet curing, the damp beans are allowed to ferment to develop the volatile aromatic oils, a process needing experience and skill, before drying and dehusking. This process gives high-quality mild coffees.

The selection and purchase of coffee are an exacting task and to maintain a standard flavor from spot purchases is by no means easy. For this reason, many confectionery manufacturers have adopted the use of extracts, relying on the manufacturers of these, who, because of their large production, are specialists in purchasing, blending, and roasting.

Soluble coffees, now often sold as instant coffees, have undoubtedly increased the consumption of the beverage and modern techniques of extraction with low-temperature concentration or freeze drying have resulted in a high standard of flavor. The true connoisseur, though, will still have nothing to do with extracts when it comes to the beverage but for bakery and confectionery they have proved invaluable.

*Use of coffee extracts in confectionery.* For the flavoring of coffee cremes, instant coffees may be added at the remelting stage, no flavor damage being done at this temperature. With higher-temperature confectionery, such as caramel or fudge, coffee extracts may develop an unpleasant burnt flavor.

The craft confectioner likes to roast the beans and make the extract just as the connoisseur will prepare the beverage. The fresh extract is used immediately in the fondant, but this is obviously applicable only to luxury assortments.

*Coffee "liquors."* Another method of incorporating coffee flavor is to grind the freshly roasted beans in fat, which may be hardened vegetable fat or cocoa butter. It is still necessary to use good roasted beans and the selection of these is an expert's job. The beans are pasted in a melangeur with the fat and then refined over chocolate rolls to a fine particle size. This fat paste may be used in a variety of products, but if added to fondant creme, it gives a stodgy texture due to the setting of the fat. A good coffee chocolate may be prepared with a liquor made with cocoa butter and some manufacturers favor the use of coffee chocolate for covering their coffee cremes as well as using some extract in the centers.

Coffee flavored chocolate has had periods of popularity. Coffee liquor in milk chocolate has a strong appeal initially but to many it is a very sickly product and the palate quickly tires of it.

The stimulating effect of coffee as a drink is well known and is attributed to the alkaloid caffeine, but is doubtful whether there is any effect from the quantity used in confectionery.

**Balsams** Commercial balsams are viscous liquids or pastes and are exudations from plants. There are three balsams that find some use in flavoring of dark chocolate—Peru balsam, Tolu balsam, and Storax preparations. They are not used in confectionery except in small amounts in the compounding of some essences.

***Peru Balsam*** This balsam is a dark viscous liquid obtained from the trunk of Myroxylon pereirae (Leguminosae). It has a pleasant odor resembling vanilla but a bitter acrid flavor and persistent aftertaste. It is partly soluble in alcohol, insoluble in water, and can be dispersed in vegetable fats.

It consists mainly of complex mixtures of compounds of cinnamic and benzoic acids. The following constants pertain to oil of Peru balsam.

| | | |
|---|---|---|
| Specific gravity | (15.5°C/15.5°C) | 1.100 to 1.125 |
| Refractive index | (20°C) | 1.571 to 1.580 |
| Optical rotation | (20°C) | 0° to +3° |
| Acid value | | 25 to 48 |

***Tolu Balsam*** Balsam of Tolu is obtained from Myroxylon toluifera (Leguminosae) and is a soft adhesive solid originally but hardens with age. Its properties are somewhat similar to Peru balsam:

Acid value      97 to 160

It yields 2 to 7 percent of a balsamic oil with the following characteristics:

| | | |
|---|---|---|
| Specific gravity | (15.5°C/15.5°C) | 0.949 to 1.080 |
| Refractive index | (20°C) | 1.544 to 1.560 |
| Optical rotation | (20°C) | −2° to +1° |
| Acid value | | 5 to 30 |

***Storax*** Storax is obtained from the trunk of Liquidambar orientalis (Hamameliaceae) and has a fragrance that makes it useful in small amounts in synthetic flavors.

Acid value      64 to 80

The balsam yields about 1 percent of an oil with the following constants:

| | | |
|---|---|---|
| Specific gravity | (15.5°C/15.5°C) | 0.950 to 1.050 |
| Refractive index | (20°C) | 1.5395 to 1.5653 |
| Optical rotation | (20°C) | −35° to +1° |
| Acid value | | 1 to 26 |

Considerable variation of this oil arises because of differences in methods of distillation.

All these balsams are used as perfumery fixatives and presumably they have similar action in flavor compounding.

## Vanilla, Vanillin, Ethyl Vanillin

Vanilla flavor is used in a great variety of foods and is invaluable in the chocolate and drinking-chocolate industry.

While natural vanilla with its incomparable bouquet is used for the highest-class confectionery, the majority of products are now flavored with the synthetic vanillin and ethyl vanillin.

**Natural Vanilla** Vanilla is obtained from the pods of a species of tropical orchid known as Vanilla planifolia. The plant is a vine that needs the support of trees or poles and in these circumstances will grow to a height of 15 ft. It is native of Central America but it is mainly cultivated in Madagascar, Seychelles, Reunion, and Tahiti. In these areas, artificial pollination of the flowers is necessary as there is none of the special kind of bee found where the plant grows naturally. The flowers have a narrow bell surrounded by thin petals that develop slowly over several months into long narrow pods about 6 to 9 in. long. Vanilla needs a process of curing similar to that for cacao to develop its characteristic aroma, and various processes are used. The pods are picked green and steeped in nut oil until they are black when they become fully ripened in about a month.

During the curing process, the flavor precursors, which are glucosides, are broken down into vanillin and glucose, and although the chemical substance, vanillin, is the main product, a number of other minor aromatics are produced. These contribute to the true aroma of natural vanilla, which is lacking in the synthetic product.

The selection of good vanilla beans needs experience. At one time, the presence of vanillin crystals on the surface was regarded as significant, but a nefarious practice of sprinkling the pods with benzoic acid crystals was uncovered by the author on one occasion. The only true way to judge is to make an alcohol extract and flavor some fondant creme.

Vanilla extract is made by cutting the beans into small pieces and soaking in successive quantities of hot 65 to 70 percent alcohol, and to prepare it this way is the best means of ensuring that a true natural vanilla flavor is obtained. There are many vanilla extracts on the market that are blends of the natural and synthetic.

**Synthetic Vanillin** The manufacture of synthetic vanillin ($C_6H_3OH \cdot OCH_3 \cdot CHO$) was one of the earliest achievements in the field of flavors and arose from the study of vanilla beans by Tiemann and Haarmann (1876).

Production of vanillin from clove oil eugenol and later from guaiacol was established and the purity of the vanillin was related to the final processes of separation, which involved vaccum distillation. For a long time, clove oil vanillin had the mark of superiority.

Today, virtually all vanillin is produced from wood lignin as a by-product of the paper pulp industry and is now accepted as equal in quality to vanillin produced from other sources.

*Constants*

| Melting point | 81 to 82.5°C (177.8 to 180.5°F) | |
| Solubility | Water | 0.5 percent |
| | Alcohol 90 percent | 40 percent |
| | Isopropyl alcohol 95 percent | 80 percent |
| | Glycerol 50 percent | 15 percent |

**Ethyl Vanillin, (Bourbonal, Vanillose)** In ethyl vanillin ($C_6H_3OH \cdot OC_2H_5 \cdot CHO$), the methyl group of vanillin has been replaced by ethyl. Although known for many years, difficulty in preparing it free from objectionable impurities retarded its sale, but today a satisfactory product is available and it is widely used. Economically, it is an advantage as it is claimed to be five times the strength but only four times the cost of vanillin. Flavorwise, it is slightly different from vanillin but in the majority of tasting tests carried out on the basis of the relative strengths quoted above it cannot be detected.

*Constants*

| Melting point | 76 to 78°C (168.8 to 172.4°F) | |
| Solubility | Water | 0.4 percent |
| | Alcohol 90 percent | 20 percent |
| | Isopropyl alcohol | A 5 percent solution of ethyl vanillin in equal parts of isopropyl alcohol and water is used for flavoring purposes |

**Use of Vanilla Flavors** As with other flavors, the method of use and quantities required rest with the experimentalist and depend on the nature of the product. Only flavor profile tests will decide the right combinations.

The best vanilla bean extracts are generally used for high-quality chocolate creme centers or uncovered fondants. Vanillin or eth-vanillin are invariably used for chocolate and drinking-chocolate powders.

Correct dispersion is very important. Originally, both vanillin and ethyl vanillin were supplied in crystal form, and this either had to be dissolved in a suitable solvent or finely ground, or isolated crystals gave flavor "hot spots." Pulverized vanillins are now available.

## Synthetic Flavorings

Although it is accepted that natural materials provide a wide spectrum of pleasurable flavor experiences, their use in confectionery products is limited as their properties have proved inadequate to meet the needs of current technology. Of necessity, confectionery manufacturers have to depend on the use of compounded flavorings, the flavor strength, composition, and nature of which have been specifically designed to satisfy the conditions involved. Such flavorings are composed of:

1. Natural flavors and flavoring substances: Preparations or substances acceptable for human consumption, obtained by physical means from vegetable, sometimes animal, raw materials either in their natural state or processed for human consumption.

2. Nature-identical flavoring substances: Substances chemically isolated from aromatic raw materials or obtained synthetically; they are chemically identical to substances present in the natural product.

3. Artificial flavoring substances: Substances that have not yet been identified in nature.

4. Appropriate permitted solvent(s) or carrier(s).

Current legislative opinion favors the control of artificial substances while allowing freedom to use natural and nature-identical materials within acceptable use limits (IOFI; Code of Practice).

Whatever the nature of the chemicals used in creating an imitation flavoring, the ultimate purpose is to reproduce, as closely as possible with the raw materials available or permitted, the odor and flavor effects of the natural flavoring in the end product. Modern instrumental techniques used in flavor research have resulted in the isolation and characterization of hundreds of organic chemicals responsible for the distinctive aromas and flavors of almost all natural flavoring materials. A bibliography of the available literature has been compiled by Heath (1981).

The preparation of imitation flavorings is not one for the amateur. The many, long-established, flavor manufacturers have the detailed knowledge and full facilities for carrying out this work and users can be assured that the imitation flavorings currently available to confectionery manufacturers are of the highest quality, are safe to use, and comply with the existing legislation of the country in which the end product is to be offered for sale to the public.

The list of available flavorings is comprehensive and there are excellent imitations of most natural flavors with a wide choice of variants to give any desired profile. Most flavor houses readily supply samples and use data applicable to the various end-product groups concerned (i.e. high boiled confectionery (hard candy), low boiled confectionery (chewy caramels), starch-deposited confectionery (pastilles, fondants), chewing gum, and chocolate. It is advisable to discuss flavoring problems with the experts.

# HANDLING AND STORAGE OF FLAVORINGS

A visit to the experimental department of any food company, and this includes confectionery, will almost certainly reveal a storage cupboard containing hundreds of bottles of flavors of all kinds, from many different flavor companies.

Most of them will be undated, some have been there for years and their contents have probably oxidized, polymerized, or partly evaporated. Nothing can be more distressing to flavor suppliers than to see their samples treated this way, but to some extent they are to blame.

A visit by a flavor company representative is often followed up by an array of samples in small bottles and the potential user probably does not have time to test them—so into the cupboard they go. At a later date, sometimes after many months, some experimental samples of confectionery are required and some of these flavors may be used. If the bottles were full, unopened, and kept cool and in the dark, they may still be satisfactory, but the risk is there to the detriment of the supplier and user. No flavors keep indefinitely; some, like vanillin, will keep for a very long time, but a system of storage should always be maintained.

1. All samples should be dated—many suppliers now do this.
2. Bottles should always be full.
3. Storage should be cool 4°C (40°F) and in the dark.
4. A record card should be kept of all samples, with the keeping period noted.

5. Once a month, all samples should be inspected and those over date discarded or replaced.

The above is intended to refer to samples but the same system must apply to stocks of flavors in production use.

Flavors are often purchased without proper consideration of usage and stocks may remain for long periods in partly filled bottles in unsuitable storage conditions. It is the responsibility of the quality control personnel to see that this does not happen. A publication by Oosterhuis (1977) outlines precisely the keeping properties of various flavors—the same publication gives a very useful summary of methods of testing and using flavors. As as general rule, adhering to the conditions outlined above, most flavors will keep from three to six months. Some encapsulated flavors, where small particles of flavor are surrounded by an inert substance, will keep for a year or more.

In relation to this, the keeping property of a flavor must also be considered in the final product. In dense confectionery, such as hard candy or caramels, the flavor once distributed is protected from the air. In aerated products, such as marshmallows, or in powders where air is present, the flavor, if susceptible to oxidation, will rapidly deteriorate. The answer is to use stable or encapsulated flavors. Flavors that are dispersed, without encapsulation, on dextrose or other base will probably have very short shelf life and are best avoided.

## THE APPLICATION OF FLAVORINGS

### Hard Candy

The specific selection and application of flavoring in confectionery and chocolate manufacture are determined by the nature of the end product and its method of manufacture. In modern factories, continuous and automated processing is replacing the long-established batch methods. The new parameters imposed by this technology place particular constraints on the composition and method of incorporation of flavoring materials. In sugar boilings, the liquid flavoring is injected into the cooked sugar mass as it passes through the system. The mixing time is limited so that ease of dispersion is an important characteristic of the flavoring used. Flavorings must be able to withstand 154°C (309°F) for a relatively short period but much longer dwell times at 140°C (284°F) are quite common. The change in profile under such conditions should have been established as part of the development program. Vacuum is frequently used, and this, too, can lead to an unacceptable level of flavor loss or a change in the flavor profile. The exposure of highly volatile flavorings to a hot sugar mass or syrup may cause the evaporation of components having a low boiling point and such fractionation will result in an unbalanced flavor effect. Again, the required additional flavoring must be established experimentally and this may be as much as 25 percent more than would be necessary under normal batch-processing conditions.

## Soft Caramels

The dairy-type soft caramels and fudge, containing butter, whole milk, and brown sugar, require little by way of added flavorings other than vanillin to accentuate a basic creaminess. Where formulations have to be modified as a cost-cutting operation, then some of the natural materials may be replaced by imitation butter, cream, caramel, or vanilla flavorings. Various fat-based flavorings are also available for caramels. These may be easily incorporated after any cooking stage and have good flavoring and storage properties.

## Deposited Articles

The flavoring of deposited articles calls for particular attention to the following factors:

*Cooking.* Usually continuous cooking and depositing affect the texture of the end-product and call for some modification in the basic formulation. The maximum cooking temperature is about 140°C so that the flavoring is not too adversely affected unless the processing time is extended.

*pH and acidity.* Most fruit flavors depend for their appeal on their

TABLE 18.4. GELLING AGENTS—ACCEPTABLE LIMITS OF ACIDITY

| Gelling agent | Added acid (as citric acid), % | pH of finished product |
|---|---|---|
| Agar-agar | 0.2–0.3 | 4.8–5.6 |
| Pectin | 0.5–0.7 | 3.2–3.5 |
| Pectin—low methoxyl | 0.4–0.7 | 4.0–5.0 |
| Gelatin | 0.2–0.3 | 4.5–5.0 |
| Starch | 0.2–0.3 | 4.2–5.0 |
| Gum arabic (acacia) | 0.3–0.4 | 4.2–5.0 |

refreshing, sharply acidic taste. As pH is critical to gel setting and stability, it may not be possible to add a sufficiently high level of acid to produce the desired flavor effect. It is usually necessary to compromise within the limits delineated in Table 18.4.

## Chocolate

Chocolate manufacture calls for quite a different approach and with few exceptions very little flavor is added directly to the chocolate mass, which has a high level of intrinsic flavor. Any flavorings are used:

1. To modify the flavor of the basic chocolate mass to give a rounded smoothness to the profile; vanilla extract, vanillin, ethyl vanillin, cassia or cinnamon oils and so on are widely used for this purpose.
2. To impose an overriding but compatible flavor; orange oil, peppermint oil, and rum flavoring are good examples of this.
3. As a flavoring in fondant-based or jelly centers.

## Chewing Gum

The flavoring of chewing gum poses quite different problems. The base itself usually has a high content of gum chicle, which has the ability to absorb very large quantities of flavor with little noticeable effect on the mouth's flavor sensibility. Dosage rates of up to tenfold those used in normal candy work are quite normal. The actual physical state of the gum and its chewing properties can be radically altered by certain flavoring materials, particularly solvents used in flavoring manufacture. To avoid this, flavor houses formulate very strong flavoring products specifically for use in chewing gum. These are relatively cheap materials and are devised to be completely free of deplasticizing agents. Alternatively, one can use encapsulated flavorings, as these have good persistency and last through an extended chewing period.

## Usage levels

There is always a risk that imitation flavorings may be used in excess and this is particularly true with the soft fruit flavors, such as strawberry and raspberry. In nature, these are very delicate flavors, but the reproduction of the natural flavor level is not successful in sugar confectionery. A much higher impact is necessary for acceptance. However a level of flavor that is too high defeats the purpose and such products are likely to be judged as synthetic or chemical in character. Manufacturers are stongly advised to seek and follow the advice of the flavor manufacturer and to ask for a demonstration sample of any particular flavor in a typical end-product application at the recommended dose rate.

## ACKNOWLEDGMENT

The author is indebted to Henry Heath for his expert assistance in compiling Chapter 18. Further information on references quoted in the chapter may be obtained from the publications by H. B. Heath 1978, 1981.

## REFERENCES

Amerine, M. A., et al. 1965. *Principles of Sensory Evaluation of Food.* Academic Press, New York and London.

ASTM 1968a. *Basic Principles of Sensory Evaluation.* STP 433. American Society for Testing and Materials. Philadelphia, Pa.

ASTM 1968b. *Manual on Sensory Testing Methods.* STP 434. American Society for Testing and Materials, Philadelphia, Pa.

Clarke, K. J. 1972. Selection and evaluation of flavorings for foods. *Proc. IFST* (U.K.) Symposium, Birmingham, pp. 32–36.

Goodall, H., and Colquhoun, J. M. 1967. *Sensory Testing of Flavours and Aromas.* Scientific and Technical Surveys No. 49. BFMIRA, Leatherhead, Surrey, England.

Heath, H. B. 1978. *Flavor Technology: Profiles, Products, Applications.* AVI Publishing Co., Westport, Conn.

Heath, H. B. 1981. *Source Book of Flavors.* AVI Publishing Co., Westport, Conn.

Kafka, W. A. 1975. Energy transfers and odor recognition. In *Structure-Activity Relationships in Chemoreception.* Information Retrieval Ltd., London., pp. 123–135.

Laffort, P. 1975. A model of olfactory mechanism based on chromatographic data. In *Structure-Activity Relationships in Chemoreception.* Information Retrieval Ltd., London. pp. 185–196.

Larmond, E. 1970. *Methods for Sensory Evaluation of Food.* Canada Department Agriculture, Ottawa.

Larmond, E. 1973. Physical requirements for sensory testing. *Fd. Technol.* 27(11), 28, 30, 32.

A. D. Little Inc. *The Flavor Profile.* A. D. Little Inc., Cambridge, Mass.

Oosterhuis, P. (1977) *Review Chocolate, Confectionery Bakery.* Beckmann, Germany.

Prell, P. A. 1976. Preparation of reports and manuscripts which contain sensory evaluation data. *Fd. Technol.* 30(11), 40–44, 48.

Schneider, D. 1968. Basic problems of olfactory research. In *Theories of Odors and Odor Measurement,* N. Tanyolac, Editor. Technivision Ltd., London., pp. 201–211.

Sjostrom, L. M., et al.   1957.   Methodology of the flavor profile. *Fd. Technol.* 11(9), 20–25.

Spencer, H. W.   1971.   Techniques in the sensory analysis of flavours. *Flav. Ind.* 2, 293–302.

Stewart, R. A.   1971.   Sensory evaluation and quality assurance. *Fd. Technol.* 25(4), 103–106.

Stone, H., et al.   1974.   Sensory evaluation by quantitative descriptive analysis. *Fd. Technol.* 28(11), 24–34.

Wright, R. H.   1971.   Steriochemical and Vibrational theories. In *Gustation and Olfaction*, G. Ohloff and A. F. Thomas, Editors. Academic Press, London and New York, pp. 161–163.

# 19

# Confectionery Processes and Formulations

The confectionery industry, like many others, has seen great changes in recent years. Basic formulations have altered little but the processes of manufacture have undergone many developments, particularly in the methods of forming small pieces and bars and their packaging.

It seems appropriate to summarize these changes at the beginning of this section and to discuss each application later with regard to particular formulations.

Certain groups of confectionery require particular methods of shaping so that they can be wrapped and packed in a manner most suitable for sale. Probably the most successful development has been the confectionery bar (also called candy bar). These bars lend themselves to economical methods of production, packaging, and presentation at the point of sale.

Concurrently, the improvement in packaging has been noticeable, with regard to both the material used and the method of sealing. Most candy bars require protective packaging to ensure good shelf life and to guard against damage by insects and extraneous contaminants. These factors are described in separate chapters.

Chocolate manufacture has already been discussed and it should be realized that chocolate and compound coatings are essentially fat based and any moisture present is very small—generally less than 1 percent. The ingredients are not in water solution.

Many confectionery processes utilize the special solubility properties of sugar (sucrose), alone or combined with others "sugars", such as glucose syrup (corn syrup) and invert sugar. There are basically two groups of sugar confectionery products: (1) those in which the sugars are wholly in solution, and (2) those in which the sugars are partly in solution and partly in the form of minute solid sugar crystals suspended in the solution. Other ingredients, such as milk and fats, may modify these products.

Group 1 includes hard-boiled sweets (hard candy), hard and soft caramels and toffees, and most jellies. Group 2 consists of such products as fondants, fudge, grained marshmallows, and grained nougats.

## SUMMARY OF CONFECTIONERY PROCESSES

From the foregoing description, it can be seen that a variety of textures is obtained as a result of the various processes and formulations and each requires a particular method of forming into pieces or bars. These methods are summarized below but specific applications are described in other parts of the book.

### Rolling and Cutting

This is probably the oldest method of producing bars and pieces, mostly from plastic products like caramel, fudge, nougat, and various pastes. The confection, in the right plastic condition, because of either its moisture, its fat content, or its temperature, is first fed through rollers to produce a slab of the required thickness. This slab is then fed to knife cutters to produce wide strips that are subsequently cut into narrow bars or small units. In a modern development of this principle, the hot product is fed to "iced" rollers, enabling the production of multiple-layered slabs. The slabs are continuously cut into strips that pass over a spreader and are then cut into bars or small pieces. An example of this process is the Sollich Conbar system (Fig. 19.1).

### Casting or Depositing

This method is applied to hard candy, fondants, jellies, some caramels and fudge, marshmallows, and other products that can be obtained in a liquid state.

### Hard Candy

Certain types may be deposited as liquid at around 150°C (302°F) into metal molds the surfaces of which are coated with a "release agent."

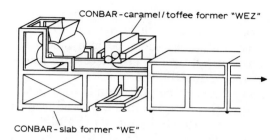

CONBAR - caramel / toffee former "WEZ"

CONBAR - slab former "WE"

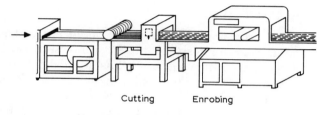

Cutting          Enrobing

Fig. 19.1. Conbar Sheeting and Cutting System (Design Is Regularly Updated)
The machine is provided with chilled rollers that permit direct feeding from the cooking unit to the slab forming units. The slabs, after forming and layering, are conveyed to a cooler and conventional cutters and spreaders followed by enrobing.

*Sollich GmbH, Bad Salzuflen, West Germany*

## Fondants, Jellies, Marshmallows

These are usually cast into starch molds. The principle is described under "Fondants." More recent developments have been the automatic depositing and release of fondants and some other confections from metal molds (Cadbury-Baker Perkins) and the depositing of caramels and toffees into silicone rubber molds (Baker Perkins). Silicone rubber has unique nonstick properties as well as being resistant to the relatively high temperatures of high boiled confections.

## Die Forming

This method is applied almost exclusively to hard candy and some caramels and toffee. It includes normal, flavored pieces, filled pieces (bonbons), and "pulled" candy.

The principle is to cool the boiled syrup under controlled conditions until it is plastic. In this state, it is reduced to a "rope," which is fed to machined dies that press the rope into pieces that usually have some special form or pattern. These pieces are immediately fed to a cooler and a wrapping machine.

A modification of this principle is applied to certain caramels and chewy candy. Here the rope is similarly produced but it is fed to a cut-and-wrap machine, which, by means of a high-speed rotary knife, cuts small pieces off the rope. These are fed to the wrapping machine.

### Extrusion and Bar Forming

The principle of extrusion, developed for many nonfood products, has been applied very successfuly in the confectionery industry. Altvater (1974) (Bepex-Hutt, Germany), has studied the application of the process to a variety of products, ranging from soft materials such as marshmallows and fondant to very plastic nougats and caramels. In the process, the material to be extruded is fed to the orifice by means of multiple rollers or screws. The cross-sectional design of the orifice determines the shape of the final unit and many extruders have a series of orifices producing ropes that can be cut into either bars or small pieces. Figure 19.2 shows the various designs used for extruders, Fig. 19.3 shows the cross section of a Werner-Lehara extruder with a mechanism for cutting small pieces from the ropes, and Fig. 19.4 shows a design by Weisert-Loser for the extrusion of chewing gum and chewable candies.

A somewhat different form of extruder is the N.I.D. Bar Former (Fig. 19.5). The fluted roller feed is similar to other extruders but the material to be formed is delivered to a second, channeled roller. The channels are Teflon coated to ensure easy release and are shaped to give the form of bar required. Fingers help to release the ropes onto a continuous belt.

Multiple bars can be made in which two layers are extruded simultaneously. Alternatively, a bar with a center of a different confection can be made. Extrusion allies itself to other continuous methods of cooking and enrobing.

Certain precautions must be observed in the extrusion of confectionery:

1. *Temperature.* The temperature of extrusion is critical, particularly with caramel and nougat where texture is closely related to temperature, and quite small changes can cause large differences in extrusion pressures. For a fairly soft caramel,

**Toothed roller press**

**Smooth roller press**

**Grooved roller press**

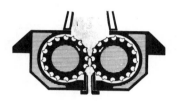

**Rotary bar roll press**

**Double rotary bar roll press**

Fig. 19.2.   Rollers and Orifices for Extruded Products

*Bepex–Hutt, Leingarten, West Germany*

temperatures ranging between 35 and 38°C (95 and 100°F) are usual but obviously a lot depends on the type of product and the exact conditions must be determined by trial. Low temperatures and high pressures will cause the safety plugs, usually provided on extrusion machines, to be ejected.

Werner Candy Extruder · Side View

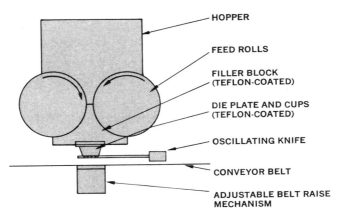

HOPPER

FEED ROLLS

FILLER BLOCK
(TEFLON-COATED)

DIE PLATE AND CUPS
(TEFLON-COATED)

OSCILLATING KNIFE

CONVEYOR BELT

ADJUSTABLE BELT RAISE
MECHANISM

Fig. 19.3.   Werner Candy Extruder—Side View

*Werner-Lehara (Baker-Perkins), Grand Rapids, Mich.*

2. *Fat separation.* If fat separates during extrusion, it is an indication of poor emulsification. Improvement is obtained by the inclusion of an emulsifier, such as lecithin or preferably glyceryl monostearate, in the recipe.
3. *Collapse after extrusion.* Some products lose their shape after extrusion and a strip of near-cylindrical cross sections may flatten appreciably after it has been on the belt for a short time. There are several reasons for this:

a. The moisture content is too high.

Fig. 19.4.   Weisert, Loser Screw-Type Extruding Machine

*Weisert, Loser, Karlsruhe, W. Germany*

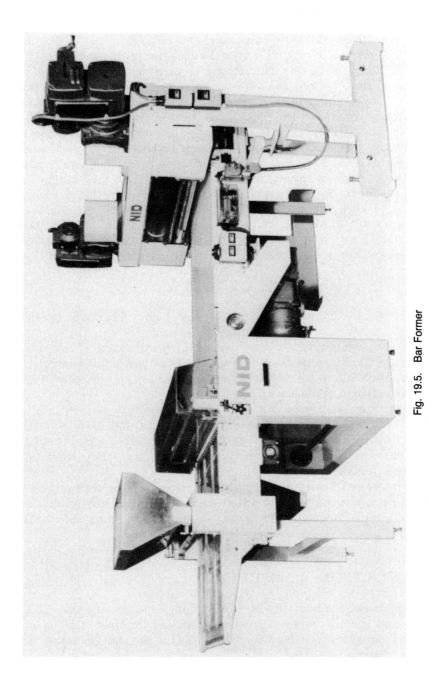

Fig. 19.5.   Bar Former

*N.I.D. Pty Ltd., Alexandria, Australia*

505

b. The fat has not been emulsified and possibly is too soft.

c. The protein of any milk ingredient is not properly dispersed.

d. In fudge or pastes, the crystal structure has not formed or may have broken down by excessive mixing after forming the crystals. This can be a very elusive defect and in continuous fudge manufacture is related to the time at which the crystallizing fondant is added prior to extrusion. Only experiment will determine the optimum conditions for particular equipment. Companies specializing in extrusion equipment have done a great deal of research into the design of machines for extrusion of confections of different consistencies. For some products, such as chewable candies and chewing gum, the roller extruder has given way to the multiple screw extruder. These processes are described elsewhere in the book.

## Panning

The principle of panning has already been described in the "Chocolate" section.

There are two kinds of sugar panning, hard panning and soft panning. In hard panning, successive layers of powdered sugar and syrup are built up on a suitable center (e.g., nuts) and dried off with warm air between applications.

Soft panning is carried out in a similar manner but is a cold process. Soft centers are used (e.g., pastes, jellies, soft caramel) and coating is with sugar/glucose syrup and powdered sugar. After coating to the correct size and weight, the pieces are partly dried and coated with a glaze.

The process has been fully mechanized, including the charging and discharging of the the pans, the automatic spraying of the syrups (or chocolate), and control of the drying or cooling air supplied to the pans.

## FONDANT CREME—BASE CREME

Confectioner's fondant made from sugar, glucose syrup (corn syrup), and invert sugar is usually spelled "creme" but some traditional products use the form "cream." This description would be incorrect in some countries.

Fondant is prepared by dissolving sugar and glucose syrup (or invert) in water and concentrating by boiling to a solution containing about 88 percent solids. At ambient temperatures, this solution is supersaturated with respect to the sugar and is unstable, and if it is rapidly agitated and cooled, the excess sugar comes out of solution in the form of minute crystals. Thus, fondant creme has a solid phase of sugar crystals suspended in a liquid phase consisting of a saturated solution of "sugars."

Without agitation or cooling, large crystals will be formed. Beating alone is unsatisfactory as a large amount of latent heat of crystallization is evolved. If the beaten syrup remains hot, crystallization is retarded and subsequent slow cooling will result in the formation of large crystals. Good-quality fondant creme should be smooth in texture.

Originally, fondant was made using sugar only, which was dissolved in water and concentrated to about 88 to 90 percent solids by boiling. Since sugar has a solubility of only 67 percent at normal temperatures, a syrup at 90 percent was highly unstable, and when cooled rapidly, it crystallized, resulting in very coarse crystals.

To overcome this, a substance called a "doctor" was used, which caused inversion of some of the sugar. This increased the overall solubility and enabled the syrup to be beaten into a fondant creme.

The substances used as doctors were citric or tartaric acids or preferably cream of tartar (potassium hydrogen tartrate). The formation of invert sugar from sugar is discussed elsewhere in the book but, briefly, the doctors decompose part of the sugar (sucrose) —which, chemically, is a disaccharide—into two monosaccharides, dextrose and fructose. The fructose has a much greater solubility (approximately 80 percent at 20°C) than sucrose, and its presence with the dextrose enables the concentrated syrup to be beaten into a fondant.

This procedure of using a doctor is very unreliable as the amount of invert sugar produced varies greatly according to the purity of the sugar, the time of boiling, and the "hardness" of the water used. An improvement is possible by adding a controlled amount of invert sugar (prepared separately) to the sugar syrup.

Fondant prepared from sugar and invert sugar has a grained texture and is very sweet to the taste; it is rarely used in modern confectionery.

Glucose syrup (corn syrup) has replaced invert sugar in fondant recipes; it is less sweet and the presence of complex carbohydrates controls crystallization and gives a more viscous fondant without the short texture.

## MANUFACTURE OF FONDANT

In the construction of a recipe for fondant creme, it is essential to include sufficient glucose syrup (corn syrup) to ensure that the final fondant has a syrup phase with a soluble solids concentration of not less than 75 percent at ambient temperatures, or microbiological troubles may arise. This condition is achieved by using a sugar/ glucose syrup ratio of 80/20, and with a moisture of 12 percent in the fondant, the syrup phase concentration is over 75 percent.

Increased glucose in the recipe will give higher concentrations but it is not usual to increase the glucose ratio above 75/25 as crystallization is retarded during beating and the texture of the fondant suffers. Fondants with higher sugar/glucose ratios are manufactured for special purposes, such as the "graining" of fudge. Occasionally, fondant creams are made with up to 8/1 sugar/glucose ratio. These, because of the crystal structure, have a very short texture but also a limited shelf life.

The original hand method for making fondant consisted of dissolving sugar and glucose in water to produce a 75 to 78 percent concentration solution, for example,

| Sugar | 3.6 kg | (8 lb) |
|---|---|---|
| Glucose syrup | 1.0 kg | (2.2 lb) |
| Water | 1.27 kg | (2.8 lb) |

and this would have a boiling temperature of 107 to 109°C (225 to 228°F). This solution, which had to be free from all traces of undissolved sugar, was then boiled to 88 percent concentration at a temperature of approximately 117°C (243°F). This syrup was then poured onto a large, cold, marble slab and rapidly turned over and over, spreading at the same time. The cooling and agitation induced rapid crystallization and, understandably, the quality of the fondant depended on the skill and energy of the operator.

There are now two basic types of mechanical equipment for fondant manufacture. The first, which works on the batch principle, consists of a shallow pan with a water-cooled flat base. Rotating within the pan is a cross-arm fitted with plow-shaped paddles that turn over the concentrated syrup while it is being cooled on the base of the pan. This rapidly brings about the conditions for crystallization and a smooth fondant with fine crystals is produced. This type of machine is still used by many confectionery companies in the United States, where it is called a "Ball" beater. Although basically a batch machine, it has one advantage in that after the fondant is made recipe

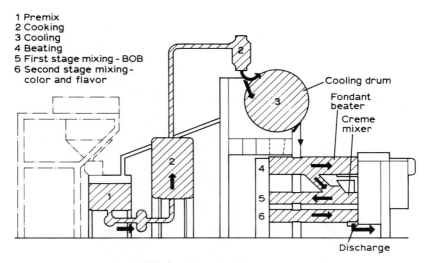

1 Premix
2 Cooking
3 Cooling
4 Beating
5 First stage mixing - BOB
6 Second stage mixing - color and flavor

Cooling drum

Fondant beater

Creme mixer

Discharge

Fig. 19.6.   Continuous Fondant and Creme Making

*Baker Perkins, Peterborough, England*

quantities of other ingredients can be added and mixed in the same machine.

The second type of machine is continuous with outputs of 453 to 635 kg (1000 to 1400 lb)/hr. In Figs. 19.6 and 19.7, two separate parts of the process are shown: *fondant* making and *creme* making. The latter provides for the inclusion of "bob" syrup, flavors, and colors, as is explained in the description of remelting.

Syrup is prepared according to one of the recipes previously mentioned. With large-scale modern production, the solution of the sugar and glucose is usually done by means of a machine called a continuous dissolver that takes the sugar and glucose from bulk storage, apportions them in recipe quantities with water, and dissolves and discharges to a storage tank that acts as a reservoir for feeding the cooker.

Continuous dissolvers work either on a volumetric or a weighing principle. The weighing method is considered more reliable, as with the volumetric method accuracy is affected by the bulk density of the sugar, and particularly by any lumps that may occur, although the latter is unlikely with bulk sugar supply.

The prepared syrup is then supplied to the fondant-making plant

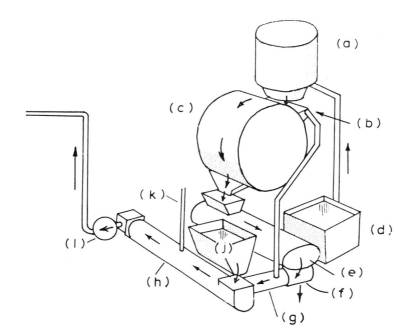

Fig. 19.7.   Principle of Continuous Fondant and Creme Making

a. Microfilm Cooker
b. Bob Syrup
c. Cooling Drum
d. Syrup Storage
e. Creme Beater
f. Fondant Discharge
for Remelting by Other Methods

g. Mixing Tube 1
h. Mixing Tube 2
j. Frappe
k. Flavor
l. Discharge Pump

*Baker Perkins, Peterborough, England*

through a superheater that raises the temperature of the syrup before it is fed to a continuous cooker. The cooker may be of two types. One consists of a spiral into which the syrup is fed from the top and this spiral is surrounded by a high-pressure steam jacket that transfers heat to the descending syrup and evaporates the water that escapes as steam from the top of the cooker. The second type uses the principle of evaporating a thin film of syrup spread over a heated surface inside a cylinder, and the microfilm cooker is an example (See "Hard Candies.")

The concentration of the syrup is measured by the temperature of the syrup issuing from the cooker, and in modern installations automatic controllers are used where signals of temperature fluctuations are fed back to adjust steam pressure and syrup pump speeds. By this means, much more consistent results are obtained than by hand adjustment based on visual observation of a thermometer placed in the syrup at the cooker exit. The syrup at about 117°C (242°F) now has to be cooled continuously and this is achieved by dropping it from the cooker onto a large, slowly rotating metal drum, cooled internally by water sprays. The syrup is cooled from 117°C to approximately 38°C (100°F) during rotation through about 270 degrees, and by means of a scraper knife, this supercooled syrup is removed from the drum and discharged into a beater. No crystallization must occur on the drum surface during cooling and after the scraper fine jets of steam are applied to the drum surface to prevent any crystallization in the very thin residual film of syrup before it again meets more syrup from the cooker.

The beater consists of a square or cylindrical casing about 3 ft long and 1 ft 6 in. in diameter, fitted internally with metal pegs and provided with a cooling jacket with water circulation. Spindles, also with pegs, rotate in the casing and provide the beating action against the fixed pegs in the jacket. The internal design is such that the crystallizing fondant moves from the syrup inlet to the fondant exit, and an important point is to keep the beater full to get maximum effect from the beaters. This is obtained by means of an adjustable slide in the exit.

The quality of the fondant is largely controlled by the efficiency of the beater, which, in addition to bringing about rapid crystallization, must remove the latent heat by sufficient flow of water through the jacket.

The temperature of the fondant flowing from the beater must be less than 43.3°C (110°F) and maximum crystal formation should have occurred in the beater by mechanical action and cooling. If appreciable crystal formation occurs after beating, coarser crystals will form and the fondant will be rough.

Examination of fondant under the microscope should show an even distribution of sugar crystals, the larger ones ranging between 10 $\mu$m and 15 $\mu$m (0.0004 in. and 0.0006 in.). The presence of a proportion of larger crystals or uneven size distribution indicates inefficient operation of the beater—mechanical or cooling. Occasionally, the presence of small amounts of colloids, such as starch, gelatin, or egg, will retard crystallization. These may be present if syrup prepared

Fig. 19.8.   Fondant Machine

*Otto Hänsel, Hannover, West Germany*

from scrap is included with the sugar and glucose ingredients. Syrup from scrap may give off-white colors, and in some circumstances, partial inversion of the sucrose—so special care in scrap reclaiming is required where fondant is concerned ( see "Reclaiming").

Fondant machines are also manufactured as smaller single units. Syrup to be fed to these machines may be made separately in a kettle or by means of a continuous dissolver. This syrup is fed to a cooling tube followed by a beater provided with a cooling jacket. A machine of this type is made by Otto Hänsel (Fig. 19.8).

## The Remelting of Fondant

Much argument has arisen in the past about the necessity of "ripening" fondant and the older confectionery makers always set aside their newly manufactured fondant in pans to mature. The result was that a great deal of manual work was entailed in getting the fondant out of the pans again and into the remelting kettles. This practice persisted even when large continuous fondant equipment was introduced and much diplomacy was required on the part of the technologists to break down the prejudice associated with this practice.

The only support for maturing is probably when products known in the United States as hand-rolled cremes are required. This involves the process of extrusion or roller depositing where a particular texture is needed.

For the mass production of fondant centers, and particularly for chocolate shell centers, this break in the sequence of operations just cannot be tolerated. Much research has been carried out on the process of "remelting," which is necessary to bring the fondant into a fluid condition for mixing with flavors, colors, and other ingredients and to enable it to be poured into molds or chocolate shells.

The traditional process was to charge the fondant from the pans into steam-jacketed kettles provided with paddles and to heat the fondant to a temperature ranging between 57°C and 66°C (135°F and 150°F) with the addition of a syrup "bob." The bob was prepared as a syrup from sugar and glucose to the same recipe as the fondant, or occasionally with a higher proportion of glucose. At one time, it was the practice to make a low-concentration bob even down to 50 percent but this is a dangerous procedure because, if the syrup is used to excess, the final fondant may become vulnerable to fermentation. A syrup bob should have a syrup concentration of 75 percent minimum and is best made with a sugar/glucose ratio the same as the fondant.

The temperature of remelt depends on the moisture content of the fondant, the ingredients, and the fluidity required at the pouring stage.

The process of remelting brings about a greater fluidity due to an increase in the proportion of syrup phase. This extra syrup phase is formed by some of the sugar crystals in the fondant going into solution or being partly dissolved and reduced in size.

The hot fluid fondant mixture when poured into molds or shells will cool and set. This is due to the growth of crystals of sugar from the syrup phase and an increase in size of crystals already there, the reverse of what has happened during melting. This increase in the number and size of the crystals results in some interlacing and gives the molded fondant a certain amount of rigidity, and with lower moisture contents a fracture appearance when broken.

When a fondant is remelted, the crystal size increase depends partly on the speed of remelting and cooling and on the level of temperature reached. Temperatures over 65.5°C (150°F) will result in an appreciable increase in the syrup phase, and when this crystallizes, it will tend to give coarse crystals. However, if the final stages of remelting are carried out through a hot nozzle, higher remelting temperatures can be tolerated. Occasionally, a rougher, short fudge-like texture fondant may be required and remelting temperatures may be taken up to 74°C (165°F), but this is an exception.

The crystal size of a good assortment creme should range between 20 $\mu$m and 30 $\mu$m. Any significant proportion over 30 $\mu$m will make the creme taste rough.

Whatever the process for remelting fondant, it has now been shown to be quite unnecessary to mature it beforehand. Freshly made fondant may be taken and heated with flavors, bob syrup, and any other ingredients to remelting temperature and poured into molds or shells without any adverse effect on the quality of the product.

Continuous mixers have been devised to take fondant straight from the plant but this method is best used when high outputs of one recipe are required (Figs. 19.6, and 19.7). It is preferable in many instances to store the base fondant from the plant with or without bob syrup and for this purpose large water-jacketed tanks may be used with slowly moving mixing arms. The water in the tank jacket is thermostatically controlled at 49 to 54°C (120 to 130°F). The tank is covered and preferably provided with some ventilation to prevent excessive condensation on the lid. Under these conditions, storage periods up to 12 hr are possible with negligible crystal growth and the fondant may be drawn upon for kettle mixing with syrup, flavors, and other ingredients. For multicenter filled blocks made on shell equipment, the fondant from the tank is sufficiently fluid to be pumped through pipelines to a battery of kettles. Sometimes continuous circulation through pipelines between the storage tank and the kettles is used but some moisture loss from the fondant will occur with this system and this must be watched carefully otherwise hard-texture creme will result in the finished product and tails will form at the nozzles depositing the creme. The additional flow diagram (Fig. 19.9) is complementary to Figs. 19.6 and 19.7.

Frappé or whip is prepared by dissolving egg albumen or a substitute in water and then mixing with sugar/glucose syrup. This mixture is then beaten to a foam by means of a high-speed whisk, under either normal or increased pressures.

A great number of recipes have been proposed using different strengths of syrup and various quantities of whipping agent. It is left

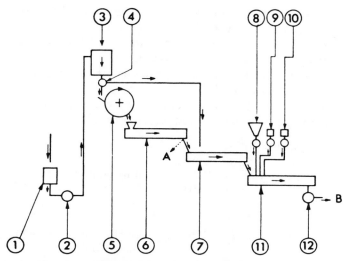

Fig. 19.9.  Flow Diagram of Continuous Fondant Maker

1. Reservoir Tank
2. Pump
3. Coil Cooker
4. Proportioning Value
5. Cooling Drum
6. Beater
7. First Mixing Worm
8. Frappe Container and Pump
9. Color Container and Pump
10. Flavor Container and Pump
11. Second Mixing Worm
12. Discharge Pump

Product removed at A requires further processing.
Product removed at B is ready for immediate use.
   In this diagram it should be noted that uncrystallized bob syrup from the cooker (4) is by-passed to the first mixer (7).
   Use of frappe: In the manufacture of creme centers for assortments or filled shells, certain recognized basic recipes are used in conjunction with the remelting procedure described above.
   If the remelted fondant, with flavors and colors, is allowed to set, the result is normally a rather dense product and a lighter texture is obtained by the inclusion of "frappe."

*Baker Perkins, Peterborough, England*

to the technologist to work out recipes to give the required texture in the finished product and the following are typical examples of an egg and substitute frappé.

*Egg Frappé*

| Egg albumen | 113 g (4 oz) | Soak in a cool place for 24 hr and pass through a fine sieve. |
| Water | 213 g (7½ oz) | |
| Sugar | 2.26 kg ( 5 lb) | Dissolve and boil to 107.2°C (225°F). |
| Glucose syrup | 3.17 kg ( 7 lb) | |
| Water | 1.12 kg (18 oz) | |

Allow the syrup to cool to 60°C (140°F) and then add the egg albumen solution and whisk. The density of this foam or frappé is of the order of 0.35 to 0.5, and as it has limited keeping properties, it should be used in fondant within a few hours.

It is advisable to pasteurize egg syrups before whipping, details are given under "Egg Albumen."

Hyfoama is a widely used proprietary whipping agent and the following recipes are given in the technical handbook.

*Frappé for Standard Beater*

| | | |
|---|---|---|
| Hyfoama DS | 99 g (3½ oz) | Mix and beat to a stiff foam. |
| Water | 1.36 kg (3 lb) | |
| Icing sugar | 2.26 kg (5 lb) | |
| Sugar | 6.35 kg (14 lb) | Boil to 111°C (232°F). |
| Water | 1.80 kg (4 lb) | |
| Glucose | 12.7 kg (28 lb) | Add to the boiled sugar batch and then add this mixture to the foam and beat again to a stiff foam. |

*Frappé for Pressure Beater*

| | | |
|---|---|---|
| Hyfoama DS | 99 g (3½ oz) | Mix Hyfoama and water. |
| Water | 198 g (7 oz) | Add glucose, mix well. |
| Glucose | 0.45 kg (1 lb) | |
| Sugar | 8.6 kg (19 lb) | Boil to 110°C (230°F). |
| Water | 2.72 kg (6 lb) | |
| Glucose | 12.2 kg (27 lb) | Add to the boiled sugar batch and then add this mixture to the Hyfoama dispersion and beat for 3 min at 30 lb/sq in. |

Hyfoama frappés are stated to have good stability but they are best used within 24 hr of making. More recently, developments in the United States with soya protein isolates have resulted in the production of whipping proteins with certain special advantages over egg and milk protein products. Details of these are given under "Whipping Agents." Some confectioners use these newly developed whipping agents as partial replacement of egg albumen.

The terms "whip" and "frappé" are sometimes used to qualify the density of the product, whips usually being of lighter density with short keeping properties, whereas frappés are of greater density with more syrup and will keep longer.

Whip and frappé may be used in fondant bases in varying quantity depending on the ultimate density required. 7 to 10 lb per 100 lb of fondant base will give a good "fluffy" texture without causing difficulty with depositing in mechanical equipment. Up to 25 lb per 100 lb can be used for very light confections but casting difficulties may arise. Extrusion methods are applicable with this formulation.

## Machinery for the Manufacture of Frappé

Frappé is not only used in fondant, it is also used in nougat, fudge, marshmallow, and wherever a lighter texture is required.

**Normal Beater**   For small batches, the vertical planetary beater such as the Hobart whisk may be used, but for larger batches, "U" troughs with horizontal beaters are sometimes employed.

**Pressure Whisks**   These make use of the principle of whisking the syrup and aerating agent in a hermetically sealed bowl, the ultimate density of the frappé depending on the quantity of syrup and the air pressure in the bowl. In this type of machine, the air pressure is used to discharge the frappé that is ejected through a valve and pipe at the bottom of the bowl. An example of this machine is the Morton Pressure Whisk ( see "Marshmallows").

This is necessarily a batch process but for very large production continuous pressure beaters are available. A solution of the whipping agent in syrup is prepared batchwise or by metering the ingredients into continuous dissolvers, and this is brought to a standard temperature in a storage tank prior to the beater. From this tank the syrup is fed into the pressure whisk at a standard rate, together with a measured supply of air or an inert gas such as nitrogen under a given pressure. The aerated batch is discharged continuously through a back pressure valve and its density is controlled by the rate of feed of syrup and air. The Oakes machine works on this principle. In continuous equipment using egg or gelatin syrups plant sanitation is of paramount importance. Residues in pipelines, pumps, etc., can be breeding grounds for all types of microflora. Planned cleaning and sterilization are essential and detachable pipelines are advantageous. This is discussed in the microbiological section.

## Fondant Remelting

Some reference has already been made to fondant remelting. The most flexible method uses steam-jacketed pans working on low steam pressures (20 psi). These pans may be up to 500 lb in capacity and must be provided with slowly moving mixing paddles that sweep the sides of the pan continuously. The motion of the paddles must also give some vertical movement within the mass. A number of manufacturers of confectionery machinery supply pans for fondant remelting.

It is important in this process to ensure that thorough mixing is obtained in the minimum time and this applies particularly when frappé is used. Prolonged mixing or mixing at high speed will cause aeration loss and variation in density.

## Kettle Process for Fondant Remelting for Cast Cremes

Cremes to a variety of recipes may be prepared for casting into starch molds, chocolate shells, or, more recently, into metal, plastic, or rubber molds.

When using fondant from a storage tank, the following is a typical process. The recipe quantity of frappé is added to the remelting pan. Fondant from the storage tank is then added, with intermittent mixing, until the recipe amount is reached. About half of the syrup bob is then mixed in and the temperature of the mix raised to the remelting temperature: 60°C, 63°C (140°F, 145°F) or as required. Further bob syrup is then added to give the right fluidity, followed by flavors, colors, and acid solution. If invertase is used, this must be added as late as possible prior to casting the batch and the fondant temperature must not exceed 65°C (150°F). (See "Invertase.")

If jam or concentrated fruit pulps are used as ingredients, these will cause appreciable thinning and may increase the syrup phase to such an extent that coarse crystallization (graining) may occur during cooling after casting. In such cases, the syrup bob may not be required, or at least be reduced in quantity. When fruits or pulps are added, care must be taken to see that the syrup phase concentration is not reduced to below 75 percent and it is always safer to make these into a jam or conserve with sugar and glucose syrup, also to a minimum concentration of 75 percent.[1]

The same applies when adding chopped, preserved fruit or candied peel to a fondant. Frequently, preserved peels and fruit as purchased have a syrup phase concentration of about 70 percent and if used in this form, the pieces can provide sites for microbiological activity in

[1] *Note*—75 percent concentration means the syrup phase concentration of soluble solids as determined by refractometer. This is discussed under "Microbiological Problems."

the finished product. Such fruit should be heat treated in higher concentration syrups (see "Preserved Fruit").

From time to time, proposals are made to include various gelatinous or colloidal ingredients in fondant creme, either at the remelting stage or in the syrup before boiling and beating. The author's opinion is that claims for these additions are usually exaggerated. Colloidal substances added at the syrup stage retard crystallization in the beater. Gelatin, agar, or starch syrup included during remelting will produce a "set" in the cooled creme and this is not necessarily an improvement. Certain essential oil flavors may have a destructive effect on the aeration introduced by the frappé and some may deteriorate from oxidation (see "Marshmallows").

## Casting Cremes in Starch

The basic process for many years for the manufacture of cremes for chocolate centers or for crystallizing has been the depositing of liquid fondant (prepared as previously described) into hollow impressions in trays of cornstarch. These trays, which are normally 32 in. × 16 in. and about 2 in. deep, are filled loosely with cornstarch, known commercially as molding starch. Although pure cornstarch can be used, better impressions are obtained with "oiled" starch and this consists of the same cornstarch impregnated with about 0.05 to 0.10 percent of purified mineral oil (medicinal white oil). It is also a fact that "used" starch will take an impression better than "new" starch, due to sugar residues from confectionery causing the starch particles to cling, and most factories use unoiled starch to make up production losses until complete replacement is required.

The filled starch tray is "printed" by means of a rigid flat board or metal plate to which are fixed rows of protrusions that represent the shape of the hollow impression to be made in the starch and ultimately the shape of the cast creme. The protrusions are often called "pips" and are made of plaster, wood, or metal. To prevent starch from adhering to them when pressed into the starch, they may be polished with graphite. The spacing of the pips and the quantity of starch in the tray are such that when the printing takes place displacement of the starch is just sufficient to give clean impressions and a level top surface. An important factor in this process is the moisture content of the starch, and in order that the cast cremes after cooling can be removed from the trays of starch, the outer layer of the creme must be fairly rigid and this is brought about by the passage of moisture from the creme to the starch.

Under normal factory conditions, the equilibrium moisture content of molding starch is between 12 percent and 14 percent and this will give bad surfaces to cremes that are poured into it—it may even result in starch adhering to the surface of each unit, which greatly detracts from eating quality. For pouring of most fondant cremes, the moisture content of the starch should be between 6 percent and 8 percent and under these conditions the cremes are removed from the starch in 5 to 8 hr provided they are cooled adequately during that period. These conditions give a firm surface to the cremes.

It is also possible to pour cremes into starch with a moisture content of 9 to 11 percent where overnight storage (16 hr) is required, but this gives a slightly different texture. The former method is preferred, particularly with cremes containing a high proportion of frappé or jam, pulp, and other hygroscopic ingredients. As a result of considerable experimental work, the conditions shown in Table 19.1 have been utilized successfully.

The alternative conditions were used depending on the circumstances of labor, planning, and the nature of the center. Item 1 in the table is interesting. Here the cast fondant was raised to a higher

TABLE 19.1. EFFECT OF MOISTURE CONTENT OF STARCH WHEN CASTING CREMES

| Type of confection | Starch moisture content, % | Air conditions | Time of removal after casting, hr |
|---|---|---|---|
| 1. Straight fondant center, No frappé, essential oil flavors only | 10–11 | 55–60°F 55–60% RH Moving air between trays | 3–4 |
| 2. Fondant with frappé, essential oil, or synthetic flavor. Some creme containing jam or pulps | 6–8 | Ambient temperature 55–60% RH air movement by convection only | 5–8 |
| 3. Fondant with or without frappé, essential oil, or synthetic flavors. Some cremes containing jam | 9–11 | Ambient | 16–24 Have been left over weekends |
| 4. Very light fondants, marshmallows | 4–6 | Ambient | 6–16 depending on recipe |

remelting temperature—74°C (165°F). The cremes in starch were cooled with moving air and setting took place with little loss of moisture. The cremes had a short texture with a larger sugar crystal than normal.

The temperature of the starch into which the cremes are cast is important. The starch must be correctly dried but not hot or the deposit may sink into the starch, causing starch crusts. The optimum temperature for cremes is 30 to 32°C (85 to 90°F). This does not apply to certain gums or jellies of lower moisture content where hot starch and heating in a stove are used.

It is also possible to store certain cremes for a limited period, using method 2, after they have been removed from the starch. The conditions found most suitable, with the cremes stored in shallow layers in ventilated trays, were 18.3°C (65°F) and 65 percent relative humidity. Under these conditions, slight drying occurred on the surface and this prevented the cremes from sticking together.

Control of the condition of cremes and other centers poured into starch for supply to enrobers that are also receiving a variety of other centers, such as caramels, jellies, and fudges, is no easy task. Neither is it a simple matter to control the moisture content of the starch into which they are poured. These problems are discussed in the following section on machinery.

## Machinery for the Casting of Confections in Starch

It was not many years ago when the production of all confectionery centers was by hand or semi-mechanical methods. They are worth description because they are still widely used in experimental confectionery workshops. The sequence of diagrams (Fig. 19.10) illustrates the methods employed and which have now become mechanized.

The first useful advance in mechanization was a machine that printed the tray of starch and moved this, on a short chain conveyor, to a multiple depositor fed from a hopper that was kept supplied with hot remelted creme. The depositor, working on the piston principle, supplied precise quantities of liquid creme to each starch impression. The starch trays were fed by hand to one end of the machine and, after filling, the rows of impressions from the depositor were removed manually from the other end and stacked to allow the cremes to cool and set. The cremes in starch, after setting, were removed and brushed free of loose starch in another machine, which also refilled the empty trays with starch. A rotary sieve was sometimes fitted to

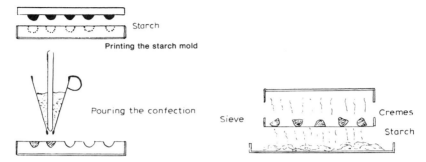

Fig. 19.10.   Hand Casting Cremes in Starch
For hand pouring, the deposit is obtained by raising and lowering the tapered stick. After cooling and setting, the cremes are removed by sieving.

this machine to remove coarse tailings of creme from the starch before filling into trays.

To dry the starch, the filled trays were stacked in a hot room until required for use. They were removed from the hot room some time before they received creme deposits. Dry, but not hot, starch is required for creme centers.

In some factories the method just described is still used, but in large plants, the sequence of events has been fully mechanized in an ingenious machine known as a Mogul (Figs. 19.11 and 19.12).

This machine accomplishes the following.

1. It receives a stack of trays containing starch with its content of cooled and set centers—deposited previously.
2. It de-stacks the trays one by one and feeds them into the first section of the machine, which inverts them over a vibratory sieve.
3. The vibratory sieve allows the starch to pass through the mesh and the cremes travel along the sieve where they meet oscillating brushes that remove most of the adhering starch. These cremes are then fed out of the machine into trays or onto conveyors where they are taken to the enrobers for chocolate covering or crystallizing. Before the enrober, the remaining starch dust adhering to the cremes is removed by an air-blowing and extraction device. (A modern development in which the cremes are fed directly to the enrober is described below.)
4. The starch is then conveyed automatically to the recirculating plant where it first passes through a sieve to remove creme particles and then to the drier (described later).

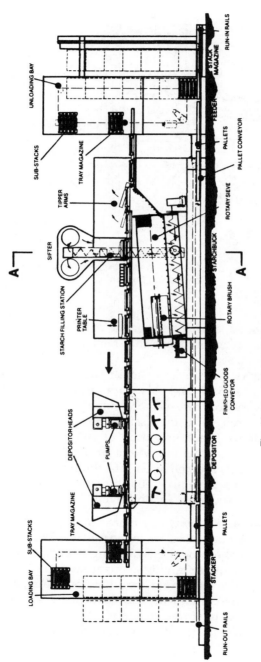

Fig. 19.11. NID 301S Starch Molding Machine

*N.I.D. Pty., Alexandria, Australia*

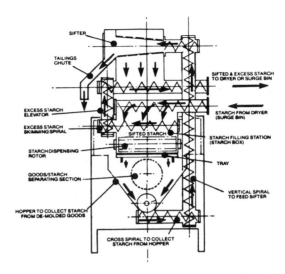

Fig. 19.12.   Cross Section A–A with External Starch
Conditioning System

5. The reconditioned starch returns to the Mogul by conveyor where it refills the empty trays that have been inverted in step 2 and levels off the starch.

6. These filled trays then pass beneath the printer, which is the board with pips affixed, and this works in unison with the filled trays moving forward toward the creme depositor.

7. The trays of starch with impressions move beneath the row of depositors and by synchronization of motion a row of creme deposits, which can be as many as thirty to forty depending on the size of the unit, are delivered into the impressions.

8. These trays of warm liquid deposits are then stacked automatically and taken to the cooling room.

The starch conditioner may serve more than one Mogul and the dried starch is sometimes used as a partial makeup of the total starch going through the Mogul circuit, but the latter procedure is best avoided unless adequate mixing is ensured.

To heat and dry the starch at regular intervals is a decided advantage from a microbiological standpoint. Yeasts, molds, and bacteria can accumulate in starch repeatedly in use and the intermittent heating and drying greatly reduce the viability of these organisms.

Some Moguls have built-in driers that use higher temperatures, and an airlift and cyclone system. Starch driers constitute a considerable explosion hazard and some serious accidents have occurred. It is wise, therefore, in any starch-drying installation to observe government recommendations regarding sparkproof switches, blast walls, relief vents, and smother grit containers. Dust explosions very often start as a very minor incident that disturbs powder in trays, on ledges, and elsewhere. This initiates a major explosion and the precautions mentioned are designed to confine the explosion to the first minor event.

A modern development in starch drying is claimed to reduce the risk of explosion and economizes in heating. It is a fluid bed drier and cooler of modular design that can be integrated into any Mogul system.

Figure 19.13 is a diagram of the system.


## The Pneumatic Starch Buck (Fig. 19.14)

A new development eliminates the need to transfer manually to the enrober conveyor confectionery pieces collected in trays after sieving. The sequence of operations is as follows:

1. The tray filled with molding starch and the deposited articles is positioned beneath a tightly fitting cover.
2. The tray and its undisturbed contents are then transferred by a unique mechanism onto a mesh conveyor belt. During this process, there is no movement between the tray, its contents, or the transferring bed or conveyor.
3. The tray is lifted cleanly off the contents, reinverted, and placed at the entry of the starch-filling section. Trays are then refilled, leveled, printed, and transferred to the depositor.
4. The starch is now loosened from the articles (which remain in their same orderly position) by fast-rotating, compressed-air jets. The fluidized starch is drawn into the starch conditioning and recirculating system.
5. The product, now freed of the main body of starch, is conveyed to a final cleaning stage where any residual starch is removed from all surfaces of the articles by fine rotating jets of compressed air.
6. The goods, still in the same positions, upon passing out of the end of the buck, are turned for transfer to the enrober conveyor.

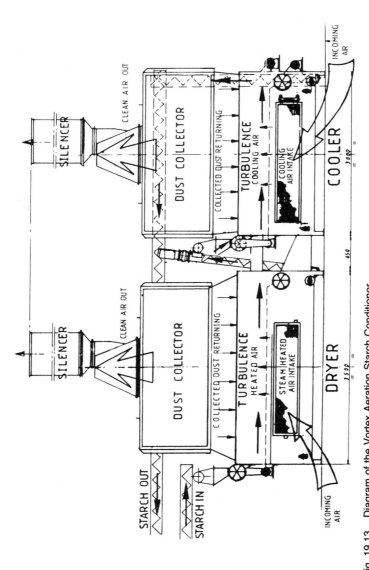

Fig. 19.13. Diagram of the Vortex Aeration Starch Conditioner
The Vortex closed circuit system can be added to any Mogul for aeration conditioning of starch at throughputs from 8,000 to 10,000 Kg per hour (17,600–22,000 lbs).

*N.I.D. Pty., Alexandria, Australia*

Unloading Bay

TO DEPOSITOR

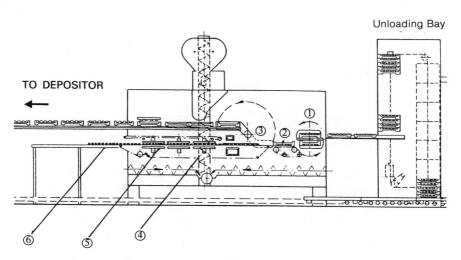

Fig. 19.14.  Pneumatic Starch "Buck"
Depositor section similar to that shown in Fig. 19.11.

*N.I.D. Pty., Alexandria, Australia*

This system permits the handling of fragile pieces such as sugar crust liqueurs that is not possible with the original starch-removal system.

Some Mogul depositors are able to deliver two types of confection, either as two separate deposits or as a multiple deposit with one material inside the other. The second system is very useful for jam-filled cremes. Nuts, cherries, or other pieces may be placed by hand or mechanicallly in the starch impressions, which then receive a deposit of creme.

The Mogul depositors consist of V-shaped hoppers provided with hot water jackets and a coarse tray sieve is fitted over the top to hold back any foreign matter or lumps.

They are kept filled from the remelting system.

Besides the large variety of fondant cremes, many other types of confectionery can be poured into starch molds provided the recipe is designed to set by cooling and partial drying. These include jellies, nut pastes, marzipans, Turkish delight, soft caramels, fudges, marshmallows, and many gums and pastilles.

With creme, texture is important and overdrying is a mistake too often made. It generally occurs as a result of leaving the pieces in starch too long due to bad production planning or perhaps because of machine breakdown.

TABLE 19.2. MOISTURE CONTENTS OF DIFFERENT FONDANT FORMULATIONS

| Conditions shown in Table 19.1 | Type of confection | Normal moisture content, % |
|---|---|---|
| 1. | Straight fondant center, no frappé | 10.5–11.5 |
| 2. | Fondant with frappé, essential oil, or synthetic flavor | 11.5–12.5 |
| 3. | Fondant with or without frappé Cremes may contain some jam | 11.0–12.0 |

Moisture contents should be maintained within fairly narrow limits—cremes much below 10.5 percent are generally too hard, and at 13 percent and over are difficult to handle mechanically and may be subject to fermentation.

Under the time/temperature/starch moisture conditions shown in Table 19.1 those shown in Table 19.2 may be expected.

## Advantages of the Mogul System

Criticism has been leveled at the system because of the quantity of starch powder in circulation, which leads to a very dusty environment. There is difficulty in maintaining good sanitation within the machine and the starch itself may develop a high microbiological count. However, improved design and good housekeeping reduce these problems to a minimum.

The Mogul and starch printing operation allow for the rapid change of product being produced. The replacement of the printer and depositor can be done very quckly. The printer is not an expensive piece of equipment.

Starch casting also produces a dry skin on the units being deposited. This helps in the handling of the pieces, gives an interesting internal texture, and, in some cases, allows a formulation to be used that is not possible when rubber or metal molds or extrusion methods are employed.

## Dextrose Fondant

The chemistry and physical properties of dextrose are described in Chapter 8.

Dextrose monohydrate has found increasing uses in confectionery processes of recent years.

Apart from its lower sweetness than sugar, its unusual crystallization properties can be of value or an embarrassment in certain processes.

Syrup mixtures containing a high proportion of dextrose will, on standing and cooling, deposit crystals. If the syrup is not seeded, these crystals will be in the form of nodules, but if seeded, they will be in orientated needle form, which gives an objectionable waxy texture.

Mechanical beating of this crystallized product will result in the breakdown of orientation and a smooth fondantlike product is formed that will deposit in chocolate shells and become only slightly more set on standing. It will not return to the waxy rigid texture.

The exact final texture of a dextrose fondant depends on the proportion of dextrose, the degree of mechanical mixing, and the amount of crystallization that takes place *after* the fondant is poured.

In some chocolate shell plant recipes, it is desirable to have a fluid recipe for pouring at a temperature below the melting point of chocolate but that has a nonfluid texture after storage for a short period when the shell is complete and demolded.

The exact proportions of dextrose depend on the nature of the confectionery center, and the best way to add the dextrose is in the form of frappé as follows:

| | |
|---|---|
| Dextrose monohydrate | 45.3 kg (100 lb) |
| Glucose syrup (low conversion) | 45.3 kg (100 lb) |
| Water | 10.4 kg (23 lb) |

Boil to 104°C (220°F). Cool to 52°C (125°F). (*Note.* This temperature is critical.)

Then add egg albumen 1.8 kg (4 lb) and water 4.5 kg (10 lb) previously soaked, dissolved and sieved.

This mixture is "seeded" by continuous stirring, followed by beating to the required density. Alternatively, it may be added to the final product after "seeding" and then beaten. In many cases it will be found that the addition of 5 to 10 percent of frappé will give the desired texture to the final fondant.

Dextrose frappé is not recommended for use in high-temperature shell plant deposits where special cooling facilities are available to prevent the remelting of the chocolate shell.

In fondants with high dextrose contents the following base syrup may be used in conjunction with frappé:

| | |
|---|---|
| Dextrose monohydrate | 20 kg (44 lb) |
| Glucose syrup (low conversion) | 11.3 kg (25 lb) |
| Sugar | 9 kg (20 lb) |
| Water | 5 kg (11 lb) |

Dissolve by bringing to the boil. Then cool to 30°C (85 to 87°F).

Cream by adding about 2 percent of a previously prepared fondant or seeded dextrose syrup, and allow to stand undisturbed for at least 16 hr.

This solution will then have set to a stiff paste and this may be beaten with dextrose frappé, flavoring materials, and some bob syrup, if necessary, until the desired fluidity is obtained.

Dextrose fondants are not easily prepared and some experimenting is necessary.

## CARAMELS, TOFFEES, BUTTERSCOTCH, FUDGE

These confections owe their character mainly to the presence of milk, butter and certain vegetable fats.

Milk solids, when heated in the presence of water and sugars (sugar, invert, glucose), develop a characteristic flavor due to the reaction between the milk proteins and the "reducing" sugars. This is known as the Maillard reaction and is described as a particular kind of "caramelization. Caramelization of a different type also occurs in sugar, glucose, and invert sugar when syrups are boiled to temperatures of 149 to 157°C (300 to 315°F). A stronger type of caramelization with yet another flavor is obtained by alkaline treatment, for example, by the reaction of sodium bicarbonate with boiling syrup at about 300°F.

The action of ammonia on certain reducing sugars also gives "caramel color."

Butter when added to high boiled syrup is subject to some decomposition and gives a characteristic and attractive flavor. No vegetable fat used in its place gives the same result, although certain fats have been developed that go some way toward attaining the butter flavor.

Brown sugars, golden syrup, and molasses have a flavor that goes well with caramelized milk and these sugars are used a great deal in caramel recipes.

The flavor produced by heating milk solids with sugars is related to the method and time of cooking and on this point great arguments have arisen with the introduction of mechanization. Continuous processes for caramel cooking invariably resulted in loss of caramel flavor compared with batch processes. However, this lack of flavor was overcome by the introduction of "caramelizers" where the continuously made caramel is held at just below cooking temperature in containers with slowly moving paddles until the extra flavor has developed. The process is still continuous, there merely is more caramel in the system.

The distinctions among caramel, toffee, and butterscotch are those of milk and fat contents, the type of fat, and the moisture content determined by degree of boiling. There are soft and hard caramels, toffees are usually hard and slightly chewy, and butterscotch is hard and brittle.

Continuous cooking and lack of flavor may be an advantage if fruit- or mint-flavored caramels are produced. With these, excessive caramelization will overrule the delicate flavors.

In all the products described above, the sugars are wholly in solution in supersaturated form with the fat and milk solids fully dispersed.

Fudge, however, and certain "grained" caramels resemble fondant and a proportion of the sugar in the form of small sugar crystals is dispersed in the remaining syrup with the fat and milk ingredients.

## The Milk Ingredient

The properties and composition of milk products are discussed in a separate chapter. Their behavior in caramel manufacture is closely related to the condition of the milk proteins and dispersion of the milk fat. Changes do occur in processing liquid milk into condensed milk or milk powder.

Liquid milk is rarely used for caramel manufacture, mainly because of the large amount of water to be removed. In the manufacture of evaporated milk, this water is more efficiently removed by multiple-effect evaporators.

If liquid or evaporated milk is used for caramels, stabilizers in the form of sodium carbonate (or, where permitted, sodium phosphate or citrate) are added. This raises the pH to a level above the coagulation point (isoelectric point) of the milk protein.

The pH of fresh milk will drop from about 6.5 to 4.5 as it ages and sours. At the lower pH, the protein rapidly precipitates on heating (curdling).

Sweetened condensed milk is favored by most caramel manufacturers and it can be either whole or skimmed.

Whole sweetened condensed milk contains the milk fat that adds to flavor, but sweetened skimmed condensed milk makes good caramel and vegetable fats with suitable emulsifiers can be used in place of the milk fat. Whole and skimmed milk powders are also used, but it is essential to make sure that the powder is properly dispersed before being incorporated in the caramel boil, or rough particles will appear in the finished product and the caramel will lose a lot of its "stand-up" properties due to incomplete dispersion of the protein.

The recipes for these reconstituted milks may be tailored to suit the caramel being made and it is advantageous to incorporate in the milk the entire vegetable fat ingredient of the caramel followed by good emulsification. It is useful to make these reconstituted milks with a higher moisture content than standard condensed milk. They emulsify better, and it aids the solution of milk powder and assists the caramel boiling.

Reconstituted milks with high moisture content must not be stored as they will be susceptible to microbiological deterioration.

There are various procedures for milk powder reconstitution but a spray-dried powder of good solubility always must be used, whether whole milk or nonfat milk. Roller process powder is not satisfactory.

## Reconstitution of Milk Powder

The following is a typical recipe and process:

| | |
|---|---|
| Water | 25 kg (55 lb) |
| | (35 kg (77 lb) alternative) |
| Nonfat-milk powder (spray process) | 22 kg (48.5 lb) |
| Sugar | 45 kg (99 lb) |
| Vegetable fat (melting point 32°C approx.) | 8 kg (17.5 lb) or increased amount depending on caramel recipe |
| Lecithin | 400 g (14 oz) |
| Sodium bicarbonate (or equivalent sodium phosphate) | 100 g (3.5 oz) (or an amount to give pH 6 to 6.5) |

Mix vigorously the water (cold), milk powder, and sodium bicarbonate. Then add the sugar and continue to mix while heating to not more than 70°C (158°F).

The fat is melted and lecithin dispersed in it. This is then added to the milk/sugar portion and again well mixed to a temperature of 70°C.

The mixture is then put through an emulsifier or colloid mill to ensure complete dispersion. This mixture contains a high proportion of water, especially if the higher recipe amount is used. It should be used within 24 hr. Utensils or equipment used for this product must be thoroughly washed and sterilized after use.

## Fats

The true confectioner maintains that there is no replacement for butter in toffees and caramels, and from a flavor standpoint this is certainly the case. It will also emulsify more readily than vegetable fats.

Nevertheless many good caramels are made with vegetable fats and for many years the recognized toffee butter was hardened palm kernel oil, but with the uncertainties of fat supplies and prices, many other vegetable oils are now used. These are described under "Confectionery Fats." Most fats now purchased are well refined and it rests with the user not to spoil them by overheating during melting. This reduces their stability and may cause oxidative rancidity later. Heating the fat alone in the presence of copper will accelerate rancidity yet it is a remarkable fact that in caramel boiling with the sugars present copper pans are used successfully. However, carefully controlled tests using copper and stainless-steel pans with the same formulation do show that stainless-steel gives a product with superior shelf life.

## Sugars

The properties of the various types of sugar are described under "Confectionery-Sugars." The brown sugars and syrups are used in caramels to give additional flavor and can be added to replace some or all of the white granulated sugar as required.

## RECIPES, PROCESSES, AND EQUIPMENT FOR CARAMEL, FUDGE, AND TOFFEE MANUFACTURE

There are numerous variations in caramel formulation and these are determined by cost and quality requirements. The best-quality caramels have, as a rule, higher milk solids and fat content.

The following experimental recipe is an example of a good quality soft caramel. For students of confectionery, the preparation of this caramel in a gas-fired pan with mechanical stirrer is a good introduction to the manufacture of this very popular confection. The experiment can be extended to include variations in some of the ingredients and to the making of fudge.

| | |
|---|---|
| Water | 3 kg (6.5 lb) |
| Sugar, white, granulated | 4.5 kg (10 lb) |
| Sugar, brown | 4.5 kg (10 lb) |
| Glucose syrup (42 DE) | 7.7 kg (17 lb) |
| Full fat sweetened condensed milk | 8.2 kg (18 lb) |
| Hardened vegetable fat (melting point/90°F, 32°C) | 3.6 kg (8 lb) |
| Glyceryl monostearate | 227 g (8 oz) |
| Salt | 142 g (5 oz) |

All the ingredients are placed in the pan and the mixer set in motion. The gas fire is lighted and heating continued on a *low flame* until the sugar is dissolved and the ingredients are completely mixed. Any sugar or other solids that may have accumulated on the sides of the pan above the liquid level are removed by means of a wet brush (after stopping the mixer). Heating and mixing proceed with the heat increased and the mixture boiling steadily. The level of heating will be obtained by experience, fierce heat will produce scorching on the pan surface and cause dark particles to appear in the mixture. The degree of boil is determined by hand thermometer, which should be kept in hot water before use. The heat is lowered, the mixer stopped, and the thermometer moved quickly through the caramel until the temperature is constant. Boiling is continued and the testing repeated until the thermometer registers 118°C (245°F). The fire is turned off, mixing continued for a few minutes, and then the caramel is di᷍ ᷍harged on to a cooling table.

N᷍᷍ the following: Change of color during heating.
Change of color on cooling table.

*Experiment 1.* Sometimes caramels become overcooked and it will be noted how rapidly the temperature rises after 245°F. In the first boiling above, a part of the batch can be left in the pan and heating continued until it reaches 135°C (275°F). It will become very dark. It can be "reclaimed" by adding water and reboiling to 245°F. The final result will be quite different—a strong, possibly burnt flavor and dark color. Correcting an error this way is bad. A sample may also be taken when the caramel reaches 127°C (261°F)—this is a hard caramel.

*Experiments 2, 3, 4.* In these experiments the glucose syrup is replaced by

1. Low DE glucose syrup
2. High DE glucose syrup
3. Invert sugar syrup

In each case, the caramel will be boiled to 118°C (245°F) but there will be appreciable differences in the final product. The caramel with low DE glucose will have increased viscosity (noted at pouring), be tougher finally, and be less sweet. The caramel with high DE glucose will be more fluid and sweeter.

Invert sugar gives greater fluidity, and a darker color in the finished product—it will also be noted that the caramel darkens appreciably on the cooling table. The flavor is changed too, with some loss of milkiness and a tendency toward bitterness.

*Experiment 5.* This is the manufacture of fudge. More will be said about this later. The basic caramel formulation is the same as Experiment 1, using 42 DE glucose but additionally 7 lb of fondant is added as follows:

The caramel is boiled to 118°C (245°F) as previously. Half the batch is discharged into another pan and cooled to 82°C (180°F) by immersing the pan in water. Half the fondant is added to the part batch at 118°C and the other half to that at 82°C. In each case, the fondant is well mixed into the caramel and then both are discharged on to a cooling table.

The caramel with the fondant added at the lower temperature will start to "set" fairly quickly while the other will remain soft for a long time. After 20 hr the first fudge will have quite a short texture while the other will still be soft although showing some signs of "setting." When the fondant is added to hot caramel, the sugar crystals in the fondant are almost completely dissolved. In the cooled batch, they remain and encourage the crystallization of the sugar in the original caramel.

## Caramel Texture

There are three consistencies of caramel—soft, medium, and hard—and they have boiling temperatures and moisture ranges approximately as follows:

|        | *Boiling range*              | *Moisture*      |
|--------|------------------------------|-----------------|
| Soft   | 118 to 120°C (245 to 248°F)  | 9 to 10 percent |
| Medium | 121 to 124°C (250 to 255°F)  | 7 to 8 percent  |
| Hard   | 128 to 131°C (262 to 267°F)  | 5 to 6 percent  |

The softer caramels are normally used for coating confectionery bars and for extrusion of layers.

The control of moisture content of caramel in open-pan cooking is difficult by hand thermometer because the mixer must be stopped and the delay in getting a constant reading not only gives wrong results but will cause overheating on the pan surface.

Some caramel cookers have built-in thermometers that are so constructed that they protrude between the revolving paddles. These are often so protected by masses of metal that they also give incorrect results.

The thermocouple or electric resistance thermometer is the best method of recording temperature. These devices are built with robust probes and may be attached to a portable lead for pan boilings. For continuous cooking the probe is inserted in the stream of caramel from the cooker and should operate a controller in the same way as that described under "Fondant Manufacture."

A skilled confectioner is able to determine hardness by taking a small sample from the boiling mixture and immediately immersing it in water. This method is of doubtful value.

The above figures are influenced by the proportion of milk solids and fat and the best way to assess texture is by means of a "penetrometer."

The standard petroleum technology penetrometer used for bitumen may be adapted for caramel, and by the use of a cone or blunt needle, consistent readings are obtained after maturing for 1 to 2 hr. A modification of this apparatus uses a spring-loaded plunger operating a dial and corrections for ambient temperature can be applied. This instrument is used on the plant and figures are correlated with the temperature of boiling. A sample from the boil may be taken, cooled in water, and the reading taken after a minute or so.

There are many variations of caramel recipes using less milk, reconstituted milks, different fats without butter, invert sugar instead of glucose. Caramels also may contain nuts, usually chopped; coconut; chocolate; or raisins. Reclaimed syrup from scrap is frequently used to replace some of the sugar and glucose in the recipe, as caramel is one of the confections that can incorporate quantities of this without degrading the product.

It is not the purpose of this book to include large numbers of recipes and the reader is referred to other literature (see References) and one can make creative experiments once learning the general principles of manufacture. A series of articles by Lees (1976) gives a great deal of information on "caramel formulations." Mention should be made of substances that are added to modify the normal caramel texture. Chewing caramels are made by the inclusion of gelatin. The gelatin is included in a proportion of about 4 oz per 100 lb of caramel and is soaked and dissolved in water before adding at the end of the batch boil.

Corn starch or a modified starch such as Amaizo also produces a different texture and is added early in the boil in the form of a water slurry. It helps to prevent caramel from losing shape.

A third modification is to add frappé and a typical recipe uses 6 lb of Hyfoama type frappé to a 100-lb batch of caramel. Frappé must be incorporated without excessive mixing to obtain a low-density aerated product.

Much which has been said about caramels applies to the manufacture of toffee but toffee generally has a much lower moisture content and less milk and fat in the recipe.

Hard toffee is usually boiled to 149 to 152°C (300 to 305°F), which gives a moisture content of 2 to 3 percent.

Butterscotch is a particular type of toffee using butter as the only ingredient other than sugar and glucose. Some lemon flavor is usually added. A typical recipe and process is:

| | | |
|---|---|---|
| Granulated sugar | 45.3 kg (100 lb) | Dissolve and boil to |
| Glucose syrup | 11.3 kg (25 lb) | 143 to 145°C (290 to 293°F) |
| Water | 18 kg (40 lb) | |

Then stir in 8 lb butter (salted), and lemon oil (1 fl oz). the butter should disperse completely in the hot syrup. The partly cooled plastic butterscotch may be handled in the same way as toffee or caramel.

## "English" Toffee, Nut Brittles

These are high boiled confections usually with much lower fat and milk contents than the caramels previously described. They are generally made from sugar only or with very low glucose content. They are particularly popular in the United States, where peanut brittle is sold in vast quantities. The nuts are, in effect, roasted as the syrup is cooked, the final temperature being 152 to 155°C (305 to 310°F). The following is an example of the process. This is a high-quality product.

| | |
|---|---|
| Sugar (white granulated) | 11.3 kg (25 lb) |
| *Salted* butter | 9 kg (20 lb) |

| Added salt | 71 g ($2\frac{1}{2}$ oz) |
|---|---|
| Lecithin | 28.3 g  (1 oz) |
| Chopped raw almonds | 2.26 kg  (5 lb) |

Use, preferably, a stainless-steel pan. Melt the butter, add the water, sugar, salt, and lecithin, and mix very thoroughly using low heat until the temperature reaches 127°C (260°F). Then add the chopped almonds and continue heating until the temperature reaches 152°C (305°F). At this stage there is a visible darkening between 152°C and 155°C (310°F) and the mixture becomes more fluid. Discharge quickly onto a cold table and spread rapidly into layers about 3/16 to 1/4 in. thick. This thickness is critical, otherwise the toffee becomes overcooked and develops a burnt flavor. A dividing frame may be pressed into the hot soft toffee to form bars. Variations with reduced butter and other nuts are possible. In such cases, a proportion of glucose syrup may be used with roasted or unroasted nuts. The degree of "roast" obtained by adding the nuts at different stages of the cooking process permits variation in flavor.

## FUDGE (JERSEY OR ITALIAN CREME)

Experimental preparation of fudge has already been described. The origin of the candy called "fudge" was probably the accidental graining of caramel prepared with a high sugar content; in fact, if a high-sugar soft caramel is mixed vigorously while it is cooling, it is likely that crystallization will occur. This type of crystallization is uncontrolled, and gives a coarse grain, and, after a period, a spotty appearance. More reliable results are obtained by causing the crystallization by the addition of fondant. Nevertheless, the fudge made by the process of rapid mixing is the "home-made" product and has it devotees.

The flavor and texture of fudge are determined by the degree of boil of the original caramel base and the proportion of fondant. The proportion of sugar to glucose in the caramel recipe will also influence the crystallization of the fudge—the greater the sugar proportion, the quicker crystallization will occur.

The crystallization of fudge can also be obtained by the addition of finely powdered sugar to the partially cooled caramel base. This gives another texture, probably with more "fracture."

As with caramels, many other ingredients can be added to fudge to give very attractive candies. There is one important difference between caramel and fudge—caramel is basically a fat emulsion in an amorphous syrup with the milk protein dispersed; fudge has a solid sugar crystal phase dispersed with the fat and milk protein in a syrup phase of a saturated solution of sugar and glucose. As a result, the fudge has a much higher water activity (ERH) and this must be borne in mind when packing or wrapping, or when it is used as a constituent of multiple confections.

Chocolate fudge is very popular in the United States. It is prepared by adding from 5 to 8 percent of cocoa liquor to the caramel batch before cooling. The addition of fondant follows during the cooling process.

## Equipment for Caramel and Fudge Manufacture

The original equipment for the manufacture of caramel and toffee was a simple pan on a gas or coke fire. The process was as described

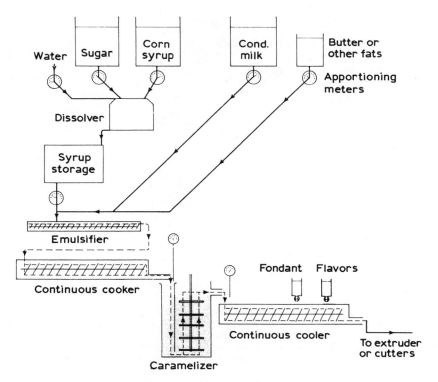

Fig. 19.15. Diagram of Continuous Caramel or Fudge Manufacture

in the previous experimental batches. Many confectioners say that gas or fire heating is the only way to get the right caramel flavor.

Fire heating was replaced by steam-jacketed pans. This was still a batch process and caramelization was still quite good. Some companies mechanized the batch system with a multiplicity of pans and pipelines, having been disillusioned with continuous processes.

However, study of the original batch process indicated the importance of the time factor in the boiling. The Maillard reaction—which takes place between the milk protein, reducing sugars, and water—is responsible for the final flavor, and the level of flavor depends on the

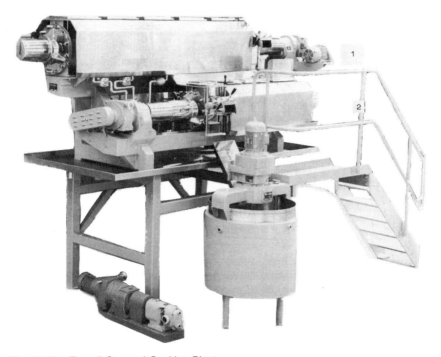

Fig. 19.16. Tourell Caramel Cooking Plant
1. Cooker
2. Blender
The cooker is of tubular trough construction with five separate steam-jacketed sections plus a weir discharge. There is a special mixing scroll with variable-speed drive. The second section is used as a caramelizer and for the mixing in of flavors, fat, etc. It is constructed to the same design as the cooker but with three sections of smaller diameter. Normal throughput is 1,000 lb/hr (450 kg) but machines are available up to 1,500 lb/hr.

*Tourell-Gardner, Cornwall, England*

time of heating, the proportion of reducing sugars, and the water present.

Continuous processes now include a caramelizer which allows the continuously produced caramel to pass through a heater during a period of about 20 min at a temperature near the final boiling point of the caramel. This gives the correct caramelization.

The type of continuous process varies somewhat and a general principle is shown in Fig. 19.15. The trough cookers are of various types, for example, internally heated rotating "scrolls" in a heated trough or a mixer in a steam-heated trough. The base caramel mix is carried along the trough as it is cooked (Fig. 19.16). A similar arrangement can be used for cooling.

The trough cooler is particularly suitable for fudge manufacture. The fondant is extruded into the cooling trough. A different product is obtained by this method, and it is also important to realize that fudge pieces, cut from a slab prepared by the pan method described previously, give a material with a definite fracture, whereas the continuous process, particularly if extruded, gives a softer paste.

With the continuous process, it has been found advantageous to use a fondant with a higher sugar content—a ratio of 10 sugar to 1 glucose syrup instead of the usual 4:1 or 3:1.

There are also various designs of thin film and scraped surface heat exchangers (discussed under Hard Candy).

In considering the use of these cookers for caramel, the possibility of burnt films of milk product accumulating on the surfaces must be realized. If this happens, heat transfer is seriously reduced and dark particles may appear in the end product. The removal of these films is a difficult process and entails the filling of the cooker with strong alkali solution and thorough washing.

## CROQUANTE (KROKANT), PRALINE (NUGAT), BRITTLES

These products bear some relationship to toffees but usually there is no milk ingredient. The descriptions cause considerable confusion in different countries. Praline in English-speaking countries means a nut paste prepared by roasting nuts in high boiled sugar syrup followed by grinding. This is Nugat in Germany. Pralines in Germany refer to all chocolates with centers. Croquant (Krokant) usually means nut pieces roasted in high boiled syrup, also called Brittles. (*Note*. International terms that apply to numerous confectionery products, processes, and machinery are explained in *Silesia Confiserie Manual No. 3,* Volume 1/11, German/English/French/Spanish.)

*Praline, Nugat*

| | |
|---|---|
| Almonds, blanched | 4.53 kg (10 lb) |
| Hazelnuts | 4.53 kg (10 lb) |
| Sugar | 8.16 kg (18 lb) |

The almonds and hazels are roasted at a high temperature—143 to 149°C (290 to 300°F). The sugar, dissolved in the minimum of water, is cooked to 157°C (315°F) and the roasted nuts added to the hot syrup and reheated to 157°C (315°F).

Alternatively, the nuts can be added to the syrup as soon as the sugar is dissolved and the mixture taken up to 157°C (315°F).

The hot mixture is poured onto a water-cooled table, and when

Fig. 19.17.  Croquant Cooker
Continuous Automatic Cooking of:
—Praline, Krokant
—Melted and Caramelized Sugar
—Brittles
—Products Containing Fats—High Boiled Toffee, Butterscotch, etc.

*Tourell-Gardner, Cornwall, England*

cold, the hard mass is broken up, pulverized in a suitable mill, and then put through refiner rolls. The nut oil content is about 30 percent and when ground is an easily workable paste that can be used for flavoring and mixing with other ingredients.

### Croquant (Krokant)

| | | |
|---|---|---|
| Chopped hazelnuts, almonds, or peanuts | 2.26 kg | (5 lb) |
| Sugar | 4.53 kg | (10 lb) |

The sugar is melted in a pan over a low fire and it is very important that all the sugar is contained in the melted mass. Slight residues on the side of the pan must be included and none of the sugar must be charred during the melting process.

When melting is complete, the chopped nuts are added and mixed in well. It is best to warm the nuts before adding to the syrup. The hot mixture is poured onto cooling tables, taking care to spread it out thinly to prevent overheating. The mixture passes through a plastic state, and in this condition can be formed into bars or pieces by pressing in a frame cutter or passing through drop rolls.

This process, using melted sugar, may also be applied to the manufacture of praline.

Traditionally, these products were made on a batch principle using special pans mounted over gas fires and provided with planetary mixers closely sweeping the sides of the pan. Recently, continuous cooking equipment has been developed that can be used for praline, croquant, and similar products (Figs. 19.17 and 19.18).

## NOISETTE, CHOCOLATE PASTE, TRUFFLE PASTE

The nugat formulation may be modified by the addition of liquor, chocolate, or cocoa powder and vegetable fat.

| | | |
|---|---|---|
| Hazelnuts (or other nuts) | 4.53 kg | (10 lb) |
| Sugar | 2.7 kg | (6 lb) |
| Cocoa liquor | 3.6 kg | (8 lb) |

The nuts are roasted and boiled in sugar syrup as with praline. After cooling and refining, the liquor or other ingredients are mixed in, with added flavors if necessary. It is then poured into trays to set. This formulation may also be termed a truffle base.

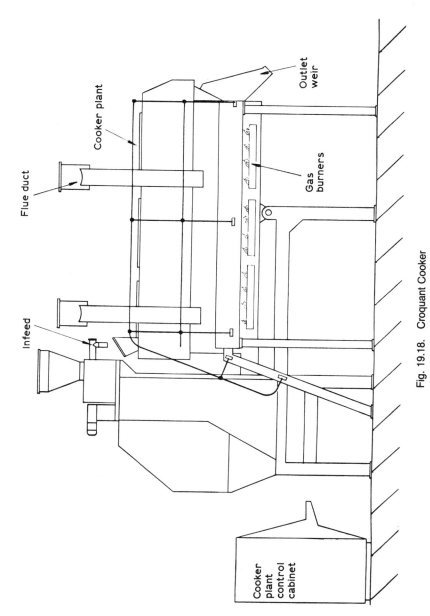

Fig. 19.18. Croquant Cooker

*Tourell-Gardner, Cornwall, England*

Labels in figure: Flue duct, Cooker plant, Infeed, Outlet weir, Gas burners, Cooker plant control cabinet

544

## Truffles

Of recent years, these have become a very popular confection, particularly in the United States. According to Richardson (1984), there are three main types—American, European, and Swiss.

The American truffle is usually a mixture of dark or milk chocolate with butter fat and hardened coconut oil, the texture being adjusted by variation of the quantity of added fats. The mixture must be subjected to a type of tempering before forming, which can be slabbing and cutting or, in certain circumstances, extrusion. This truffle, being virtually free of moisture, has a good shelf life.

The European truffle combines syrup with a chocolate base of similar ingredients (i.e., cocoa powder, milk powder, fats, sugars, glucose syrup, and invert). The final truffle is an oil-in-water emulsion with syrup phase adjusted to give a water activity of 0.7 or below and a syrup phase concentration of 75 percent plus. Invertase may be added to assist in these respects. Assuming that these conditions are met and that the fat is well emulsified, shelf life is good.

The Swiss-type truffle is prepared from dairy cream, dark chocolate, and butter. The method of manufacture is to bring the cream and butter to a boil and then to add the melted chocolate, the proportions being approximately 60 percent chocolate, 10 percent butter, and 30 percent dairy cream. Sometimes egg yolks are added.

The ingredients are mixed in a whisking machine and then poured into trays to set. Aeration in the whisk helps to give some rigidity to the paste, which, after setting, is still very soft. It is frequently dusted with cocoa powder and cooled to aid handling, when it can be rolled into balls or piped into shapes. The formed pieces may be chocolate covered or rolled in vermicelli.

These truffles are very delectable but have a shelf life of only a few days. Freezing may extend this to a few weeks.

This formulation is suitable for the small candy shops that specialize in short-life confections such as chocolate-covered strawberries.

In Europe, particularly Germany, truffles invariably contain liqueurs such as brandy, Cointreau, and rum, and this considerably extends shelf life. Alcohol is a good preservative. A typical recipe using condensed milk instead of cream is as follows:

| | |
|---|---|
| Dark or milk chocolate | 500 g |
| Sweetened condensed milk | 500 g |

The chocolate is melted and the condensed milk warmed to about the same temperature. It is mixed in a planetary beating machine for 3 to 5 min and then, with *slow* mixing, 75 g Cointreau added. The mix is poured into a tray to set, and then formed or piped as noted previously. Truffle filling is also described under "Pulled Sugar."

## BOILED SWEETS, HARD CANDY

Boiled sweets may be defined as highly concentrated solutions of sugar, glucose syrup, and sometimes invert sugar, with flavor added as required. The method of production is such that a glassy mass is formed in which the sugars are dissolved in less than 2 percent of moisture. The viscosity of the mass is so high that it is stable at normal temperatures provided that it is not allowed to pick up moisture. It is very hygroscopic.

The basic formulation is simple but great care is necessary in preparation, handling, and wrapping, or crystallization and moisture absorption will take place.

In hard candy sugar boiling, sugar alone cannot be used as it rapidly crystallizes while cooling, particularly if it is stirred. To prevent crystallization, the traditional sugar boiler used a process called "doctoring," and this meant the addition of a substance that inverted a proportion of the sugar. The substance that was used for many years was potassium hydrogen tartrate or cream of tartar, and this was added to the syrup at the beginning of the boilings. The quantity used was of the order of 1 oz to 25 lb (11.2 kg) of syrup and this produced sufficient invert sugar when the boiling was taken to 149 to 154°C (300 to 309°F).

Doctoring was an uncertain procedure and the quantity of invert sugar produced varied appreciably according to the time of boiling, purity of sugar, and hardness of the water used to dissolve the sugar.

A more certain process is to incorporate a proportion of invert sugar or glucose syrup, and by so doing much more consistent results are obtained. The texture given by invert sugar addition is shorter and less chewy than glucose syrup and glucose imparts less sweetness.

There are some manufacturers of pulled boilings and Edinburgh rock who claim that the correct texture can only be obtained by the use of cream of tartar.

## Process and Recipe for Sugar Boilings

The original sugar boiler used gas- or coke-fired boiling pans. These were superseded by high-pressure steam pans and now by continuous and vacuum cookers.

Much can be learned by the student or a newcomer to the industry by making some boilings by the pan method and a typical procedure is as follows:

| | | |
|---|---|---|
| Sugar (high-grade granulated) | 14 kg (6.35 lb) | proportions |
| Glucose syrup (42 DE) | 3 kg (1.16 lb) | may vary |
| Water | 5 kg (2.27 lb) | |

Dissolve the sugar in the water and bring to the boil—then add the glucose and bring to the boil again. An essential procedure is to see that any traces of sugar crystals around the pan above the level of the boiling syrup are washed down and dissolved completely. The old sugar boiler used a wet brush for this but a steam jet will also prove satisfactory. Traces of residual solid sugar will cause grain to develop and this will be very detrimental to shelf life of a finished product. When the boiling has reached 143 to 146°C (290 to 295°F), a solution of color may be added and this must also be free from solid particles. Boiling is continued to 150°C (302°F) and the syrup is then poured out onto an oiled table and allowed to cool until it is plastic but still soft and mobile. The edges of the mass are folded inward and a depression made in the center of the mass. Into this are placed finely powered citric acid (monohydrate) 0.5 to 1.0 percent and the flavoring essence and the folding and kneading are continued until the ingredients are dispersed throughout the plastic mass. When cool enough, the batch is cut into sections and fed into "drop rolls," which press the plastic sugar into shapes. On a large scale, the plastic sugar is supplied to a batch roller that forms it into a "rope," which may be pressed into shapes in a continuous former.

## LARGE-SCALE PRODUCTION

### Syrup Dissolving and Cooking

The sugar, glucose, invert (and sometimes "scrap" syrup) may be metered into continuous dissolvers. These provide a supply of syrup of constant composition to the cookers (Fig. 19.19).

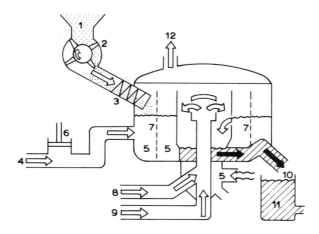

Fig. 19.19.   Continuous Dissolver

| | |
|---|---|
| 1. Granulated Sugar Feed | 7. Mixture of Sugar and Water |
| 2. Metering Wheel | 8. Glucose Feed |
| 3. Worm | 9. Feed of Other Ingredients |
| 4. Water Feed | 10. Preboiled Solution of Sugar and Glucose |
| 5. Steam | 11. Intermediate Container |
| 6. Water Pump | 12. Boiling-Vapor Discharge |

*Hamac-Holler, Viersen, West Germany*

The cookers may be thin-film batch or vacuum. The Microfilm cooker (Fig. 19.26) has already been mentioned in the production of fondant. It operates with a steam pressure of 120 to 150 lb/psi. The principle of this cooker is for a thin film of syrup to be spread mechanically over the inner surface of a cylinder heated by the high-pressure steam. The rapid heat exchange to the thin film evaporates the water very quickly from the syrup and the cooked syrup is then discharged onto a water-cooled rotating table with an aperture in the center that removes the partly cooled syrup. This cooled mass is conveyed on steel belts and through automatic kneaders where the acid, flavors, and colors are continuously incorporated. The flavored plastic mass is then fed to forming machinery as follows:

*Batch former.* This consists of ribbed conical rollers that revolve and oscillate. The plastic sugar mass is placed in the top end of the machine where the rollers are widest. By means of the movement of these rollers, the mass is reduced to a rope 2 to 3 cm (0.8 to 1.2 in.) in diameter (Fig. 19.20).

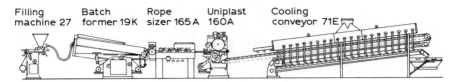

Filling | Batch | Rope | Uniplast | Cooling
machine 27 | former 19K | sizer 165A | 160A | conveyor 71E

Fig. 19.20.   Hard Candy Production Line

*Hamac-Holler, Viersen, West Germany*

*Rope sizer.* This takes the rope from the former. The sizer consists of four pairs of channeled "wheels" through which the rope must pass, thereby making the rope of precise diameter. In both the former and rope sizer, it is obvious that the temperature and plasticity of the sugar mass must be correct and even throughout. The forming rollers may be of variable dimensions according to the size of candy required.

*Sweet forming die head.* The rope, still in a plastic state, is fed into this machine and is subjected to a series of die plungers that press the rope into individual pieces. The pattern of these pieces is determined by the impressions on the surface of the die (Fig. 19.21).

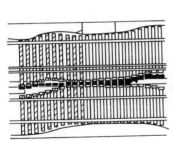

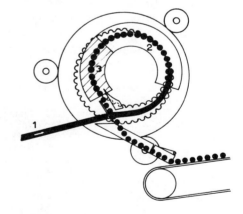

1 Sweet rope,
2 Forming, stamping, releasing
3 Ejection

Fig. 19.21.   "Uniplast" Sweet Former
*Method of operation*: The sugar rope arriving from the batch former and rope sizer is fed into a pair of sizing rollers and then runs on into a rotating die head fitted with plungers and guiding cams for the stamping and forming of the individual candies. The formed candies are then delivered onto the narrow top belt of the 71E cooler, which is driven by the Uniplast.

*Hamac-Holler, Viersen, West Germany*

*Cooling conveyor/wrapping machine.* The candies from the die head are then delivered to cooling conveyors to prevent deformation. These are of metal netting, which, with air circulation, provide the necessary dissipation of heat (Fig. 19.20).

*Center filling.* It is possible to deliver a filling to the center of the rope by means of a hopper and feed pipe. The filling may be fatty pastes or concentrated fruit preserves, and these are introduced into the rope while forming in the batch rollers (see later "Soft-Center Bonbons").

*Packaging.* There are numerous types of packaging machines but it is always desirable to wrap the pieces while they are still slightly warm and preferably the packing room should be air conditioned to a relative humidity of 45 percent. This prevents sticky surfaces and subsequent "graining."

## Vacuum Cooking

There are several advantages of cooking the syrup for boiled sweets under vacuum:

- The color of the syrup is better; fire cooking causes some browning.
- The temperature of boiling is lower, which reduces the inversion of the sucrose ingredient. This helps to retain a light color in the syrup.
- there is a saving in evaporation costs; cooking time is shorter.

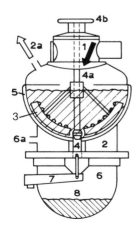

Fig. 19.22.  Vacuum Boiling Pan
1. Filling (Water, Sugar, Glucose, and Possibly Milk and Fat)
2. Batch Cooker
2a. Vapor Exhaust
3. Mixer
4. Valve
4a. Valve rod
4b. Valve Operating Wheel
5. Steam Heating
6. Vacuum Chamber
6a. Vacuum Connection
7. Swivel Device
8. Delivery Pan with Ready Boiled Sugar Mass

*Hamac-Holler, Viersen, West Germany*

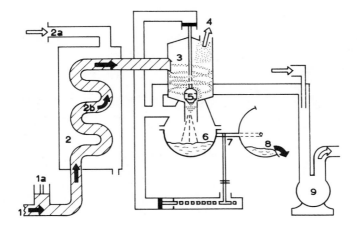

Fig. 19.23.  Continuous Vacuum Boiling System
1. Precooked Sugar-Glucose Solution
1a. Feed Pump
2. Steam Chamber
2a. Steam Supply
2b. Cooking Coil
3. Vapor Space
4. Extraction of Vapors
5. Valve
6. Vacuum Chamber
7. Pan-Swiveling Device
8. Discharge Pan
9. Vacuum Pump

*Hamac-Holler, Viersen, West Germany*

There are batch, semicontinuous, and continuous vacuum cookers. Brief descriptions and diagrams are given in Figs. 19.22 and 19.23).

Normally, in vacuum cooking the proportion of glucose syrup is higher and in a batch cooker there is a syrup cooking pan mounted above a vacuum pan. After cooking the syrup to a given temperature in the upper pan, it is drawn by the vacuum into the lower pan. A valve, which can be controlled, delivers the syrup at a standard rate. Acid, flavor, and color are added to the syrup in the pan and stirred. This is followed by discharge to a cooling table, and when in the correct plastic condition, it is formed into a rope and then pressed into shaped pieces as described previously.

An example of a formulation and cooking conditions is as follows:

| | |
|---|---|
| Glucose syrup (42 DE) | 15 kg (33 lb) |
| Sugar (high-grade granulated) | 15 kg (33 lb) |
| Water | 5 kg (11 lb) |

Cook sugar and water to 110°C (230°F). Add glucose and cook to

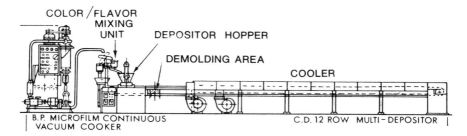

Fig. 19.24. Continuous Candy Molding Plant

*Baker Perkins, Peterborough, England*

138°C (280°F). Vacuum is drawn in the lower pan at 620 mm (24.5 in.). The discharge time of the syrup to the lower pan is 3 min.

Another method of continuous production is used for making clear fruit drops. The high boiled syrup is deposited into impressions in metal molds that are coated internally with a "release agent." The molds are cooled and the units are then easily removed.

This process necessitates the addition of flavoring, color, and acid to the high boiled syrup and special types of flavors are required to reduce loss of volatile constituents. Buffered lactic or malic acids are used in the process in place of citric acid (Fig. 19.24).

## PULLED SUGAR, SOFT-CENTER BONBONS, AERATED BOILINGS, FOURRÉS

The "pulling" of sugar is another example of the sugar boiler's craft. The process consists of cooling the boiled sugar mass until it reaches the plastic state and then stretching it, folding over, and stretching again until it becomes translucent and develops a sheen. Continued pulling increases the opacity and eventually, if the sugar/glucose proportions are correct, the mass will "grain off" and become quite short in texture. Air is also incorporated during this process. The graining may be accelerated by storage under warm humid conditions at about 38°C (100°F).

This method is frequently used for after-dinner mints, one of the few formulations that uses cream of tartar (see later).

The sugar boilers who carried out this process by hand used a large

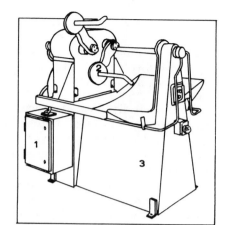

Fig. 19.25. Sugar Pulling Machine
1. Control Box
2. Pulling Arms
3. Frame

*Hamac-Holler, Viersen, West Germany*

hook about 6 in. long and $\frac{3}{4}$ in. in section fixed about 6 ft high on the wall adjacent to the cooling table. When the mass of boiled sugar was sufficiently plastic, it was slung on to the hook stretched downward, and then over the hook again, and this process was repeated until the desired condition was obtained. Gloves were necessary to handle the hot mass repeatedly and the final condition of the product depended entirely on the sugar boiler's judgment.

This process has been mechanized by the use of oscillating cranks that fold and pull plastic sugar batches of 56 lb. This operation can be timed precisely and a more constant product is obtained (Fig. 19.25).

Pulled sugar can be used as a casing for centers such as soft caramel, jam paste, chocolate paste, truffle, or sherbet, and these are often called bonbons.

The following are typical processes and recipes for this class of confectionery but many others are possible, depending on the skill and ingenuity of the experimental confectioner.

*Edinburgh Rock*

| | |
|---|---|
| Sugar | 11.3 kg (25 lb) |
| Glucose syrup | 3.6 kg (8 lb) |
| Water | 3.85 kg (8$\frac{1}{2}$ lb) |

Dissolve the sugar and glucose in the water and boil to 138 to 139°C (280 to 282°F). Pour onto the cooling table, fold in from the edges, and when plastic knead in

| | |
|---|---|
| Citric acid power | 113 g (4 oz) |
| Lemon flavor | |
| Yellow color solution | as required |

When well distributed, place the mass on the pulling machine and keep running until the batch has developed a spongy texture. The mass is put through formers to obtain a diameter of 2 to 3 cm (1 in.) and then cut into bars. On cooling, the bars will grain and become quite short.

This class of product contains air in finely divided form and is in intimate contact with the flavor ingredient. This fact is also mentioned in the section on marshmallows and certain flavors, particularly essential oils, are very susceptible to oxidative deterioration under these conditions.

Flavors resistant to oxidation must be used; reputable flavor manufacturers can give advice on the most suitable type and synthetic citrus oils are usually better than the natural oils for this kind of confectionery. This is very important—there are many instances of rock being sold in an unpalatable condition because of this alone.

### Mintoe-Type Hard Candy

This is a very popular class of sweet and aeration is obtained by the inclusion of frappé.

*Frappé*

| | |
|---|---|
| Egg albumen (or substitute) | 142 g (5 oz) |
| Water | 284 g (10 oz) |
| Soak with occasional stirring until dissolved. Mix, and then add | |
| Glucose syrup | 1.81 kg (4 lb) |
| Whip to a stiff foam. | |

*Syrup*

| | |
|---|---|
| Sugar | 9 kg (20 lb) |
| Glucose syrup | 6.8 kg (15 lb) |
| Water | 3.1 kg (7 lb) |
| Dissolve and boil to 138 to 139°C (280 to 282°F) | |

Add the frappé and beat into the syrup, allow to cool to about 121°C (250°F). Then quickly mix in the following.

| Butter | 0.90 kg (2 lb) |
|---|---|
| Peppermint oil | 14 g ($\frac{1}{2}$ oz) |

*Note*: Peppermint oil is a very volatile flavor and some will rapidly evaporate off the surface of a high boiling. This will not only reduce the strength of the flavor but will change its character as some fractions of the oil will distill off more readily. An emulsion of the flavor can be prepared, but encapsulated flavors are also available.

## Soft-Center Bonbons

These confections are made by feeding the liquid center composition from a hopper through a flexible tube into the plastic boiled sugar mass. This is done in the batch roller and the soft mixture becomes a core in a high boiled casing and this is carried away to the forming rollers that press the plastic rope into individual pieces with a soft center.(See previous descriptions and diagrams.)

This is also a skilled procedure, as the casing must be handled as hot as possible and the center should be about the same temperature. To obtain a satinlike sheen, the boiled sugar casing is pulled for a short period only and is generally colored and flavored in a minor way. The soft center must contribute most of the flavor.

*Casing recipe*

| Sugar | 11.3 kg (25 lb) |
|---|---|
| Glucose syrup | 3.6 kg (8 lb) |
| Water | 3.85 kg (8$\frac{1}{2}$ lb) |

Dissolve the sugar and glucose in the water and boil to 146°C (295°F). The boiled sugar is poured onto the table and flavored and colored as previously described, and, if desired, pulled to give the required appearance.

*Center Recipes*
*Jam Center*

| Raspberry jam | 11.3 kg (25 lb) |
|---|---|
| Glucose syrup | 5.5 kg (12 lb) |
| Water | 0.90 kg (2 lb) |
| Citric acid | 21.2 g ($\frac{3}{4}$ oz) dissolved in 28 g (1 oz) water |

Heat the jam, glucose and water and then boil to 117°C (242°F).

Cool slightly, then add the citric acid solution, flavor and color as desired. This mixture is best prepared freshly before use and kept warm.

*Truffle center*

| | |
|---|---|
| Fondant creme | 11.3 kg (25 lb) |
| Cocoa liquor | 1.36 kg (3 lb) |
| Sweetened condensed milk (full cream) | 2.72 kg (6 lb) |
| Vanillin, color | as required |

Melt the fondant at 60 to 63°C (140 to 145°F) and stir in the liquor. Heat the condensed milk to 93°C (200°F) with careful stirring for 15 min and then add to the fondant and stir in well.

Nougat, soft toffee, and various paste centers can be made. It is well to avoid "cold process" centers in which mixtures are made without heating to a temperature that will destroy microorganisms and lipolytic activity. Alternatively, ingredients may be sterilized by heating before use. Milk products, some preserved fruits, and ground nuts may cause trouble, and it is always necessary to maintain a minimum syrup phase concentration of 75 percent in the centers.

*After-Dinner Mints*

| | |
|---|---|
| Sugar | 15 kg (33 lb) |
| Cream of tartar | 45 g (1.6 oz) |
| Water | 6 kg (13.2 lb) |

Boil to 135°C (275°F). Pour onto table to cool, then fold into center.

| | |
|---|---|
| Icing sugar | 750 g (26.4 oz) |
| Peppermint oil | 7 ml (0.25 oz) |

Pull until spongy consistency is obtained, then put through formers and store the pieces in warm, humid conditions until they are short in texture.

These mints can be made without the icing sugar, the graining taking place during pulling and subsequent storage. Graining is slower and texture of the finished mints is different.

## True Bonbons, Fourrés

The original bonbon consisted of a center of nut paste, marzipan, preserved fruit, or truffle dipped in fondant creme.

Fondant creme is melted at 60 to 63°C (140 to 145°F) and

transferred to shallow, hot-water-jacketed pans. These are usually fitted into tables so that the rim is level with the table surface and the water circulation is thermostatically controlled. The centers are coated by the method of fork dipping as used for hand covering with chocolate and for this purpose a two or three-pronged wire fork is used. The center, balanced on this, is immersed in the liquid fondant and removed with the fondant coating and then inverted on to a waxed paper layer in a tray. This is a hand process, highly skilled if uniformity is to be obtained, and expensive.

Fondant recipes that have low glucose syrup content are often used for this process in order to give quick setting of the fondant. These should not be used if good keeping qualitites are required. Some confectioners advise the inclusion of 0.05 percent of acetic acid as a preservative but it is possible to use the conventional recipe fondant if sufficient time for drying is allowed after coating.

These bonbons can be crystallized or covered with pastel or white chocolate coating.

Mechanical methods have been devised for fondant coating, but one of the problems is constant evaporation of moisture from the fondant itself, giving gradually increasing thickness to the fondant layer.

Syrup of 75 percent concentration and the same sugar/glucose ratio as the fondant may be mixed in judiciously to decrease viscosity but water must never be used.

## SPECIAL PROPERTIES OF BOILED SUGAR CONFECTIONS

While boiled sweets are relatively simple in formulation, they are subject to physical changes that greatly affect their shelf life. Chemical changes, inversion, and "caramelization," also have a marked effect on their properties.

The following scientific data should provide some understanding of the problems associated with high boilings.

### Microstructure

High boiled sugar has been likened to glass, being in a metastable state between the completely amorphous and crystalline state.

Maintenance of a condition close to the amorphous state is a result of the very high viscosity of the boiled sugar at ambient temperatures. A change towards the crystalline state is brought about through reduction of viscosity by either raising the temperature or increasing the moisture content. Hence, warm storage or damp conditions encourage the graining of high boilings and caramels.

A publication by Andersen (1968) describes many interesting features of hard boilings.

The following are the main observations.

**Strain** Strain striae observed optically in freshly deposited high boilings disappear fairly quickly, showing that the sugar is still in a mobile condition although of very high viscosity. Similar striae in glass are practically permanent.

**Air Occlusions** Vacuum cooking, kneading, and pulling produce air bubbles of different shapes and sizes and these affect the smoothness of the product.

**Flavor Substances** These are frequently very badly dispersed and appear as large droplets unevenly distributed throughout the mass. Citric acid particularly is mentioned and unmelted crystals have been found throughout the mass of hard candy. The melting point of citric acid crystals is given as 126°C (259°F) or anhydrous 153°C (307°F) so that good dispersion would be expected only with the crystalline acid if it is incorporated in the very hot or liquid melt. Acid particles give an unsatisfactory flavor sensation.

## Water Activity (ERH)

There is a great difference between the water activity of boiled sugar "glass" and the same composition when crystallized (0.25 compared with 0.75).

The noncrystalline sugar is very hygroscopic and will readily pick up surface moisture. If this is allowed to happen, a relatively dilute syrup is formed on the surface. This will crystallize and the dilute syrup generated will set up a type of chain reaction, and eventually the whole sweet will crystallize. Hence, it is important to wrap the candies immediately after forming.

## Composition

Most boiled sweets are now made with sugar and glucose syrup. Invert sugar is rarely used, except in countries where sugar is grown and glucose is not readily available.

The ratio of sugar to glucose varies according to the method of boiling—open pan, thin film, or vacuum. However, it is also related to the type of glucose used (see "Glucose Syrup"), for example, the low conversion glucoses are very viscous, less hygroscopic, and less sweet.

In any composition, the sugar ingredient is subject to breakdown into the two monosaccharides, dextrose and fructose. The fructose causes the candy to be more hygroscopic and liable to become sticky when exposed to the air.

Vacuum cooking reduces inversion because temperatures are lower and boiling time shorter. The inclusion of "acid scrap" will bring about considerable inversion. Such scrap should be dissolved, neutralized, and decolorized before use (see "Reclaiming").

## Color

The longer the boiling and higher the temperature, the more the color changes toward yellow or light brown. Fruit drops, clear mints, and similar products need to be free from this defect.

## Faults

These are summarized, with possible causes, as follows:

*The candies have become sticky.* They have picked up moisture from the air and causes are

1. Too much "invert" content (see above).
2. Relative humidity of forming and packing room too high. Packing room should be 45 percent relative humidity or below. Candy should be wrapped warm at about 32°C (90°F).
3. Unsuitable packaging material—moisture vapor transfer of candy wrap and bag wrap too high.
4. Humid conditions at point of sale [also related to (3)]

*The candies have "grained"*

1. Sugar/glucose ratio too high.
2. Moisture content too high.
3. Addition of solid "scrap."
4. Humid storage.
5. Warm storage.

With reference to (5), it should be understood that warm conditions have the effect of softening the hard amorphous candy. This will make the very concentrated syrup more mobile and liable to crystallize.

## AERATED CONFECTIONERY—PROBLEMS AND CONTROL

The incorporation of air (or other gases) in foodstuffs has been practiced for many centuries. The leavening of dough by fermentation is an example of forming a spongy texture from a dense mixture.

There are various methods of aeration that increase bulk for a given weight, thus improving texture, possibly digestibility, and generally appearance of better value because of the increased size of the product.

### Mechanical

1. Air or an inert gas such as nitrogen may be introduced into a syrup or fat by mechanical beating under normal or increased pressure. To prevent premature collapse of the foam, stabilizers are usually added. These may be gelatin, egg albumen, hydrolyzed milk protein, or edible gums. Marshmallows and nougat are typical examples of confections made by this method.

2. Air or gas may be dissolved under pressure in a syrup, fat mixture, or chocolate, and when this mixture is allowed to exude from a nozzle in the pressure vessel, it will expand by reason of the gas coming out of solution in the form of small bubbles.

   Aerated chocolate can be made by this method. Carbon dioxide will dissolve quite readily in warm liquid cocoa butter under pressure and liquid chocolate can be subjected to this treatment. When it is extruded through nozzles from the pressure mixer, the gas comes out of solution and a viscous chocolate foam is formed. This can be deposited into molds or formed into slabs and solidified by cooling.

3. Pulling and kneading—are used for hard boilings when in the warm plastic condition and air is entrained during the process of folding and pulling. Satin hard boilings and Edinburgh rock are examples of this type of confectionery.

4. In extrusion, plastic high boiled sugar is forced through a plate with multiple star-shaped perforations, the threads joining up loosely after extrusion to form channeled strips.

   There are several machines that extrude high boiled sugar

and the result is a type of honeycomb that gives a crisp texture. The voids can be filled with a soft paste such as peanut butter.

5. Flaking off rollers and compacting the flakes to form a channeled bar are used for chocolate, fat pastes, and some boiled sugar products. It gives a bar with structure somewhat similar to (4) but less regular. (See "Chocolate Flake.")

Some confections that are already aerated in a minor way by one of the previous methods can be considerably expanded by treatment in a vacuum vessel or oven. If heat is applied with the vacuum, cooling is required before the vacuum is released or the aeration will collapse. If drying occurs with the application of vacuum, cooling may be unnecessary.

Chocolate may be aerated by vacuum treatment. The liquid chocolate is first tempered; it is then rapidly beaten (without destruction of the temper). This introduces small air bubbles throughout the mass, and when subjected to vacuum, these will expand. The chocolate is then allowed to set, which will happen fairly rapidly under moderate cooling conditions and if the chocolate is well tempered. Setting must take place under vacuum.

Pulled boiled sugar will expand a great deal if the warm plastic product is subjected to vacuum but the expanded article must be cooled quickly to prevent collapse.

Under these methods, a reference should be made to pressure gun puffing, which is used for cereals and some other products. The cereal grains containing their natural water content are heated under pressure in a special cylinder to a temperature well above the boiling point of water. The cylinder lid is suddenly released and the moisture in the cereal cells is instantaneously transformed to steam and the cellular volume is greatly increased.

## Chemical Methods—Surface Activity—Stability of Foams

Certain chemical substances decompose under heat; sodium bicarbonate is the most important example, yielding carbon dioxide. Others react without heat but in the presence of moisture; sodium bicarbonate plus citric or tartaric acid are examples. It is generally necessary to control the degree of aeration and bubble size by means of stabilizers.

The following examples explain the problems involved:

In the days of the individual candy makers using batch processes, aeration could be controlled, as thought fit, by adjustment of the quantity of aerating agent or modification of the process.

With modern continuous processes, it is necessary to use a precise recipe and to understand the properties of the ingredients that are responsible for giving a standard volume to the aerated product with consistent gas bubble size.

The surface activity of ingredients of any particular confectionery product must be considered if control of the foam responsible for aeration is to be obtained.

A simple example is the expansion of a high boiled syrup by the addition of sodium bicarbonate, which decomposes by heat into sodium carbonate, steam, and carbon dioxide ($2NaHCO_3 \rightarrow Na_2CO_3 + H_2O + CO_2$).

A similar reaction occurs if an organic acid ($\bar{A}$) such as citric or tartaric is added ($H\bar{A} + NaHCO_3 \rightarrow Na\bar{A} + H_2O + CO_2$).

In each case, the carbon dioxide is evolved in the form of small bubbles, which are dependent on the particle size of the bicarbonate. The strength of these bubbles and their ability to coalesce into larger bubbles are dependent on surface activity. The structure of the resultant honeycomb may vary greatly because of this factor alone.

Hard candy foam, known in the trade as "cinder taffy," "honeycomb," or "crunch," has been made for many years and sold in lumps in fair grounds. To make it in bar form with consistent density and aeration needed study of the surface properties of the ingredients.

In the development of a continuous process for the manufacture of this bar, the properties that controlled the bubble size and stability of the aeration were determined. Two factors were shown to be responsible.

1. The particle size distribution of the sodium bicarbonate.
2. The foam-forming properties of the sugar and glucose syrup ingredients.

The particle size distribution of the bicarbonate was determined, first, by microscope and elutriation measurements in coordination with grinding experiments on a micropulverizer. These measurements were related to performance in experimental and factory production using the same syrup. Ultimately, a quick sediment test was developed that proved equally satisfactory in giving an empirical judgement of size distribution.

When investigating the factors that determined whether the

bubbles coalesced, it was found that certain "surface active" substances in the sugar and glucose could give variations in density and bubble size, resulting in considerable variation in the candy product.

The quantities of these substances were so small that they were not readily detected by quantitative analytical methods and physical tests were therefore applied. Surface tension measurements were tried but the practical difficulties were so great, particularly if they had to be applied to routine testing, that they were abandoned in favor of foaming tests.

Two methods were used to investigate foam-forming tendencies in the sugar ingredients. The first consisted of a "foaming stone" used for oil testing, and this was connected to an air supply and meter that delivered air at a constant rate into a syrup held in a graduated cylinder.

This served admirably to show up gross variations in sugar and glucose ingredients, and in the early days of research variations were very great, particularly in glucose syrup and some beet sugars.

It showed that foam values obtained by this method gave a close correlation with the nature of the aeration in the aerated high boiling.

As development of the continuous process proceeded, greater accuracy was needed and an improved method for foam value was devised. This method is similar to that proposed by Bikerman and modified later by Clark and Ross (1940)—it is the determination of the "dynamic foam" value. Details of this method are given in the Appendix.

At the time of the original investigations, sugar and glucose manufacturers had little knowledge of the foaming properties of their products. It was possible to show that there were significant differences between some high-grade sugars prepared from cane or beet raws. Beet sugars had high foam values whereas cane sugars had low values and in some cases showed no foaming tendencies.

In low-grade or badly refined sugars, these differences were very marked, and as the refining improved, the foam values of sugars of cane or beet origin became much closer.

Beet sugars may contain minute amounts of protein matter, mucilage, saponins, and similar foam-producing substances. Cane sugar, on the other hand, may contain small amounts of cane wax, which has some antifoam properties. High ash contents in sugar also contributed to foam formation.

In the examination of glucose syrup it was found that significant amounts of protein residues may be present that produce very marked foaming but no direct connection between nitrogen content and foam values was found.

Some of the glucose syrup manufacturers recognized that these residues were liable to cause foaming, and to avoid trouble in the evaporators, antifoam substances were sometimes added. This had the disastrous effect of causing very coarse aeration in the aerated product, and in some cases almost complete coalescence of the gas bubbles. At the same time, the density increased greatly.

At the present time, most manufacturers of sugar and glucose syrup are well aware of these problems and the degree of refining is to a very high standard. However, some sugars and glucose syrups of foreign origin still exhibit the difficulties mentioned.

These special properties of sugar and glucose syrup are not only significant in aerated boilings but may be responsible for troubles in clear boilings where foaming may cause difficulty at the boiling stage or produces unsightly bubbles in the finished product.

## Stability of Foams

The stability of a foam, whether in hard boilings, marshmallows, nougats, or other confections, is important if constant density and good keeping properties are to be achieved.

Destruction of foam structure is readily brought about by the presence of fatty material, fatty acids, essential oils, and various other flavors. In an aerated high boiled sugar product without a stabilizer, very small amounts of fat are damaging. Even when stabilizers such as egg albumen or gelatin are present, fat can cause considerable variation in low-density aerated confections.

In experiments made to assess the effect of stabilizers in aerated boiled sugar, the additions of gelatin, albumen, gums, and hydrolyzed milk protein were investigated. The presence of 10 parts per million of gelatin was shown to produce a very fine aeration but addition of these substances to the syrups to try to control foam led to difficulties in the syrup cookers. Very small additions of protein material to the cooked syrup or with the aerating chemical gave better control. These same products, however, showed a completely changed aeration structure if a fat or fatty acid was added to the extent of 100 to 500 parts per million, and in some cases most of the aeration was lost completely. The stability of the foam in a confection is also affected by the extent of mixing. Overmixing has the effect of producing a "sponge" where some of the bubbles are connected but have not coalesced into larger bubbles.

Foam collapse, whether due to a fatty oil or overmixing, will occur in high boiled products only when the magma is hot, but with higher moisture confections such as marshmallows the collapse is progressive over a long period.

There are several high boiled confections on the market that have had considerable additions of gelatin, albumen, or milk protein. They are generally not crisp, and have a higher moisture content (5 to 7 percent) and a very close aeration.

The crisp aerated high boilings without stabilizer addition have a coarser aeration and lower moisture content (3 to 4.5 percent).

When it is necessary to add a fat or essential oil to an aerated product, it is possible to reduce its antifoam properties by making an emulsion in syrup but the syrup, not the fat, must be the continuous phase.

## Shelf Life of Aerated Confectionery

Reference to shelf life is made elsewhere in the description of particular types of confectionery but some general observations are given here as this is a very important aspect of these lines.

## Oxidative Rancidity

Because of the intimate contact of air with the other ingredients in an aerated product, oxygen may cause deterioration of flavoring materials and fats. The stability of such ingredients should be checked by accelerated oxygen absorption tests. Some essential oils are very badly affected in a product such as marshmallows and some natural fruit flavors lose their strength.

Fats in nougats and light texture fudges may develop rancidity but fat deterioration is more likely to be due to lipolytic activity.

## Microbiological Spoilage

Aerating agents, particularly egg albumen, are heat sensitive and there is a tendency to keep processing temperatures as low as possible to preserve the properties of the whip; marshmallow is an example. Pasteurization temperatures in the region of 74°C (165°F) for 15 to 20 min will destroy most organisms liable to cause spoilage if the heating is carried out in water solutions or dilute syrups, but with more concentrated substrates these organisms are resistant to higher temperatures.

The maintenance of a syrup phase concentration of 75 percent minimum will avoid most trouble but occasionally osmophilic yeasts may cause fermentation, and liquefying bacteria can produce local breakdown of the foam.

A more serious problem of spoilage arises from fat-splitting enzymes that are associated with the microorganisms. Esterases and lipases are very resistant to destruction by heat in low moisture substrates. Temperatures of 93°C (200°F) for 20 min will destroy these enzymes in 75 percent concentration syrup but in dry powders such as cocoa, temperatures up to 115°C (239°F) have been found necessary.

Fat-splitting enxymes are responsible for "soapy rancidity" in nougats and fudges, especially where lauric glyceride fats are used.

Ingredients and processes should be checked at all stages of manufacture, particularly in continuous production, where process temperatures may be of short duration.

Aeration, by reason of the very fine division of air cells in a confection, will promote the deterioration of many ingredients— flavors, fats, milk solids, nuts.

The phenomenon of "microclimates" should also be mentioned. This concerns the minute atmosphere within a bubble that may, locally, have a very high relative humidity. In such circumstances, a localized deterioration may occur. This condition often arises due to poor mixing or emulsification.

## Cold Cracking

This is an interesting phenomenon of light, aerated, low-moisture products when covered with chocolate. Chocolate-covered sugar "honeycomb" or light cookies will, if subjected to temperatures below 40°F, develop cracks in the covering, and if temperatures reach 20°F or below, they may shatter completely. Milk chocolate shows this defect much more than dark chocolate.

This cracking will accelerate moisture gain in the center. With aerated boiled sugar, this will cause collapse of the center to, in the worst circumstances, a syrup. With wafer cookies expansion occurs and the chocolate will split off. The wafers become tough.

The reaons for this cold cracking have been investigated and they may be attributed to the following.

1. *Mechanical strength of the centers.* A light crisp "honeycomb" or cookie will break up under stress. Soft centers are able to adjust

their shape slightly if subjected to outside forces. Similarly, dense solid centers will not break or change shape under ordinary stresses.

2. *Composition of the chocolate covering.* Milk-chocolate-covered centers are most susceptible to damage under cold conditions. From dilatation measurements the fat portion of many milk chocolates under ambient conditions contains about 20 percent of liquid fat phase. When this is cooled, for example to 20°F, this liquid fat portion becomes solid, and in so doing sets up sufficient stresses in the covering to break the center. Dark chocolate with much less liquid phase at normal temperatures does not give this trouble.

3. *Tempering and cooling.* During enrobing and subsequent cooling similar stresses will be set up in the chocolate if it is undertempered and subjected to severe cooling. Low temperatures and rapid cooling must always be avoided with chocolate-covered aerated centers.

## MARSHMALLOWS

Marsh mallow (Althaea officinalis) is of the hollyhock family of plants and its root yields a mucilaginous juice that has demulcent properties. The juice was originally compounded with egg and sugar to produce an aerated product. It was a pharmaceutical medicament used for treatment of chest complaints and eventually was recognized as an attractive candy.

Marshmallows today do not contain the root juice and aeration is obtained by the inclusion of egg albumen, gelatin, hydrolyzed milk protein, gum arabic, or other whipping agents in a sugar/glucose syrup.

The texture depends on the type of aerating agent and the final moisture content, which may range between 12 percent and 18 percent. Additionally, particularly in the United States, there are "stoved" marshmallows that may be dry and fairly hard. They are used in cookery recipes and flour confectionery.

The formulation and process of some mallows are such that the syrup phase concentration may not reach the minimum of 75 percent. This, together with the fact that ingredients such as egg albumen and gelatin may have significant microbiological counts, renders the confection liable to fermentation, mold growth, or infection by some liquefying bacteria. Another factor, mentioned earlier in the chapter, is possible flavor oxidation due to the intimate contact of air cells with the substance.

The manufacture of marshmallows is frequently a cold process after preparation of the base syrup. Even when cast into starch, which gives superficial drying, the remaining center of pieces may have too high a moisture content.

Marshmallows can be made in batches by whipping under normal atmospheric pressure or under increased pressure. They may be pressure whipped in continuous machines.

Batch processing may take place in two ways: In the one-step process, the syrups and solutions of the whipping agents are mixed and beaten to a foam of required density. In the two-step process, the whipping agent is beaten separately to a light frappé and the syrup, prepared separately, is added to this, followed by beating to the required density.

The aerated product may be cast into starch on a Mogul machine or extruded as a rope onto a long bed of starch. The extruded rope is chopped into lengths, the formula being adjusted to give rapid setting.

The following formulations and processes are examples only. The variations possible and, in fact, employed in the manufacture of marshmallows, are very large indeed and detailed information is given by Jackson (1973) and Minifie (1971).

A typical method for the manufacture of marshmallows is as follows:

Gelatin after soaking is dissolved in hot water.

Egg albumen is soaked in cold water to dissolve, strained through a fine sieve, and added to the gelatin solution.

Sugar, glucose and invert sugar are dissolved in water to a syrup and boiled to the appropriate degree.

The gelatin/egg solutions are placed in a beater, which can be of the planetary or horizontal type, and the syrup added after cooling to 71°C (160°F). The mixture is then beaten to a fluid foam and the density checked. The foam, which should be at about 50°C (120°F), is then ready for casting into starch. For marshmallows, the starch must be dried to a moisture content of 4 to 6 percent and the temperature of the starch should preferably be below 100°F, and if these conditions are not observed, the marshmallow foam may partially soak into the starch and produce starch crusts on the pieces.

Marshmallows may be made in a pressure whisk using the same procedure as above but the density is controlled by beating at a given

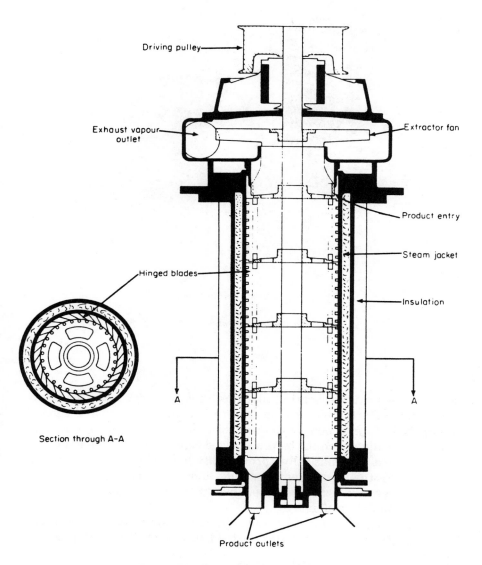

Driving pulley

Exhaust vapour
outlet

Extractor fan

Product entry

Steam jacket

Insulation

Hinged blades

A

A

Section through A-A

Product outlets

Fig. 19.26. Microfilm Continuous Cooker

*Baker Perkins, Peterborough, England*

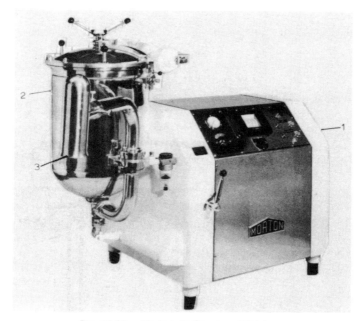

Fig. 19.27.   Morton Air Pressure Whisk

*Morton Machine Co., Motherwell, Scotland*

pressure. A pressure beating machine that has been used for many years is the Morton whisk (Fig. 19.27). The air pressure and time of whipping is determined by the control unit (1). The syrup/whipping agent is rapidly mixed in the pressurized container (2). The foam mix is discharged automatically through (3) by means of the internal pressure in the container. Beating under pressure, followed by release to atmospheric pressure, gives maximum expansion of the foam and minimum density.

It is also possible to use a continuous pressure beater where the syrup/gelatin/albumen mixture is held in a tank for feeding to the beater.

A machine that has been used very successfully is the Oakes continuous automatic mixer (Fig. 19.28). The essential part of this machine is the high shear mixing head, which consists of a rear stator, a rotor, and a front stator. The delivery of syrup mixture, air,

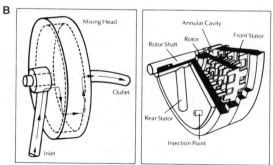

Fig. 19.28.  A. Oakes Continuous Mixer
B. Diagram of Mixing Head

*Vicars Group Ltd., Newton-le-Willows, England*

and pressure is controlled automatically and the aerated product is delivered continuously through a back pressure valve.

Other continuous machines are the Turbomat (Otto Hänsel), Air-O-Matic (Weesp Holland), and the Whizolator. This last machine, developed by the late J. Alikonis and the Beich Company, Bloomington, Ill., is interesting because it contains no moving parts at the mixing stage. Air under pressure and the whipping syrup are fed through spiral jet valves that give all the dispersion needed for production of a stable foam.

A typical recipe for marshmallow that can be poured into starch and then enrobed or sugar/starch dusted, is as follows;

| 1. | Gelatin | 340 g (12 oz) | soak and dissolve by |
| | Water | 1.58 kg (3½ lb) | warming. |
| | Egg albumen | 113 g (4 oz) | soak, dissolve, and mix |
| | Water | 0.68 kg (132 lb) | these two solutions. |

| 2. | Sugar | 6.35 kg (14 lb) | Heat to dissolve, then |
| | Glucose syrup | 2.72 kg (6 lb) | boil to 112°C (233°F). |
| | Water | 2.26 kg (5 lb) | |
| 3. | Invert sugar | 2.7 kg (6 lb) | Add to boiled syrup 2. |
| 4. | Flavor as required. | | |

Cool the mixed syrups (2 and 3) to 71°C (160°F) and add these to the gelatin/egg solutions (1) and beat to the required density, which should be 0.40 to 0.50.

Pour at about 49°C (120°F) into starch at 4 to 6 percent moisture content and leave to dry and set for 16 to 24 hr in a warm, dry atmosphere (27°C/80°F).

The texture and density of marshmallow can be varied greatly by adjustment of the quantity of egg albumen and gelatin and by the inclusion of other gelatinizing agents or gums. The quantity of water affects the density of the whip, as will the pressure applied in the machines described above. Mallows may be made with gelatin alone provided setting of the gelatin is controlled in the process.

The following gelatinizing agents may be used to impart different textures to marshmallows:

*Agar.* This is dissolved in boiling water, the solution strained and cooled, and then added to the albumen solution. This takes the place of the gelatin and is used in the proportion of about 1 to 2 oz per 10 lb of marshmallow. This gives a firm set to the mallow.

*Gum arabic.* This gum is dissolved in cold water by soaking and occasional stirring, then strained and added in the place of the gelatin solution. Gum arabic is usually added in fairly strong solution (25 to 50 percent) and produces a tenacious, stringy confection.

*Pectin.* Pectin is an ingredient not often used in mallows but it gives an unusual texture and has been applied to recipes that include fruit pulp. The syrup of sugar and glucose is prepared as before. The fruit pulp is added to the hot syrup followed by the liquid pectin, which is high methoxyl pectin solution prepared from powdered pectin or a pasteurized liquid pectin (not preserved with sulfur dioxide). This mixture is then added to the gelatin solution and the whole beaten to a foam. This pectin/gelatin/syrup must be at a temperature of 71 to 82°C (160 to 180°F) for beating and pouring so that premature setting of the pectin does not occur (see "Pectin"), and it may be necessary to adjust the pH for the optimum gel formation. This recipe gives a short jellylike texture with pectin about 1 percent of the mallow.

## Short Mallows

A short texture mallow can be made by the addition of powdered (icing) sugar to the beaten foam. In this recipe, the syrup base is prepared with a higher sugar content, for example

| | |
|---|---|
| Sugar | 6.35 kg (14 lb) |
| Glucose syrup | 1.36 kg (3 lb) |
| Invert sugar | 0.9 kg (2 lb) |
| Water | 2.26 kg (5 lb) |

The mallow is otherwise made as for the typical recipe previously given and the icing sugar (2 lb) mixed into the foam for just long enough to give complete dispersion. Over mixing must be avoided.

## Kisses, Angel Kisses

These confections consist of a very light density mallow (0.25) extruded through nozzles onto a wafer cooky. The mallow so prepared is piped warm and will stand up as a dome $1\frac{1}{2}$ to 2 in. high and this is covered with chocolate or coating, but the bottom is usually left uncovered.

The method for preparation of this mallow is to beat the albumen (or substitute) solution with gelatin solution and icing sugar to a light stiff foam and then add to the sugar/glucose syrup and mix *slowly* until completely incorporated. Overmixing will increase the density.

| | | |
|---|---|---|
| 1. Egg albumen | 85 g (3 oz) | Dissolve the egg and |
| Gelatin | 43 g (1½ oz) | gelatin in part of the |
| Water | 0.9 kg (2 lb) | water and mix. |
| Then add icing sugar | 0.9 kg (2 lb) | |

Beat at high speed until the lowest density is obtained.

---

| | | |
|---|---|---|
| 2. Sugar | 1.8 kg (4 lb) | Dissolve and boil to |
| Glucose syrup | 1.36 kg (3 lb) | 112°C (233°F). |
| Water | 0.68 kg (1½ lb) | |

Cool to 71°C (160°F) and add to 1 with slow mixing.

Marshmallows on cooky bases covered in chocolate, or more frequently compound coatings, have become a very popular confectionery piece. Often they are combined with a fruit jelly deposit before coating. They do have a limited shelf life, which is considerably longer than most flour confectionery items but less than sugar or chocolate confectionery such as caramels and soft nougat bars. The mallow deposits are denser than with the Angel kisses mentioned above. Often they are combined with a fruit jelly deposit before coating.

## Sanitation in the Production of Marshmallows

The subject of sanitation and spoilage is frequently mentioned throughout this book. The importance of this in relation to good shelf life and quality cannot be overemphasized.

The freshly made marshmallow is a delectable sweet, but careful examination of many receipes given in textbooks will reveal that they must have a short shelf life. In constructing a recipe for marshmallow, the following points should be checked:

**Moisture Content and Concentration of the Syrup Phase** It will be noted from a rough calculation of the moisture content of a mallow syrup before beating that it is often over 30 percent. This will give a syrup phase concentration well below 75 percent in the whipping mixture when it is poured into starch. The effect of the dry starch and low-humidity storage will reduce the moisture considerably and may even bring the syrup phase concentration to 75 percent or above. This may not apply to large mallow pieces, and particularly to the center portion. Although after a period moisture will distribute itself, the center may be at a low concentration long enough to bring about microbiological deterioration. In most mallow recipes, the sugar/glucose/invert ratio is such that a sugar solubility can be obtained that will give a 75 percent minimum soluble solids if the moisture content is low enough. It is particularly important, therefore, to check moisture and soluble solids through the process. Where no starch drying or stoving is used, as would be the case if poured into a chocolate shell, the soluble solids of the mallow must be over 75 percent at this stage.

Some mallow recipes such as kisses and those used for fillings and topping for flour confectionery are not intended for long shelf life and are made to very light densities and high moisture contents.

**Microbiological Deterioration**  Apart from any defects arising from incorrect syrup phase concentrations, the ingredients and process invite microbiological action. Both egg albumen and gelatin may be sources of microorganisms and the low temperatures used in many of the processes do not destroy these.

Reference has been made elsewhere to the pasteurization process for egg albumen. This has been usefully applied to egg/gelatin syrups in continuous mallow manufacture for depositing into chocolate shells. Egg albumen is not subject to coagulation at so low a temperature in syrup as it is in water and the egg/gelatin/syrup can thefore be pasteurized by holding it at 71 to 74°C (160 to 165°F) for 15 to 20 min with *very slow* stirring. Local overheating must be avoided as coagulated egg films appear on the surface of tanks and pipes. Rapid stirring will create an aeration in the syrup that affects the syrup density sufficiently to upset metering accuracy in continuous plant.

The introduction of ferments into mallows must be avoided at all cost and scrupulous plant hygiene is very necessary. The need for good plant hygiene is again emphasized. Egg and gelatin syrups are very vulnerable to microbiological attack. *A regular system of cleaning and sterilization of containers, tanks, pipelines, and machines must be operated. With egg syrup, which may coagulate and form layers on the interior surfaces of plant equipment, washing out first with tepid water is necessary. This is followed by steaming and circulation of a bactericidal detergent.*

**Flavors—Antifoam Effect—Deterioration**  In a marshmallow, the air in the bubbles is in intimate contact with the syrup film around the bubbles, and this oxygen in the air bubbles has a maximum opportunity of reacting with ingredients in the syrup film.

Certain flavors and essential oils, lemon and orange oil particularly, are very susceptible to oxidative rancidity when in thin layers and develop objectionable resinous flavors. Careful selection of flavors for mallows, in fact, for any aerated confection, is very important. It is useful to carry out oxygen-absorption tests on natural essential oils before using them for aerated products. Terpeneless oils and synthetic flavors usually keep better than the original essential oil and natural fruit concentrates are generally satisfactory.

Most essential oils and many other flavors that contain oily or fatty acid components have distinct antifoam properties and this can have a marked destructive effect on the aeration. In the worst cases, the mallow center will separate into a syrup layer and a coarse foam with little substance after several weeks. Where flavors that have an oily base must be used, they are better emulsified. Alternatively, powder "locked" flavors may be included. It is now a practice in the manufacture of some aerated foods to use an inert gas (nitrogen) instead of air for preparation of the foam. This is readily applied in the continuous-pressure machines previously described.

**Graining—Drying Out**  With the exception of short mallows where crystallization has been intentionally encouraged by the addition of icing sugar, the syrup phase composition should be such that no sugar crystals form on storage.

Marshmallows that are not chocolate coated will dry out under most conditions of storage, and in severe cases the drying will start crystal formation because the syrup phase becomes more concentrated. Drying out will also make the mallow tough.

Uncoated mallows are now generally wrapped in moistureproof film bags. Various proposals have been made to retard drying without making them hygroscopic. Glycerol and sorbitol solution (70 percent) are used and these retard crystallization as well but appreciable amounts have to be included to be effective, and this is expensive. The pectin mallows mentioned previously show remarkably good keeping properties as pectin gels generally retain moisture well.

A proportion of invert sugar in the syrup phase helps, but an excess may bring about dextrose crystallization.

## NOUGAT, SOFT NOUGAT (MONTELIMART, NOUGATINE)

Nougat, traditionally, is a French product made from honey and egg white, beaten to a frappé. Nuts and dried fruit are added to give the product called montelimart.

Originally, nougats were hard and chewy but of recent years soft nougats (sometimes called nougatines) have become very popular. They are higher in moisture content, lighter in density, and contain such powder ingredients as cocoa, milk powder, lactose, malt, and icing sugar, which have a shortening effect. Fat and emulsifiers are also included to overcome stickiness, and these aid cutting and forming.

In the hard nougats, the sugars are totally in solution, whereas in the short, softer nougats, there is a syrup phase interspersed with the solid substances (above) and the fat. Compare this with caramels and fudge.

In the production of the short nougats, the proportion of sugar to glucose syrup is higher and icing sugar is added to promote crystallization in the mixture. Alternatively, fondant may be added in the later stages of mixing.

There are numerous formulations possible and they may be made by a batch or continuous process and, as with marshmallows, mixing of the whip and syrup may be done under pressure.

The following describes batch, continuous, and pressure systems but there are many variations.

Certain essential factors must be observed and some of these have been mentioned in connection with marshmallows. They apply particularly to continuous process.

## Egg Syrups

These should be pasteurized and used within a short period after making—a few hours maximum. *Never* leave overnight in a warm place. Vessels, pipelines, and pumps must be sterilized after each production period. Wash first with tepid water, followed by boiling water or steam—this avoids films of coagulated egg coating the equipment. If the equipment is to be out of action for a period, wash out with bactericidal detergent.

## Cocoa Powder, Milk Powder, Spice Flavors

These should be free from lipase and have low bacteria counts. Cocoa and spices particularly may be suspect. The temperatures of processing may not be sufficient to destroy lipase. A method that has been used to treat cocoa and spices is to make a slurry in a heat-stable vegetable oil (Durkee 500) and heat to 110°C (230°F). This procedure also aids dispersion of the cocoa powder and prevents lumping.

## Fats

Lauric fats (coconut and palm kernel oil), which are often used, are vulnerable to soapy rancidity, from lipase action, and it is better to use nonlauric fat if the texture is acceptable. When adding fats to aerated products, use the minimum mixing time in keeping with dispersion. Fat destroys aeration.

## Examples of Formulation and Processes

*Soft Nougat*

Using planetary beater:
1. Egg albumen (or substitute)    0.25 lb
   Water                          3.0 lb
   Icing sugar                    5.0 lb

Dissolve egg in water. Add icing sugar. Beat at high speed with wire whisk.

2. Sugar           13.0 lb
   Glucose syrup    20.0 lb
   Water            4.0 lb

Dissolve sugar in water. Add glucose syrup. Boil to 127°C (260°F).

Add syrup in a thin stream to the whip using low speed and the flat beater.

3. Cocoa powder (10 to 12 percent fat)    2.0 lb
   Malt powder (diastase free)            2.0 lb
   Nonfat-milk powder                     2.0 lb
   Icing sugar                            1.5 lb

Mix powders dry and stir gradually into 1 plus 2.

4. Fat    1.0 lb

Melt at low temperature and add to mixture with slow mixing and minimum time for dispersion. Spread onto cooling table and cut when set.

Cocoa may be dispersed in fat as mentioned previously.

*Hard Chewy Nougat*

Using planetary beater.
1. Egg albumen (or substitute)    0.25 lb
   Sugar                          4.5 lb
   Water                          3.5 lb

Dissolve egg in 2 lb water. Dissolve sugar in 1.5 lb water (hot). Cool and add to egg solution. Beat at high speed using wire whisk.

2. Sugar                              25.0 lb

   Glucose syrup (low conversion)    25.0 lb

   Water                          10.0 lb

Dissolve sugar in water. Add glucose syrup. Boil to 141°C (286°F). Add to 1 in thin stream. Fat and fillings (chopped nuts) may be added if required. Spread onto cooling table and cut when set.

With the ever increasing popularity of the soft nougats, mostly sold in bar form and chocolate covered, large-scale batch and continuous equipment has been constructed.

A typical example is the twin batch machine Ter Braak Preswhip (Fig. 19.29). This machine provides a metered delivery of sugar and glucose followed by dissolving and cooking to a given temperature. The solution of whipping agent is prepared separately. The syrup and whipping agent are delivered to a mixing vessel, which is pressurized. Mixing is standardized to a given pressure time and temperature.

After beating, the aerated mixture is delivered to a mixer in which the fat and other ingredients (flavor, cocoa powder, milk powder, icing sugar, preserved fruit) are incorporated. This *must* be a short

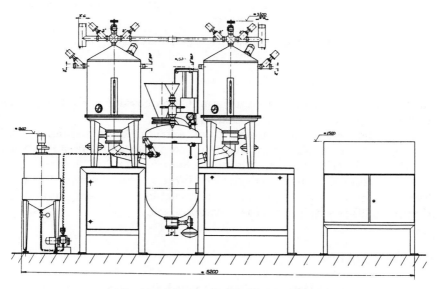

Fig. 19.29. Preswhip Air Pressure Whisk

*Bepex-Ter Braak, Rotterdam, Holland*

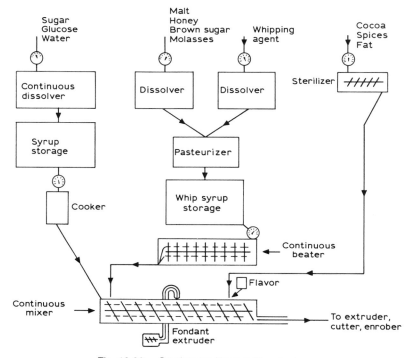

Fig. 19.30. Continuous Nougat Manufacture

mixing as the fat acts as an antifoam. Cocoa powder is best dispersed in the fat as previously mentioned.

The nougat can be fed directly to an extruder or to a slabbing and cutting system. Another continuous process uses the Oakes machine, which was described under "Marshmallows."

Some companies that produce very large quantities of soft nougat bars for chocolate covering have constructed their own equipment. Figure 19.30 is a diagram of the system used.

## FRUIT CHEWS

This type of confectionery has achieved much popularity of recent years—particularly as summer candies, because they are mostly fruit flavored and acidulated.

The basis of this product is a sugar/glucose syrup combination and a gelatinizing agent, which may be gelatin, gum arabic, maltodextrin, or a modified starch. A proportion of fat is included to improve texture and prevent excessive stickiness in the mouth.

The mixture after compounding is pulled, which again contributes to the texture. These candies have a lasting effect in the mouth, somewhat like chewing gum but without the presence of the insoluble gum base. Texture is also modified by the inclusion of fondant, which has a shortening effect.

The following formulation and batch process are representative.

*Base Syrup*

| | |
|---|---|
| Sugar | 14 lb |
| Water | 5.0 lb |
| Glucose syrup (low conversion) | 20 lb |
| Vegetable fat (melting point 33°C approx.) | 26 oz |
| Lecithin | 3 oz |

Dissolve sugar in water and add glucose syrup. Cook to 120°C (248°F). Disperse lecithin in melted fat, add to syrup, and mix thoroughly in a planetary mixing machine.

*Gelatin Solution*

| | |
|---|---|
| Gelatin (130 Bloom) | 4.5 oz |
| Water | 9.0 oz |

Soak gelatin until swollen, then warm to dissolve (70 to 80°C). Cool syrup to 90°C (194°F) and then add the gelatin solution. Mix slowly and pour onto cooling table, where it is "tempered" by folding as with hard candies. Add, during the mixing process or on the table, citric acid (monohydrate), 3.5 oz, for flavor and color.

If shorter candies are required, fondant is added at the same time as the syrup is mixed with the gelatin solution (fondant, 1 lb 2 oz).

The tempered, flavored mass is pulled for from 3 to 10 min according to the final texture required, and then formed into pieces by pressing or extrusion.

As with many other confection, variations are large.

Table 19.3 gives the ranges of ingredient quantities. Popular flavors are the essential oils—lemon, orange, lime, peppermint—and combinations of natural and synthetic fruit flavors.

Since these candies are partly aerated, the flavors used must be resistant to oxidation (see "Marshmallows").

Large-scale equipment is now available for making this type of

TABLE 19.3. CHEWING SWEETS COMPOSITION

| Percent | Range | Average |
|---|---|---|
| Moisture content | 5.3–7.3 | 6.5 |
| Acid (citric) | 1.0–1.6 | 1.3 |
| Fat | 2.8–10.4 | 5.8 |
| Gelatin (or other agent) | 0.5–2.5 | 1.4 |
| Sugar | 36.7–46.1 | 42.1 |
| Glucose | 45.0–60.0 | 49.5 |
| Proportion sugar/glucose syrup | 1/0.9–1/1.7 | 1/1.2 |

sweet—including cooking and kneading, conveyor cooling, and extrusion.

## JELLIES, GUMS, PASTILLES, TURKISH DELIGHT

The group of confections ranging from hard gums to soft jellies is a very large one and the nature of each product is largely determined by the gelatinizing agent and the moisture content.

The main gelatinizing agents used are summarized in Table 19.4

TABLE 19 4. GELLING AGENTS—ORIGIN AND USES

| Type of gelling agent | Origin | Use |
|---|---|---|
| Gelatin | A protein of animal origin extracted from bones and skins | General. Must not be boiled. To be added to warm syrups for setting on cooling. |
| Agar <br> Alginates | Extracted from various seaweeds | Various. Produce short neutral jellies. Weakened by boiling in acid solutions. |
| Gum arabic or acacia | Exudation from trees | Used to produce hard gums and as an extender and thickener in products such as marshmallows. |
| Starch and modified starches | Seeds and various roots | These have completely or partly replaced other gelling agents in gums, Turkish delight, glazes. |
| Pectin | Fruit residues, particularly citrus and apple pomace | Used largely in acid fruit jellies but low methoxyl pectin is used in neutral jellies. |

and these are described in greater detail in the ingredient section of the book.

Various other gums are used in the food industry but to a lesser extent in confectionery. Among these are guar and carob (locust bean) gums—which are seed products. Gum tragacanth is an exudation from the Astragalus bush. These gums are used as mucilages for thickening and for stabilizing emulsions. Some are used in chocolate spreads or syrups. Xanthan gum a natural gum prepared by biosynthesis is similarly used.

Then there are the methyl celluloses that produce great bulk when water is absorbed—they find application in certain dietetic and slimming confectionery. Chicle gum is a natural latex used in chewing gum. Modified starches are being used more and more in the manufacture of jellies. High amylose starch combined with pressure cooking will produce jellies with much reduced setting time.

In the manufacture of jellies, particularly the softer jellies, certain general precautions must be observed.

## Solution of the Gelling Agent

Obviously, the gelatinizing material must be properly dissolved and, if necessary, the solution strained to remove any extraneous material. Some, like gelatin, agar, and gum arabic, need soaking in cold water first and when this is done care must be taken to see that the solid does not consolidate at the bottom of the soaking vessel. Careful stirring during soaking is necessary and this applies particularly to powdered gelatin and agar. Gelatin solutions must be warmed and not boiled to dissolve. Gum arabic needs slow dissolving in warm conditions and excessive stirring or boiling results in an unmanageable foam. Agar must be boiled to dissolve, but this must not be prolonged. Unmodified starches require boiling to dissolve but first must be dispersed as a thin slurry with cold water. Some modified starches are soluble in cold water. Amylose starch requires pressure cooking to dissolve. Pectin needs special care in dissolving— the powder must be carefully dispersed through the solution and the pectin/sugar/acid balance must be correct (see "Pectin").

## Syneresis, pH, Gel Breakdown

Syneresis is the property some gels have of exuding syrup (sweating) after a period of storage—it is detrimental to sales of the product as it not only spoils the eating quality but causes sticky syrup to adhere to the wrappings. This defect arises in agar jellies from excess acid addition and in pectin gels by incomplete solution of the pectin, overacidification, or as a result of depositing below the setting temperature.

Some starch gels are prone to syneresis and it is customary to include another gelling agent as a stabilizer. High amylose starch gels generally are not affected.

All jellies may exhibit this defect as well as granulation if the finished confectionery product is mixed after the gel has started to form. Granulation occurs if a jelly confection is poured into starch, chocolate shells, or other molds below the setting point of the mixture and in any mixture, before depositing, the setting point temperature must be known.

## Gums

**Hard Gums** This is another confectionery product that has been handed down from the pharmacist who combined medicaments with gum arabic, sugar syrups, and honey. The presence of gum ensured slow dissolving, a feature particularly useful for treatment of throat infections.

Most recipes for gums rely on gum arabic or a mixture with gelatin as the gelatinizing agents but certain modified starches are now also used.

A typical recipe and process is as follows:

| | | |
|---|---|---|
| Gum arabic | 12.7 kg (28 lb) | Soak with gentle warming and stirring until the gum is dissolved. Strain through a fine sieve to remove foreign matter. |
| Water | 11.3 kg (25 lb) | |
| Sugar | 6.8 kg (15 lb) | Dissolve and boil to 124°C (255°F) |
| Glucose syrup | 1.8 kg (4 lb) | |
| Water | 2.26 kg (5 lb) | |
| Glycerol | 0.45 to 0.68 kg (1 to 1½ lb) | May be added to prevent overdrying in the hot room. |

The syrup mixture is poured into the gum solution and gently mixed. A certain amount of scum will rise on standing and this should be skimmed off. A second heating will cause further scum and this is also removed and the clear gum solution drawn off for deposition in starch which should be dried to 4 to 5 percent moisture. The gums are then dried for six to ten days in a dry, hot room at 49°C (120°F) until the required texture is obtained. They are then removed from the starch by sifting and well brushed to remove as much starch as possible. The gums are then subjected to a glazing process by placing them on wire sieves and steaming, which gives a glossy surface, but this must not be overdone or the surface will become unduly softened. They are then subjected to a further drying process. A continuous machine is manufactured for this process and the original practice of "oiling" the gums is not necessary.

The above basic recipe requires flavor and acid additions and a variety of substances may be used. Lemon, orange, and lime oils give good citrus flavors and concentrated juices are very popular for other fruit flavors. Licorice juice, honey, and various substances such as menthol, eucalyptus, and aniseed are employed for throat gums.

**Soft Gums and Pastilles**  With the softer gums and pastilles it is usual to include gelatin as well as gum arabic and the glucose syrup content is higher; otherwise, the process is very similar. A recipe used for fruit pastilles is as follows:

| | | |
|---|---|---|
| Sugar | 4.1 kg (9 lb) | |
| Glucose | 4.1 kg (9 lb) | |
| Fruit juice concentrate or fruit pulp | 3.1 kg (7 lb) | Dissolve and boil to 121°C (250°F). |
| Water as required depending on concentration of juice or pulp and some citric acid may be used. | | |
| Gum arabic | 3.1 kg (7 lb) | Dissolve and strain. |
| Water | 3.1 kg (7 lb) | |
| Gelatin | 0.45 kg (1 lb) | Soak and dissolve. Then add to gum solution. |

Add the gum/gelatin solution to the syrup and mix well, then pour into dry starch and dry in a hot room until the the correct texture is obtained.

Pastilles are normally sanded with granulated or graded sugar. This process starts with steaming the units after removal from starch as for hard gums, but before drying they are sugar coated in a rotary drum and the excess sugar removed by sieving. The sugared pastilles are then dried off on a wire mesh, the sugar forming a layer over the surface. After sugaring, a second steaming is given before drying and this gives a more continuous sugar coating.

## Starch Jellies

Starch, in its various forms, has been used for many years for making jellies. In its original unmodified form (e.g., corn flour, wheat flour), it was used for making Turkish delight. This entailed long boiling in water or dilute syrup (4 to 5 hr) in order to burst the starch grains.

The first special starches to be produced were the boiling starches. These are chemically softened by acid treatment and are still used a great deal in confectionery. They are satisfactory for open-pan boiling

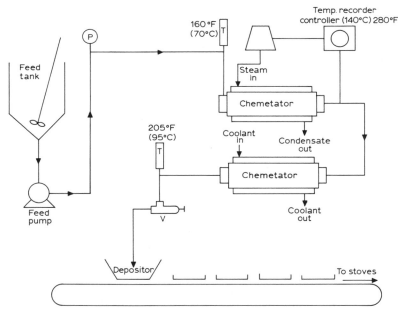

Fig. 19.31. Continuous Starch Jelly Production
P = Pressure Gauge
V = Back Pressure Valve
T = Thermometer

*Chemtech, Reading, England*

as well as for pressure or jet cooking. Genetically modified starches have been developed of recent years. High amylose starch is the most important, and this requires high-pressure/temperature cooking. This results in a great saving in time and heat and jellies prepared this way have become very popular, particularly in the United States.

They are produced in a variety of flavors and textures. They may be sugar sanded, steamed, dried, or oiled, and used as centers for panning, as in jelly beans.

These different starches are described elsewhere (see "Starches").

The equipment for processing may be the steam-jacketed pan, the Chemetator, or the jet cooker (see Figs. 19.31 and 19.32).

**Chemetator®** A premix is made that contains just sufficient water to

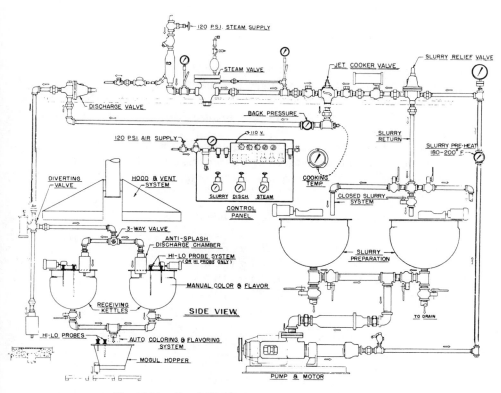

Fig. 19.32.  Typical Flow Diagram Jet Cooking System

*A.E. Staley Mfg. Co., Decatur, Ill.*

cook and dissolve the ingredients and is only 1 to 2 percent above the moisture content of the final cooked jelly. This premix is heated to 70°C (160°F) and then pumped to the Chemetator unit, where it is heated to 140°C (284°F) under pressure. The bladed shaft, which spreads the slurry over the internal heating surface, ensures even cooking.

From the cooker the slurry is fed to the cooling Chemetator and then through a back pressure valve to the depositing system.

**Jet Cooking** A premix slurry is prepared according to the formulation required. This slurry is preheated to 82°C (180°F) and fed to a jet chamber, where it is cooked continuously at about 80 lb/in² pressure at 140°C (284°F). The discharged cooked product is flavored and colored, either in a separate pan or continuously in a pipeline.

## Process and Formulation Details

**Open Pan Boiling** Open kettle processing has been used for many years and the starch/sugar slurry is cooked in steam-jacketed pans with mechanical stirrers. These must sweep the surface thoroughly so that no film accumulates, thereby forming an insulating layer that retards boiling.

In this method of cooking, thin boiling starch is used with sufficient water to ensure that the starch grains are burst. If insufficient water is used and sugar is present, the sugar retards the starch gelatinization.

*Open Pan Boiling (typical recipe and process)*

| | |
|---|---|
| Sugar | 22.6 kg (50 lb) |
| Glucose syrup 42 DE | 28 kg (62 lb) |
| Invert syrup | 5.4 kg (12 lb) |
| Thin boiling starch | 7.25 kg (16 lb) |
| Water | 56.7 kg (125 lb) |
| Citric acid | 14 g ($\frac{1}{2}$ oz) |
| Flavor ⎱ as required | |
| Color ⎰ | |

Dissolve sugar, corn syrup, and invert sugar in half the water and bring to boil. Prepare a slurry of the starch and the remainder of the water (cold). Pour slurry in a thin stream into boiling syrup. Boil to 76/78 percent solids by refractometer. Deposit in starch.

**Pressure Cooking** The first developments in the pressure cooking process used scraped surface heat exchangers of the Chemetator type. Later, the steam injection principle was invented where steam under pressure is continuously injected into the starch/sugar slurry through radially aligned nozzles. This gives very rapid mixing and cooking. The disadvantages of open-pan cooking are overcome by pressure cooking. A large excess of water must be used to cook starch in an open pan, with a consequent long cooking period.

With pressure cooking at a high temperature, the starch can be gelatinized with much lower water contents and in the presence of sugar. With the steam jet method, small amounts of slurry are cooked continuously in the jet chamber in a matter of seconds. A diagram of the system is shown in Fig. 19.32.

*Pressure-Cooked Starch Jellies*

Formula A—(Chemetator principle) using thin boiling starch;

| | |
|---|---|
| Granulated sugar | 18.1 kg (40 lb) |
| 64 DE corn syrup | 27.2 kg (60 lb) |
| Thin boiling starch | 5.9 kg (13 lb) |
| Water | 7.7 kg (17 lb) |

Put water in the slurry kettle. Then add the corn syrup followed by the other ingredients. Mix well and preheat to 82°C (180°F), being careful not to boil off moisture at the pan edges. The sugar should be in the solution and the slurry is ready for pressure cooking at 138°C (280°F).

Formula B—(Jet cooker) using high amylose starch;

| | |
|---|---|
| Granulated sugar | 18.1 kg (40 lb) |
| 64 DE corn syrup | 27.2 kg (60 lb) |
| Thin boiling starch | 1.8 kg (4 lb) |
| High amylose starch | 27 kg (60 lb) |
| Water | 5.4 kg (12 lb) |

Prepare slurry as for formula A but preheat to 93°C (200°F) and pressure cook at 168°C (335°F).

In each case, the cooked product is run into a suitable mixer for incorporating the flavor and color. It is also possible to feed these continuously into the exit pipeline. Note the higher cooking temperature for the high amylose starch.

**Turkish Delight**  This very popular candy has a very long history originating in the East where it was made from honey and flour and flavored with rose otto, which is still the accepted flavor for this product.

Turkish delight recipes are of two types—one is poured onto tables to cool and set, after which it is cut up into cubes and dusted with fine sugar. It is sold in this fashion, usually in exotically designed boxes. The second type is poured into starch and after removal is covered in milk chocolate.

Starch is an essential ingredient of Turkish delight as it gives the opacity characteristic of the traditional article.

Turkish delight prepared from starch alone does not have a very good shelf life. It may suffer from syneresis and will dry out if not chocolate covered. Gelatin or agar may be used to improve the properties of the jelly.

Low methoxyl pectin with boiling starch has proved very successful with Turkish delight for starch casting.

Typical recipes are given below, but that containing LM pectin is more appropriately given in the section on "Pectins."

*1. Starch Only*

| *Thin* boiling starch | 2.26 kg | (5 lb) |
|---|---|---|
| Water | 13.6 kg | (30 lb) |

Mix cold to a slurry and then bring to the boil with continuous mixing and boil for 2 min. Then add:

| Sugar | 13.6 kg | (30 lb) |
|---|---|---|
| Glucose | 3.6 kg | (8 lb) |
| Invert sugar | 0.90 kg | (2 lb) |

Continue boiling until the solution reaches a soluble solids of 78 to 80 percent when it is poured onto cooling tables. Flavors, which should include rose otto or synthetic rose flavor and a small amount of citric acid, must be added after boiling has ceased.

*2. Starch with Additional Gelatinizing Agents.* Use procedure as recipe 1, and then add, after boiling has ceased:

Gelatin    340 g (120 oz)    (previously soaked in water)

Alternatively, agar may be used and a solution prepared by boiling as previously described and added to the base syrup before boiling is completed. This recipe may be boiled to a lower soluble solids content (75 to 76 percent).

## Agar and Gelatin Jellies

Both agar and gelatin are used a great deal in the production of confectionery jellies. The final jelly is short if produced from agar and rather tough and rubbery from gelatin, but the texture from gelatin depends much on the quantity used. Pectin jellies, on the other hand, are tender and readily dispersed in the mouth.

The properties of agar and gelatin are described in the ingredient section of this book.

**Preparation of Gelatin Jellies** The type of gelatin used for confectionery jellies may be powder, sheet, or granule. The powder can be dissolved fairly quickly in warm water but care is necessary to prevent if from consolidating on the bottom of the vessel. Gentle stirring is necessary. Sheet or granule gelatin must be soaked in cold water and it will absorb as much as ten times its weight, but it is usual to soak in an equal amount of water. Care in this process is necessary; prolonged soaking in warm conditions will encourage microbiological spoilage. Use clean, sterilized covered vessels and soak not more than 12 to 18 hr. The strength of gelatin is recorded by its Bloom value, which may vary from 50 to 300. In confectionery, that usually employed for jellies ranges between 180 and 220. The proportion of gelatin in a jelly varies from 4 to 12 percent, depending on the texture required. Gelatin solutions must never be boiled, and the gelatin solution must be added at the end of the syrup boil after cooling to below 80°C (176°F). Similarly, any acid *must* be added at the latest stage. Heat and acidity rapidly degrade jelly strength.

The following is an example of the formulation and manufacture of a gelatin jelly. Many variations are possible. In the manufacture of fruit jellies, the mixing of some fresh fruits with gelatin syrups must be avoided. Pineapple and papaya, for example, contain proteolytic enzymes that rapidly destroy gelatinizing powder. While this problem is most likely to occur in domestic use, it is a wise precaution to use only canned fruits or pulp in confectionery or to boil fresh fruit syrups to destroy these enzymes.

*Gelatin Jelly (Acid Jelly)*

| | |
|---|---|
| Sugar | 25 lb |
| Glucose syrup (42 DE) | 18 lb |
| Water | 10 lb |

Dissolve sugar in water, add glucose syrup, and boil to 115°C (239°F).

| | |
|---|---|
| Gelatin | 3.2 lb ⎱ Soak, warm to dissolve. |
| Water | 3.2 lb ⎰ |

Cool the above syrup to below 80°C (176°F), and then add the gelatin solution and mix.

Immediately before casting, add

| | |
|---|---|
| Citric acid | 8 oz dissolved in |
| Water | 8 oz |
| Essential oil of orange (or other) | 18 ml |
| Orange color | as required |

Cast in dry starch impressions; allow to set for 12 hr. Remove by sieving. The jellies may be sugar sanded or chocolate covered.

**Agar Jellies**  Agar-agar is a word derived from an Eastern expression meaning seaweed, and commercially the product is now shortened to "agar." (See "Gelatinizing Agents.")

It is a useful substance for the production of confectionery jellies. These tend to be short in texture and find particular use in the preparation of fruit slices, shaped, colored, and flavored to resemble sections of oranges, lemons, or grapefruit.

Agar was originally available only in strips, somewhat resembling dried seaweed, but the powdered product is now mostly used.

It is not soluble in cold water but swells slighly on soaking. To dissolve, a large volume of water at boiling temperature is required, usually thirty or forty parts of water to one of agar. Agar is not destroyed by boiling water unless acid is present, when it is rapidly degraded. Therefore, any acid added must be after the syrup boil is completed.

It is advisable to add a buffer salt, such as sodium citrate, to offset the action of acid addition.

Agar jellies may be subject to syneresis and a general improvement in texture and shelf life may be obtained by combination with other gelatinizing agents, particularly pectin and starch.

### Preparation of Agar Jellies, Fruit Slices

*Jelly*

| | |
|---|---|
| Sugar | 25 lb |
| Glucose syrup (42 DE) | 20 lb |
| Agar | 1 lb |
| Water | 40 lb |
| Sodium citrate | 5 oz |

Soak the agar in the water for 2 to 4 h, and then add the citrate. Bring to a boil and keep at this temperature for 5 to 10 min (simmer) until the agar is in solution. Fibrous material is usually present, especially in the strip agar, so it is necessary to pass the solution through a fine sieve.

Add the sugar and, when dissolved, the glucose. Boil to 107°C (225°F). Pour into another pan to cool to 75°C (167°F) and remove any scum. Mix in

| | |
|---|---|
| Citric acid | 3 oz dissolved in 3 oz water |
| Essential oil, orange or lemon | 20 ml |
| Color | as required |

Cast without delay into dry starch (6 to 8 percent moisture). Do *not* hold the hot mixture for a long period after acid addition. The jellies should remain in starch overnight in a dry, warm room.

Small jelly pieces may be made in this way, or for fruit slices, the liquid mixture is deposited into semicylindrical impressions. These, when set, are removed, cut into slices, moistened, rolled in caster sugar, and dried.

To imitate the outer "rind" of the the fruit slices, a paste may be prepared of the following composition

| | |
|---|---|
| Glucose syrup (42 DE) | 8 lb |
| Gelatin | 4 oz |
| Icing sugar | 12 lb |
| Cornstarch | $1\frac{1}{2}$ lb |
| Citric acid | 1 oz dissolved in 1 oz water |

The gelatin must be soaked and dissolved in 12 oz of water and then mixed into the heated glucose. After cooling, add citric acid. Mix the icing sugar and cornstarch dry, and then add the syrup gradually, kneading the mixture continuously.

At this stage, color and flavor can be added as required. The paste should now have a consistency such that it can be rolled into thin sheets. These sheets may be cut and applied to the round surface of the pieces after removal from the starch and cleaning. The coated jelly may then be sliced, moistened, icing sugar applied, and dried.

***Special Note on the Preparation of Jellies***  When jelly mixes containing the gelatinizing agent are boiled (e.g., starch, agar, pectin), thermometer readings to determine the end point are unreliable. A refractometer should be used and the concentration of the syrup phase of the final product should always be greater than 75 percent.

## MARZIPAN, ALMOND AND OTHER NUT PASTES

A great deal of argument has arisen from time to time regarding the definition of marzipan. Dictionary descriptions say that it is a paste composed of crushed almonds and sugar, but because of the high price of almonds, many substitutes have appeared on the market. These have been prepared from apricot kernels, soya flour, and various other ingredients, and are often much overflavored with synthetic almond essence.

Composition standards are, however, recognized in some countries and a distinction is also made between "raw marzipan" or "base almond paste" and the finished article, which is called "marzipan" or "almond paste."

A recognized formulation in the United Kingdom at one time for raw marzipan was a mixture of two parts of almonds to one of sugar. According to Jackson (1973), the United Kingdom "agreed" standard for marzipan is one part of ground almonds to three parts of sugar.

There are also the legal compositions quoted in the *Silesia Manual* (Vol. 2, No. 3, 1984), which are as follows:

1. *Raw Marzipan, Base Almond Paste.* Is a mass produced from blanched almonds, containing a maximum of 17 percent moisture and 35 percent sugar, and 10 percent of the whole mass may be invert sugar. Almond oil content minimum 28 percent. May contain up to 12 percent of the almonds as blanched bitter almonds. "Debittered" bitter almonds not permitted.

2. *Marzipan, Almond Paste.* Is a mixture of raw marzipan plus not more than an equal amount of sugar. The sugar may be partly replaced by glucose syrup and/or sorbitol. Up to 3.5 percent of the total weight of the marzipan may be glucose syrup and/or up to 5 percent of 70 percent sorbitol solution.

It is debatable whether much is achieved by all these complications. The public will decide whether it likes the product, and if not, will not buy it. That is just what has happened with the overflavored pastes mentioned previously.

### Manufacture of Marzipan

Raw almonds are cleaned by sieving, air elutriation, and other electronic and mechanical devices. They are then immersed, usually in wire baskets, in water just below boiling point for about 5 min. This loosens the skins and the soaked nuts are then put through a blanching machine consisting of rotating rubber rollers that squeeze the skins from the kernels. There are also continuous blanching machines.

The blanched almonds are then washed with cold water and inspected on a conveyor, where any residual skins, foreign matter, or defective nuts are removed.

From this stage different methods for paste manufacture may be used. Some manufacturers advocate drying the almonds, which, they claim, gives a better flavor—possibly it has the effect of a light roast, but slight loss of color results from this.

Whether they are dried or not, the next part of the process is to coarsely grind the almonds, and this is done by passing them through kibbling rolls or other grinding machine and then through a triple roll refiner.

This refined paste is mixed with sugar and a small proportion of glucose or invert sugar, and then transferred to a paster, which is usually a heated kneader. This gives the mixture a further cooking and reduces the moisture content.

After cooking, the paste is spread on tables to cool. This base is often refined again before making into bars and softened by the addition of invert sugar.

Another method for the preparation of almond paste consists of placing the refined blanched almonds in a trough mixer and pouring into this a syrup consisting of sugar and glucose syrup boiled to 250°F (121°C). Fondant creme is then added and mixing continued until the creme is fully dispersed, when the mixture is cooled and again passed through a refiner.

A typical recipe for this second process is:

| | | |
|---|---|---|
| Sugar | 22.6 kg (50 lb) | |
| Glucose syrup | 6.35 kg (14 lb) | Boiled to 250°F |
| Water | 9.1 kg (20 lb) | (121°C) |
| Ground almonds | 15.9 kg (35 lb) | |
| Fondant creme | 6.8 kg (15 lb) | |
| (3 sugar 1 glucose syrup) | | |

## The Stephan Method

The Stephan machine is useful for making marzipan and other nut pastes. Its action is such that it will comminute fibrous materials not readily handled by other refining methods.

The basic machine consists of a bowl with an internal rotating cutter that operates at speeds between 1,500 and 3,000 rpm. The bowl can be heated, cooled, and provided with steam injection. It can withstand pressure and vacuum.

Raw materials may be fed directly into the machine and are then

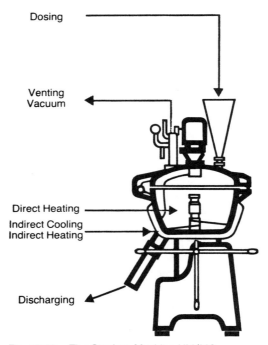

Fig. 19.33.  The Stephan Machine UM/MC

The bowl is mounted in a machine stand and counterbalanced with its geared-down motor, making it easy to tilt. Once tilted, it can be locked at any angle in the range.

Bowl, lid, motor shaft, and all processing tools are made of stainless steel. The motor shaft extends into the bowl and carries the processing insert with its sturdy cutter blades (or the mixing and kneading insert). Bowl and lid are interlocked by means of accidentproof quick-release clamps.

The dosing of liquid and fluid raw materials can be carried out via the dosing valves without opening the lid and discharge is effected via a bowl outlet valve. As with all other functions, both of these valves are controlled from the separate switch cabinet.

*Stephan u. Sohne, Hameln, West Germany*

subjected to pulverizing, mixing, and sterilizing. In the case of marzipan, blanched almonds, sugar, glucose syrup, and any of the other ingredients, such as invert sugar and sorbitol, are charged into the bowl. In a short time, the mixture is pulverized, mixed, and sterilized. Marzipan made this way has a soft texture.

Various models are available and a popular machine is the series UM/MC: see Fig. 19.33.

## Shelf Life of Marzipan and Nut Pastes

Much has been written about fermentation in marzipan and many are the proposals to prevent it, including the addition of glycerol, acetic acid and bitter almonds. None of these additions is essential if the fundamental principles of confectionery technology are observed.

The first concerns the elimination of microorganisms in the raw material combined with good plant hygiene. The second is to make sure that the sugar/glucose/invert sugar ratios are correct, which, in conjunction with the moisture in the recipe, maintains a syrup phase concentration of 75 percent minimum after maturing.

Nuts in the natural state contain microorganisms, including osmophilic yeasts as well as active enzymes. Unless destroyed, these will cause slow fermentation and off-flavors in the confectionery prepared from them. These organisms are resistant to destruction by heat in high sugar concentration media unless subjected to temperatures of 93 to 100°C (200 to 212°F) for 10 to 20 min. It is better to destroy these organisms during the blanching process by immersion in boiling water just long enough to ensure sterility. Three minutes should be long enough if the water surges through the nuts. Some manufacturers claim that blanching for longer than an instantaneous immersion is detrimental to flavor but this is doubtful.

Plant hygiene has been mentioned in many processes. With marzipan manufacture utensils used for raw nuts must never be used for finished products unless cleaned and sterilized. Dusts from nut cleaning are particularly dangerous and this operation should be done away from the paste department. Marzipan scrap should be sterilized by heating if it is more than 24 hr old.

The practice of covering pastes and fondant with wet cloths to prevent drying out is entirely wrong and unfortunately is still recommended in some textbooks. The same applies to wet cloths used for hand wiping, and bacterial examination of specimens found occasionally in the factory have revealed a very high population of microorganisms.

Concerning the soluble solids concentration, base marzipan consisting of almonds and sugar alone cannot attain a concentration higher than 67 percent because this is the maximum solubility of sucrose and the soluble part of the nut contributes very little. Therefore, if this is made, it should be kept for the minimum time. A better procedure is to use the sugar/glucose/invert mixtures as the ingredients of the syrup base with a minimum of 20 percent glucose syrup. This will give a satisfactory syrup concentration when the moisture content of the marzipan is 8 percent. A check can be made by extracting some of the syrup phase from matured paste by means of a "syrup press" (Fig. 19.34) and determining the concentration by refractometer.

Since the inhibiting action of certain chemical substances has been mentioned, some idea of the quantities required should be given.

The introduction of an organic acid is effective if, for any reason, low soluble solids have to be tolerated. Acetic acid seems the best, but at the really effective levels of 0.1 to 0.2 percent an objectionable flavor is detectable. Even at the minimum of 0.05 percent some

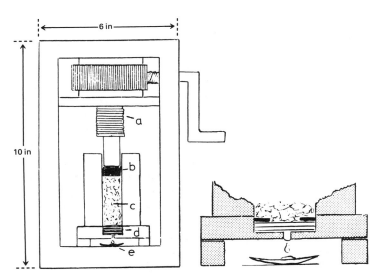

Fig. 19.34.   Diagram of Syrup Press
a. Geared Ram          d. Filter and Gauze
b. Leather Washer      e. Syrup Drop
c. Fondant or Marzipan
Section of filter (d), (e) showing upper fiber washer, upper wire gauze, filter paper, lower wire gauze. Beneath the lower gauze are grooves leading to the central orifice (approx. dimensions shown).

people can taste the acid. Lactic acid is less effective and at higher levels the taste is not so easily detected. Glycerol and sorbitol syrup have an inhibiting effect and will help to keep the paste soft and retard crystallization, usually 2 to 4 percent is used in the recipe. Bitter almonds have also been shown to inhibit fermentation, possibly due to the glucoside amygdalin, and 2 to 3 percent of bitter almonds are often added to almond paste to enhance flavor and these help with preservation.

Sorbic acid where permitted, is an effective preservative used at the rate of 0.1 to 0.2 percent.

### Persipan

This is an "official" name for a paste made from debittered apricot or peach kernels. Correctly made, it bears a very close resemblance to the true almond paste.

There are two methods of debittering. One is to subject the kernels to washing with cold water until the bitter glucoside (amygdalin) is removed. This is wasteful as other useful soluble solids are lost. The second method is to make use of the natural enzyme, emulsin in the outer layers of the kernel. This will decompose the amygdalin into hydrogen cyanide, benzaldehyde, D-glucose. The kernels are chopped finely and water added until a slightly wet mash is obtained, and this is incubated at 30°C (86°F) for 24 hr. Ventilation and extraction must be provided to remove the hydrogen cyanide evolved. If the kernels have been blanched by a scalding process, emulsin enzyme must be added to bring about the reaction. The only loss in this process is the cyanide, and the wet nuts can be used direct to make the persipan.

The preparation of persipan from the debittered kernels is the same as for marzipan.

## COCONUT PASTE, COCONUT ICE

Coconut is a very popular ingredient in confectionery. Coconut ice is one of the traditional sweets made at home.

The main ingredient of most coconut lines is desiccated coconut, although some confectioners claim that only fresh coconut gives the true flavor and a juicy tender texture. Coconut is also available in other forms, such as sweetened desiccated, tenderized, and canned in syrup.

Desiccated coconut, being prepared from a natural product, may contain microorganisms. Unfortunately, it has had some reputation in the past for being responsible for outbreaks of salmonella poisoning and typhoid, due to unhygienic methods of preparation in some parts of the world. Unsatisfactory drying procedures also have caused the presence of lipase, resulting in soapy rancidity developing in the final confections.

Because of these dangers, methods of sterilization of all supplies were introduced and at the same time delegations from user countries advised on improved methods of processing in the growing areas. The production of desiccated coconut is described in the chapter on "Nuts," and today supplies are of much higher quality.

## Coconut Ice

Coconut ice was originally prepared from boiled sugar syrup to which coconut was added together with some color and flavor, the sugar being allowed to crystallize with the coconut as a result of stirring.

This home-made sweet had a relatively short shelf life—it dried out, developed a coarse grain, and sometimes fermented because, with sugar alone, the minimum syrup phase concentration of 75 percent could not be obtained.

Modern recipes contain glucose and often condensed milk, honey, gelatin, and various flavors. Texture varies from soft pastes grained off with fondant creme to semihard caramel-like Japanese desserts.

Some typical recipes are given below:

*Coconut Paste*

| | | |
|---|---|---|
| Sugar | 4.53 kg (10 lb) | Dissolve and boil to 121°C (250°F). |
| Glucose syrup | 1.36 kg (3 lb) | |
| Water | 1.81 kg (4 lb) | |
| Sweetened condensed milk | 0.68 kg ($1\frac{1}{2}$ lb) | |
| Fondant creme (80/20) | 0.90 kg (2 lb) | |
| Fine desiccated coconut | 1.80 kg (4 lb) | |

Add the condensed milk to the hot syrup and mix well. Then stir in the desiccated coconut and follow this with the fondant creme, which should assist the crystallization of some of the sugar in the syrup.

Flavor and color can be added as required. Texture may also be adjusted by the addition of a small amount of honey or invert sugar.

The paste is cooled on a table, rolled, and cut into bars, or it may be extruded. This type of paste has the best texture when the moisture content is 11 to 12 percent and the coconut then appears less tough and fibrous. Adjustment of moisture content can be obtained by altering the syrup's boiling temperature or by adding invert sugar. By using sweetened condensed milk, some milk fat will be included, which may affect shelf life. Nonfat milk is preferable.

*Japanese Dessert*

| | | |
|---|---|---|
| Sugar | 2.26 kg (5 lb) | Dissolve and boil to |
| Glucose syrup | 3.17 kg (7 lb) | 115°C (240°F). |
| Water | 0.68 kg (1½ lb) | |
| Fine dessicated coconut | 2.7 kg (6 lb) | |

Stir the coconut into the hot syrup and add flavors as desired. Dried or preserved fruit pieces may be added to improve the character of this sweet. It is cooled on a table, rolled and cut.

## Tenderizing Coconut

In many coconut confections, the coconut exhibits a fibrous texture and may leave a somewhat objectionable residue in the mouth. It is claimed that the use of fresh coconut will avoid this. This is generally not a practical proposition for most manufacturers, who normally use desiccated coconut.

As mentioned previously, processed coconut in cans, or sweetened and tenderized, is available and the confectioner must judge for himself whether the extra cost of these is justifiable.

Various methods for tenderizing desiccated coconut have been proposed. These include soaking in hot invert syrup with or without a proportion of sorbitol.

Important factors to recognize are the moisture content and syrup phase concentration of the final confection and the proportion of syrup to nut.

High syrup concentrations have a toughening effect on the nut fiber. A moisture content of about 11 percent with a syrup phase concentration of 76 to 78 percent is optimum.

## CREME AND LOZENGE PASTES, CACHOUS, TABLETS

### Lozenges, Cachous

Lozenge pastes are used a great deal for the manufacture of medicated sweets, cachous, and the like. They are relatively simple recipes consisting of icing sugar kneaded into a paste with a gum mucilage, gelatin solution, or both, and then rolled, shaped, and cut, followed by drying.

Since these are made from gum solutions and sugar only, the formulation of the mucilage is important. Gum arabic, gum tragacanth, and gelatin are used.

For machine-made lozenges, some manufacturers say that mucilages prepared from a mixture of gum tragacanth and gelatin are more easily managed than gum arabic solution, which is reserved for hand-made recipes.

The following is a typical formula for mucilage:

1. Gum arabic      1.81 kg (4 lb)
   Water (cold)    4.1 kg (9 lb)

Soak with occasional slow stirring for 24 hr. Sieve through a fine mesh to remove foreign matter.

2. Gum tragacanth    141 g (5 oz)    Soak first using 5 lb of
   Water (cold)       3.17 kg (7 lb)   water for 6 hr and then add the remainder of the water and soak for a further 18 hr.

   Gelatin           198 g (7 oz)    Soak until the gelatin is
   Water (cold)       2.26 kg (5 lb)   soft, then warm to dissolve, and cool.

The mucilage consists of a mixture of equal amounts of these solutions. The gum solutions must be used fresh as they will deteriorate on keeping, due to bacterial action, and they must be used cold.

The mucilage is mixed with powdered sugar to give the required texture, flavor and color being added as necessary.

With larger-scale batches, a "Z"-arm kneader is used. Judgment of the texture of the mixed paste is visual unless the particle size distribution of the sugar can be guaranteed. The condition of the sugar dough must be correct for the stamping machinery to give good shapes without sticking or collapse of the pieces.

After forming, the pieces must be dried slowly, first with slightly warm air, 24°C (75°F), preferably humidity controlled at about 50 percent relative humidity. This can be done on wide, moving belts. The lozenges are turned over automatically by transfer from one belt to another, and this process may take up to 4 hr to ensure even drying.

The lozenges are then sufficiently firm for drying in a stove or hot room, which is usually done in trays. Stoving is at 38°C (100°F) for 24 hr.

**Flavor**  The procedure of drying is responsible for the loss of flavor in some formulations. Some flavors are particularly volatile, peppermint, for example.

Loss of flavor can be reduced by a premix with fine sugar or cornstarch or by the use of encapsulated flavors.

## Creme Pastes

Creme pastes are made in a manner similar to lozenges but with the inclusion of glucose syrup to give a softer finished product. The following is a representative formula and process.

| | |
|---|---|
| Fine sugar | 4.53 kg (10 lb) |
| (icing, caster or mixed) | |
| Glucose syrup | 1.58 kg ($3\frac{1}{2}$ lb) |
| Gelatin | 56.7 g (2 oz) |

The gelatin is soaked until soft and then mixed into the glucose syrup, which has been warmed until fluid. A little water may be added to increase fluidity. The sugar is placed in a mixer, usually of the trough type with horizontal spindle and carrying paddles, and while in motion the glucose/gelatin syrup is added slowly until a fairly stiff paste is produced. This is flavored, colored, rolled, and shaped and the finished pieces are allowed to dry and set.

Lozenges and pastes have many applications in the confectionery and pharmaceutical industries. The formulation is simple but the production needs care and understanding of the kneading and drying processes. Incorrectly prepared paste will not form correctly; it will stick to the cutters or deform after cutting. Uneven drying will also give poor shapes and uneven texture.

## Tablets

Compressed tablets are mixtures of base materials, binders, flavors, and a lubricant. The base material may be icing sugar, dextrose, or lactose. The binder is gelatin or gum arabic. Lubricants are necessary to give good release from the press and consist of magnesium and calcium stearate or stearic acid. Flavors and colors must be powders.

Special machinery is used for tablet making and consists of compression punches that close over the mixture of ingredients. These ingredients are in a granulated form produced either by wet granulation or "slugging."

In wet granulation, the mixture of ingredients is wetted to a dough, granulated, sieved, and dried. This dry material is flavored, and colored, and the lubricant added in a separate mixer. It is then ready for feeding to the tablet press.

Slugging is used when the tablet mixture consists of substances of a hygroscopic nature or is volatile (e.g., some flavors and medicaments).

The slugs are prepared from the mixed powders in a tablet machine of large bore, which produces pellets suitable for breaking and granulating. They do not need drying and can be fed directly to the tablet press.

Compressed tablets are particularly popular for sweets sold in roll or stick packs. Mint-flavored tablets are very well known.

Compressed tablets containing dextrose give a cooling sensation in the mouth. Sugarless tablets have been sold in large quantities in the United States and consist of powdered sorbitol and flavor, usually mint. Sorbitol is of a plastic nature and will consolidate under pressure without a binder.

Effervescent tablets are possible. Icing sugar, sodium bicarbonate, color, and mucilage are first mixed, granulated, and dried. Citric acid is then added in a tumbler mixer, with dry flavor and lubricant. This is fed to the tablet press.

Jackson (1973) describes tabletting in considerable detail and specialists in tabletting machinery will provide information.

As with all confectionery products, many formulations exist. As an example, the following describes a compressed peppermint tablet:

| | |
|---|---|
| Icing sugar (free flowing) | 50 lb |
| Peppermint oil | 5 oz |
| | (up to twice this amount for very strong mint) |

| *Powdered* gum arabic | 1 lb |
| Water | up to 3 lb |
| Magnesium stearate (or stearic acid) | 8 oz |

Mix the powdered sugar and gum arabic thoroughly. Add water sufficient to make granules. Dry the granules at about 60°C (140°F). Mix in the peppermint oil and lubricant and subject this to tabletting. The compressed tablets should be dried at a low temperature, 38°C (100°F).

## LIQUEURS

There are several types of liqueur chocolates.

### Chocolate Liqueurs

Chocolate shells are preformed and the alcoholic liqueur is deposited into the shells. For this type, the shells are usually bottle shaped.

Another type of chocolate shell, used for assortments, resembles the shape of an enrobed chocolate. The "liqueur," not necessarily alcoholic, is deposited into the shell, followed by spraying with cocoa butter and backing with chocolate (see "Chocolate Shells").

### Sugar Crust Liqueurs

This liqueur piece consists of a crystalline sugar shell containing a liquid center and is chocolate covered. The liqueur in the center may be alcoholic or merely a flavored syrup.

The principle of preparation, whether or not alcohol is involved, is to deposit the syrup mixture into starch impressions at a concentration that will form a crystal layer at the surface in contact with the starch.

The interior must remain liquid. Crystal formation is aided by the cooling and slight loss of moisture to the starch, which should have a moisture content of 5 to 7 percent and a temperature of about 30°C (85°F). After depositing is completed, a starch layer is dusted over the surface. The syrup remains in the starch for $2\frac{1}{2}$ to 3 hr during which time crystallization occurs, mostly on the lower surface.

At this stage, the semicrystallized pieces are turned over. This requires skill and the operation involves passing a shaped wire through the starch and beneath the crusted shapes.

In the process of manufacture, great care must be taken to avoid unwanted or poor crystallization. High-grade sugar must be used and vibration avoided at all stages.

The following are typical formulations. They all need practice before good results are obtained.

*Sugar Crust Liqueurs with Alcohol*

| | |
|---|---|
| Sugar, high grade | 100 lb |
| Water | 40 lb |

Cook to 113°C (236°F). Cool to 70°C (158°F). Add the following alternative liqueurs by very carefully pouring them, in a thin stream, into the syrup at a temperature between 60 and 70°C:

1. Proprietary liqueur, 60 percent alcohol, 25 lb
2. Rectified spirit, 96 percent alcohol, 15.6 lb
   Water, 9.4 lb
   Flavor as required.
3. With red wine, 12 to 14 percent alcohol
   Cook syrup above to 118°C (245°F)
   $Add$ "red wine," 28.5 lb ⎫ warm to dissolve.
   Fine sugar, 8.5 lb ⎭
   Then deposit in starch [5 to 7 percent moisture, 30°C (85°F)].

*Sugar Crust Liqueurs Without Alcohol*

| | | |
|---|---|---|
| Sugar, high grade | 100 lb ⎫ | Cook to 110°C (230°F). |
| Water | 40 lb ⎭ | |
| | Cool to 60°C (140°F). | |
| Add citric acid | 0.5 lb (in 0.5 lb water) | |
| Flavor as required | | |

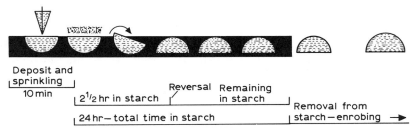

Deposit and
sprinkling
10 min
2½ hr in starch    Reversal  Remaining
                             in starch
24 hr—total time in starch    Removal from
                              starch—enrobing →

Fig. 19.35.  Sugar Crust Liqueurs with Alcohol

*Central College of Confectionery, Solingen, Germany*

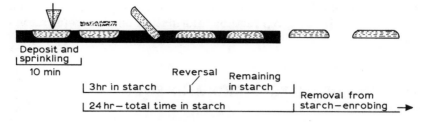

Fig. 19.36. Sugar Crust Liqueurs without Alcohol

*Central College of Confectionery, Solingen, Germany*

Deposit in starch at this temperature.

Figures 19.35 and 19.36 illustrate the sequence of operations.

In the preparation of sugar crust liqueurs with alcohol, two factors must be realized:

1. Sugar solubility is depressed by the presence of alcohol as is shown in Table 19.5.
2. Alcohol and water contract in volume on mixing, as shown in Table 19.6.

TABLE 19.5. PERCENT ALCOHOL IN SOLUTION BY WEIGHT

| °C | 0 | 6 | 12 | 20 |
|----|------|------|------|------|
| 20 | 66.9 | 61.0 | 55.1 | 47.4 |
| 40 | 70.6 | 65.2 | 59.6 | 52.0 |
| 55 | 73.7 | 68.6 | 63.2 | 55.5 |
| 70 | 77.1 | 72.4 | 67.1 | 59.1 |

TABLE 19.6. MIXTURE OF WATER WITH ALCOHOL (96%). PERCENTAGE BY VOLUME

| Alcohol | Contraction | Alcohol | Contraction |
|---------|-------------|---------|-------------|
| 15 | 0.223 | 54 | 2.754 max. |
| 25 | 1.257 | 60 | 2.696 |
| 40 | 2.438 | 70 | 2.410 |
| 50 | 2.732 | 80 | 1.854 |
|    |       | 90 | 0.908 |

Complete tables are available in handbooks of physics and chemistry. Jackson (1973) gives further information on the preparation of liqueurs.


## PANWORK, DRAGEES—HARD AND SOFT PANNING

"Dragees" is the name given to sugar-coated confectionery where the sugar has been applied as successive layers of syrup in a rotating pan. Reference has already been made to panning under "Chocolate Processes" where a similar procedure is used to coat nuts and other confectionery pieces with chocolate.

There are basically two types of sugar panning:

1. *Hard Panning*

    (a)  Hard texture, smooth, round or ovate, opaque appearance.

    (b)  Pearl—hard with a rough surface, opaque.

With hard panning, successive layers of syrup mixture are applied. Each layer is dried by injection of warm air into the pan. Pans may be provided with heating coils attached to the exterior of the pans or, in some instances, the pan is heated while rotating by means of gas flames beneath the pans.

Traditionally, all pans were made of copper but stainless steel is gradually replacing copper. Pans are of various sizes and shapes and the speed of rotation and angle of tilt can be varied.

A basic requirement in all panning processes is to ensure that the articles in the pan are rolling and not sliding. This is usually achieved by careful intermittent addition of the syrup and powdered sugar.

Occasionally, ribs may be attached to the interior surface of the pan. Often a roughened layer of confection is built up on the inside of the pan to encourage rolling.

If the pieces in the pan slide and the syrup mixture is applied erratically, bad shapes occur and the articles will conglomerate.

2. *Soft Panning*

Soft confectionery centers, (e.g., coconut pastes, soft caramels) are coated in successive layers of sugar/glucose syrup. They are "dried" by application of fine sugar without air blowing. Smaller pans and higher speeds are usual for soft panning.

Examples of various panning processes are given below but it must be realized that this is a confectionery operation that is probably subject to more variations and personal opinions than any other.

The panning process with all its ramifications is a highly skilled operation when a variety of products is being manufactured with a small number of pans. Addition of the syrups and powders and application of the air to the pan becomes a matter of judgment and experience.

On the other hand, when large quantities of the same kind of product are being made, the process can be mechanized and standardized.

A description of mass production machinery is given later.

## Sugared Almonds

With nuts it is advisable to preglaze by coating the nuts with a gum arabic solution or gum/syrup mixture.

| | |
|---|---|
| Gum arabic | 0.90 kg (2 lb) |
| Water | 1.36 kg (3 lb) |

Dissolve and sieve as described in previous recipes.

The almonds, which have been previously dried or lightly roasted, are placed in a revolving pan and the gum solution applied in small quantites so that thin layers are distributed evenly over the surface of the nuts and any wrinkles filled. The pan and gum solution are cold but slightly warmed dry air may be directed into the pan. Fine sugar is sifted over the rotating nuts, which helps to dry the gum film and to separate the nuts. More gum is applied followed by sugar until an even coating is obtained when the nuts are discharged into wire mesh trays and dried in a warm room (29°C/85°F) for 24 hr.

It is also possible to dry using the air blown into the pan if time is available.

The nuts are then subjected to the "engrossing" process and sugar/gum syrup is required for this:

| | |
|---|---|
| Sugar | 4.53 kg (10 lb) |
| Water | 2.04 kg ($4\frac{1}{2}$ lb) |

Dissolve by boiling and adjust the concentration to 67 percent determined by refractometer or hydrometer. Then add 12 oz of gum solution as used for the almond coating and mix well.

The almonds with the gum/sugar coating are placed in the rotating pan with slight heat applied and a small amount of gum syrup added to give a continuous thin layer over the surface. As the layer approaches dryness, icing sugar is dusted on, followed by further applications of gum/syrup. The nuts rotating in the pan will gradually enlarge with each application of gum syrup and icing sugar and this is continued until the correct unit weight is obtained. During the last stages of building up, color may be added to the syrup, and to obtain opacity and bright colors or pure white the addition of some titanium dioxide (Pure-food grade) is permissible.

Lake colors are preferable for coloring as they are insoluble. Soluble colors may be used but these are prone to cause staining in the mouth.

After coloring, a finishing glaze is applied, which consists of a sugar syrup of 60 percent concentration.

In some panning operations, cornstarch is used for building up as well as fine sugar. This is cheaper, but uncooked starch in quantity gives a mealy texture and is not readily digested.

Roast almonds, hazelnuts, and peanuts can be coated by the method just described and flavors can be added during the building-up process.

A very popular product is made from small chocolate pellets as the center. These are made on a Roller depositing machine and they are coated with a thin sugar layer, multicolored, and polished. They are essentially a children's line.

## Nonpareils, Comfits, Aniseed Balls

These are much smaller panned articles but the process of manufacture is very similar and needs only adjustment of the syrup strengths to suit the size and allow the pieces to tumble over one another. Syrups of the following concentration may be used:

First coating foundation     40 to 45 percent
Building up                       65 percent
Finishing                    50 to 55 percent

For nonpareils, free-flowing, fairly coarse crystal granulated sugar is used. It is sifted free from dust and the coarse grain charged into a warm pan. Fifty percent syrup is poured on until well distributed over the grains, a process that must be assisted by running the hands through the mass. Small amounts of cornstarch are used to promote free flowing and the addition of syrup is continued alternately with cornstarch until sugar grains have built up to the required size. Coloring and flavoring are added in the final syrup additions.

Seeds of caraway, coriander, and aniseed are processed as nonpareils, the original seeds being well dried and sifted free from dust before panning commences.

## Pearling

This is a process of sugar panning where syrup with a small addition of gum arabic is allowed to drip in a controlled flow over the confectionery revolving in a pan until the required size is obtained. The centers may consist of pieces of candied peel or other fruits, dried fruits, angelica, sugar boilings, or pastes. Before pearling they are coated with gum solution and dusting sugar as described previously.

## Polishing Sugar-Panned Products

The gloss obtained by the above process can be enhanced by wax glazing, using beeswax, carnauba wax, or spermaceti. The best way to apply this is to coat the interior of a pan by pouring in sufficient molten wax to cover it completely with a thin layer. The surface must be quite smooth and this is attained by polishing with a fiber-free cloth, preferably coated in talc. The use of talc, however, is not permitted in some countries.

The units to be polished must roll and not slide in the pan and an oscillating pan is better for this part of the process.

Cool air should be applied to the pan as frictional heat will cause the wax on the pan to become too soft.

Other proprietary materials are available for polishing.

These consist of acetylated monoglycerides and are described under "Chemical and Allied substances" and "Glazes."

## Soft-Centered Pan Work

A large variety of confectionery centers can be used for this process. Coconut and other pastes, jellies, marshmallows, Turkish delight, Japanese desserts, and caramels may be coated.

It is essentially a cold process in which the centers are rotated in a pan coated internally with a rough sugar layer. During normal sugar panning a layer of this nature is obtained.

The centers, which are usually pressed or cut into shapes, are charged into the pan and while rotating, glucose syrup solutions are applied alternately with caster sugar until the correct size is obtained. Syrup of a concentration of 60 to 65 percent is used but this may be varied according to the texture of the center. The final sugar dustings are with icing sugar, which gives a smooth surface, and when this has been obtained, the units are discharged onto a tray and allowed to set in a dry, but not hot, atmosphere for 16 to 24 hr. Glaze is then applied by the same process as for dragees.

Jelly beans, birds' eggs, and many candies attractive to children are prepared by this method and as with all soft confectionery great care must be taken to ensure that a minimum syrup phase concentration of 75 percent is obtained in the center.

The speckled appearance of birds' eggs may be obtained by subjecting the unglazed centers to a coarse spray of dye solution, which is allowed to dry before glazing. For small batches, a stiff brush with long bristles may be dipped in the solution and droplets flicked off by bending the bristles with a rod and releasing.

Sometimes soft panning is used as a preliminary to enrobing. Preserved ginger is an example; this tends to exude syrup and chocolate covering by enrober is difficult in these circumstances.

The syrup may be absorbed by panning with the addition of sugar or cocoa powder, followed by a chocolate layer. The units are then suitable for enrobing.

When egg, gelatin, or milk products are used, the same care is needed in observing hygienic methods and to destroy microorganisms and lipolytic activity by pasteurization.

## Silver Cachous

The manufacture of these needs skill and experience and the final sugar coating is carried out in a glass or glass lined pan.

The center is built up from granulated sugar graded by sifting, and the application of syrups as for nonpareils. Some gum arabic solution can be incorporated in the syrup and when the correct size of sphere is obtained with a smooth surface the silver coating is applied as follows:

*Adhesive Solution*
High-grade gelatin    226.8 g (8 oz)

Soak in water until, soft, then dissolve in a warm mixture of

Glacial acetic acid    397 g (14 oz)
Water                  795 g (28 oz)

Acetic acid has the effect of hardening the gelatin jelly.

The cachou centers are placed in a separate bowl, some of the gelatin solution added, and the pan oscillated and rotated to give a very thin film spread completely over the surface of the centers.

This operation is very important and the exact conditions and quantity of syrup can only be found by trial. Too much will cause the silver leaf to soak into the syrup and it will not spread out as a layer. Too little syrup will give patchy results where the silver leaf has not adhered properly.

Some of the leaf is placed in the glass pan and this is set in motion. The moistened cachous are added in a thin stream, generally through a funnel, and will pick up the silver leaf and gradually become completely coated. No heat is used. When coated, it is best to transfer them to a separate glass pan or bowl where they are rotated at high speed for polishing. When manufactured, the silver cachous must be transferred quickly to closed, preferably glass, containers to prevent them from tarnishing. When used on soft confections or cakes, they tend to have a short shelf life and lose their luster.

Silver leaf must be food grade, and foreign metals, particularly lead, must be absent.

## Modern Machinery for Panning

In recent years, there has been a great deal of development in the mechanization of the panning process. The following methods are in use:

1. Mechanization of the standard rotating pan system. Here progress has been made by
   a. Automatic filling and discharge of the pans by tilting.
   b. Automatic control of the application of chocolate or syrups using spraying and air-drying techniques controlled by electronic programming and instrumental measurement of relative humidity within the pans.
      By this automation one man may handle production from a number of pans but the system is more suitable for the manufacture of large quantities of the same line.
2. New principles
   a. Redesign of the rotating "pan"—the cylindrical machine by Dumoulin is an example (Fig. 19.37). Another example is the Driacoater, which uses a perforated drum; the process is computer controlled.
   b. Coating on a continuous belt. The principle is to do the coating in the loop of a continuous belt. Several machines have appeared and there are problems with this method.

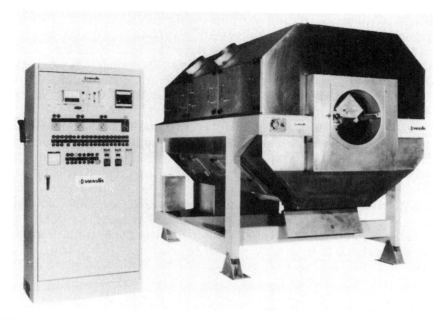

Fig. 19.37.   Cylindrical Rotating Pan I.D.A.-X
The interior of the cylinder is provided with baffles to ensure tumbling of the product. Sprayers extend the length of the cylinder. Discharge is automatic. Controlled air drying with twin circuit air distribution

*Dumoulin et Cie, La Varenne, France*

## CRYSTALLIZATION OF FONDANTS AND OTHER CONFECTIONS

Cremes and other centres that have been cast in starch will dry out under average ambient conditions unless they are given a protective coating. Normally, they would be covered with chocolate or a compound or pastel coating. These are fat based and retard the loss of moisture from the centre.

Protection will also be given by forming a sugar crystal layer over the surface. Formation of this layer by steaming and sanding has already been described under gums and pastilles.

This method is not suitable for cremes or pastes, such as marzipan.

Wet crystallization is used in these circumstances and the process consists of growing a continuous sugar crystal layer over the surface of the confections by immersion, in single layers, in a solution of sugar slightly supersaturated. This not only gives protection but greatly enhances the appearance, giving a sparkle to the surface.

The preparation of the syrup must be carefully done to avoid premature crystallization, and this means boiling quickly to dissolve all solid sugar, and to prevent inversion, cooling without movement, and filling the crystallizing bath or trays with as little disturbance as possible.

Vibration from machinery or other causes in the department must be absolutely avoided.

The solubility of sugar at 20°C (68°F) is 67 percent and to promote controlled crystallization on the surface of fondants the syrup concentration should be just above this.

## Preparation of Syrup

The highest quality white sugar is required to make the syrup:

| | |
|---|---|
| Sugar | 45.3 kg (100 lb) |
| Water | 24.5 kg (54 lb) |

The sugar is dissolved and heated rapidly to boiling, when any scum is removed and crystals adhering to the inside of the pan dissolved by washing down or steaming. Boiling is continued until a concentration (determined by refractometer) of 68 to 70 percent is reached. The higher the concentration, the coarser is the crystal layer. The layer is also formed more quickly but the appearance suffers and the layer may contain agglomerates of sugar.

When the correct concentration is reached, the hot syrup is passed through a fine sieve into a cylindrical pan and allowed to cool in the crystallizing room without disturbing. The pans are covered with lint-free cloth—not lids, which cause condensation and a weaker solution on the surface of the syrup.

## Crystallizing

There are two methods of crystallizing—the tray and tank systems.

**Tray Crystallizing** Rectangular trays and wire mesh inserts are used; the trays have sloping sides and both the mesh and trays are heavily tin plated or are stainless steel.

Dimensions are approximately 15 in. by 9 in.

The cremes are arranged on the wire mesh at the bottom of the tray.

The trays are placed on open racks in a room held at a constant temperature between 21°C and 22°C (70 to 72°F). The prepared syrup is poured over the fondants until they are covered to a depth of a $\frac{1}{4}$ to $\frac{1}{2}$ in.: a second wire mesh is then placed over the top of units. The trays of syrup and fondants are allowed to stand undisturbed for up to 16 hr depending on the thickness of crystal layer required.

At the end of the crystallizing period, the units should be coated with a continuous layer of crystal and the trays are then tilted to drain off the syrup from the corners. The tray racks are provided with slots to hold the trays in a sloping position and troughs to collect the syrup. When draining is complete, usually 4 to 5 hr, the trays are placed flat again, the top mesh removed, and left to dry for 16 to 24 hr. During the drying process, it is customary to tap each tray periodically to release any cremes adhering to the mesh, and it will also help to make the crystal layer more continuous over the surface in contact with the wire.

**Tank Crystallizing** The second method of crystallizing is more in keeping with modern methods of production and uses a tank system. The principle is to use a nest of wire baskets held in a large cage, the baskets being packed in single layers with the fondants to be crystallized. The cage is controlled by a hoist that enables it to be let down into a tank containing the crystallizing syrup. The syrup for the tank is prepared in a separate pan with the same meticulous care as described above and is fed through a fine sieve from the pan to the tank. After the crystallizing process, the cage of baskets is lifted out of the tank and allowed to drain and dry on a large tray adjacent to the tank. The crystallized fondants should be ready for removal after 24 hr but good air circulation is necessary.

The syrup from the tank and draining trays is discharged for reboiling or reclaiming. The crystallizing tank must be steamed out and no crystal residue whatever must remain or trouble with the next batch will arise.

The process of crystallizing can give endless trouble if methods are slipshod, and before leaving the subject, the following points should be emphasized:

*Syrup* This must be prepared with great care as described and must not be allowed to come in contact with any sugar crystals while cooling and must never be stirred or vibrated.

It is best not to use the syrup more than once although there are manufacturers who reboil the syrup and add small amounts of sodium citrate buffer salt to reduce inversion. The syrup should have a pH in the region of 6.0 (±0.2) for best results and this gives good color retention and minimum inversion.

Repeated use gradually darkens the syrup, and where this has occurred, it can be used for crystallizing colored fondants or jellies. New syrup must be used for white fondants such as mints or other high-class pale-colored fondant.

By far the best procedure is to use freshly prepared syrup. The larger manufacturers usually have reclaiming equipment and very little cost is involved in disposing of spent crystallizing syrup with solutions of other confectionery scrap.

*The Crystallizing Department* Rarely is enough care taken with the construction of the crystallizing room and too often it is part of another confectionery department. This is false economy and results in poor work and a lot of waste. The essential points are:

1. The room must be free from vibration.
2. The syrup boiling section must have good extraction ducts to remove steam and preferably be partitioned from the crystallizing room.
3. The crystallizing room must have an even temperature of between 21 and 22°C (70 and 72°F) and it is preferable to have it thermostatically controlled. High humidites are most undesirable and a range of 50 to 60 percent should be maintained.

*Cleanliness* Without clean working success will never be achieved. All tanks, trays, or utensils used for crystallizing or for containing the new syrup must be absolutely free of sugar residues.

Dust from starch rooms or elsewhere must not settle on the crystallizing syrup or equipment.

## LICORICE

Licorice confections are very popular, particularly the "Allsorts" type where licorice pastes are sandwiched with other confectionery pastes. Many other products make use of the medicinal properties of licorice, as in throat lozenges and gums and for aperients.

The licorice plant, botanically, is Glycyrrhiza and there are many species, Glycyrrhiza glabra being the most well known. It grows wild in the subtropical areas of Europe and Asia and is a leguminous weed. The root will, in favorable circumstances, extend to 25 ft below ground and varies in thickness from thin fibers to several inches in diameter.

It is mostly harvested in the autumn by peasant populations by the process of hand digging, and delivered to central collecting stations and dried to about 10 percent moisture.

Licorice is unique as it is the only known plant that contains considerable quantities of the glycoside called glycyrrhizin, which varies from 6 to 14 percent according to the origin of the root. This is the sweetest chemical substance found in nature, being about fifty times as sweet as sugar.

All licorice products are prepared from an extract of the root and this is made by grinding the root, leaching with hot water, and evaporating the solution. Modern processes use counter current extraction methods and multiple-effect evaporators for maximum economy. The concentrated extract is then dried to produce block or granulated licorice.

Spray-dried licorice powder is also available. It is free flowing with a moisture content of 3 to 5 percent. It is easier to dissolve than block juice that requires breaking and soaking.

A typical composition of block licorice is, in percent:

| | |
|---|---|
| Moisture | 18 |
| Glycyrrhizin | 18 |
| Natural sugars | 11 |
| Gums, starch | 28 |
| Color and other extracts | 20 |
| Ash | 5 |

A licorice derivative known as ammoniated glycyrrhizin is finding application as a special sweetener in baking and confectionery products. It is stated to enhance cocoa and chocolate flavors and has the following approximate composition, in percent:

| | |
|---|---|
| Moisture | 10 |
| Glycyrrhizin | 81 |
| Natural sugars | trace |
| Gums, starch | 1 |
| Ash | 0.5 |
| Other extracts | 7.5 |

## Manufacture of Licorice Confectionery

**Ingredients**  The important ingredients of licorice paste are:

1. Licorice block juice (or other prepared licorice as mentioned previously).

2. Flour—preferably a strong wheat flour.
3. Brown sugar—the various brown grades described under confectionery ingredients are all suitable. Molasses, treacle, and golden syrup may also be included.
4. Glucose syrup. High DE glucose is preferable or dextrose hydrate may be used.
5. Gelatin. This gives body to the final product and acts as a binder. A medium-grade gelatin is satisfactory and no advantage is obtained by using high-grade, light-colored material.
6. Color. Caramel color is generally used in addition to food-grade black dye, as it softens the intensity of the dye and imparts luster to the product.
7. Emulsifier. Glyceryl monostearate is usually added to prevent the licorice from sticking to the teeth. About 0.1 percent is adequate and sometimes 2 to 3 percent of fat (hardened palm kernel oil) is emulsified into the mix. This acts as lubricant and will aid extrusion.
8. Flavor. Aniseed oil is regarded as the flavor for licorice.

**Manufacture** The fundamental part of the process is to gelatinize completely the starch in the flour ingredient—this applies to all confectionery—but in licorice bad dispersion and cooking will show as white flecks or lumps in the end product.

**Recipes** Published recipes for licorice pastes vary considerably, some are deficient in licorice content and overburdened with starchy material. Those shown in Table 14.7 are typical of the proportions of various ingredients used.

TABLE 19.7. LICORICE—PROPORTIONS OF INGREDIENTS

| | Normal range, % | Published recipes, parts by weight | | |
|---|---|---|---|---|
| | | 1 | 2 | 3 |
| Flour | 30–40 | 24 | 12 | 30 |
| Sugar (various types) | 50–60 | 16 | 6 | 22 |
| Glucose (corn syrup) | | 8 | 3 | — |
| Treacle | | — | — | 14 |
| Caramel | 6–25 | 12 | 6 | 3 |
| Block licorice juice | 3–6 | 4 | 2 | 2 |
| Gelatin (150 Bloom) | 0.5–4 | 8 | 2 | 1 |
| Residual moisture | 17–18 | | | |
| Water | | 30 | 18 | 30 |

## Process

**Licorice Base for "Allsorts" and Similar Products as Recipe 1** The gelatin is soaked overnight in some of the water. The sugar, glucose, and block licorice are dissolved in the remainder of the water by heating. This solution should be sieved and cooled to about 37.7°C (100°F) and the soaked gelatin and the flour added, a portion at a time, so that a smooth slurry is obtained.

Cooking is then commenced and continued with stirring until the flour is completely cooked and the correct final moisture achieved.

Modern methods of cooking starch/sugar slurries employ heat exchangers, which give complete gelatinization with a lower proportion of water. The licorice produced may be poured onto tables and cut or extruded and in some cases it is dried off further in a hot room.

Licorice production, in its various forms, is specialized in certain companies and there are very many processes and formulations.

The plastic nature of licorice confectionery lends itself to the manufacture of tubes, strips, and other shapes particular to children's lines.

Jackson (1973) gives more detailed description of these products.

## CHOCOLATE AND CONFECTIONERY SPREADS—CHOCOLATE SYRUPS

These spreads may be considered confectionery or bakery products. The composition and processes are closely associated with confectionery but the uses are almost always related to cakes, sponges, and bread, and occasionally drinks.

Chocolate spreads became popular during World War II and in the immediate postwar period when there were still food shortages in Europe.

### Water-Based Spreads

The most popular were those made from cocoa powder dispersed in a syrup of sugar, invert sugar, and sometimes glucose syrup. The composition of these spreads is similar today.

The proportion of cocoa powder is usually between 18 and 22 percent, and is low fat (10 to 12 percent cocoa butter). The concentration of the syrup is important and ranges between 75 and 77 percent—syrups below 75 percent are vulnerable to microbiological attack. High-concentration syrups are too viscous for spreading and the same applies if the cocoa content is increased.

The syrup phase concentration depends on the proportion of invert and glucose syrup in the mixed syrup.

Natural cocoas are normally used in these spreads but alkalized cocoas may also be included to give darker or redder shades. Occasionally, light-colored cocoas derived from Criollo beans are used.

Manufacture of these spreads is a simple process.

The cocoa powder is placed in a low-pressure steam, stainless-steel pan with contrarotating scraper paddles, and a small amount of syrup added. This mixture is run out to a smooth stiff paste at a temperature not exceeding 49°C (120°F). The remainder of the syrup is added slowly until the cocoa paste is dispersed completely. It is general, though not essential, to heat the mixture to 180 to 190°F to sterilize and develop flavor, but if this is done, some moisture loss will occur and this must be allowed for in the recipe. The heating time must also be strictly controlled. It is usual to add flavors such as vanilla at the end of the process. The mixture is cooled to about 43°C (110°F) and deposited into waxed cartons with screw top lids, which allows a certain amount of ventilation. If jars with sealed lids are used, these should be sterilized or there is a risk of mold growth.

In some spread recipes, modified starch, alginates, or low methoxyl pectin are added, usually in conjunction with a reduced cocoa content. These tend to produce a paste rather than a viscous syrup—they may spread better, but lack flavor.

Milk chocolate spreads may be manufactured by including milk powder as partial replacement for the cocoa. Nonfat powders are preferable as the inclusion of milk fat in a water-based spread may lead to rancidity or other off-flavors.

## Fat-Based Spreads

Fat-based spreads may be made where the continuous syrup phase in the water-based spreads is replaced by a soft vegetable fat.

The fat must have properties that give good spreading qualities at normal temperatures. It must not show oiling out or separation of the liquid fractions and there must be no shrinkage due to excessive solidification of the higher melting components.

The most successful fatty spreads of recent years have been those containing nut ingredients. All nuts contain natural oils and those most applicable to spreads have the following approximate contents:

Hazelnuts 64 percent
Almonds 55 percent
Peanuts 45 percent

Probably the original nut spread would be peanut butter, which is consumed in vast quantities in the United States.

The product made from the nut alone is subject to oil separation. Physical modification of the oil has resulted in a more stable material suitable as a spread for crackers, and cakes and as a filling for confectionery items.

Hazlenut/cocoa spread has become very popular and proprietary brands on the market now sell very well. Fatty spreads are also manufactured using milk powder, without nuts or cocoa, and with cocoa powder alone.

According to Heemskerk (1981), the following compositions of fatty spreads are representative (Table 19.8). The fat ingredient is mostly responsible for the properties of the final spread and must consist of:

1. A liquid oil stabilized by the inclusion of a small percentage of a high melting component; or
2. A natural or tailormade oil, which has the appropriate stabilizing fractions so that it has satisfactory physical properties under ambient temperatures.

## Manufacture of Fatty Spreads

In some respects, processing resembles that applied to chocolate or compound coatings. The solid ingredients are mixed with a propor-

TABLE 19.8. COMPOSITION OF FATTY SPREADS

| | Type of spread | | |
|---|---|---|---|
| Ingredient | Nut | Milk | Cocoa or chocolate |
| Fats % | 25–40 | 30–40 | 30–40 |
| Nuts % | 5–15 | — | — |
| Cocoa powder % | 5–10 | — | 10–18 |
| Milk solids % | 0–10 | 18–25 | 0–5 |
| Sugars % | 40–55 | 40–55 | 40–55 |
| Lecithin % | 0.3–0.6 | 0.4–0.6 | 0.3–0.6 |
| Flavor | as required | as required | as required |

Source: Haemskerk (1981), Friwessa Zaandam, Holland.

tion of the fat to give a paste of consistency suitable for roll refining. After refining, the remainder of the fat is added and the mixture conched or otherwise suitably mixed to give a smooth paste. The nuts are normally blanched and roasted to a degree sufficient to enhance the nutty flavor.

Other mills, such as the Stephan or MacIntyre, may be used, when all the ingredients are charged in at the outset of processing. These machines are described in other parts of the book.

## Chocolate Syrups

These are basically similar to syrup spreads with a much higher water content. Consequently, they are sold in sterile cans or jars. Superfine cocoa is used and stabilizers are usually added to prevent separation on storage.

## RECLAIMING OF CHOCOLATE AND CONFECTIONERY "SCRAP" (REWORK)

In all manufacturing processes, material is produced that cannot be included as part of the final article for sale. Food, including chocolate and confectionery, is no exception.

In mass production food processes, particularly those involving cutting and stamping, offcuts are produced in considerable quantity. This material is quite wholesome and should be returned to a previous stage in the process in a constant proportion so that the quality of the saleable product does not suffer in any way.

Psychology plays a part in the handling of this material. If it is consigned to containers, even temporarily, which are in poor condition, dented, with ill-fitting lids, or perhaps dirty, the contents are immediately downgraded in the eyes of the workers. In the worst circumstances, other trash will be deposited on what is a perfectly edible product.

Assuming care is taken, reincorporation of substances such as chocolate bars, cooky dough, or marzipan is merely a process of blending with new production in a suitable mixer.

When it comes to chocolate-covered confectionery, aerated products, high boilings, and highly flavored and colored cremes, the problem becomes more complicated and special processes have to be used.

The word "scrap" is now considered a bad description and the same translated into other languages implies material that is waste, unwholesome, and not suitable for consumption. The EEC directive states that the word "rework" should be used and this causes no problems with translation.

However, in this chapter the word "scrap" is perpetuated. By some reclaiming methods, a good-quality material is obtained by filtering and decolorizing a product that otherwise would be unfit for reincorporation direct.

## Foreign Matter, Sanitation

As suggested above, the handling of scrap will involve a risk of contamination with foreign matter or microorganisms however well operators are trained in sanitation.

Some publications recommend segregation of different types of scrap so that each can be returned to its appropriate formulation. Experience has shown that this procedure is generally unworkable unless there is a very small number of products.

The following methods of reclamation have been proved and should be followed by all companies concerned with good manufacturing practice and quality control.

All processes that do not allow for an immediate return of scrap to manufacture should include filtration or screening and sterilization by heating.

In the chapter on "Hygiene," various devices for foreign matter prevention for inclusion in standard equipment and processes are described and some of these are applicable to reclaiming also.

## Reclaiming of Scrap as Syrup

Confectionery fondant centers, hard candy, aerated confections, and jellies that are not chocolate covered can be returned to any process as a basic raw material if converted to a colorless, flavorless neutral syrup.

The basic process of reclaiming as syrup consists of dissolving the scrap in water, screening to remove foreign matter and pieces of nut and fruit, adjusting pH, and then adding decolorizing carbon and filtering. With fondants this a straightforward procedure but if the scrap contains egg albumen, other whipping agents, or fruit pulp, a modified procedure is necessary to aid filtration. With some confections such as acid-flavored hard candies, fruit-filled fondants, or fruit jellies, the proportion of invert sugar present or formed in the reclaiming process is considerable and it is preferable to convert the scrap to an invert syrup that contains invert sugar and any glucose syrup in the original recipe, but no sucrose. This syrup can be used in the original recipe to replace invert sugar and, if necessary, the glucose can also be allowed for but this usually can be ignored if the reclaim syrup used is not more than 10 percent of the basic recipe.

## Enzyme Filter Aids

Protein matter, particularly egg albumen, is a great nuisance in causing syrup foaming and retarding filtration in the reclaiming process.

Trypsin and pepsin (peptide peptido-hydrolases) can help considerably in making a syrup filterable, and the choice depends on whether acid or alkaline conditions are most suitable for the process. The optimum conditions for trypsin are pH 8.5 at 40 to 46°C (104 to 115°F) but good results can be obtained at pH 7.5 to 8.0 over a period of 48 hr at the temperatures quoted, with a 50 percent syrup concentration. The lower pH reduces alkaline caramelization, which can be serious at pH 8.5. With pepsin, a low pH is required (1.8 to 2.0), and in a syrup of 50 percent concentration at 40 to 46°C (104 to 115°F) a digestion period of 48 hr should break down all protein matter. This latter process is preferable provided inverted syrups are required. A check on the end point of digestion in both cases can be made by adjusting a sample of the syrup to pH 5.0, adding 1 percent activated carbon, filtering, and the filtered syrup subjected to gentle boiling. No persistent foam should be produced, and the test filtration should have taken place without clogging of the filter paper. When protein has been removed, the batch of syrup is neutralized, active carbon added, and filtered.

Pectinaceous material causes filtration difficulties, particularly if the proportion of jam confections is high. Pectin enzymes are a useful aid. The natural enzymes that degrade the various forms of pectin in fruit pulps and juices are essentially pectinesterases and polygalacturonases but when it is necessary to assist the removal of pectin, pectolytic enzymes produced synthetically by the action of molds are used. Commercial pectinase preparations vary appreciably and the efficiency depends upon the exact nature of the enzymes present. For reclaiming scrap, the commercial preparation has a high polygalacturonase content and some pectin methyl esterase activity. Pectin methyl esterase hydrolyzes the methyl ester groups in pectin giving low methoxyl pectin. Pectin polygalcturonase hydrolyzes the 1.4 glycosidic link in the polygalacturonic acid molecule with the formation of polygalacturonic acids of lower molecular weight and some galacturonic acid. Activity is greatest in the normal pH range for fruit (3.3 to 5.0) at temperatures of 57 to 63°C (135 to 145°F) and 30 to 60 min are required for completion of pectin degradation in a 50 percent syrup. The quantity of enzyme required depends on the fruit substance present and can be found by experiment and reference to the enzyme manufacturers' literature.

Enzyme application has made many advances in recent years and specific enzymes that work in near-neutral solutions are now available for the treatment of scrap containing starch, gelatin, and pectin.

For gumdrops and starch jellies 1 percent of a bacterial alpha-amylase is used with processing at 77 to 80°C (170 to 175°F), pH 6.7 to 7.0, for 20 min. Scrap containing gelatin, is processed with 0.25 percent bacterial protease pH 7 at 44 to 52°C (111 to 125°F) for 20 min. Various pectinase enzymes are also available. (Novo Enzymes, Denmark).

## Invert Syrups—Use of Invertase Enzyme

The traditional method for the manufacture of invert sugar from pure sugar (see "Invert Sugar") is by acid hydrolysis, and with pure sucrose the quantity of acid required is quite small. When invert syrups are made from scrap confectionery, inversion by acid is greatly retarded by the presence of small amounts of mineral salts, and the use of invertase is preferable because it avoids excessive amounts of inverting acid. The technologist will require to work out the optimum conditions for dealing with a particular type of scrap and the effects of concentration, pH, and temperature are given in the sections on "Invertase" and "Invert Sugar." The economics of using syrups of higher or lower concentration depends on how the scrap is dissolved or used after inversion. Similarly, the actual quantity of invertase used in relation to the time of inversion needs careful costing. Equipment for the inversion of syrups must be designed to have thermostatic control, adequate mechanical stirring to prevent local overheating, and automatic pH control. After inversion, the syrup must be heated to inactivate the enzyme; this usually occurs automatically when the syrup is used in subsequent processes or during concentration.

## Process and Equipment

The equipment for reclaiming scrap as syrup need not be complicated or expensive, but economies can be realized by installing

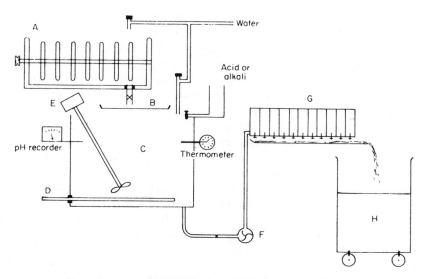

Fig. 19.38. Diagram of Plant for Reclaiming Scrap

ancillary equipment for hoisting drums of scrap and for pumping the reclaimed syrup. A diagram of the type of equipment that has been used for many years is given in Fig. 19.38.

(A) is a horizontal steam-jacketed mixer with rotating paddles. Water is added to the mixer to a predetermined level followed by a weighed quantity of scrap. Since the scrap may be in large drums, a hoist is necessary and a steam jet will assist in obtaining complete discharge. Steam is applied to the jacket, the mixer set in motion, and when the scrap is dissolved, the syrup should have a concentration of 50 to 55 percent, which is checked by refractometer.

The hot syrup is then discharged to tank (C) through a fine mesh sieve (B) that removes foreign matter, pieces of nut, fruit, and other solids. When the tank is full, the pH is adjusted to 5.0 after the mixer (E) has been set in motion and the concentration adjusted to 50 percent. If enzyme treatment is necessary, acid or alkali is added to obtain the pH appropriate for the enzyme being employed, but after enzyme treatment the pH is readjusted to 5.0.

(D) is an electric heater thermostatically controlled to obtain the correct temperature conditions for enzyme treatment or for decolorization.

When the enzyme treatment (if necessary) is completed and the pH adjusted, decolorization is obtained by the addition of powdered active carbon; there are various brands on the market.

The quantity required depends on the amount of color to be removed and is usually between 0.5 percent and 1.0 percent of the syrup by weight, but the most economical and effective quantity can be determined by experiment.

There is also some latitude allowable in the temperature of the syrup at the decolorizing stage and this usually varies between 49°C (120°F) and 71°C (160°F), the optimum temperature being found by experiment for any particular mixture of scrap. At the higher temperatures, decolorization will occur more rapidly and at all temperatures the reaction should be complete in 15 to 30 min but a check on a filtered sample of the suspension will show how the process is proceeding. The syrup with the carbon is then pumped (F) to the filter press(G) and the clear syrup collected in the trolley tank (H). Filtration is greatly assisted by mixing a filter aid with the carbon in the decolorizing tank in a quantity equal to the amount of carbon.

Filter aids are diatomaceous earths, and to effect an economy they may be mixed with a proportion of pumice powder.

Sometimes these filter aids are suspended in water or clear syrup and pumped to the filter press so that a bed of filtering material is laid down on the filter pads before the carbon suspension is filtered.

An alternative procedure is to add the majority of the filter aid with the carbon in the first batch of syrup pumped to the clean press. Subsequent batches of syrup will require much smaller quantities of filter aid until the press chambers have become filled. The same also applies to the quantity of decoloring carbon required, because the layer of carbon in the press still retains a lot of its decolorizing power.

Some syrups containing caramelized sugar, or high boiled sugars that have been aerated with sodium bicarbonate, are difficult to decolorize completely without using large quantities of active carbon. In such cases, it is possible to use the double decolorizing principle. This makes use of the carbon/filter aid paste discharged from the filter press for the first syrup treatment and the filtered syrup from this first treatment receives a second decolorizing with new carbon. A somewhat different arrangement from that shown, with extra treatment tanks, is required for this process and the economics of extra decolorizing must be studied closely.

When sodium bicarbonate has been used in a confectionery product, much effervescence will take place when acid is added for neutralization or pH adjustment and it is very important then to ensure that an unmanageable foam is not produced. Enzyme treatment or the addition of antifoam will help and a small amount of fat addition is usually effective for this purpose. Laboratory experimentation before adopting a reclaiming process on a commercial scale must be emphasized. The type of scrap, whether it is to be pulverized, and how it shall be dissolved can all be worked out on a small scale and the economics studied.

## Scrap Containing Fat

Caramel, fudge, nougat, and other confectionery containing a high proportion of fat are not satisfactorily treated by the above method as the fat will partially separate and cause embarrassing scums on the surface of the tanks. In such cases two methods can be adopted: (1) The syrup at 50 percent concentration is heated and allowed to stand in tanks or drums until fat has risen to the surface, when it is skimmed off. This process is costly because of the handling and loss of fat, as reuse of this fat is not recommended after this treatment and it can only be sold for refining. (2) The caramel or other material is dissolved hot at about 60 percent concentration, screened, and instead of going through the decolorizing process, is used in this form as a base ingredient for certain specific recipes of the caramel and fudge type. This process should apply only to fresh scrap.

It is possible to dissolve solid scrap in a new boiling of caramel but screening to remove foreign matter is more difficult in a finished boiling and it is not easily applicable to a continuous boiling process.

## The Use of Reclaimed Syrup

The decolorized syrup must be used within two or three days as it will not keep at 50 percent concentration, and where there is large fondant creme manufacture, this presents no problem.

The process can be operated at 60 percent concentration but decolorizing and filtration are more difficult. Normally, the filtered decolorized syrup will have a pH somewhat higher than 5.0 and can be used without further neutralization for most confectionery if the proportions are not too high. If the syrup is to be stored, it is much better to convert it to an invert syrup and concentrate to 75 to 78 percent, and to do this the filtered decolorized syrup is treated with invertase enzyme in thermostatically controlled tanks followed by vacuum concentration. This is described under "Invert Sugar."

## Reclaiming Chocolate-Covered Scrap

The return of chocolate-covered confectionery scrap to the original product can be brought about by several methods but some manufacturers of high-quality products prefer to make a separate line from the scrap and sell this through cheap stores or markets under a name that does not identify the origin. This is not a profitable method as a rule.

The types of line that will accept a variety of scrap in the confectionery center are limited and usually they are confined to caramels, fudges, pastes, and dark-colored nougats or nougatines. It is better from a quality standpoint and more economical to return the processed scrap to the chocolate if legislation permits.

## Reclaiming of Scrap as Refiner Paste

Chocolate-covered cremes and confectionery units are reduced to a paste in a warm melangeur or edge runner with the addition of either fine sugar, chocolate refiner paste, or cocoa butter, or a combination of these, depending on the nature of the confectionery scrap. This process removes most of the moisture in the confectionery centers, and pulverizes any nut or similar ingredient as well as ensuring a good mixing. This paste is then put through a roll refiner. This refined paste can be mixed with a confectionery center base where light colors are not required or added to chocolate in quantities not exceeding 5 percent. In this proportion, the presence of small amounts of other sugars (invert, glucose), and perhaps foreign oils and fats, will have negligible effect on chocolate viscosity, flavor, or other properties.

But this process has two failings: (1) it does not take into account foreign matter that may be present, and can even grind it into the paste; and (2) it does not sterilize the scrap.

Since most of this foreign matter is likely to be metallic, this can be mostly removed after passing the scrap through an electronic metal detector before the melangeur process.

It is always possible that microorganisms will contaminate the paste. Although they will not propagate, they will certainly not be destroyed by the low process temperature. Subsequently, if the paste is incorporated in a higher-moisture confection, they may become active.

## Reclaiming of Scrap as "Crumb"

A method that has been used very successfully employs the milk crumb process described under "Chocolate" and this has distinct advantages over the paste process because it:

1. Removes foreign matter.
2. Sterilizes the scrap.
3. Removes the majority of flavors by steam distillation.

This process is available to those manufacturers who make their own milk crumb by the vacuum process. It is expensive to install equipment specially for this method of reclaiming but some factories have done so.

The principle of the process is to replace the fat-free cocoa content of some of the liquor used in the crumb process by the fat-free cocoa present in the scrap. The sugars in the scrap (calculated as sucrose) similarly replace some of the sugar in the crumb recipe. By this process a crumb can be produced that contains as much as 15 to 20 percent of chocolate-covered scrap and in the chocolate-making process may be blended with crumb containing no scrap so that the finished chocolate contains a maximum of 5 percent.

## Process

In the manufacture of milk crumb, the milk is first evaporated to about 37 percent total solids and the sugar is then added, followed by condensation. A weighed quantity of scrap that is calculated to replace certain quantities of liquor and sugar in the crumb batch is dispersed with water in a separate kettle, raised quickly to the boil, and this slurry is then sucked or pumped through a fine-mesh screen to the condensing kettle. Condensing proceeds as for normal crumb and the volatile flavors are removed with the evaporated steam. Foreign matter, nuts, and other solids are extracted on the screen and the preboiling and condensation are sufficient to destroy all microorganisms. The crumb process then proceeds normally through the condensing, kneading, and drying stages and a crumb is produced that is practically indistinguishable from that made without scrap. It is sterile, has good keeping properties with very little variation in composition, and can be used without difficulty in the normal milk chocolate manufacturing process.

A similar process can be adopted without the milk ingredient and a dark crumb is produced that can be used in dark chocolate manufacture.

In this method of reclaiming, some sorting of the constituents of the scrap is desirable, for example, milk-chocolate and dark-chocolate-covered cremes should be kept separate or blended together in some definite proportion so that the fat-free cocoa content is known. Filled blocks or cremes made on shell equipment usually have a higher chocolate content and these are best kept separate from enrober covered chocolates. By attention to these points, it is possible to make a crumb with very constant composition.

It is surprising how effectively flavors are removed in this process, including peppermint, which by other reclaiming methods is impossible to eliminate and necessitates its use only in peppermint confectionery.

Some flavors are not removed and these include caramelized sugar, but because some caramelization takes place in the crumb process itself this extra flavor may be tolerated to a limited extent. A few synthetic flavors are not volatile in steam and tests can be made to prove this. In many cases, a volatile flavor equally good and not fugitive in the confectionery product can be found.

## Legislation

Some countries that have strict laws regarding chocolate composition do not permit the return of rework as an ingredient of the chocolate but insist that it must be used only in centers. It is regarded as an adulterant of chocolate, which must be made from pure cocoa nib, sugar, and, if milk chocolate, milk solids in addition.

While there may be some reason for such argument in the case of solid chocolate bars or blocks, it is difficult to substantiate if the chocolate is mixed with nuts, raisins, cookies, or cereals, or is used as a covering for various confectionery centers.

This is a useful cost saving by the inclusion of rework and quality control checks have shown that no deterioration is detectable when rework not exceeding 5 percent is incorporated and the chocolate used for covering or recipe mixtures.

## MULTIPLE CONFECTIONERY BARS

Many confectionery bars have appeared on the market that consist of layers of different types of product in contact with one another. Many of these are attractive but often the nontechnical members of a company are unaware of the limitations that must be placed on the combinations that may be used. Many skilled sugar confectioners also are not at all sure of permissible combinations and lack of knowledge of the theory behind the problem can lead to very unsatisfactory shelf life in the end product.

The shelf life of a multiple confection is related to the equilibrium relative humidity (water activity) of the constituent layers. Equilibrium relative humidity (ERH) and its significance are explained in several sections of the book.

Confectionery products with similar ERH may be placed in contact without significant moisture transfer. If there is a wide difference, then moisture will transfer from the confection with the high ERH to that with the lower. In the worst cases, this can have disastrous effects on quality and shelf life. The following examples, which were actually suggested by a marketing organization, illustrate what might happen.

1. Wafer cookies are layered with fondant creme. After a short period of storage, the wafers became sodden and tough. The creme became dry and short. Wafers have an ERH of about 20 percent and fondant creme about 70 percent.
2. A honeycombed boiled sugar is layered with a soft jelly. The high ERH jelly will transfer moisture to the boiled sugar. The honeycomb structure will collapse completely to a syrup. If it was a chocolate-covered bar, it would become hollow. Boiled sugar has an ERH of about 25 percent, and the jelly, 70 percent.

The equilibrium humidity of a confection is controlled primarily by the concentration of the syrup phase, and to a lesser extent by its composition. Sucrose crystals in fondants and fudges, and fat in fudges and caramels, play no active part. The higher the syrup concentration, the lower is the equilibrium humidity and the more hygroscopic the confection. The *proportion* of each confection must also be considered in relation to moisture transfer, and sometimes the transfer of moisture from one layer to another does not detract from the quality of the whole bar. An example is a thin layer of caramel on top of a bar of soft nougat. The nougat will have an ERH of about 65 percent and the caramel 50 to 55 percent. Moisture will transfer from the nougat to the caramel, raising the ERH and making the caramel softer. Moisture will be lost by the nougat but because of the small proportion of caramel layer, the amount of moisture lost by the nougat is small and its texture is not changed significantly. If, however, the proportions of caramel to nougat were equal, it is likely that the caramel would become liquid and the nougat hard.

The low equilibrium humidity type of confection is generally brittle or hard (butterscotch, hard candy), whereas the medium ERH type is chewy (caramel, nougat). The higher ERH type contains crystals (fudges and fondants), or may be noncrystalline but high in moisture (jellies, Turkish delight).

Moisture content alone, therefore is not a reliable guide to the ERH but an increase in moisture in any type of confection will raise the ERH and vice versa.

A list of the ERH values of various confections is given in other sections of the book.

## Inclusion of Other Ingredients

Nuts and cereals (e.g., corn flakes, rice krispies) must be used in combination with confections of low equilibrium humidity or they lose crispness and are more prone to flavor deterioration. Nuts can tolerate a higher humidity than cereals, but when the nuts are roasted, a low-humidity confection is desirable.

Cookies are similar to cereals and should be combined with low ERH confections although some fatty-type cookies can be combined with some of the medium ERH type without becoming objectionably soft.

Dried fruits, such as raisins or preserved peels, are sometimes mixed with low-moisture confections. This leads to the transfer of moisture from the fruit pieces to the confection. In some instances, the fruit pieces become objectionably hard.

## Insulating Layers

In some instances, a fatty layer will provide a barrier to retard, if not prevent, moisture transfer. The effectiveness of this barrier will depend on its continuity, thickness, type of fat, and probably the presence of other materials, such as sugar or milk powder in the fatty layer.

There is some evidence to suggest that a high-melting-point fat is more effective than one of low melting point, although there are obviously limits on the score of palatability.

# REFERENCES

A list of textbooks containing formulations and other information is given in the Appendix.

Aerated confections—*Hyfoama Manual.* Lenderink & Co., Schiedam, Holland.
Altvater, F. 1974. Candy Extrusion. Manf. Conf. June, U.S.A.
Anderson, G. 1968. Manf. Conf. 39, U.S.A.
Clark, G. L., and Ross, S. 1940. *Ind. Eng. Chem.* 1954.
Heemskerk, R. 1981. Speciality Fats Seminar. Friwessa Zaandam, Holland.
Jackson, E. B., and Lees, R. 1973. *Sugar Confectionery and Chocolate Manufacture,* pp. 299–315.
Lees, R. 1976. Conf. Prod., London.
Minifie, B. W. 1970. Marshmallow—technology and methods of manufacture. Manf. Conf. 31, U.S.A.
Novo enzymes—Novo Industri A/S, Bagsvaerd, Denmark.
Panning machinery—Dumoulin, La Varenne, France; Driam GmbH, Eriskirch, Germany.
Richardson, T. 1984. Manf. Conf. 47, U.S.A.
Smith, O. B. 1979. Advantages and future trends in food extrusion. *Proc. IFST* London.
Tabletting machines—Manesty Machines, Liverpool, England; Bramigk Ltd., London, England.

## Confectionery Plant and Machinery Information

*Directory of Equipment and Supplies.* Manf. Conf. U.S.A. (July 1986 and annual).
*Food Processing Industry Directory,* IPC Consumer Industries Press Ltd., London.
*Silesia Manual,* Vol. 1/1 No. 3. Silesia, Essenzenfabrik, Neuss, Germany.

# General Technology

Part **3**

# 20

# Science and Technology of Chocolate and Confectionery

There are various terms associated with confectionery technology that are used in different parts of this book. Suggestions have been made by readers that it would be useful if these terms could be explained in one place, in addition to the references elsewhere.

Therefore, the part of the second edition of this book that dealt with spoilage problems has been extended to explain the technical expressions so often used in the industry. It is hoped that better understanding of the terms in relation to various processes will result.

## SUGAR CONFECTIONERY

### Solubility, Saturated and Supersaturated Solutions

Most solids will dissolve in water to some extent. The limit to which each will dissolve is called the solubility. Since most solids dissolve to a greater extent when the temperature is raised, temperature must also be quoted when defining solubility.

**Example**  Sugar has a solubility of 67.1 percent by weight at 20°C (68°F). At this temperature with this quantity dissolved, the solution is said to be saturated. No more sugar will dissolve at 20°C.

Some substances will produce supersaturated solutions and sugar will do this.

If a saturated solution of sugar at 20°C is raised in temperature, it will dissolve more sugar until it becomes saturated at the higher temperature.

If this solution is allowed to cool without movement, the sugar will remain in solution and it is then stated to be supersaturated. Supersaturated solutions are unstable and will rapidly deposit the excess sugar when agitated. This is the principle of fondant and fudge manufacture. Each of the sugars described in Chapter 8 has a different solubility and combined sugars have a greater solubility than sugar alone.

## Syrup Phase Concentration

In many confections, there are different "phases." The solid phase is usually sugar crystals, but may also be milk solids, cocoa powder, and so on. These solids are dispersed in the liquid phase, which is a saturated solution of mixed sugars. There may also be small amounts of other soluble ingredients, such as fruit substance. An important factor is the concentration of this liquid phase and it is now universally accepted that it must be a solution of sugars and other minor substances with a minimum concentration of 75 percent as determined by refractometer at 20°C. This induces conditions that will resist most microbiological deterioration and give good shelf life. However, certain enzymes will react under these conditions. Lipase (fat-splitting enzyme) is an example, and some ingredients such as cocoa powders, nuts and nut pastes, and egg albumen may contain active enzymes. It is necessary to ensure when purchasing ingredients that they are free from lipase or, alternatively, the lipase is destroyed. On this point, it must be understood that destruction of lipase in dry powders requires high temperatures, up to 110°C (230°F). Other phases may be present in a confection—fat and air are likely. Some fats are very prone to rancidity and air in finely divided form can accelerate rancidity and cause bad flavors by oxidation.

## RELATIVE HUMIDITY, DEW POINT, VAPOR PRESSURE, WATER ACTIVITY—EQUILIBRIUM RELATIVE HUMIDITY

All these factors have a significant bearing on the successful production and storage of chocolate and confectionery.

## Relative Humidity, Dew Point

Relative humidity is an expression that compares the amount of water vapor present in the air with the amount of water vapor required to saturate the air at the same temperature. Air saturated with water vapor is at 100 percent relative humidity. If air is 60 percent saturated, that means the relative humidity is 60 percent.

Relative humidity is a very important factor in confectionery technology. It is necessary to know and control the relative humidity in packing and stock rooms, hot rooms, and particularly in chocolate coolers.

As the temperature of air is raised, so will the relative humidity be reduced, but the air at the higher temperature will also hold more moisture.

Therefore, in comparing air at 20°C and 50°C at 70 percent relative humidity, the *quantity* of moisture held in the air at 50°C is considerably greater.

Now consider the reverse: As air is cooled, the relative humidity increases and a temperature will be reached when it is 100 percent at which moisture will be deposited. This is known as the dew point.

The practical side of this arises in chocolate coolers, because, if the chocolates, in passing through the cooler, become overcooled, they will be below the dew point of the air in the packing room. This results in moisture depositing on the chocolates and "sugar bloom" will form later.

## Vapor Pressure, Equilibrium Relative Humidity, Water Activity

All water solutions, and water itself, exhibit a vapor pressure. The vapor pressure of a solution depends on the substance dissolved and the quantity in solution.

Saturated solutions of different salts have different vapor pressures, which means that in a closed container the relative humidity above these solutions is constant for each salt at a given temperature.

With confectionery, the vapor pressure depends on the amount of various sugars in solution and bears some relationship to the syrup phase concentration and moisture content.

## Equilibrium Relative Humidity, Water Activity

These expressions mean the same; water activity is expressed as a unit (1.0) whereas equilibrium relative humidity (ERH) is expressed as a percentage (100 percent).

As mentioned above, sugar solutions have a vapor pressure and "solutions" will include low moisture products such as hard candies. As with the salt solutions, these confections, if put in a closed container, will initiate a relative humidity in the air surrounding them. The candy will eventually set up an equilibrium with this surrounding air and the actual relative humidity of that air is known as the ERH.

Alternatively, a series of small, enclosed containers giving a range of relative humidities may be devised using special salt solutions. By exposing small samples of confectionery to the air inside the container, it is soon noted which of these gains or loses weight in the different humidities. By plotting a graph of the weight changes, the ERH can be calculated.

*By definition the ERH of a confection is that relative humidity at which it will neither gain nor lose weight.*

Knowledge of this property is essential in deciding the style of package needed and storage conditions.

The acceptability of different combinations of confections in contact with one another depends on their ERHs.

A summary of ERH/water activity properties of various products is given in Chapter 22. Multiple confectionery bars are discussed in the "Confectionery" section.

Methods of determining ERH are given in the Appendix.

## pH, HYDROGEN ION CONCENTRATION

The definition of pH is baffling to many people. In confectionery, it is probably best described as "active acidity" and in products such as pectin jellies an understanding of pH is very necessary.

The pH range covers acidity and alkalinity, although rarely are alkaline conditions met in confectionery. Alkalized cocoa is an exception, and possibly also are aerated products using sodium bicarbonate—these are slightly alkaline.

Scientifically, pH value is the negative logarithm of the concentration of hydrogen ions. Water, chemical formula $H_2O$, will "dissociate" in the presence of acids into positive hydrogen ions and negative hydroxyl ions and the product of the concentration of $H^+$ and $OH^-$ ions is always $10^{-14}$.

This figure gives the well-known pH range of 14 that is, 0 to 7 acidic, 7 to 14 alkaline. pH 7 is neutral.

There are strong and weak acids. In confectionery, weak acids such as citric and tartaric acids are used for flavouring. Strong acids, for example, hydrochloric acid, are used for special purposes only, such as sugar inversion.

There is a relationship between the pH and concentration of these

TABLE 20.1. ACIDITY OF VARIOUS CANDIES

|  | pH | % Citric acid |
|---|---|---|
| Acid drops | 2.2 | 1.8 |
| Pectin jellies | 3.3 | 0.8–1.0 |
| Fruit jellies | 4.2 | 0.5 |
| Nut pastes, marzipan | 6.0 | — |

TABLE 20.2. RELATIONSHIP BETWEEN pH AND TASTE

|  | Parts by weight to give | |
|---|---|---|
|  | Equal pH drop | Equal acid taste |
| Citric acid | 1.00 | 1.00 |
| Tartaric acid | 0.56 | 1.00 |
| Malic acid | 1.00 | 0.80 |
| Lactic acid | 1.00 | 1.25 |

acids, and to obtain an acidic solution of pH 2.0, citric acid must have a concentration of 2.4 percent whereas with hydrochloric acid only a concentration of 0.03 percent is required. The natural acid of vinegar, acetic acid (another weak acid), requires a concentration of 2.0 percent. In confectionery products, the approximate relationships shown in Table 20.1 are worth noting.

In addition to the actual acidity, the flavor factor of each particular acid must be recognized. If citric acid, which is the acid normally used, is replaced by another acid, a different acid taste will result from the same pH.

According to Buckle (1980), Table 20.2 indicates an approximate relationship between pH and taste.

The taste factor depends to some extent on the type of confectionery product.

## Measurement of pH

Although pH can be measured by the use of colored test papers, these are unreliable if the product is highly colored or very viscous.

The only method that gives consistent results is an electric meter with immersion electrode. A 50 percent dispersion of the confection should be used and measurement should always be at 20°C (68°F).

## Buffer Salts

In many food processes, it is necessary to control the active acidity (pH), and to do this, use is made of salts of strong alkalies and weak acids. These are buffer salts. The substance used mostly in confectionery is sodium citrate but occasionally phosphate compounds are used.

The best example of buffer action is the control of the rate of setting of pectin jellies. With the acid addition alone, setting is too rapid to control the casting process. The addition of a buffer delays this action without changing the character of the jelly.

## Reducing Sugars

This is a term that has an application in the determination of the composition of sugar syrups. Invert sugar in hard candy, for example, is estimated by the presence of reducing sugar. The description arises from the fact that the copper salt in the reagent Fehling's solution is reduced to cuprous oxide.

The common reducing sugars are the monosaccharides (invert sugar), dextrose and fructose, and disaccharides, lactose and maltose.

Sucrose, which is a disaccharide, is not a reducing sugar. Reducing sugars have an aldehyde group in their molecule. This chemistry is explained in detail in books on sugar analysis.

## OPTICAL ACTIVITY

This is another description applied to analytical procedures. It is a property possessed by many substances and solutions (including sugars) whereby a beam of polarized light, in passing through them, is subject to the rotation of its plane. The strength and composition of sugar syrups may be determined in this way, as can the purity of essential oils. Polarized light is light in which the waves are vibrating in a single plane.

Some sugar solutions turn the plane one way, and others in the reverse direction. For example,

Dextrose—dextrorotatory (to the right)
Levulose (fructose)—levorotatory (to the left)

## Specific Rotation—Dextrose Equivalent

Each sugar has an optical rotation, and when this is related to the specific gravity, and concentration of the solution, it is defined as the specific rotation.

In the analysis of sugar syrups, the reducing sugars are calculated as dextrose even though the actual reducing sugars may be other than dextrose. This is called the dextrose equivalent (DE).

Full definitions of these terms are given under "Confectionery Sugars".

## Scientific Instruments

Scientific instruments are used to an increasing extent in factory control; see Table 20.3. They are described in the Appendix and in other chapters where they are applied to particular processes.

TABLE 20.3. SCIENTIFIC INSTRUMENTS—APPLICATION

| Instrument | Applications |
|---|---|
| Refractometer | Determines the concentration of syrups and end points of confectionary boilings, particularly jellies. |
| Viscometer | Control of chocolate and coatings fluidity, also some syrups and slurries (see "Lecithin, Emulsifiers, Chocolate"). |
| Hygrometer | For measuring relative humidity in stores, production rooms, and coolers. |
| Hydrometer | For measuring density of syrups. Now often replaced by refractometers. |
| Thermometer | The original mercury-in-glass thermometer is a reliable instrument but is inclined to be slow in recording. Electronic probe thermometers are now used both for small-scale recording and continuous process. |
| Particle-size instruments | Micrometers, microscopes (visual and projection), and special counting instruments (Coulter, laser beam). |
| pH | Many meters are available, portable and continuous recording. |

# SPOILAGE PROBLEMS

Most foods are subject to spoilage. Fresh foods such as meat and fish are recognized as having a short shelf life, and, if necessary, are preserved by canning, freezing, or drying.

Other foods should have good keeping qualities because they are processed in sugar syrups or salt, or are low in moisture content. These products may develop faults in appearance or flavor either through poor storage or some failure in production control.

Chocolate and confectionery, because they are usually regarded as luxury foods, must meet the highest quality standards.

## Chocolate Bloom

Bloom on chocolate has been the subject of much argument about its cause, composition, and prevention. There are two types of bloom—fat bloom, arising from changes in the fat in the chocolate and sugar bloom formed by the action of moisture on the sugar ingredients.

Although this bloom is detrimental to the appearance of the chocolate, it does not harm its eating qualities unless there have been very bad storage conditions. In such cases, the chocolate may have a stale taste, and if it has been subject to excessive dampness, surface mold may have developed.

## Fat Bloom

Fat bloom is recognized as a grayish coating on the surface of chocolate, milk or dark, but is more visible on dark. It looks like the bloom on some ripe fruits such as plums and grapes, and when touched lightly with the finger, it has a greasy appearance and is easily removed. Under the microscope minute fat crystals are visible.

It is caused by:

1. Bad tempering of the chocolate in that stage of the process.
2. Incorrect cooling methods, including covering cold centers.
3. The presence of soft fat in the centers of chocolate-covered units.
4. Warm storage conditions.
5. The addition to chocolate of fats incompatible with cocoa butter.
6. Abrasion and finger marking, particularly under warm conditions.

**The Theory of Fat Bloom—Causes and Methods of Prevention**  A great deal has been written about chocolate fat bloom. Much of this arose from the very great losses suffered by the chocolate manufacturers when little was understood about the polymorphism of cocoa butter and the effect of added fats other than cocoa butter. Even when the science of fats was beginning to be understood, engineers often chose to avoid the issue by installing machinery that gave speed, but completely ignored the need to prepare and set cocoa butter in a stable condition.

In England, bloom troubles were magnified in the early days by some abnormally hot summers of which 1921 was an example, and it was also during those years that extra mechanization was being introduced.

At the same time, fats other than cocoa butter were being incorporated into chocolate. Subsequently, many of these were shown to be incompatible and fat bloom and discoloration were the result. The original hand methods, particularly of tempering, allowed for the production of stable forms of cocoa butter in the chocolate with, consequently, greater chocolate stability.

A great deal of research work on bloom, its causes, and prevention, has been carried out by various workers.

A summary of their conclusions is given later.

The formation of fat bloom is closely related to the polymorphism of cocoa butter and this has been discussed in Chapter 3.

There are four essential polymorphic forms. They are tabulated below with their melting points.

$\gamma$ form 17°C (63°F)—very short "life" at all temperatures.
$\alpha$ form 21°–24°C (70°–75°F)—short "life" at all temperatures.
$\beta'$ form 27°–29°C (81°–84°F)—gradual transition at ordinary temperatures;
to $\beta$ form 34°–35°C (95°F)—the stable form.

Various workers have stated that other forms exist but from a practical point of view the four mentioned above are universally accepted.

Bloom is formed by transition of the lower melting polymorphs to the stable $\beta$ form. The technology of good chocolate production depends on ensuring that only the stable form of the cocoa butter ingredient exists in the final product.

In practice, full transformation to the stable form is rarely achieved but modern methods of tempering and cooling go a long way towards this.

In relation to chocolate processing, the following is again emphasized:

1. The lower melting and unstable forms of cocoa butter are formed by using too low temperatures for tempering and cooling of the liquid chocolate.

2. "Seed" crystals of the stable form of cocoa butter should be dispersed in the liquid chocolate. This promotes the formation of the stable form in the liquid cocoa butter, which has still to be solidified in the cooling process after enrobing or molding.

Therefore, correct seeding and moderate cooling go a long way toward the prevention of fat bloom.

Work carried out by the author and co-workers was reported in the second edition of this book. Since this has a bearing on the findings of other workers, summarized below, it is again included later in this chapter.

The following publications are among those recorded in the references at the end of this chapter. The main conclusions drawn by the authors and related to fat bloom are as follows:

**Easton and Moler** (1952) Detailed study of the composition of fat bloom has indicated that it is composed of the higher melting fractions of cocoa butter with lower iodine value.

**Vaeck** (1960) The summarized conclusions presented by this worker in a detailed report include the following.

1. Bloom results from the growth of *large* cocoa butter crystals out of the surface of the chocolate, originating from the unstable forms still in the chocolate.
2. The first conditions for producing bloom-resistant chocolate is correct tempering. Nuclei of the stable form of cocoa butter must be present. These may be formed by the addition of chocolate shavings, previously tempered and solidified, or by mixing and cooling.
3. Tempering by mixing and cooling must be adjusted according to the melting characteristics of the fat. Milk chocolate needs lower temperatures because of the milk fat present. If illipe fat is present, higher temperatures are required.
4. In properly tempered chocolate, the cocoa butter first exists in the $\beta'$ form but transformation to the $\beta$ form starts immediately. This is a controversial conclusion [author].
5. With properly tempered chocolate, the cooling rate is not very critical. This can be challenged [author]. Contraction and possible formation of unstable forms are factors; these are discussed elsewhere.
6. The assumption that fat bloom may be formed as a result of heat evolution by transformation from unstable to stable forms of cocoa butter only applies if the chocolate product is packed closely in boxes before it is completely set [author].

7. Foreign fats such as nut oils, which lower the melting range of cocoa butter, promote fat bloom. Milk fat is the exception. Fats that elevate the melting range generally reduce bloom formation. If the fat is not compatible, this may not be the case [author].

***Merken and Vaeck*** (1980)  These workers have carried out further investigations of polymorphism of cocoa butter by means of the differential scanning calorimeter.

***Kleinert*** (1961)  In a very comprehensive paper that describes the causes and prevention of bloom, the effects of moisture and temperature on the promotion of bloom development are given. Experiments with numerous additives to chocolate that may prevent or delay bloom are described. Butter fat, now well accepted in the industry, is one of those shown to be effective. Contrary to findings by some workers, Span 60, Tween 60, and glyceryl monostearate were found to be ineffective.

The value of heat treatment, described later in this chapter, is demonstrated.

***Duck*** (1957, 1961, 1963, 1964, 1965, 1967)  Many investigations have been carried out by this worker. They are largely concerned with chocolate tempering and the characteristics of cocoa butter during this process.

***Errboe*** (1981)  This worker describes the value of certain cocoa butter equivalents in reducing fat bloom. He also claims that the retarding effect of butter fat is due to the delaying of crystal transformation from the unstable $\beta'$ to the stable $\beta$ form of cocoa butter.

Many of the observations stated above have been confirmed by the author and co-workers. As noted, some have been disputed.

Other experiments on the behavior of cocoa butter, illipe butter, and butter fat, carried out some years ago, are again worth including in this chapter. As far as is known, no similar information has been reported elsewhere.

It is believed that bloom formation is related to the properties of these fats. The experiments concern:

1. Changes in the melting point of tempered and untempered cocoa butter when stored at different temperatures for periods up to six months.
2. Fractionation of cocoa butter, butter fat, and illipe butter. Determination of fractions and their melting points.

## Crystallization of Cocoa Butter Under Different Ambient Conditions—Melting Point Changes

Filtered untempered cocoa butter and well-tempered cocoa butter were poured into shallow covered molds and stored in incubators, without any movement, at 18°C (65°F), 23°C (73°F), 27°C (80°F), and 29.5°C (85°F). After three months at these temperatures, storage of all samples was continued at 65°F. "Complete melting points" by the capillary tube method were determined at intervals and recorded graphically as in Figs. 20.1 and 20.2.

These graphs show that both tempered and untempered cocoa butters take a considerable time to reach melting point stability, and the untempered samples take longer at the lower temperatures of storage. The untempered samples finally attain a melting point higher than the tempered.

Some of the cocoa butter used for this experiment was mixed with carbon black and tempered and untempered portions subjected to the same storage conditions as those without the pigment.

On the untempered samples, white nodules were visible on the surface of all samples after 1 hr, and after three months storage at the various temperatures followed by three months at 65°F the appearance of the samples was as shown in Table 20.4.

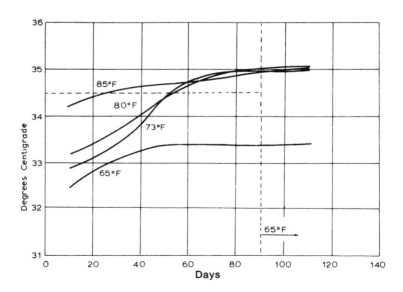

Fig. 20.1. Melting Points Tempered Cocoa Butter After Storage

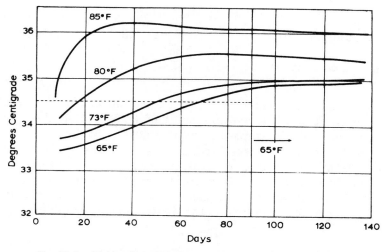

Fig. 20.2.  Melting Points Untempered Cocoa Butter After Storage

Some of the bloom from the tempered samples that had been stored at 27°C (80°F) and 29.5°C (85°F) and as free as possible from underlying surface fat was transferred to a capillary tube and the melting point determined. The results were:

From 80°F sample    Melting point 34.6°C (94.3°F).
From 85°F sample    Melting point 34.2°C (93.5°F).

Suppose a figure of 34.5°C is taken as the true melting point; this is very close to the melting point of the stable β form of cocoa butter.

On the graphs of melting points (Figs. 20.1 and 20.2) it is interesting to note the periods of storage before the melting point of

TABLE  20.4. TEMPERED  AND  UNTEMPERED  CHOCOLATE—APPEARANCE AFTER STORAGE

|  | 18°C (65°F) | 23°C (73°F) | 27°C (80°F) | 29.5°C (85°F) |
|---|---|---|---|---|
| Tempered | Glossy surface. No bloom. | A few white specks. Surface dull. | Surface covered with white specks. | Whole surface covered with white bloom. |
| Untempered | Mottled surface but glossy. | Mottled surface dull. | Very mottled. | Very discolored large white blotches. |

the sample reaches 34.5°C (94°F), they are:

| Storage temperature | 29.4°C (85°F) | 26.7°C (80°C) | 22.8°C (73°F) | 18.3°C (65°F) |
|---|---|---|---|---|
| Tempered cocoa butter (days) | 24 | 52 | 52 | ∞ |
| Untempered cocoa butter (days) | 8 | 18 | 48 | 68 |

It will be observed that the tempered cocoa butter stored at 18.3°C (65°F) never attained a melting point of 34.5°C, and after seven months' storage of the samples, there was still no bloom.

A separate experiment was carried out with (a) pure cocoa butter and (b) the same cocoa butter to which had been added 12 percent of butter fat (≡4 percent in chocolate). Samples were tempered as before and stored at 80°F. Melting points were determined at intervals up to seven months and the figures plotted graphically (Fig. 20.3).

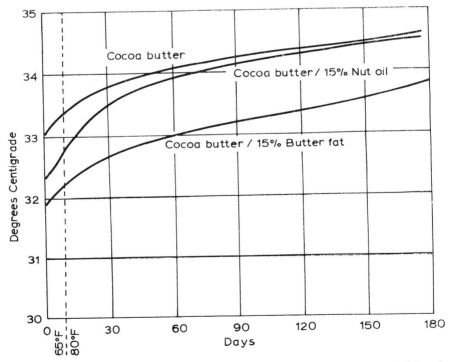

Fig. 20.3.  Melting Points of Pure Cocoa Butter, Cocoa Butter with 15 Percent Nut Oil, and with 15 Percent Butter Fat after Storage

This type of experiment indicates what happens when chocolate (tempered, badly tempered, or with butter fat addition) is stored at temperatures ranging from "temperate" to "tropical" and some idea may be obtained of the time required to produce bloom at a particular storage temperature and to convert the unstable forms of cocoa butter to the stable form.

But this alone does not explain why the stable $\beta$ form comes to the surface. Let it be assumed that chocolate is badly tempered and rapidly cooled; then the unstable forms ($\alpha$ and $\beta'$) of the cocoa butter constituent will be formed in considerable quantity and will be present in the solid chocolate when it is first set.

When this chocolate is subjected to storage under ambient conditions, there will be a gradual transition of these unstable to stable forms and stable crystals will be formed throughout the chocolate mass. This in itself will not account for bloom, and a second factor is concerned. From dilatometer and adiabatic calorimeter measurements, cocoa butter at ambient temperatures contains a proportion of liquid fat, and as the temperature is raised, the proportion of liquid phase increases with progressive softening of the fat (or chocolate). The proportion of liquid phase varies according to the previous history of the cocoa butter (or chocolate). Cocoa butter (or chocolate) set by a cooling procedure that might be expected in a commercial cooler with normally tempered chocolate may contain 20 percent liquid phase, decreasing to 15 percent after several hours at 18°C (65°F). It is also estimated that about 25 percent of the cocoa butter may still be in the unstable state ($\alpha + \beta'$). Obviously these figures may vary appreciably according to the tempering and cooling procedure.

Taking the above figures as an example, there remains 25 percent of the cocoa butter to be converted to the stable form and of this 15 percent is still to be solidified.

During the process of conversion of the unstable to stable form and solidification of the liquid phase, cocoa butter, or more especially chocolate, may be considered as a semimobile matrix within which cocoa butter is crystallizing over a lengthy period of time.

If the temperature rises, say, to 24°–27°C (75°–80°F) during storage, the $\alpha$ and $\beta'$ forms will mostly melt and the percentage of liquid phase increases. Therefore, the fat within the matrix becomes more mobile, and as crystals of the stable $\beta$ form arise, they will tend to grow out between the surface particles of cocoa butter, sugar, and cocoa particles, as well as inside the chocolate.

At lower storage temperatures, there is less mobility within the matrix and there will be no crystals growing outwards, or at least their growth will be restricted.

To prevent or delay bloom, therefore, it is necessary to prevent the growth of large $\beta$ stable crystals, and this can be encouraged by setting as much as possible of the cocoa butter in its stable form during the tempering and cooling process. If small, rapidly grown $\beta$ crystals are formed, the slow growth of large $\beta$ crystals is avoided.

The question will now be asked—How does butter fat prevent bloom? It is not only a question of adding a proportion of a more liquid fat, as other liquid fats tend to promote bloom and will still allow the slow crystallization of the large stable $\beta$ cocoa butter crystals. In fact, if this crystallization can still occur, the liquid fat will increase the liquid phase and hence the mobility of the matrix, resulting in easier migration of the large $\beta$ crystals to the surface. An experiment with hazelnut oil confirmed this and also failed to prolong the suppression of melting point as was the case with butter fat (Fig. 20.3).

The melting point experiment (Fig. 20.3) suggests that butter fat prevents or delays the formation of the normal $\beta$ forms of cocoa butter crystals. Vaeck suggests that it prevents the formation of *large* $\beta$ crystals.

**The Fractionation of Cocoa Butter, Illipe Butter and Butter Fat**  Samples of these three fats were liquefied at 110°F, filtered, and then stored without movement at temperatures of: 35°C (95°F), 29.4°C (85°F), 26.7°C (80°F), 22.8°C (73°F), 15.6°C (60°F). A sample was also stored at a temperature ranging between 15.6°C (60°F) and 10°C (50°F).

Slow crystallization occurred and crystals were carefully removed from the fat at each storage temperature until no more crystallization was apparent. This operation was not quantitative but five fractions were collected (Illipe butter—4) and their melting points determined.

At 95°F no crystals were collected from cocoa butter, 3 percent from illipe butter, and 12 percent from butterfat.

Results were as given in Table 20.5.

These figures show the complexity of the crystalline forms of these three fats and the variation of melting points of the different fractions.

Other observations from this fractionation are the wide ranges of melting points of the illipe and butter fat fractions. The iodine value of the cocoa and illipe butter fractions vary inversely with the melting points.

TABLE 20.5. MELTING POINTS OF FRACTIONATED FATS

|  | Fraction, % | Melting point after 2 days | Melting point after 16 weeks | Iodine value |
|---|---|---|---|---|
| Cocoa butter | A 12 | 25.9°C (79°F) | 28.2°C (83°F) | 48.8 |
|  | B 16 | 31.5°C (89°F) | 32.7°C (91°F) |  |
|  | C 42 | 33.5°C (92°F) | 33.9°C (93°F) |  |
|  | D 21 | 34.0°C (93°F) | 34.0°C (93°F) |  |
|  | E 9 | 33.2°C (92°F) | 37.0°C (99°F) | 33.2 |
| Illipe butter | A 44 | 33.5°C (92°F) | 34.5°C (94°F) | 31.0 |
|  | B 22 | 34.0°C (93°F) | 35.0°C (95°F) |  |
|  | C 31 | 35.8°C (96°F) | 36.8°C (98°F) |  |
|  | D 3 | 48.5°C (119°F) | 55.0°C (131°F) | 21.0 |
| Butter fat | A 29% below | 17°C (63°F) | — |  |
|  | B 26 | 24°C (75°F) | — |  |
|  | C 15 | 28°C (82°F) | — |  |
|  | D 18 | 34°C (93°F) | — |  |
|  | E 12 | 47°C (117°F) | — |  |

The small proportion of a very high melting fraction of illipe butter is significant in supporting the need for higher tempering temperatures of chocolate with a proportion of this fat. If poor tempering and low temperature cooling are used for this type of chocolate, it is readily understood that these higher melting fractions will cause bloom troubles if there are quantities of them still in the unstable condition and that will crystallize slowly on storage. The extreme range of melting points of butter fat suggests that some of the fractions may affect the crystallization of the forms of cocoa butter including the $\beta$ stable form and prevent growth of this in large crystals. Fractionation of butter fat has recently been done commercially (see Chapter 10).

**Other Observations** Research on the tempering of chocolate indicates that if, during the cooling of chocolate, the crystals of stable fat being formed are subjected to rapid shearing action, the "seed" crystals are distributed in very fine form throughout the mass. This prevents the development of large crystals and the chocolate so tempered is more resistant to bloom.

Some tempering machines have been constructed on this principle. The chocolate is distributed in a thin film against the cooling surface by means of a rapidly moving scraper blade.

Relatively low coolant temperatures can be used but because of the movement the chocolate never reaches the temperature of the coolant. At the same time, only fine cocoa butter crystals are formed.

There is very little in the published literature on the reasons for the marked differences in susceptibility to bloom of molded and enrober covered chocolates.

Chocolate molded in polished molds, even when the chocolate has a low degree of temper and is rapidly cooled, shows no bloom development on the glossy molded surface. Under storage conditions that promote bloom, the glossy surface of a chocolate bar or shell remains free from bloom whereas the back that has been cooled out of contact with any surface develops severe bloom.

Microscopic examination of the surface of molded and enrober chocolates shows a considerable difference in their structure. The molded surface is compacted and continuous whereas the enrobered surface is relatively rough and uneven.

The violent agitation of the molds containing the liquid chocolate forces the sugar and cocoa constituents, together with liquid fat, against the mold surface forming a continuous layer and it is cooled and set in this condition free from fissures.

The enrober chocolate coating when setting is able to move according to the needs of the cocoa butter crystallization and, depending on the condition of temper in the chocolate and the cooling, will set in an uneven condition. Kleinert has shown that badly cooled chocolate coating, or coating that has been put on cold centers, contains minute fissures.

This difference between molded and enrober chocolates lends considerable weight to the argument that the physical condition of the chocolate has a bearing on its susceptibility to bloom as well as the crystal form of the cocoa butter. In molded chocolate, this physical strength of the surface layer is enhanced by the presence of the sugar, cocoa and milk solids but in an enrober coating these may act in the reverse way and encourage the development of fissures.

It is well known that centers that contain low melting fat or oil (apart from butter fat) will promote bloom formation on enrobered chocolates but this is not the case with molded shell plant chocolates unless the chocolate coating is very thin (this would be a defect in a shell unit).

It seems that fissures that may be formed in chocolate enrober coatings allow the diffusion of fat from the centers and also promote mobility of the liquid unstable fractions while they are converting to the stable $\beta$ form.

**"Rubbed" Chocolates** This was a term used at one time to describe chocolate assortment units that had a bad shape or appearance but otherwise were quite edible. They were sold to employees or to the public in the cheaper food stores.

Examination of these chocolates often showed that fat bloom had developed in patches or on the bottoms.

This exemplifies another cause of fat bloom and possibly again supports the theory of the unbroken surface on chocolate preventing bloom. Enrober covered chocolates may have damaged surfaces either through finger marking, abrasion, or where they have been pulled off the enrober belt before they are completely set. Bloom will develop on the damaged surfaces after storage under conditions of temperature shown previously to promote blooming.

An interesting observation is that molded chocolate surfaces will also develop bloom on abraded patches and it is suggested that the crystallizing stable $\beta$ form gains access to the surface through the damaged area.

**Bloom on Milk Chocolate**  Practically nothing has been published about bloom on milk chocolate and it is probable that because of the natural butter fat content it is assumed that it does not bloom. This is incorrect, as milk chocolate will bloom if held for long periods at about 18°C (65°F), the actual temperature being dependent on the milk fat content. The time required for bloom to develop may be from six months to nine months, and although chocolates would not be expected to be kept so long, these conditions occasionally prevail in warehouses and shops.

Bloom on milk chocolate will arise basically from bad tempering and cooling and the risk of bloom occurring on milk chocolate if good techniques are used is remote.

Milk chocolate will not bloom if stored under warm conditions but it may show patchy bloom from abrasion on the bottoms or from finger marking when stored for long periods at temperatures around 18°C (65°F).

Bloom on milk chocolate can be avoided by heat treatment.

## Heat Treatment of Chocolate

A useful method of preventing or delaying bloom is by heat treatment of the chocolate shortly after it comes from the enrober. Heat treatment of molded chocolate is unnecessary for reasons previously explained.

There are two methods of heat treatment:

1. Heating of dark chocolate covered centers at 32.2°C (90°F ± 1°F)

for periods up to 2 hr and it is essential that the chocolates are treated as soon as they come from the enrober cooler.

This process gives a brilliant glossy surface that is very resistant to bloom development—it has been called the "super-glossing process." It has not been commercially developed to any great extent because it does not lend itself to continuous operation in conjunction with the enrobing process. In some companies there was some resistance to the glossy appearance as it was maintained that the sheen of the traditional enrober covered chocolate was a sign of quality. The gradual increase of shell plant chocolates in assortments has changed this opinion. Kleinert has proposed a continuous method with infrared heating at 32°C (89.6°F).

2. Heating of covered chocolates from the enrober cooler for 48 hr:
   (a) dark-covered chocolates at 26.7 to 29.4°C (80 to 85°F) in trays;
   (b) milk-covered chocolates at 22.8° to 25°C (73 to 77°F).

Milk-covered chocolates can be heat treated in the packed boxes in a warm room without damage provided the boxes are stacked in open form to allow access of the warm air. This is preferably fan circulated and thermostatically controlled.

Method 2 does not give a high gloss appearance; in fact little change is noted. Method 2(a) has the disadvantage of not conforming to continuous layout operation.

Method 2(b) for milk chocolate is quite practicable and is not very costly except for the provision of what amounts to an intermediate stockroom. It acts as a useful safeguard when very viscous chocolate is used for enrobing and where the enrober operator may work near the upper limit of chocolate temperature. With both of these methods of heat treatment, there should be as little delay as possible between enrobing and heat treatment.

Methods 1 and 2(a) need moderated cooling after heat treatment with air at about 14.4 to 16.7°C (58 to 62°F) but method 2(b) needs no special cooling and the packed boxes can be moved to the normal stockroom.

There are possibly two explanations for heat treatment being a means of bloom prevention. With the high-temperature method it may be related to the formation of a continuous surface layer as with molded chocolate. With the lower temperature of heat treatment it is probable that the unstable forms of cocoa butter are still in the very early stages of transition to the stable forms. These are partly remelted and transformed to the stable $\beta$ form by the subsequent moderate cooling, thereby reducing the amount of unstable cocoa butter to be transformed over a long period of storage. Another feature of heat treatment is the relaxation of stress within the chocolate itself—this could cause compacting of the chocolate and prevent movement of the stable form of cocoa butter during crystallization. With dark chocolate, in addition to good tempering and cooling techniques butter fat is used as an anti-bloom additive and this makes heat treatment unnecessary.

### Fat Bloom—Summary

To a reader who may be new to the industry the foregoing dissertation on fat bloom could well result in confusion and despair at the lack of precise knowledge of the crystal structure of cocoa butter and associated fats. Scientists and technologists have now developed machines and processes that take into account the special requirements needed for handling chocolate and other coatings.

Sometimes these machines are worked incorrectly or above the rated throughputs.

The following is a summary of the important points to be observed by the chocolate technologist:

1. Chocolate must be well refined and conched to give good distribution of solids (sugar, milk solids, cocoa matter) throughout the cocoa butter.
2. Correct tempering of the chocolate is of prime importance. The methods used are discussed under "Chocolate Manufacture" and "Enrobing."

   The main points to watch during production processes are:

   a. If viscous chocolate is used to save cocoa butter or give thick coatings on enrobered units, make sure the machine operator does not raise the chocolate temperature excessively to obtain extra fluidity and thereby reduce temper to a dangerous level.

   b. If autotemperers are used, work them at the correct output whatever the demands of the production personnel. If extra-tempered chocolate is required, install an extra temperer and work them in unison.

   c. Use a "tempermeter" to check the degree of seed periodically—the instruments on a temperer may be giving incorrect readings.
3. Moderate cooling is necessary particularly with enrobed chocolates. Details of cooling methods are given under "Chocolate

Processes" but it is essential to avoid low-temperature air in the first stages of cooling. Confectionery centers must be warmed before covering and this is usually done by storing them in the enrober room for a period. Their temperature should never be less than 24°C (75°F) but depending on chocolate viscosity and the size of center the temperature may be raised to 29.4°C (85°F). Above this temperature, chocolate is likely to run off the units and produce flanges on the bottoms.

Molds for chocolate blocks must be similarly warmed before receiving chocolate deposits and if nuts, raisins, cookies, or similar materials are mixed with the chocolate, these should be warmed to the chocolate temperature.

4. Do not add foreign fats to chocolate unless they are compatible. Where soft fats or oils may migrate from the center, use an antibloom additive such as butter fat.

5. Do not allow freshly covered chocolates to be handled with warm hands and make sure bottom cooling is adequate so that chocolates detach readily from the cooler band. Abrasions or finger marks promote bloom.

6. Store finished chocolates at cool temperatures, 10 to 13°C (50 to 55°F), is adequate for most purposes; 7 to 10°C (45 to 50°F) is probably better for chocolate cookies with fatty fillings.

"Deep freeze" using temperatures of the order of −10°C (14°F) is sometimes used for very long storage. This method is useful for chocolate centers containing butter or cream that may develop rancidity under the above cool storage conditions.

Care must be taken when removing these goods from storage as much condensation will occur on the outer surfaces of the boxes. Shrink-wrap plastic covers for platform loads can be used so that there is adequate protection until the loads have reached ambient temperatures.

Compound coatings containing fats other than cocoa butter require different tempering, cooling, and storage conditions. These depend on the nature of the fat ingredient and are discussed in Chapter 6 on "Confectionery Coatings."

## Sugar Bloom—Causes and Methods of Prevention

Sugar bloom has a grayish appearance and in a mild form resembles fat bloom but when touched with the finger it is not removed and has no greasy feeling. In a more severe form it has a crystalline appearance (frosting), is quite rough to the touch, and when examined under the microscope small sugar crystals can be seen. It will form on milk and dark chocolate. In the early stages of sugar bloom formation the chocolate may be coated with a very thin layer of sugar syrup and in severe cases this is quite sticky. In due course this layer deposits crystals of sugar.

It is caused by:

1. Storage of chocolates in damp conditions or against damp walls.
2. Deposit of "dew" during manufacture from damp cooler air or allowing chocolates to enter a packing room at a temperature below the dew point of that room.
3. Use of hygroscopic ingredients (e.g., low-grade or brown sugars).
4. Removal of chocolate from cold storage without adequate wrapping protection.
5. Use of damp packing materials.
6. High-temperature storage conditions of chocolate covered confectionery where centers have a high equilibrium relative humidity (e.g., fondants) and the water vapor given off is trapped in impervious wrappings.

**Storage of Chocolates in Damp Conditions** Chocolate will absorb moisture on the surface if stored in air at a relative humidity above 82 to 85 percent for dark chocolate or above 78 percent for milk chocolate. These relative humidity values are not precise as the proportion of milk solids, total fat, and small proportions of other sugars have an effect.

The time of storage under damp conditions affects the final appearance of the chocolate surface—minor sugar bloom gives only a very slight dullness to the surface. A long exposure allows the moisture to penetrate more and the first effect is a sticky surface layer not necessarily greatly affecting the general appearance. When the damp conditions cease, the film of syrup will dry and sugar crystals form giving a gray appearance. Sometimes the crystals are visible to the naked eye but they can readily be seen under a low power microscope.

Where chocolates have been wrapped or boxed, different effects appear. Obviously an impervious heat-sealed wrap gives full protection but overlap wraps or boxes with waxed linings allow penetration at the folds or corners and sugar bloom appears on the chocolate near these points. Boxes stored against a damp wall will cause the chocolates on the side near the wall to be most affected.

With some of the modern methods of display for sale, particularly chocolate bars and hard candy in bags, the ambient conditions are virtually those of the open air with protection only from a canopy or kiosk roof. Unless there is a very quick turnover, such conditions demand a protective wrap or deterioration will result.

**Sugar Bloom from Deposit of "Dew"**  This is mentioned under "Chocolate Cooling" and it can be formed in several ways.

If chocolate is cooled under conditions that allow the units to emerge from the cooler at a temperature below the dew point of the packing room, moisture will deposit and ultimately form sugar bloom.

At one time this was a common occurrence in summer months but better cooler design and the air conditioning of packing rooms have helped to eliminate this trouble.

Occasionally, through bad cooler design or during defrosting of cooling coils, air at high humidities may pass through the coolers and the result will be the same as with damp storage.

Checking of unit surface temperatures with a thermocouple needle and insertion of a hygrometer in the cooler will show up defective cooling conditions.

**Damp Packing Materials**  This is most unlikely in a well-managed factory but the packing of cool chocolates in a damp carton is not unknown. Excessive glue or undried lined strawboard are possible causes and the effect is similar to damp storage.

**Removal from Cold Storage**  Where low-temperature storage has been used below 10°C (50°F) for boxed chocolates, considerable deposition of moisture can form on the outside of the boxes if they are brought out into normal atmospheres. Either an intermediate room with fairly dry air is required or the loads can be enveloped in polythene sheeting until they have attained the temperature of the outside air.

Complaints are received occasionally from tropical or subtropical areas where it is the habit to keep boxes of chocolates in a refrigerator. If the box is opened before it has reached room temperature, dew will deposit and the result is sugar bloom.

**Tropical Conditions**  Chocolates with fondant or semiliquid centers when contained in impervious wraps will, under warm storage conditions, result in high humidities inside the wrap and diffusion of the syrup from the center into the covering. This results in sugar bloom and in some cases mold on the chocolate surface.

## Other Faults

**Microbiological Problems, Rancidity** Chocolate and confectionery are foods that are not prone to attack by spoilage organisms in the same way as meat and fish, nor do the types of deterioration mentioned under this heading have any pathogenic effect other than, possibly, nausea through the sight of a badly discolored bar or a rancid flavor.

Some ingredients, however, may have had microbiological history that can result in some toxins being present.

Chocolate and confectionery are not manufactured or packed under aseptic conditions but they preserve their edible condition by reason of low moisture or high soluble solids contents.

The main causes of deterioration when they do occur are:

*Fermentation* Ferments can arise from raw materials or bad plant hygiene and mostly they will remain inactive if the soluble solids content is above 75 percent or the moisture content is very low (as with chocolate).

*Rancidity* This can arise from the action of the air and is catalyzed by the action of light, heat, and some metals. This is oxidative rancidity and can give objectionable flavors in nut oils and other vegetable fats.

Soapy rancidity arises from the action of fat splitting enzymes and is called hydrolytic rancidity—its development sometimes is greatly delayed and off-flavors arise that are not readily recognized as associated with enzyme action.

*Molds* Molds in themselves are readily seen in the advanced state of growth by the presence of masses of hyphae that cover the food, giving it an offensive odor and making it unpalatable.

Molds that have affected a small part of a raw material, for example, adjacent to a damp spot on a sack or the inside of a nut or cocoa bean, may pass unnoticed and be dispersed in the mass of foodstuff.

Such mold will leave in the food mold spores, fat-splitting enzymes, or toxins. Aflatoxin in peanuts is a typical example of the latter.

Molds and ferments may occur in plant and pipelines if hygiene is at fault and these can subsequently cause some of the defects mentioned above.

**Fermentation** There was a time when fermentation of chocolate centers in boxes of assortments was a common occurrence and one defective unit could ruin the entire contents by bursting the chocolate covering through the pressure of the carbon dioxide evolved and distributing sticky, beery syrup throughout the box.

Fortunately, with better understanding of the nature of the organisms causing fermentation and careful attention to the proportion of different sugars in the recipe, fermentation is now rare, and when it does occur, it is usually due to some deviation in the recipe or process.

Fermentation in confectionery is largely caused by osmophilic yeasts (Zygosaccharomyces, Torulopsis). These microorganisms are capable of growing in sugar syrups of high soluble solids content. They are associated with the spoilage of honey, golden syrups, raw sugar, fruit concentrates, and dried and preserved fruit. In the early days, cases of fermented confectionery were attributed to the action of bacteria, as these were the microorganisms isolated from fermented units. The osmophilic yeasts failed to grow on the medium used. There are two courses of action that can be taken to avoid fermentation: (1) to eliminate, as far as possible, the sources of osmophilic yeasts, and (2) to provide concentration in the syrup phase of the confection sufficiently high to prevent the yeasts from acting. Sporadic outbreaks of fermentation may arise from fruits, nuts, cocoa, certain cereal flours, and other natural products that are prime sources of these yeasts and they also provide protein matter as additional nutrient. It is always desirable to heat any confection containing these ingredients to 82°C (180°F) or above for 15 to 20 min and sterilization is only effectively achieved in a moist medium, e.g., by heating the ingredient in syrup before incorporating it into the confection.

Osmophilic yeasts are also liable to occur in plant pipelines and valves where residues of confectionery may remain undisturbed, particularly when production is intermittent. These organisms exhibit increasing virility, and, therefore, plant hygiene is of great importance and all pipelines, valves, mixers and mincers must be regularly sterilized by live steam or bactericidal detergents. To prevent fermentation it is usually sufficient to provide a syrup phase concentration over 75 percent by incorporating liquid glucose (corn syrup) or invert sugar equivalent to two and a half to three times the moisture content of the confection, e.g., approximately 25 to 33 percent glucose or invert solids in the confectionery recipe. However, isolated instances have been recorded where osmophilic yeasts have caused fermentation of syrups above 75 percent concentration. The cause has usually been due to one of the ingredients previously mentioned, when sterilization has not been carried out, or to contaminated equipment.

Fermentation will also be retarded by reduction of pH or the inclusion of permitted preservatives.

Jam or fruit paste of pH 3.0 to 3.5 will be resistant to most fermenting organisms down to 72 percent soluble solids and similar pH values will inhibit mold growth. Certain organic acids such as acetic, lactic and sorbic acids have inhibiting effects on fermentation but sorbic acid is not universally accepted as an additive. Acetic and lactic acids are acceptable because they are naturally occurring substances but their flavor effect is appreciable at quite low levels, particularly acetic acid.

Sorbic acid is effective at 0.1 to 0.2 percent levels and acetic acid at 0.05 to 0.1 percent.

Sulfur dioxide and benzoic acid are permissible as preservatives in fruit pulps, preserved and dried fruit, and some other commodities, and reference should be made to the food laws of a particular country to ascertain the quantities permitted.

Summarizing, therefore, fermentation is prevented by:

1. Maintaining all soluble solids concentrations above 75 percent.
2. Testing suspect raw materials and sterilizing them before or during manufacturing of the confection.
3. Observing strict plant hygiene and testing residues for fermenting organisms.

Preservatives are unnecessary except in special circumstances.

**Rancidity** The type of rancidity caused by the action of air, oxidative rancidity, is widespread in foods generally but in confectionery hydrolytic rancidity is more often the cause of bad flavors. This arises usually from enzymic (lipolytic) action on lauric fat ingredients, so often used in formulations.

*Oxidative rancidity* Fats and oils containing unsaturated fatty acids and esters are oxidized when exposed to the action of air and the rancid flavor condition that arises is due to the formation of aldehydes and ketones that have characteristic odors and flavors.

Oils and fats can also become increasingly rancid through auto-oxidation. It must be understood that rancidity is a collective description for all sorts of "off-flavors" and the bad flavor is often described as tallowy, fishy, or metallic. The cause may be wrongly diagnosed as some foreign contaminant when it is merely rancidity in a mild form. The natural flavors associated with some of our foods when isolated and analyzed have aldehydic and ketonic compositions very similar to those separated from rancid foods.

Some people do not detect rancidity easily by taste, whereas others are extremely sensitive, and this applies particularly to hydrolytic rancidity. Oxidation of fats is a very complex chemical reaction and must fats go through an induction period where little rancid flavor is detectable but after a certain period of time oxidative rancidity starts to develop and intensifies with time. This phenomenon is related to the formation of "peroxides" and a method of determining rancidity or incipient rancidity of a fat is to estimate its peroxide value. The unsaturation of a fat or oil is determined by means of the iodine value—the higher the iodine value, the more susceptible the fat is to oxidation. A method for determining the length of "induction period" uses a constant air or oxygen stream blown through the oil at constant temperature, plotting the increase of weight against time. An accelerated weight increase coincides with the end of the induction period. Another method using an oxygen "bomb" is mentioned under "Peanuts." Many fats and oils contain natural antioxidants, notably lecithin and the tocopherols, and these prevent or delay oxidation and rancidity, but synthetic and other natural antioxidants may be added. BHT (butylated hydroxy toluene) and Sesamol (in Sesame oil) are examples and further information is given under "Antioxidants."

Another interesting fact is that certain substances act as synergists to the antioxidants—these are citric and phosphoric acids, various amino acids, and other substances present in some vegetable and animal foods; see Abrahams and Naismith (1968).

The reader who wishes to pursue this chemistry is referred to the many books on oils of fats.

The food technologist is more concerned with the methods of prevention of the bad effects of oxidation.

First of importance is the purchase of good-quality fats and fat-containing raw materials.

Next it is necessary to ensure that good raw materials are not spoiled in processing. Oxidation of fats is accelerated by the action of heat, light, and the presence of certain metals.

*Heating of Fats* Good fats are frequently spoiled by bad melting. Fats today are often supplied in bulk by tanker, and if this results in quick use and no overheating, so much the better. Where fats are to be melted from solid blocks this should be done in hot-water-jacketed, stainless-steel vessels—steam-heated kettles should be avoided as the operator who wishes to be always ready will not only raise the temperature too high but will hold the liquid fat for long periods at high temperatures. Also, when the kettle is discharged, a thin layer of fat is exposed to the air at the temperature of the pan surface, which is, in effect, the steam temperature in the kettle jacket at the pressure used. This thin fat layer will deteriorate very rapidly.

Steam- or hot-water-heated grids are often used. These consist of a number of tubes through which the heating medium circulates and the blocks of fat are placed on the tubes. The fat melts where the blocks touch the tubes and runs off into a container below. This system is better than a steam-jacketed kettle but checks should be made on the temperatures attained in the liquid fat.

*Metals* Iron and particularly copper, aid fat oxidation and many other metals are undesirable. Aluminum alloy or heavily tinned copper are poor alternatives to stainless steel, and mistakes are often made by installing brass or bronze valves between stainless kettles and pipelines.

The melted fat should preferably be kept below 60°C (140°F) and never raised above 71°C (160°F) or if storage is anticipated, the maximum should be 49°C (120°F).

*Light* Bright light greatly accelerates fat deterioration and blocks of confectionery fat exposed to sunlight will soon develop a rancid layer on the surface. If trolleys are used to transport blocks or shredded fats to production departments, they must be regularly and completely emptied and cleaned and never stored near bright windows.

*Hydrolytic Rancidity* Fats can be broken down into their constituent fatty acids and glycerol by various means and the process is known as hydrolysis. A small amount of hydrolysis may take place when a wet food is heated in an oil, but mostly hydrolysis is brought about in foods by enzyme action. Lipases change fats into glycerol and fatty acids but some lipases decompose not only true fats but other fatty compounds and esters that may be water soluble. This must be borne in mind when investigating problems of rancidity.

Hydrolytic or "soapy rancidity" can be a most objectionable defect in a confection containing fat. It arises from the introduction of fat-splitting enzymes present in ingredients such as coconut, milk products, egg albumen, and cocoa. Lipolytic activity can arise from the activity of molds during the preparation or storage of foodstuffs. These molds can act in isolated damp spots and be distributed throughout the food unnoticed.

Fat splitting by lipolytic enzymes does not always give soapy flavors but where fats of the lauric type are present very bad soapy rancidity will develop. Lauric acid ($C_{11}H_{23}COOH$) is a saturated fatty acid, the name being derived from the plant family Lauracaeae, some members of which contain over 90 percent lauric acid in their natural glycerides. The common fats containing lauric glycerides are coconut oil, and palm kernel oil with 40 to 50 percent and butter fat 2 to 6 percent. Cocoa butter, illipe fat, palm oil, and peanut oil do not contain lauric acid. Only traces of free lauric acid are required to give a soapy flavor.

Coconut confectionery has been noted for its susceptibility to rancidity. This problem was investigated some years ago (Minifie & Carpenter, 1968), and the information gained then has been invaluable in dealing with isolated instances of hydrolytic rancidity that have occurred since that time. It was shown that coconut or similar fatty confectionery inoculated with lipase prepared from castor seed must be heated to 88 to 93°C (190 to 200°F) to ensure inactivation of the enzyme. To achieve this in practice it was found convenient to sterilize the coconut before it was used in confectionery and a continuous plant was constructed. This used live steam that impinged on a thin layer of coconut on a moving belt and also served the purpose of destroying salmonella and other organisms that have been found in desiccated coconut. Kiskova and Rasper (1965) have shown that lipolytic rancidity in coconut is associated with Micrococcus candidus, M. luteus, M. flavus, Achromabacter lipolyticum, and Bacillus subtilis. Since that date, many improvements have been made in coconut-growing areas to ensure that lipolytic enzymes do not exist in the final coconut product.

Milk fat emulsions prepared from milk powder, or condensed milk, sugar, and fat are used in caramels, fudges, and other paste confections. Rancidity in some of these confections has been traced to the milk emulsions that had become rancid in a very short time while awaiting use. Hydrolytic rancidity has also occurred in confectionery containing fat to which egg albumen in the form of a frappé has been added. The origin of lipase in egg albumen may be due to contamination of egg white by egg yolk during the preparation of egg albumen and the use of pancreatic lipase to remove traces of yolk fat (Lineweaver, 1948; Frazier, 1967).

Some authorities state that lipolytic organisms are inactivated in albumen syrup mixtures by pasteurization treatment at 71°C (160°F) for 20 min, and evidence from tests carried out using methods mentioned later suggests that lipolytic action may have been eliminated. However, storage tests on fatty confections that have used pasteurized albumen frappé have shown that the treatment may be inadequate and soapy rancidity will develop on occasions. The temperature required to inactivate lipases depends greatly on the nature of the substrate. Strong sugar syrups, fats, and absence of water exert a protective influence and in normal confections temperatures of 88 to 93°C (190 to 200°F) are necessary for inactivation; with dry powders survival at temperatures over 104°C (220°F) has been recorded.

Certain cocoa (e.g., from expeller cake) may cause serious rancidity trouble, and when this ingredient is used in low-moisture confections, it is often impracticable to sterilize in syrup media. In these circumstances, it has been found possible to destroy enzyme activity by making a slurry of the cocoa in fat and heating for 5 min at 107°C (225°F).

A stable fat or heat-resistant frying oil must be used. In recent years, much research has been carried out on the deterioration of low-moisture foods. Microbial and enzyme activity associated with these products is more related to water activity than moisture content.

Acker (1962, 1965, 1968) studied the influence of water activity on enzymic reactions and found that the most widespread enzymes, amylases, phenoloxidases, and peroxidases, were inactive when the water activity was below 0.85. In contrast, the lipases remained active at values below 0.3 or as low as 0.1, and under such conditions there was extraordinary resistance to thermal destruction.

At low water activities, lipases react very slowly.

Loncin et al. (1962) states that bacterial growth was unlikely at water activities below 0.90; mold and yeasts were inhibited between 0.88 and 0.80, except for some osmophilic yeasts that were active down to 0.6. Thermal destruction became more difficult as water activity decreased.

*Detection of Lipolytic Activity* The detection of lipase presents some difficulty especially when activity is very low; also, enzymes of bacterial origin may remain long after the bacteria themselves have been destroyed. Numerous methods have been proposed, and while some have useful applications for specific food raw materials and recipes, they can be very unreliable.

One of the chemical tests applied is the indoxyl acetate test. Indoxyl acetate is hydrolyzed at pH 7.2 by esterases of various origins and the indoxyl so formed is rapidly converted by atmospheric oxygen to indigo blue (Purr, 1965). Another, the tributyrin agar test, provides a medium that may be used for plate counts of lipolytic organisms or for plating out colonies (Harrigan et al., 1966). Oterholm and Ordal (1966) have modified this method to increase sensitivity. In the sigma test, a 1-g sample is incubated for 24 hr at 37°C with an olive oil emulsion buffered at pH 7.0 and the free acidity determined by titration with standard alkali (Sigma, 1970).

Although somewhat time consuming, the most reliable is the sensory method, and this should always be used as a confirmation of chemical tests.

Meursing (1983) describes a procedure in which an intimate mixture of the sample, palm kernel oil, and sugar syrup is incubated at 36°C for periods up to three months. Organoleptic tests are made regularly during that period.

**Molds**  In addition to molds causing the introduction of microorganisms, enzymes, and toxins into foodstuffs, they are able themselves to produce taint and objectionable flavors.

If the water activity of a confection is reduced to a level that will prevent mold growth at normal temperatures, the same immunity may not apply at tropical temperatures.

Where closed packs are sent to tropical markets, the water activities at high temperatures must be determined and occasionally, where violent fluctuations of temperature are experienced, local condensation may occur in the head space of a package and this may promote mold growth.

In closely wrapped confections (crystallized fondants, soft-center chocolates) and under warm conditions, molds will cause liquefaction of small areas on the surface of these units. Mold spores (and bacteria) present in the starch into which the confections are deposited are often the source of the infection. This type of defect can be reduced by regularly cycling the molding starch through a heater and drier, but because of the low moisture content, sterility cannot be attained and in any case aseptic conditions do not prevail in the process of confectionery casting.

Very soft and moist confectionery is best avoided for tropical export.

**Graining, Crystallization**  Crystallization, or "graining" as it is known to the confectioner, is a defect of chocolate confectionery centers, hard candies, caramels, toffees, preserved fruits, and jams. It does not occur in chocolate and any coarse crystal present in chocolate is due to bad refining, or if superficial, to sugar bloom. Graining gives a gritty, unpleasant texture and sometimes causes other deterioration.

Graining is caused by the following.

1. Incorrect balance of the proportion of different sugars in a confectionery formulation. Usually sucrose crystals are deposited, and if growth is slow, they can be large and gritty. In some instances, dextrose crystals will be deposited and these may be in the form of large nodules consisting of agglomerates of crystals. Lactose will sometimes crystallize from milky products, such as fudge. They are slow forming and are very hard in texture.

2. Bad storage conditions. In hard candies, toffees, and caramels, the dissolved sugars are in a supersaturated condition and the extremely high viscosity of the sugar "glass" retards crystallization. If moisture is picked up on the surface or the temperature is raised, the viscosity is reduced and crystallization commences. Graining will also occur if storage results in drying out of the syrup phase in a confection so that coarse sucrose crystals slowly form.

3. Incomplete impregnation of a fruit, peel, or ginger during the preservation process resulting in an incorrectly balanced syrup in the interior of the fruit. Coarse gritty crystals appear inside the preserved fruit in due course; they may be sucrose or dextrose.

4. Breakdown of protective colloids such as gelatin or egg albumen, which retard crystallization.

## Solubility of Mixed Sugars

The solubility of the various sugars has been noted, and in many instances, the saturation solubility at ambient temperatures is insufficient to prevent microbiological deterioration; sucrose, for example. When considering mixtures of the sugars, incorrect proportions will also cause unwanted crystallization as well as microbiological troubles.

Glucose syrup has special significance in the prevention of crystallization due to the presence of higher-molecular-weight sugars and a complex variety of oligosaccharides. These increase the viscosity of the syrup and at the same time result in greater solubility.

The combined properties of solubility and viscosity are made use of in candy technology. In considering the microbiological aspect, it is fortunate that stable, mixed-sugar syrups can be made that are above the minimum of 75 percent. Invert sugar, glucose syrup, and dextrose

TABLE 20.6. SOLUBILITIES OF SUCROSE/GLUCOSE SYRUP MIXTURES

| Percentage of solids present | | Solids content of solution saturated at 20°C (68°F) with sucrose (%wt/wt) |
|---|---|---|
| Sucrose | Glucose syrup solids | |
| 100 | 0 | 67.1 |
| 78.6 | 21.4 | 70.0 |
| 67.6 | 32.4 | 72.0 |
| 57.6 | 42.4 | 74.0 |
| 53.0 | 47.0 | 75.0 |
| 48.8 | 51.2 | 76.0 |
| 47.5 | 52.5 | 76.1 |
| 40.9 | 59.1 | 78.0 |
| 34.1 | 65.9 | 80.0 |
| 28.4 | 71.6 | 82.0 |
| 23.7 | 76.3 | 84.0 |

have similar effects on the solubility of sucrose. Tables 20.6 and 20.7 by Jackson (1973) illustrate this fact.

**Storage, Packing and Transport** Bad storage will cause the defects already mentioned—it can also result in bad flavors. The value of wrapping for protection and preservation of the original quality of confectionery is dealt with under "Wrapping and Packing Materials," as also are the taints that may arise from bad wrappings.

TABLE 20.7. SOLUBILITIES OF SUCROSE/INVERT MIXTURES

| Percentage of solids present | | Solids content of saturated solution at 20°C (68°F)% wt/wt | |
|---|---|---|---|
| Sucrose | Invert sugar | Saturated with sucrose | Saturated with dextrose |
| 100 | 0 | 67.1 | — |
| 78.6 | 21.4 | 70.0 | — |
| 67.6 | 32.4 | 72.0 | — |
| 57.6 | 42.4 | 74.0 | — |
| 48.8 | 51.2 | 76.0 | — |
| 47.5 | 52.5 | 76.1 | 76.1 |
| 40.0 | 60.0 | — | 73.6 |
| 30.0 | 70.0 | — | 70.5 |
| 20.0 | 80.0 | — | 67.7 |

Inadequate packing followed by bad transport arrangements will result in broken and abraded units and fluff from the paper packings through constant movement inside the pack. Boxes that are not strong enough will give crushed chocolates and make them unsaleable. Methods of travel testing and simulating conditions experienced in transport are discussed under "Quality Control." Export of chocolates and confectionery presents special problems as movement of chocolates in warm conditions will cause distortion, fat seepage, and fat bloom. In completely enclosed wraps, mold problems can arise.

Infestation dangers are dealt with under "Pest Control."

## REFERENCES

Abrahams, N., and Naismith, D. J. 1968. Dehydration of foods in edible oil in vacua. *J. Fd. Technol.* 3, 55–68.

Acker, I. (1962, 1965, 1968). Biochemical and microbiological aspects of low water activities in dehydrated foods. Institut für Lebensmittelchemie der Universitat Munster/W.f. Germany.

Buckle, F. J. 1980. *Pectins.* H. P. Bulmer Ltd., Hereford, England.

Duck, W. N. 1957–67. Many publications, e.g., PMCA projects, Franklin and Marshall College, Lancaser, Pa.; Manf. Conf., U.S.A.; *Twenty Years of Confectionery and Chocolate Progress.* AVI Publishing Co., Westport, Conn.

Easton, N. R., and Moler, E. S. 1952. Composition of cocoa butter. In *Twenty Years of Confectionery and Chocolate Progress.* AVI Publishing Co., Westport, Conn.

Errboe, J. 1981. Confectionery fats. ISCMA/IOCC Report, General Assembly, Hershey, Pa.

Frazier, W. C. 1967. *Food Microbiology* (2nd ed.). McGraw-Hill Book Co., New York.

Harrigan, W. F., and McCance, M. E. 1966. *Laboratory Methods in Microbiology.* Academic Press, New York.

Kiskova, R., and Rasper, V. 1965. Listy Cukrova, 81. 214 (1966). *JSFA* 1, 38. England.

Kleinert, J. 1961. Studies on the formation of fat bloom and methods of delaying it. *Int. Choc. Rev.* Switzerland.

Lees, R., and Jackson, B. 1973. *Sugar Confectionery and Chocolate Manufacture.* Specialized Publications, Surbiton, England.

Lineweaver, H., et al. 1948. *Arch. Biochem.* 16(3), 443.

Loncin, M., Bimbanet, J. J., and Langes, J. 1962. *Fd. Technol.* 3, 131–142.

Merken, G. V., and Vaeck, S. V. 1980. Study of polymorphism in cocoa butter by differential scanning calorimetry. *Lebensmittel. Wiss. Techn.* 13(6) 314–317. Germany.

Minifie, B. W., and Carpenter, W. J. 1966. Microorganisms in the confectionery industry. *Proc. Biochem.* England.

Meursing, E. H. 1983. Detection of lipase activity in cocoa powder (sensory method). In *Cocoa Powders for Industrial Processing*. Cacaofabriek de Zaan, the Netherlands.

Oterholm, A., and Ordal, Z. J. 1966. *J. Dairy Sci.* 49(10). 1281. Technical Bulletin no. 800. Sigma London Chemical Co. Ltd., Poole, England.

Purr, A. 1965. *Nahrung,* 9(4), 445.

Vaeck, S. V. 1960. Cacao butter and fat bloom. In *Twenty Years of Confectionery and Chocolate Progress*. AVI Publishing Co., Westport, Conn.

# 21

# Pest Control

A number of the comments in this chapter refer to procedures in the United Kingdom but it is felt that many of these apply worldwide. The technical material, identification of insects and rodents, and methods of control are generally applicable in every country. Life cycles, particularly of some insects, vary with the climate, and this will affect methods of control to some extent.

## PEST CONTROL

The scientific control of pests in food factories and warehouses is now an accepted procedure. Before World War II, losses resulting from the consumption and fouling of foodstuffs by insects and rodents had become a very serious problem.

Although the larger manufacturers operated their own method of control, there was virtually no supervision of most warehouse and dock storage. The same applied to the transportation of food and very poor conditions prevailed in ships and railroad cars.

In order to conserve food supplies and prevent damage, the United Kingdom established an Infestation Branch of the existing Ministry of Food. As a result, training courses for new personnel were organized with the objective of improving sanitation and pest control on all premises where food was prepared, stored, or sold.

In the early days of these courses, students were taken to see some of the worst cases of infestation and younger readers will be incredulous that such conditions could have existed in food premises. Examples are worth quoting:

1. In a warehouse storing cocoa beans (which had been there for several years), the degree of infestation by the cocoa moth had reached such proportions that layer upon layer of webbing spun by the larvae had formed a continuous mat resembling a fabric over all the bags of beans. Ephestia moths were flying everywhere, almost like a snowstorm.

2. Bags of cocoa beans were unloaded from a boat from West Africa into railroad cars that stood in a siding overnight. It was late autumn and the drop in temperature overnight started migration and next morning hundreds of thousands of ephestia larvae were crawling from the wagons over the ground and up adjacent walls. Migration had also commenced from stacks in a warehouse and larvae were so numerous that from one corner of the stack they were crawling over one another in a seething mass some 3 in. deep.

3. Rats were a serious problem. In a London warehouse in which cattle meal was stored, rat "smears" at the top of the roof supports had reached a depth of an inch or so. Smears are formed by rats that have inhabited dirty sewers and other filthy buildings, rubbing against the pillars in passing, in this case by swinging under the cross-rafters against the roof-supporting pillars.

Conditions such as these have mostly disappeared. In the enlightened countries of the world, pest control and good sanitation are now accepted and this is reflected in a healthier population.

Continued vigilance must be maintained because insects and rodents are capable of very rapid increase if uncontrolled. Where there have been political and economic problems in some African countries, cocoa beans, for example, have remained too long in storage without proper supervision. In such instances, quality deteriorates and insect infestation increases rapidly.

## Pests Associated with Cocoa, Chocolate, and Confectionery

The pests that cause damage and loss to these products can be divided into (1) those that are indigenous, such as rats, mice, some birds, cockroaches, and certain moths and beetles that are native to warmer climates, but that have since become acclimatized to survive a mild winter, and (2) those that are tropical in origin, mainly insects, and that cannot survive the winter unless introduced into buildings that are permanently heated.

## INSECT PESTS

### Identification

Jan.: "What's got thur, you?"
Will: "A blastun Straddlebob crawlun about in the bag."

Jan: "Straddlebob! where didst larn t'call'n that?"
Will: "Why, what should e caal'n then—tes the right naam."
Jan: .: "Right naam no—yer gurt vool casn't zee tes a Dumbledore."
Will: "I know tes; but vur aal that, Straddlebob's so right a naam
    vorn as Dumbledore es."

(With apologies to Haliwell *History of the English Language*—Latham)
This is English West Country jargon—Author.

The control of insect pests in foods necessitates a knowledge of
their habits and especially their life cycle, and to obtain this
knowledge requires correct identification of insects found. Most big
food companies employ an entomologist whose duty is to act as
hygiene and pest control officer and obviously that person's identification of insects would be better than that of Jan or Will.

Fortunately, the variety of insect pests associated with chocolate
and confectionery products and the raw materials used for their
manufacture are relatively small compared with the vast number of
species distributed throughout the world.

The experienced factory entomologist can readily identify a particular member of this select group and will know its habits and
methods to be used for its control. Occasionally, a strange insect may
appear with a customer's complaint and identification usually suggests its origin, which may be some other manufacturer's product or
premises. This is useful to provide evidence for a public health officer
and absolves the manufacturer from blame.

All food manufacturers receive complaints from time to time, and
although the proportion is small compared with total production,
publicity is damaging and everything should be done to prevent this.

At one time, most complaints arose from insect attacks that had
occurred in shops and warehouses selling and storing finished
products. Many of these premises, on inspection, revealed considerable infestation and the causes of damage to the foods were not
understood by the shop owners. Stock control was often bad and
sanitation lax. In some cases, there was the opinion that chocolate,
for example, and insect infestation were inseparable and the manufacturer was usually blamed.

With the development of supermarket chains stock control improved. A very big change has occurred in packaging, particularly
with confectionery bars that are now generally in heat-sealed
wrappings, which make access by insect larvae or beetles much more
difficult. Because of this, the responsibility is transferred to the
manufacturer, who must ensure that infestation of the product does
not occur before it is wrapped.

Insect control and sanitation are costly and often it is difficult to convince senior management that these expenses are justifiable. As a result of improved control over the years, infestation is probably not so obvious so the cost of preventative measures may be even more difficult to vindicate.

## Descriptions of Insects

The following insects have been found associated with raw cocoa, chocolate and confectionery ingredients, and factory plant. Brief descriptions are given of the "order" and main "species" associated with a particular product or factory location. Further descriptions and means of precise identification are given in references at the end of the chapter.

**Order Lepidoptera** This includes butterflies and moths, and certain members of the moth species are the greatest problem in the chocolate industry.

**Species** *Ephestia elutella* is of subtropical origin; now called the "warehouse moth," it was formerly known as the "cocoa moth." It has become established in warehouses and shops throughout the country and is capable of wintering in the larval form in unheated premises.

*Ephestia cautella,* the tropical warehouse moth that comes in cocoa beans, does not survive the winter in adult or larval form but has become established in warmer factories or warehouses where it continues to breed profusely if control measures are not taken.

Both moths are small, about $\frac{3}{8}$ in. long when wings are folded, of a mottled gray color, and the adults of the two species are indistinguishable to the casual observer [Fig. 21.1(a)]. For correct identification, dissection is necessary. They take no food, but water is required for fecundity. Both go through the same stages of metamorphosis, the female moth lays up to 300 eggs but from observation the number is usually much less than this. The eggs are about 1/50 in. in diameter, are deposited indiscriminately, and will hatch in ten to fourteen days in average warmth. The emerging larva is a minute caterpillar capable of entering an orifice greater than 7/1000 inch. It will wander until food is found, which then becomes the larval home until fully grown, but if it should be disturbed by outside happenings, it may crawl to other food.

The duration of this stage can be variable but usually occupies about eight weeks when the larva will be about $\frac{3}{8}$ in. long. It will then crawl upwards if possible to a crevice in which it will pupate. The

Fig. 21.1a.  Ephestia Moth

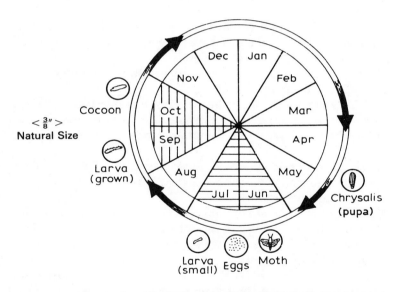

Fig. 21.1b.  Life Cycle of the Ephestia Moth Under Unheated
Conditions in a Warehouse or Shop in the United Kingdom

pupation results in the formation of a chrysalis that is the resting
stage, and for the warehouse moth (E. elutella) this normally lasts
until the following year in unheated shops or warehouses. Occasionally, where there is continuous warmth, more than one life cycle
a year occurs.

The life cycle may be represented diagrammatically [Fig. 21.1(b)]. The life cycle of E. cautella is similar to E. elutella with the exception of the pupal stage, which is much shorter in the ambient conditions of the factory. Two life cycles per year are usual and in certain circumstances three are possible.

Insects of four other species may occur occasionally, they are:

*Ephestia kühniella*—the Mediterranean flour moth, which is found infesting cereals, cookies and flour.

*Plodia interpunctella*—mostly found associated with dried fruits.

*Endrosis lactella* (Sarcitrella)—white-shouldered house moth, found in cereals and dried pulses.

*Paralipsa gularis*—now recognized as the "nut moth."

*Corcyra cephalomica* (rice moth)—is becoming a pest of increasing importance. In Ghana it has been more prevalent than Ephestia.

**Order Coleoptera**   These are the beetles and there are in all about a quarter of a million described species and according to Hinton (1943) some 400 are associated with stored foods.

*Species*   Those associated with cocoa, chocolate, and confectionery are mostly confined to the following.

*Araecerus fasciculatus.* A small, brown beetle about $\frac{1}{4}$ in. in length infesting tropical warehouses—occasionally arrives in warehouses with cocoa but the insects are in a comatose condition and soon die in cold weather.

*Carpophilus dimidiatus.* A small, dark-brown, oval beetle about $\frac{1}{8}$ in. long with some yellow markings. Occasionally found in warehouses, rarely a factory pest.

(*Sylvanus*) *Oryzaephilus surinamensis* (saw-toothed grain beetle). A small, narrow, brown beetle ($\frac{1}{8}$ in.) of cosmopolitan distribution. Pockets of infestation have been found in factory locations where conditions are warm and damp.

*Dermestes lardarius* (larder beetle, $\frac{3}{8}$ in.). This is a fairly large oval beetle with conspicuous band marking on the back. Feeds on dried animal products such as skins, furs, dried fish. Occasionally returned as "complaints," having been found in a chocolate package. The origin of such complaints is normally other material at the point of sale and it is not a factory pest.

*Necrobia rufipes* (copra beetle). Red-legged ham beetle ($\frac{3}{16}$ in.) bright bluish-black, almost luminescent beetles sometimes found infesting raw cocoa in considerable numbers but they do not multiply in a moderate climate. Not a factory pest.

*Lasioderma serricorne* (cigarette beetle, $\frac{1}{16}$ in. +). This is a notorious tobacco pest. Squat brown beetles with head bent downward. Found infesting raw cocoa in considerable numbers but die out in cool weather—not a factory pest.

*Stegobium (Sitodrepa) paniceum* (drugstore beetle, $\frac{1}{8}$ in.). A small, brown beetle. Pockets of these insects have been found occasionally in the factory and sometimes in cocoa stores and other warehouses. Will live in stale foods such as bread and in seeds and spices.

*Tribolium confusum* (confused flour beetle).

*Tribolium castaneum* (rust-red flour beetle $\frac{3}{16}$ in.). Reddish-brown elongated beetles. Can be a serious pest in the factory and will breed in warm locations in great numbers if undisturbed. Heavy infestations have occurred in milk crumb stores and silos. Is found from time to time in raw cocoa.

*Tenebroides mauretanicus* (cadelle beetle). A black-brown insect ($\frac{3}{8}$ in.) occasionally found in cocoa. It is a predator of moth larvae and its presence should be regarded as an indication of moth infestation.

*Tenebrio molitor* (meal worm). This is a similar black-brown beetle well known as a pest of cereals but will also infest other food products.

*Ptinus tectus, Ptinus fur, Niptus hololeucos* (spider beetles, $\frac{1}{8}$ to $\frac{3}{16}$ in.). These have long legs and stout hairy bodies and resemble spiders. They will breed in a moderate climate and have been found in great numbers in the crevices and floor boards of older warehouses and shops. In recent years, they have become a major pest. They are omnivorous and mobile and will infest any premises that handle foodstuffs. The adult often enters packages without attacking the contents but the larvae, which can gain access through very small apertures, may start a high-density infestation.

**Order Orthoptera** This order includes the cockroaches and the jumping insects, crickets, and grasshoppers. Cockroaches can be serious and offensive pests in a food factory. They are now regarded as a separate order, i.e., Dictyoptera.

**Species** *Blatta orientalis* (common or oriental cockroach, 1 in.). In the past, these insects were a major domestic pest but better sanitation and building construction have eliminated them from the better-quality housing. In some densely populated city areas, however, particularly in warm climates, they are still a problem almost beyond control.

Cockroaches like warm, damp locations and will feed on moist decaying food and refuse that gets beneath stoves and cupboards. They are nocturnal and if infestation is suspected, night inspection with sudden lighting of the area can be very revealing.

*Blattella germanica* (German cockroach, steam fly, $\frac{1}{2}$ in.). These are light-brown, quick moving insects that have been a much more serious pest in chocolate and confectionery factories than the common cockroach. They also are nocturnal and feed on decaying matter in warm, wet places and in these conditions breed very rapidly. They do not attack the finished product, but when numbers are large, they may fall into liquid chocolate in a molding plant, and if undetected and wrapped with the chocolate, can have a distressing effect on the purchaser.

The most serious aspect of cockroach infestation is that they distribute pathogenic organisms. These are present in their bodies and in their droppings and the following have been identified: Staphyloccus aureus, Salmonella typhimurium, Salmonella typhi.

General procedure for insect control is given later but it seems appropriate here to mention the special methods used to eradicate cockroaches.

*Lacquers.* Persistent insecticides are available in the form of lacquers or varnishes that are applied to surfaces in the infested area. The insecticides used are fairly long-lived, the actual principle being contained in the lacquer.

*Residual insecticidal sprays.* A concentrated weekly treatment is best. On food premises, a synthetic pyrethroid of the resmethrin or deltamethrin type, giving some persistency as well as "knock down", should be used. Regular treatment using ultra low-volume sprayers with resmethrin/deltamethrin will eventually establish control.

*Wettable powders.* These are mixed with water but should not be sprayed onto surfaces that may come into contact with food. These powders contain residual contact insecticides that may be toxic and great care should be exercised during the mixing process. They persist for a period and are useful when heavy infestations exist. Bendicarb is one such insecticide; fenithrothion is another.

*Dry powders.* These can be used in food factories if synthetic or natural pyrethroids are the active chemicals. They are blown into cracks and crevices or other locations not easily accessible by other methods.

*Acheta domesticus* (house cricket, $\frac{3}{4}$ in.). This is mentioned because they may occur in considerable numbers in canteens and the odd insect may get into an employee's food. They are easily controlled.

**Order Arachnida**   Spiders and mites are included in this group. Spiders, which are eaters of other insects, seem to be accepted as useful citizens although looked upon with horror by some people.

*Tyroglyphus* (various). These are mites and quite serious infestations have been found in flour and powdered sugar, usually when damp conditions prevail. They have also been found in crevices under machinery when floor washing has left moist sugar residues. These pests are mostly very small (less than $\frac{1}{32}$ in.) and in certain circumstances the infected food can become a seething mass before it is detected.

**Order Diptera**   The common housefly and relatives are in this order and can be a serious pest in food factories. More are likely to enter the factory from outside sources than breed in the factory, although bad hygiene, particularly in canteens, will encourage breeding. They are likely to contaminate by alighting on liquid chocolate, particularly enrober covered units. A chocolate bar with a fly attached can have serious repercussions on the manufacturer if it gets into the hands of a public health officer.

*Drosophila*. This is a small fly often called the "fruit" or "vinegar" fly. It is an indication of fermentation and can multiply in huge numbers if a pan of syrup is left unattended.

*Musca autumnalis* (cluster flies). These do not feed on stored products but, like the housefly, can be a pest because of the very large numbers that assemble in light, warm window areas in the autumn.

**Order Hymenoptera**   This includes bees, wasps, and ants, and sometimes these are serious pests that contaminate confectionery products and constitute a foreign matter hazard.

*Vespa* (*vulgaris*). The common wasp will enter a sweet or jam factory in vast numbers in the early autumn. They need no description for identification.

*Microbracon hebetor*. This is worth mentioning as it is a parasitic wasp found in cocoa bean stores. The adult wasp kills ephestia larvae by stinging and the microbracon larvae are parasitic on the ephestia larvae.

Another parasitic hymenoptera is *Nemeritus*.

*Apis* (hive bee). This valuable insect can also be a pest in a food factory and like wasps, will, late in the year, arrive in their thousands to devour syrup or soft confectionery. The only answer is to keep them out by means of window screens. Bees are particularly partial to ginger syrup.

*Monomorium pharaonis* (Pharaoh's ant). This very small ant (less than $\frac{1}{10}$ in.) is reddish brown and nests in building structures, usually in very inaccessible places and will emerge in thin lines from crevices often a long way from the nest. With food available they will multiply in great numbers and are very difficult to eradicate.

Ants have been discovered in purchased foods. On one occasion boxes of crystallized violet and rose petals used for decoration evolved vast numbers of these insects when delivered to the workrooms.

**Orders Thysanura (Silver Fish), Psocoptera (Book Lice)**  These are frequently associated with paper and carton board packing materials and thrive where humidity is high. They feed on starchy products such as dried adhesive. They are a nuisance if their numbers are large as they may be present in chocolate boxes or corrugated paper and pass unnoticed when the candies are packed. They cease to exist when drier conditions prevail.

A serious development of this insect has arisen, particularly in England, due to extension of the wooden pallet system. Loads of packed boxes of chocolates and other candies are moved on these pallets by fork-lift truck. When not in use, the pallets are frequently stored in the open and become wet. When brought inside for use, even if they are dried, there is residual moisture in the crevices, which harbors the insects.

## THE CONTROL OF INSECT PESTS

There are two methods of approach to the control of insect pests—the first is good housekeeping, one of prevention, making it impossible for them to develop and increase. The second is extermination by means of insecticides, fumigants, or heat sterilization. The first makes the second less necessary in a factory. Raw materials may need chemical methods occasionally.

### Raw Materials

The first precaution against infestation is to see that the raw materials brought into the factory are free of insects. Some indication of the trouble of many years ago has been given and fortunately these have mostly disappeared, but vigilance is still very necessary.

**Raw Cocoa Beans** This is a major source of insect infestation, the trouble generally being the moth Ephestia cautella. Beans brought into West African stores from the farms may be held in storage for a period awaiting shipment, and insect development may occur there. In recent years, great improvements in warehousing have been made and, when necessary, this improved housekeeping is aided by the use of insecticides or fumigants. During shipment of the beans, which are contained in sacks stacked in the holds, latent infestation can develop and multiply, so immediate inspection of the beans in the boat on arrival is necessary. This is usually carried out by the factory sanitation officer. The contents of each hold are examined, the extent of infestation assessed, and any insects found identified.

The sanitation officer decides whether to:

1. Fumigate the whole consignment.
2. Put into warehouse with, possibly, fumigation later.
3. Store without fumigation.

The large port warehouses hold many thousands of tons of beans, often for different companies and agencies, and in some circumstances beans are put into a selected store straight from the boat until it is full. No infestation will develop during the winter months as the stores are unheated, and at a later date the whole store may be sealed and fumigated. This is usually done with beans that will be stored until the following summer; other cocoas that are required during the winter months would be delivered in small quantities to the user factory at a rate to meet production requirements. In this way, a serious potential source of infestation is kept out of the factory.

Some users of cocoa beans fumigate all deliveries immediately on receipt into the factory, and this is fairly general practice in the United States where there are different arrangements for port control and purchase of beans.

Methods of fumigation and the use of insecticides are discussed later.

## Container Transport

Large quantities of raw cocoa are now shipped by container and generally it is not possible to determine levels of infestation or to assess quality until the shipment arrives at the manufacturing premises. It is important, therefore, that some system be introduced to ensure that infested material is prevented from entering the factory. In continental Europe, most containers are unloaded at the port and quality surveillance can be carried out there. In the U.K., in order to obtain maximum financial benefit from the door-to-door transport, most containers are delivered directly to the factory.

It is important that there be some prior knowledge of the condition of a consignment before arrival at the factory. If containers, on opening, are found to be infested and cannot be delivered to production departments, manufacturing will stop—an inconvenient and expensive occurrence.

Inspection or sampling at the port of loading has been investigated but reliability of these procedures is doubtful. Samples have to be air freighted to the user to be of any value.

Where fumigation of all consignments takes place on arrival, as in many U.S. factories, treatment in the container or immediately after unloading seems possible.

**Nuts**  Nuts are probably the most vulnerable raw material, particularly walnuts, peanuts, and hazelnuts; almonds are less affected and with brazil nuts, infestation is practically unknown.

The only safe way to keep nuts is in cool storage at a temperature that will not promote insect development. This is usually 7°C (45°F) (range 43 to 48°F) and some control of humidity is necessary unless the nuts are in moisture-proof packing. A relative humidity of 55 to 65 percent is generally satisfactory and this is particularly important for walnuts, which will soon become sour if they absorb moisture.

Another method tried successfully for hazelnuts was to give a very light roast on receipt of the new crop nuts, reducing the moisture content from about 5 percent to 2 percent and store in cool store in well-lidded drums. Apart from controlling infestation, this keeps the nuts sweet for at least twelve months.

Many makers of walnut-topped confectionery or chocolate blocks fumigate immediately before use because walnuts are not roasted whereas almonds, hazels, and peanuts are, and this destroys infestation. In addition, the many crevices in a walnut provide ideal conditions for tiny larvae to find protection and feed.

**Raisins, Sultanas, and Currants**  At one time, raisins were a serious source of infestation but they are now fumigated at the packing station and little trouble is experienced unless they are stored for long periods in unsatisfactory warehouses. If the humidity in the storage area is high, the fruit may become infested with mold feeding beetles.

## Other Raw Materials

Flour may become infested but usually stocks are kept for quite short periods. Care must be taken to see that the flour store remains free from infestation as dust will accumulate in crevices.

Other dried fruits such as dates and figs may occasionally be infested but these are now usually treated with the same care as raisins.

Packing materials, particularly strawboards, may be contaminated with insects, usually through bad storage on boat or in warehouse, and china novelties for gift packs sometimes received from the pottery packed in straw of doubtful origin have been known to contain numerous beetles (Ptinus, Trigonogenius).

Railway cars that have been used for other materials will occasionally introduce pests into raw foods. Close scrutiny by the receivals department and claims against the railway authority are usually effective in preventing recurrence.

Container transport has already been mentioned in relation to incoming cocoa beans. There are also problems with transport of other raw materials or finished products.

Containers are often issued from a common depot and may have been used previously for materials completely incompatible with foods.

Empty containers for food use should be inspected at the depot and also on receipt at the loading point to ensure they have been adequately cleaned.

They should be examined for:

1. Condensation or dampness on *all* interior surfaces.
2. Physical damage that would result in entry of rain or other extraneous materials.
3. Foreign odors that could cause taint.
4. Infestation, rodent excreta, and any other substances remaining from previous use.

On receipt of a loaded container, a similar inspection should be made, and obviously an examination of the contents for damage.

## SANITATION—THE HUMAN FACTOR

Factory, warehouse, and office sanitation with a realization of personal responsibility is of prime importance. Unless the management and supervision are alive to the need for tidiness and cleanliness, little progress will be made and it is surprising how many people do not notice untidiness.

The sanitation officer and staff should hold meetings with supervisory staff and trade union representatives. At these meetings, examples and photographs of dirty, untidy, and infested locations may be shown, together with samples of foreign matter or infested goods that have been returned by the public, but unless the impact is such that the audience transmits the talks into deeds in their departments, no use is served.

The sanitation officer and staff must see that regular inspection of departments is carried out by the supervisors, and from time to time the officer or a senior member of the staff should accompany them.

## Cleanliness

The following are the main points to watch concerning tidiness and cleanliness:

**Floor Cleaning**   Much of this is done mechanically but often splashes of dirty water containing sugar and pieces of chocolate are driven into crevices and beneath machines. These dry out slowly, become moldy, and will support the development of mold-feeding beetles and mites. Mechanical cleaning must be supported by some hand cleaning, frequent changing of the cleaning water, and the use of mild, odorless detergents in some cases. Structural simplicity and good machine design assist cleaning.

**Refuse Bins**   Containers for refuse must be provided at strategic points but these must be watched carefully and emptied and cleaned regularly, or they will become reservoirs of infestations. Many types have been tried and those which have a removable inner receptacle or destructible waterproof bag are preferred.

**Cupboards, Drawers**   These can become dumping grounds for the human squirrels. Food and chocolate residues encourage mice as well as insects; wet brushes and mops breed cockroaches. Only regular inspections cure these troubles.

**Machinery**   Many tradespeople are not tidy minded and when machines are dismantled it seems customary to leave the fittings in a pile in a corner of the production room where chocolate dust accumulates and becomes infested. These places also become dumping ground for other refuse. Such locations can lead to heated arguments because the engineers will say that the fittings will be required again when the machine is reerected or replaced—but that may be months hence. Some machines have dead spaces that harbor refuse or become places to deposit rubbish; this is dealt with under machinery design.

**Utensil Cleaning**   A department that cleans and sterilizes molds, trays, drums, trollies, and platforms is an essential part of any food factory. For trays and molds, washing machines are used that carry the articles through boiling water jets with detergent, followed by rinsing and drying.

The larger pieces of equipment are cleaned manually with high-pressure water and steam jets and dried in a hot room, and repair and maintenance of equipment are coincident with cleaning. Some of the new high-velocity, low-volume pumps are very useful for removing fatty deposits.

## Special Cleaning

The foregoing has dealt with visible dirt and untidiness but equally important is that which cannot be seen and which accumulates in ducts, coolers, elevator boots, support girders, sheet metal safety covers, and the hosts of machinery crevices. These accumulations of foods, often consolidated and undisturbed, provide ideal conditions for the growth of larvae and it is in these locations that the Ephestia larvae are found. This insect is the major pest.

In spite of efforts to improve the construction of plants and machinery, such crevices still exist. Special cleaners directly supervised by the sanitation officer may be used to deal with this problem, using special tools and aided by the engineering staff to dismantle guards and ducts. This cleaning is carried out on a rotation system and when production is stopped on weekends and holidays. Such a rotation is particularly important for wrapping machines. Dismantling of plant for cleaning reveals faults in design and these can often be eliminated or provision made for easy access—doors in large air ducts are a typical example.

Certain materials are much more prone to infestation than others. Milk crumb dust accumulating on ledges and in crevices will tend to cake, refiner flake chocolate is similar, and harbors a great deal of infestation if not removed. Liquid chocolate kills all forms of insect life and little trouble occurs in the conches, enrobers, or storage kettles except where liquid chocolate has splashed and solidified.

Food drinks containing milk powder, malt, sugar, and cocoa are very bad sources of infestation as any dust allowed to accumulate will pick up moisture from the air and consolidate. In such a condition, it cannot be brushed away from girders and ledges as is the case with crumb dust and chocolate flake.

Any plant handling nuts, whole or chopped, is vulnerable, as infestation once started develops rapidly and a strict program of special cleaning is essential.

During the special cleaning operation an obvious precaution is the disposal of infested material. Sometimes hundreds of larvae may be present in a few pounds of material removed from a space normally inaccessible and this refuse must immediately be put in a tightly lidded drum and the accumulation destroyed by burning.

An area of the factory that needs special attention is the bean-cleaning and -roasting sections, for although the inspection of raw cocoa beans should prevent them from coming into the factory infested, this is not infallible. This part of the factory is best built away from the main production areas and no insects will survive the roasting process.

The special cleaning-staff should be properly trained and supervised by the sanitation officer. The personnel should not be changed frequently as the work requires a certain amount of skill and experience. However, good housekeeping will not totally control infestation and the special cleaning should be supported by spraying, and possibly fumigation. These procedures are described later.

**The Detection of Factory Infestation**  In a well-managed factory it would be hoped that insect infestation is not obvious in the form of numbers of flying moths or crawling larvae.

The special cleaning staff must be directed to positions where their efforts are most effective, and a system that operated well was as follows: In departments where infestation could develop, moth trays made of aluminum or plastic about 12 in. by 8 in. by $1\frac{1}{2}$ in., were placed on brackets about 8 ft from the floor. Small departments would have one tray only but large departments that extend the length of the building might have three or four trays. The trays were filled with water to a depth of an inch and this was replaced on inspection once weekly, and at the same time the number of moths caught in the water recorded. Some of these would be dissected for identification. The number caught in each tray throughout the factory was charted and this would show up a sudden increase in a particular tray. Often none would appear in the tray, or at most two or three, but if this jumped to a dozen or more, a detailed inspection of the area would be made and invariably some pocket of infestation would be found. When this was removed, the moth catch returned to normal although the catches in all trays showed a seasonal pattern with increases in the warmer months.

When infested material was removed by special cleaning, an assessment of the density of insects was always made as this could give some indication of the age of infestation. The method used was to place a layer of the debris, about $\frac{1}{2}$ in. thick, on a tray over a hot plate. After a short period, the larvae would rise out of the layer away from the heat. These were removed by forceps, counted, and the density calculated. A more recent method is to use sexual attraction. Female moths are confined in a small cage and this is set up on a plate coated with horticultural adhesive (Bandite). Alternatively, a synthetic sexual hormone (pheromone) may be used.

## Design of Machinery, Equipment, and Buildings

Great efforts have been made to influence manufacturers of food machinery in the design of their plant and machines so that they will reduce to a minimum inaccessible places that harbor insects. At the same time, the construction of simple safety covers for moving parts and their easy removal by those authorized to do so must be considered. Publications by the British Food Research Association with the cooperation of leading food manufacturers have given valuable advice to machinery constructors.

It is not possible to quote other than the important items from these publications but a few examples are given to enable the reader to understand the enormity of the problem.

**Buildings** *Tongued and grooved* boards must not be used in production departments.

*Inside brickwork.* All pointings should be flush and smoothly finished. Holes must be filled with hard setting cement. Hard glazed surfaces to be used up to the window sill for general factory buildings, but in the case of cocoa, crumb, and starch rooms, these to be continued to ceiling level.

*Floors.* These must be constructed with a minimum of joints that may open and become crevices into which water and food materials penetrate. Trolleyways need particular care.

*Pits in floor.* All pits should be avoided, but when essential they must be large enough to allow room for cleaning. Glazed surfaces and rounded corners must be specified.

*Partitions.* For partitions, plywood or compressed fiberboard must be used. The underside of the partitions must be kept 4 in. off the floor, and only where heat loss is to be prevented, a hinged flap provided.

**Plant** *Purchased machinery.* All machinery manufacturers for the food industry should be guided by the publications mentioned above. Many of these are published in German, French, and Italian as it must be realized that much machinery, particularly in the chocolate and confectionery industry, is produced in these countries.

*Side frames and general castings.* (a) Ribbed castings. The top side of horizontal ribs must slope at 45° or alternatively have a large radius between vertical and horizontal surfaces. Corners and pockets should be avoided and ribs should always be in the position that allows the greatest accessibility. (b) Hollow castings. If hollow castings have to be used, it is important that the inside be completely sealed. All unused holes or openings must be filled in.

*Bolts.* These should be reduced to a minimum; welding is preferred. Screw threads of bolts must not be left protruding through nuts. Such screw threads are difficult and expensive to clean. Holding-down bolts should always be fixed on the outside of a casting. Holding-down bolts for heavy machinery must give at least a 2-in. clearance between the nut and the vertical face of the machine. Where holding-down bolts are in a slotted hole, the slot should be filled with grease. Counter-sunk socket headed bolts should have the socket filled with grease.

*Steel platforms, gantries, etc.* Supporting structures should be of welded tubes.

*Motors.* Fixed on floor. For all motors up to $7\frac{1}{2}$ hp brackets should be made of flat iron to give at least 4 in. clearance. For all motors over $7\frac{1}{2}$ hp, channel supports with flanges turned outward should be used. Channels should not be less than 6 in. apart. If a motor is fixed in the corner of a department, there should be at least 12-in. clearance all round.

*Enclosed metal conveyors.* (a) Where conveyors pass through walls or partitions, an easily detachable sheet-metal filling-in piece should be fitted tightly around the conveyor, particularly where the conveyor passes through from a dusty room, or into a finished product room. Conveyors should be fixed at least 12 in away from the ceiling. (b)

Conveyors carrying raw material should not pass through production rooms handling finished goods.

*Sheet-metal racks.* Tubular supports must be used, the tops and bottoms to be sealed. Lugs for carrying shelves should be welded on. A space of 1 in. must be left between shelves and a wall or partition. The bottom shelf must be at least 9 in. off the floor. The lower portion of the vertical edges of the shelves must be folded back tightly against itself. Complete racks must be portable and they must also be open all round, that is, have no backs or sides.

*Electrical work.* All conduit pipes must be positioned at least $\frac{1}{2}$ in. clear of all walls or partitions by the use of spacing saddles. All junctions, bends, etc., must have cover plates properly fixed. Conduits should be sealed and should not be supported through ferrule holes.

*General.* Trolleys, trays, and load platforms are inclined to be forgotten but are none the less important. Trays should never be made of wood but from metal pressed from a single piece, and wood is best avoided for trolleys and load platforms, or if it is used, precautions already mentioned must be taken to avoid crevices and inaccessible places. A marking system to ensure that all this loose equipment goes through a rotation of washing is of great value in preventing infestation and in rejecting damaged articles.

The use of stainless-steel equipment should be encouraged. Initially, it is more expensive but maintenance costs are less—it requires no painting and cleaning is easy.

Windows and doors that can be opened should be screened to prevent the entry of insects and birds.

## Heat Sterilization, Fumigation, Insecticides

It has already been indicated that cleanliness and tidiness are the first line of defense against insects, but it is virtually impossible for all machinery and plant to be insectproof or for all raw materials to be insect-free.

The development of automation and increased use of intricate electronic equipment has made it difficult to adopt all the constructions outlined previously, although from a mechanical efficiency standpoint exclusion of dust is important.

The use of heat to destroy insect life is the method that leaves no after effects, as both fumigants and insecticides may leave trace residues. The application of heat is not always possible or completely effective and requires specialized equipment and a precise program of operation. Fumigants are very lethal to insects and people and, because of this, security precautions must be very strict. It is best to employ an experienced company for large-scale work, but batch fumigators that take up to 1 ton of goods can be operated by the sanitation officer's staff. Insecticides have limited use and do not penetrate large stacks of material—they can be used to attack large concentrations of flying insects or larvae, but the result is only temporary.

**Methods of Heat "Sterilization"** Heat can be used to destroy insect life in isolated plant or machines or in large rooms. In some cereal mills, steam radiators built into the construction are used and periodically the plant is stopped, and the whole area sealed and raised to a lethal temperature. In some factories, heat sterilization and fumigation are used, but there has been a gradual trend toward fumigation as it is easier to operate. Heat is applied by the use of very large electric fan heaters and the design of these has been improved over many years of operation. Originally, experiments were conducted by erecting tents of fireproof fabric over refining rolls and wrapping machines and heating the internal air electrically. Modern equipment consists of portable 25-kW fan heaters that direct large volumes of hot air in any desired direction.

The destruction of insect life by heat is largely due to desiccation and provided a temperature of 120°F is reached for several hours *in all positions* where infestation is likely, no life stage (adult, egg, larva, pupa) will survive. A sufficient number of heaters is required to raise the temperature to 120°F and hold it for four hours. The total period should not exceed 24 hr.

To ensure that these conditions are met, considerable preparation of a department is required:

*Cleaning.* Heat will not penetrate thick layers of powder—a certain amount of special cleaning is required before commencing the heating.

*Ducts, safety covers.* These must be opened or removed to allow hot air to penetrate.

*Sealing.* Drafts from badly fitting windows or doors cause a big heat loss and, locally, lethal temperatures may not be reached. Sealing of cracks and crevices with adhesive paper is necessary.

*Machinery.* Without precautions, machinery may be damaged even at 120°F and locally temperatures may rise higher. Refiners and other precision machinery should be slackened and, after heating, replacement of grease and oil is required.

It might seem that this procedure is complicated and expensive but over a period an efficient team organization can be built up that is able to sterilize a large department and return to production in 48 hr. A disadvantage of this system is that the time taken to build up a lethal temperature may allow mobile insects to escape from the heated area.

Electronic apparatus, such as electric eyes, foreign-matter detectors, computers, punched card systems, and production control consoles may be damaged by high temperatures. Advice must be sought from the manufacturers of such equipment.

## Fumigation

**Warehouses and Factory Departments** Fumigation has been used for many years to exterminate insects and rodents in ships, dock stores, barns, and mills. Hydrogen cyanide gas was employed originally but owing to its extreme toxicity it has been superseded by ethylene oxide or methyl bromide and more recently by aluminum phosphide (see later). Ethylene oxide or a combination with carbon dioxide was used extensively at one time but was also found to have toxic effects. It is now banned as a fumigant in most countries but is still used for sterilizing medical instruments and certain spices.

Methyl bromide is now frequently used and can be distributed readily from cylinders through narrow tubing to all parts of the area being fumigated. Concentration checks are made in different parts of the area by a simple recorder and small, gauze-covered tubes containing specimen insects are placed in various locations. To aid detection in very minute concentration, a lachrymator may be added—usually chlorpicrin. Methyl bromide is a very penetrating gas and will destroy insect life in fairly thick layers of powder, and although it is preferable to remove as much residue as possible, as with heat sterilization, gas fumigation is very suitable for treatment of departments containing complicated equipment where dismantling for special cleaning is difficult and expensive.

The preparation of a department or warehouse for fumigation is somewhat similar to that for heat sterilization. Sealing of door and window crevices and pipe channels is important as, in addition to prevention of gas loss, seepage into adjacent rooms must be avoided.

In carrying out these treatments, the sanitation officer's staff must work closely with the fumigation contractor in the provision of warning notices, door shutters, and locks, and in making security arrangements with factory patrols.

Fumigation of a large department can be carried out in somewhat less than 48 hr, allowing 24-hr exposure to the gas.

The concentration of gas normally used for methyl bromide is 0.68 kg (24 oz) per 1,000 ft$^3$. The temperature of the area to be fumigated should be above 15.5°C (60°F) for best results.

In addition to treatment of factory departments and warehouses as described above, stacks of raw materials are fumigated under large plastic sheets. The sheets at the base of the stacks are kept flat on the ground with heavy weights and the gas distributed through narrow tubing to various extremities under the sheets. Wrapping machines, sets of refining rolls, and other isolated equipment can be dealt with by this sheeting method but the fumigation of one piece of plant in a department is of value mainly when it is to be removed to stores or outside the factory. If it remains in the same department where some infestation is general the machine will soon become reinfested.

Cocoa beans are sometimes fumigated in barges at port. Two hundred to three hundred tons are transferred from the ship's hold to a barge alongside, which is sheeted and gassed. This method is particularly suitable when the beans are to be transported to a warehouse or factory by waterway.

**Fumigation Chambers** A chamber that can take up to 1 ton of material is a very useful piece of equipment in a large good factory. It is an advantage to use vacuum fumigation where the chamber is evacuated before the gas is applied. This ensures very good penetration of the fumigant.

A vacuum oven with the internal heating shelves removed is an ideal arrangement. It has the vacuum pumping equipment, and the strength to withstand evacuation, and it can easily be arranged by means of a ramp to charge and discharge the material to be fumigated on trollies or platforms.

The discharge from the vacuum pump is taken to the open air so that it can be used to flush out the gas from the chamber after fumigation. Figure 21.2 is a diagram of a fumigator.

The material to be fumigated is loaded into the chamber and vacuum applied until the gauge reaches full vacuum (25 in. is usually sufficient). The gas is then let in from cylinders until the vacuum gauge drops to a reading (found by experiment) that gives the correct dosage of gas. The chamber remains closed for 24 hr, during

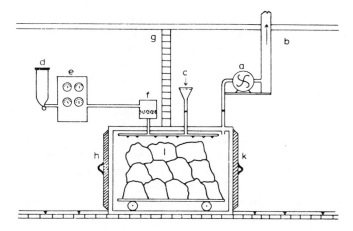

Fig. 21.2. Fumigation Chamber
a. Vacuum Pump    f. Heater
b. Vent to Outside Air    g. Partition Wall
c. Air Inlet    h. Intake Door
d. Fumigant    k. Exit Door
e. Control Panel    l. Material for Fumigation

which time most of the vacuum will have been lost. The vacuum is then raised, exhausting to open air, after which the inlet valve is opened while the vacuum pump is running. The fumigant, by this means, is flushed to the open air in an hour or so. The gas should never be ventilated into a room.

This type of plant has proved very useful for dealing with infested "returns," and small quantities of raw materials, such as walnuts and dates. Utensils such as trays and molds that may have been out of use for some time, sacks and bags, and infested material removed during special cleaning, could also be handled.

Similar chambers, usually larger, can be used without vacuum, especially when using methyl bromide. A forced ventilation and circulatory system is necessary for reliable results.

**Recent Developments in Fumigation** One of the disadvantages of fumigation is that trace residues may be left in the foodstuffs and occasionally slight taint occurs in the food or wrapping materials. One example of this concerned a polythene laminate that developed a very bad odor when exposed to methyl bromide; since that time, all packing materials are removed from a department under fumigation.

This particular case was investigated in detail and the evidence suggested the effect was caused by an impurity in either the chlorpicrin or methyl bromide, because a different batch of fumigant did not produce the same results. Chlorpicrin is a lachrymator that indicates the presence of traces of fumigant should it escape in working areas.

A fumigant developed in the United States, stated to have no taint problem, is Phostoxin. It has the approval of the FDA and its use has increased greatly of recent years, particularly for dried fruit.

Phostoxin is prepared in tablet or pellet form and the active ingredient is aluminum phosphide. The commercial tablets are 3 g in weight and the pellets are 0.6 g, and they will release 1 g and 0.2 g of phosphine gas respectively. The composition of the tablets is aluminum phosphide (55 percent) ammonium carbamate (41 percent), and purified paraffin wax (4 percent).

The aluminum phosphide slowly evolves phosphine, the active fumigant, after about 2 hr exposure to the atmosphere.

The chemical reactions occurring are as follows:

1. Ammonium carbamate decomposes in the presence of moist air into ammonia gas, carbon dioxide, and water

$$NH_2COONH_4 + H_2O \rightarrow (NH_4)_2CO_3$$

$$(NH_4)_2CO_3 \rightarrow 2NH_3 + H_2O + CO_2$$

2. Aluminum phosphide reacts with moisture to form phosphine and aluminum oxide

$$2AlP + 3H_2O \rightarrow 2PH_3 + Al_2O_3$$

The ammonia and carbon dioxide gases act as damping agents to prevent spontaneous ignition of the phosphine.

Phostoxin is very useful for the fumigation of freight cars, containers, and stacks of raw material under tarpaulins.

Pellets of the chemical are easily distributed throughout the product to be fumigated.

When using Phostoxin, in order to ensure the safety of operators, it is important that the spent tablets and sachets are collected and disposed of in water.

There has been much publicity in recent years on the subject of pesticide residues. To date, ethylene oxide has been virtually lost, and ethylene dibromide, not previously mentioned but a useful grain fumigant, can no longer be used. Methyl bromide has been subject to much criticism although evidence against it is scanty. Phosphine, to date, is satisfactory.

It is well known that fumigants and insecticides have been used indiscriminately in the past, resulting in appreciable residues in foods. Much has been published on this topic and it only remains to be said that the technologists and sanitation officers should constantly review the literature. Care and restraint in the use of these substances must always be heeded.

## Insecticides, Biological Control

The value of insecticides has been proved beyond doubt and many animal and human parasites and disease-carrying insects have been eliminated to the advantage of the community.

When DDT and BHC (Lindane) were discovered, this was undoubtedly a great advance in the technology but unfortunately these insecticides have been shown to be toxic at quite low concentrations. DDT will accumulate in human and animal organs. These chemicals and some others are called persistent insecticides because, after application, they remain effective for a considerable time. Many other insecticides have appeared since. It seems to be a human failing to think that better results are obtained from the use of quantities of medicines, vitamins, and antibiotics beyond the normal dose. This has certainly applied to insecticides. Plants and their products have been subjected to successive doses in quantity far in excess of that necessary to be effective and this has led to high pesticide residues being found in a number of foods.

Another factor resulting from erratic application is the production of insect strains that are resistant to a particular insecticide. The search for safer insecticides continues.

A valuable insecticide for food use is pyrethrum extract as it is nontoxic to humans and animals. The original preparations had a very short life and thin films rapidly lost their toxicity, particularly when exposed to bright daylight. The discovery of certain synergistic substances (e.g., piperonyl butoxide), which increased shelf life and potency, has made this insecticide a valuable aid in combatting infestation. The original was known commercially as "pybuthrin" and sprays of this substance have very quick "knock-down" action against most flying insects and will also destroy crawling larvae by contact action. Its residual effects are limited but are claimed to be about one month in dark, sheltered positions.

Pyrethrum extract is a natural product, and because of its unique value as a nontoxic insecticide, much research has taken place to produce synthetic equivalents.

Improvements continue with these substances and the latest are more effective, for example, deltamethrin and resmethrin.

Pyrethrin insecticides can be used with safety as sprays where they are liable to settle on foods or bags of raw materials. In large stores, high-pressure sprayers are used, which produce a fog of very small droplets. These sprayers can be directed against undersurfaces whereas with some fogging machines droplets fall only on upper parts of the materials and are therefore less effective.

Spraying will destroy flying insects and crawling larvae but is not effective against internal infestation, such as in cocoa beans inside a sack. Fumigation is necessary to deal with this. Spraying is a temporary means of preventing the spread of infestation. It does *not* replace special cleaning or fumigation.

**Insecticidal Emulsions and Paints** The principle of these is to incorporate insecticide in the formulation, and when applied to a wall or machine, the active substance will crystallize, forming a lethal layer on the surface of the film.

Substances used in these paints were persistent insecticides such as gammexane, aldrin, and dieldrin. Many of these chemicals are now prohibited and alternative substances are being used. It is claimed that chemicals that would not be suitable in sprays may be used in these paints.

These paints are very effective against houseflies and many crawling insects. They are particularly valuable in shops, stockrooms, and canteens where other methods of treatment are not applicable.

**Insecticidal Smokes** Pyrotechnic candles, which evolve gammexane (BHC) smoke, are used a lot in horticulture, for greenhouses. They have been used with some success in small stores, silos, and railway cars when they are empty—the smoke deposits as a toxic film and will eliminate flying insects and any beetles or larvae in crevices. Obviously, caution is necessary in the use of these smokes as residual films of insecticide will remain for a period.

**Repellents** Insect repellents are frequently employed outside the factory or warehouse, particularly if there are areas of grassland or rough terrain. It is assumed that refuse dumps are nonexistent or reduced to a minimum.

Gammexane sprays in conjunction with substances like ortho- or paradichlorobenzene are useful for this purpose and can be applied to refuse bins outside the factory particularly in hot weather.

**Insecticide Vaporizers** An invention for the dispersal of persistent insecticides by incorporating a small electric heating device around a cell of BHC has been used in shops troubled with flies, but from the results of experiment, it has little application in factories or warehouses against Ephestia. It is claimed to deposit a "molecular" layer on surfaces in the room containing the apparatus, but this might include foods in these areas.

Another invention is the vaporizer strip, which consists of a card impregnated with an insecticide and these are hung in the area to be treated. Vapona and Dichloros are examples and the active substance is an organophosphorus compound (2.2-dichlorovinyl dimethyl phosphate).

The "strips" are impregnated cards of dimensions 10 in. by 2.5 in. and this size is designed for the treatment of 1,000-ft$^3$. The active substance has a high volatility and the presence in the atmosphere of concentrations as low as 0.02 μg/l will give insect control. The strips are designed to give a continuous emission of vapor over a period of ten to fifteen weeks.

**Baits** Some insects can only be exterminated by means of poison bait, and ants are the most likely invaders requiring this treatment. Monomorium pharaonis, which has been recorded in great numbers in certain locations in confectionery factories, is difficult to eradicate, and a government publication recommends thallium sulfate or sodium fluoride baits. Since these chemicals are extremely poisonous, numbered bait boxes must be used in concealed positions. If the nest can be located, BHC is effective.

## Biological Control

The name "insect growth regulators" has been applied to the substances used for this type of control. The principle is to use a chemical that attacks a specific stage of development, thus preventing the adult insect from maturing.

A substance known as methoprene has been used for this purpose, it attacks the larval stage of most insects. Sex attractors are being applied, mostly experimentally, at the present time. Chemicals called pheromones have found use in traps and in assessing the density of infestation in certain areas of storage or production.

Recent publications by Jacobson (1984) and Wilbur (1984) give good summaries of potential methods of insect control.

## RODENT CONTROL

Rats and mice mice multiply in dirty and untidy conditions and factory sanitation is the first priority in combatting these pests. Rats will also live in wasteland or rubbish in areas around the factory where they may breed undisturbed and invade the factory for food. Rats, and more often mice, may be introduced into factories in packing cases, particularly where there is loose paper or shavings.

Rats are renowned as carriers of disease and were responsible in the Middle Ages for the spread of plague in its various forms, which devastated the populations of Europe and Asia. The disease was carried from the rat to the human by the plague-infested rat flea.

Other diseases attributed to rats are typhus, infectious jaundice (Weil's disease), trichinosis, and rabies.

The organisms of food poisoning, poliomyelitis, and types of meningitis can be communicated to humans by eating food contaminated with the excreta of rats and mice, and also merely from rats' contact with food.

Rats use sewers as a means of traveling from one place to another and there they may pick up polio infection from human excreta and transmit this to food without themselves being infected.

Many other examples of horrors arising from rodent infestation could be quoted but it is hoped that sufficient has been said to impress on the management of food factories the necessity of strict hygiene and the provision of funds to implement it.

### Rats

There are two main species of rat, Rattus rattus, the black or ship rat, which is mainly confined to ships and port areas; and Rattus norvegicus, the brown rat, which is the main pest associated with domestic and factory surroundings.

The distinguishing features, according to Mallis (1964), are shown in Table 21.1.

**Breeding Places**   In the published literature on rats it is emphasized that they will nest in undisturbed rubbish if this is left for more than a month, so it is vital to remove such dumps quickly or never make

TABLE 21.1. SOME CHARACTERISTICS SEPARATING THE ADULT OF THE BLACK RAT FROM THE ADULT OF THE BROWN RAT

| Character | Black rat | Brown rat |
|---|---|---|
| Size | Body and head approximately 200 mm | Body and head approximately 250 mm |
| Muzzle | Sharp | Blunt |
| Ears | Large and almost naked | Small and densely covered with short hairs |
| Tail | Slender, and often longer than head and body | Stout, usually not as long as head and body |
| Mammary glands | Usually 10 in number | Usually 12 in number |
| Weight | 8 to 12 oz | 16 or more oz |
| Fur | Somewhat stiff | Soft |

them. Among other places favored are sheltered areas beneath concrete foundations, sheds, poultry houses, and pig-sties. Water is essential to rats and they must live near a supply, however dirty, and they are able to swim.

**Food** Rats eat a large amount of food for their size and an adult rat consumes about 1 oz of food and 2 oz of water per day. In so doing, it tries to select the most palatable, but at the same time fouls large quantities of other food. In circumstances where food is short, they will eat almost anything, such as plastic buttons, bones, leather, and wood.

**Breeding** It is well known that rats are extremely prolific in a suitable environment. The brown rat may have six litters a year with six young as an average litter, and can breed any time of the year. There is a tendency to migration in the spring and autumn but this varies with the location and the country and no doubt accounts for the tales of armies of rats being seen by late-night revellers.

## Mice

The house mouse is principally Mus musculus domesticus, and it is this species that is the principal pest with which one must contend.

Although much has been said about the scourge of the rat, it must be admitted that the mouse in many food factories is a bigger problem than the rat.

**Breeding Places**  Mice, unlike rats, require little water, and they can obtain sufficient for their needs from the natural moisture content of many foods and can therefore live unobserved in stacks of boxes, sacks, or piles of raw materials or finished goods. Their presence may only be detected when their numbers have increased to such an extent that their faeces become obvious or the typical musty odor associated with mice is detected.

They make nests of shreds of paper inside boxes, and in the center of sacks of any dry commodities, and, like rats, they breed throughout the year with six to ten litters per year of some six to eight young. Another observation by students of this rodent is that some nests may be communal, used by several families.

**Food**  Like rats, they can damage a great deal more food than they consume but will eat or gnaw almost anything available.

They are disease carriers through their faeces, urine, or deposit of body hair, and by transmission of body insects and mites. Food-poisoning organisms (salmonella) can arise in foods from mouse contamination.

## Control of Rats and Mice

Several methods are available for rodent destruction—trapping, baiting, gassing, and the use of poisoned tracking powder.

**Rats**  Modern food factories are practically rat proof, but when evidence of rats appears it must be dealt with vigorously because, if left unheeded, they multiply at an alarming rate.

In dealing with any rat infestation, the first operation is a survey of the surroundings, which includes not only the condition of the premises, but the position of runs, harborage, and food supplies.

Older buildings are much more prone to rat infestation than new concrete structures that are devoid of hollow floors and entries near beams and pipelines. The survey should lead to the adoption of ratproofing measures and the elimination of harborage from refuse and rubble dumps.

*Trapping.*  Control of rats by trapping can be very unreliable as these rodents are very suspicious of new objects, particularly if they have a human odor. Traps may be put into positions where the rat frequents, with attractive food and the trap not set. When the food is taken regularly, the trap is set; this is the only procedure that gives results.

Adhesive-based traps, which are put down in the runs, are

sometimes effective if the rodent is forced to cross them to get its foods.

*Baiting.* Unless a precise system of baiting is adhered to no catches will be obtained. Prebaiting with unpoisoned material is essential to allow the rats to gain confidence and to record where baits are being taken. Bait boxes are always used, or, in open sites, the bait is placed in narrow drain pipes that will allow a rat to enter but not dogs or cats.

The site where rodents are suspected is baited in this way and unpoisoned baits put down until "takes" are being recorded regularly, which may be a week or so. Then the same bait is introduced with the poison added and a similar number of "takes" should occur. Later, dead rats may be found in the area although they may travel some way toward water before dying. The procedure of prebaiting followed by poison baiting is repeated if rat activity reappears. However, it is stated that with the modern anticoagulant baits prebaiting is not necessary.

The reader is referred to publications on rodent control for details of design of bait boxes and bait formulations. Generally, rats prefer cereal-based baits such as wheat, oatmeal, corn flakes, and bread crumbs, but this depends a lot on local conditions and competition from other food. Fish and meat meals are favored in some areas.

*Rodenticides.* The selection of the poison to include in the bait depends a lot on the location. Inside a food factory, it is always best to avoid the baits that are poisonous to human beings and animals, but on outside sites suitably concealed baits with more lethal poisons may be used.

*In all cases baits must be numbered and their location recorded so that those not taken by rodents can be removed and destroyed.*

Many bait poisons have been proposed over the years; some were extremely poisonous to humans, for example, phosphorus and strychnine. The use of such poisons should be avoided.

The anticoagulant bait warfarin has been used successfully in recent years and seems least likely to cause bait shyness but resistance has developed in some areas.

*Biological methods.* Certain virus preparations have been proposed that will impart diseases to rats, but so far these have not proved reliable, their safety is disputed, and their use is not recommended by many government authorities.

**Mice** It has been mentioned that mice can live concealed in stocks of foodstuffs without access to water but they do have their "runs" and rarely travel more than 20 to 30 ft from their nest. They are not normally bait or trap shy, and provided their location is known, an intensive baiting or trapping campaign will be successful.

An important precaution that applies to rodent as well as insect control is never to stack foods or packing materials against walls, but to allow access to all sides. Boxes containing poisoned bait can be placed close to and even built into stacks. Traps should be placed close to walls and set very finely as mice are very light in weight and can readily walk over a badly set trap.

Baits to be used on traps vary according to the nature of surrounding food but nuts, bacon, and cereals are generally popular. Baits may contain the same poisons as used for rats and again it is essential to record the position of baits and traps and note the number of takes.

Warfarin is the most effective bait for mice but because some resistance has been shown other baits are now in use. These are:

Alphachlorolose (Alphakil, Rentokil)
Coumatetralyl ⎱ Used in tracking dust 0.75 percent or in baits
Chlorophacinone ⎰ at 0.25 to 0.4 percent.

Two other more recently developed anticoagulant baits are Calciferol and Brodifacoum. These can be used without prebaiting.

Tracking dusts are useful and these are picked up on the fur in the runs and the rodents are poisoned by licking it off.

When trapping is used, large numbers of traps should be set. The setting should be very light or the bait will be taken without releasing the trap. A new type of trap with a very light setting has recently come on to the market.

Fumigation is being used increasingly for the extermination of rodents, particularly mice, and carbon dioxide is usually employed.

To destroy rats in burrows in sites outside the factory, calcium cyanide dust is injected into the holes with a powder blower. The moisture in the soil decomposes the cyanide with the evolution of hydrogen cyanide.

This method of treatment is best entrusted to expert fumigation companies.

## BIRDS

Birds can be a serious menace in large warehouses, manufacturing departments, and canteens. They foul foods with their droppings and because of this are a potential source of food poisoning. Unfortunately, they are frequently encouraged by employees, who feed them and object to their extermination.

Pigeons and sparrows are the main offenders and the only satisfactory answer is to proof the building by netting windows and spaces under the eaves. Pigeons can be repelled by placing plastic jelly compounds on their usual perches—proprietary materials are available for this purpose.

In very large buildings, sparrows will nest, breed, and feed without ever going outside, and in these circumstances the nest must be located and destroyed, often a very difficult task.

Other methods of combatting the bird problem that have been used include:

1. Use of prerecorded distress calls, flashing lights, and any method to prevent roosting.
2. Use of narcotic poisons. This has great value, as expected birds can be revived and released to the wild. This technique usually requires a license and involves some on-site training of the operator. Prebaiting is required to obtain successful results.
3. Use of real or imitation falcons to prevent large numbers of birds roosting in the factory environment. When trained, live birds are used. It is not always necessary to fly them as their presence may be enough.

The indiscriminate killing and poisoning of birds are prohibited by law and certain birds are protected entirely.

## BUILDING CONSTRUCTION AND THE PREVENTION OF INFESTATION

Much has already been said about the internal construction of machines, plant, walls, and partitions, but many modern factories built in the United States and Europe are without windows and internally are artificially lighted and air conditioned or otherwise ventilated.

This may have psychological disadvantages for the employees, but it is of great value in keeping out flying insects, birds, and rodents. It also helps considerably in maintaining suitable temperature and humidity conditions inside the factory. The traditional factory may have acres of windows and it is very expensive and not entirely effective to protect windows and doors with wire mesh guards.

## ACKNOWLEDGMENT

I am indebted to Barry Penney (BAP Services) for help in updating this chapter on pest control. Barry Penney was entomologist at Cadbury in England, and is now a consultant to the food industry on infestation in storage and shipping.

## REFERENCES

*Ants*. British Museum Aconomic Leaflet No. 9.

Busvine, J. R. 1980. *Insects and Hygiene* (3rd ed.). Chapman & Hall, London; Methuen, New York.

*The Design of Machinery and Plant in Relation to the Control of Insect Pests*. 1956. British Food Research Association, Leatherhead, England.

Graham, J. R. 1977. *Training Manual for Analytical Entomology in the Food Industry*. Food and Drug Administration, U.S. Department of Health, Washington, D.C. [An excellent manual.]

*Handbook of Rat and Mouse Control*. Ministry of Agriculture, Fisheries and Food, London.

Hayhurst, H. 1937. Insect infestation of stored products. *Ann. Appl. Biol.* 24(4), 797–807.

Hinton H. E. and Corbet A. S. 1943. *Common Insect Pests of Stored Products*. British Museum, Natural History Series, London.

*Infestation Control in the Cocoa, Chocolate and Confectionery Industry*. 1984. Joint publication by British Food Research Association and Cocoa, Chocolate and Confectionery Alliance. Obtainable from British Food Research Association, Leatherhead, England.

*Insect Travellers*. Vol. 1 1975. Coleoptera. Tech. Bulletin No. 31, Her Majesty's Stationery Office, London.

Jacobson, F. B. 1984. Cocoa bean pests and problems of control. *Manuf. Confec.*

Janes, N. F. 1985. *Recent Advances in the Chemistry of Insect Control*. Royal Society of Chemistry, London.

Mallis, A. 1964. *Handbook of Pest Control*. Macnair Dorland Co., New York.

Pickett, J. 1984. Prospects for new chemical approaches to insect control. *Chem. Ind.* 657, London.

*The Rentokil Library*. Cornwell, P. B. The Cockroach. Munro, J. W. Pests in stored products. Lucus, C. M. Hygiene in buildings. Hutchinson, London.

*Review of the Persistent Organo-Chlorine Pesticides*. 1965. Her Majesty's Stationery Office, London.

Richards, O. W., and Herford, G. V. B. 1930. Insects found associated with cacao, spices and dried fruits in London warehouses. *Ann. Appl. Biol.* 17(2).

*Rules and Regulations for Buildings, Machinery etc.* 1961. Cadbury Ltd., Birmingham, England.

Solomon, M. E. 1943. *Tyroglyphid Mites in Stored Products*. Her Majesty's Stationery Office, London.

Wilbur, D. A. 1984. Sanitation and pest control. *Manuf. Confect.*

# 22

# Packaging in the Confectionery Industry

## TRENDS AND DEVELOPMENTS

Since World War II, developments in packaging generally have been phenomenal, and this applies to both materials and machinery. Changes have occurred with all commodities, large and small, edible and non-edible.

In the food industry, which includes chocolate and confectionery, special factors arise. The package and packaging operation must conform to rigid standards of sanitation, toxicity, and odor, and must provide protection of the contents. Also, modern distribution and sales methods have influenced the type of package. Supermarkets and garage areas, for example, require unit packs to encourage impulse buying as well as ease of display. Examples include candy bars, which now mostly have heat-sealed, protective wraps, and pouch packs, which are heat-sealed films or laminates giving adequate protection in what are often hostile atmospheric conditions.

Cost is obviously important, and this applies both to the packaging material and the packaging operation. An interesting aspect of this is the adoption of containers made of plastic instead of metal and glass. This has also applied to flexible laminate aseptic packs, which are replacing cans and bottles of juices at about one-third the cost.

Gas packing using inert nitrogen gas is being used for bulk and individual packs to avoid rancidity in such products as peanuts and butter toffee toppings.

## THE PURPOSE OF A PACKAGE

The main reasons for using a package on any food may be summarized simply:

It serves as a "container," for distribution and sales.
It protects the contents.
It describes and advertises the product in the package.

It may occur to the reader that these three factors apply only to the finished product, but the food technologist is also faced with the problem of the packing used for the raw materials received into the factory. It is regrettable that the standard of packing of many of these raw materials is lamentably low and packings are frequently made worse by bad transport and storage before receipt on the factory premises.

Reference to bulk materials is also made under entomology, rodent control, and storage.

## The Container

Accepting the fact that the prime function of the package is to be a container, special needs arise when the material inside is a food, whether it is in bulk or a small retail pack. The container must necessarily be strong enough to carry the weight of material as any damage will result in foreign matter in the food, or loss, and it must be easily filled and emptied. The material used for the container must not contaminate the contents either through foreign flavors, odors, or toxic substances, and must not be corroded or weakened by the foodstuff.

**Appeal** Unfortunately, there is still an opinion among many of the sales and advertising fraternity that the design and appeal of a wrapper or label on a box are the only things that matter and as a result costly designs and multicolored printing are used at the expense of the quality of the product. The heavy printing can cause serious contamination. Sometimes the actual construction of the retail package is so complicated that this also adds greatly to the cost.

In the chocolate and confectionery industry, there are packings of every type. Probably the most elaborate and expensive are those used for fancy assortments and Easter eggs, and the public generally is prepared to pay for a chocolate box of attractive shape and design. When the box is opened, customers expect to see bright linings and foils and glossy chocolates and other candies free from blemish.

The result is that the packing becomes as expensive as the contents and it is a fact with some Easter eggs and novelties that the pack has become a work of art in printing and carton construction and the confectionery item contained therein is almost a nonentity lost in cardboard.

This is not true with chocolate and confectionery bars and hard candies, where the trend is toward simple wrappers and bags, usually glossy paper or transparent or printed film with a compromise in printing design—fewer colors but adequate and still attractive if designed well.

These bars and bags are displayed in special racks in candy shops, on the shelves of supermarkets, and even in the open air with minimum overhead protection.

Then the protective quality of the wrap becomes important.

**Protection** With any foodstuff, including chocolate and confectionery, protection can mean either preservation of the condition of the unit inside the wrap, such as prevention of drying out, or action as a barrier against exterior contaminants.

The second is probably the more important for the following reasons:

1. It *excludes light*. Light can promote rancidity and cause fading.
2. It *prevents access of water vapor, oxygen, and contaminating odors*. Damp air can be most damaging, causing stale flavors, surface mold growth, surface crystallization or graining in hard candy, and sugar bloom formation on chocolate.

Oxygen, obtained as a continuous supply due to a wrapper being permeable to air, will cause staleness and rancidity, particularly if combined with humid conditions.

Contaminating odors will ruin any food and the worst offenders are naphthalene (moth balls), paint, particularly with naphtha solvents, printing inks incompletely dried after printing the wrappers, perfumed soap, and various strongly flavored foods like cheese. Unseasoned and resinous woods used for cupboards can cause contamination and damp musty storerooms can affect the flavor and appearance.

Strawboard and chipboard used for chocolate boxes and outer cartons for bars may contribute to musty flavor if stored damp, and a relative humidity as low as 70 percent can give this effect over a prolonged period.

Recycled board may be a hazard as the fibre slurry often is treated with odorous fungicides.

3. It *keeps out insects*. The majority of food manufacturers take great precautions to ensure that their premises are free from insect and rodent infestation but there are still shops and warehouses receiving the manufactured goods that are not so free. An insectproof wrap, either on the confectionery unit or

over the carton containing the product, does a great deal to prevent public complaints of insect damage and can save the manufacturer the ignominy of a court case, even though it may not be the manufacturer's fault. Fortunately public health officers have become much more vigilant toward retail premises.

Under this heading, something might be said about packaging for export, particularly to tropical climates. Protective wraps are very important as it must be realized that there are usually long transport periods as well as adverse conditions under which the goods are sold.

It is surprising how many nontechnical people think that a protective wrap protects a food from heat damage. In fact, even quite heavy padding of a case of chocolate gives only temporary protection against direct sun or storage near boilers or radiators, and melting and bad discoloration can arise from not heeding the storage location. What is worse, a good protective wrap in these circumstances gives a false impression and often no visible signs of deterioration are apparent before the package is opened.

High temperatures and an impervious wrap can result in evaporation of moisture from a fondant or similar center, causing high humidities and condensation inside the wrapper. Mold will soon develop under these conditions.

**Shelf Life** The foregoing remarks are related to the expected shelf life of the food; if the product is eaten a few days after it is manufactured, the degree of protection required is greatly reduced and is confined to the need to see that it travels well and gives the sales information required. Shelf life is related to sales policy, the size of the company, method of distribution, and zeal, or lack of it, on the part of the retailer.

Some companies spread their production throughout the year by holding large stocks of both raw materials and finished goods. For seasonal lines sold at Christmas, Easter, or on Mothers' and Fathers' days, this entails costly low-temperature storage, with a measure of control over humidity, if nine to twelve months' stock is to be held.

Some raw materials, such as nuts, also need cold or cool storage, particularly if the policy is to buy fresh crops to spread over a year or at least six months' production. The conditions required and methods of storage are discussed elsewhere.

Other companies work on a day-to-day production basis, holding negligible stocks of either raw materials or finished goods, and make use of daily public or private transport. This method usually ensures that the goods get to the retailer in a fresh condition but they are still at the mercy of the shopkeepers' stock rotation unless a system of merchandizing is used where the salesperson controls the stock in the shop.

Assuming the second method is used successfully, savings can be made on wrappers, which, in many instances, do not need to be protective.

## MATERIALS

### Metal Cans

Most chocolate and confectionery products are not perishable and therefore sterile packs are not required except, perhaps, for chocolate syrups and similar low-concentration mixtures. However, metal cans are often used for cocoa and food drink powders.

Many granular food drink products are very hygroscopic, and if not adequately protected, will pick up moisture and consolidate. For these, lever-lid cans with latex lined bottom and side seams are required.

For sterile cans, it has been the practice to use soldered seams, but in the United States, the FDA has been pressing for the reduction of lead levels in foods. This has resulted in increasing production of welded and "drawn" cans. (See "Metals Cans" later.)

Slip-on lid tins are not completely moisture- or insectproof unless the lids are sealed with adhesive tape, and even this does not guarantee that insects are excluded, as minute larvae have been known to crawl up the tiny channel along the side seam and beneath the tape. Lever-lid tins are also used for export packs of panned nuts and twist-wrapped chocolates and are often made from elaborately colored printed tinplate.

### Paper and Associated Materials

Paper is the most widely used packaging material and there are many types in industry. It can be just a simple wrap or lining, a bag, the base of a label, or a stuffing in the form of shavings.

**Manufacture**   Paper and carton board are produced from natural cellulose wood fiber. The wood is pulped and the natural binding material is removed by chemical dissolution. This is known as the

Kraft process and produces cellulose fibre of maximum length. It gives final papers with greatest strength.

Other processes are mechanical methods to tear the fibers apart, and although many improvements have been made, fibers are shorter and the final paper or board has less strength.

The paper-making operation consists of running a suspension of the processed fiber, containing 99 percent of water, onto endless moving belts. Thus, a layer of matted fiber is laid down.

On these belts, most of the water is removed by draining, vacuum suction, and pressing. Then the remaining moisture is evaporated by means of internally heated drums. Surface treatment (calendering) is often applied to give a smooth surface.

Modifications of these manufacturing methods enable other substances to be incorporated into the pulp. These may be starch, china clay, or resinous materials that impart gloss, opacity, waterproof properties, and wet strength.

Most paper is made from virgin pulp but the recycling of waste is now a very important part of the industry. Newspapers, cartons of all types, and wrapping papers are pulped and usually made into "chipboard."

It has been mentioned previously that there is always a danger of contaminating residues if recycled waste is used. This must be carefully checked where food packaging is concerned.

## Types of Paper

**Unbleached Kraft**  This is commonly known as "brown paper" and is the most used and cheapest packaging paper available. It is manufactured in vast quantities from long-fiber soft woods to which its strength is attributed. Suitably calendered, it can be printed and it also serves as a base for coated papers and laminates. It is used as a support for polythene and wax and laminated to itself with asphalt—these combinations are employed as strong moistureproof bags for bulk materials. Kraft paper can be modified mechanically in process by the addition of size or resins and its extensibility can be increased by the process of "creping."

**Bleached Sulfite Papers**  These are prepared from chemical wood pulps, usually from a mixture of long-fiber soft woods and the shorter-fiber hardwoods. They are used where brightness and whiteness are required and are used for food wrapping, interleaving and laminating. They are reasonably strong and water resistant.

For high-class printing, they are glazed by special calendering or coating. These papers are defined as MG (machine glazed) or MF (machine finished).

**Glassine and Greaseproof (G&G) Papers** Glassine is a heavily calendered paper with a shiny smooth surface and is transparent. Greaseproof is similar but is less calendered. These papers are produced by processes that beat the pulp until the cellulose fibers become highly hydrated and this has the effect of rendering them resistant to the penetration of oil and grease. Glassine is used extensively for wrapping confectionery bars. Its greaseproof properties are very valuable in this respect and it prints well. Because of its dense structure, it gives a measure of protection against extraneous flavors. When special protection against moisture is required, glassine may be coated or laminated using wax, lacquer, polythene and other plastic films. With plastic films it may be heat sealable.

**Vegetable Parchment** Parchment paper is manufactured by immersing unsized paper in a solution of sulfuric acid, followed by neutralizing, washing, and drying. This gives a tough, fiber-free, greaseproof paper with good wet strength. It is used frequently for box linings and for "platforms" for confectionery bars inside glassine wraps. It is a "sterile" paper. It can be coated or laminated to other substances to give an impermeable material.

**Waxed Paper** Almost any paper can be coated or impregnated with wax. The wax may be on the surface (wet waxed) or in the paper substance (dry waxed) and the latter gives more protection. Wax on the surface tends to crack and lose its protective quality. Waxed paper is used a great deal for all food packing as the purified wax is odorless, tasteless, and nontoxic. It can be heat sealed.

**Tissue Paper** Fine tissue paper is in the same category as the MG and MF bleached sulfite papers. It is used for interleaving as protective layers and for padding to prevent movement of articles or confectionery items inside a box. Cellulose wadding is used similarly and is "creped," formed into pads that may be printed and used as padding in the tops of chocolate boxes.

**Carton Board, Paper Board**  This is the material used for making carton outers, boxes, trays, and liners. It is made from various combinations of cellulose pulp and generally has a thickness greater than 0.25 mm (0.01 in.).

These boards have somewhat different descriptions in the trade in different countries but are generally known as strawboard, chipboard, manilla board, and Scandinavian board. Many boards have combination layers of good white facing on a cheaper, poor-color base.

The following are some of the more important types of board.

*Strawboard.* Strawboard, as the name suggest, is made from straw, has long fibers, is used for making rigid boxes, and is frequently paper lined. Under damp conditions, it has a strong odor and may contaminate foodstuffs.

*Plain chipboard.* Is made entirely from mixed paper waste and, like strawboard, is used mainly for boxes. Odor originating from the makeup materials can be a problem.

*Manilla board.* Is made from chemical pulp with long fibers. It bends well and presents no odor problem. May be lined with reclaimed material.

*"Scandinavian"* board (uncoated and clay coated). Wholly sulfate wood pulp. Expensive pure fiber board, no odor problems. Very good color. Suitable for high-quality food packing.

## Metal Foil

Metal foils have long been used for protecting foods and the time is not long past when lead foil was to be found lining bulk cases of certain foods.

Tinfoil was used for many years, particularly for wrapping chocolates, and older employees of the chocolate industry claim that tinfoil clings more readily than aluminum foil and is therefore more suitable for Easter eggs and novelties.

Improvements in the metallurgy of aluminum and the soaring price of tin have caused a complete change to aluminum for wrapping and as a constituent of laminates.

Foil has many advantages as a packing material—it has a most attractive appearance and still carries with it an image of quality of the product wrapped in it. It is odorless, tasteless, nontoxic, easily printed, and not troublesome with static electric charge as are many plastic films, and it is a very good barrier against moisture, gases, and light. It is nonmagnetic.

Its disadvantages are lack of strength, liability to porosity, and cost.

The strength factor can be overcome by lamination with plastic materials, but in spite of its lack of strength, foil wrapping by machine is a feature of the confectionery industry. Individual close wrapping of assortment chocolates for tropical markets not only gives a good appearance but retains the shape of soft chocolate. Foil is also manufactured with a heat seal lacquer coating that is used with reasonable success for moisture- and insectproof wraps. It still has the fragility of the unlacquered foil and wrapping machines using this material require very careful adjustment and maintenance.

**Gauge**   By definition (United States), "foil" is less than 0.006 in. (0.15 mm) thick but commercially it is rolled to a thickness as little as 0.00024 in. (0.006 mm).

**Porosity**   The porosity varies with the thickness and Table 22.1 gives some average values.

Pores are easily visible if the foil is held before a light in a darkened room although the average size may be only 0.001 mm. Even with the 0.009-mm foil, these pores have but little effect on its barrier properties compared with the effects of small tears and the overlaps.

If foil is laminated with polythene or similar film, these pores are closed and permeability reduced.

TABLE 22.1. POROSITY

| Gauge mm | in. | Pores per in.$^2$ |
|---|---|---|
| 0.03 | 0.0012 | 0 |
| 0.012 | 0.00047 | 80 |
| 0.009 | 0.00035 | 400 |

*Source*: Aluminium Development Association, London.

**Printing**   Aluminum foil can be printed by multicolor processes giving attractive, bright reproductions. Foil of this type is used to wrap round hollow chocolate figures such as gnomes, rabbits, and

Santa Clauses, the printing being made to coincide with the features and clothing of the figures (see "Hollow Molding").

**Foil Containers**  Vast numbers of heavier-gauge, rigid foil containers are now used in the food industry for pies, tarts, frozen foods, and many other products.

In the confectionery industry, they are used for fudge and similar pastes. Smaller cups are filled with chocolate and truffle pastes for inclusion in boxed chocolates.

Fudge, for example, may be deposited in shallow trays of 4 oz (114 g) capacity. A metal "lid" is crimped on, or the closure may be a heat-sealed laminate.

Products with a high water activity may develop mold within the head space. The container must then be sterilized or the tray, with deposit, passed beneath a sterilizing ultraviolet lamp immediately before closure.

## Transparent Films

The manufacture of transparent films has followed closely the development of synthetic fibers. The chemical processes are similar but the engineering is different. With synthetic fibers, the base polymer is extruded through minute orifices followed by spinning, whereas films are made by forcing the material through slits machined to great accuracy. Films are frequently coated or laminated to give resistance to water vapor and gas transmission or to improve heat sealing. The plastic base of the film may have plasticizers incorporated to improve flexibility or the molecular structure of the "polymer" may be modified by the addition of different "monomers." The plastic film is then called a copolymer.

Among the types of film available at the present time are regenerated cellulose film (viscose, cellophane, and Sidac, etc.) and there are many proprietary names for this film. It is made by first reacting high-grade wood pulp with caustic soda (sodium hydroxide) solution. This prepares the cellulose for the next treatment with carbon disulfide which forms sodium cellulose xanthate.

This product is aged to bring about a molecular change to aid the next reaction with dilute sodium hydroxide. Thus, a heavy syrupy liquid is formed that is termed "viscose." This is filtered and passed through a horizontal slit into a sulfuric acid bath.

In this bath, a cellulose gel is regenerated and this product is then passed through a series of acid and washing baths to produce a film of great clarity. The final baths may incorporate softening agents to give flexibility. Glycerol and various glycols are currently used in quantities of the order of 15 percent.

The film is finally dried on steam-heated rollers. This film has a brilliant appearance but little protective qualities—none against moisture penetration.

An interesting property of this film is its expansion and contraction under humidity change. A chocolate box wrapped in a packing room at average humidity will, under changing atmospheric conditions, exhibit a tightening of the film over the box, enhancing its appearance. This effect, however, must be used with care as the initial expansion under humid conditions is less than the subsequent shrinkage and the box may distort and the film burst. The film is readily sealed with a simple, nonodorous adhesive such as gum arabic or dextrin, which is an advantage in food wrapping.

Much has been done to improve the protective qualities and the first "moistureproof" film produced was solvent coated with wax-impregnated nitrocellulose. Nitrocellulose-coated films are now made in heat-sealing and non-heat-sealing forms. There is also a film of this type that gives partial protection where it is necessary for the food contents to "breathe," thereby preventing internal moisture condensation.

Another coating is polyvinylidene chloride (PVDC or Saran) and this gives to the film excellent barrier properties against both moisture and gases. Cellophane also provides the base for many laminates with foil, paper, polythene, and PVDC (see "Laminates").

**Polythene (Polyethylene) Film**    Polythene film, in fact polythene in all its forms, has become an article of commerce of enormous importance and constitutes about 30 percent of all plastics produced. In the packing of chocolate and confectionery it has found numerous uses in bags for hard candies, outer envelopes for cartons of confectionery, coating of carton board to reduce water vapor transfer, and as a constituent layer in many types of laminate. It is low in cost.

Polyethylene is manufactured by heating ethylene gas at 200°C under very high pressures ranging from 500 to 3,000 atmospheres. The higher the pressure, the greater is the molecular weight of the product. The resultant polymer is a translucent waxy solid that melts around 115°C (239°F) and softens considerably below this at about 70°C (158°F). The molecular weight of the polythene used for film is of the order of 15,000 to 20,000 and it is not necessary to add any plasticizer to give flexibility.

Polythene is produced to a range of densities: low, 0.915 to 0.925; medium, 0.93 to 0.94; and high, 0.94 to 0.97.

The low-density film, which is very flexible, is used for a variety of food wraps and for shrink wrapping. Medium-density film has better barrier properties and is somewhat stiffer. It is used for outside wraps and some pouches. High-density film is much stiffer and is more often employed as box and carton liners.

The film is manufactured by two processes—blown and cast. Blown film is made by extruding the molten polymer through a circular orifice and expanding the tube so formed by compressed air. The gauge of the film may be controlled by the degree of expansion. The tube is air cooled and removed by a pair of rollers.

Cast film is made by extruding the molten polythene through a slit onto chilled rolls. This method gives a high-clarity film but the molecular orientation is uniaxial (single direction) whereas the blown film has random orientation. Because of this, the cast film is less flexible.

Polythene film has a low value for water vapor transfer but it has a relatively high permeability to gases such as carbon dioxide and oxygen. It is probably the easiest film to heat seal, and has great flexibility and considerable strength, which are both useful when the film is used in laminates with paper and foil. Oils, particularly essential oils, will diffuse through polythene and these can cause "stress cracking" of the film or container. They will also impart an odor that cannot be removed.

In the early days of laminate production, the absorption of odors by the polythene layer was a problem to the manufacturers. Residual solvents from the inks on the printed paper layer were readily taken up by the polythene and no means of subsequent treatment would remove them, although they were sufficiently volatile to contaminate any wrapped food.

Another problem arose from the process of extrusion—some polythenes developed a "burnt wax" odor during heating of the polymer in the extrusion machine. This was equally difficult to remove and could also cause contamination of food in contact. Modern production techniques seem to have eliminated most of these problems.

Polythene film is now available in thicknesses ranging from 0.0004 in. (0.01 mm) to 0.010 in. (0.25 mm).

**Polypropylene Film** This is a substance closely related to polythene made by a catalyzed low-pressure polymerization of propylene. It has a much higher softening point (150°C/302°F) and melting point (170°C/338°F) than polythene and this makes it very suitable for foods which are to be cooked in a bag.

The film has very good optical properties; it is brilliantly clear, transmits color well, is resistant to scuffing, and has good moisture barrier properties. It is favored for display purposes in the packing of various garments. In the confectionery industry, it is used for twist wrapping.

Polypropylene became a commercial article in the 1950s as a nonoriented film, and later, in the 1960s, oriented film was produced.

**Oriented Film** Orientation means the mechanical alignment of the polypropylene molecules. Like polythene, it can be made by the blown or cast method, and in both instances, the plastic film is stretched by the take-off rollers to produce the orientation.

Various types of oriented film are now produced:

*Heat stabilized.* This is a nonshrinking film in which the stresses set up by orientation are released by an annealing process.

*Non-heat stabilized.* This is used for shrink wrapping.

*Coated and coextruded heat stabilized.* The original film can be combined on one or both sides with PVDC, cellophane, or paper, or it can be coextruded with another polymer. Coextrusion, in which two plastic substances are extruded through one orifice, has obvious commercial advantages over solvent or lamination.

**Nonoriented Film** The properties of this film lie between those of blown polythene and oriented polypropylene. It is less costly and it does not shrink, and it has the optical brilliance previously mentioned.

**Polyvinyl Chloride (PVC)** The monomer, vinyl chloride, may be obtained from acetylene and polymerization is brought about by oxidation. The basic polymer is a horny substance and to make into flexible film requires the addition of plasticizers.

PVC packaging films may contain 25 percent or over of plasticizers. For food packaging, these must be innocuous and not leached out by the constituents of the food. The FDA in the United States and similar bodies in other countries have strict regulations on the use of plasticizers. Stabilizers are also used, some of which are not suitable for food purposes. However, PVC is capable of being formulated with such a variety of additives that it has become a very important packaging film.

PVC films find great use in the wrapping of fresh produce such as meat and vegetables as they have medium water vapor and gas transmission rates, allowing the product to breathe. Perforation, as with polythene, is not necessary.

In the confectionery industry, an important use is vacuum forming (see later), which provides very attractive liners for boxed chocolates. Certain PVC films are very suited to all forms of stretch wrapping (as opposed to shrink wrapping). These possess exceptional toughness and heat-sealing properties as well as the stretch characteristics.

Boxed chocolates wrapped with PVC have an attractive appearance and the film is more durable and protective than viscose film. PVC films can be oriented uniaxially and biaxially and the properties obtained applied to all types of packaging.

**Polyvinylidene Chloride (PVDC)** This polymer is used extensively as a constituent of food wrappings and has already been mentioned under cellophane. It has exceptionally good barrier properties. In its original form as a single polymer, it was of little interest as it was inflexible, but later copolymerization with vinyl chloride produced satisfactory films. These have been on the market for some years under various names (Saran and Cryowrap). They are very resistant to the passage of water vapor and most gases, even when of thin gauge. The film is a valuable constituent of laminates, or in solution or emulsion form as an impervious coating material. PVDC-coated cellulose film and laminates with polythene paper and foil have been widely used in recent years. Some problems have arisen with heat sealing. Partial decomposition of the PVDC on the hot platen surface evolves hydrochloric acid, and this can cause corrosion. PVDC itself is not readily heat sealed but lamination with polythene overcomes this problem. However, it has now been found possible to seal PVDC using electronic methods.

In the confectionery industry, as a constituent of laminates, coated papers, and films, PVDC has proved exceptionally useful for the protection of hygroscopic hard candies, wafer sandwiches, and similar products.

**Polyamides and Polyesters** Nylon is the general name given to polyamides as a class. They can be converted into films that are very heat resistant [140°C (284°F)]. Because of this, they are used for sterilizable packs and cooking bags.

These films are permeable to water vapor but resist the passage of most gases and so are good barriers to odor. This property is useful in packing fresh meat.

The various nylons can be produced by many extrusion processes—

blown, cast, and coextruded. They may be oriented or unoriented and used as a constituent of laminates.

Polyester film, known commercially as terylene, is produced from dimethylterephthalate and glycol, and the polymer itself is polyethylene terephthalate, a brilliant film of exceptional strength, free from plasticizers, and nontoxic.

It will not heat-seal unless very high temperatures are reached or it is laminated to polythene. Its water vapor transfer values are similar to those of polythene, and barrier properties against gases and odors are good.

When used as a wrapping material, its great strength is a disadvantage for simple packages but is valuable in laminates used for vacuum packing of meats, cheese, etc. Metalized polyester film is finding increased applications. It has superior barrier properties and is very attractive in appearance. Polyester film is used for conveyor belts, usually as the top layer of a laminate, and can be embossed for enrober use.

**Polystyrene**   This plastic produces bright, clear films and in its rigid form is used for small food containers of all types. Polystyrene is readily obtained from the monomer, which itself is obtained by reacting ethylene or vinyl chloride with benzene.

It is not suitable for flexible films but because of its brilliance and clarity it has found application in the manufacture of decorative candy containers and as "windows" in cartons to display the contents.

It is also used to produce trays by vacuum forming and these are used instead of paper cups for packing chocolates and fondants into boxes. Thin-gauge material is used, but polystyrene obtained a bad reputation at one time because it introduced a taint to the foodstuffs. This trouble was traced to free monomer and the plastic manufacturers now claim to have overcome this defect. Additives in the form of plasticizers or stabilizers can still cause tainting, as with other plastics.

**Rubber Hydrochloride (Pliofilm)**   This is made by treating crepe rubber in organic solvent with hydrochloric acid, antioxidants, and plasticizers being added. A number of grades with different quantities of plasticizer are manufactured and the permeability to water vapor and oxygen increases with the quantity of plasticizer. It is a clear, transparent, elastic film that has the property of conforming to the shape of an article wrapped, and because of this, its application to mechanical handling and printing is limited.

## Metallized Films

Metallizing is a process that coats the surface of a film or other material with a "molecular" layer of metal, giving it a brilliant sheen resembling foil.

The metallizing takes place in a very high vacuum chamber (approximately one-millionth atmosphere). The metal, usually aluminum, is heated to a point where its vapor pressure exceeds that of the chamber.

The band of film being treated passes over an internally cooled drum within the chamber and the metal vapor deposits on the cooled surface of the film.

Metallized films have found increasing uses, and mostly they are polyester, polypropylene, and nylon, but to date they have not replaced foil. Metallized paper has been developed but there are problems because of the presence of moisture and volatiles and also the rougher surface.

## Shrink and Stretch Films

*Shrink packing* is a process where the wrapping material is tightened over the article to be wrapped. It can be applied to small articles on display cards or to loads of boxes on a platform.

The principle of shrink wrapping is first to orient the film by stretching the semimolten polymer, followed by cooling. When this film is wrapped over an article and heat applied, the tension in the film is released and the film shrinks tightly over the article. The degree of shrink may be from 30 to 70 percent and is related to the original orientation.

Many types of film can be used for shrink wrapping but polyvinyl chloride and polypropylene are probably the main products in use.

*Stretch wrapping* is an alternative to shrink wrapping and is cheaper in material and capital costs. It is a process where an elastic film is stretched round a load to prevent collapse and to give protection.

Rougher materials such as bricks, wooden articles, and glass bottles are usually stretched wrapped.

## Laminates

Reference was previously made to the combination of various materials in layers by adhesive or pressure and heat. Extrusion lamination is also used in which polymers, particularly polythene and polypropylene, in the semiliquid state are forced through orifices onto a web of paper, foil, or film. The materials are then pressed into contact to give a very firm bond.

Coextrusion is also used where two polymers are fed through a common orifice.

Laminates may consist of more than two materials, for example, paper/foil/polythene. The paper is for printing and to give ridigity

TABLE 22.2. LAMINATES AND THEIR USES

| Laminate | Uses | Comments |
|---|---|---|
| 1. Aluminum foil/ heat-seal "plastic" or polythene | Moistureproof, insectproof wrap for chocolate blocks and covered cookies where close wrap is required. | Low-water-vapor permeability when undamaged but fragile. Polythene laminate rather stronger. Good odor barrier. |
| 2. Aluminum foil/ waxed tissue or glassine | Close wrapping of chocolate blocks without heat sealing. | Suitable for warm climates, prevents fat staining of outer wrapping, and also protects against odor contamination. |
| 3. (a) Aluminum foil/paper/ polythene or (b) Paper/ aluminum foil/polythene | Strong wrap for very hygroscopic products—for heat sealing. Suitable for pouches for "complete drinks." (b) is stronger. | With paper on outside, gives well-"plimmed" package. Good odor barrier. |
| 4. Paper/PVDC/ polythene or paper/polythene/ PVDC | Similar to 3 but cheaper. Used for outer wrap of larger packs, e.g., biscuits. | |
| 5. Regenerated cellulose/ polythene | Transparent heat-seal wraps where visibility and tight wrap required. | Fairly good water vapor barrier. Not a good odor barrier. |
| 6. Regenerated cellulose/ PVDC | Transparent wrap heat-sealable at high temperatures. Wrappings and bags for hard candy, outside wrapping of cartons. | Very good water vapor, gas, and odor barrier properties. |
| 7. Regenerated cellulose/ PVDC/ polythene | As 6 above But heat-sealable at lower temperatures. | As 6 above. Probably a little stronger. |
| 8. Paper/ polythene | Heat-sealed, insectproof wraps. | Moderate water vapor barrier properties. Poor odor barrier. |

and a neat appearance; the foil is for maximum barrier properties to moisture and odor; and the polythene for heat sealing and to give flexibility and strength to the whole laminate.

Where adhesives are used for lamination, waxes and thermosetting resins are preferred to any containing solvent as there is always a problem of evaporating the solvent out of the laminate, and if trapped, it will cause odor problems. Another type of laminate is formed by applying several coatings of an emulsion of a plastic. A continuous film is built up that will have very intimate bonding with the supporting layer. This method can be used for application to paper where the fiber assists bonding.

With some film laminates, the surface of one film is printed and then a second film is applied over this after complete drying so that the print is sandwiched between the two film layers. This enhances the appearance and prevents the print from being in contact with the food.

Some examples of laminates and their uses in the chocolate and confectionery industry are given in Table 22.2.

## SELECTION AND USE OF WRAPPING MATERIALS FOR CHOCOLATE AND CONFECTIONERY

### The Machinery

It is not proposed to give details of wrapping machines or their maintenance as this is the prerogative of the engineering profession but some acquaintance with their performance is essential to the technician. It is necessary to know about adhesives to be applied to the various wrapping materials, and whether these adhesives have the right tack and drying speed and are compatible with the confection being wrapped. With heat-sealing, the technician must know the temperature of the sealing irons, wheels, or crimpers. The scientific person is responsible (or should be) for negotiating with the wrapping material suppliers on such points as the melting point of the sealing layer and the nature of the print, which must not be odorous or pick off on the sealing irons.

Some knowledge of paper and film webs is desirable in relation to their behavior under variable humidities.

The quality control duties of the technician include checking the perfection of the sealing, as a heat-seal wrapping machine can be performing well according to the maintenance fitter but careful observation of seams often reveals imperfect joints. This can occur during machine start-up when the sealing irons have not become fully heated.

Invisible faults such as broken bars inside a good wrapper or heat damage of the chocolate can be found only by inspection methods that are discussed in another chapter.

## The Type of Wrap

The method of wrapping a confectionery product is determined by:

1. Appearance required.
2. Degree of protection.
3. Economy of wrapping material.
4. Speed of wrapping, which is related to the size of unit to be wrapped.

The methods of wrapping described below are "basic." Obviously, there are variations of type and methods of sealing, which can be adhesive, heat under pressure, welding, cold seal, and in some cases, as with metal foils, the mechanical fold is sufficient to hold the wrap in place.

**Parcel Wrap** This is probably the best known wrap used for hand and machine, particularly suitable for larger packs. It has a single or double overlap on the long seam and tucked-in folds on the ends (Fig. 22.1).

There are various machine modifications of this type of wrap. With a moistureproof wrapping material, it gives protection against intermittent high humidities but is not suitable for long storage of hygroscopic products due to access of air through seams and overlaps. For the same reason, it is not insectproof. It has a neat appearance.

**Bias or Oblique Overlap** This gives maximum economy of material but very little protection unless the overlap is considerable. It is used

Fig. 22.1. Parcel Wrap

Fig. 22.2.　Bias or Oblique Wrap

for foil wrapping of chocolate blocks, often in conjunction with an outer sleeve wrap. The foil/sleeve wrap is "traditional" for many chocolate lines and gives a pleasing appearance. Paper bias wraps are likely to create the impression of poor quality and are liable to unseal when handled for display (Fig. 22.2).

**Sleeve Wrap** This is a simple one-way band wrap with a single overlap and ends of the unit exposed. It is rarely used on its own but serves the purpose of a descriptive label. It gives no protection on its own but is often used in conjunction with inner foil or glassine bias wrap (Fig. 22.3).

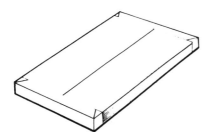

Fig. 22.3.　Sleeve Wrap

**Twist Wrap** This involves a single sheet of paper or film formed into a cylinder around the confection with each end then twisted. This is always a mechanical operation and the wrapping material must be limp or the twist will unwind. It is used for small hard candies and some bars and gives moderate protection only, as the overlap is not sealed—no adhesive is used on these machines. Twist-wrapped small candies have an attractive appearance if colored film is used and the "twists" are useful to give padding and bulk in a loose carton pack. A twist-wrapped candy is easily unwrapped by pulling at each twist (Fig. 22.4).

With hard candies and caramels, the individual pieces are first wrapped in wax-impregnated paper or protective film to prevent the

Fig. 22.4.    Twist Wrap

candy from sticking to the wrapper. Twist-wrapped candies are often contained in pouch bags (see below).

**Bag and Pouch (Pillow) Wraps.** These are generally used for packing small candies, dragees and some hard candies, and transparent film is usual. The bags may be preformed and the units filled into the mouth of the bag by hand or a mechanical method and the bag then sealed by twist, metal strip or ribbon, or sealed overlap.

More often today pouches are formed from continuous traveling films from reels and sealed by clamping jaws at predetermined distances followed by weighed or measured deposits of candies, and further seal. These pouches and contents, sealed at both ends, are then cut off at each crimping. Such bags, made from heavy-gauge polythene, PVDC-coated film, or other laminates, have become a standard pack for hard candies, which, being heat-sealed in material with good barrier properties, will withstand prolonged storage in humid conditions (Figs 22.5 and 22.6).

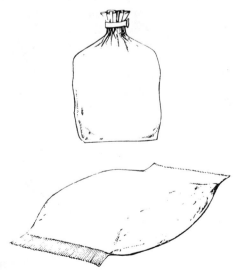

Fig. 22.5.    Bags, Pouch and Pillow Packs

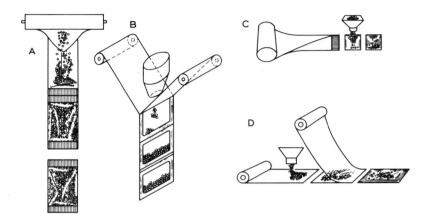

Fig. 22.6.   Methods of Filling Pouches
A. Pillow Type
   One Web—Vertical Seam, Top and Bottom Seals
B. Two Webs—Four-Side Seal
C. One Web—Bottom Fold, Three-Side Seal
D. Two Webs—Four-Side Seal

*Modern Packaging Encyclopedia, McGraw-Hill, New York*

**Roll and Stick Wraps**   These have become very popular for wrapping filled molded chocolates or individually wrapped hard candies.

The chocolates often have a foil wrapper inside with an outer paper sleeve. Assembly of the separate pieces is done mechanically as is the final wrapping, and a complicated machine is used (Fig. 22.7).

**Heat-Seal Wraps**   Heat sealing is used for a variety of purposes, usually with laminates with one layer of heat-seal material.

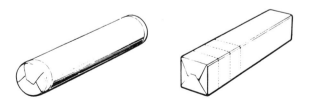

Fig. 22.7.   Roll and Stick Wraps

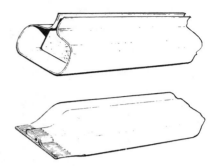

Fig. 22.8.   Heat-Seal Wraps

Hand methods can be employed for larger boxes using an electrically heated sealing iron; machines are available for smaller cartons. Parcel wraps are general for these boxes.

Confectionery bars, which are wrapped at high speed, use laminate that, coming flat from a reel, is bent to form a trough that accepts the bars fed continuously to the machines. The inner surfaces of the heat-seal material at the top edges of the trough are clamped together either with heated serrated wheels or clamping jaws closing intermittently over each bar. This closes the side seam and the ends are then similarly treated and are usually bent over to aid packing (Fig. 22.8).

Fig. 22.9.   Bunch and Fold Wraps

This method gives, with a triple laminate, a highly protective pack suitable for very hygroscopic confections.

Machines may also operate with the fin seal at the bottom.

**Cold-Seal Wraps**   A later development is the cold seal, which is intended to give the same result as heat sealing. The wrapping material is coated with a layer of latex compound, which when pressed in contact with itself, will give a hermetic seal. Experience suggests that this was introduced to the food industry prematurely as some of the coatings used were liable to detach or stick to the wrapped confection. An improvement was "printing to register" of the

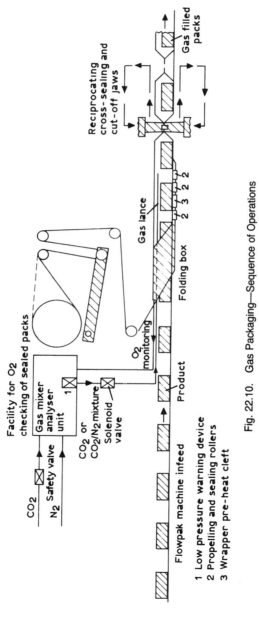

Fig. 22.10. Gas Packaging—Sequence of Operations

*Reproduced by the permission of A.P.V. Baker, Leeds, England*

cold-seal strip, which avoids the need to have the cold-seal material over the whole wrapper surface. Very high machine speeds are possible with this type of wrap.

**Bunch Fold Wraps** Foil is generally used with this method for wrapping small chocolates in export boxes, Easter eggs and other molded novelties, and sometimes for fancy inserts in boxes of assortments (Fig. 22.9).

**Vacuum and Gas Packing in Flexible Materials** This type of packing is used mostly for food other than confectionery, but roasted peanuts and hazelnuts show considerably increased shelf life if gas or vacuum packed. Drink mixes containing full-cream milk powder and other fatty products where the fat is on the surface and in thin layers benefit from the exclusion of oxygen.

Roasted peanuts are packed by this method probably more than any other confections but these are also often vacuum packed in cans. Peanuts that have developed oxidative rancidity are particularly distasteful. It is usually beneficial to exclude light as well as oxygen, and foil laminates are to be preferred. PVDC is a favored film constituent of the laminates because of its exceptionally low permeability to oxygen. Machinery is available for gas and vacuum packing using flexible wrapping material, and the technician must decide whether the expense of this process is justified by the extended shelf life given to the products (Fig. 22.10).

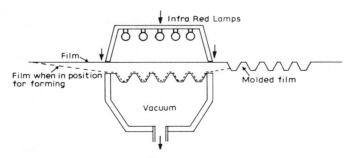

Fig. 22.11. Vacuum Forming

**Vacuum Forming** This is used to produce shaped inserts for boxes of chocolates, cookies, and cakes, which can replace paper cups and dividers.

The principle of vacuum forming is to draw, by vacuum, a plastic film softened by radiant heat into a perforated mold made to the shape required for the packing insert. Thin-gauge, colored film, either polystyrene or polyvinylchloride, is used as both of these soften at about 70°C (160°F).

As the candies are in direct contact with the film, it is essential that there be no odor problem or transfer of plasticizer. Figure 22.11 shows the method of vacuum forming.

**Other Forms of Wrap** Shaped novelties and Easter eggs molded in chocolate are generally closely wrapped in foil, and this is hand operation. A development to avoid hand wrapping, which can be applied to some molded products, is to press the wrapping foil into a metal mold, which will receive liquid chocolate, so that the foil takes on the shape and impressions of the mold. The chocolate is then deposited into the foil, the molds are cooled, and the solid chocolate, with foil adhering, removed. To provide for overlaps at the back of the articles, the mold impressions must be widely spaced and the foil cut round each impression. (See "Hollow Chocolates, Molding").

## PHYSICAL PROPERTIES OF WRAPPING MATERIALS—TESTING METHODS

The suitability of a wrapping material depends on the physical properties of the material as well as the type of substance. Classes of wrapping material have already been described but the performance of these materials when used on a product must be assessed.

It is not possible to describe the tests in detail but the following summary gives the nature of the various methods and the official American Society for Testing and Materials ASTM and British Standards tests.

### Strength

It is little use having a material that is impervious to moisture and gases if it is easily torn or damaged during machine wrapping, transport, or handling.

**Tensile Test** Material of standard dimensions is clamped between jaws that can be pulled apart until breaking occurs. Force and elongation are measured (ASTM D828, 882; BS 1133).

**Impact**   A sample of the material in a taut condition is subjected to the action of a pendulum knob swung from a given height. It is an indication of damage that might arise from dropping (ASTM D3420).

**Tear Strength**   This is a very important factor. It is often observed, with films particularly, that they seem to be very strong until a nick appears. Then they tear very easily. On the other hand a high tear value may mean the package is difficult to open by hand. This is a modern "fault." The test measures the force required to continue tearing after a nick made (ASTM D689, 1922).

**Bursting Strength**   The Mullen tester is the recognized apparatus. It consists of a means of clamping a sample of the material over a rubber diaphragm. The diaphragm can be expanded to a bubble by pumping a liquid beneath it until the test material bursts. The force is measured hydrostatically (ASTM D774, D2529, D2738; BS 1133).

## Permeability

The degree of permeability is probably the most important property of a wrapping and as a rule is related to the rate of diffusion of gas, water vapor, or liquid through the material.

**Gas Permeability**   Barrer (1941) produced an equation of the permeability of gases and vapors as follows:

$$\text{Permeability (P)} = \frac{qx}{At(p1 - p2)}$$

where $q$ is quantity of gas diffusing through film of area $A$ and thickness $x$ in time $t$, and $p1$ and $p2$ are the pressures on either side of the film.

This equation has been confirmed by various experimental workers and is useful for calculating permeability under different conditions of temperature, pressure, and film thickness, having established experimentally one set of results. The permeability of a material is also affected by its crystalline nature, the presence of plasticizers, and its porosity. Diffusion, where solution of the gas occurs on one side of the film and in this state passes through the film to be evolved on the other side, is also a factor.

The practical significance of gas permeability lies in the rate of passage of air, oxygen, nitrogen, and carbon dioxide through a material used for inert gas or for vacuum packing. This can be measured by simple manometer methods where a pressure is maintained on one side of the material with the gas being investigated or where the gas is continuously swept over the surface at normal pressures.

Gas transmission may be measured by the ASTM apparatus, which consists of two cells with the sample of material clamped between them. The test gas ($O_2$, $N_2$, $CO_2$) is flushed through one compartment and the other is evacuated. The amount of gas that passes the film in a given time is estimated (ASTM D1434; BS 1133).

When protective wraps are required to keep out odors from confectionery or chocolate, the best practical test is organoleptic, although gas-chromatography is also now used, Sealed wrappers containing, for preference, milk chocolate are placed in closed bottles with a small amount of moisture, which must not wet the wrapper. To different bottles are added traces of naphthalene, peppermint oil, and acetic acid; to others, large pieces of damp strawboard and some foodstuff affected by molds. The closed bottles are incubated at 24°C (75°F) for 24 hr and then the chocolate tasted by a selected panel. This range of substances has been found very effective in forming a judgment of odor barrier and it is important to test a number of packages, as the efficiency of seal as well as the wrapping material is significant in such a test.

**Water Vapor Permeability**  The pleasure of eating many foods is closely associated with texture as well as flavor, as, for example, prepared cereals, wafer biscuits, aerated hard candy. These, which should be crisp, rapidly become soft or tough when exposed to average humidities. Other products such as hard candy and caramels, will become sticky and "grain" while fondants and icing will dry out under normal atmospheric conditions and their original quality is lost. Chocolate will deteriorate by being exposed to damp air and a sugar crystal "bloom" will form. The water vapor permeability determines the protective quality of a wrapping material against moist air. A simple method of determination is described later (BS1133; also ASTM E96).

**Water Activity**  The property of any foodstuff to pick up or lose moisture is defined by its water activity (A/W). Water activity is also termed equilibrium relative humidity (ERH). ERH is expressed as a percentage whereas A/W is unity, that is,

$$A/W \ 1.0 \equiv ERH \ 100 \ percent$$

TABLE 22.3. WATER ACTIVITY OF CONFECTIONERY PRODUCTS

| Class | Class of confectionery products or raw material | A/W | ERH, % |
|---|---|---|---|
| 1 | Wafer cookies (freshly baked), roast nuts, fresh cereals (corn flakes) | 0.15–0.25 | 15–25 |
| 2 | Hard candy, aerated high boilings, hard toffee, butterscotch | 0.25–0.30 | 25–30 |
| 3 | Hard caramels, hard nougats (ungrained), soft cookies, milk crumb | 0.35–0.50 | 35–50 |
| 4 | Gums, pastilles, low-moisture jellies, licorice pastes, soft caramels | 0.50–0.60 | 50–60 |
| 5 | Soft marshmallows, Turkish delight, some fondants, fruit jellies, soft nougat (grained) | 0.65–0.75 | 65–75 |
| 6 | Soft fondants, marzipan, pastes, fudge | 0.60–0.85 | 60–65 |
| 7 | Chocolate coatings, compound coatings, lozenge pastes | see note below | |

See also Appendix I, method of determination. Figures quoted are approximate but are comparable. Some variation occurs with different formulations.

The ERH of a substance is the relative humidity of the ambient air at which that substance will neither gain nor lose moisture. The A/W of confectionery products and ingredients varies greatly and this must be taken into account when choosing a wrapping material or making a line with multiple layers of different confections. Table 22.3 gives a list of confectionery products and ingredients and their approximate A/W. (See also "Multiple Confectionery Bars".) This will give a good idea of the variation and an indication of the care needed in the selection of a wrapping material to meet the conditions that prevail in commercial storage, transport, and sales outlets.

*Note*: Chocolate, compound coatings, tablets, and dragees need a special comment. The A/W of these is related to the hygroscopicity of the ingredients. The moisture content of all of these products is low, 1 percent or less, and the main ingredients are probably sugar, milk solids, and cocoa material, with very small amounts of glucose syrup or invert. Fat is also present, which is moisture-free.

Dark chocolate will start to absorb moisture approximately at 85 percent relative humidity, and milk chocolate at about 78 percent. With these chocolates, the surface starts to become sticky or discolored at about these relative humidities (see "Sugar Bloom"). Tablets or lozenges will soften. At the relative humidities quoted, or slightly above, moisture will be absorbed until equilibrium is reached with the solution of the solids. Theoretically, absorption will continue until a solution of all the solids is obtained. This depends on the relative humidity and the amount of moisture available in the surrounding atmosphere. In an enclosed space, this is limited, and in the cell of an A/W determination apparatus, it is probable that the value obtained for a normal chocolate will be of the order of 0.1 or 0.2.

If chocolate or the other products described above have absorbed moisture, the A/W will be higher. The A/W of low-moisture substances, including powders, depends on the absorption properties of the constituents and the actual moisture content. The A/W is influenced by the protective nature of fats, colloids, and emulsifiers, and their physical condition.

In considering the figures in Table 22.3, it should be recognized that the relative humidity in temperate climates, including most of northern Europe and parts of North America, is within the range of 55 to 70 percent. It is frequently higher and rarely lower, and these figures apply also during summer months at temperatures ranging between 16 and 27°C (60 and 80°F). During the summer period, these humidities prevail inside many shops and stores whereas in winter they are lower because of artificial heating. In winter, unheated buildings can have humidities of 80 to 90 percent. It is realized that different countries have extreme variations in temperature and humidity and the technician must understand the need for different types of packaging.

Air conditioning in supermarkets and large stores has increased, which is advantageous in most cases. However, often low humidities are maintained, which encourage drying out of some products.

## Physical Structure

The physical structure of a confection must also be considered in relation to the wrapping material required. For example, a hard candy is dense and moisture picked up on the surface will temporarily raise the ERH and this moisture will have to diffuse inward before more is picked up from the surrounding air. Graining may also occur on the surface. In contrast to this, aerated high boilings, light cookies, or prepared cereals, being cellular in nature, will allow the surrounding atmosphere to diffuse rapidly through the whole of their substance and moisture is absorbed quickly if unprotected.

Deterioration of hard candy begins by softening of the outside layer (stickiness) followed by crystallization (graining), and it is a considerable time before the whole candy is affected, and even then it is still edible. The same happens with a caramel and this, when it has picked up moisture throughout, changes from a hardy chewy texture to one resembling a soft fudge or paste.

In aerated or cellular products, the deterioration is much more serious. With cookies and cereals, the absorption of moisture makes then either soft or tough and at the same time accelerates staling. Cookies that have lost crispness are disliked by most people. With aerated high boiled confections, the cellular structure will collapse on absorbing moisture and in the worst cases will transform into a layer of syrup or sticky crystal. If these are chocolate covered, and the covering becomes damaged or is defective, the center can apparently "disappear" because the volume of the syrup is much less than the aerated product.

The protective qualities of a chocolate covering are almost entirely related to its perfection or continuity over the center and with a very hygroscopic center the covering must be complete if no protective wrapper is used. In experimental work carried out on the enrober covering of a honeycombed center, it was never possible to get higher than 98 percent perfectly covered units and this high figure was obtained by means of a specially constructed enrober that immersed the center in chocolate as well as covering it.

In normal enrober practice, the standard is much lower and during the period between covering and sale more defects will arise in the form of cracks or breaks; a survey of bars at the point of sale showed 6 to 10 percent had defective covering.

Good protection of a hard candy or caramel can be obtained by means of a close overlap wrap of low permeability and the actual area exposed to the air is confined to a very small part of the surface near the overlap and a long exposure to humid air is required to affect the candy. However, even if small amounts of moisture are absorbed, the surface will become sticky and there will be difficulty in removing the wrapper so it is usual to provide an outer bag or other container for bulk protection as well.

With aerated products and cookies, a wrap with overlap is of little value even if the wrapping material has a very low permeability. The diffusive properties of the center have the effect almost of drawing in moisture through folds in the wrapper and the only sure protection is a strong heat-sealed wrapping. There is another factor not often realized with the hard candy confectionery in classes 2 and 3 (Table 22.3. It is that the sugars are in the amorphous or vitreous form, and as long as they are in this form, their A/W will be low, but should they crystallize or "grain," their A/W becomes that of the syrup phase, which is probably a relatively dilute solution of the mixed sugars with A/W approaching 0.70 to 0.80.

Confectionery in class 4 is all of a stiff or rubberlike consistency and changes little in texture unless it becomes very damp—the gelatinizing ingredients—gelatin, gum arabic, modified starch, or pectin—have a stabilizing effect and some can be protected from stickiness by sugar coating. The gelatinizing substances give stability as long as the ratio of crystallizable to noncrystallizable sugars is correct and no graining occurs.

Class 5 and 6 confections tend to give up moisture and dry out. Their physical structure consists of a syrup phase maintained at a soluble solids concentration of about 75 to 78 percent and a solids phase, which may be sugar crystals, ground nuts, or fruit substance. Fondants, marzipan, and pastes during part of their time in shops or storage will be exposed to drying conditions and will not regain the loss completely if more humid conditions are encountered. With wraps for these lines some "breathing" is necessary, because under warm storage conditions moisture will tend to deposit on the inside of impervious wraps with risk of mold growth, or if they are chocolate covered, moisture may transfer from the center to the covering. Milk chocolate particularly will develop a cheesy flavor as a result of this transfer.

Class 7—chocolate (and coatings)—are subject to surface attack by humid air. The moisture in the air will have no effect on the fat ingredient but will react with the sugar and milk solids in the surface layer only. (See note below Table 22.3).

In the confectionery section of the book, the relative A/W values of ingredients and different confections in contact with one another are discussed and the effect of mixtures of ingredients in a bar or chocolate block does have some bearing on the wrapper to be used.

A chocolate bar may contain raisins, nuts, small cookies, pieces of hard candy (perhaps aerated), nougat, or cereal such as Rice Krispies. Confectionery bars may also be coated with chocolate containing nuts or cereal and rough surface particles may protrude from the chocolate covering.

The protection to be given by the wrapper is related to the ingredient that has the lowest A/W because this will deteriorate where it is exposed or only light coated with chocolate—wafer, and honeycomb hard-candy pieces are typical examples. When moisture is absorbed, another factor must be considered. It has previously been noted that staleness of cereal products will develop but the moisture can also accelerate rancidity of fats in neighboring nuts or other fatty materials. It will be seen, therefore, that the permeability of the material used for the wrapping of a confection is only part of the information required. The nature of the recipe must be considered as well and it is necessary to carry out storage and travel tests on wrapped and boxed goods before the whole packing can be judged satisfactory.

**Water and Gas Permeability of Various Packaging Materials**   Table 22.4 and Tables 22.5 and 22.6 (Haendler, 1978, 1983) give figures that are representative of the large range of wrapping materials. Particular note should be taken in Table 22.4 of the effect of creasing or folding and of the difference between temperate and tropical conditions (see "Testing Methods" later and references).

TABLE 22.4. WATER VAPOR PERMEABILITY, GRAMS PER SQUARE METER PER 24 HOURS

| Material | Temperature conditions 25°C/77°F 75% relative humidity | | Tropical conditions 38°C/100°F 90% relative humidity | |
|---|---|---|---|---|
| | Uncreased | Creased | Uncreased | Creased |
| Metal foil/film Metal foil/film/papers PVDC/polythene* Laminates | 0.1–0.2 | no change | 0.2–0.5 | very slight increase |
| Moistureproof cellulose film | — | — | 6–10 | 12–20 |
| Polythene films (varies with gauge) Polythene/ paper laminates | 1–3 | no change | 3–7 | no change |
| Paraffin waxed paper | 4–10 | 90+ | 30–50 | 200+ |
| Paper impregnated with microcrystalline or plastic wax Bitumen/Kraft laminates | — | — | 10 approx. | moderate increase only |
| Plain cellulose film Glassine Vegetable parchment | 500–700 very little protection | no change | 1400+ very little protection | no change |

*Polyvinylidene chloride laminates with other films and paper have similar properties. The low water-vapor permeability is related to the PVDC, and strength to the other constituents of the laminate.

TABLE 22.5. WATER VAPOR PERMEABILITY—VARIOUS PACKAGING MATERIALS

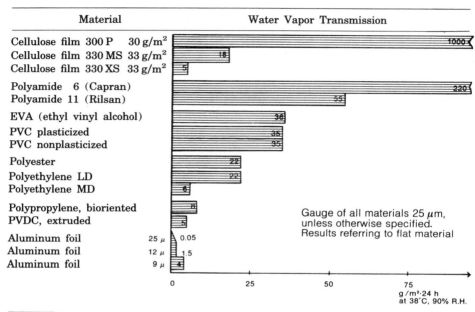

| Material | Water Vapor Transmission |
|---|---|
| Cellulose film 300 P 30 g/m² | 1000 |
| Cellulose film 330 MS 33 g/m² | 18 |
| Cellulose film 330 XS 33 g/m² | 5 |
| Polyamide 6 (Capran) | 220 |
| Polyamide 11 (Rilsan) | 55 |
| EVA (ethyl vinyl alcohol) | 36 |
| PVC plasticized | 35 |
| PVC nonplasticized | 35 |
| Polyester | 22 |
| Polyethylene LD | 22 |
| Polyethylene MD | 6 |
| Polypropylene, bioriented | 8 |
| PVDC, extruded | 5 |
| Aluminum foil   25 μ | 0.05 |
| Aluminum foil   12 μ | 1.5 |
| Aluminum foil    9 μ | 4 |

Gauge of all materials 25 μm, unless otherwise specified. Results referring to flat material

0    25    50    75

g/m²·24 h
at 38°C, 90% R.H.

## Printing Odors in Food Wrappers

All food manufacturers receive occasional deliveries of printed wrappers that smell of printing ink, which, if used in that condition, will contaminate the food wrapped in them. Chocolate and fatty foods are particularly prone to pick up these odors.

Odorous wrappings have been experienced for many years, but the problem seemed to become worse when the speed of printing machinery increased. This involved modification of ink composition and often insufficient drying was permitted before the printed material was reeled or bundled.

The odor residues in wrappings are of two types:

1. *Residual solvents.* Gravure printing was once the worst in this respect because a low-viscosity ink was used with solvents such as toluene and xylene, which are highly aromatic, and xylene is

TABLE 22.6. GAS PERMEABILITY—VARIOUS PACKAGING MATERIALS

| Material | Gas Transmission Rate—$O_2$ |
|---|---|

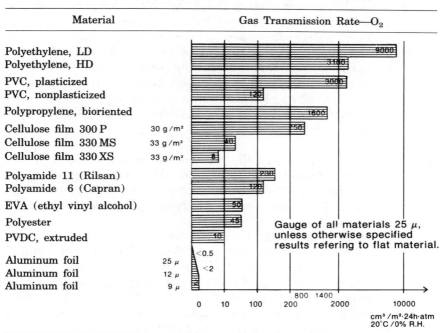

not very volatile. Toluene, if pure, is readily volatile, but low-grade toluene may contain small amounts of nonvolatile residue with very objectionable odor. Alcohol can be used in some inks and this gives little trouble from odor but has technical difficulties.

Small amounts of plasticizer are also used in printing inks and these can delay solvent evaporation and in themselves be odorous. Improvements have been made to these inks and less trouble is now experienced.

2. *Drying oils.* Letterpress and lithographic printing use high-viscosity inks and their drying depends partly on absorption into the material accepting the print but mostly on oxidation. The latter process gives rise to a characteristic oily odor that disperses with difficulty and only when oxidation is complete. Chemical "driers" are used to speed up this action but these may increase the odor given off.

**Prevention of Odor**   The prevention of odor in printed material rests jointly with the printer, the printing machine designer, and the manufacturer of printing inks, particularly the last.

The author's early experience of the printing inks and paint industries indicated that few companies constructed their formulations on a scientific basis and the printing processes were considered solely on an engineering and trade union basis. Fortunately, there have been improvements of recent years and formulations, purity of solvents, and residues in the printed material are checked, aided by modern analytical techniques.

**Removal of Odor**   Although the incidence of odor decreased, it was not entirely eliminated, and provision for deodorizing offending deliveries had to be adopted. It was not always practical to return consignments to the printer and generally the printer would bear the cost of treatment.

Procedures for deodorizing are as follows:

*Airing*   Offending reels or opened packets of labels (or printed cartons) are stacked on platforms in a manner that allows plenty of air circulation through them. Air is circulated by fan and some heat to give a temperature of 27 to 32°C (80° to 90°F) is desirable. This method may require several days of airing to disperse the odor but a great deal of printed matter was treated this way. It takes up a lot of space.

*Vacuum Treatment*   Large vacuum chambers are available in many food factories and these have been used successfully without heat, but better results are obtained if some leakage is allowed to give a purging effect. The use of vacuum with charging and discharging is expensive.

*Hot Oven Treatment*   Large stoves were constructed that contained circulatory fans drawing in hot air from a steam heater in the roof. Ventilators at floor and roof level provided air change and control of temperature. The reels or open packets of labels were open stacked on platforms but because of the better air circulation they needed less space between than with the airing method.

These stoves held up to 200 reels and could operate between room temperature and 65°C (150°F). The time and temperature required were determined for each type of wrapping material—gravure needed temperatures below 49°C (120°F) to prevent "blocking" of the labels by softening of the ink, and cellulose film required still lower temperatures with a maximum of 38°C (100°F). A diagram of the stove is shown in Fig. 22.12.

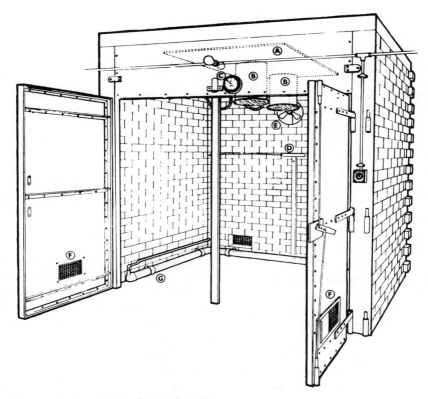

Fig. 22.12.   Deodorizing Stove—Front View
A. Steam Heater
B. Air Circulating Fans
C. Thermometer
D. Thermostat for Steam Valve
E. Vent and Sliding Shutter
F. Adjustable Fresh Air Inlet
G. Pipe Heater

Difficult wraps. Some wrappers were found to have persistent odor incapable of removal and there was no alternative but to destroy them—usually the printer took responsibility for the cost of this. Certain laminates were the greatest offenders and, as an example, if polythene is laminated to a printed paper before the ink has dried completely, solvents can be absorbed by the polythene and no treatment will remove them. Drying oil ink odors, previously mentioned, were sometimes impossible to remove and also some inks with cellulose solvent odors.

*Cadbury-Schweppes, England*

**Testing Wrappers for Printing Ink Odor**   The level of odor at which a wrapper is acceptable can be a subject of contention between supplier and user as it must be recognized that the staff at the printing factory is likely to be constantly in an atmosphere of aromatic substances and it is virtually impossible to detect the low levels required by the food manufacturer in such surroundings. Some companies make use of administration centers away from the factory for testing.

The following methods of checking may be used:

*Bottle Test* 1   Shredded pieces of wrapper are placed in a glass-stoppered bottle of 500 ml capacity and allowed to stand for 16 to 24 hr. A selected panel of five will decide from smelling the opened bottle whether deodorizing is needed. In sampling for this test, a reel is taken from the center of the consignment and the outer six layers removed—the sample is then taken from the next six layers for the bottle test.

*Bottle Test* 2   A sample similarly taken is placed in a glass-stoppered bottle with a piece of milk chocolate and incubated at about 21°C (70°F) for 24 hr. The chocolate is then tasted by a trained panel.

*Gas Chromatograph*   Both the above methods depend on an arbitrary standard and at low odor levels can result in argument. A more positive method uses the gas chromatograph and this is now in use by many large printing firms as a quality-control instrument to determine traces of residual solvents in their own printed material and to check the purity of the solvents or inks supplied to them. To volatilize the traces of solvent or other odorous substances in the wrapper, special cells with heated jackets are used and the vapors evolved injected into the chromatograph column for analysis. The peaks on the chromatograph charts are compared with known controls and the analysis is most effective when the types of inks used by the printer are known.

A modern development of this procedure is known as head space analysis. Stoppered bottles of 1 oz capacity are used and $100\,cm^3$ of the wrapper to be tested is shredded and placed in the bottle. The closed bottle is then subjected to a temperature of 105°C for 10 min. At the end of this time, 1 ml of the "air" in the bottle is extracted by means of a gas syringe and injected into the chromatograph.

In one large food company in the United States, chromatograph charts were seen that were used as a means of rejection or acceptance of a consignment of wrappers and an agreement had been worked out between the printer and the food company on the basis of certain peaks on the charts. Nevertheless, the bottle test was still used as a preliminary.

## Testing of Wrappers for Various Other Properties

Many detailed publications are concerned with the testing of wrapping materials and the reader is referred particularly to British

Standard Packaging Code BS 1133 Section 7—Paper and Board Wrappers, Bags and Containers Appendices A to W, which cover most of the methods required by a food technologist and to appropriate U.S. agencies. They are:

BS 1133 (1949–64) Revised 1967 (with amendments to 1985).

| | |
|---|---|
| Appendix A | Method of conditioning paper and board samples for testing. |
| Appendix B | Methods for the determination of the bursting strength of paper. |
| Appendix C | Method for the determination of the melting point of wax and/or other extractives (waxed papers). |
| Appendix D | Method for determining resistance to oil or grease (waxed papers). |
| Appendix E | Methods for the determination of water-vapor permeability. (See later.) |
| Appendix F | An accelerated oil resistance test for greaseproof paper. |
| Appendix G | Method for the determination of moisture in vegetable parchment. |
| Appendix H | Method for the determination of ash. |
| Appendix J | Method for the determination of acidity. |
| Appendix K | Method for the determination of total water-soluble extract. |
| Appendix L | Method for the determination of reducing matter in the water soluble extract. |
| Appendix M | Method for the determination of water-soluble copper and iron. |
| Appendix N | Method for the determination of bursting strength of wet paper. |
| Appendix O | Determination of chemical impurities in cellulose film. |
| Appendix P | Heat sealing test for cellulose film. |
| Appendix Q | Waterproofness test for cellulose film. |
| Appendix R | Oilproofness test for cellulose film. |
| Appendix S | Method for the determination of bursting strength of fiberboard. |
| Appendix T | Method for the determination of weight of fiberboard. |
| Appendix U | Method for the determination of thickness of fiberboard. |
| Appendix V | Method for the determination of waterproofness of outer surface of fiberboard. |

Appendix W   Immersion method for the determination of the water resistance of solid fiberboard.

Further sections of British Standard Packaging Code are listed in the References.

**Determination of Water Vapour Permeability**   The method quoted in Appendix E of BS1133 (1949–69) uses two atmospheric conditions:

1. Temperate conditions: 25°C (77°F) and 75 percent relative humidity for 24 hr.
2. Tropical conditions: 38°C (100°F) and 90 percent relative humidity for 24 hr.

A simpler modification of this method has been used for routine testing and is sufficiently accurate to give the information required on most wrappings used for confectionery purposes. It uses one set of atmospheric conditions and results obtained are closer to the tropical test above. The conditions are 25°C (77°F), 100 percent relative humidity for 24 hr and the detailed method is as follows:

The apparatus consists of a heavily tinned mild steel container, 2.5 in. in internal diameter and 0.90 in. deep. The top of the cup is fitted with a $\frac{3}{8}$-in. flange, making the total diameter 3.25 in. (Fig. 22.13).

A straight-sided metal dish 2.4 in. in external diameter and 0.75 in. deep is placed inside the container. The dish is fitted with a cover to prevent any increase in weight during weighing when it contains the desiccant.

The wrapping to be tested is placed on a clean flat surface and a 3-in diameter thin metal disk is placed on top and inscribed with a pencil. The circle of wrapping is cut with scissors for the test. The wrapping must not be creased or the surface damaged in any way.

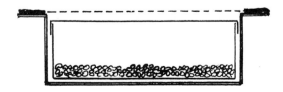

Fig. 22.13.   Determination of Water Vapor Permeability

Then, $10 \pm 0.1$ g of calcium chloride graded between 8 to 16 mesh is weighed into the dish covered with the lid. The lid is removed just before the dish in inserted in the container and the circle of wrapper placed on the flange. The disk is fastened in three or four places, using a brush dipped in molten beeswax, and then the perimeter is waxed to the width of the flange.

The apparatus is placed on wire gauze in a trough containing water,[1] arranged so that the level of the test material is $2\frac{1}{2}$ in. above the water surface. The trough is closed with a well-fitting glass plate with greased edges, and is kept in an incubator at 25°C (77°F). After 24 hr, it is removed and the wrapper cut away. The internal dish is removed, covered with its own lid, and weighed. The increase in weight in g $\times$ 316 represents the amount of moisture passing through 1 m$^2$ of material in 24 hr and is not influenced by moisture gain in the material itself. The result is termed the moisture vapor transfer (MVT) value or water vapor permeability (WVP). Materials with very low permeability may be tested for longer than 24 hr to give greater accuracy but the result is always calculated on a 24-hr basis.

During machine packaging, the wrapping material is likely to be creased or crumpled. With some material, this seriously reduces its resistance to water vapour and gas permeability. Waxed paper is an example. The following tests are designed to estimate the resistance to damage.

**Creasing Test**  A modification of the British Standards procedure is employed to suit the apparatus used in the permeability test. The procedure is as follows:

The British Standards method states that the length of the creases in centimeters in the paper shall be equal to the area of the paper in square centimeters. The containers used in the test previously described are 2.5 in. in effective diameter, that is, 6.3 cm so that the exposed area is 31 cm$^2$. A length of 31 cm is obtained by folding the paper across the center at right angles and having four other folds 2 cm away from the center folds. This gives two folds of 6.3 cm and four of 4.6 cm. The creasing also includes nine places where the folds cross at right angles.

The template for marking the paper comprises a flat piece of metal plate 8.8 cm square. Each side is notched in the middle and also at 2 cm on each side of this center notch. The template is placed on the

---

[1] A desiccator is suitable.

paper to be tested and the twelve folding positions (i.e., notches) are marked with a pencil. The wrapper is trimmed to approximately $\frac{1}{8}$ in. each way larger than the template and then taking one side as the top is folded across the center and gently creased by finger pressure. (No hard pressure must be given.) The two side folds are made in the opposite direction from the first fold. The paper is then opened and, keeping the original top surface underneath, is folded across the center at right angles to the first three folds. The other folds are then made in the opposite direction. The four quadrants of a 3-in. diameter template are marked permanently on the circumference. This template is placed on the creased paper so that the four marks correspond to the folds on the major axes. The paper is then marked round the template and cut with a pair of scissors. The paper is arranged centrally on top of the prepared container and coated with hot beeswax. The test is then completed as described previously.

**Crumpling Test** In addition to the creasing test, it was found useful to test certain specially strong laminates by determining the water vapor permeability after severe crumpling. The following procedure was used.

An 8-in. square from a representative sample of the material to be tested is cut out. This is pushed gently into the upright cylinder of the apparatus illustrated (Fig. 22.14). The piston is positioned in the mouth of the cylinder and allowed to fall under its own weight on to the sample. It must remain stationary for 1 min when the piston is

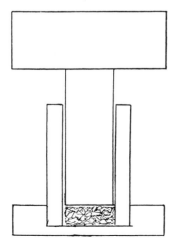

Fig. 22.14.   Crumpling Test Apparatus
Weight of Plunger—10 lb (4.53 kg)
Diameter of Plunger—4 cm
Distance of Fall of Plunger—10 cm

removed, the cylinder opened, and the paper smoothed flat. The MVT is then determined on the crumpled sample.

**Testing the Efficiency of Heat Sealing on a Wrapped Confectionery Bar**   Whether a heat seal wrap is required for insectproofing or for preventing water vapor or gas penetration it is necessary to know whether the wrapping machines are performing well. Visual inspection alone will only detect major faults such as a completely unsealed seam. With a high-speed, bar-wrapping machine, many bars may be wrapped before a minor sealing fault is detected visually. It is important, therefore, to have a rapid test that can be made adjacent to the machine so that as soon as any trouble occurs corrective action can be taken.

A method found to be very effective uses a hypodermic needle connected by flexible tubing to a compressed air line conveying air at 5 lb/in.$^2$ pressure. The heat-sealed package is tested by holding it lightly in the left hand and inserting the tip of the needle through the wrapper using the right hand.

With a good seal the wrapper quickly inflates quite hard. There are stages between this and no inflation at all indicating a minor defect in the heat sealing, and careful inspection will usually reveal where the bad seal is.

If repeat tests show similar defects, the reason is sought on the wrapping machine by the maintenance fitter.

The incidence of sample testing depends on how the machine is performing and very little experience is required to judge the degree of inflation. This method is simpler and just as effective as immersion in a water tank and the samples tested are not wasted—the puncture can be quickly resealed with a hot iron.

Although the needle system provides a quick answer to the performance of the heat-sealing machine, it must be realized that there is also an "aging" factor. Seals that may appear satisfactory immediately after leaving the wrapping machine may open after a period of storage.

In the heat-sealing operation, there should be complete fusion of the two surfaces. Microscopic examination of the seals will reveal this, but to apply this method on a continuous basis is hardly possible.

There have been laboratory machines that estimate the heat and pressure required to give the perfect seal but often the results do not correlate with machine performance. According to Young (1982), U.S.

Army-Natick Laboratories have developed an infrared system for evaluating seals. Perfect sealing is essential for sterile food packaging.

## Resistance of Printing Ink and Varnish to Tropical Conditions

Two other tests not covered by BS 1133 are particularly valuable when wrapped chocolate or other fatty confections are sent to tropical countries.

**Effect of Liquid Fats**  Under warm storage conditions, the fat in chocolate or compound coating will partly melt and transfer to the wrapper. Oil from nut ingredients has a similar effect. In such circumstances, unsightly stains may appear on the wrapper or carton. Glassine (greaseproof) paper is used as an inner wrap for fatty confections but many transparent films now have grease-resistant properties.

Some printed wrappers are varnished so that they do not show fat stains but an essential test with export wrappers is to see that the inks do not "bleed" by being fat soluble or that the ink varnish does not soften by being in contact with fat. The effect of the latter can be disastrous and there have been instances where whole consignments have been returned from warm climates with all the wrapped bars stuck together as complete blocks in the carton. At the same time, the chocolate inside was not unduly discolored, showing that this trouble can arise without excessive heat.

**Grease Staining Test**  The wrappers are stained by allowing small cubes of milk chocolate to remain in contact with the inside of the wrapper at 32 to 35°C (90 to 95°F) for 6 hr. The position of the chocolate should be so adjusted that a portion of each color on the wrapper is in contact with seeping fat. The degree of staining may be classified as follows:

Wrappers that show no change in color and little change in appearance when fat is absorbed—Excellent.

Certain colored nongreaseproof wrappers where the color is not dissolved by fat and where the fat stain is masked by the color of the paper—Good.

Wrappers with colored printing on a plain background where the color is not dissolved but the fat stain is very obvious—Fair to Poor depending on the extent of the staining.

Wrappers with coloring that is soluble in fat and bleeds into surrounding areas and would cause under wrappers and chocolate to be stained. The appearance of the wrapper may be altered considerably—Poor to Very Poor. Unsuitable for use.

Printed wrappers that are quite suitable for temperate climates sometimes react badly to these tests.

**Tropical Test for Determining the Adhesion of Printed or Varnished Surfaces of Wrappers** A number of the printed surfaces of the wrapper are arranged face to face and placed on a piece of plate glass. A sheet of $\frac{1}{8}$-in. thick aluminum 4 in. by 3 in. is laid on top followed by a 4-lb weight applied centrally. Two similar tests are prepared, one for incubation at 18°C (65°F) and the other at 32°C (90°F).

Parallel tests are carried out with a smear of liquid cocoa butter between the printed surfaces of the wrappers.

After an incubation period of 48 hr, the printed surfaces are pulled apart. Any signs of adhesion or damage are noted, especially tearing or marking of the wrapper.

## Toxicity

Descriptive matter so far will have given a good guide to the selection of a wrapping material for a particular product but it must be realized that plastics are used for a large variety of purposes besides making film for foods. The fabrication of many other articles has led to the inclusion of numerous pigments, plasticizers, stabilizers, and other additions, some of which, if used in plastics for food purposes, can create a toxicity hazard. Similarly, the adhesives used in laminates may be unsuitable in contact with food. Many of these additives are not copolymers or chemically combined with the plastic and can be readily leached out by moist or fatty foods.

For some years this problem was studied in the United States and the Lehman reports (1951–1956) were accepted as the basis of U.S. legislation and the most reliable guide to manufacturers and users of plastics for food purposes.

In the United Kingdom, the British Plastics Federation (1958) issued a list of substances used in plastics and classified them according to toxicity. Later, the Food Standards Committee of the British Ministry of Agriculture, Fisheries and Food collected information from manufacturers and users of plastics for the purpose of formulating regulations. Similar activity was initiated by the U.S. FDA. In 1976 an EEC directive was adopted to explore the laws of member states relating to articles that come into contact with foods. The materials to be considered were ceramics, glass, plastics generally, and particularly those plastics containing vinyl chloride. With ceramics, for example, the likelihood of extraction of lead and cadmium was considered and a simulated method of test proposed extraction with 4 percent acetic acid over 24 hr. Considerable study of the implications of legislation and its application suggests that not only should plastic wrappings be considered but food containers and plant as well. It is often the case that large containers, tank linings, packings, and piping which are made for general purposes are used in food factories, and the plastics used may be totally unsuitable for food. While the "resin" manufacturers are usually well acquainted with food requirements, the fabricators of the articles may not be, and unsuitable pigments, plasticizers, and additives may be included. All food utensils should be identified in such a way that they are recognizable as suitable for food purposes.

**Printing** The printing on a wrapper must be nontoxic. Pigments based on lead chromate are undesirable and many heavily coated prints have been found to have a very high lead content. Arsenical pigments are not unknown.

From time to time, reports appear in the press of children being poisoned by sucking lead-painted toys; contaminated print can do the same.

The problem of printing ink composition is similar to that of plastics, as the majority of printing is done on materials that have no contact with food, and solvent and pigment composition is therefore of little importance from a purity standpoint. With the enormous increase in prepackaged foods, the quantity of packaging materials required for these is no longer such a small proportion of the total.

Some printing requires the use of "gilt powder." This can cause trouble if not fast and can distribute itself throughout a box of chocolates or on the surface of a chocolate bar. Gilt print is usually varnished to fix it and the copper in the gilt has been known to react with constituents in the varnish to produce contaminating odors. Such reactions are accelerated if the wrapper becomes damp, in which case discoloration also occurs, with the formation of verdigris.

The printing trade[1] is now more aware of the problems with food packages and trouble is less frequent.

---

[1] Brochures by Colorcon, Inc. describe No Tox inks that are prepared entirely from FDA-approved ingredients (see References).

**Toxicity "Standards"**   Although no legal limits are universal, arbitrary standards have been agreed upon among some food companies and suppliers of printed matter. Examples of limits are given in Table 22.7 and in some circumstances other metals, such as copper, are considered.

TABLE   22.7. PACKAGING MATERIALS—POISONOUS   METAL CONTAMINATION

|  | Arsenic and antimony $mg/m^2$ | Lead $mg/m^2$ |
|---|---|---|
| Paper, film, laminate, carton board, foil, in direct contact with food | 14 | 70 |
| As above, not in direct contact, e.g., outer box labels | 28 | 140 |

## Wrapping Materials in Display and Advertising

Some reference has already been made to the various methods used for display and sale of chocolate and confectionery. Many changes have occurred in the past fifty years. At one time, huge displays of confectionery were seen in shop windows, some candies wrapped but mostly unwrapped. The conditions to which these goods were exposed varied enormously—bright light, sometimes sunlight, and high and low temperatures and humidities.

As more prepacked goods came into the shops, these were mostly contained in fancy boxes, printed foils, and wrappers. Many were exposed to the same conditions in shop windows and a high proportion of the printing faded and became unsightly. A faded wrapper always gives an impression that the contents are old and stale—very often due to bad stock control.

The next phase came when window displays were largely composed of dummies. These consisted of wax or painted plaster imitations of chocolates and confectionery bars and assortments, usually combined with elaborately printed advertising material.

Because these did not deteriorate in the same way as the edible product did, they were left far too long in the shop window and also often became faded, warped, and dusty.

Much more emphasis today is placed on counter and shelf display particularly with the advent of self-service stores and supermarkets. Here the wrapping and packing material is designed to attract attention and create impulse purchase. The wrapping must also be reasonably strong, and for most lines a protective wrap is desirable.

Much thought has been given to the design of racks used for display purposes both for counter use in shops and kiosks and in self-service stores. The first need is to be able to display as large a number and variety of bars and cartons as possible and to make it easy for the customer to choose and remove just what is wanted.

Unfortunately, this very fact makes it difficult for the shopkeeper to rotate stocks properly and often the slower selling lines remain for long periods in the racks. Sometimes the occasional bar gets "lost" and may become insect infested. In the bottom of the racks, small pieces of chocolate accumulate, and these also will become infested in time. The development of infestation is encouraged by the temperatures that prevail in supermarkets and large stores.

The cure for this trouble is meticulous care in stock rotation and cleanliness.

Fading of the wrappings is much less likely with interior display but outside stands and trolleys have been appearing where they are exposed to all the elements except direct rain.

## Testing of Wrapping and Advertising Material for Fading

In the early days of prepacking, many of the printing inks used were very susceptible to fading. Much investigation was carried out in conjunction with the printers and there was a gradual improvement, accelerated when the ink manufacturers and printers started to employ their own technicians.

In both the textile and the paper trades, dyes that are much less fugitive are now available and fading troubles are far less than they were.

Several methods have been employed to check wrapping materials—one uses the mercury vapor lamp, which gives a light very rich in ultraviolet rays; the other uses the Kelvin Fugitometer, which has a carbon arc lamp and gives a light corresponding more closely to daylight. The fugitometer also includes a humidifying device, as fading is accelerated by high-humidity conditions.

The mercury vapor lamp gives quicker fading than the fugitometer and results are obtained in 2 to 3 hr, whereas the fugitometer takes up to 48 hr.

More recently the xenon lamp has been employed, which, in addition to the standard light source, controls temperature and humidity and can provide intermittent water sprays. The intensity of radiation on the sample under test is slightly higher than the maximum intensity of the sun.

Whatever method is used to check fading, it is advisable to establish controls by exposing test strips to daylight at different times of the year and comparing these with the results obtained with the artificial light source. In addition to wrappings, advertising display cards and posters should undergo vigorous testing.

## ADHESIVES

A good knowledge of adhesives is an asset to anyone concerned with packaging, and although heat-sealing is employed a great deal in wrapping confectionery, many adhesives are also used on wrapping machines, and for hand and semimechanical wrappings of larger packages.

Adhesives are also an essential part of the makeup of laminates, paperlined strawboard, cartons, and boxes. Special glues are often required for joints in conveyor belts.

**Odor**  As with printing inks a vital factor in the use of adhesives for food packing is that they shall be noncontaminating both from an odor and toxicity standpoint. Some of the new wrapping materials, particularly coated film and metallized paper, are difficult to stick and many of the adhesives proposed have quite unsuitable solvents that, although they may lose their odor on drying, have vapors that may percolate into the inside of the pack and contaminate the food. The bottle test described under "Odorous Wraps" can be applied in the examination of adhesives.

Like plastics, adhesives are used in the packing of a great number of commodities besides food where sticking the wrapper together is the only criterion, and the special requirements of the food manufacturers are not always realized by the adhesive supplier.

Preservatives are frequently added to adhesives and many are badly contaminating—chlorophenol derivatives and similar substances can affect food flavor more than the odor would suggest, and some are toxic.

**Discoloration**  Discoloration of wrappings can arise from excessive acidity or alkalinity of an adhesive, from the solvent partly dissolving printing inks, or merely from soaking into the paper and causing stains.

Some glues darken appreciably on drying and exposure to air, and this can lead to unsightly uneven patches on the wrapper.

**Storage** Many troubles arise from incorrect storage. It is necessary, particularly with proprietary adhesives, to obtain from the suppliers the keeping limit and storage conditions necessary.

Starch- and dextrin-based adhesives change in viscosity on prolonged storage, and under extremes of temperature, syneresis may occur and no amount of mixing will bring them back to normal.

Gelatin glues, supplied in jelly form, may degrade by bacterial action unless heavily preserved, which in itself can be undesirable.

Other adhesives contain solvents that are highly inflammable or have toxic vapors, and these should be kept in "solvents" store. Dermatitis can be caused by handling concentrated adhesives, but these risks are usually well noted in the instructions for use. When dilution of an adhesive is required to make it suitable for use, a frequent mistake is to make too much dilute material, and this usually deteriorates rapidly. The excess must be discarded and this includes what is left in the glue pots on machines.

## Physical Properties of Adhesives

Certain terms used in relation to adhesives should be understood:

*Tack.* This describes the property of immediate adhesion. Two pieces of material are brought together with glue applied, and even if the glue is not dried or set, it will hold firm and prevent the pieces from moving in relation to one another. Hot glue, molten waxes, and resins have this property because they will cool rapidly in thin films.

Water pastes and gums rely on their viscosity and rate at which they dry or soak into the substances to which they are applied.

*Slip.* This is the opposite of tack and denotes the tendency for the materials to slide after the adhesive is applied. This gives untidy and irregular wrapping. Absorbency of the surfaces to be joined is a prime factor.

*Viscosity.* This is a loose description because solutions of adhesives do not necessarily conform to the normal laws of viscosity. The range of consistencies varies from stiff pastes with no flow properties to viscous solutions of hot animal glue. Checks on particular adhesives are always made by performance tests and moisture content, or an animal glue by determination of the Bloom value.

*Spinning*. Some glues show this fault when used on machines with the usual gluepot and pickup mechanism. The pickup does not give a clean break-off on deposit of a blob of glue but produces a filament that can wind round all local moving parts and cause mechanical trouble as well as fouling the reels of wrapper.

*Foaming*. This is often related to spinning. Foaming in the glue pot or in the preparation of the glue makes the process unmanageable and the whole gluing equipment must be cleaned out and fresh material prepared. Both these troubles can be reduced by adding small amounts of oleic acid, silicones, or other antifoam. If the glue is too concentrated, these troubles are accentuated.

*Thermosetting adhesives*. This type of adhesive may be used hot but after application will not melt again on heating, or it may be a cold liquid adhesive that will set by the addition of a chemical hardener. An irreversible chemical action takes place that gives a permanent bond.

*Thermoplastic adhesives*. These are resins that are applied as hot melts and set on cooling. They can also be applied in solution in organic solvents. Although theoretically they can be remelted and the joint pulled apart, in fact the bond is extremely strong.

## Adhesive Groups

There are various types of adhesives and a brief summary of these, their properties, and uses in relation to chocolate and confectionery are given. Greater detail may be obtained from publications specializing in the subject (see References).

**Vegetable Adhesives**   This is the most important group as far as food packaging is concerned and there are basically two types: (1) that made from starch or starch containing material, and (2) vegetable mucilage.

**Starch Adhesives**   The simplest is flour paste made by mixing boiling water with a cereal flour, usually wheat; it has little tack and it is suitable only for absorbent paper where its moisture can soak into the fiber and dry out through the paper. Starch pastes are made by heating the purified starch (which has been freed from protein and fiber) with water or dilute alkali, sometimes under pressure. The result is a viscous, translucent paste with rather more tack than flour paste—very little color or odor and pH close to neutral. The starch in these pastes is gelatinized or partially converted into dextrin. Dextrin can be made by the dry process of heating starch with nitric acid, by acid hydrolysis of slurries under pressure, or by acid hydrolysis and enzyme treatment. The treatment of these dextrins with alkali gives increased tack and adhesive strength. Dextrin gums have a slight but characteristic odor and are usually light brown or off-white in color. Dextrins are frequently compounded with borax to give improved adhesion and rate of set.

*Vegetable Mucilage* Gum arabic is probably the best known of these. It gives a very pale solution with considerable tack but its properties are variable.

Guar gum has become an important product of recent years and is used for paper sizing and as a stabilizer and thickener in liquid foods. Gum tragacanth is used in conjunction with gum arabic but not very frequently as a gum.

Many natural gums are used as ingredients in confectionery and they are described in greater detail under "Gelatinizing Agents".

While the simple pastes and mucilages are used for sticking paper and uncoated cellulose film, the dextrins are employed for many machine purposes and are made to a variety of viscosities and tack. The adhesive companies usually provide grades for particular machinery and it is essential to use these as prescribed and within proper keeping periods.

Some of these gums will become brittle and flake off under very dry conditions and plasticizers or humectants such as glycerol may be added to prevent this.

**Animal Glues**  These are used for carton and cardboard box making, for gummed paper strapping tape, and for all types of wood gluing. They are prepared by the extraction of bones and skins, subsequently purified to remove objectionable odor.

These glues are used generally as hot solutions, have very good tack, and give strong joints, but these joints will come apart and promote mold under damp conditions. The addition of certain hardeners such as potassium dichromate will help to prevent this.

These glues are sold commercially in three forms:

Solid: as granular powder, sheet or "pearls"—This form should be soaked in cold water for a period before it is used. This swells the pieces and makes solution on warming easier.

Jelly: this is a strong solution that has set and needs warming only for use.

Liquid: this is a fluid solution ready for use. It can be applied to surfaces that are absorbent and where there is time for drying. Like vegetable glues, these sometimes require the addition of plasticizers to prevent embrittlement and flaking when dry.

Fish glues are included in the animal group but would rarely be used for the food industry, because of the odor of both the glue and added preservatives.

The strength and character of animal glue are affected by prolonged heating and temperatures of 60°C (140°F) should not be exceeded and minimum quantities should be kept hot. The traditional glue pot with boiling water jacket is about the worst treatment animal glue can have.

Casein glue is strictly of animal origin as it is prepared from milk protein (curd). It is used to bond-paper and wood and can be used for foil but it is usually alkaline and may cause staining and corrosion. It has no tack and may be used cold under pressure when a good bond is given by penetration of the surfaces. Casein adhesives are frequently combined with latex and resin emulsions. Compared with casein alone, they have improved setting speed, flexibility, and tack.

**Synthetic Resin Adhesives**  These adhesives are particularly useful where highly glossy and metallized papers and laminates are used on chocolate boxes. They are also suitable for use on some coated cellulose films.

The vinyl resins are used mostly—they are polyvinyl acetate in organic solvent or emulsion and polyvinyl alcohol, which is soluble in water.

The resins and solvents have a distinct odor but provided the solvent is not trapped or an excessive amount of adhesive used, they are noncontaminating.

The cellulose esters have found increasing use for paper bonding, particularly the carboxyl methyl cellulose derivatives. They are water soluble with a very high viscosity for dilute solutions.

Most of the other synthetic resin adhesives, such as phenol and urea formaldehyde, are used for bonding wood.

**Sodium Silicate**  Sodium silicate solutions of various strengths are used extensively in the paper and cardboard industry, particularly for corrugated board. Being inorganic, solutions will keep well, are not affected by microorganisms, and are resistant to high temperatures but less so to dampness. They are of low viscosity and have little tack. Neutral grades are sold for adhesive purposes and this term is misleading because, although less alkaline than other grades, they have a pH of 10 +. Sodium silicate adhesives are useful for sticking paper labels on tins but inks must be resistant to alkali. If exposed to the air, these adhesives will solidify and will not redissolve in water.

The viscosity of sodium silicate solutions is increased by adding inorganic fillers such as china clay, and this retards the rapid absorption by porous surfaces.

**Bitumen**  Bitumen is rarely used as an adhesive in the food industry but it appears as a very useful bonding material for making waterproof paper. Two sheets of heavy kraft paper bonded together with bitumen form a waterproof laminate very suitable for lining cases or parcel wrapping large cartons for export. This material is robust and retains its waterproof properties well on creasing.

**Microcrystalline Waxes, Rosin, Mineral Jelly Mixtures**  These are hot-melt adhesives and are very rarely used in solvent solution. They are applied at temperatures of 100 to 140°C (212 to 284°F) using a heated glue box and roller, but prolonged heating must be avoided as partial decomposition occurs and the glue loses its plasticity. Small blobs set immediately by rapid cooling and this type of adhesive is very useful for high speed tin labelling. Some are used for making foil, film, and paper laminates.

**Rubber Solution and Latex**  Rubber adhesives have the unique property of retaining "tack" almost indefinitely and because of this are used on self-seal envelopes, packets and labels. They have limited use in the confectionery industry but latex adhesives will bond paper and foil.

Solvent solutions of rubber are not satisfactory for food purposes on account of odor and are often highly inflammable.

One fault of latex adhesives and self-seal packets is that threads and small pieces of rubber may get attached to the sweet inside the wrap.

Latex performs a very useful function for lining the seams of cans to render them moistureproof.

## Mechanical Sealing Methods

Heat sealing has been mentioned previously to provide wrappers that are completely closed to prevent access of moisture, odor, or insects.

There are several methods of achieving this mechanically.

**Bar Sealing**   This is the simplest and least expensive. One or two heated bars press the films together, thereby forming a weld by partial melting of the plastic.

Rotary sealing bars are also used to avoid intermittent action.

**Impulse Sealing**   These machines have a construction similar to bar sealers except that the bars are silicone rubber coated and a nichrome ribbon is stretched over one of the sealing surfaces. The ribbon is covered with teflon impregnated fiberglass. When the bars are closed over the article to be sealed, an electric current is applied through the ribbon for a short period of time to give the necessary heating and welding. This method is used for sealing "tacky" unsupported films.

**Hot Wire, Knife Sealing**   This is a very rapid method used for sealing or cutting bags (polythene).

**Other Methods**   There are many other methods of sealing used for larger articles. These are described by Young (1982).

## Adhesive Tapes

Adhesive tapes have become an important adjunct to the packaging industry not only as a convenient and secure means of sealing a package but to provide protection against moisture and insects.

**Gummed Paper Tape and Cotton Fabric Tape**   Gummed paper tape is a sized kraft paper coated with animal glue and is usually applied to larger packs for chocolate or confectionery to secure the lids and outer wrapping paper. It is supplied in reels and for use is moistened mechanically and applied immediately. Storage of the reels must be watched as damp conditions cause solid blocking—it is usually supplied in waterproof wrappers.

Paper tapes are available of various strengths, but when extra strength is required, cotton fabric tapes are used. These are coated with animal glue and are used and stored in a manner similar to that for gummed paper tape.

**Self-Adhesive Tapes**   Although these tapes have largely taken the place of string for securing parcels, their primary use is to seal container lids and prevent the contents from picking up moisture and becoming contaminated with dust and dirt.

The basic type consists of cotton fabric, paper, and various plastic films, and all may be colored or plain. The adhesive coating is usually derived from rubber, neoprene, butadiene–acrylonitrile copolymer (nitrile rubber), or polyvinyl ethers.

The following properties of these tapes must be carefully assessed as many types on the market are by no means perfect.

**Keeping Properties of Adhesive Tapes**  The adhesive coating must remain permanently adhesive and a period of at least twelve months should be guaranteed as moistureproof tapes are used particularly to ensure protection over considerable periods of storage. Experience has shown that some tapes become loose or even fall off due to perishing of the adhesive layer under tropical storage conditions.

It is useful to carry out accelerated keeping tests on taped articles using the BS tropical conditions 37.7°C (100°F)—90 percent relative humidity for periods of at least fourteen days, and for up to three months at 29°C (85°F) and average humidities.

**Strength**  This seems to vary greatly with cellulose film tapes and a breaking strength should be specified.

**Storage of Reels of Tape**  Sometimes when tape is unwound some of the adhesive comes away on the back of the tape. This may be due to ageing or bad storage and all adhesive tapes should be stored in moistureproof packages or cans.

**Water Vapor Permeability**  This is obviously an important property of waterproof tapes and the method described under "Water Vapor Permeability of Wrappers" can be applied to these tapes by using a modified container.

The container flange is fitted with bolts with wing nuts, rubber gasket, and a brass disk $3\frac{1}{4}$ in. in diameter, drilled to fit the bolts on the flange, and having three slits cut in it. For wide tape, two of the slits are $1\frac{1}{2}$ in. $\times \frac{3}{8}$ in. and the third 2 in. by $\frac{3}{8}$ in., while for narrow tape the slits are the same by length but $\frac{1}{4}$ in width.

A suitable disk is chosen and the tape is pressed firmly into position over each of the slits; an overlap of $\frac{1}{16}$ in. being made all round the slit.

The inner dish and calcium chloride are weighed, the rubber gasket fitted, and the disk screwed down on top of the gasket.

An exposure of 48 hr is given, after which the dish and contents are reweighed and the permeability over 24 hr calculated from the area of the slits in the disk.

**The Application of Adhesive Tapes**   Self-adhesive tapes are used to advantage for slip-on lid tins, but frequently the tapes are applied badly. On a cylindrical tin with sealed joints, the ultimate degree of protection depends on the quality and method of application of the sealing tape. This should be in contact with the lid and body surface over the whole circumference with good overlap and also pressed into the seam recess of the body. Storage under severe conditions of temperature and humidity should be applied to filled taped tins to decide if this method of packing is adequate, or whether seamless bodies and/or lever lids should be used.

## METAL CONTAINERS

Metal containers are used in the chocolate and confectionery industry for packing cocoa, food drinks, and chocolates for export to hot and damp climates. Often the protection value of the tin is coupled with decorative multicolored tin printing for sales promotion.

The materials used for metal containers are tinplate and aluminum. Tinplate is rolled mild steel plate coated in tin of varying thickness, either electrolytically or by dipping in liquid metal. The thickness of the tinplate and the tin coating is related to a "basis box," which is 112 sheets 20 in. by 14 in. (31,360 in$^2$.) and the thickness of the plate is defined by the weight of material in this basis box. Similarly, the thickness of the tin coating is specified by the weight of tin coating on a basis box and this may vary from 8 oz to 46 oz depending whether it is electrolytically coated or dipped. Metric standards are now being applied.

It must be understood that tin coating is not a complete protection against rusting under damp conditions as the tin coating is noticeably porous. A test with ferrocyanide jelly will show up these imperfections and a surprising variation exists in any particular grade of tinplate. Where adverse storage conditions are expected or if the contents of the tin are likely to be discolored by traces of iron, lacquering of the tinplate is necessary and this can also eliminate the necessity for a paper lining.

*Aluminum*   (99 percent   purity)   or   aluminum   alloy   (with 1.25 percent manganese) is used for "aluminum" containers, and this metal has specific uses and is not just an alternative to tinplate.

It can be extruded to give seamless bodies, is relatively incorrodible, and can easily be given decorative finishes, but it cannot be soldered and is not as strong as tinplate.

## Types of Cans

For details of the various types of can bodies, reference should be made to other literature Sullivan (1982) but a few are described below.

### Built-Up Body

1. Where the cylindrical body is seamed, the top rim presents a ridge at the end of the seam and slip-on lid is necessary. The tightness of this lid is determined by the depth of the side of the lid, the relative dimensions of the lid and body (a function of the stamping and forming machines), and the gauge of the metal. This type of closure is not water-vapor- or insectproof but can be greatly improved in this respect by tape sealing.
2. To the top of the seamed body a continuous rim can be rolled that can be flanged to take a slip-on lid, lever lid, screw-on lid, or vacuum cap. These give, progressively, a more secure and vaporproof closure.

**Seamless Body** Slip-on lids are more effective on these bodies as the edge of the top of the body is continuous, but the vacuum pressure lid is particularly applicable to this type of body.

**Vacuum Sealing** This usually refers to partial vacuum in the can. The interior of the lid is provided with a soft rubber or compound gasket that rests on the upper seamless rim of the body. Evacuation of the cans containing the food product is carried out in cabinets and on release of the vacuum the lid is forced down on to the gasket giving a permanent seal. For release of the lid, prising or slight distortion of the body is required.

**Gas Packing** Lever-lid or screw-top cans are required for this. Milk powders and drinks containing them, roasted nuts, and some fatty products have much longer shelf-lives if oxygen is excluded. The cans are evacuated in a cabinet under reduced pressure and the vacuum then released in an atmosphere of nitrogen or carbon dioxide. Alternatively, the tins are individually "sparged" by blowing the gas through nozzles into the cans. The last is more applicable to larger products such as nuts.

Cans have a much wider application in food industries other than confectionery. They have particular use in canning of meat, fish, fruit, and vegetables where a hermetic seal of sterile contents is required. Such conditions are rarely required for confectionery—an exception is chocolate syrup, which is prepared with solutions of sugar of low concentration.

The history of can development is interesting, dating from the Napoleonic wars, and some of the original cans have become collectors' items. The earliest can makers made about ten per day compared with modern "drawn" cans where machines can produce twelve cans per stroke.

In 1979, it was estimated that about 90 billion cans were made in the United States that year.

Originally, all cans were made by the three-piece system—top, bottom, and body. The seams have been variously sealed by solder, cement, or welding.

In the 1960s, a revolutionary development was the two-piece drawn and ironed can, which eliminated the side seams and separate bottom.

## DESSICANT POUCHES

Some reference should be made to methods of retention of low humidity inside packs. This is used for the bulk packing of dehydrated fruits, vegetables, and egg and milk powders. The packages are usually cans or drums.

In addition to the food, a pouch made of moisture-permeable material containing a desiccant is included in the container. By this means, dehydration of the occluded air and the food continues after closure and the moisture content is reduced below that obtained by commercial drying.

The common desiccants used are calcium oxide, silica gel, or calcined alumina, which retain their "dry" condition after absorbing moisture.

## REFERENCES

Angel, T. H. 1957. *Aluminium in Packaging*. Aluminium Development Assoc., London.

Barrer, R. M. 1941. *Diffusion In and Through Solids.* Cambridge Press, London.

Haendler, H. 1978. Practical aspects of use P.V.D.C. as a component of flexible packaging materials. *International Review. Chocolate, Confectionery, Bakery.* Beckmann, Hannover, Germany.

Haendler, H. 1983. Rugenberger Grossbäckereien, Moers, Germany.

Hurd, J. *Adhesive Guide.* British Scientific Instrument Research Association, Chislehurst, Kent, England.

Jones, O., and Jones, T. W. 1941. *Canning Practice and Control.* 124 et seq. London, U.K.

Lehman, A. J. 1951, 1954, 1955, 1956. *Q. Bull. Assoc. Fd. Drug Off.,* U.S.A. *Report of Toxicity.* Sub-Committee British Plastics Federation (1958), London.

*Modern Packaging Encyclopedia.* McGraw Hill, New York.

*No-Tox Inks.* Colorcon, Inc., West Point, Pa.

Oswin, C. R. 1954. *Protective Wrappings.* Cam Publications, Cambridge, England.

*Printing Odors in Food Wrappers* (Extracts from). Publication Dept., Cadbury Schweppes, Bournville, Birmingham, England.

Printing, Packaging and Allied Trades Research Association, Leatherhead, England.

Sullivan, C. 1982. Cans Metal. *Package Engineering* (163) (*Packaging Encyclopedia*). Cahners Publishing, New York.

Young, W. E. 1982. Machine sealing. *Package Engineering* (304) (*Packaging Encyclopedia*). Cahners Publishing, New York.

Wake, W. C. 1966. *Adhesives.* Royal Society of Chemistry Lecture Series No. 4. London.

The following sections of the British Standards Packaging Code (BS 1133) provide extensive information on the subject matter of this chapter. All sections are regularly updated.

British Standards Institution, London.

*Sections 1–3: 1967*
Introduction to packaging.
Factors influencing selection of packaging: contents, marketing, distribution regulations, costs, packaging methods.

*Section 5: 1985*
Protection against spoilage of packages and their contents by microorganisms, insects, mites, and rodents.

*Section 7: 1967*
Paper and board wrappers, bags and containers.
Materials, uses, and methods of test.

*Section 10: 1966*
Metal containers. Tins, cans, drums, crates, collapsible tubes, Materials, construction, uses.

*Section 14: 1961*
Adhesive closing and sealing tapes. Uses and preparation of gummed tape and self-adhesive tape. Illustrations.

*Section 16: 1968*
Adhesives for packaging.
Types, factors influencing choice, precautions in use of bituminous, inorganic, natural rubber and latex, vegetable, waxes, wax-based adhesives.

*Section 21: 1978*
Regenerated cellulose film plastics film, aluminum foil, and flexible laminates. Nomenclature, availability, properties and uses, methods of test.

# 23

# Quality Control

The general principles of quality control have remained the same over the years but there is a tendency today to use the term "assurance" instead of "control."

In the dictionary, "to assure" means to make certain, whereas "to control" is to govern with authority. So perhaps assurance is the better word as authority often does not guarantee good results. Quality control managers—take note!!

During the past sixteen years, and, in fact, for many years previously, the author has visited factories, large and small, in different countries. The approach to quality maintenance has varied greatly; in some places, control hardly exists, and in others there are laboratories carrying out many analytical tests, often meaningless in relation to the product.

Frequently, there is little knowledge of basic science and technology or the understanding of such elementary principles as relative humidity, water activity, cooling, solubility, and pH is minimal.

In the bigger factories, "in-line" control has developed and this results in continuous monitoring of manufacture. Signals of deviations in the final product are fed back to an earlier stage of production where corrections are made automatically. An example is continuous fondant manufacture where quite small changes in moisture content produce marked differences in texture. Thermometer probes at the syrup boiling stage transmit information that adjusts syrup flow and steam pressure.

Much greater attention is now given to microbiological standards as there have been instances of food poisoning from salmonella, Escherichia coli and other pathogenic organisms.

Packaging has seen many changes. Whereas at one time many packages had simple paper overlaps, now most confectionery and chocolate are in heat-sealed protective wraps. The packing material, whether film or laminate, is resistant to the passage of water vapor and gases. It also keeps out most insects.

Much of the material in this chapter is similar to that in previous editions. The principles of quality control have not changed much but the methods of application have been adjusted in line with modern processes. Many of the views expressed are results of personal experience but nevertheless are believed to be valid.

## PRINCIPLES OF QUALITY CONTROL

Many books on quality control abound with mathematical formulas and we are indebted to the statisticians and mathematicians who are able to guide the manufacturer in the acceptance or rejection of raw materials and finished products, but there is a great deal more to quality control than just statistics.

The statistician must be provided with facts, usually a large number of them, before being able to give sound advice, and often the scientific knowledge of the technologist makes it possible to eliminate much of this mathematical treatment.

In the confectionery industry, the past sixty to seventy years have seen a change from the craft of the single worker to mass production methods with much mechanization and relatively few operators who need have no knowledge of confectionery. The technologists have developed the product, and the engineers have constructed machinery to make the product consistently to standards laid down by the technologists and marketing personnel.

Quality and production to a standard then become a combination of very close control of raw materials, instrumentation at all stages of the process, and statistical examination of the finished product. The last involves weight control, tasting tests, and package examination.

## Who Decides the Standard of Quality?

In the development of a new product or in changing the process for an existing product, the quality standard is set by the sales and marketing executive. It will obviously be as high as possible but should be compatible with the market for which it is intended, and at the right price. Many instances have arisen where the quality has been aimed too high at a price that only appeals to a small minority of the public. In the confectionery industry, quality will also include the packing. At one end of the scale there are the elaborate boxes used for chocolate assortments and at the other the simple printed wrapper of a chocolate-covered bar that just tells the buyer what is inside and gives some protection to the product.

The modern thought is for the quality standard to be fixed fairly early in the development of a product, and this includes the type of packing. Highly mechanized equipment for mass production of a confectionery bar may involve a very large capital expenditure and it is essential to establish the precise nature of the line, including quality standards, before machinery is ordered. Alterations at a later date can be very costly.

## Who Controls Quality?

The responsibility for manufacture to a quality standard rests entirely with the production departments, which today include the engineers and technologists. It is quite erroneous to believe that the quality control department is responsible for the quality of a product. If this is the case, a mentality will develop among the process personnel where they will consider quantity only and not quality, and it will be left to the inspection departments to reject the bad work at the end of the production line—a highly wasteful operation. Provision of information on deviation from the quality standard is the real duty of the quality control department, and it must be provided quickly.

## How Should Quality Control Be Organized?

There are three main spheres of operation in the control of quality: (1) raw ingredients, (2) process of manufacture, and (3) inspection of the finished product.

The approval of raw materials for use in production and the inspection of the finished product are the prime responsibilities of quality control departments. There is, of necessity, liaison with the process staff, but considerable diplomacy is required for the most effective results. Inspection, strictly speaking, is not quality control, as it works only on the basis of acceptance or rejection of the finished product. The only value of rejection is that advice may be forthcoming on how to prevent further occurences.

Inspection of finished product should also include examination of stock in warehouses and depots. Many companies include the investigation of complaints and the inspection of retail outlets as duties of the quality control department.

## RAW MATERIALS

The acceptance or rejection of raw materials is normally the duty of the analytical chemists' department which should be a section of quality control. Whereas the chemist's analysis is important to decide whether the material meets the agreed-upon specification, it is also important to make a visual inspection of the consignments as a whole. Very quickly, damaged bags or cases will be noticed and it should be possible to trace manufacturers' batch numbers, which will considerably reduce the incidence of sampling.

Inspection is usually combined with sampling and this enables the sampler to make a visual examination of the material in the open bags or cases. A good idea of variation, cleanliness, or infestation can then be obtained.

Sampling is the most important part of raw material inspection and an incorrect sample, not representative of the bulk, will jeopardize all subsequent analysis, and may even result in spoilage of large quantities of product. A sampler must be reliable and given implicit instructions.

A statistician will advise on the best method of sampling a consignment purely from a mathematical standpoint, but the technologist or analyst usually has an advantage in knowing the product, its origin, its susceptibility to variation, and the effect of the variations on the final product. This knowledge is usually able to reduce sampling and analysis considerably.

The following procedure is a commonsense approach to raw material inspection and it has been used successfully with very few errors. Rarely is it necessary to sample and examine on a fully statistical basis.

## Type of Raw Material

Some raw materials are practically pure substances. Sugar, for example, is, 99.9 percent plus the pure chemical substance saccharose (sucrose), and if the source of supply and manufacture is known, it is senseless to do a detailed analysis. It is usually sufficient to do a superficial inspection of the consignment (assuming it is in bags) for damage or local contamination, together with making a syrup from a single representative sample for color examination.

Glucose syrup, starch, and fats are in a similar category but with one difference. Glucose may be of different degrees of conversion, fats may have different melting points, and starch may be for creme depositing or "thin boiling" for use as an ingredient. In such instances, *all* packages, drums, or bags must be marked adequately to describe the product. (Bulk deliveries are considered later.) A visual inspection of the whole consignment is made and limited analysis done on a representative sample. As an example, this may be the melting point of a fat. Essential oils, other flavors and spices, cocoa beans, nuts, dried fruits, egg albumen, and similar materials need a different treatment. The essential oils may be sophisticated if the origin is not very definite, therefore it is necessary to do more sampling and to carry out flavor tests and certain analyses such as specific gravity, optical rotation, and refractive index. Purity tests are also important as contamination with metals such as lead and copper is not uncommon. Spices, in addition to flavor tests, need microbiological testing and examination for extraneous "filth." Cocoa beans require the "cut test" to ascertain whether they are correctly fermented. Nuts and dried fruits require careful examination for flavor, foreign matter, and moisture content.

The above serves to show the pattern of examination—more detailed examinations, where necessary, can be applied and references should be made to raw materials described elsewhere in the book and the many analytical textbooks.

## The Supplier

Knowing the supplier of a raw material and the supplier's method of manufacture and quality control greatly influences the degree of inspection to be carried out at the receiving end.

The manufacturers of most raw materials, particularly the bulk ingredients, have improved and mechanized their methods of manufacture in the same manner as the confectionery industry. They have also improved their quality control. All that is necessary, therefore, is for a senior quality control chemist and a purchasing executive to visit the supplier and, through discussion, agree on specifications, type of packing, batch signs, and keeping periods. The quality control chemist will also discuss methods of testing used by the supplier's chemist.

This greatly simplifies the receiver's task, and reduces inspection to the bare minimum. In many cases, the supplier will give a certificate of analysis with each delivery, and with bulk tanker deliveries this is essential.

With any new supplier or new ingredient, the above procedure is the only one to produce good liaison between supplier and user.

## Receipt of Raw Materials and Preparation for Production Use

**Bulk Delivery** Sugar, glucose syrup, mixed syrups, and fats are delivered to large factories in tank truck or rail tank cars. Similar bulk transport is used for liquid chocolate, coatings, cocoa liquor, and milk crumb, when carrying these products between factories of the same company, or when a factory specializing in the manufacture of covering chocolate is supplying another not equipped to make chocolate.

A tank car may hold 10,000 to 20,000 kg (22,000 to 44,000 lb) and when it arrives at the receiving factory, the contents are normally pumped to storage containers that may hold several hundred tons. It would be disastrous if a tanker load of contaminated material or the wrong substance were delivered into the storage container, so it is essential that a rapid inspection be made of a sample drawn from the tanker. Usually the tanker cannot be held while any chemical analysis is being made but some physical tests such as specific gravity or moisture by an electronic method are possible. The sample must *always* be tasted and examined visually for foreign matter.

It has previously been mentioned that most raw material suppliers will give a "certificate of analysis" with each bulk delivery and it is then necessary only to taste and make a visual inspection of a sample before discharging to storage. At the supplier's factory, it is likely that chemical analysis and other quality control tests will have been made in the course of production and a tanker delivery would be made only from approved batches.

**Delivery in Separate Containers**   Many food raw materials are still delivered in hessian or paper sacks, wood or fiber cases, or drums. A more recent improvement is to use heavy-gauge polythene bags or to line the containers mentioned with thinner polythene bags.

A superficial inspection of a consignment is made and a representative sample obtained for analysis. If prior knowledge of the consignment (batch numbers, supplier's analysis) is not available, the sample may be obtained by the method known as random sampling. This consists of numbering each package or group of packages and then referring to tables of random numbers that appear in many statistical publications. Having obtained the sample or samples, it is necessary to decide whether to reject or accept the consignment. The supplier should have used a quality control system for the manufacture of a product but is bound to accept a degree of variation. Similarly, the consumer has fixed a standard of quality and is prepared to accept a certain tolerance.

It is possible, if these ranges are known, to calculate the number of samples to be taken to make sure there is a very low risk of acceptance of a poor, or rejection of a good, consignment.

Analytical chemists will usually have more time to conduct tests on consignments delivered in separate containers. Such tests are useful in checking doubtful material during visual inspection of the consignment.

Assuming a delivery is accepted, it has then to be transferred to production, and this means opening containers and weighing up for production batches. It is generally most undesirable to handle containers in production departments with, perhaps, the exception of drums, as foreign matter in the form of nails, wood splinters, staples, and pieces of paper or polythene can find their way into the product. The tipping of sacks may also deliver fiber or dirt from the outside of the bag into the raw material—bags should always be brushed off before emptying.

Containers should be opened in batch rooms or in a section of the raw material stores where special care can be taken to ensure that no contamination takes place. In many instances, it is a good plan to transfer the contents of packages to metal bunkers for use in production departments where they may be discharged mechanically to metering or autoweighing equipment.

In the operation of opening the containers, every opportunity is presented for detecting defective material, and the value of training the personnel to report anything unusual is obvious. The same applies in the batch rooms and in the production departments, and this emphasizes the point previously mentioned that production personnel are responsible for quality control.

**Alternative Raw Materials**  Another quality factor in a recipe is the use of alternative materials. The quality control analyst will be able to recommend a new source of supply of a raw material following chemical examination and some trial factory batches. Occasionally a cheaper source of supply is available and at the same time some aspect of the quality may be below standard. Yet some special treatment or minor processing may render the inferior material suitable for use without affecting the quality of the final product. In such cases, the cost of the extra process must be balanced against the lower cost of the material—very often the additional process shows a useful net profit.

Another, more difficult decision to make is the use of a substitute material—for example, the replacement of egg albumen with another whipping agent. This is not the responsibility of the control analyst, except from a purity aspect. It is mainly a job for the development staff members, who will make small-scale experiments and, if these show promise, production batches will be made and the resultant product submitted to control tasting panels. It these samples prove acceptable, they will still have to be subjected to shelf-life tests.

## PROCESS CONTROL

A great change in process methods has occurred in the confectionery industry in recent years. In various parts of the book, mention has been made of the adoption of continuous processes in place of a multiplicity of batch makings.

In the batch process, control is largely in the hands of two persons—one who weighs out the ingredients according to the recorded recipe and the other who makes the batch of product. The first is aided by scales and volumetric measures, and the second is armed with a thermometer or, if very up to date, a refractometer as well! Sometimes both jobs are done by the same person. It can be easily understood that quality greatly depends on the reliability and understanding of these workers.

The larger the quantity of product manufactured, the greater will be the number of batches, and the more likely will be the risk of variation between the batches, and also the people making them.

To check the quality of a large number of small batches is a very big task, and if treated on a statistical basis will mean the examination of many samples. Each should be checked for flavor and at least one analytical determination may be made on many of them, usually moisture content. In actual fact, sufficient checking is never done with this sort of production and the result is a very variable product.

The alternative to many small batches and a lot of checking is either to greatly increase the batch size or to make the product continuously. Unfortunately, so frequently in the confectionery industry, insufficient study is given to the effect on product quality, particularly flavor, before new processes and equipment are installed. As a result, continuous processes have been looked upon with suspicion as always giving an inferior product. In many factories, when increased production is required, instead of using a continuous manufacturing process an automated batch system is installed where, perhaps, rows of steam kettles of the type used for hand-made batches are automatically filled with ingredients, boiled for a precise period, and then discharged automatically. While this system may be highly ingenious, it is not scientific and the more sensible and rewarding method is first to study the batch process. This consists of recording cooking time, storage periods in kettles with temperature gradients, effect of the sequence of ingredient addition, and timing and examination of the reactions that occur when a batch is spread on a table. The data to formulate a continuous process are then available.

In batch processes, the times of cooking and cooling are usually longer than in continuous processes.

The effects of the difference are exemplified by the following.

*Caramel, fudge.* Flavor and color are related to the Maillard reaction between the milk protein and the sugar ingredients. The longer the cooking time, the more the flavor and color are developed.

With short-duration continuous processes, additional caramelizers are necessary to develop flavor (see "Caramels").

*Pectin jellies.* Acidity is necessary to set pectin jellies. Batch boiling will cause appreciable sugar inversion but in a continuous process of short duration it will be much less.

*Microbiological.* Raw materials such as cocoa powder, nuts, preserved fruits, some milk products, and egg albumen contain ferments, molds, and enzymes. These may not be destroyed in a continuous process, particularly if added toward the end of the process.

Sometimes the sheer mechanics of a continuous process may cause trouble. Fudge manufacture is a typical example. When fudge is made batchwise, the caramel is boiled to about 118 to 121°C (245 to 250°F), cooled in the pan to around 82°C (180°F), and fondant stirred in quickly. The mixture is then poured onto a table, where it cools without movement and develops the characteristic crystal structure. This fudge has a degree of rigidity and can be rolled, cut, and enrobed without showing signs of collapse.

If, however, the fudge is made and cooled in the pan with continuous mixing during cooling, the crystal structure is broken down and the fudge becomes a soft paste and difficult to handle. Continuous processes are prone to overmixing, and if a continuous caramel cooker is used for fudge making, followed by a continuous cooler (both of which agitate the product), it is essential to add the fondant near the discharge point of the cooler, allowing just sufficient time for proper dispersion of the fondant in the caramel. It is better to cool the mixture finally on a continuous band where crystal growth results in a more rigid product. This may then be shaped by rolling and cutting, or by extrusion, as with the batch process, and no loss of shape should occur.

## In-Line Process Control

A great deal of attention is being paid to this system in the chocolate and confectionery industry. It was developed in the chemical and oil industries where it is used extensively. In principle, the system makes use of certain physical or chemical characteristics such as boiling temperature, density, viscosity, refractive index, pH, color, or even the thickness or width of a slab. In batch production, these constants are measured manually, usually on a proportion of the batches only, by means of laboratory instruments or an adaptation to make them more robust for factory use. With "in line" control, continuously recording instruments are used. These feed signals to controllers, which, in turn, operate valves adjusting steam pressure or flows of liquids or solids, thereby quickly correcting any defect appearing in the end product.

In the planning of in-line control, it is first necessary to establish the critical points in the process and the appropriate methods of control are focused on all these points.

The principle is best explained by giving an example. This concerned the development of a continuous process for an aerated boiled sugar bar that had been made for many years on the batch principle. The basic process was simple—merely adding sodium bicarbonate to a high boiled syrup, stirring well, and tipping the magma onto tables to cool and mature, after which it was cut into bars. The original process produced small batches and slabs that had to be cut, and those that were not the right thickness were wasted and had to be reclaimed by an expensive decolorizing and filtration process.

After a great deal of experiment, a continuous process was devised and Fig. 23.1 is a diagram of the flow path of the plant showing the critical points of control.

The two most important points in the system are (1) at (H) where the boiling temperature of the syrup is recorded, which determines the final texture of the confection and influences the density and type of aeration, and (2) at (K), which controls the quantity of sodium bicarbonate delivered to the mixing pot. This affects the thickness of the layer on the belt, the degree of aeration, and the flavor and color of the finished product.

The pump supplying the bicarbonate suspension must be a precision instrument as the flow is very small compared with that of the syrup, and a slight error in the quantity delivered shows immediately in the thickness and color of the material on the band. The effect is so

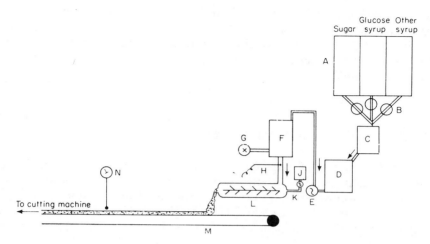

Fig. 23.1.   Flow Diagram of "In Line" Plant

A. Ingredient supplies—quality accurately controlled as previously described. The sodium bicarbonate suspension at (J) is similarly checked.

B. Automatic apportioning, volumetric or weighing.                        (Critical point 1)

C. Continuous dissolver supplying a syrup of constant recipe to

D. Syrup storage tank.

E. Variable pump supplying syrup to                                       (Critical point 2)

F. Continuous cooker. This is supplied with high-pressure steam through control valve (G)                                                           (Critical point 3)

H. This is an electric resistance thermometer with a sensitive probe in syrup stream issuing from the cooker.                                          (Critical point 4)
   This syrup is fed into

L. A continuous mixer also supplied with a suspension of aerating agent from a small tank (J) through

K. A variable-flow peristaltic pump.                                      (Critical point 5)
   The hot magma discharges from (L) onto a conveyor (M) where the magma continues to react and ultimately attains a uniform thickness.

N. Is a recording device that monitors the thickness.

marked that correction is easily made, manually, from the reading on the thickness recording device, but the safer way is for the thickness gauge to adjust the bicarbonate pump electronically.

   In addition to the major points of control shown there are various other mechanical devices controlling the temperature, width, and uniformity of the mixture on the belt. The cooled product arriving at the cutting machines is therefore of constant quality and only occasional analytical checks are required. These are confined to a determination of the sugars ratio in the original syrup and moisture content and density.

The latest development in this system is computer control whereby the critical point conditions may be obtained immediately on the computer screen.

## FINISHED PRODUCT INSPECTION

If quality control of raw materials and process has been carried out correctly, no inspection should be necessary. Such a state of affairs in any factory would be Utopia indeed, but with the continued development of in-line control it should be possible to reduce the incidence of inspection. At the same time, rejection of the finished product should attain a very low level.

In the best factories with the most conscientious workers things still go wrong occasionally, so finished product inspection is necessary to cross-check the previous operations. The problem is to decide the minimum level of inspection. Inspection is costly and it is foolish to concentrate on labor saving at all stages of the process and then to offset these savings by an army of inspectors. This can happen and not only does it cost money but it is bad for the morale of the process workers.

At the end of a confectionery line, probably hundreds of wrapped pieces appear every minute and it must be decided what checks are to be made. The following are the most important.

1. Is the appearance satisfactory? (This includes the product and the packing.)
2. Is the taste correct?
3. Is the weight within the agreed-upon tolerance?

### Appearance

The inspection staff in the factory is provided with a "pattern," which consists of a packed box of the product being manufactured. A breakdown of the packed box should also be available on trays to show the wrapped and unwrapped units.

The pattern has already been approved by the marketing and quality control departments.

## Incidence of Sampling—Taste Checks

The samples taken for weight control (see below) may be judged for appearance and also set aside for taste tests.

Taste checks will be made by the methods outlined in Chapter 18, on flavor.

Often taste checks are made on a sample from the original bulk batch. A storage tank of chocolate, for example, may contain 5 tons, which would produce about 90,000 2-oz bars. A sample from the tank is molded into bars and a rapid taste check made. This will indicate whether the chocolate deviates greatly from standard, and will at least avoid molding and wrapping a product that would eventually be rejected.

Inspection staff members who are responsible mainly for weight and appearance may also make periodic taste checks. They would not be responsible for making rejection decisions but they could alert others to anything seriously wrong.

## Weight Control[1]

Frequent production of overweight pieces obviously reduces profit. Likewise, the making of many underweight products cheats the consumer and constitutes a legal offense. A weight control chart becomes essential. Some plants may use the under-over types of system, with the zero line being the exact weight. The whole numbers that indicate fractional ounces or grams depend upon the product. At a glance, a supervisor can get an idea of how the weights are lying in relation to the distribution chart. If shifting takes place toward the upper control limit, adjustments are made by the operators, and a similar action is taken when shifting occurs toward the lower control limit.

A distribution curve makes it possible to ascertain the cost weight of the product. This is the average weight. Also, it shows the characteristics of the plant and points out spots where more control is necessary. It enables the operator to know when the product is "out of control," and adjustments must be made to correct that situation. Adequate sampling is essential to develop a distribution curve, and

---

[1] The section on weight control is based on a presentation on quality control by the late Wesley Childs (formerly of Curtis Candy Div. Standard Brands) at the University of Wisconsin NCA candy course.

from it, control charts. Unless a bell-shaped curve is obtained, the process is out of control. When a bell-shaped curve is obtained, the lower "sigma" is the lower control limit and the upper "sigma" the upper control limit of the process. The following is a method used to prepare control and distribution charts:

The weights of confectionery bars are tabulated as on the control chart (Fig. 23.3) over a period of time, usually an 8 hr shift. Each column is totalized, and each horizontal space is totalized, e.g., one at 0.82 oz, four at 0.81 oz, twenty-three at 0.75 oz, etc. (Note that Fig. 23.2 covers 4 hr only.) From data so obtained, the distribution chart is made (Fig. 23.3). The frequency, f, is the number of different ounce fractions found by weighings, as given above. Because 0.75 oz was guessed to be the average of weights taken over two 8 hr shifts, this group was designated as the 0 group. Weight over 0.75 oz were marked *plus*—e.g., 1, 2, 3, etc. Weights below 0.75 oz were marked *minus*—e.g., 1, 2, 3 etc. The reason for this is to simplify calculations.

Frequency (f) multiplied by the class deviation (markings plus or minus) gives a figure as fd. Multiplying this fd by d results in $fd^2$.

The formula for standard deviation (sigma) is

$$A\sqrt{\frac{fd^2}{n-1} - \left(\frac{fd}{n}\right)^2}$$

A is the actual group interval (1 in above).
n is the number of samples or total of frequencies.

The standard deviation ("sigma") is the most useful means of measuring dispersion of weights. It is valuable when trying to determine how great the variations are within the samples taken. The larger the value of the standard deviation, the less closely grouped are the samples. What do we mean when we say the standard deviation of the bars was 3.04? It means that the bars varied in weights; some were above and some were below the average or mean. Standard deviation is defined as the square root of the average of the squared deviations from the mean. When the distribution of weights taken enables a normal bell-shaped "curve" to be drawn, then the standard deviation can be interpreted in relation to the mean. Between 1 sigma minus and 1 sigma plus or 72.3 to 78.4, 68.3 percent of the bars should be included. Between 2 sigma minus and 2 sigma plus or 69.3 to 81.4, 95.3 percent of the bars should be included. Between 3 sigma minus and 3 sigma plus or 66.3 to 84.5, 99.7 percent of the bars should be included.

A control chart cannot be made until upper and lower control limits

**TIME** | 7:00 | 7:15 | 7:30 | 7:45 | 8:00 | 8:15 | 8:30 | 8:45 | 9:00 | 9:15 | 9:30 | 9:45 | 10:00 | 10:15 | 10:30 | 10:45 | 11:00 | **ACTUAL WEIGHT OZ.**

| | | | | | | | | | | | | | | | | | | Weight |
|---|---|---|---|---|---|---|---|---|---|---|---|---|---|---|---|---|---|---|
| 14 | | | | | | | | | | | | | | | | | | 0.89 |
| 13 | | | | | | | | | | | | | | | | | | 0.88 |
| 12 | | | | | | | | | | | | | | | | | | 0.87 |
| 11 | | | | | | | | | | | | | | | | | | 0.86 |
| 10 | | | | | | | | | | | | | | | | | | 0.85 |
| 9 | | | | | | | | | | | | | | | | | | 0.84 |
| 8 | | | | | | | | | | | | | | | | | | 0.83 |
| 7 | | | | | | | | | | | | | | | | | | 0.82 |
| 6 | = | − | | | − | − | − | | ≢ | | | | ≡ | | | | | 0.81 |
| 5 | − | | − | | ≢ | ≡ | = | − | = | − | − | | − | ≡ | ≡ | | − | 0.80 |
| 4 | ≡ | | = | = | − | = | − | = | − | − | ≡ | − | − | − | ≢⁄ | = | ≡ | 0.79 |
| 3 | | − | = | ≡ | ≡ | = | ≡ | = | − | ≡ | | ≡ | ≡ | − | | = | ≡ | 0.78 |
| 2 | = | = | = | | − | = | − | = | = | − | ≡− | − | − | | | ≡ | − | 0.77 |
| 1 | | − | = | = | − | − | − | − | | − | − | | | ≡ | ≡≢⁄ | ≡ | ≡ | 0.76 |
| + | | | | | | | | | | | | | | ≡ | | | ≡ | 0.75 |
| 1 | = | = | = | | | | = | = | | ≡ | ≡ | = | − | | | ≡ | − | 0.74 |
| 2 | | − | = | = | − | = | = | = | − | − | | | | − | ≡ | | 0.73 |
| 3 | | | | | | | − | − | − | = | ≡− | | | | = | | 0.72 |
| 4 | | − | = | | ≡ | | | | | | | = | | − | | | 0.71 |
| 5 | | | | | | | | | | | | | | | | | | 0.70 |
| 6 | | | | | | | | | | | − | = | | | | | 0.69 |
| 7 | | | | | | | | | | | − | | | | | | 0.68 |
| 8 | | | | | | | | | | | | | | | | | 0.67 |
| 9 | | | | | | | | | | | | | | | | | 0.66 |
| 10 | | | | | | | | | | | | | | | | | 0.65 |
| 11 | | | | | | | | | | | | | | | | | 0.64 |
| 12 | | | | | | | | | | | | | | | | | 0.63 |

Fig. 23.2. Weight Control Chart of Confectionery Bars

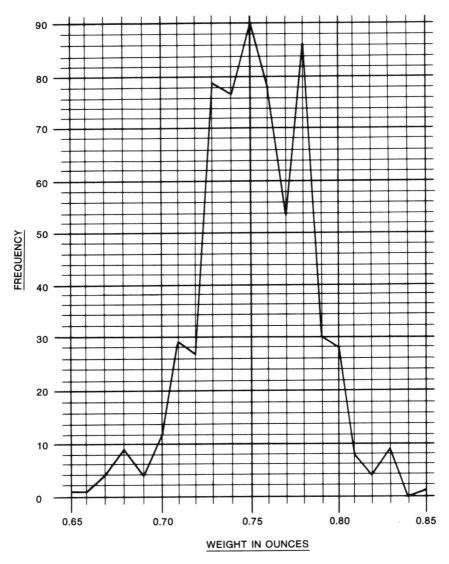

Fig. 23.3. Weight Distribution Chart of Confectionery Bars

have been determined along with average weight. This must be based on the distribution curve, and the data derived from it. For the upper control limit, we would have the 3 sigma or three standard deviations mark of 84.5, with the average weight of 75.4, and a lower control limit of 66.3, three standard deviations below the average. Hence, one item at 0.85 oz and two items (one at 0.66 oz and one at 0.65 oz) are rejects, out of control. This calculates to be 0.48 percent. The following example shows the actual weights of the confectionery bar taken over two 8-hr shifts.

These weights are arranged in groups and calculations made as indicated in Table 23.1.

The average was assumed to be 75. Then, 227 divided by 620 equals 0.366 to be added to assumed average giving a true average of 75.366.

TABLE 23.1. TABLE OF CONFECTIONERY BAR WEIGHTS

| G | d | f | fd | $fd^2$ |
|---|---|---|---|---|
| 85 | 10 | 1 | 10 | 100 |
| 84 | 9 | 0 | 0 | 0 |
| 83 | 8 | 9 | 72 | 576 |
| 82 | 7 | 4 | 28 | 196 |
| 81 | 6 | 8 | 48 | 288 |
| 80 | 5 | 28 | 140 | 700 |
| 79 | 4 | 30 | 120 | 480 |
| 78 | 3 | 86 | 258 | 774 |
| 77 | 2 | 54 | 108 | 216 |
| 76 | 1 | 68 | 68 | 68 |
| 75 | 0 | 90 | 0 | 0 |
| 74 | −1 | 77 | −77 | 77 |
| 73 | −2 | 79 | −158 | 316 |
| 72 | −3 | 27 | −81 | 243 |
| 71 | −4 | 29 | −116 | 464 |
| 70 | −5 | 11 | −55 | 275 |
| 69 | −6 | 4 | −24 | 144 |
| 68 | −7 | 9 | −63 | 441 |
| 67 | −8 | 4 | −32 | 256 |
| 66 | −9 | 1 | −9 | 81 |
| 65 | −10 | 1 | −10 | 100 |
| Totals | | 620 | 227 | 5,795 |

G—group (weights, reported in 0.01 oz)
d—class deviation
f—frequency

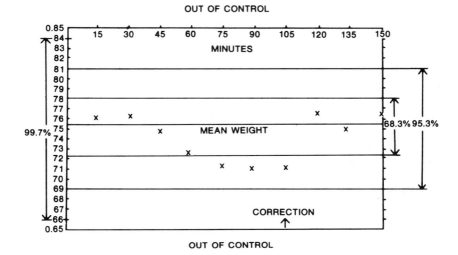

Fig. 23.4. Production Control Chart

Standard deviation (sigma) then equals

$$1*\sqrt{\frac{5795}{619} - \left(\frac{227}{620}\right)^2} \quad \text{or} \quad 3.037$$

A control chart for production use can now be made (Fig. 23.4).

During production, check weighings are made every 15 min and hypothetical examples are shown by the crosses (x) on the chart of Fig. 23.4. There is an obvious trend to low weights after 60 min, necessitating corrective action. The decision regarding permitted tolerance from both legal and cost standpoints can be determined from the control charts.

## Shelf Life—Keeping Limits and Keeping Tests

**Keeping Limits** With any foodstuff deterioration of quality occurs with time and eventually a condition will arise that makes the product unacceptable and therefore unsaleable. The rate of deterioration depends on a combination of factors such as temperature of storage, the recipe, the quality of ingredients, and the nature of the wrapping material. It is necessary to try to assess a period described as the "keeping limit," which is the time after which the product,

* Group width or range 1.

stored under average shop temperature and humidities, arrives at a condition when it is just unsaleable. In view of the variable conditions that are likely to prevail in shops, a somewhat pessimistic attitude must be adopted.

In the calculation of the keeping limit for a particular product, both its residence time in the shop and the conditions to which it will be exposed must be considered. If average values were used, then only half the items sold would be better than the minimum quality laid down and half would be worse. The "90 percent points" of the two distributions are therefore used so that only 10 percent of all items will experience worse shop conditions and 10 percent will have a longer residence time in the shop. This procedure ensures that approximately 98 percent of the items will be of better quality than the set minimum.

Many foods, including confectionery, are now labeled with the statement "best before ...." Some of the periods quoted are somewhat optimistic and to maintain maximum quality of a food for periods of more than a year entails perfect storage conditions and packaging.

A formulation with low water activity and one that contains no vulnerable ingredients, such as some milk products, egg albumen, some fats, and natural flavors, would be desirable.

**Shelf-Life Tests**  To obtain absolutely true values for keeping limits, prolonged storage tests under average shop conditions are required. This is very time consuming and is usually applied only to verify the results obtained with accelerated tests.

Accelerated keeping tests are carried out using the following procedure. Representative samples of the product to be tested are taken and a number of tastings of fresh material made. These samples are subjected to different storage conditions and possibly in several types of packaging depending on the nature of the product. Samples are then brought out for tasting at appropriate intervals. All tastings are carried out by a particular panel and samples are scored on a "saleability scale." Thermostatically controlled storage rooms or large incubators are used for the tests and the temperatures are preferably 18°C (65°F), 23°C (73°F), 27°C (80°F), and 29.5°C (85°F). Samples are normally examined after exposure to these temperatures for one and two months. Without special humidity control, the samples in the 80°F and 85°F incubators become subjected to drier conditions than those normally experienced, and this is allowed for in assessment of the results.

Other tests are designed to examine the effect of special storage conditions, for example,

| | |
|---|---|
| Tropical conditions | 29°C (85°F) 85 to 90 percent relative humidity |
| Cool storage conditions | 7°C (45°F) and 10°C (50°F) |
| Cold storage conditions | −7°C (20°F) |

A special comment must be made about low-temperature storage. Although low temperatures will preserve flavors, chocolate-covered confectionery bars containing wafer cookies, honeycomb centers, or similar fillings will split, or sometimes shatter, under very-low-temperature storage. Milk-chocolate-covered pieces fare worse in this respect.

**"Destructive" Testing** It is the practice in some confectionery companies to subject samples to conditions of fluctuating temperatures and humidity. The maximum and minimum conditions used are frequently such that they imitate only the most severe tropical climates. For example, the temperature of test is likely to vary every 24 hr between 15.5°C (60°F) and 32°C (90°F). At the same time, relative humidity can rise to 85 to 90 percent. This method of accelerated testing can be very misleading and should never replace the incubation tests mentioned previously. It has some limited value in the testing of protective wrappings but should never be applied to unwrapped confectionery pieces.

The exact conditions required for accelerated shelf-life tests depend on the product, areas of sale, and the type of sales outlet (e.g., shop, supermarket, kiosk) and these must be determined by the quality controller in conjunction with the marketing department.

The slope ratio method may be used for the analysis of keeping tests. Estimates may be obtained of rates of deterioration for use in the calculation of keeping limits or for comparison between types of packaging or storage conditons.

A representation of a set of experimental data is shown in Fig. 23.5. The four straight lines are taken to correspond to four different sets of storage conditions.

It can be seen that the material is assumed to start off fresh (A) and deteriorate linearly from that point at a rate that is determined by the packaging and storage conditions.

When applying a slope ratio analysis, a mathematical model that is a precise description of this situation is used. In practice, the results are not expected to conform to it exactly. Several factors, such as tasting error, unstable storage conditions, or non-homogeneous

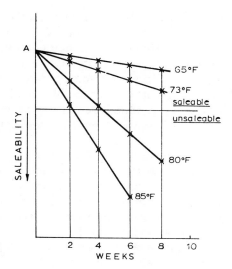

Fig. 23.5. Keeping Qualities Under Various Conditions and Lengths of Storage

materials, could affect the results. The model is, however, expected to fit the data within certain limits as estimated by statistical test. A number of different tests are therefore applied to the particular set of data to determine the limitations of the application of the model to the data. A series of curves will usually be produced from tasting test and from these "average slopes" are obtained from which keeping limits may be assessed.

## Microbiological Quality Control

Microbiological problems are discussed in detail in several parts of the book and the reader of this chapter should refer to these. The reasons for microbiological control are essentially as follows:

1. To prevent food-borne disease organisms from being present in the product.
2. To prevent microbial spoilage.

Contamination may arise

1. from purchased raw materials
2. from microbial growth during processing
3. through unsatisfactory handling or storage of the finished product.

It should also be recognized that if the final product is contaminated microbiologically, it may be quite satisfactory in taste and appearance. However, the microorganisms present will multiply rapidly in a suitable environment such as the human system or in warm moist conditions during further preparation of a food. Unlike meat, fish, and other perishable foods, microorganisms in most confectionery products remain dormant at ambient temperatures, as a result of low water activity and high syrup phase concentration.

**Raw Materials** The factory microbiologist has the responsibility first of all to see that the raw materials entering the factory are free of pathogenic organisms. For example, in the past salmonella has been detected in coconut, certain milk products, and egg albumen. Mycotoxins have been detected in peanuts and various spices.

In raw materials for confectionery processing, microbiological examination should show: salmonella, negative; E coli and Enterobacteriaceae, negative in 1 g.

These latter two organisms indicate unsanitary preparation of the food.

Spoilage organisms, molds, and yeasts (ferments) that can cause deterioration of the product by producing rancidity, fermentation, and other off-flavors must be at a low level, for example: a total plate count of less than 5,000; and molds and yeasts, maximum 50 per gram.

Enzymes, especially lipase, will cause soapy flavors in fatty confectionery. In cocoa powder, and egg albumen, lipolytic activity should be negative.

**Process Control** While most confectionery boiling processes will destroy all organisms, in the manufacture of certain pastes, fondants, and aerated products such as marshmallows, this is not the case.

Intermediate processes such as the soaking of gelatin or egg albumen are very vulnerable and must be carefully controlled. Solutions must be freshly prepared and equipment and utensils used for these solutions must always be washed and sterilized.

**Finished Products** With proper control of raw materials and processing, the finished products should be free of criticism. Handling during packaging can be a hazard and exposure to contaminating dust is always a possibility. Therefore, some checks on the finished product are always necessary and these will be a double check on the raw materials and the process.

**Complaints** A final aspect of microbiological quality control is the investigation of complaints, which all manufacturers hope to avoid. However, complaints do occur, and in confectionery, those arising from microbiological causes may concern fermentation, rancidity, or other off-flavors, and occasionally microbial liquefaction. The last defect may promote mold growth. Analytical examination of complaint samples should reveal the cause, which may have been an error in formulation or a breakdown in process control.

## SANITATION, HYGIENE, FOOD POISIONING, FOREIGN MATTER

*Author's Note.* The first and second editions of this book have had worldwide circulation. Much of the information in this chapter is based on United Kingdom experience and practice and it is hoped that readers from other countries will benefit.

Because of climate variations and different populations, the problems may be greater, but knowledge of the general methods of quality control, including sanitation, is useful everywhere.

Parts of this section are based on publications by the well-known British store Marks and Spencer—"Hygienic Food Handling" and "Hygiene in Marks and Spencer Staff Kitchen and Dining Rooms."

Hygiene is not instinctive in humans and historically their record has been one of indifference, arising from ignorance, and positive neglect.

The present attitude of many people is, unfortunately, still depressing to the sanitation officer and as is mentioned in this book under "Pest Control." Constant pressure is needed to maintain the necessary standards in a food factory.

The following story, based on an English west country experience, is indicative of the attitude in some areas of civilized countries.

A traveler, wishing to escape from the civilization of the city to the peace and beauty of the countryside, chose one of those remote farmhouses where they add to their income by taking in paying guests.

Arriving late at night and tired, he went straight to bed. Rising early, he searched the house for the toilet, and not finding it, asked a farmhand in the yard below.

"Privy—aye zur—'e be up thur behind they trees."

Our guest, after finding it, returned to the house and again met his informant in the yard.

"Did'st find 'un, zur?"

"Yes, but I didn't like the hundreds of blow flies there."

"Flies, zur, aye–they be bad. Pity tho', thee's should'st 'ave gone thur ha'f hour later, they'd be in the kitchen then."

## Historical

Certain eras of civilization, for example, the Babylonians and the Romans, practiced standards of hygiene equal to the best today but not, it seems, with the understanding of the real causes of disease and discomfort.

After the Roman occupation of Europe, standards fell to a very low level with people herded into cities with inadequate and impure water, no sanitation, and putrefied and adulterated food. These conditions prevailed for over 1,000 years and it is understandable that visitations of the plague, cholera, fevers, and other diseases destroyed large sections of the population from time to time. These diseases, not being understood, were supposed to be the work of the devil and other supernatural beings.

The knowledge of hygiene and practice of sanitation as it is understood today is not much more than 100 years old, although the discovery of "germs" attributed to the Dutchman Leeuwanhoek (1632–1723) had inspired the beginning of a worldwide interest in the subject of the cause of disease.

The great names associated with this development may be noted: Louis Pasteur (1822–1895) made bacteriological discoveries that opened the way to the development of antiseptic surgery by Joseph Lister, the famous Scottish surgeon (1827–1912).

Robert Koch (1843–1910) identified certain diseases with specific germs and discovered the bacillus responsible for tuberculosis. Semmelweiss (1818–1865), the Austrian obstetrician, discovered the cause of child bed fever responsible for many deaths at childbirth.

By the middle of the nineteenth century, medical science had advanced considerably but a great deal of social change had to be brought about before much improvement could be made.

The conditions of living in the industrial areas, which were growing rapidly in the early nineteenth century, were appalling in the extreme and this, with the cholera epidemics of the 1830s, led to an outcry for action to improve public health legislation. The names of Sir Edwin Chadwick (1801–1890) and Sir John Simon (1816–1904) are historic in the pioneering work for public health and it was mainly through their efforts that a Board of Health was established and the Public Health Act in 1875.

Along with the bad living conditions, the public had to endure poor-quality and adulterated food, but this could not be improved until medical and scientific knowledge would ensure satisfactory inspection and analysis. The British Food and Drugs Act, also established in 1875, laid the responsibility on the local authorities for seeing that food was wholesome and unadulterated, and eventually specific standards were enforced. Milk and its products received special attention as it had been shown that they were prime factors in the spread of infection.

The enforcement of the act through the local courts, backed by the efficient working of the public analysts, made a tremendous contribution toward improvement of food quality, and the Public Health Act gave powers for the destruction of unsafe food.

As time went on, and methods of food manufacture and preparation changed, it became clear that the original act was not sufficiently comprehensive, but sixty years had to elapse before stricter legislation came into being.

The British Public Health Act of 1936 and the Food and Drugs Acts of 1938, 1944, and 1950, laid greater stress on the importance of the conditions of premises used for storage and preparation of food and for its transport, and most important, the personal hygiene of people who handled food.

In the United States, federal, state, and local community laws have been established to protect the population from the hazards to human health.

The Federal Food, Drug and Cosmetic Act of 1938, Sec. 402, may be quoted. This states that a food product shall be deemed to be adulterated—(a) (3) "if it consists in whole or in part of any filthy, putrid, or decomposed substance, or if it is otherwise unfit for food; or (4) if it has been prepared, packed, or held under unsanitary conditions whereby it *may have become* contaminated with filth, or whereby it may have been rendered injurious to health; or (5) deals with diseased animals; (6) refers to poisonous container rendering contents injurious to health, etc., etc."

Another, more recent aspect of food hygiene, if hygiene is the correct word, is related to food additives, intentional and accidental. To prevent food from becoming inedible or developing bad flavors, preservatives may be added. Some of the original compounds proposed have subsequently been found to be hazardous to health.

## Food poisoning

Although the possibility of contamination by chemical or metallic substances has been indicated, by far the majority of food poisoning incidents of recent years have been attributable to bacterial causes. Food today is manufactured or prepared in large quantities, and communal eating is much more common. This means that contaminated food is more likely to result in a large number of people being affected.

Food poisoning is generally caused by one of two types of bacteria—salmonella and staphylococcus, both resulting in severe inflammation of the digestive system with vomiting, diarrhea, abdominal pain, and prostration.

The degree of poisoning varies from slight discomfort to distressing illness, and occasionally death. There is also the possibility that continued consumption of mildly contaminated food results in chronic debility.

Mention also should be made of other organisms which, while not directly concerning confectionery, may occasionally cause cross-contamination.

**Botulism (Clostridium botulinum)** Serious outbreaks of botulism have occurred from time to time resulting in many deaths. The cause is invariably inadequately processed food that has been stored and then eaten cold. Home-processed vegetables are particularly vulnerable because the heat treatment is inadequate. Autoclaving is necessary.

With fruits, the lower pH is a safeguard.

**Clostridium perfringens, Bacillus cereus** These exist in the excrement of animals, insects, and human beings. Contamination arises from bad sanitation followed by unsatisfactory cooking and storage.

**Vibrio parahae molyticus** An organism at one time limited to the Far East, it is now more widely distributed. It has been associated with the practice of eating raw fish.

At one time, bad sanitation in food handling led to the transmission of serious infectious diseases such as typhoid, cholera, and tuberculosis. Improved methods of handling, pure water supplies, and better toilet facilities in food factories have practically eliminated this problem. The control of insects and rodents is an important factor in preventing food contamination (see "Pest Control").

In the chocolate and confectionery factory, the product being handled is generally of low moisture content, which is not conducive to bacterial growth. As a result, the standards of hygiene may slip well behind those of a factory handling milk, egg, or meat products, where bad hygiene results in loss of product as well as the rapid increase of any organisms introduced.

In the confectionery factory, there are, however, intermediate products and ingredients that may contain or cultivate food-poisoning organisms. These are milk products, egg albumen, gelatin, nuts, and some dried fruits. Egg, milk, and gelatin are usually made into solutions before incorporation in a confectionery mix and these are often allowed to stand in warm places where microorganisms will multiply profusely. Fortunately, the solutions are generally added to a batch at a temperature high enough to destroy the organisms but in the meantime they may have contaminated the hands of factory personnel and sometimes the utensils are used again, without proper cleaning, for bulk finished products before transferring to shaping and cutting machinery.

Some processes do not destroy microorganisms—pastes, marzipan, and chocolate may remain infected after manufacture if an ingredient is contaminated or the mixture handled by a "carrier" during the process. Because of the low moisture content of these products, they will not cause microbes to multiply, but if eaten while still infected, may cause illness as a result of their rapid development in the digestive system.

In suitable surroundings, food poisoning organisms will multiply at an alarming rate; for example, in moist conditions at 38°C (100°F) the bacterial count may increase 100,000 times in 12 hr.

## Factory Hygiene and Sanitation

In any food factory the maintenance of good hygiene is related to several factors (1) persons, (2) ingredients, (3) equipment and premises, (4) control of insects and rodents.

*Persons.* Unless employees are versed in the rudiments of hygiene, the provision of the best ingredients, equipment, and premises will be wasted. The responsibility for personal hygiene rests primarily with the medical department. Following medical examination, which approves a person for work with food, an intensive training in personal hygiene should be given. This involves emphasis on clean hands, overalls, hats to cover hair, use of toilets, and the reporting of illness to the supervisor, particularly intestinal troubles. The showing of films and talks to new employees and supervisors have been used to advantage.

*Food ingredients.* Materials should never be handled with the bare hands. Unfortunately, this rule is frequently broken in a confectionery factory but every effort must be made to use scoops and paddles for raw materials, and gloves to handle finished goods. Dust from raw materials should not settle on processed work. Bulk handling is useful.

*Premises and equipment.* Nothing encourages slipshod methods more than untidy surroundings. Tidiness is the responsibility of each room supervisor. Equipment must be washed regularly with hot water and steam jets, adequate to sterilize, and sufficient spare equipment must be available to allow this to be done on a rotation basis. Damaged equipment must be taken out of service. Apart from cleaning facilities within each department, a separate department should be provided for the cleaning and sterilizing of large equipment, bunkers, and trolleys. Disinfectants should be used with caution as many are odorous.

*Insects and vermin.* This is a very important factor and is dealt with in detail under "Pest Control."

*Storage.* The premises used for storing raw materials and finished goods must have a very high standard of cleanliness. An essential feature of these stores is correct stock rotation.

## Prevention of Foreign Matter Contamination

It is very damaging to the reputation of a food manufacturer to have the publicity of a court case where a member of the public has found a screw, piece of metal, or glass in a product. Probably more nauseating to the customer is to find a human hair, a housefly, or rodent excreta.

The prevention of foreign matter of all types is closely linked with hygiene. The following summarizes the precautions that should be taken at all stages of manufacture:

**Raw Materials** The quality control chemist approves raw materials for use in the factory. In addition to making sure that the composition is according to specification, the chemist must ensure that the product is clean and wholesome. Analytically, the material must be checked for rodent contamination and the presence of insect fragments. Solution or sieving tests will reveal any foreign matter, such as pieces of wire, nail, splinters of wood, or bag fiber.

Raw materials received in sacks, wooden cases, or paper-lined cartons may easily become contaminated during opening. Quality control inspectors must see that packages are dealt with correctly and always in a department away from production. Nuts, raisins, desiccated coconut, and similar materials require mechanical and hand sorting to remove stones, nut shells, twigs, and other foreign matter. Machines that operate on the principle of air elutriation or other means of classifying from the different densities of stones and metal are available from specialist firms.

**Electronic Sorting**   This is used for nuts and beans and works on the basis of rejection of any unit that is the incorrect color. Each nut is examined very rapidly and individually by photoelectric cell when passing down a shute, those with defective color being rejected. This process has its limitations.

**Electromagnets**   These are very necessary to extract metal such as bolts, screws, and metal implements from raw materials or partly processed products. This is particularly important when materials are being sent to pulverizing or refining machinery where serious damage may occur.

With these magnets it is very necessary to remove the adhering metal at frequent intervals and *not* to release the whole accumulation into the product.

**Vibratory and Pipeline Sieves**   When processing involves dissolving a raw material or grinding to a liquid as with chocolate, sieving provides a valuable method of eliminating foreign matter. With modern sieving machinery, fine meshes are possible without a reduction in throughput.

Whenever a solution or slurry of raw material is prepared, for example, prior to a caramel or fondant creme boil, it should be passed through a sieve with mesh as fine as possible in keeping with output required.

Chocolate may be passed through vibratory sieves inserted in the production line at the latest stage prior to molding or enrobing.

Vibratory sieves of various sizes are available for this purpose and good outputs through a 30 or 40 mesh are obtainable even with viscous chocolate.

Pipeline sieves in enrobers or molding equipment are a final safeguard against stray pieces of metal or nuts and bolts that may detach during the working of the machine.

**Electronic Metal Detectors, X-Ray Detectors**  Whatever the precautions taken in the production processes, some foreign matter will get through to the finished product, and additional means are required to detect this and eliminate the offending bar or assortment unit.

The *electronic metal detector* is a device that may be fitted over a conveyor band and will detect particles of metal present in confectionery pieces passing through it. In recent years, great improvements have been made in the sensitivity of the detector and it is also less sensitive to extraneous influence, which at one time was a great nuisance, and greatly impaired the value of the apparatus.

The modern detector consists of a bridge placed over the conveyor carrying the bars, small candies, or packed boxes. Immediately beneath the bridge is a "balanced electronic field" and a particle of metal passing through this field will disturb the balance and set up an electrical impulse, which operates a warning system. This warning may take the form of a flashing light or bell combined with the release of counters that mark the place on the conveyor belt where the contaminated candy is to be found. This method necessitates the removal of the units between the counters by an employee who may not always be reliable.

A much more positive procedure is automatic rejection whereby the offending units are mechanically swept off the band or the conveyor may be split and the confectionery pieces dropped into a container below. Pipelines may also be fitted with detectors that operate a diversion mechanism to discharge contaminated product.

Figures 23.6 and 23.7 show the principles of operation of the detectors and rejection devices. Sensitivity is shown in Fig. 23.6. It is now possible to detect ferrous or magnetic stainless-steel particles in aluminum-foiled packets. Selective apparatus will record particles in a package of cookies, for example, down to 1.5 mm in diameter. To obtain maximum sensivitivity, the packets should be examined in a single line.

Confectionery units that are rejected may be examined individually by passing through the detector at times when production is not taking place. Some factories use a separate detector unit for this purpose, which is a safer procedure. Occasionally, an accident occurs when a metal implement breaks and the pieces are distributed unnoticed in a batch. This will cause the production line detector to operate with great frequency and it is much better to examine the products from such a batch on a separate detector.

The units that have been finally rejected are often worth examining to find the pieces of metal, particularly if there is an abnormal

**Metalchek 9** sensitivity guide.

| Aperture Size | | Steel, Iron, Copper, Brass, Aluminium | Non-Magnetic Stainless Steel | |
|---|---|---|---|---|
| Height | Width | | EN58J 316 | EN58E 304L |
| 4in (101mm) | 4in × (101mm) | 0.5mm | 0.6mm | 0.7mm |
| 14in (355mm) | 5in × (127mm) | 1.0mm | 1.2mm | 1.4mm |
| 30in (762mm) | 8in × (203mm) | 2.0mm | 2.4mm | 2.8mm |
| 26in (660mm) | 14in × (355mm) | 2.5mm | 3.0mm | 3.5mm |

Minimum diameter of metal sphere detected

Fig. 23.6.   Metal Detector and Sensitivity Table

*By permission of Lock Metal Detectors, Oldham, Manchester, England*

frequency in one particular line. To do this, a slurry is made with hot water and this is passed through a sieve that will remove the foreign matter. Detection is not so easy if nuts or fruits are present, but usually close inspection with the help of a magnet will find the offending metal particles. Ferrous metal particles are generally much more frequent than nonferrous.

When metal pieces of a similar nature are found frequently, it may be necessary to examine partially processed or raw materials.

Magnetic extraction and sieving cannot be used in some processes and metal particles then get through to the finished product. Examples are fine pieces of wire found in desiccated coconut and lead shot in raisins arising from the shooting of birds in vineyards.

X-ray detectors have the advantage, theoretically, of detecting stones, glass, and other dense particles besides metal. To use X-ray detection in large factories is a very expensive and tedious procedure and the human factor is involved to such as extent that only a

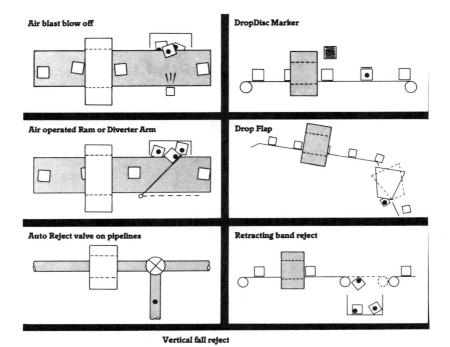

**Air blast blow off**

**DropDisc Marker**

**Air operated Ram or Diverter Arm**

**Drop Flap**

**Auto Reject valve on pipelines**

**Retracting band reject**

**Vertical fall reject**

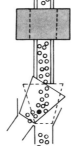

Fig. 23.7.   Typical Automatic Rejection Systems
Metal detector installations:
Metal detectors are fitted to conveyors, chutes, and pipelines as well as to packaging and processing equipment or other bulk flow-handling equipment to provide completely unattended inspection. When metal is detected, the metal detector relay operates so as to actuate alarms, stop the conveyor motor, or operate an auto reject mechanism to divert the contaminated product.

*By permission of Lock Metal Detectors, Oldham, Manchester, England*

relatively low level of detection can be expected. Since the method involves visual examination on a screen, the rate at which examination takes place must be slow compared with an electronic detector. However conscientious the observer is, fatigue is likely after a very short period and many particles on the screen have an indistinct outline.

To cope with the examination of hundreds of thousands of packages is an enormous task, and even with semiautomatic operation, which has been tried, it is generally not a commercial proposition. Fully automatic X-ray detection is the ultimate answer.

**Other Causes of Foreign Matter** Foreign matter arising from personal belongings, stationery, and small ancillary equipment can be quite a serious factor. Pins, hair clips, and buttons are among the articles that may fall into batches in the course of production—only personal care and strict supervision will prevent his. Even with smoking prohibited, which is the rule in all food factories, cigarette ends and matches may be found occasionally. These are often carried into factory production areas on the soles of boots and shoes. Paper clips, staples, and drawing pins must not be allowed in production areas. Personal pen knives and similar small tools should not be required if production personnel are provided with adequate equipment—even safety razor blades have found their way into products, used by fitters for some purposes.

Food and drink should never be taken into a production department—most factories now provide areas adjacent to workrooms where refreshments may be taken.

Broken windows or electric lamps may be the cause of glass foreign matter. All broken or cracked windows must be reported and repaired immediately and all electric lighting near bulk foods should be protected.

The possibility of criminal activity must not be overlooked. Personnel in a factory are sometimes motivated, because of a grudge against the company or for political reasons, to introduce dangerous articles or substances into the product. Descriptions of such occurrences appear in the press from time to time.

Large companies usually have a medical department and a psychologist is sometimes helpful in the screening of employees who may have the potential for mental imbalance.

**Implementation of Hygiene and Foreign Matter Control** To implement the measures outlined above effectively, top management interest is essential and managers must make periodic inspections, without prior notice. The detailed work is carried out by the sanitation officer and staff. They will cover insect and rodent control and arrange for plant and equipment swabs to be taken for checks on pathogenic organisms. Management must support action required by the sanitation officer; no other method will give proper control.

## Travel Tests

The protective qualities of the package with respect to the type of material used and method of sealing have been considered elsewhere in the book.

The construction of the outer case is a factor that often is not given sufficient attention. It should be the responsibility of the quality control department to provide information on the strength required of the outer package that will ensure protection during transport and storage.

Some companies transport loads of finished product on pallets from the production departments to distribution depots situated in various parts of the country. These depots are located in areas suitable for road distribution and the pallet loads are then made up into individual orders for delivery to wholesalers or retailers.

Other companies distribute directly from factory to retailer in corrugated cartons, making use of public or other hired transport. The strength of the carton containing the goods must be sufficient to resist the weight of other filled cartons stacked above. With pallet loads, subject to movement on trucks and in vans, it has been found that a single carton must be able to withstand a pressure equivalent to five times the total weight of filled cartons stacked above it in a load. This pressure can be calculated from the height of the load on the pallet, which may be from 3 ft to 5 ft.

To test cartons, a laboratory hydraulic press[1] fitted with flat platforms is used. The box to be tested is placed between the platforms, which gradually close in on the box by steady application of hydraulic pressure that is recorded continuously by a gauge. The pressure at which the box just begins to collapse is recorded. Various box constructions using different weights and type of carton board, with and without internal supporting collars, may be tested and the type of confectionery in the box has some bearing on the strength of box required. For example, solid chocolate blocks packed closely in a

[1] Carver Laboratory Press.

box are able to support a considerable weight without damage, whereas with soft-centered assortments or marshmallows, the entire external weight must be taken by the outer case. Gross negligence sometimes is found in stockrooms and instances of very high stacking have been seen where the bottom cases have been crushed to half their normal size! The pressure test gives a good guide to the nature of the box required but travel tests or simulated tests provide the final answer. For this purpose, pallet loads or fiber cartons are made up and sent on cross-country journeys, usually to several distribution depots where they are examined on arrival.

Complicated "travel testing" equipment has been designed to simulate the movements and pressures that might be expected on a journey, but results obtained must be interpreted with some caution.

## REFERENCES

Birch, G. G., and Parker, K. J.   1984.   *Control of Food Quality and Food Analysis.* Elsevier Publishers, Barking, England.

Butcher, D. W.   1983.   *On-line Monitoring of Continuous Process Plant.* Ed. North East London Polytechnic, Dagenham, England.

Gould, W. A.   1977.   *Food Quality Assurance.* AVI Publishing Co., Westport, Conn.

Institute of Quality Assurance, U.K.   1974.   Quality control in the food industry conference papers. *Conference Communications.* Holly Tree House, Farnham, Surrey, England.

Kramer, A., and Twigg, B. A.   1970.   *Quality Control for the Food Industry.* AVI Publishing Co., Westport, Conn.

McFarlane, I.   1983.   *Automatic Control of Food Manufacturing Processes.* Elsevier Publishers, Barking, England.

Pearson, E. S., and Hartley, H. O.   *Biometric Tables for Statisticians.* Cambridge University Press, London.

Steiner, E. H.   1967.   *Statistical Methods of Quality Control.* Academic Press, London.

Stiles, E. M.   1968.   *Handbook of Total Quality Assurance.* Prentice-Hall, Englewood Cliffs, N.J.

24

# Food Value of Chocolate and Confectionery

Rarely do people who eat chocolate and confectionery consider its food value—they eat candy because they like it. Sometimes it is called a "fun food."

The variety of confectionery available is probably greater than any other foods. Attractive presentation and packaging and its gift value are additional factors in its favor. Confectionery, in recent years, has also become a snack food, particularly where nuts, cookies, wafers, and cereals are included.

Because confections are pleasant to eat, some people consume large quantities, and so these substances are blamed for causing obesity, dental caries, and other serious problems. As a consequence; sweets receive adverse publicity, often based on medical evidence related to abnormal intake.

Obesity results when more nutrients are absorbed than the body requires, and this applies to food generally and not only to confectionery. The consumption of chocolate and confectionery in different countries as tabulated in Chapter 1 showed that the British, West Germans, and Swiss eat 11 to 12 kg per person per year. Even so, confectionery accounts for only 2 to $2\frac{1}{2}$ percent of the average diet, or on an energy basis about 5 percent.

## FOOD VALUE AND THE COMPOSITION OF FOODS

The constituents of all foods may be divided into five classes—carbohydrates, fats, proteins, minerals, and vitamins. These are absorbed into the body to promote growth, energy, and good health. Additionally, there are the nondigestible constituents that provide bulk and help the foods to pass through the bodily system. Fiber, largely composed of cellulosic substances and natural to many foods, has achieved much publicity of recent years.

TABLE 24.1. RECOMMENDED DAILY CALORIE INTAKES FOR MEN

| Age | Occupation | Calorie requirements |
|---|---|---|
| 18 to 35 years | Sedentary | 2,700 |
| Weight about 150 lb | Moderately active | 3,000 |
| | Very active, hard working | 3,600 |

"Calorie" is the term used to measure the energy produced when a food is absorbed. In some countries, calories are regarded almost as a food ingredient.

The calorie requirements of a person depend on age, sex, and the amount of exercise taken, and calorie-conscious people must understand this when they are considering the caloric value of the food they eat.

Recommended daily calorie intakes for men, for example, are as shown in Table 24.1.

## Carbohydrates

These generally provide the main source of energy and are quickly absorbed into the system. They consist of the sugars (sugar, glucose syrup, honey, invert sugar, lactose), starches (wheat flour, corn-starch, potato starch, tapioca, and the natural starch of cocoa), and the various celluloses, gums, and pectinaceous substances. The celluloses are generally indigestible.

In most confectionery and chocolate products, "sugars" are the main ingredient. Starches are used to a lesser extent for cooky ingredients and certain jellies and gums. Cellulosic substances exist as fibrous material in very small amounts in nuts, cocoa, fruits, and similar natural ingredients.

## Fats

These provide more than twice the calorie value of carbohydrates and are the main source of body fat. Fat is the energy reserve of a person in sound health and should not be associated with obesity.

In confectionery and chocolate, fats are provided by the natural cocoa butter of the cocoa bean; by milk fat in milk chocolate and confectionery centers such as caramels; and by vegetable fats in centers and ingredients such as nuts.

## Proteins

Proteins are generally regarded as the most important nutritional constituent of food as they provide the material for growth and replacement of body tissue and muscle. Proteins consist of a number of substances known as amino acids and some twenty or more different amino acids may go toward making up a protein. It is often not realized by the average person that there are first- and second-class proteins and that a proportion of first-class proteins is necessary for proper body building. Animal products generally contain adequate first-class proteins while many vegetable proteins are second class or contain small amounts of the first-class proteins. It therefore is necessary to consume a greater quantity of a vegetarian diet to get the necessary quality proteins, in which case some of the second-quality proteins will be converted to fat. Some authorities consider this a waste of valuable protein foods.

Milk, in the form of milk solids in milk chocolate, caramels, and fudge, is the most valuable source of animal protein in confectionery. Nuts, which are an ingredient of many sweets, are one of the best sources of vegetable protein. Wheat flour in cookies or wafers constitutes another protein supply, and cocoa matter itself contains some vegetable protein.

Soya flour is becoming increasingly important as a source of good protein. Soya flours and soy protein concentrates had a reputation in the past of having an earthy flavor and poor textures. Great improvements have been made and flours, concentrates, and isolates are now available to enhance the protein content of many foods, including confectionery.

## Mineral Matter

Mineral substances composed of inorganic compounds are essential in the human diet, although required in very small amounts. They go toward making up bone structure, and are trace constituents of organic tissue and fluid such as blood hemoglobin. In a properly constructed diet, all the required mineral salts are provided, but iron and calcium are occasionally deficient.

Fruits, milk solids, cocoa beans, and many natural products provide the mineral substances, but highly refined ingredients such as white sugar, which is practically a pure carbohydrate, have much reduced inorganic constituents. It must be recognized, however, that when a pure ingredient such as white sugar is used, it is no way detrimental if sufficient other ingredients are present to provide the necessary trace substances.

"Raw" sugar, which contains minerals, is now used in many foods.

## Vitamins

The existence of vitamins in food was not recognized until comparatively recent times (1912–1915) although it was known that certain foods, such as fresh fruit or vegetables, were essential for the maintenance of good health.

The vitamins are complex organic compounds vital to life and the regulation of body processes. They cannot be formed from the classes of nutrients already described.

Chocolate and confectionery cannot be regarded as rich in vitamins, but small quantities occur in the milk, cocoa, nut, and fruit ingredients.

Since there has been a tendency to fortify foods with vitamin concentrates, some knowledge of the subject by the food technologist is desirable.

Although it is not the practice to fortify confectionery or chocolate for general sale to the public, food drinks, particularly for export to underdeveloped countries, have been manufactured with the addition of vitamin concentrates. Certain cereal components are useful sources of vitamins as well as protein. Defatted wheat germ is used for the fortification of breakfast cereals, which, in turn, are popular ingredients of chocolate bars (Vitamins Inc., 1978–81).

In the immediate postwar period, "vitamin relief chocolate" was manufactured to send to distressed areas in Europe where malnutrition had occurred. This was usually dark chocolate with vitamin concentrate added.

The development of a fortified product is best done in conjunction with a company specializing in the manufacture of vitamin concentrates. Such companies have the means of checking the potency of the constituents and will supply analysis certificates with each delivery of concentrate. They will also advise on the probable loss of potency in the processing or storage of a fortified product.

The vitamins are listed in Table 24.2.

**Vitamin A**   Vitamin A in the diet is derived from previously formed retinol and from carotenoid precursors that are converted to vitamin A in the intestinal tract. This vitamin is vital for correct growth, and prevention of eye diseases and congenital malformation. Vitamin A is

TABLE 24.2. VITAMINS AND THEIR MAIN FUNCTIONS

| | | |
|---|---|---|
| Fat-soluble vitamins | Vitamin A | Retinol |
| | Vitamin D | Calciferol |
| | Vitamin E | Tocopherol |
| | Vitamin K | Phylloquinone |
| Water-soluble vitamins | Vitamin $B_1$ | Thiamin |
| | Vitamin $B_2$ | Riboflavin |
| | Niacin | |
| | Folacin | |
| | Biotin | |
| | Pantothenic acid $B_5$ | |
| | Pyridoxine $B_6$ | |
| | Cobalamin $B_{12}$ | |
| | Vitamin C—ascorbic acid | |
| | Choline | |

widely distributed in fish liver oils, eggs, butter, milk, and animal livers. The precursor carotene exists in many yellow vegetables and fruits.

**Vitamin D** Several substances have vitamin D activity. The two most important are cholecalciferol ($D_3$) and ergocalciferol ($D_2$). $D_3$ is formed from animal sources by irradiation and $D_2$ similarly from plant sources. Vitamin D is essential for correct bone calcification; its deficiency causes "rickets." The vitamin is found in eggs, liver, and some fish—it is synthesized in the body and in the skin from solar radiation.

**Vitamin E (Tocopherol)** Vitamin E exists in several forms. It has many important functions in the proper working of the blood, muscle, nerve, and reproduction systems. It is present in adequate amounts in whole grains and natural vegetable oils.

**Vitamin K** Vitamin K is now recognized as a group of quinone compounds. Essential in maintaining the function of the blood coagulation system, it is widely distributed in nature in all green leafy growth and in potatoes, rose hips, and some seed oils.

The chemical forms are $K_1$ (phylloquinone, phytomenadione, phytonadione) in green plant tissue; $K_2$ (menaquinone) produced in food and animals by microbial action; and $K_3$ (menadione).

**Vitamin $B_1$ (Thiamin)** Vitamin $B_1$ is known as the antineuritic or anti-beri-beri vitamin. It is effective in nervous disorders and loss of appetite.

$B_1$ is well distributed naturally in whole-grain cereals, wheat germ, beans and peas, egg yolk, nuts, brewer's yeast, and, to a lesser extent, in various meats, especially offal. Chemically, it can be synthesized as thiazolium hydrochloride.

**Vitamin $B_2$ (Riboflavin)**   This vitamin occurs as flavin mononucleotide and dinucleotide, and is involved in tissue oxidation and respiration. Riboflavin is widely distributed in foods; animal muscle, fish, milk, and eggs.

Vitamin $B_2$ deficiency is associated with inflammation of the mucous membrane, some types of dermatitis, and itching.

**Vitamin $B_3$ (Niacin)**   A derivative of pyridine, niacin functions as a component of two enzymes, nicotinamide adenine dinucleotide and its phosphates. It is necessary for cell respiration in using up fat, protein, and carbohydrate. Deficiency causes pellagra, a severe form of dermatitis.

Niacin is present in animal tissue such as kidney and liver. Whole cereal grains and yeast are rich sources.

**Vitamin $B_6$ (Pyridoxine)**   This vitamin includes the chemical complexes pyridoxine, pyridoxal, and pyridoxamine. Deficiency causes types of dermatitis, skin erosions, and convulsions in infants.

It is widely distributed in cereals, milk, vegetables, and various meats, such as liver.

**Vitamin $B_5$ (Pantothenic Acid)**   This is an important vitamin widely distributed in ordinary foods. Chemically, it is a derivative of $\beta$-alanine. Its deficiency results in weight loss, insomnia, fatigue, cramps, and certain nervous disorders. Its action promotes the biosynthesis of cholesterol, steroids, and various fatty acids.

**Vitamin H (Biotin)**   Biotin functions in the synthesis of fatty acids and proteins and in carbohydrate metabolism. Its deficiency causes dermatitis and reduction in hemoglobin. It is rarely deficient in normal diets but can be synthesized for fortification purposes.

Chemically, it has several isomers but only d. biotin is physiologically active.

**Folic Acid**   (Folic acid, folacin, pteroylglutamic acid) is a derivative of glutamic acid, amino-benzoic acid, and pteridine. Its deficiency causes anemia, and this may arise in times of severe stress. It is present in green vegetables and liver, but is susceptible to destruction by heat.

More recently, folic acid deficiency has been recognized during pregnancy, in alcoholics, and in low-income families with poor diets.

**Vitamin $B_{12}$ (Cyanocobalmin)**   Vitamin $B_{12}$ is essential for cell function, particularly bone marrow, intestines, and the central nervous system. Its deficiency causes anemia and spinal cord degeneration.

It is present in most foods of animal origin and it has been recorded that deficiency arises among vegetarians who refuse to eat animal proteins.

**Choline (Neurine)**   Choline is essential in the synthesis of cell membranes. It is a constituent of egg and soya lecithins and chemically is a quaternary ammonium compound.

Choline deficiency results in fatty degeneration of the liver and certain kidney diseases.

**Vitamin C (Ascorbic Acid)**   Vitamin C has a number of functions, including regulation of respiratory cycle and maintaining blood vessel strength. Its deficiency causes anemia, hemorrhage, swollen gums, and scurvy. It is readily available in fresh foods but is produced biologically and used for fortifying foods. Vitamin C is widely available in fresh fruits and vegetables.

## The Labeling of Foods Containing Vitamins

Strict regulations control the claims that may be made for the curative effects of vitamins occurring naturally in or added to a food.

To advertise vitamins at all necessitates a normal daily requirement being present in a reasonable quantity of food. The regulations for any country where a fortified food is sold must be consulted.

## FOOD VALUES OF CHOCOLATE AND CONFECTIONERY

Table 24.3 gives the food values of a selection of chocolate and confectionery products. Items 1 through 8 are from McCance and

TABLE 24.3. FOOD VALUES OF CHOCOLATE AND CONFECTIONERY

| Product | g per 100 g | | | per 100 g | mg per 100 g | | | | | | | | |
|---|---|---|---|---|---|---|---|---|---|---|---|---|---|
| | Protein | Fat | Available carbohydrate | Calories | Na | K | Ca | Mg | Fe | Cu | P | S | Cl |
| 1. Dark chocolate | 5.6 | 35.2 | 52.5 | 544 | (143) | 257 | 63 | 131 | 2.9 | 0.8 | 138 | — | 4.8 |
| 2. Milk chocolate | 8.7 | 37.6 | 54.5 | 588 | (275) | 349 | 246 | 59 | 1.7 | 0.5 | 218 | — | 170 |
| 3. Chocolate assortment | 4.1 | 18.8 | 73.3 | 467 | 60 | 243 | 92 | 51 | 1.8 | 0.5 | 121 | — | 177 |
| 4. Milk-chocolate-covered bar | 5.3 | 18.9 | 66.5 | 447 | 145 | 249 | 163 | 35 | 1.1 | 0.3 | 154 | — | 295 |
| 5. Hard candies | — | — | 87.3 | 327 | 25 | 8 | 4.8 | 2.4 | 0.4 | 0.1 | 11.6 | — | 68 |
| 6. Cocoa powder | 20.4 | 25.6 | 35.0 | 452 | (650) | 534 | 51 | 192 | 14.3 | 3.4 | 685 | 160 | 199 |
| 7. Toffee | 0.2 | 6.2 | 90.8 | 399 | 115 | 91 | 11 | 4.0 | 0.6 | 0.04 | 9.7 | 20.7 | 40 |
| 8. Malt food drink | 11.4 | 7.5 | 67.6 | 370 | 360 | 660 | 89 | 170 | 3.3 | 1.0 | 411 | 243 | 185 |
| 9. Chocolate creme bar | 2.5 | 15.5 | 75.1 | 438 | | | | | | | | | |
| 10. Aerated candy, milk chocolate covered | 5.3 | 22.9 | 71.6 | 501 | | | | | | | | | |
| 11. Milk chocolate bar with cookies and raisins | 7.1 | 28.1 | 58.3 | 508 | | | | | | | | | |
| 12. Cocoa | 20.4 | 21.8 | 34.9 | 417 | | | | | | | | | |

Widdowson (1960) and 9–12 are the author's figures. McCance and Widdowson's figures for chocolate and cocoa show high fat contents compared with present-day recipes. Chocolate and cocoa may have added alkali or salt, hence the figures in brackets.

Food factories get regular requests from hospitals, nutritionists, general practitioners, the general public, and food faddists to give details of the composition of their products. It is as well to build up a file of information on all lines sold so that inquiries may be answered promptly.

## Calculation of Caloric Value of a Food

Often it is necessary to calculate the caloric value of a food. This is a relatively simple matter and is based on the fact that foods consumed in the body are burned up in the same way as they would be in a physical calorimeter. Allowing for small losses due to indigestibility, the caloric equivalents are as follows:

| | |
|---|---|
| Carbohydrates | 4.0 calories per gram |
| Fat | 9.0 calories per gram |
| Protein | 4.0 calories per gram. |

Thus, if the composition of a food is known or obtained by analysis, the caloric value may be calculated.

As an example, consider liquid milk consisting of protein, 3.3 percent; fat, 4.0 percent; carbohydrate, 5.0 percent; and water, 87.7 percent. It is found that 100 g of milk will provide:

| | |
|---|---|
| Protein | $3.3 \times 4.0 = 13.2$ calories |
| Fat | $4.0 \times 9.0 = 36.0$ calories |
| Carbohydrate | $5.0 \times 4.0 = 20.0$ calories |
| Total | 69.2 calories per 100 g |

The water is merely a "carrier" and provides no fuel.

## Supposed Harmful Effects of Confectionery

Much publicity has been given to the harmful effects of confectionery in the modern diet. But how can this be so if it constitutes only 4 percent of the total calorie intake? Such a general statement is misleading. It is certain that some people eat far more than the average, and often these same people will also consume excessive quantities of other food.

**Obesity** A British National Food Survey estimated an average calorie intake from all foods of 2,900 to 3,300 per day against a requirement of 2,100 to 3,000 per day. This is a reason for obesity that is related to overeating, that is, consuming food beyond one's actual calorie requirements, with the excess going to reserve fat.

The consumption of excessive carbohydrates was advanced by Yudkin and his team at Queen Elizabeth College, in London as a cause of atherosclerosis, work that was substantiated by McDonald and Butterfield of Guy's Hospital and Medical School. T. L. Cleave in *The Saccharine Disease* provided further evidence against sweet and starchy foods, suggesting they were responsible for ulcers and diabetes as well as atherosclerosis.

In contrast, the Royal College of Physicians, London, in its 1976 report on the prevention of coronary heart disease, concluded that there was no evidence to link the disease with the consumption of sugar.

If the evidence of medical researchers is examined, it will often be found that many of their conclusions are derived from tests that involve abnormal intakes. Periodically, other foods come under suspicion—"saturated" fats, smoked foods, and even cranberries. The more research that is carried out, the more things are supposed to be wrong with our food, and this includes many of the natural foods that food faddists insist on eating.

This is not to decry research, as much good comes out of investigations on food and nutrition, but sensational statements resulting from evidence based on abnormal food intakes are very misleading. Because sweets have a high caloric value and are often eaten between meals, they are frequently blamed for obesity. Evidence may be produced, however, to show that sweets eaten before a meal may actually reduce the appetite and prevent the consumption of extra food during meals. Sweets provide a rapid increase in the blood sugar level which helps to reduce the desire for food.

**Dental Caries** The allegation that sweets are the cause of dental decay has been given much publicity by the medical and dental professions. But dental evidence points to excessive consumption of all soft, sweet foods as a main cause. Starch is less damaging than the sugars. Sweet foods that cling to the teeth and stay in the crevices produce the substrate and anaerobic conditions necessary to form what are now termed the cariogenic acids. However, of the annual consumption of carbohydrates, confectionery accounts for only 7 percent, so why make it the sole culprit?

Statistical evidence comparing countries with lower confectionery consumption against those with higher shows no difference in tooth decay. In a special survey on 2,905 Scottish children (McHugh, McEwan, & Hitchens, 1964) there was no evidence to suggest that a high consumption of sweets (25 oz per week) resulted in any more decay than a small consumption (5 oz per week).

**Other Publications**   Grenby (1968) produced a survey of causes and prevention of dental decay and his recommendations were:

1. Control and reduce the consumption of sweets, ice cream, soft drinks, cakes, and cookies, and replace them with savory foods.
2. Practice rigid oral hygiene by eating cleaning foods such as applies, celery, and other fruit and vegetables, and by regular tooth brushing and cleaning.
3. Add fluoride to water. Fluoridation has been shown conclusively to make teeth more resistant to decay.

Mention is made elsewhere in this book of anticariogenic substances. Xylitol has come into prominence in this respect.

Karger (1975) in the *World Review of Nutrition and Dietetics* quotes a series of papers on "Sugar in the Diet of Man." These reports discuss the role of sugar in nutrition and it's relationship to various diseases (see References).

## THE VIRTUES OF CONFECTIONERY

Much has been said about the harmful effects of overindulgence in sweets but these are probably much less than many other habits about which the public is warned with monotonous regularity.

What are the virtues of eating confectionery? First of all, it is enjoyable to eat, and has a pleasant flavor. Because of its high sugar content and physical structure, it is quickly digestible and provides a rapid source of supply of blood sugar—this means a quick replenishment of energy.

It is useful as a snack food and many are the occasions when a confectionery or chocolate bar and a cup of coffee or tea are a welcome interlude.

Chocolate particularly, is much in demand for survival rations or as a constituent of food packs for use during feats of endurance such as mountain climbing and rescue.

The role of confectionery as a gift is not to be overlooked. There is great pleasure in receiving a box of chocolates. Holidays such as

Easter and Christmas are enhanced by gifts of attractive foil-wrapped candies—Easter eggs, rabbits, and Santa Clauses.

## REFERENCES

Clydesdale, F. M., and Francis, F. J. 1983. *Food, Nutrition, Health.* AVI Publishing Co., Westport, Conn.

*Confectionery in Perspective.* 1982. The Cocoa, Chocolate and Confectionery Alliance, London.

Grenby, T. H. 1968. Some Aspects of Food and Dental Caries. *Chem. Ind.*

Marks, J. 1975. *A Guide to the Vitamins.* Medical and Technical Publishing Co., Lancaster, England.

McCance, R. A., and Widdowson, E. M. 1960. *The Composition of Foods.* Her Majesty's Stationery Office, London. New Edition (1985) Pave, A. A., Southgate, D. A. T.

McHugh, W. D., McEwan, J. D., and Hitchen, A. D. 1964. *Br. Med. J.,* London.

Peterson, M. S., and Johnson, A. H. 1978. *Encyclopedia of Food Science,* Vol. 3, AVI Publishing Co., Westport, Conn.

Vitamins Inc., Chicago, Ill. 1978–81. U.S. Patent No. 4,256,769.

Also see:

Karger, S. 1975. *World Review of Nutrition and Dietetics,* Vol. 22 (U.S.A.)

including the following:

Stare, F. J. (Boston, Mass.) Role of sugar in modern nutrition.

Grande, F. (Minneapolis, Minn.) Sugar and cardiovascular disease.

Danowski, T. S., Nolan, S., and Stephan, T. (Pittsburgh, Pa.) Obesity.

Bierman, E. L. (Seattle, Wash.) and Nelson, R. (Rochester, Minn.) Carbohydrates, diabetes, and blood lipids.

Finn, S. B. (Birmingham, Ala.) and Glass, R. B. (Boston, Mass.) Sugar and dental decay.

Danowski, T. S., Nolan, S., and Stephan, T. (Pittsburgh, Pa.) Hypoglycemia.

# 25

# Research and Development in the Confectionery Industry

Much has been published about research and development in all industries. Confectionery is a part of the food industry, and food as a whole has been subjected to research in every form.

Starting with food production, basic raw materials, such as sugar, fats, flours, and starches, nuts, fruits, and the cocoa bean, are all involved in research to produce the best-quality material at minimum cost.

The aim also is good nutritional value and food safety. Included in the latter is freedom from pesticide residues, noxious additives, and microbiological contamination. Maintenance of flavor and keeping qualities during storage, transport, and sale are of increasing importance with modern methods of marketing and extended stock periods. Wastage between manufacture and consumption must be avoided. Utilization of the most economical methods of production and modern machinery is another aspect of research and development.

Last, there is market research. However clever the development of formulation, process, and machinery might be, the product must repeatedly sell well.

Test marketing in selected areas will provide useful information before large sums are spent on equipment and its installation. If the information thus obtained is favorable, skillful advertising is then necessary.

## RESEARCH IN THE CONFECTIONERY INDUSTRY

The size of sales outlets varies enormously in the confectionery industry. At the lower end of the scale there is the small candy shop where a small range of products is made by hand or with a minimum of mechanical aids. The products are sold in the immediate locality.

They may have a short shelf life, and under these circumstances, exotic sweets such as chocolate-covered strawberries are possible. Research, if it can be called that, consists of making some samples, tasting them, then putting some on sale. Under these circumstances, much expansion of sales is not possible.

At the opposite end is the very large company where products are made by the million using fully mechanized methods of production and packing. Candy bars lend themselves admirably to such methods.

## R&D in Large Companies

It is within the large companies that the methods of research and development (R & D) mentioned at the beginning of the chapter become necessary. The results of such research are rarely published, and certainly nothing if it will affect the sales or profitability of the company. Exceptions are analytical procedures or microbiological investigations that are common to the food industry as a whole.

Large companies are likely to have the advantage of financial reserves but often lack coordination of their research information. It is essential to have persons at the level of director who can understand and streamline R&D information.

In the development of a new line, team work must be fostered from the beginning and engineers, production personnel, and scientists must be present at all stages. Table 25.1 shows the responsibilities that each department might be expected to assume throughout the project.

The terms "engineers, production, and scientists" are used, but in large organizations, R&D departments have personnel representing all these operations. They work as a team and in liaison with their respective factory departments.

## R&D in Small Companies

Small companies are likely to have meager resources in scientific and engineering staff and experimental equipment. Some are content to make small quantities of a limited number of lines for local distribution and by so doing are able to make reasonable profits even when obsolete methods of production and plant are used.

In such circumstances, very little research or development is necessary and probably one chemist covers quality control, trouble-shooting, and any necessary investigations. Sometimes a particular line will grow to a stage where new machinery capable of larger

TABLE 25.1. COORDINATION OF RESEARCH AND DEVELOPMENT

| | Sales/ marketing | Engineers/ production | Scientists including experimental confectioner |
|---|---|---|---|
| Conception of line | Will suggest type of line and presentation | Will comment on feasibility from machinery and process angle, particularly in relation to mass production. | Will decide on the suitability of recipe in relation to shelf life, nature of package required. Will make experimental samples for submission to marketing and sales members. Many large companies employ confectioners who have had "continental" training. These people are craftsmen who may design formulations that become the basis of a new product and feasibility study. |
| Experimental production stage 1 | Will examine samples of line that might represent final mass-produced article. | All will work jointly, probably as members of a project team in the erection of experimental and pilot production plant. This will lead to complete production plans and plant layout from which costs can be obtained. | |
| Experimental production stage 2 | Will be shown an example of the final product that will include expected size and wrapping. | Will continue to examine details of layout, looking at possible economies in methods of production and in the use of alternative raw materials without affecting quality of product. | |
| Test marketing | At this stage it may be appropriate to conduct a test market exercise. If pilot plant is adequate, samples may be made on this. Sometimes "co-packers" who have suitable plant are used but that is a security risk. Certain consulting companies undertake this type of work and produce samples under strict security. | | |
| Erection of plant | Assuming test marketing gives a satisfactory result, final plans for production will be made: Mainly engineers' responsibility, but production and scientists must see that process conditions in the original plan are being achieved. | | |
| Commissioning of plant | This must be the joint responsibility of a small team, usually one engineer, and one each from the production and scientific departments. Data of all operations *must* be recorded so that ultimately in production causes of troubles can be quickly recognized. | | |
| Production | When the commissioning team is satisfied, the plant and process are handed over to the production and quality control departments. | | |

output and more efficient production becomes necessary. Development then needs careful consideration before such machinery is installed and there are several ways to decide what is best.

*Make inquiries directly to machinery companies.* This necessitates considerable technical knowledge on the part of someone in the company because, however honest a supplier may be, there is always the tendency to oversell. The confectionery manufacturer is then faced with the necessity to increase sales considerably to meet the output of the plant and to pay for it. Writing off capital and interest can have a very adverse effect on profits.

Where single machines are involved, it is generally possible to negotiate a sale or return agreement, with the user paying rental on the machine for experimental purposes until satisfied with its performance.

*Attend trade shows.* Certain exhibitions specialize in the display of confectionery machinery and much information can be obtained by visiting the various stands. This again, however, implies some knowledge of the technology of the industry and the warnings outlined above must be recognized.

*Use a consultant.* In many cases, this is probably best for a small company. Consultants are available for product and process development and some specialize in plant and machinery installations (see later).

## Research Facilities

There are many research associations, each with its own special expertise—for example, baking, meat processing, fruit and vegetables, food preservation, chocolate and confectionery, metals, building and construction, and many others.

Often companies become members of the association most appropriate to their business. That association will provide current information regularly, and will also undertake contract work.

With contract work, the client must give a fairly precise account of the problem and at the outset negotiate the method of payment. Also, some idea of the time involved must be agreed. Codes of practice, including preservation of confidential information, must be observed.

Many universities and technical institutes have sections that undertake contract or consulting research.

Their value is generally related to special scientific problems that need detailed discussion between the university and client before work is undertaken. They are not often concerned with investigating problems involving production.

There are also government research organizations that are not generally involved in work for private companies but do publish brochures and handbooks for public consumption. In most countries, registers of consulting scientists are published.

## CONSULTANTS AND CONSULTANCY

Consultation consists of a meeting between an expert in a particular field and persons requiring the help of that expert. Most industries use consultants but some companies will never use them!

There are large consulting associations and smaller companies that will carry out investigations under contract in the same way as mentioned previously.

In the confectionery industry, there are relatively few companies that specialize. These have laboratories or "kitchens" equipped with small-scale chocolate and confectionery machinery and are usually operated by people who have had long experience in the industry.

There are also individual consultants who may have had particular experience in a section of the industry—for example, chocolate, cocoa powder, or hard candy.

It may have been implied that consultants are mainly used by smaller companies where technical staff is limited. This is not the case, and many large companies use outside consultants to develop new products and to verify work that has been carried out by their own technical staff. Mistakes are sometimes made in large companies and consultants are hired to correct them. Confectionery consultants with their own laboratories are often better equipped for certain development work than those of the larger companies. These consultants are also able to develop new plant and machinery with companies that normally make equipment for other food manufacture. This is a way in which the business of an engineering company may be extended with minimum risk of costly mistakes in design. Companies that develop new machines invariably find the greatest difficulty in obtaining a valuation of their machine from the user.

The same applies to new raw materials where a consultant is able to give an unbiased evaluation.

Another development in consultancy of recent years is the provision of confectionery courses, run either in the consultant's laboratories or in conjunction with the department of food technology at a university. There are also specialist schools that are concerned with confectionery and probably baking as well. A list of some of these schools and consulting companies is given in the Appendix.

It has been mentioned that some companies will not use consultants; their reason is that "secrets" are obtained and communicated to other clients. In fact, often the "secrets" exist only in the minds of certain of that factory's personnel. In any case, a consultant will recognize a process or machine that *is* unique and will never give details to others—the consultant's reputation is at stake. Many companies ask consultants to sign a security agreement.

Some companies believe they have special processes and formulations that the directors say "must never be changed". When sales and production increase, such attitudes lead to complicated multiplication of batch processes whereas technical investigation and advice from a consultant would have led to more efficient continuous processes without alteration of the product.

Cooperative effort over a period between the consultant and client is best. In this way, problems may be solved amicably and development undertaken in stages.

Consultancy has reached very large proportions in the United States and a survey some years ago indicated that four of the large organizations—Battelle, Arthur D. Little, Stanford, and Illinois—among them employed about 12,000 persons and earned $150 million a year.

In Europe, private contract research units are generally smaller but research associations have been in operation for over fifty years.

# Appendix I

# Special Methods of Analysis

There are many publications available on food analysis and some of these give details of methods that can be applied to confectionery. All chocolate and confectionery manufacturers should be acquainted with The International Office of Cocoa and Chocolate (IOCC), 172, Avenue de Cortenbergh, 1040 Bruxelles, Belgium; and The International Sugar Confectionery Manufacturers' Association (ISCMA), 194, rue de Rivoli, 75001 Paris, France. Both of these organizations convene a general assembly every five years and monitor all aspects of the industry.

The IOCC publishes "official" methods of analysis that are regularly updated and accepted internationally as reference methods. Periodic bulletins are also published that describe the activities of these organizations.

Analytical methods special to the industry are described in some manufacturers' brochures. Trade journals frequently contain articles concerning the application of these methods and the significance of the results obtained.

Analytical procedures were mentioned when necessary in earlier chapters. Some of the following methods are repeated from the second edition of this book with updated insertions where considered necessary.

## PARTICLE-SIZE DETERMINATION

The particle size of cocoa, chocolate, and confectionery products is a fundamental property. It has importance both organoleptically and in manufacturing processes.

In chocolate, the smoothness on the palate is related to the absence of coarse particles of sugar and cocoa material. Fondant creme is similar. In cocoa powder, the fineness of particles is related to suspension in a liquid, for example, milk and water. Particle size *distribution*—that is, the *proportion* of particles of different sizes—is important economically.

In chocolate, the use of cocoa butter to obtain a given fluidity is linked with the proportion of very fine particles. Because of greater total surface area, these require more cocoa butter to "wet" them.

## Particle Variation

The type of particle in the various chocolate and confectionery products varies considerably in appearance, shape and size.

In cocoa liquor, the particles, when examined under the microscope, are irregular but without sharp edges, and range in color from a transparent light brown through reddish brown to opaque dark brown.

In dark chocolate, the cocoa particles are similar but less conspicuous because of the presence of sugar. The sugar crystals are readily visible as clear plates usually of irregular shape because they have been shattered during the grinding process.

Milk chocolate contains cocoa particles, sugar crystals, and milk solids particles, but if milk chocolate is made from crumb, only one type of particle is visible. These are aggregates of milk solids, sugar, and cocoa liquor that have been formed during the crumb process by the final crystallization and drying processes, and these conglomerates are broken down in the chocolate-refining process but not separated into their constituent particles.

These aggregates are a light-brown color and in some of them sugar crystals and cocoa particles can be seen cemented together.

In the manufacture of milk crumb, it is important to ensure that, at the crystallizing stage, the sugar crystals formed are small and in the microscopic examination the aggregates are separated by the use of chemical mounting liquid so that the sugar crystals can be measured independently. In a milk chocolate made from milk powder, the milk powder particles are clearly visible as separate, pale-yellow, irregular shapes.

In fondant creme the visible particles are sugar crystals only and as most of these have grown out of a syrup without much hindrance they are mainly cubic and present a very regular appearance under the microscope.

There is a difference between a freshly prepared "base" fondant and a fondant that has been remelted and cast into molding starch. The latter has a proportion of large crystals due to growth at the remelting and casting stage with a background of small crystals as present in the "base fondant." Large crystals may also be present in "base" fondant due to bad beating or cooling during manufacture, and this is discussed under "Fondant Manufacture." The size distribution

TABLE A.1.  PARTICLE SIZE OF CONFECTIONERY PRODUCTS

| Base product | Manufactued product | Particle size | |
|---|---|---|---|
| *Cocoa liquor* | | | |
| Cocoa particle size | Superfine cocoa or chocolate manufacture | 100 μm | 0.0040 in. |
| | Drinking cocoas | 200 μm | 0.0080 in. |
| *Chocolate—dark* | | | |
| Cocoa particle size | Fine eating chocolate or high-quality covering | 30–50 μm | 0.0012–0.0020 in. |
| | Average covering chocolate | 75–100 μm | 0.0030–0.0040 in. |
| Sugar crystal size | Fine eating chocolate or high-quality covering | 25–35 μm | 0.0010–0.0014 in. |
| | Average covering chocolate | 50 μm | 0.0020 in. |
| *Chocolate—milk* | | | |
| Crumb aggregate size | Fine eating chocolate | 35 μm | 0.0014 in. |
| | Average covering chocolate | 50 μm | 0.0020 in. |
| *Fondant creme* | | | |
| | "Base" fondant from machine | 10–15 μm | 0.0004–0.0006 in. |
| | Fondant after casting (assortments) | 25–30 μm | 0.0010–0.0012 in. |

of the sugar crystals in fondant has an important bearing on smoothness on the palate.

An approximate guide to the particle sizes that may be expected in these products is shown in Table A.1. The figures are "averages of the larger particles."

*It must be understood that these figures cannot be compared with micrometer readings* (see later).

## Methods of Determination

Methods of particle size determination may be summarized:

1. Micrometer and modifications.
2. Microscope, including computer scanning.
3. Wet sieving, using either water or petroleum solvent.
4. Sedimentation.
5. Electronic counting, Coulter counter, laser beam.

## Micrometer

Over the years, the standard engineer's micrometer has been used more than any other instrument for determining the fineness of chocolate paste from the refining rolls.

The method utilizes a small amount of the refiner paste mixed with an equal amount of a liquid oil. One drop of the mixture is applied to the lower face of the micrometer jaws and the jaws closed, using the spring-loaded control, until movement ceases. The value is read on the micrometer scale.

It is a "figure" only and does not represent the size of any particles or take their shape into account. Isolated large sugar crystals are most likely crushed. The result, in the hands of a skilled person, is meaningful but really represents the thickness of a layer of particles compressed between the jaws of the micrometer using a standard force. The weakness of the instrument is in the variability of micrometers, the people using them, the size of the sample, and the fact that there is no precise information about the actual size of the particles or the proportion of the different sizes.

An improvement of recent years is the electronic micrometer in which the jaws are larger in area and are closed under precise pressure. The reading is digital but the figure obtained is in the same category. The value of the micrometer method as a means of monitoring the refining process led to research on how its value might be improved.

## Metriscope

This apparatus was devised by Lockwood (1958) of the Cadbury Research Laboratories, and is an ingenious mechanical device for measuring the efficiency of chocolate or liquor grinding. The method is really a refinement of the micrometer method and uses a much larger sample for test. Further details of the apparatus are available from: Confectionery Division Research Laboratories, Cadbury/Schweppes Ltd., Bournville, Birmingham, England.

The principle of the method is the support of a tapered steel "stopper" in a socket by a film of the chocolate or liquor to be measured for fineness.

The stopper and socket have a taper of exactly one in ten and the trapped film of cocoa and sugar particles causes the stopper to be raised twenty times the thickness of the film by reason of the taper slope. The protrusion of the stopper is measured accurately by micrometer and this can be related to the average size of the larger particles in the chocolate or liquor.

The Metriscope is fully automatic and will give several repeat readings in a short space of time. The machine uses a 1 oz (28.4 g) sample dispersed in 5 oz of lecithinated cocoa butter and this gives a

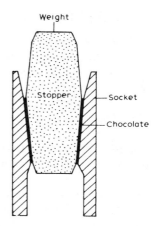

Fig. A.1.  Principle of the Metriscope

much more representative result than that obtained from the very small sample taken from micrometer assessment.

Figure A.1 is a diagram of the stopper and socket and Figs. A.2 and A.3 are illustrations of the machine in use in a number of factories.

Fig. A.2.  Metriscope—Front

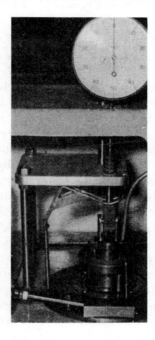

Fig. A.3.  Metriscope—Rear

## Method of Determination

*Materials normally tested.* Paste from refiners, chocolate from conche, finished chocolate, liquor.

*Apparatus required.* Metriscope, electric Vibro-mixer, 5-fl-oz measure (for cocoa butter), 1-oz dispenser (for refiner paste), stainless-steel container (approx 10 fl oz capacity), palate knife, oven at 45°C (113°F).

**Cocoa Butter for Dispersion** This is cocoa butter in which is dissolved 0.25 percent of soya lecithin.

**Preparation of Sample for Testing** One ounce of refiner paste is measured out using the "dispenser" and the plug of chocolate placed on the perforated disk of the Vibro-mixer and mixed for 1 min with 5 fl oz of the cocoa butter containing 0.25 percent lecithin. The sample is then ready for immediate test.

The other materials tested are usually weighted into small beakers, mixed with a small amount of the 5 fl oz cocoa butter using the palate knife, and then transferred to the stainless-steel container for mixing.

**Operation of the Metriscope** The instrument heater is switched on for 2 hr before using the Metriscope to ensure that the instrument will operate on liquid cocoa butter. The operating temperature range is 40 to 45°C (104 to 113°F) and this can be maintained by the use of the thermostat.

When the Metriscope is required for measurement, the "motor" switch is turned on, cocoa butter is run through the wire gauze sieve into the hopper, the handle is turned to "Test," and the butter allowed to fill the inner "moat." Any excess butter entering the inner moat overflows into the outer moat, from where it runs into the collecting bin in the bottom of the Metriscope.

The instrument is fully automatic from this stage and the stopper rises and falls to allow a certain amount of the cocoa butter to remain between the faces of the stopper and socket. The vertical beam is released and the reading given on the dial is an indication of the thickness of the film. This operation is repeated very 20 sec and after three or four complete cycles the inner moat should be empty. With the cocoa butter, the dial readings should be 25 or less, and when this has been achieved, the test suspension is run in through the wire gauze sieve. When the inner moat is full, the handle is turned to

"Waste" allowing the rest of the sample to run into the collecting bin. The first dial reading is ignored and the average of the next three recorded as being the Metriscope value. This result is in 1/1000th of a centimeter.

**Calibration of Metriscope**  To ensure that all Metriscopes give the same readings in the operating range (30 to 150), a standard chocolate of known Metriscope value is issued from the central laboratory. The standard chocolate is made up, as previously described, and the dial of the instrument set by loosening the bolt at the back and moving the dial up or down the column until the required reading is obtained when the bolt is tightened. A small final adjustment can be made by turning the front of the dial but this should not exceed two or three units. This operation is carried out with the zero adjustment removed. The Metriscope is then allowed to run and the beam should bring the dial to a negative reading of about 10 units. The zero adjustment is then replaced so that the knife edge fixes the outer ring of the dial and the zero screw is adjusted so that the pointer of the dial returns to zero after each reading. The accuracy of the instrument is checked daily by the scientific staff and no adjustment by the factory operatives is necessary.

**Standard Chocolate/Cocoa Butter Mixture**  Two ounces of chocolate are weighed into a beaker, melted, mixed with 10 fl oz cocoa butter, and the mixture poured into a labeled, stoppered bottle. This mixture is kept in the Metriscope ready for confirmatory tests that are carried out by shaking the bottle thoroughly and pouring about 1 fl oz into the inner moat of the instrument.

## Microscopes

In the early days of refining control, sugar crystal and cocoa particle measurement was carried out with standard microscopes fitted with transparent scales in the eyepiece and a mechanical stage to hold the microscope slide. The mechanical stage enabled the material on the slide to be examined methodically and the particles were measured by means of the eyepiece scale.

This microscope method was very tedious and caused eyestrain over long periods of working, but even so it was used for many years. Hand microscopes have been superseded generally by projection microscopes and these have also greatly improved in design, par-

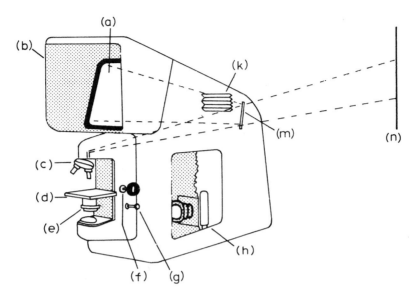

Fig. A.4. Projection Microscope

a. Viewing Screen
b. Hood
c. Double Objective
d. Stage
e. Condenser
f. Coarse Adjustment

g. Fine Adjustment
h. Lamp and Condenser
k. Ventilator
m. Mirror (Hinged to Allow Rear Projection onto a Screen)
n. Wall Screen

*Elcometer Instruments Ltd., Manchester, England*

ticularly in the brilliance of illumination and the size of the viewing screen.

A projection microscope is shown in Fig. A.4. This instrument is not currently available but the illustration shows the principle of operation, which may be applied generally. Illumination is by a high-powered xenon lamp and the image of the particles is projected through a prism onto a mirror from where it is reflected onto a 10-in. screen.

The primary microscope objective is low powered but by means of a lever a supplementary lens is interposed and an enlarged image of the center of the screen is obtained. With this instrument, the image is sufficiently bright to enable measurement in subdued daylight. A lattice is provided on the screen that is calibrated for easy measurement of the particles.

Microscopic measurements are empirical and their value depends to some extent on a precise method of preparing the slide and a systematic method of measuring the particles. With the system already described, the larger particles only are measured (10 $\mu$m to 50 $\mu$m). That is usually adequate for refiner control and fondant manufacture.

For assessment of the very fine particles and size distribution other methods are used (see later).

It is often said that one advantage of the microscope is that the particles are "seen" whereas with other methods they are not. With many products, this qualitative aspect is a great advantage.

**Methods of Microscopic Measurement**   The following are simple practical methods. The actual details may be modified to suit the user.

**Cocoa Liquor—Preparation of Sample**   A representative sample from the refiners is obtained and well mixed. With liquor the larger particles tend to settle on standing and if the test is not to be made immediately the liquor should be tempered and molded.

**Preparation of Microscope Slide**   The amount of liquor used on the slide is controlled by using a stainless-steel plate 2 in. by 1 in. and 0.048 in. in thickness. The plate has a $\frac{1}{8}$-in.-diameter hole in the center.

This plate is placed on a microscope slide and packed with solid or pasty liquor using the blade of a scalpel. The excess liquor is removed by drawing the edge of the knife across the hole. The liquor is expelled with a blunt metal poker that is just able to pass through the hole. The pellet of liquor is deposited in the center of the slide.

One drop of thin oil is added from a micropipette marked to deliver the necessary quantity for a ground liquor. For coarse liquor, slightly more oil is required. The slide is warmed and the oil and liquor mixed with the point of the scalpel, keeping the mixture in a circle less than $\frac{1}{2}$ in. diameter. The small amount of oily mixture on the end of the scalpel is disregarded and must be the minimum possible. When thoroughly mixed, a $\frac{7}{8}$-in.-diameter cover glass is placed on top and allowed to sink by its own weight. No pressure must be applied and the amount of oil used should be sufficient just to fill the cover slip without oozing beyond the edge.

It is important that no material is removed from the slide, and the details given must be strictly followed so that slides of standard density are prepared.

**Measurement by Projection Microscope**  The slide is placed on the mechanical stage and focused on the screen. The slide is first examined methodically to see that particles are evenly distributed and to assess approximately the size of the larger particles. It is then examined a second time and the large particles measured and recorded.

When ten to fifteen measurements have been made it can be readily assessed whether they are representative of the majority of large particles on the slide. The average of the first ten is then taken.

**Chocolate, Refiner Paste, and Milk Crumb Cocoa Particle Size**  To determine cocoa particle size in dark or milk chocolate, refiner paste or milk crumb, it is necessary to make up the slides so that the density of cocoa particles on the slide is the same as that when examining cocoa liquor. If this is not done, results are not strictly comparable. Stainless-steel plates with larger holes and poker rods will be required in keeping with the cocoa liquor content of the chocolate.

Chocolates and crumb vary somewhat in composition but examples of the cocoa liquor content of these are, in percent:

Dark eating chocolate      30
Milk eating chocolate      11
Milk refiner paste         13
Milk crumb                 14

The calculation of the size of the hole and rod must be done according to recipe.

With milk crumb it is best to weigh the amount required (approx. 15 mg) onto the slide and it must be softened to enable the slide to be prepared. This can be done by using a mountant consisting of ethylene glycol saturated in ammonia or with dichlorhydrin.

The measurement of the particles on the slide is accomplished by using the same procedure as for cocoa liquor.

**Sugar Crystals and Crumb Aggregates**  The procedure for slide preparation is similar to that used for cocoa particle measurement except that the mounting oil is colored red by the addition of an oil-soluble dye. The density of crystals and aggregates on the slide is such that it is not usually necessary to make adjustments for the sugar content of the chocolate.

First inspection of the slide may reveal occasional "abnormal" crystals of a size well above the average of the larger crystals. It is usual not to include these in the measurements.

In dark chocolate, the sugar crystals have well-defined outlines and are easily measured, but in milk chocolate and refiner pastes it is the ground crumb aggregates that will be measured and these are less distinct. Occasionally, the grinding process separates some of the sugar crystals from the aggregates and these will be smaller in size than the aggregates.

If during the manufacture of crumb the crystallization has proceeded too slowly, the sugar crystals will have grown too large and aggregates containing these large crystals will be formed. These make grinding more difficult and output will be reduced.

The measurement of sugar crystals in crumb is done using the glycol or dichlorhydrin mountant but this test is rarely required when chocolate manufacturers' make their own crumb, as they should check the sugar crystal size at the kneader stage. (See "Milk Crumb Manufacture".)

In milk chocolate made with milk powder, the sugar crystals are distinct as in dark chocolate. Milk powder particles are light yellow and irregular in shape.

**Fondant Creme** It has already been mentioned that fondant creme may contain two ranges of crystal size due to bad beating conditions in the fondant machine or bad remelting techniques. Large crystals may also arise from the inclusion of rework but this is rarely done today and reincorporation of rework is best done by syruping and decolorizing (see "Reclaiming").

*Preparation of Slide* A plate similar to that used for cocoa liquor is packed with fondant and discharged onto a slide where it is mixed with glycerol mountant colored red or green. The slide must be read immediately and not warmed or some solution of the crystals may occur.

A well-prepared base fondant should have a size distribution similar to the following: 10 $\mu$m and 15 $\mu$m—a very large number of crystals; 20 $\mu$m—none. The average crystal size would be assessed at 12.5 $\mu$m = 0.0005 in.

**Measurement of Cast Cremes** This is complicated if on inspection of a number of fields there appears to be two distinct ranges of sizes.

TABLE A.2. CAST (ASSORTMENT) CREMES

| | | Size | | | | |
|---|---|---|---|---|---|---|
| | 20 $\mu$m and below | 25 $\mu$m | 30 $\mu$m | 35 $\mu$m | 40 $\mu$m | 45 $\mu$m |
| *Example 1* | | | | | | |
| Number of | large number | xxx | xxx | xxx | | |
| crystals | | xxx | xxx | | | |
| *Example 2* | | | | | | |
| Number of | large number | xxx | | x | | |
| crystals | | x | | | | |
| *Example 3* | | | | | | |
| Number of | large number | | | xx | xxx | xxx |
| crystals | | | | | xxx | |
| | | | | | xxx | |
| | | | | | xxx | |

The method then involves the counting of crystals in the larger ranges and recording them as shown in Table A.2.

The average size of the "larger" crystals would then be assessed as:

Example 1    28 $\mu$m = 0.0011 in.
Example 2    20 $\mu$m = 0.0008 in.
Example 3    40 $\mu$m = 0.0016 in.

Examples 1 and 2 would be considered average size distributions but Example 3 shows two distinct size ranges and indicates bad remelting and casting techniques or poor base fondant, but if Examples 1, 2, and 3 are made from the same fondant, bad remelting must have been the cause of Example 3.

## Sieving Methods for Determining Particle Size

Dry sieving is not often used for particle-size determination except occasionally for low fat cocoa powder and fine sugars.

For this purpose, banks of sieves of varying mesh size are subjected to automatic vibration and the amount of material collected on each sieve is weighed together with the total amount passing the finest mesh. Thus, some idea of the particle size distribution is obtained. Dry sieving is more suitable for nonclogging powders—fatty powders tend to blind the sieve.

## Petroleum (Wet) Sieving

The principle is to suspend the powder or liquor in a petroleum solvent and wash it (with clean solvent) through a fine mesh. The method below uses a 325 mesh woven wire sieve, which, for most practical purposes, is adequate; in fact, 200 mesh is more frequently quoted for cocoa powders [aperture size British Standard: 200 mesh—0.0030 in. (75 $\mu$m); 325 mesh—0.0018 in. (45 $\mu$m)].

Micromesh sieves are now available and particle sizes down to 8 $\mu$m may be measured.

**Wet Sieving Method (Petroleum)**  The sieve comprises a 3 in. diameter hollow cylinder of tinned copper $2\frac{1}{2}$ in. high to which is attached a handle in the form of a long hoop. The base of the sieve is flanged and covered with standard wire gauze, 325 mesh, and soldered all round the edge. A metal ring is soldered on the lower side of the gauze to protect the sieve when standing on the bench.

Twenty-five grams[1] of cocoa are placed in the sieve and lowered gently into a vessel containing petroleum ether. 60 to 90°C (140 to 194°F). The sieve is held with the gauze inclined at about 30 degrees to the horizontal and given a rotary movement throughout the test. When the sieve is something less than a quarter full, it is slightly withdrawn so that about half the gauze is below the surface of the petroleum ether in the vessel. When the volume in the sieve has been reduced appreciably, the sieve is lowered until it is again something less than a quarter full and drained as before. The process is repeated as required, generally for about 5 min, the gentle swirling motion being maintained throughout. At the end of this time, the residue in the sieve will appear appreciably coarser than the original cocoa.

The sieve is then transferred to another vessel containing clean petroleum ether. Any cocoa clinging to the side is washed down using a wash bottle of petroleum ether. After flooding the sieve several times by dipping followed by draining, the sieving should be complete.

The sieve is dried in an air oven for 10 min. The cocoa tailings are transferred to a tared dish, using a small brush, weighed, and the percentage calculated.

---

[1] For routine cocoa-testing, 10 g may be used.

*Note*—A fine-grade cocoa powder will have less than 2 percent rejects by this test. Liquor can also be tested by this method and a fine liquor will have less than 1 percent rejects.

Desiccators, approximately 8 in. internal diameter and having well-fitting lids, make good washing vessels for the foregoing method.

After a number of determinations, the sieve will have an appreciable number of clogged holes. The sieve is cleaned in a large beaker containing a half-inch layer of boiling 3N sodium hydroxide. The sieve is boiled for about 2 min by which time the holes should be clear. Without delay the sieve is washed throughly with water, then alcohol, and dried in an oven.

**Wet Sieving Method (Water)** In this method, the residue on a 200 mesh is determined as below. Full details are given in the brochure *Cocoa Powders for Industrial Processing* (3rd ed.). Meursing (1983).

***Procedure*** Weigh approximately 5 g of cocoa powder into a 400-ml beaker to an accuracy of 10 mg. Add 20 ml of distilled water and stir carefully with a glass stirring rod until all lumps have disappeared. Then add 280 ml of hot distilled water ($75 \pm 5°C$) and stir vigorously for 2 min using a mechanical stirrer but taking care not to create a vortex (about 300 rpm). Pour the suspension through a 200 mesh sieve, simultaneously rotating the sieve horizontally.

Rinse the beaker and sieve with hot distilled water ($75 \pm 5°C$). If the suspension does not run smoothly through the sieve, tap it gently. Attach a Buchner funnel (diameter about 7 cm) to a 500-ml suction flask.

A glass-fiber filter, previously dried for 30 min at 103° to 105°C, is cooled and immediately weighed to an accuracy of 1 mg. It is then moistened with water and pressed firmly against the perforated base of the Buchner funnel. Wash the residue from the sieve on to the filter and apply the vacuum. The top part of the sieve is washed first and then the bottom. Wash the filter with about 15 ml of acetone and dry it for 60 min at 103° to 105°C. Cool and immediately weigh the filter plus residue to an accuracy of 1 mg.

Calculate percent residue as "rejects 200 mesh sieve."

**Microsieving Methods** Microsieving down to particle size 8 to 10 $\mu$m has already been mentioned.

To obtain satisfactory results, Niediek (1978) has shown that controlled vibration of the sieve is necessary. To obtain information on smaller particle sizes (4 $\mu$m and below), the electron microscope is used and Sutjiadi (1974) describes a method using a spray system to give even distribution on the microscope slide.

## Elutriation (Sedimentation) Methods

There are numerous elutriation methods for obtaining information on particle-size distribution. If a powder is suspended in water or some other liquid, the rate at which the particles settle depends on the size of the particle and the density of the liquid.

With elutriation the suspension may be partly separated after a period and filtered and weighed. Alternatively, the sediment may be measured volumetrically after a given time.

A standard piece of equipment for elutriation tests is the Andreasen pipette. A method was described in the first edition of this book called the Residimeter Value. This was used for many years to determine a selected particle-size range in milk chocolates. A valuable water sedimentaion test for cocoa powder is described below. Cocoa containing shell and germ particles will form sediments rapidly. This is objectionable in chocolate drinks.

**Imhoff Test** The Imhoff tube is that designed by Imhoff for sewage effluent and sludge control, but for cocoa sediment testing it is further tapered to enable sediment readings down to 0.05 ml to be obtained (Fig. A.5).

The powder to be tested should be free from compressed flakes, which may occur in packing and transport, and to remove these the powder is passed through a 100 mesh sieve. Weigh 2.5 g of cocoa and transfer to a 1-liter beaker. Rinse a 500-ml cylinder with hot distilled water. Then fill the cylinder to the 500-ml mark with hot distilled water. Allow the water in the cylinder to cool to 82°C (180°F).

In the meantime, fill the Imhoff tube to the 500-ml mark with hot distilled water (82°C) (180°F approx.).

Make a thin paste of the cocoa with a small amount of water at 180°F from the measuring cylinder by rotating the beaker to give a swirling motion to the water. Add about 200 ml water slowly, swirling the contents. Add the remainder of the 500 ml water at 180°F without swirling the contents.

Stir the mixture with an electric stirrer for exactly 30 sec. and during this period drain the Imhoff tube *completely* by inverting and tapping the tapered end to remove the last small quantities of water.

Place the Imhoff tube in the stand and immediately transfer the cocoa suspension to the tube and allow to stand for exactly 5 min.

Sedimentations may be watched by placing a shaded light at the side of the Imhoff tube and a magnifying glass can be used to assist measuring the sediment.

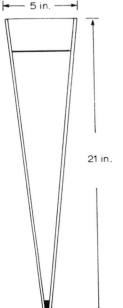

5 in.

21 in.

Fig. A.5.   Imhoff Tube

The sediment line should be well defined, but if the line is irregular or sloping the test is repeated.

A good cocoa suitable for the preparation of drinks should, by this method, have a sediment figure of less than 0.25 ml.

Sedimentation methods are empirical, and some are very slow to perform. They tell nothing about the shape of the particles.

### Particle Sizing by Laser Beam

This method has been applied successfully for determining the particle size and distribution in chocolate pastes and powders.

**Principle**   A low-power laser beam illuminates a cell containing a suspension of the particles that can be dispersed in a range of organic or aqueous liquids in either small- or large-volume cells. The light is scattered and focused by means of a convergent optical system onto a multielement ring detector. The detector ring position is selected for each particular size distribution and measurement.

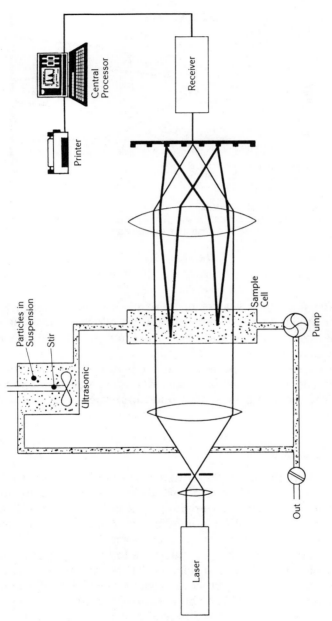

**Fig. A.6. Principle of Diffraction Scattering**

*Malvern Instruments Ltd., Malvern, England*

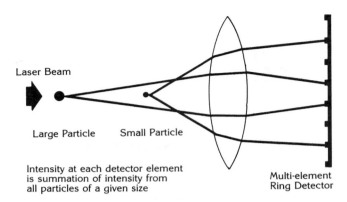

Laser Beam

Large Particle     Small Particle

Intensity at each detector element
is summation of intensity from
all particles of a given size

Multi-element
Ring Detector

Fig. A.7.   3600E Detector
Intensity at each detector element is summation of intensity from all
particles of a given size.

*Malvern Instruments Ltd., Malvern, England*

The degree of light scattering depends on the particles size—the smaller the particle, the wider is the angle of scattering. The detector is scanned continuously and the output amplified and programmed on a microcomputer.

Tabulated results of particle size distribution with histogram are shown on the visual display unit and may also be recorded on a printer. The principle is shown diagrammatically in Figs. A.6 and A.7.

## The Coulter Counter

Developed in the late 1950s, this principle has been used for the measurement of particle-size distribution. The instrument, known as the Coulter Counter,® was applied originally to blood cell counting but has since been used for a wide range of particulate materials, in powder, paste, suspension, or emulsion form.

The method requires that the particles to be analyzed be suspended in a suitable electrolyte solution, and this suspension is caused to flow through a narrow aperture that has an electrode on each side. As each particle passes through the aperture, it replaces its own volume of electrolyte solution, momentarily changing the impedance between the two electrodes, and producing a modulation of the current path (e.g., as a change in resistance) in the form of a pulse of magnitude closely proportional to the volume of the particle. The series of pulses arising from the stream of particles is amplified electronically and scaled to provide particle count and/or volume (mass) above or between a series of known size levels.

A special electrolyte solution is necessary for chocolate, and the most suitable found was a 5 percent w/v solution of ammonium thiocyanate in either industrial alcohol or isopropanol, with the chocolate particles being dispersed into either a fractionated (alcohol-soluble) lecithin or "Span 80" (sorbitan oleate, Honeywill Atlas) solution. To analyze dark chocolate, the electrolyte solution is presaturated with sucrose, and to analyze milk chocolate, the electrolyte is presaturated with defatted milk chocolate solids. By the use of these electrolyte solutions and various apertures, particle-size distribution in chocolate, cocoa, or cocoa mass, by number and weight (mass or volume) percentage, has been obtained down to 0.5 $\mu$m. Analyses of chocolate prepared by different processes can be made and the proportion of fine to coarse particles determined. It is not proposed to describe the operation of the various models of the instrument or the methods used, as these are covered in detail elsewhere, but it will interest the technologist to note the particle-size distribution of a popular milk chocolate determined by Coulter Counter (Figs. A.8 and A.9).

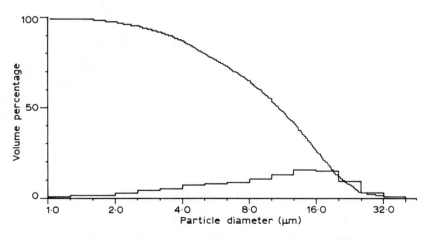

Fig. A.8.  Particle-Size Distribution, Popular Chocolate, Differential and Cumulative Volume (Mass or "Weight") Percentage Against Size ($\mu$m).

Coulter Electronics Ltd., Luton, England

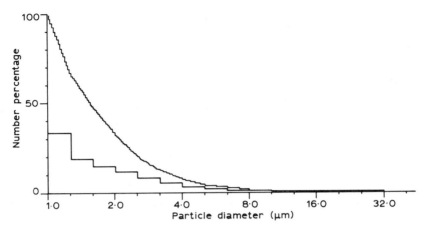

Fig. A.9. Particle-Size Distribution, Popular Chocolate, Differential and Cumulative Number Oversize Against Size (μm)

*Coulter Electronics Ltd., Luton, England*

## DETERMINATION OF FAT IN COCOA AND CHOCOLATE PRODUCTS

Analytical textbooks describe the standard methods in detail:

*Soxhlet extractor*⎫
*Bolton extractor* ⎭ Continuous extraction with petroleum ether.

*Werner-Schmid method* Used where protein interferes with extraction. Product heated with hydrochloric acid before extraction.

*Mojonnier and Rose Gottlieb methods* Product treated with ammonium hydroxide and alcohol, which dissolve the protein. Suitable for milk products, caramels.

*Gerber* A volumetric method used for routine testing of milk. Product treated with sulfuric acid followed by centrifuging.

### Rapid Methods using Trichlorethylene Extraction

This is a simple, quick, inexpensive method that can be used for routine testing.

**Preparation of Sample** Where possible, samples of chocolate and liquor should be properly tempered and molded. Alternatively, a homogeneous sample is obtained by finely grating a large sample and mixing.

Except as stated below, 10 g are taken for the estimation. Samples containing large amounts of shell, such as whole beans and winnowing products, should be examined by the Soxhlet method as complete extraction does not occur with the rapid method.

*Cocoa nib—winnowing products*: *use Soxhlet method.*

*Cocoa liquor.* Take 5 g in thin shavings for this estimation. If weighed while still fluid, mix well.

*Chocolate.* Use 10 g. Grate finely. If weighed while still fluid, mix well.

*Cocoa press cake.* Use 10 g. Grind until it will all pass a 30 mesh sieve.

*Cocoa powder.* Use 10 g. Mix sample well.

*Milk crumb.* Use 10 g. Grind and sieve as for cocoa cake. Checks should be made periodically by Soxhlet method. In some milk crumbs, some of the fat may be "locked" in protein or sugar agglomerates.

**Weighing Out of Sample** Some 5 or 10 g of the sample, as the case may be, are weighed in a counterpoised metal scoop. An accuracy of ±0.01 g is sufficient when weighing 10 g but an accuracy of ±0.005 g is necessary when weighing 5 g. The weighed sample is carefully emptied into a 6-oz narrow stoppered bottle, the last traces being removed by tapping the metal scoop or by use of a camel hair brush. Then 100 ml of pure distilled trichlorethylene are added from an automatic pipette supplied from a reservoir of solvent. Two pipettes are provided so that one may empty while the other fills. After replacing the glass stopper, the bottle is well shaken and allowed to stand for at least 30 min. The bottle is shaken at intervals during the 30 min.

**Filtration** A no. 5 Whatman filter paper (18.5 cm) is fluted by folding over a test tube (closed by a rubber stopper) to form a thimble, and inserted into the bottle so that the filter is held under the shoulder (Fig. A.10). The fact that the filter paper tends to spring open keeps it in place quite easily. The stopper is replaced. Ten minutes are usually enough to allow for sufficient filtrate to accumulate by filtering to the inside of the thimble.

**Pipetting** Twenty millilitres are now pipetted from the interior of the filter thimble into a weighed 100-ml wide-mouthed flask. Before pipetting, the pipette should be rinsed with a little of the filtrate. In

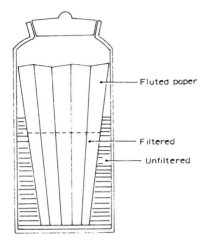

Fig. A.10.  Filtration Method

routine work, when examing a large number of samples, this rinsing procedure may be omitted after the first one. The 20-ml pipettes must be calibrated against the automatic 100-ml pipettes to deliver exactly one-fifth. *Note*: "Safety" pipettes should always be used to avoid inhaling trichlorethylene vapor.

**Distillation**  The excess solvent in the flask is now removed by distillation in a special apparatus accommodating 8 flasks (Fig. A. 11). This apparatus consists of a water-cooled condenser, hot plate, and distillation "adapters." Care must be taken not to carry the distillation too far or there will be a risk of decomposing the fat. When most of the solvent has distilled, the flask is removed and hot air blown in to remove as much as possible of the traces remaining. The flask is then placed in an oven maintained at 90°C (194°F) for 2 hr. After cooling to room temperature, the flask is again weighed, giving the weight of extracted fat.

In cases of greater urgency, the solvent may be removed by hot air blowing alone, and this takes 20 to 30 min followed by cooling and weighing. Further blowing for a short period is desirable to check that the weight is constant.

$$\% \text{ Fat} = \frac{\text{Wt of extracted fat} \times \text{factor} \times 100}{\text{Wt of sample taken}}$$

*Factor*. The factor compensates for the volume increase due to the

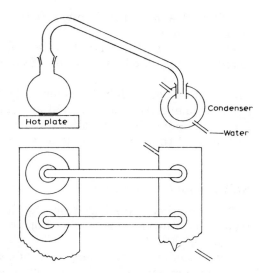

Fig. A.11.   Distillation Process and Adapters

fat extracted from the sample and is derived from the equation

$$\% \text{ Fat} = x(4.988 + 0.28\,x)\,\frac{100}{W}$$

where $x$ = weight of fat in 20 ml of solution
   $W$ = weight of sample taken.

For routine work it is customary to produce tables from this equation corresponding to the weight of sample taken (5 g or 10 g). These tables can be either "factors" or longer lists of fat contents equated to weights of fat obtained from the 20 ml of solution.

**Correction for Temperature Due to Solvent Expansion**   Should an appreciable change in the room temperature take place between the times the solvent is measured out and the 20 ml pipetted, a temperature correction must be applied to compensate for any expansion or contraction in the volume of the solvent.

**Corrected Factor**   If the aliquot part is measured out at a temperature of $T°C$ and the temperature of the solvent originally added was $T_1°C$, then the true factor = Factor + 0.0055 $(T - T_1)$.

**Recovery of Solvent Residues**   The residues from fat determinations are filtered and distilled into a bottle containing anhydrous potassium carbonate.

The dried distilled solvent is filtered and stored in amber bottles. Residue, determined on 50 ml must not exceed 0.0010 g.

**Preservation of Trichlorethylene** Trichlorethylene is subject to slight decomposition on repeated use and distillation. The addition of 2 percent industrial alcohol (ethanol) to each new supply of solvent will prevent this and will have no effect on the determination.

## Refractometer Methods

These methods are based on the determination of the refractive index of solutions of cocoa butter in the nonvolatile solvent l. chloronaphthalene.

The procedure is to weight 2.5 g of cocoa powder (or liquor) into a small beaker and add 5.0 g of l. chloronaphthalene. The mixture is heated to 70°C with stirring followed by filtration. The refractive index of the filtrate is measured using a high-precision refractometer with accurate temperature control and compared with the refractive index of the pure solvent (Meursing 1976).

## Determination of Fat Using Nuclear Magnetic Resonance

The system of low-resolution nuclear magnetic resonance (NMR) has been developed very successfully in recent years. By this method, it is possible to measure the solids/liquid ratio in fats and to determine the fat content of chocolate products, nuts, seeds, and so on. It can also be applied to the determination of moisture.

The technique is now well established as a means of quality control. Monitoring the fat content of chocolate during the process of manufacture is used in many large factories. Operation of the instrument is simple and unskilled staff may be used. Very little sample preparation is required.

The initial cost of the apparatus is high by comparison with the analytical methods previously mentioned. However, where frequent estimations are to be made, the saving in personnel and time by the use of NMR is significant.

Further information is available from Newport Instruments Ltd., Milton Keynes, England.

## Determination of Cooling Curve of Cocoa Butter and Similar Fats

The meaning of "cooling curve" has been explained in Chapter 3 on cocoa butter and other fats. The method described below indicates the need for precise control of all operations.

Most of the large fat manufacturing companies have fully mechanized the method and have a number of units working simultaneously. They are used as part of the control of production and quality control.

The Shukoff-De Zaan modification uses a vacuum-jacketed tube, recording thermometer, and printout chart. A diagram of the Shukoff tube is shown in Fig. A.12. The method is described by Meursing.

### Method—Apparatus (Fig. A.13)
1. Sample tube: 6 in. by 1 in. light-walled, rimmed Pyrex test tube conforming with BS 3218: 1960 and closed with a stopper

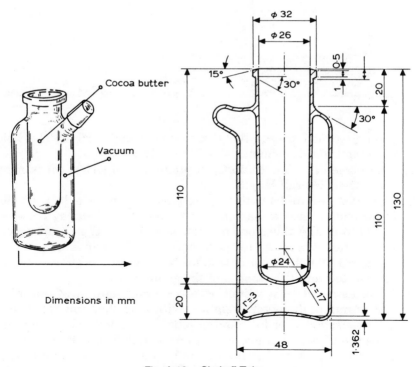

Fig. A.12.   Shukoff Tube

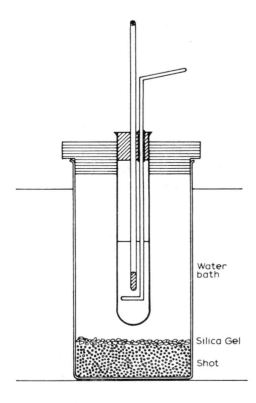

Fig. A.13.  Cooling Curve Apparatus

Water bath

Silica Gel

Shot

through which passes the thermometer and stirrer. The hole carrying the thermometer should be a loose fit.

2. Thermometer: 0.5°C to 50°C, graduated in 0.1°C, 6-cm immersion, 35-cm overall length.

3. Stirrer: A 4-mm-diameter Pyrex glass rod with a loop at the lower end of such a diameter as to fall about halfway between the bulb of the thermometer and the side of the sample tube when the stirrer is in position. Overall length of stirrer, 190 mm ± 10 mm. A suitable glass bearing for the stirrer should be inserted into the stopper of the sample tube.

4. Air jacket: Tall-form 1-liter beaker, approximately 190 mm high and 88 mm in diameter, weighted with lead shot and fitted with a lid constructed of five laminations of $\frac{1}{4}$-in. felt. Three laminations are of such a diameter as to be a snug fit inside the beaker and two to rest on the top of the beaker. The five laminations should be glued together and bored to hold the sample tube

firmly and centrally in the air jacket. The inclusion of dried silica gel in the air jacket prevents misting during the test. The prepared air jacket should be immersed in a water bath to within 2 cm of the top, temperature controlled at $17 \pm 0.2°C$, and allowed to reach equilibrium by standing for 30 min before carrying out the test. The water bath should be made of glass or have a glass window so that the contents of the tube can be observed.

## Procedure

1. *Preparation of the fat for seeding.* Melt a representative portion of the sample in an oven at a temperature between 55 and 60°C, (131 and 140°F), and filter at the temperature through a dry filter paper. Cool the filtered fat, with occasional stirring, until the temperature falls to within 32 to 34°C (90 to 93°F) and then stir continuously either by hand or mechanically until the fat has the consistency of a paste. Immediately transfer to a vessel previously kept at a temperature between 15 to 22°C (59 to 72°F) and allow to stand at this temperature for not less than 24 hr before use as the seed in the determination.

2. *Cooling curve determination.* Transfer $15 \pm 0.1$ g of a representative sample of previously filtered fat into the sample tube, stopper the tube, and completely melt the contents in a separate waterbath at 50°C (122°F). Replace the stopper carrying the stirrer and thermometer and leave in the water bath at 50°C for not less than 15 min, stirring occasionally. Remove the tube, plus stirrer and thermometer, from the water bath, wipe dry the outside of the tube, and clamp it in air. Gently stir the sample until 40°C (104°F) is reached, at which temperature transfer the tube to the air jacket. Clamp the thermometer so that the bulb is central in the fat, and adjust the stopper so that it can slide up the thermometer and be raised sufficiently from the tube to insert the seed at a later stage without disturbing the siting of the thermometer. The stirrer can, if facilities are available, at this stage be attached to any suitable mechanical device adjusted to the required stirring rate.

   Stir the fat occasionally until 35°C (95°F) is reached. From this point note, and record, the temperature at 1-min intervals and stir the fat by two gentle strokes of the stirrer at each 15 sec interval, taking care not to break the surface of the fat with the stirrer loop. At 28°C (82°F) (see Note) quickly add 0.03 to 0.04 g of finely shaved flakes obtained by lightly scraping a well-grained sample of the fat prepared as described in procedure 1, and continue recording the temperature and stirring at the same rate as before, and additionally note the temperature at which the first definite sign of crystallization occurs. Finally, stop stirring when the increase in temperature per minute has just passed its maximum, but continue recording the temperature until five successive readings are the same. Plot the time/temperature curve on $\frac{1}{10}$ in. graph paper with time (1 min. = $\frac{1}{10}$ in.) along the horizontal axis and temperature (1°C = $\frac{1}{2}$ in.) along the vertical. At least two runs on any one sample should be carried out.

*Note* Should the fat under examination have solidification properties differing widely from those normal to cocoa butter, some modifications to the method may be desirable. Thus, if crystallization occurs at a temperature above 28°C, the seed should be added 2 to 3°C above the crystallization temperature; also, the temperature at which regular stirring and readings are commenced may need to be altered.

## The Melting Point of Fats

The methods of determining the melting point are many and varied. The results obtained depend on close adherence to detail in the method used. There are recognized standard methods in different countries, for example, the Wiley melting point in the United States. When comparisons are required, the method used must always be stated.

The capillary tube method below is widely accepted. Preparation of the sample is of great importance in order to obtain the fat in a stable condition before the melting point determination.

## Determination of Fusion, Slip and Clear points

**Preparation of Sample** Transfer 30 to 50 g of a representative sample of the fat to a small, clean, dry beaker. Melt the fat by warming in an oven until the temperature is 55 to 60°C (131 to 140°F). Filter the fat through a dried Whatman no. 41 paper, keeping the temperature during the filtration within 55 to 60°C. Cool the fat, with occasional stirring, until its temperature is 32 to 34°C (90 to 93°F) and then stir continuously until the first signs of cloudiness appear. When a pasty consistency is achieved, transfer the fat quickly to a mold that has previously been stored at 15 to 22°C (59 to 72°F). Store the fat at 15 to 22°C for not less than 24 hr before testing.

**Determination**

*Apparatus* Beaker of 400 ml capacity. Thermometer, reading to 0.1°C (calibrated against a standard thermometer). Capillary tubes, 5 to 6 cm long, 1.1 to 1.3 mm internal diameter, 1.4 to 1.7 mm outside diameter, washed in chromic acid, distilled water, and then dried. Mechancial stirrer. Cotton wool. Rubber bands.

*Procedure.* Roll a wisp of cotton wool between the thumb and forefinger and introduce it into the capillary. Push it along with a piece of wire until it is 2 cm from the end. Lightly compress it in this position by simultaneously pressing with a piece of wire from the other end. The function of the cotton wool plug is to retain the fat (after it has "slipped"), below the level of the water bath, so that the clear point can be determined on the same sample.

Press the capillary (cotton wool end) into the fat to obtain a plug of fat 1 cm long in the tube. Attach the capillary to the thermometer with two small rubber bands so that the plug of fat is coincident with the bulb of the thermometer.

Into the beaker pour previously boiled and cooled distilled water to a depth of not less than 6 cm and clamp the thermometer centrally in the beaker so that the lower end of the capillary is 4 cm below the surface of the water. Fit the mechanical stirrer and heat the water so as to obtain a heating rate of 0.5°C per minute as the slip point is approached, that is, from about 5° before the slip point. Note and record:

1. The point at which softening is first seen = fusion point.
2. The point at which the fat starts to rise in the tube = slip point
3. The point at which the fat clarifies = clear point.

## Softening Point of Fats

**Barnicoat Method[1]**

*Principle.* The procedure, which is based on the ring and ball technique for testing bitumen, consists of recording the temperature at which a 3-mm ($\frac{1}{8}$-in.) steel ball penetrates halfway down a column of fat the temperature of which is being raised at the rate of 0.5°C per minute.

[1] Analyst, 69, 176.

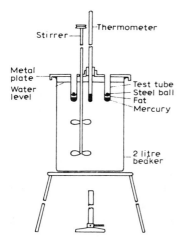

Fig. A.14. Barnicoat Apparatus

*Apparatus.* Beaker—2 liter squat. Metal plate, cut circular to fit over the top of the beaker, is provided with a central aperture for a thermometer, and has a number of symmetrically drilled holes in which test tubes containing fat are suspended by their lips. A special hole is drilled for the shaft of a mechanical stirrer. Lipped test tubes—thin walled, 5 cm long, 1 cm external diameter, 3-mm ($\frac{1}{8}$-in) steel balls (ball bearings). Standard titer thermometer. Mercury. Mechanical stirrer.

*Assembly of Apparatus.* Place the beaker on a wire gauze on a tripod, fill with distilled water at 20°C, and put the thermometer, plate, and stirrer in position. For details refer to Fig. A.14.

*Process.* Place 0.5 ml of mercury in a 5 by 1-cm test tube, cool the tube and contents for 5 min in crushed ice and water, pour onto the mercury 1 ml of the melted fat, leave the tube in icewater for 15 min.[1] A number of determinations equal to the number of test tubes that the cover of the apparatus will hold may be carried out simultaneously. Place a 3 mm ($\frac{1}{8}$ = in.) ball bearing in each tube in the depression of the fat surface that forms when the fat cools. Put the tubes in position in the holes in the plate and adjust the height of the thermometer bulb so that it is level with the fat column in each tube. Start the determination with the temperature of the bath at

---

[1] In the original Barnicoat method, the tube was held in ice water for 30 min and then placed in a refrigerator overnight. It was found that this procedure made the test too long for routine purposes and that the test could be modified as described.

20°C (68°F), and maintain it at this temperature for 15 min. (This was reduced to 15 min from the original time of 30 min.) Then raise the bath temperature at the rate of 0.5°C per minute, stirring vigorously. Record the temperature at which the steel ball has fallen halfway through the fat column. This is the softening point.

## Hardness of Fats

**Penetrometer Method**   The penetrometer (Fig. A.15) is that used for the testing of bitumen and waxes. It can be applied to fats, chocolate, and coatings. The degree of penetration of a needle or cone is measured at different temperatures and the results are considered in conjunction with melting point, dilatation, and NMR.

## Foam Test—Modified Bickerman Method

The foam test is useful for checking the foaming properties of syrups made from sugar, glucose syrup, and invert sugar when used for hard candies and is particularly useful if a raising agent such as sodium bicarbonate is added to the high boiled syrup.

High foam values indicate the presence of traces of foam-forming

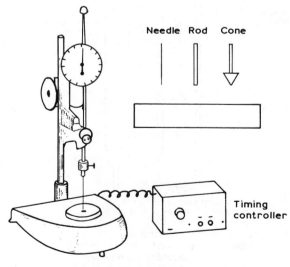

Fig. A.15.   "Seta" Penetrometer Showing Needle, Cone, and Timing Controller

substances such as protein, mucilage, and saponins—they can be present in badly refined beet sugar and some low-grade glucose syrups. High ash contents in sugar promote foam formation. High foam values in the ingredients give dense aeration with small bubbles and low bulk density of an aerated product.

Low foam values can be given by the presence of traces of fat or fatty acids, usually some form of contamination from machinery or containers. Cane sugar, if badly refined, may contain traces of cane wax.

These "antifoams" are very detrimental in the manufacture of aerated confections and they either cause complete destruction of the aeration or give coarse aeration with large bubbles (see "Aerated Confectionery").

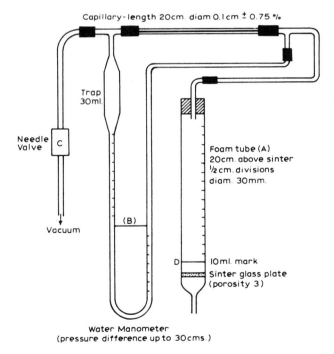

Capillary-length 20cm. diam 0.1cm ± 0.75 %

Trap
30ml.

Needle
Valve C

Vacuum

(B)

Foam tube (A)
20cm. above sinter
½ cm. divisions
diam. 30mm.

D — 10ml. mark
Sinter glass plate
(porosity 3)

Water Manometer
(pressure difference up to 30cms.)

Fig. A.16.   Foam Test Apparatus
Unclip tube A, empty, and swill out with hot water before
conducting the next test.

Clark, and Ross (1940). *Ind. Eng. Chem.* (32), p. 1594

**Apparatus**   Foaming apparatus (Fig. A.16) includes 150-ml beakers and watch glasses, stirring rods, and a refractometer.

*Cleaning of Apparatus*   The beakers, watch glasses, glass rods, and sintered glass foaming tube are cleaned by immersion in strong chromic acid for at least 20 min and then rinsed with distilled water and allowed to drain on clean filter paper.

**Procedure**   Weigh 20 to 25 g of glucose or sugar (or 35 g of sugar/glucose syrup) into a beaker and add 85 ml of distilled water. Add a few clean glass beads, gently warm, and stir until the sample dissolves. Then cover the beaker with a watch glass and boil the solution for exactly 3 min. Cool the solution rapidly to 20°C (68°F) and adjust the concentration to 20 ± 0.5 percent using recently boiled and cooled distilled water.

Swill out the tube (A) with a little of the solution, and replace the tube in its retaining clips.

Set the manometer levels to zero by adjusting the sliding scale (B) and then slowly open the valve (C) controlling the vacuum until a manometer head of about 8 cm is obtained.

Remove the foaming tube and pour in the sample to the mark (D), while sealing the lower outlet from the foaming tube with a finger.

Replace the foaming tube firmly on the rubber bung and adjust the pressure until the correct manometer head (12.5 cm for glucose and 30 cm for sugars) is obtained, when the timing period is started. The introduction of the sample into the foaming tube and the pressure adjustment should be completed within 1 min.

Maintain the pressure, 12.5 cm (or 30 cm for sugar), for the duration of the test, and note the maximum head of foam after each min.

The foam value is the height in centimeters of the foam after 10 min (the record of minute readings is used only to indicate the stability of the reading).

After the test, switch off the vacuum valve, and when the bubbles in the foaming tube have subsided, remove the rubber bung.

## WATER ACTIVITY, EQUILIBRIUM RELATIVE HUMIDITY

Equilibrium relative humidity (ERH) has been mentioned in several parts of the book. The ERH or relative vapor pressure is the humidity at which a food neither gains nor loses moisture and the figure is expressed as a percentage.

"Water activity" (A/W) is the expression now more often used but the significance is no different. It is based on a unit of one instead of percentage; as an example, an A/W of 0.65 is the same as an ERH of 65 percent.

The A/W of a food, including confectionery, has a marked bearing on its shelf life and hygroscopicity. As an example, take two confections at opposite ends of the range of A/W—hard candy at 0.25 to 0.30 and fondant creme at 0.65 to 0.75. In a temperate climate, the hard candy will almost always absorb moisture whereas the fondant will most likely dry out. Confectionery products with A/W of 0.75 and over will become vulnerable to the action of microorganisms and molds. Figure A.17 is a diagram giving an approximate relationship between A/W and microactivity but formulation variations in the confections will create some differences.

Moisture content and syrup phase concentration (mentioned elsewhere) do not have a definite relationship with A/W. For example, in a cereal with a moisture content of 12 to 14 percent, the A/W will be about 0.65 whereas an oil seed of the same A/W will have a moisture content of about 8 percent. Some syrups or preserves with a moisture content of 25 to 30 percent will also have an A/W of 0.65.

As with saturated salt solutions mentioned later, the vapor pressure (and hence the A/W) of a solution of sugars depends on the actual substances in solution as well as the concentration.

## Determination of A/W

A simple practical method for the determination of A/W (Fig. A.18) makes use of saturated solutions of various salts that have different vapor pressures (see Table A.3).

Modern apparatus consists of thermohygrometers plus cells that contain the product under test. The Protimeter described later is an example.

## Calculation of A/W/ERH

For relatively simple formulations where the syrup phase contains sugar, glucose syrup, and invert sugar only, a calculation of ERH may be made using the Money and Born equation. These workers showed that if the concentrations of dissolved substances are related

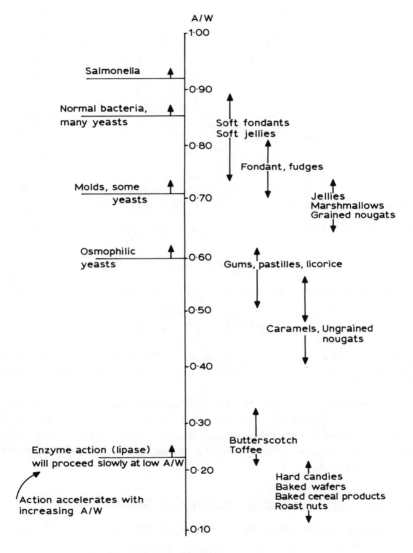

Fig. A.17.   Relationship Between Microbial Action and Water Activity

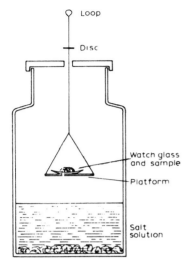

Fig. A.18. Apparatus

to 100 parts of water by weight, then the equation becomes

$$\% \text{ ERH} = \frac{100}{(1 + 0.27N)}$$

where N is the total number of mols of dissolved substance. (Mol is the unit of weight of a compound equal to its molecular weight in grams.)

Subsequently, Norrish (1964) produced a nomogram and a series of articles by Cakebread (1970) gave further explanations of the principles involved in the calculation. For complicated formulations, it is generally more satisfactory to estimate the ERH by one of the methods quoted.

## Determining A/W Using Solutions of Salts

This is a simple inexpensive means of obtaining the A/W value of confectionery.

**Apparatus** An apparatus (Fig. A.18) is used consisting of a lidded glass jar containing a saturated solution of the salt selected. A small hole is drilled in the lid and through this is passed a fine wire terminating in a small, stainless-steel platform. The top of the wire is

TABLE A.3. RELATIVE VAPOR PRESSURE OF SATURATED SOLUTION OF SALTS. EXPRESSED AS PERCENT RELATIVE HUMIDITY AT TEMPERATURES 0°C TO 50°C (32°–122°F).

| Solid salt present | | 0°C 32°F | 5°C 41°F | 10°C 50°F | 15°C 59°F | 20°C 68°F | 25°C 77°F | 30°C 86°F | 35°C 95°F | 40°C 104°F | 45°C 113°F | 50°C 122°F |
|---|---|---|---|---|---|---|---|---|---|---|---|---|
| * Lithium chloride | $LiCl \cdot H_2O$ | 14.7 | 14.0 | 13.3 | 12.8 | 12.4 | 12.0 | 11.8 | 11.7 | 11.6 | 11.5 | 11.4 |
| Potassium acetate | $CH_3COOK \cdot 1\frac{1}{2}H_2O$ | | | | 23.0 | 22.9 | 22.7 | | | | | |
| * Magnesium chloride | $MgCl_2 \cdot 6H_2O$ | 35.0 | 34.6 | 34.2 | 33.9 | 33.6 | 33.2 | 32.8 | 32.5 | 32.1 | 31.8 | 31.4 |
| Chromic acid (chromium trioxide) | $CrO_3(H_2Cr_2O_7)$ | | | | 38.7 | | 39.5 | | | | | |
| Potassium carbonate | $K_2CO_3 \cdot 2H_2O$ | | | | 44.3 | 44.0 | 43.7 | | | | | |
| Potassium nitrite | $KNO_2$ | | | | 50.0 | 49.1 | 48.2 | | | | | |
| * Magnesium nitrate | $Mg(NO_3)_2 \cdot 6H_2O$ | 60.6 | 59.2 | 57.8 | 56.3 | 54.9 | 53.4 | 52.0 | 50.6 | 49.2 | 47.7 | 46.3 |
| * Sodium dichromate | $Na_2Cr_2O_7 \cdot 2H_2O$ | 60.6 | 59.3 | 57.9 | 56.6 | 55.2 | 53.8 | 52.5 | 51.2 | 49.8 | 48.5 | 47.1 |
| Sodium bromide | $NaBr \cdot 2H_2O$ | | | | 60.5 | 59.3 | 57.8 | | | | | |
| Sodium nitrite | $NaNO_2$ | | | | 66.2 | 65.2 | 64.2 | | | | | |
| Ammonium nitrate | $NH_4NO_3$ | 77.1 | 74.0 | 71.0 | 68.0 | 64.9 | 61.8 | 58.8 | 55.9 | 53.2 | 50.5 | 47.8 |
| * Sodium chloride | $NaCl$ | 74.9 | 75.1 | 75.2 | 75.3 | 75.5 | 75.8 | 75.6 | 75.5 | 75.4 | 75.1 | 74.7 |
| * Ammonium sulfate | $(NH_4)_2SO_4$ | 83.7 | 82.6 | 81.7 | 81.1 | 80.6 | 80.3 | 80.0 | 79.8 | 79.6 | 79.3 | 79.1 |
| Potassium chromate | $K_2CrO_4$ | | | | 86.8 | 86.6 | 86.5 | | | | | |
| * Potassium nitrate | $KNO_3$ | 97.6 | 96.6 | 95.5 | 94.4 | 93.2 | 92.0 | 91.7 | 89.3 | 87.9 | 86.5 | 85.0 |
| Ammonium dihydrogen phosphate | $NH_4H_2PO_4$ | | | | 93.7 | 93.2 | 92.6 | | | | | |
| * Potassium sulfate | $K_2SO_4$ | 99.1 | 98.4 | 97.9 | 97.5 | 97.2 | 96.9 | 96.6 | 96.4 | 96.2 | 96.0 | 95.8 |

Figures marked with * are from Wexler, A., & Saburo, H. (1954). Relative humidity temperature relationships of some saturated salt solutions. *J. Res.*, National Bureau of Standards.

provided with a loop and a small flat metal disk so that the platform may be suspended close to the surface of the salt solution at the same time as the hole in the lid is covered by the disk. The loop at the top of the wire is attached to the balance pan hook for weighing when the platform (with sample) hangs free in the bottle, the suspension wire not touching the edge of the hole in the lid.

**Method**  A number of bottles are required for test purposes, and saturated salt solutions are prepared from Table A.3, according to the samples being tested. Assemble the bottles required for test, and these must be absolutely clean.

Each sample must be tested at four different A/W values, two above and two below the expected value, and the difference between any two consecutive values should not exceed 5 percent.

Into each bottle introduce 50 ml of a saturated salt solution of known A/W and adjust the wire holding the platform so that the latter is about 19 mm ($\frac{3}{4}$ in.) above the level of the liquid when the small disk rests on the lid. The platform carries a small watch glass for the sample.

Allow the apparatus to stabilize at 18°C (65°F) (or other testing temperatures) for at least 24 hr.

Immediately before testing, the sample is prepared in such a manner that equilibrium is established as quickly as possible, e.g., caramel and fudge should be shaved and hard candies quickly ground to a coarse powder.

Weigh the empty watch glass, wire support, and platform in each jar—this may be done by placing the jar on a "bridge" over the pan of a single-pan balance, and attaching the upper end of the wire support to the hook on the balance arm.

Add about 1 g of the prepared sample to the watch glass, and immediately reweigh.

Allow the apparatus to stand at 65°F (or other temperature) for 48 hr, and then reweigh the sample, watch glass, wire support, and platform in each jar, and record the change in weight.

Weigh again after further 24-hr periods of stabilization, until there is no significant change in weight.

Calculate the percentage change in weight for each A/W value, and plot a graph of percentage change in weight against A/W.

From the graph, read off the A/W corresponding to zero change in weight. This value is the A/W of the sample.

If many estimations are to be made, small cabinets containing the salt solutions may be used and small dishes to contain the confection

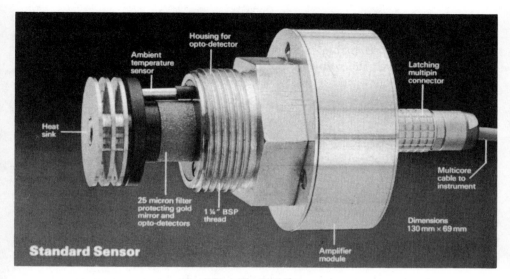

Fig. A.19.    Standard Sensor

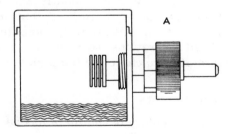

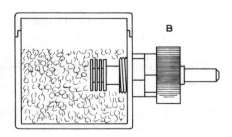

Fig. A.20.   Arrangement of Sensor in ERH
Cell for Measurement of A/W ERH.
A. With Liquid, Sticky, or Lumpy Materials
   or
B. With Clean Granular Materials for
   Quickest Results.
*The use of the ERH cell.* The instrument
may be used to measure the ERH by
placing a sample in the ERH cell or other
suitable container.

*Protimeter Ltd., Meter House, Marlow, Bucks,*
*England*

under test. They should be fitted with a lid for weighing. With cabinets it is advisable to have a small internal fan to provide air circulation over the solution.

## A/W Determination by Protimeter

This is an electronic thermohygrometer (Protimeter Ltd, Marlow, England) with digital readout. It is capable of determining water activity, percent relative humidity, dew point, and moisture content.

For determination of A/W (ERH), a specially designed sensor is placed in a cell, as shown in Figs. A.19 and A.20.

## SCIENTIFIC INSTRUMENTS EMPLOYED IN PRODUCTION DEPARTMENTS

Where results need to be obtained rapidly in order to monitor production processes, instruments are installed in the production department.

The first development employed instruments that gave visual readings and these were then used to make manual adjustments to the equipment. Subsequently, the instruments were designed to feed back signals to controllers that made the necessary adjustments. Now many plants are fully computerized with monitoring and control of all operations such as temperatures, steam pressures, time, formulation, chocolate tempering, and viscosity.

### Viscometers

The rheology of chocolate has been the subject of much research in recent years. The need to work with chocolate with low cocoa butter contents for reasons of economy has initiated a study of plastic viscosity and yield value.

The simpler forms of viscometer suitable for fluid chocolates do not provide the information necessary when working with more viscous products.

Viscometers are described in detail in Chapter 4; the following is a summary.

**Redwood type** These are flow-through instruments with the orifice dimensions similar to the no. 2 Redwood viscometer. The method is empirical and standards are prepared to suit the chocolate recipes in a particular factory. These instruments are used only for fluid chocolates. The Ford Cup viscometer employed in the paint industry has also been used for chocolate.

**Falling Ball Viscometer**   This instrument is useful for determining the consistency of thick chocolates used for molding, piping, and drops. It is less accurate than the Redwood viscometer but more precise results can be obtained by using a falling cone with ring markings similar to the ball. (See also "Mobilometer.")

**Rotational Viscometer**   The viscometer that was accepted for many years in the United States is the McMichael rotational viscometer. It was adopted by the National Confectioners Association of America as a simple, relatively inexpensive instrument, but of recent years has been the subject of much criticism for not providing all the rheological data required.

The Brookfield and Haake viscometers are being increasingly used in many companies.

## Refractometers

Refractomers are now indispensable for control in a confectionery factory. They have replaced thermometers for determining the end point of boilings of syrups, jellies, and jam, and in conjunction with the syrup press (see "Confectionery Processes"), the syrup phase concentration of fondants, pastes, and marzipan. The following is a brief description of the different types of instrument.

**"Pocket" Refractometer**   This telescopic instrument with hinged prisms at one end is portable [weight 241 g ($8\frac{1}{2}$ oz) length 6.7 in.] and relatively inexpensive.

It is made in several ranges and can be applied to a large variety of products as well as syrups. As an example, it is used for checking sugar content, and hence the ripeness, of apples, beets, and potatoes. It is particularly useful for examining the progress of sugar penetration during the preservation of fruits.

**Abbé Refractometer**   Generally regarded as laboratory instruments, they are now made with a very robust construction, and with moderate care can be used for production work. The refractive index range is 1.300 to 1.740 or sugar scale 0 to 95 percent. A flow-through model is also available. In this model, a water-jacketed compartment, which replaces the conventional hinged prism, is fitted with a funnel and antisyphon drain tube to waste. Alternatively, the funnel and drain tube can be replaced with couplings to enable the flow cell to be introduced into a laboratory pilot plant in order to carry out refractive index measurements of flowing liquids. Water circulation, providing temperature control, is taken from the fixed prism box without the need for interconnecting tubes.

**Immersion Refractometer**  This instrument may be used for laboratory or process work in the confectionery and food industries.

Since the accuracy of measurement is high, it is necessary to restrict the range of the instrument to a comparatively small span of refractive index, but by changing the prism to one of another range, the instrument can be altered with no loss of accuracy. There are several ranges of prisms available. For absolute measurements, the accuracy will depend upon the precision to which the instrument has been standardized, and for this reason the principal use of the refractometer is in differential measurements involving comparison of refractive index between known and unknown solutions where the differences are small.

The body of the instrument consists of a rigid tube covered with leather to provide a safe grip. At the upper end is an eyepiece, through which is observed an evenly divided scale. A drum enables the scale to be moved by one scale division. The drum has ten divisions, thus enabling the scale to be subdivided to one-tenth. The critical angle borderline, due to the difference in refractive index between the prism and the sample, appears against the scale and is observed as a division of the circular field of view into a light and a dark portion. The point on the scale where the demarcation line intercepts it provides the scale reading, from which the refractive index of the sample can be found from the calibration table supplied with each prism.

At the lower end of the body is the dispersion circle, which is adjusted to eliminate color from the demarcation line and below this the prism, which must be immersed in the solution to be measured. The illumination, which must be external to the instrument, may be provided by a daylight window or an electric lamp.

To protect the prism, and as a sample chamber, a metal beaker is provided that is partly filled with the sample so that the refractometer may be conveniently used in the hand and directed toward the light source.

There are two modifications of the immersion refractometer. One has a jacketed prism box mounted on the lower end of the instrument and a few drops only of the sample are required. The other has a cell attachment through which the liquid under examination flows and can therefore be continuously monitored.

**Projection Refractometer**   This refractometer has many applications in the confectionery industry and works on the principle of internal reflection. It is large and robust (length 2 ft 5 in., width 6 in., height 9 in.) and generally no cooling is required as the bulk of the instrument is sufficient to cool the small sample to ambient temperature. For certain purposes, a cooling attachment can be provided.

*Method of Measurement*   A critical angle borderline, observed as the edge of a shadow intercepting the scale, provides a reading directly in terms of sugar percentage. The illumination is provided by one of two 6-volt lamps mounted on a shaft terminating on a milled knob on the side of the instrument. Either lamp may be brought into the illumination position with full adjustment to provide (1) diffused white light or (2) filtered orange illumination; the latter to reduce coloration at the edge of the shadow.

The supply to the lamp may be provided from an external transformer or battery. The instrument is therefore used only at low voltage and is safe in damp conditions. The scale is within the body of the instrument and is viewed through a window on the top surface of the instrument.

The sugar concentration of fondants, marzipan, or gums may be determined regardless of color, optical density, pips, or other hard particles. The sample is merely spread on to a horizontal glass surface, the illumination adjusted, and the sugar percentage read off directly from a scale without the need for magnification or eyepieces. After measurement, the material is sponged off the surface, which is dried with a soft cloth and is then ready to receive the next sample.

**Process Refractometers**   Two types are available, the pan refractometer and pipeline refractometer.

*Pan Refractometers*   These are constructed for direct clamping to a boiling pan in which a cooking or evaporation process is carried out. The refractive index, expressed as percent solids, usually sugar, is continuously displayed on the screen at the front of the instrument and may be seen at a glance. A wiper mechanism is provided whereby the material in contact with the measuring prism may be changed periodically, thus bringing a fresh layer forward. The wiper is drilled for water supply in order that washing in addition to wiping of the prism surface may be carried out while the pan is in operation. The instrument employs internal reflection principles and is thus capable of measuring optically dense materials. A built-in light source is provided powered from an external supply of 6 volts at 0.5 ampere either battery or transformed mains supply; the refractometer is therefore electrically safe even under wet conditions.

*Pipeline Refractometers*  These are designed for direct installation in a stainless-steel pipeline and give continuous readings.

In one instrument, the principle employed is the same as that used in the immersion refractometer and the reading depends on light transmitted through the material in the pipe. Another instrument uses the method of reflection and this can be used for opaque liquids.

*Automatic Refractomers, Electronic Refractometers*  These instruments for continuously recording refractive index are used for special production processes. They are expensive by comparison with those already described.

## Specific Gravity, Density

The determination of density or specific gravity of syrups was a traditional method of check by the craft confectioner, and the scaled hydrometer is too well known to warrant description. Continuous methods of density checking and adjustment have been developed that rely on the change of weight of a column of the liquid in a flexible U-tube.

For checking the density of whipped confectionery products, a simple cylindrical container with a wide mesh wire bottom has proved useful. The whip is filled into the container until it exudes from the wire mesh, when it is scraped off flat at both ends with a knife, and the whole weighed. The tare of the container being known, the density is quickly determined. This container avoids the formation of voids that are liable to form in a cylinder with a solid bottom.

There are many physical methods for the determination of density. These are described in scientific textbooks.

## Temperature Measurement

**Thermometers**  Temperature measurement is still the means of controlling the concentration of sugar syrups for hard boilings and other syrups that do not contain interfering ingredients such as gelatin, pectin, or solid substances.

Accurate temperature measurement and control are very important in chocolate tempering and cooling. The following types of thermometer are in use.

**Mercury-in-Glass Thermometers**   These are reliable and generally retain their accuracy well. For production use they are usually provided with protective cases that prevent the bulb from being broken. The protection should be an open, thick wire guard at the bulb end, as a heavy metal sheath greatly reduces sensitivity and results in errors.

Some cookers have built-in thermometers that may be encased in heavy metal stirrers. Such protection can lead to greatly retarded readings and considerable inaccuracy in determining the end point.

"Standard" mercury-in-glass thermometers are available, certified by the National Physical Laboratory, England. It is good practice to have these on hand in the laboratory for checking other thermometers used in pilot plant and production. They should *not* be used for general purposes.

**Mercury-in-Steel Dial Thermometers**   These make use of the principle of expansion of mercury in a bulb immersed in the liquid of which the temperature is required. The expansion is transmitted through a steel tube to a dial indicator. With careful handling, these thermometers are reliable but may get out of adjustment and therefore need constant checking.

**Air Expansion Thermometers**   These operate dials by the expansion of air in a bulb. These can be very unreliable and should not be used for any purpose where consistent accuracy is required.

**Thermocouples and Electric Resistance Thermometers**   The use of these instruments has greatly increased and their particular value is great sensitivity due to the small size of the probe compared with the bulb of the other thermometers described. They are also very suitable for operation with recording charts.

There are many portable thermocouple thermometers that are very sensitive and reliable with careful handling. Unfortunately, they are often abused and become inaccurate. They should be regularly checked against a standard mercury-in-glass thermometer.

**Telemetric Methods**   For measuring temperature inside tunnels, coolers, etc., radio telemetric instruments are available. These emit signals that are transmitted to a receiver and no wire or tube connections are necessary. This method is very useful for monitoring the temperature on moving belts in enclosed spaces.

Several companies specialize in the manufacture of these thermometers and they will provide all the information required for any special process.

*Infrared Thermometers* These thermometers measure product temperature accurately without contact. They are particularly valuable for measuring the temperatures of very viscous substances, which tend to clog thermometer probes. A telescopic system with spotlight ensures that the point at which the temperature is measured is always the same.

**Checking Thermometers** In spite of assurance by instrument manufacturers, errors in thermometers are not uncommon. All thermometers received at the factory must be checked. For this purpose, a thermostatically controlled oil bath with stirring device may be used and it should be large enough for a number of thermometers to be tested at one time. The thermostat is reset periodically at different temperatures so that the thermometers can be checked at specific temperatures throughout the range.

The checking of "in plant" thermometers is also very important and regular inspection must be made by plant superintendents. When errors are found, the thermometer should be immediately replaced—it is very bad practice for the plant operator to have to make corrections to an observed temperature or to attach a label to the thermometer showing the correction to be applied.

"Certified" thermometers, previously mentioned, should be used for checking.

## Electronic Moisture Meters

Electronic testers are used in many industries for the determination of moisture in powders or granulated products, such as flour and grain.

In the confectionery industry, the one main application is to check the moisture content of molding starch, either in the depositing plant or from the driers.

Most instruments used are based on electrical capacitance, and the material to be tested is contained in a cell of standard dimensions. It is important, for consistent results, that the degree of packing be constant and most meters incorporate a means of achieving this.

Nevertheless, periodic checks are necessary by gravimetric methods, and if this is done, quick, accurate results are possible for one type of material. The meters must be calibrated for each product or raw material.

Other electronic methods for moisture determination during production are also used. (See References.)

## Hygrometers

It is often necessary to obtain records of relative humidity conditions in factory departments such as stockrooms, starch rooms, or crystallizing departments. The simplest reliable instrument is the whirling or sling hygrometer used in conjunction with the *correct* psychrometric tables.

The Mason type of wet and dry bulb hygrometer is notoriously unreliable unless provision is made for reasonable air movement around the bulbs. These hygrometers suspended flat against a wall in stagnant air are utterly useless.

Where continuous records are required, the recording hair hygrometer gives the best results but needs checking and adjustment regularly. The wet and dry bulb recording hygrometer is, like the Mason hygrometer, unreliable unless in good air movement. The wet bulb consists of a bimetallic coil with a wet fabric cover. The hair hygrometer is much more sensitive.

The instrument that is regarded as a standard reference hygrometer is the Assmann psychrometer. This consists of sensitive wet and dry bulb thermometers contained in a cylinder fitted with a small fan, which ensures that ambient air is drawn over the bulbs at a standard velocity.

Another type of instrument is the paper or sword hygrometer. This can be inserted between layers of paper or strawboard, and by recording the relative humidity, the equilibrium humidity, and hence the moisture content of the board, can be determined.

## Nuclear Magnetic Resonance

This instrument was described earlier in the chapter.

## REFERENCES

Apparatus for determinations of Isotherms for water vapour sorption in foods. Department of Food Science, Agriculture University, Wageningen, Holland.

British Standard BS 3406: Part 5. 1983. *British Standard Methods for Particle Size Distribution. Part 5. Recommendations for Electrical Sensing Zone Method (the Coulter Principle)*, 33 pp.

Cakebread, S. H. 1970. *Mafg. Confect.* 50(11), 36; 50(12) 42; 51(1) 25.

Chemical analysis and its application to candy technology. *Conf. Prod.* 161. London.

Coulter, W. H. 1956. *Proc. Nat. Elect. Conf.* 12, 1034 (Coulter Counter® is a registered trademark of Coulter Electronics Inc.)

*Coulter Counter* ® *Industrial Bibliography* 1986, (1521 references) Coulter Electronics Ltd., Luton, England.

*Instruments for the Food Industry,* British Food Research Association, Leatherhead, Surrey, England. (Series of leaflets on special equipment.)

Jacobson, A. *Chocolate—Use of N.M.R. in the Chocolate Industry.* Cloetta, Sweden.

Lockwood, H. C. 1958. A new method for assessing chocolate grinding. *Chem. & Ind.,* 1506–1507, Nov. 15.

Meursing, E. H. *Cocoa Butter, Quality and Analysis.* Cacaofabriek de Zaan, Holland.

Meursing, E. H. 1976. *Cocoa Powders for Industrial Processing.* Cacaofabriek de Zaan, Holland.

Money, R. W., and Born, R. 1951. *J. Sci. Fd. & Agric.* 2, 180.

Niediek, E. A. 1978. New equipment for determination of particle size. *Chocolate, Confectionery, Bakery.*

Norrish, R. S. 1964. *Conf. Prod.* 30, 769.

Reade, M G. 1971. Fat content by refractometer. *Rev. Int. Choc.* 26, 334–342.

Refractometers, electronic moisture meters. (From a series of 17 papers on Candy Analysis, B. W. Minifie.) Confectionery Production (1970 et seq) London.

Sutjiadi, I., and Niediek, E. A. 1974. Preparation methods for control of fineness (in German). *Gordian,* 284–291.

Van den Berg, C. 1983. *Description of Water Activity of Foods for Engineering Purposes.* International Congress, Dublin, Ireland.

Department of Food Science, Agriculture University, Wageningen, Holland.

Wiggins, P. H., Ince, A. D., and Walker, E., *Rapid determination of Fat in Chocolate and Related Products Using Low Resolution N.M.R.* Cadbury Schweppes, Bournville, Birmingham, England.

## Equipment

Baird and Tatlock, Romford, England (various equipment)
Bellingham and Stanley, Ltd., London (refractometers)
Raytek Inc., Mountain View, Calif. (infrared thermometers)

# Appendix II

# Resources

## BOOKS RELATED TO CHOCOLATE, COCOA, AND CONFECTIONERY SCIENCE AND TECHNOLOGY

Some of the books listed are out of print and are only available in company or municipal libraries. They are listed because they contain useful information, historic and technical, of interest to the student and technologist. Many contain formulations not available elsewhere and which serve as a basis for experimental work.

The reader is also referred to references at the end of each chapter.

*Chocolate Confectionery and Cocoa*, H. R. Jensen (1931). Blakistons, Son and Co. Ltd., Philadelphia, Pa., U.S.A.

*Chocolate and Confectionery*, C. Trevor Williams (1964). Leonard Hill Books, London, U.K.

*Chocolate Evaluation*, S. Jordan and K. E. Langwill (1934). G. Ferguson Co., New York, N.Y., U.S.A.

*The Problem of Chocolate Fat-Bloom*, R. Whymper (1933). The Manufacturing Confectioner Publishing Co., Glen Rock, N.J., U.S.A.

*Chocolate Production and Use.*, L. Russell Cook, revised by E. H. Meursing (1982). Harcourt Brace Jovanovich, New York, N.Y., U.S.A.

*Cocoa*. G. A. R. Wood and R. A. Lass (1985) (4th Ed.). Longman Group Ltd., Harlow, U.K.

*Cocao Fermentation*, A. W. Knapp (1937). J. Bale, Sons & Curnow Ltd., London, U.K.

*Candy Production*, W. L. Richmond (1948). The Manufacturing Confectioner Publishing Co., Glen Rock, N.J., U.S.A.

*Cocoa*, E. M. Chatt (1953). Interscience Publishers Inc., London, U.K.

*Cocoa and Chocolate Making*, H. C. J. Wijnoogst (1957). Mannhein, Amsterdam, Holland.

*Cocoa and Chocolate Manufacture (Hints)*, H. C. J. Wijnoogst (1938). Resultat Verlag, Amsterdam, Holland.

*Confectionery Problems,* S. Jordan (1930). The National Confectioners' Assn., Chicago, Ill., U.S.A.

*Processing of Raw Cocoa for the Market,* T. A. Rohan (1963). Food & Agriculture Organization of the U.N., Geneva, Switzerland.

*Complete Confectioner,* Skuse (1957). Bush, Boake, Allen, London, U.K.

*Sweet Manufacture,* N. F. Scarborough (1933). Leonard Hill Ltd., London, U.K.

*The Technology of Chocolate,* N. W. Kempf (1964). The Manufacturing Confectioner Publishing Co., Glen Rock, N.J., U.S.A.

*Twenty Years of Confectionery and Chocolate Progress,* C. D. Pratt et al. (1970) AVI Publishing Co., Westport, Conn., U.S.A.

*Cocoa and Chocolate Processing,* H. Wieland (1972). Noyes Data Corp., Park Ridge, N.J., U.S.A.

*Confectionery Products Manufacturing Processing,* M. Gutterson (1969). Noyes Data Corp., Park Ridge, N.J., U.S.A.

*Choice Confections.* W. Richmond, *The Manufacturing Confectioner,* Glen Rock, N.J., U.S.A. (reprint).

Pennsylania Manufacturing Confectioners Annual Production Conference Proceedings 1965 to date. *The Manufacturing Confectioner,* Glen Rock, N.J., U.S.A.

*Sugar and Chocolate Confectionery,* S. H. Cakebread (1975). Oxford University Press, London., U.K.

*Sugar Confectionery and Chocolate Manufacture,* R. Lees, and E. B. Jackson (1973). Specialized Publications Ltd., Surbition, U.K.

*A Basic Course in Confectionery,* R. Lees (1980). Specialized Publications Ltd., Surbiton, U.K.

*Faults, Causes, Remedies. Sweets and Chocolates,* R. Lees (1981). Specialized Publications Ltd., Surbiton, U.K.

*Candy Technology,* J. J. Alikonis (1979). AVI Publishing Co., Westport, Conn., U.S.A.

*The Art and Science of Candy Manufacturing,* C. D. Barnett (1978). Candy Industry, Duluth, Minn., U.S.A.

*The Complete Wilton Book of Candy,* E. T. Sullivan, and M. C. Sullivan (1983). Wilton Enterprises Inc., Woodridge, Ill., U.S.A.

*Cocoa Powders for Industrial Processing* (3rd ed.) E. H. Meursing (1983). De Zaan Co., Holland.

## Publications by Silesia-Essenzenfabrik, Gerhard Hanke K. G. Norf., West Germany

*Handbook for the Sugar Confectionery Industry,* A. Meiners and H. Joike (1969).

*Silesia Confectionery Manual No. 2,* K. W. Stock and A. Meiners (1973). (Specializing in Dragee production.)
*Silesia Confectionery Manual No. 3.*
> Volume 1/1. A. Meiners, K. Kreiten, and H. Joike (1983). Tables, bibliography, illustrations.
> Volume 1/11. A. Meiners, K. Kreiten, and H. Joike (1984). *International Dictionary of Biscuit Confectionery and Chocolate Industry Terms,* in English, French, German, Spanish.
> Volume 2. A. Meiners, K. Kreiten, and H. Joike (1984). *Handbook of the Confectionery Industry.*

## Textbooks in Other Languages

A comprehensive list of textbooks in Arabic, French, German, and Spanish is given in *Silesia Handbook No. 3,* Vol. 1/1. The following are specially mentioned.
*Guide Technologique de la Confiserie Industrielle,* Y. Fabry and P. H. Bryselbout (1985). SEPAIC, Paris, France.
*Handbuch der Kakaoerzeugrusse,* H. Fincke (1936, 1965). Springer-Verlag, Berlin, West Germany.
*Technologie Zuckerwaren,* Autoren Kollektiv. (1977). VEB Fach-buchverlag, Leipzig/G.D.R.

## Analytical

*Analytical Chemistry,* R. Belcher and C. L. Wilson (1964). Chapman & Hall, London, U.K.
*Instrumental Methods of Analysis,* H. H. Willard, L. L. Merritt, Jr., and J. A. Dean (1967). D. Van Nostrand Co. Inc., New York, N.Y., U.S.A.
*Absorption Spectrometry,* J. R. Edisbury (1966). Hilger & Watts, London, U.K.
*ICUMSA Methods of Sugar Analysis.* H. C. S. de Whalley (1964). Elsevier Publishing Co., New York, N.Y., U.S.A.
*Methods of Sugar Analysis,* C. A. Browne and F. W. Zerban (1941). John Wiley & Sons Inc., New York, N.Y., U.S.A.
*Polarimetry Saccharimetry and the Sugars,* Bates (1942). National Bureau of Standards, U.S. Dept. of Commerce, Washington, D. C., U.S.A.
*Laboratory Techniques in Chemistry and Biochemistry,* P. S. Diamond and R. F. Denman (1966). Butterworths, London, U. K.
*Laboratory Handbook of Chromatographic Methods,* O. Mikes (1961). D. Van Nostrand Co. Inc., New York., N.Y., U.S.A.

*Gas Phase Chromatography,* Vols. 1, 2, 3, R. Kaiser (1963). Butterworths, London, U.K.

*Confectionery Analysis and Composition,* S. Jordan and K. E. Langwill (1946). The Manufacturing Confectioner Publishing Co., Glen Rock, N.J., U.S.A.

*A Handbook of Laboratory Solutions,* M. H. Gabb and W. E. Latchman (1967). Andre Deutsch., London, U.K.

*Methods of Analysis* Association of Official Agricultural Analysts, Washington, and Association of Official Analytical Chemists (current editions). Washington, D. C., U.S.A.

*Quantitative Paper and Thin-Layer Chromatography,* E. J. Shellard (1968). Academic Press, London, U.K.

*Instruments for the Food Industry* (1973). Food Research Association, Leatherhead, U.K.

*Microbiological Methods,* C. H. Collins (1967). Butterworths, London, U.K.

*Recommended Methods for the Microbiological Examination of Foods,* J. M. Sharf (1966). American Public Health Assn. Inc., Washington, D.C., U.S.A.

*Food Science* (4th ed.), N. N. Potter (1986). AVI Publishing Co., Westport, Conn., U.S.A.

*Pearson's Chemical Analysis of Foods,* H. Egan, R. S. Kirk, and R. Sawyer (1981). Longman Ltd., Harlow, U.K.

*Advanced Sugar Chemistry,* R. S. Shallenberger (1982). AVI Publishing Co., Westport, Conn., U.S.A.

*Food Analysis— Theory and Practice,* Y. Pomeranz and C. E. Meloan (1978). AVI Publishing Co., Westport, Conn., U.S.A.

*Developments in Food Analysis Techniques,* R. D. King, Vol. 1 (1978), Vol. 2 (1980). Elsevier Applied Science Publishers, Barking, Essex, U.K.

## Fats and Oils

*Fats and Waxes,* T. P. Hilditch (1941). Baillière, Tindall & Cox., London. UK.

*Industrial Oil and Fat Products,* A. E. Bailey (1951). Interscience Publishers Inc., New York, U.S.A./London, U.K.

*Lipids and their Oxidation,* H. W. Schultz (1962). AVI Publishing Co., Inc., Westport, Conn., U.S.A.

*Oils, Fats and Fatty Foods,* K. A. Williams (1966). Churchill, London, U.K.

*Oil and Fat Analysis.* L. V. Cocks, and C. van Rede (1966). Academic Press, London, U.K.

*Food Oils and Their Uses,* T. J. Weiss (1982). AVI Publishing Co., Westport, Conn., U.S.A.

*Fats and Oils—Chemistry and Technology,* R. J. Hamilton, and A. Bhati (1981). Elsevier Applied Science Publishers, Barking, Essex, U.K.

*Analysis of Oils and Fats,* R. J. Hamilton, and J. B. Rossell (1986). Elsevier Applied Science Publishers, Barking, Essex, U.K.

*Hydrogenation of Fats and Oils,* H. B. W. Patterson (1983). Elsevier Applied Science Publishers, Barking, Essex, U.K.


## Milk Products

*Condensed Milk and Milk Powder,* O. F. Hunziker (1949). Hunziker. La Grange, Ill., U.S.A.

*Diary Chemistry,* J. G. Davis and F. J. MacDonald (1952). Griffin & Co. Ltd., London, U.K.

*Modern Dairy Technology,* R. K. Robinson (1986). Elsevier Applied Science Publishers, Barking, Essex, U.K.

*Fundamentals of Dairy Chemistry* (2nd ed.), B. H. Webb, A. H. Johnson, and J. A. Alford (1974). AVI Publishing Co., Westport, Conn., U.S.A.

*Chemistry and Testing of Dairy Products* (4th ed.), H. V. Atherton, and J. A. Newlander (1977). AVI Publishing Co., Westport, Conn., U.S.A.


## Hygiene, Pest Control

*Hygiene in Food Manufacturing and Handling,* B. Graham-Rack and R. Binsted (1964). Food Trade Press Ltd., London, U.K.

*Food Poisoning and Food Hygiene,* B. C. Hobbs (1968). Edward Arnold Ltd., London., U.K.

*Food Sanitation,* K. Longree (1967). Interscience Publishers Inc., New York, U.S.A./London, U.K.

*Pest Control,* W. W. Kilgore and R. L. Doutt (1967). Academic Press, London, U.K.

*Handbook of Pest Control,* Mallis (1964). MacNair-Dorland Co., New York, N.Y., U.S.A.

*Insects and Hygiene,* J. R. Busvine (1966). Methuen & Co. Ltd., London, U.K.

*Infestation Control, Cocoa, Chocolate and Confectionery Industry* (1970). Cocoa, Chocolate and Confectionery Alliance, London, U.K.

*The Lever Industrial Hygiene Plan,* Hygiene Advisory Service, Lever Industries Ltd., London.,U.K.

## Nuts

*Peanuts, Production, Processing, Products,* J. G. Woodroof (1966). AVI Publishing Co., Westport, Conn., U.S.A.

*Nuts—Their Production and Everday Uses.* F. N. Howes (1948). Faber & Faber Ltd., London, U.K.

*Tree Nuts. Production, Processing, Products* (2nd ed.), J. G. Woodroof (1979). AVI Publishing Co., Westport, Conn., U.S.A.

*Coconuts, Production, Processing, Products* (2nd ed.), J. G. Woodroof (1979). AVI Publishing Co., Westport, Conn., U.S.A.

## Food, Nutrition, Quality Control

*The Composition of Foods,* McCance and Widdowson (revised by A. A. Paul and D. A. T. Southgate) (1985). Her Majesty's Stationery Office, London, U.K.

*Enyclopedia of Food Science,* M. S. Peterson and A. H. Johnson (1978). AVI Publishing Co., Westport, Conn., U.S.A.

*Quality Control in the Food Industry,* A. Kramer and B. A. Twig (1970). AVI Publishing Co., Westport, Conn., U.S.A.

*Rancidity in Foods,* J. C. Allen, and R. J. Hamilton (1983). Elsevier Applied Science Publishers, Barking, Essex, U.K.

*Introductory Food Science,* D. B. Smith and A. H. Walters (1967). Classic Publications Ltd., U.K.

*Dictionary of Nutrition and Food Technology,* A. E. Bender (1965). Butterworths, London, U.K.

*Food Microbiology,* W. C. Frazier (1967). McGraw-Hill Book Co., New York, N.Y., U.S.A.

*Chemistry, Medicine and Nutrition,* Royal Society of Chemistry Symposium (1966). Russell Sq., London, U.K.

*Quality Control in the Food Industry,* S. M. Herschdoerfer (1967). Academic Press, London, U.K.

*Odors and Taste* (1986). American Society for Testing and Materials, Philadelphia, Pa., U.S.A.

*The Carbohydrates,* W. Pigman (1957). Academic Press Inc., London, U.K.

*Pectic Substances,* J. J. Doesburg (1965). Wageningen, Holland.

*Pectic Substances,* Z. I. Kertesz (1951). Interscience Publishers Inc., New York, N.Y., U.S.A.

*The Starch Industry,* J. W. Knight (1969). R. Maxwell Publications, London, U.K.
*Chemistry of Essential Oils,* Vols. 1 and 2, E. J. Parry (1918). Scott, Greenwood & Son, London, U.K.
*The Essential Oils,* Vols. 1 and 2, E. Guenther (1948, 1949). Van Nostrand Inc., New York, N.Y., U.S.A.
*Unit Operations in Food Processing,* R. L. Earle (1966). Pergamon Press, London, U.K.

## Packaging

*Introduction to Packaging* (current edition). British Standards Institution. (British Standards Packaging Code: 1133 Sections 1–21.) London, U.K.
*Fundamentals of Packing,* F. A. Paine (1962). Blackie & Son Ltd., London, U.K.
*Adhesives Guide,* J. Hurd (1960). British Scientific Instrument R. A., London, U.K.
*Protective Wrappings,* C. R. Oswin (1954). Cam Publications Ltd., Cambridge, U.K.
*Principles of Package Development* (2nd ed.), R. C. Griffin, S. Sacharow, and A. L. Brody (1985). AVI Publishing Co., Westport, Conn., U.S.A.
*Food Packaging Materials—Analysis and Migration of Contaminants,* N. T. Crosby (1981). Elsevier Applied Science Publishers, Barking, Essex, U.K.
*Modern Food Packaging Film Technology,* J. W. Selby (1968). British Food Manufacturing Industries Research Assoc. Leatherhead, U.K.
*Modern Packaging Encyclopedia and Planning Guide* (Annual). McGraw Hill Book Co., Inc., New York, N.Y., U.S.A.

## Various

*British Pharmaceutical Codex,* Pharmaceutical Society of Great Britain. The Pharmaceutical Press, London, U.K.
*British Pharmacopoeia,* General Medical Council. The Pharmaceutical Press, London, U.K.
*Handbook of Chemistry and Physics,* American Rubber Co. CRC Press, Cleveland, Ohio, U.S.A.
*Food Processing Industry Directory* (1979). I.P.C. Consumer Industries Press, London, U.K.

*Surface Chemistry for Industrial Research,* J. J. Bikerman (1948). Academic Press Inc., London, U.K.

*Viscometry,* A. C. Merrington (1949). Edward Arnold & Co., London, U.K.

*Hygrometry,* H. Spencer-Gregory (1957). Crosby Lockwood & Son Ltd., London, U.K.

*Rheology and Texture of Foodstuffs,* S.C.I. Monograph No. 27 (1968). Society of Chemical Industry, London, U.K.

*Aerated Confections. "Hyfoama" Manual,* Lenderink & Co., Schiedam, Holland.

## JOURNALS

| Journal | Publisher |
|---|---|
| *United States* | |
| *Candy Industry* | Harcourt, Brace, Jovanovich, 747 Third Ave., New York, N.Y. |
| *Manufacturing Confectioner* | Manufacturing Confectioner Publishing Co., 175 Rock Road, Glen Rock, N.J. |
| *The Confectioner* | American Publishing Corporation, 771 Kirkman Road, Orlando, Fla. |
| *United Kingdom* | |
| *Confectionery Production* | Specialized Publications Ltd., 5 Grove Road, Surbiton, Surrey |
| *Confectionery Manufacture and Marketing* | J. G. Kennedy & Co. Ltd., 22, Methuen Park, London |
| *West Germany* | |
| *Süsswaren* | Rhenania-Fachverlag GmbH, Possmoorweg 5, Hamburg |
| *Kakao und Zucker* | Zeitschriftenverlag R.E.D.V., Postfach 1135, Düsseldorf |

| | |
|---|---|
| *Zucker und Süsswaren Wirtschaft* | Verlag Eduard Beckmann K. G., Postfach 1120, Lehrte |

*France*

| | |
|---|---|
| *Revue Internationale des Fabricant Produits á base de farine et sucre* | La Press Corporative Francaise, 5 rue d'argout, Paris |
| *Revue des industries de la biscuiterie, chocolaterie, confiserie* | Sepaic, 42, Rue de Louvre, Paris |

## RESEARCH ASSOCIATIONS AND TRADE ASSOCIATIONS

| Association | Address |
|---|---|
| *United Kingdom* | |
| Cocoa, Chocolate and Confectionery Alliance | 11, Green St., London |
| Leatherhead Food Research Association (BFMIRA) | Randalls Rd., Leatherhead, Surrey |
| *United States* | |
| National Confectioners Association (NCA) | 7900 Westpark Drive, Suite 514, McLean, Va. |
| American Association of Candy Technologists (AACT) | 175 Rock Rd., Glen Rock, N.J. |
| Pennsylvania Manufacturing Confectioners Association (PMCA) | 3404 Verner St., Drexel Hill, Pa. |
| Retail Confectioners Institute International (RCI) | 1701 Lake Avenue, Glenview, Ill. |
| *Other* | |
| International Office of Cocoa and Chocolate (IOCC) | 172, Avenue de Cortenbergh, 1040, Bruxelles, Belgium |

International Sugar    194, Rue de Rivoli,
  Confectionery       7500, Paris,
  Manufacturers Assocn,  France
  (ISCMA)

Information on other associations, organizations, etc., is available from the above or from the journals previously quoted.

## EXHIBITIONS

The main international exhibition of the confectionery, chocolate, and packaging machinery is "Interpack" held every three years in Düsseldorf, West Germany, usually in May.

Two other exhibitions of importance are the International Sweets and Biscuits Fair in Cologne, West Germany; and the International Confectionery and Biscuits Fair, Paris, France. Both exhibitions are usually held in January.

A confectionery exhibition held in Birmingham, England, for the first time in 1986 is InterConfex.

Exhibitions on a smaller scale are often held in conjunction with conventions and seminars, for example, NCA and PMCA in the United States and the Zentralfachschule, Solingen, West Germany.

## TECHNOLOGICAL COURSES

| Establishment | Type of Course |
|---|---|
| Zentralfachschule der Deutschen Süsswaren, Wirtschaft, Solingen-Grafrath, Germany | A full-time school well equipped with process machinery. "Residential" courses for periods up to eight weeks. Theory and practical work. Courses are given in English, French, German, and Spanish. International seminars of about three days duration are held throughout the year. |
| University of Wisconsin, Madison, Wis., U.S.A. (Department of Food Science) | Three-week vocational course each year in July—theory and practical. Organized in conjunction with the National Confectioners Association. |

| | |
|---|---|
| Retail Confectioners Institute (International) Glenview, Ill., U.S.A. | Two-week course annually at Erie, Pa. |
| Richardson Researches, Hayward, Calif., U.S.A. | One-week courses in chocolate and confectionery technology. Usually three weeks in March and other courses throughout the year. |
| British Food Research Association, Leatherhead, Surrey, England | Courses in analytical techniques. Others by arrangement. |

Other short courses are held from time to time and are advertised in the trade press. Departments of food technology in various universities and colleges hold courses in bakery and some of these include confectionery technology.

A recent development by the Cocoa, Chocolate and Confectionery Alliance and the Cake and Biscuit Alliance in England is a self-study training system. Termed CABATEC, it consists of a workbook with audio and video cassettes and samples. The first module in the series was "Chocolate Enrobing and Molding."

| | |
|---|---|
| Knechtel Research Sciences Skokie, Ill., U.S.A. | Instructional Courses in Chocolate and Confectionery Technology by arrangement. |

# Index

A.A. sugars, 240
Acesulfam K, 266
Acetic acid, 410, 413, 414, 598
Acetylated monoglycerides, 426–7
*Acheta domesticus,* 682
*Achromabacter lipolyticum,* 668
Acid salts as extenders for egg albumen, 322
Acid value, 284
Acids in confectionery, 409–13
Adhesive tapes, 763–5
  application of, 765
  keeping properties of, 764
  water vapor permeability, 764–5
Adhesives, 757–65
  discoloration, 757–8
  hot-melt, 762
  odors, 757
  physical properties, 758–9
  starch, 759–60
  storage, 758
  synthetic resins, 761
  types of, 759–62
  vegetable, 759
Advertising, wrapping materials, 755–6
Advertising material, fading tests, 756
Aerated chocolate, 185
Aerated confectionery, 560–7
Aerating agents, 315–16, 565, 567
Aeration, 315–16
Aflatoxin, 663
After-dinner mints, 556
Agar, 572
  jellies, 592–3
Agar-agar, 329–30
Air scrubbing, 58
Aldrin, 700
Alflatoxin, 398
Alginates, 330–1
  as extenders for egg albumen, 322
Alkalization, 46–7, 61–7, 82
  cocoa beans, 66–7
  cocoa cake, 66
  cocoa liquor, 65

effect of quantity and concentration of alkali, 64–5
nib, 62–4
Almond oil, 475
Almond paste, 594–5
Almonds, 385–9
  blanching, 405
  sugared, 609–10
Alphachlorolose, 706
*Althaea officinalis,* 567
Aluminum containers, 765–6
Aluminum phosphide, 698
Amino acids, 22–4, 666, 809
Ammonium carbamate, 698
Analytical methods, 825–72
*Anarcardium occidentale,* 390
Angel kisses, 573–4
Aniseed balls, 610
Aniseed oil, 476
Antibloom additives, 109, 170–1, 206
Antioxidants, 419–23
  action of, 420–1
  in utensils, 423
  limitations, 420–1
  methods of incorporating, 422–3
  natural, 420, 422
  permitted, 421–2
  prohibited, 423
  properties of, 424
  wrapping materials, 423
Antitailing device, 208
Ants, 683, 684, 701
*Apis,* 683
*Apium graveolens,* 478
Apparent density, 287
Apples, 381, 422
Apricot, 380, 475
  kernels, 599
*Arachis hypoqaea,* 395
Arachis oil, 109, 281–2
*Arachnida,* 683
*Araecerus fasciculatus,* 680
Aroma, 20–3
Arrowroot starch, 360

885